An Introduction to Molecular Neurobiology

AN INTRODUCTION TO

MOLECULAR NEUROBIOLOGY

ZACH W. HALL

With 11 Contributors

SINAUER ASSOCIATES, INC. • PUBLISHERS

Sunderland, Massachusetts

THE COVER

In situ hybridization of a sympathetic ganglion cell with ^{35}S-labeled poly(U) to reveal the distribution of poly A-mRNA. Silver grains are found over the cell body and dendrites but not over the axon. From D. A. Bruckenstein, P. J. Lein, D. Higgins and R. T. Fremeau, Jr. 1990. Neuron 5: 809–819.

AN INTRODUCTION TO MOLECULAR NEUROBIOLOGY

For information address Sinauer Associates, Sunderland, MA 01375 U.S.A.

Library of Congress Cataloging-in-Publication Data

An Introduction to molecular neurobiology / edited by Zach W. Hall.
p. cm.
Includes bibliographical references and index.
ISBN 0-87893-307-7 (cloth)
1. Molecular neurobiology. I. Hall, Zach W.
[DNLM: 1. Neurobiology. 2. Neurons—physiology.
WL 102.5 I618]
QP356.2.I58 1992
599′ .0188—dc20
DNLM/DLC
for Library of Congress 91-5138
CIP

Printed in U.S.A.
6 5

For Julie

CONTENTS IN BRIEF

CONTENTS

PART TWO THE CELL BIOLOGY OF NEURONS AND GLIA

THE CONTRIBUTORS

DAVID J. ANDERSON
Division of Biology, California Institute of Technology

GARY BANKER
Department of Neuroscience, University of Virginia School of Medicine

XANDRA O. BREAKEFIELD
Molecular Neuroscience Unit, Massachusetts General Hospital

ZACH W. HALL
Department of Physiology, University of California, San Francisco

MARY B. KENNEDY
Division of Biology, California Institute of Technology

GREG LEMKE
Molecular Neurobiology Laboratory, The Salk Institute

EVE MARDER
Department of Biology, Brandeis University

PAUL H. PATTERSON
Division of Biology, California Institute of Technology

ELLIOTT M. ROSS
Department of Pharmacology, University of Texas Health Sciences Center at Dallas

PETER B. SARGENT
Department of Stomatology, University of California, San Francisco

RICHARD H. SCHELLER
Department of Molecular and Cell Physiology, Stanford University

RON D. VALE
Department of Pharmacology, University of California, San Francisco

PREFACE

Although the molecules of the nervous system and their roles in neural function have fascinated neurobiologists for over one hundred years, molecular neurobiology has only recently come into its own as a discipline. Thanks to the powerful techniques of biochemistry and molecular biology, hundreds of new proteins have recently been identified and their primary structures determined. Techniques for expressing the proteins in new cells and organisms have helped to define their roles in determining the size and shape of neurons, their synaptic connections, and how they receive, transmit, and retain information. Sophisticated biochemical and physiological techniques have revealed an almost bewildering variety of molecules that carry messages between and within cells in the nervous system. Genetic techniques allow defective genes responsible for diseases of the nervous system to be identified, and questions of neural plasticity and development can now be phrased in terms of how gene expression is controlled.

An Introduction to Molecular Neurobiology aims to use the wealth of new molecular information to examine how cells function and develop in the nervous system. It is intended as a textbook for students who are curious about the roles of molecules in cellular neurobiology. Although many who use the book will have some knowledge of neurobiology, we have attempted to make it accessible to those without previous exposure to the field. An introductory chapter outlines the basic concepts and terminology of neuroscience, and a second chapter provides instruction in basic neurophysiology. The book can thus be used in an ambitious introductory course with a strong molecular emphasis, or in an advanced course devoted specifically to molecular neurobiology. We also hope that the book will be useful for practicing neurobiologists whose primary tools are not molecular, but who wish to acquire literacy in a new and important area.

In a field as large and rapidly moving as molecular neurobiology, a text cannot hope to be comprehensive or up to the minute. For each topic, we have tried to emphasize fundamental questions, relevant experimental techniques, and future directions. A small number of references are given at the end of each chapter, with the most important ones starred. We envisage that instructors using the book will supplement these references with others from the recent literature.

Among many colleagues who gave generously of their time and advice as we prepared the book, several deserve special mention. Discussions with Reg Kelly were invaluable in formulating many of the ideas in Chapters 4 and 5; Dennis Bray made an important contribution to the introduction of Chapter 7; and Peter Sargent wrote the box in Chapter 8. Wolf Almers, Allan Basbaum, Henry Bourne, Jeremy Brockes, Thomas Carew, Martin Chalfie, Arlene Chiu, David Colman, David Copenhagen, John Dani, Charles Edwards, Stanley Froehner, A. J. Hudspeth, Steven Hyman, Katherine Jones, David Julian, Eric Kandel, Regis Kelly, Marc Kirschner,

Juan Korenbrot, Dan Madison, Mike McKeown, William Mobley, Perry Molinoff, Roger Nicoll, James Patrick, Martin Raff, James Sabry, Julie Schnapf, Tom Schwarz, Sigrid Reinsch, Babette Stewart, Peter Stewart, Michael Stryker, Elly Tanaka, Anthony Trevor, and Larry Zipursky made critical and constructive comments on individual chapters.

We are indebted to many at Sinauer Associates. The idea of an edited textbook originally arose in a conversation with Andy Sinauer; his high standards and patient encouragement have been indispensible throughout. Carol Wigg carefully and skillfully guided the editorial process and Joe Vesely provided expert design and production. We are grateful to John Woolsey and his colleagues at J/B Woolsey Associates for their imaginative illustration of the book.

I personally owe a debt to many colleagues at UCSF, expecially those in Neurobiology, and to my teachers and former colleagues at Harvard Medical School. I am particularly grateful to my advisors, Peter Stewart, Edward Kravitz, Stephen Kuffler, and I. R. Lehman. Members of my laboratory, past and present, have taught me much about molecular neurobiology. The successful examples of Alan Pearlman and Kenneth Walker inspired me to undertake this book, Henry Bourne encouraged me, and the intelligent and reliable help of Monique Piazza made it possible. Finally, I thank my mother, wife, and children for their constant support and encouragement.

ZACH W. HALL
San Francisco, California

1

The Cells of the Nervous System

Zach W. Hall

THE HUMAN NERVOUS SYSTEM is the ultimate product of biological evolution. Understanding how it functions and develops not only has the allure of self-knowledge but is an intellectual challenge of the highest order. To understand how the nervous system works requires knowledge at several levels. At one extreme we need to know how large numbers of neurons interact to produce the complex behavior of the organism. How do our eyes follow a moving object, for example? At another level we need to know the properties of individual cells and how they interact. How do photoreceptors in the eye convey information about light to the rest of the nervous system? Finally, we want to understand the molecules of the nervous system and how they determine the properties and behavior of the cell. How do ion channels work? How do growing axons move? What molecules determine how neurons differentiate and which genes they express?

The major focus of this book is to examine the properties and behavior of neurons and glia, the principal cells of the nervous system, and how they arise from the molecules that constitute them. What are the important molecules in nerve cells, how do they interact with each other, and how do these interactions lead neurons to grow, to differentiate, to make correct connections with each other, and, in the adult, to transmit signals and organize information?

These are exciting times in molecular and cellular neurobiology. New molecular, biophysical, and anatomical methods developed within the last 20 years have literally transformed the study of neurons and indeed have changed modern biology. Problems that were previously understood only at the level of cellular behavior can now be approached in terms of the responsible molecules. The resulting wealth of molecular information has revealed to us how fundamentally conservative nature is. Different cells use similar molecules or parts of molecules to accomplish diverse ends. These unexpected molecular relationships have added excitement to modern biology and have fueled its progress. For cellular and molecular neurobiologists the challenge is to understand how very common biological molecules are organized to give neurons and glia their very uncommon properties. Success in understanding the molecular basis of neuronal and glial behavior will lead to better understanding of how neurons work together in complex circuits to regulate our bodies, to coordinate our movements, to perceive our environment, and to store memories of these events. Perhaps someday these studies will even lead us to an understanding of the basis of our emotions, our creativity, and our conscious sense of ourselves.

An Introduction to Neurons

Although they are not the most abundant cells in the nervous system, neurons are the most important. Their fundamental task is signaling. Their every aspect—their size, shape, cellular organization, development, and organization within the brain—is shaped by their specific functions of receiving, organizing, and transmitting information.

Neurons communicate with each other at synapses

Neurons encode information by a combination of electrical and chemical signals. In primitive organisms, cells communicate with each other through chemical messengers that are secreted into the extracellular environment and bind to surface receptors in other cells. Neurons use a similar but highly modified form of chemical signaling, which is faster, more precise, and more flexible. Neurons secrete **neurotransmitters** and other chemical messengers at sites of functional contact called **chemical synapses.** At each chemical synapse a region of membrane in the **presynaptic** cell that is specialized for rapid secretion is closely and firmly attached to a specialized region on the **postsynaptic** cell that contains a high density of **receptors** for the neurotransmitter. In this way, chemical communication in the nervous system is specifically directed from one cell to another. Electrical signals in the presynaptic cell cause the release of neurotransmitter; its binding to surface receptors triggers an electrical signal, the **synaptic potential,** in the postsynaptic cell. Neurons thus transmit information through the nervous system by an alternating chain of electrical and chemical signals (Figure 1).

Neurons are polarized

The number of synaptic connections that a neuron makes can be extraordinarily large. A single motor neuron in the mammalian spinal cord, for example, is estimated to receive signals from over 10,000 synapses on its surface. To accommodate these many connections, neurons have elaborate networks of **dendrites,** highly branched cellular processes that are specialized to receive synapses from other cells by having receptors and other proteins necessary to convert chemical signals into electrical signals. The signals received in the dendrites are combined, or **integrated,** and then translated into another series of electrical signals that are sent out over a long, thin cellular process, the **axon.** These signals go to the **presynaptic**

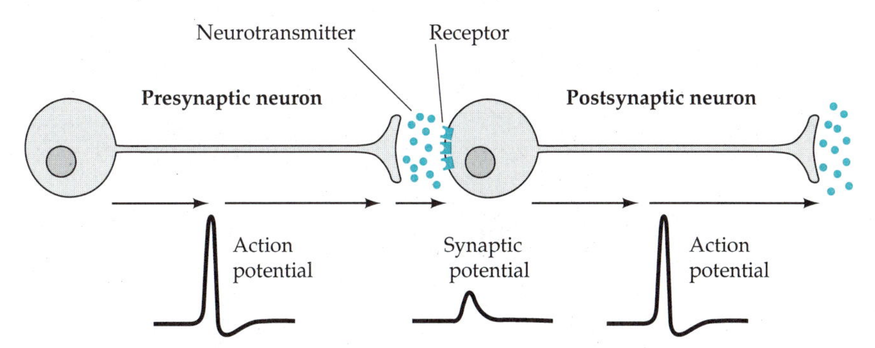

FIGURE 1. Neurons convey information by a combination of chemical and electrical signals. At chemical synapses, electrical signals in the nerve terminals cause them to secrete neurotransmitters onto the postsynaptic cell. The neurotransmitters bind to postsynaptic receptors to cause electrical signals, called synaptic potentials, in the postsynaptic cell. The synaptic potentials trigger an action potential in the axon which is conveyed to the nerve terminal and triggers secretion of neurotransmitter onto the next cell.

nerve terminal, a bulbous region at the end of the axon that releases transmitter onto the next cell. Because electrical signals and information generally flow from the dendrites to the axon to the nerve terminals, neurons are said to be functionally and anatomically **polarized.** Axons can be very long, up to two meters, and are usually branched, giving rise to multiple terminals that synapse with many different cells.

In most vertebrate neurons, the dendrites and axon extend from a central **cell body** that contains the nucleus and the membrane organelles responsible for the synthesis and processing of cellular proteins. In these neurons, synapses often terminate on the cell body as well as on the dendrites, and the cell body, along with the dendrites, participates in signal integration. In other neurons the cell body is displaced from the main routes of information flow (Figure 2). In invertebrate neurons, for example, the cell body receives no synapses and is attached by a single stalk to the dendritic tree from which the axon arises. Sensory neurons in the vertebrate peripheral nervous system represent another exception to the general scheme. In these neurons, stimuli such as pressure, temperature, and stretch generate electrical signals in specialized regions at one end of the sensory neurons. These signals then travel along a single axon to nerve terminals at the other end of the cell. The cell body, which is attached to the axon by a single process or stalk, receives no synapses, and, again, plays no immediate role in signaling.

Neurons have diverse forms

In spite of the general plan of cellular organization that most neurons share, they have an extraordinary variety of sizes and shapes (Figure 3). Estimates of the number of different classes of neurons in the vertebrate brain extend as high as 10,000. Neurons range from small bipolar cells in the retina that have a single, simple dendrite to large Purkinje cells of the

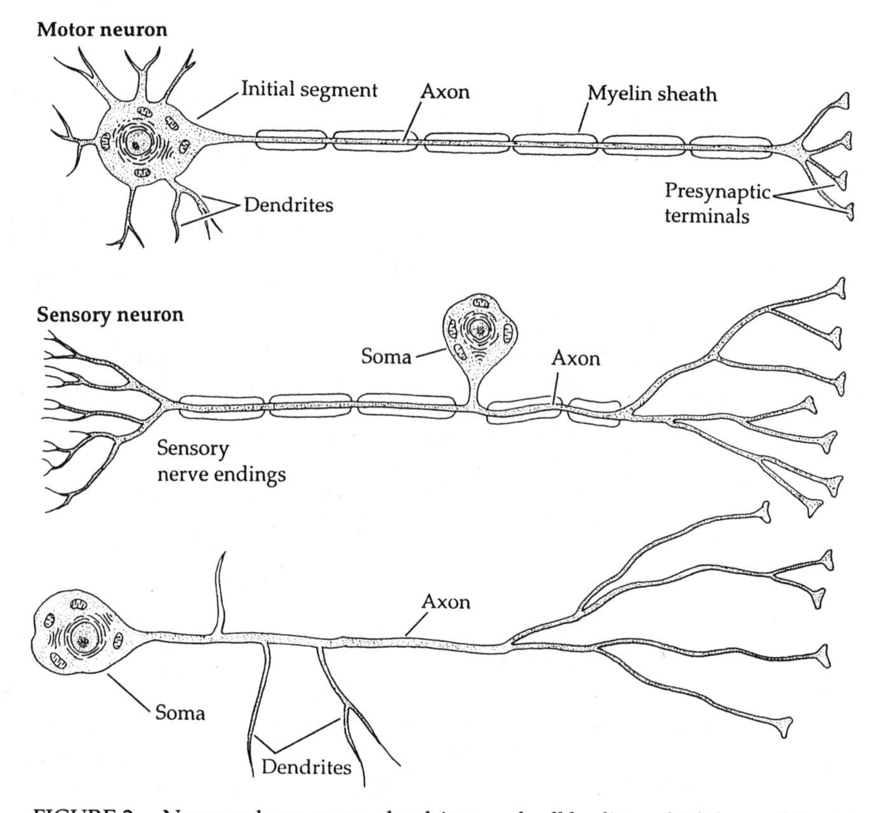

FIGURE 2. Neurons have axons, dendrites, and cell bodies, which have different arrangements. (A) A conventional neuron from the vertebrate central nervous system in which both axons and dendrites extend from the cell body. (B) A vertebrate sensory neuron in which the proximal regions of the axon and dendrite leading from the cell are fused into a single process that connects the cell body to the rest of the cell. (C) An invertebrate neuron in which the cell body is connected to an extensively branched dendritic tree from which the axon arises. Note that only the vertebrate axons have myelin sheaths.

cerebellum with large, flamboyant, and highly stylized dendritic trees. The cell bodies can be large or small and their axons short (microns) or long (meters). This enormous diversity of size and shape allows flexibility in the pattern and sites of synaptic connections that a neuron makes.

Differences in form are the most visible and dramatic expression of neuronal diversity, but variation between neurons pervades almost every aspect of neuronal function. Different neurons make different transmitters and other synaptic messengers, express different combinations of receptors and ion channels on their surfaces, use different configurations of second messengers, and have different cell surface and cytoskeletal molecules. Ultimately all of these depend on differences in gene expression. How do neurons generate such variation? One theme that will recur throughout this book is the powerful **combinatorial strategy** that neurons use to generate diversity. Combining members of a particular set of mol-

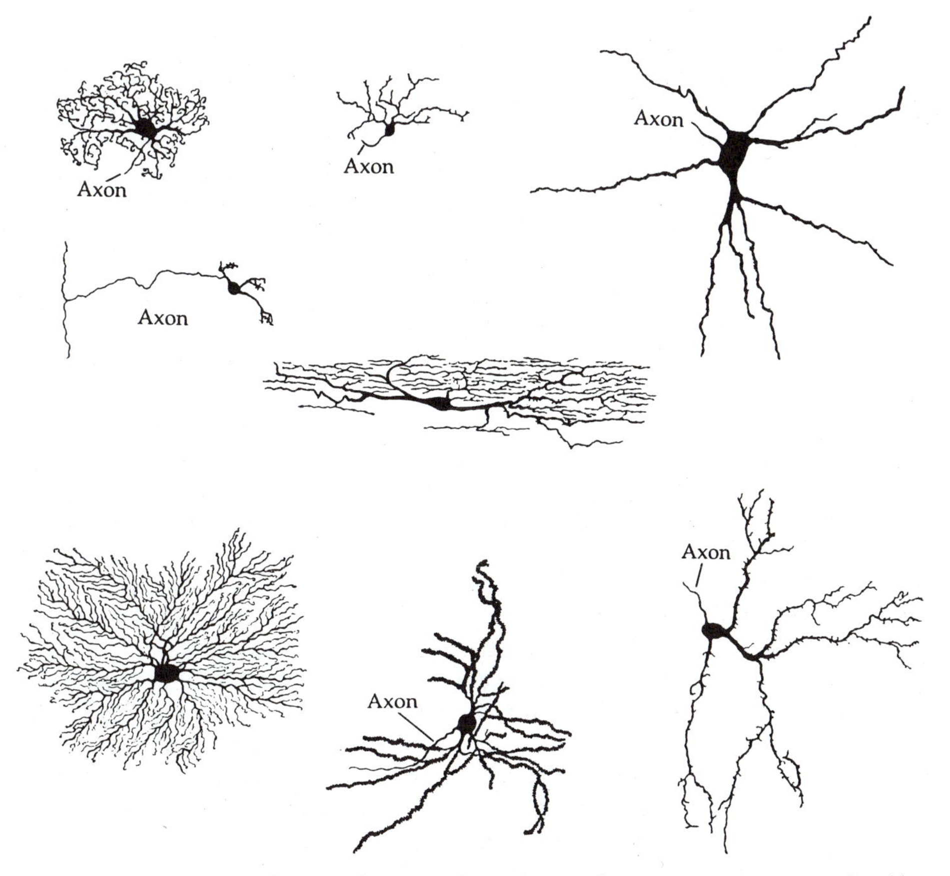

FIGURE 3. A selection of neurons from the vertebrate nervous system, stained by the Golgi method, illustrating the diversity of their sizes and shapes. (Adapted from M. Carpenter, 1976. *Human Neuroanatomy*, 7th Ed., Williams & Wilkins, Baltimore.)

ecules in different ways is a much more efficient way of achieving variation than having a unique molecule for each task. Out of many possible choices, neurons select the particular combination needed.

Signaling in Neurons

Quiescent neurons, like other cells, have a difference in electrical potential across their surface membrane, with the inside of the cell about 60 mV negative to the extracellular fluid outside. This **resting potential** arises because of the unequal distribution of ions across the membrane and because of the selective permeability of the membrane to them. These same principles underlie the complex electrical signals that neurons use. Proteins in the membrane called **transporters** use metabolic energy to actively transport ions into and out of the cell, resulting in ion concentration gradients across the membrane. These gradients represent a form of stored energy used by neurons and other cells to generate electrical signals. The signals themselves result from the opening and closing of **ion channels,** transmembrane proteins that facilitate the passive flow of specific ions across the membrane and down their electrochemical gradients. The flow of ions through the channels constitutes a current that changes the membrane potential, making it either larger or smaller. Such changes, which vary in time and in space, constitute electrical signals. Electrical signals mostly convey information within cells but can also mediate cell–cell communication at **electrical synapses,** where specialized membrane proteins form large channels that span the two membranes and connect the cytoplasm of the two cells, allowing ionic current to flow between them.

Electrical signals organize and transmit information within neurons

Neurons use several types of electrical potentials to convey information; each serves a different purpose. A brief, stereotyped signal, the **action potential** (Figure 4), conveys axonal signals rapidly and efficiently over long distances. During the action potential, the membrane potential is quickly reversed (i.e., the membrane becomes electropositive inside) and then returns to nearly the original resting potential, all within a few milliseconds. Action potentials are more or less the same in all cells and are self-propagating. They are triggered by an automatic series of changes that occurs whenever the membrane potential of axons is reduced past a **threshold** value. Because a stereotyped action potential is automatically triggered once threshold is passed, it is said to be "all-or-none." Action potentials encode information by their frequency. The higher the frequency of action potentials that propagate into nerve terminals, the higher the rate of transmitter secretion.

Other potentials, sometimes called **local potentials,** encode information by their size and shape. The synaptic potentials produced in postsynaptic

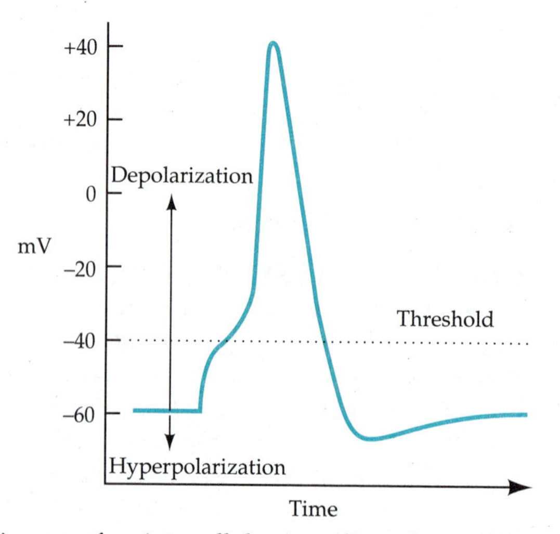

FIGURE 4. Diagram of an intracellular recording of an action potential.

neurons by the binding of neurotransmitters to their receptors are one example. The size of synaptic potentials, which varies, depends upon how much transmitter is released, which in turn depends upon the frequency of action potentials in the presynaptic nerve terminal. Synaptic potentials can either **depolarize** the membrane (bring it nearer to zero potential) or **hyperpolarize** it (take it farther away), and can thus be either excitatory or inhibitory. Synaptic currents can sum, integrating the information coming into the cell from different synapses. Unlike the action potential, synaptic potentials are not self-propagating but decrease in size as they are transmitted away from the postsynaptic membrane and die out altogether after a few millimeters. The electric currents that give rise to potentials in the dendrites and/or the cell body add and subtract, and the potentials that result are propagated to a region at the base of the axon, called the **initial segment** (Figure 2). If at any moment the local potential at the initial segment exceeds threshold, an action potential is initiated that travels down the axon. The variation of summed synaptic potentials with time at the initial segment thus determines the frequency of action potentials. The initial segment is the site of "read out" in the cell for the integration that takes place in the cell body and dendrites.

In sensory neurons the stimulus produces a graded, local electrical potential called a **sensory generator potential** that is usually translated near its site of initiation into a train of action potentials. Other neurons may have endogenous electrical activity. In these cells, a repeating, self-generating series of permeability changes in the membrane produces a steady series of action potentials. Such cells often serve as **pacemakers** for rhythmic activity in the nervous system (see Chapter 14).

Regulated ion channels are the basis of electrical signaling

Neurons produce electrical signals by controlling the flow of ions across their surface membrane. The channels through which the ions flow have two important properties: they are **ion-specific**, allowing the flow of different ions to be controlled independently, and they are **regulated.** In a simplified scheme (Figure 5), the ion channels, which are formed by transmembrane proteins, exist in either of two states, open or closed. The equilibrium between these two states can be shifted by a variety of agents: by neurotransmitters; by cytoplasmic second messengers; by metabolic alterations of the channel, such as phosphorylation; by mechanical deformation of the membrane; or by membrane potential itself.

Over 75 different types of channels are expressed in the nervous system, with more being discovered at a rapid pace. Neurons use different combinations of these channels to create a rich variety of electrical signals. By controlling the types of channels, and their density and distribution on the surface, neurons generate a varied repertoire of electrical signals in different cells and in different parts of the same cell. These signals give neurons the complexity and flexibility that they need as the cellular components of the large signaling system that is our brain.

In addition to receiving chemical signals from neurotransmitters that initiate rapid changes, neurons also respond to molecules in their extra-

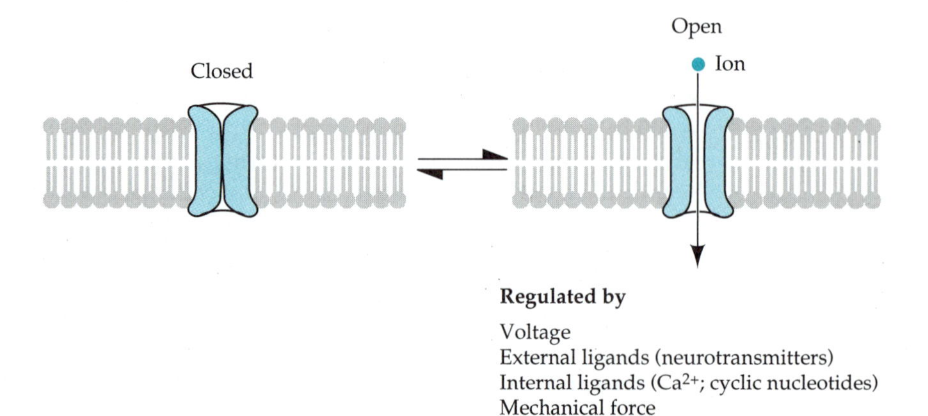

FIGURE 5. Ion channels in neurons are regulated. Ion channels can be represented as having two states, open and closed, that are in equilibrium. Ions flow through the open channel down their electrochemical gradients. The relative proportion of the two states can be regulated by diverse factors.

cellular environment that modify their behavior over an extended duration. These include peptides released by nerve terminals and circulating hormones. These molecules modify neuronal function by interacting with specific cell surface receptors, and, through intermediary proteins and second messengers, modify ion channels, metabolic reactions, cytoskeletal cell surface molecules, and gene expression.

Neuronal Cell Biology

With a profusion of processes, and a span that can reach a meter, neurons look like no other cells in biology. Their morphological variety, their ability to generate electricity, and their peculiar staining properties led earlier generations of investigators to emphasize how different neurons were from other cells. More recent study of the biochemical and molecular properties of neurons, however, has focused attention on the properties that they share with other types of cells. Their metabolic pathways, membrane and surface proteins, organelles, and cytoskeletal elements all resemble those of other cells. Even many of their unique characteristics have turned out to be variations or exaggerations of well-known themes in cellular physiology. Once recognized, these common motifs are a great boon to modern neurobiology because they allow progress in understanding basic cellular processes in other cells to be readily and profitably translated to the investigation of similar problems in neurons.

Although neurons use common cellular mechanisms, their unique functions impose particular demands that shape these processes. The use of electrical signals by nerve cells, for example, means they must have high rates of metabolism to maintain and restore the ionic gradients that provide the currents that create these signals. Neurons thus have many mitochondria and a high rate of glucose transport. The length of neuronal processes requires that they have an efficient transport system for commerce between the different parts of the cell. Neurons thus have a well-developed cytoskeleton. Neurons secrete small molecules and peptides at a furious rate, but unlike other secretory cells this activity takes place far from the site of synthesis of membrane and protein. Neurons thus recycle membranes, not only in the cell body, but also in the nerve terminal, which is maintained as an active outpost, autonomous in some activities, but ultimately dependent on the cell body for macromolecular supplies. Exploration of some peculiarly neuronal activities not only illuminates their mechanisms but also casts light on the general cellular processes of which they are special examples.

The cell body synthesizes most neuronal macromolecules

The cell body makes most of the macromolecules that neurons use. All neuronal DNA, except mitochondrial DNA, is in the nucleus. Because neurons do not divide, they do not synthesize DNA, but their DNA is actively transcribed in the nucleus to make RNA for protein synthesis. After synthesis, ribosomal and messenger RNA (mRNA) are transported through nuclear pores into the cytoplasm. Most of the mRNA remains in the cell body, although some is transported into the proximal dendrites. The organelles that carry out protein synthesis and processing, the **ribosomes,** the **endoplasmic reticulum (ER),** and the **Golgi complex** are also mostly in the cell body (Figure 6).

Messenger RNA from the nucleus attaches to ribosomes, which are the machines for protein synthesis. If the protein encoded by the RNA is soluble, the ribosomes remain free in the **cytosol** (the cytoplasm minus organelles). If the encoded protein is a secretory or membrane protein, however, it is attached to the ER and translocated into the ER lumen as it is being synthesized. As polypeptides are synthesized and translocated into the lumen, a branched chain of carbohydrates rich in mannose may be added to particular asparagine residues to form an *N*-linked glycosyl chain.

The ER is an extensive system of flattened sheets and tubules of membrane, at least part of which is continuous with the nuclear membrane. In some neurons, it is dispersed throughout the cell body and appears fragmented, whereas in others it appears as an orderly array of flattened sheets. The distribution of ER in the cytoplasm appears to result from its translocation along microtubules. In many neurons the ER extends into the proximal dendrites, and fragments of it can be seen even farther out. Microtubules presumably transport it to these locations. Because of the large amount of cytoplasm that is sustained by it, the ER in many neurons is highly expanded. Its prominence and staining properties earned it a special name from classical cytologists: the **Nissl substance.**

Membranes and secretory proteins are synthesized in the **rough ER,** so called because the membranes are studded with ribosomes attached to their cytoplasmic surface; lipids are synthesized in the **smooth ER,** which lacks ribosomes. In neurons, as in other cells, membranes are not made de

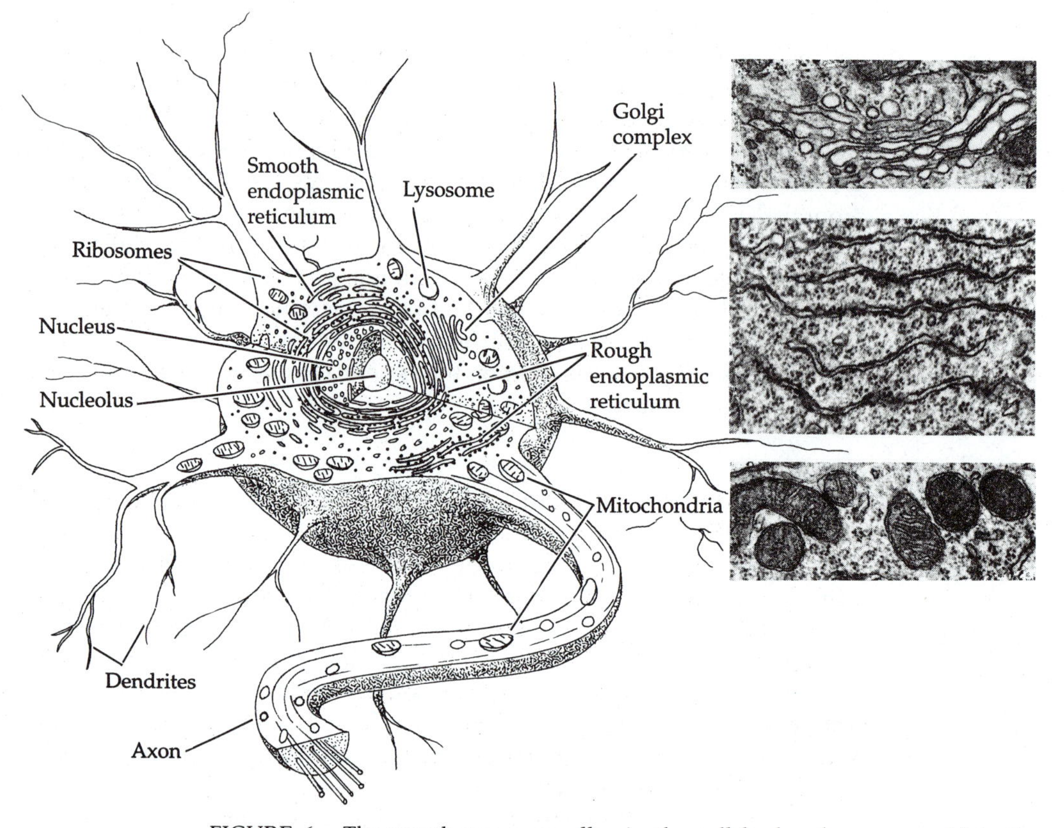

FIGURE 6. The membrane organelles in the cell body of a neuron. Note the nucleus with its nucleolus, the rough and smooth endoplasmic reticulum, the Golgi complex, mitochondria, and lysosomes. The rough endoplasmic reticulum is often very prominent in neurons. The inset electron micrograph figures are from a neuronal cell body in the cochlear nucleus. Magnification of the photomicrographs is 23,500. (Photomicrographs courtesy of D. McDonald.)

novo, but their area is increased by the addition of components to preexisting membrane. Since both phospholipid and membrane protein, the two major components, are made in the ER, most new membrane arises there.

After synthesis, membrane and secretory proteins are transported in small vesicles to the Golgi complex (sometimes called the Golgi apparatus) which consists of a series of flattened membrane sacs that appear in some cells as stacks. The Golgi complex is usually located in the cell body near the nucleus, although fragments of it also appear in dendrites in close association with the ER. The stacked membranes represent a series of compartments through which the proteins move sequentially. As the proteins move through, specific enzymes modify their *N*-linked sugar chains by trimming them back and adding new sugars. Other modifications made in the Golgi complex include the attachment of fatty acids, the addition of sugars to serine or threonine residues (*O*-linked sugars), sugar phosphorylation, and the sulfation of tyrosine residues. The modification of proteins to form proteoglycans, which have many long, highly sulfated, unbranched oligosaccharide chains, also occurs in the Golgi.

The transfer of membrane and secretory proteins from the ER to the Golgi complex and from compartment to compartment within the Golgi occurs via small transport vesicles that pinch off from the membrane of the donor compartment, and fuse with the membrane of the recipient compartment. The pinching off and fusion of the vesicles is not random. At each step proteins that are ready to be transferred are recognized and segregated into the vesicles away from resident proteins of the donor membrane. The transport vesicles are targeted to fuse correctly to the acceptor membrane compartment.

In addition to organelles concerned with macromolecule synthesis, neuronal cell bodies contain **lysosomes,** which are responsible for intracellular degradation. Lysosomes are large vesicles that contain an impressive cocktail of degradative enzymes, including proteases, nucleases, phospholipases, phosphatases, and glycosidases. Lysosomes actively transport protons into their interior, so that they have an internal pH of about 5. All of the enzymes in the lumen have acidic pH optima and are relatively inactive at neutral pH. Lysosomes are commonly recognized in neurons by histochemical stains for the enzyme acid phosphatase. Proteins and other macromolecules that are to be degraded are delivered by vesicles to lysosomes, which fuse with them.

Membrane traffic shuttles to and from the surface of neurons

In all eukaryotic cells, newly synthesized and processed membrane proteins are sorted and dispatched to their appropriate cellular compartment in a region of the Golgi complex known as the ***trans*-Golgi network (TGN).** Some proteins go to the surface membrane, others to lysosomes, and others are stored in **secretory vesicles,** until an appropriate signal is given for secretion.

In cells, there is thus an orderly pattern of traffic to and from the cell surface (Figure 7). Vesicles carrying proteins destined for the cell surface deliver them by fusing with the surface membrane, a process known as **exocytosis.** Secretory vesicles also release their contents by exocytosis at the surface membrane. Proteins can be retrieved from the surface by the reverse process of **endocytosis,** in which membrane is pinched off to form a vesicle. The budding or pinching off of small vesicles is energetically unfavorable, and can be aided by the protein **clathrin,** which forms a cage, or coat, around the vesicle. Once internalized, the **coated vesicles** shed their coats and then fuse with other vesicles to generate **endosomes,** a system of membrane vesicles and tubules with an acidic interior. From the endosomes, membrane proteins are either returned to the surface or are transported to lysosomes for degradation.

Endocytosis is an important mechanism for the internalization of receptors that bind protein carriers of nutrients used by eukaryotic cells. Iron, for example, is bound by the plasma protein transferrin, which in turn binds to transferrin receptors in the surface membrane. These are internalized by endocytosis and delivered to the endosomes. At the acidic

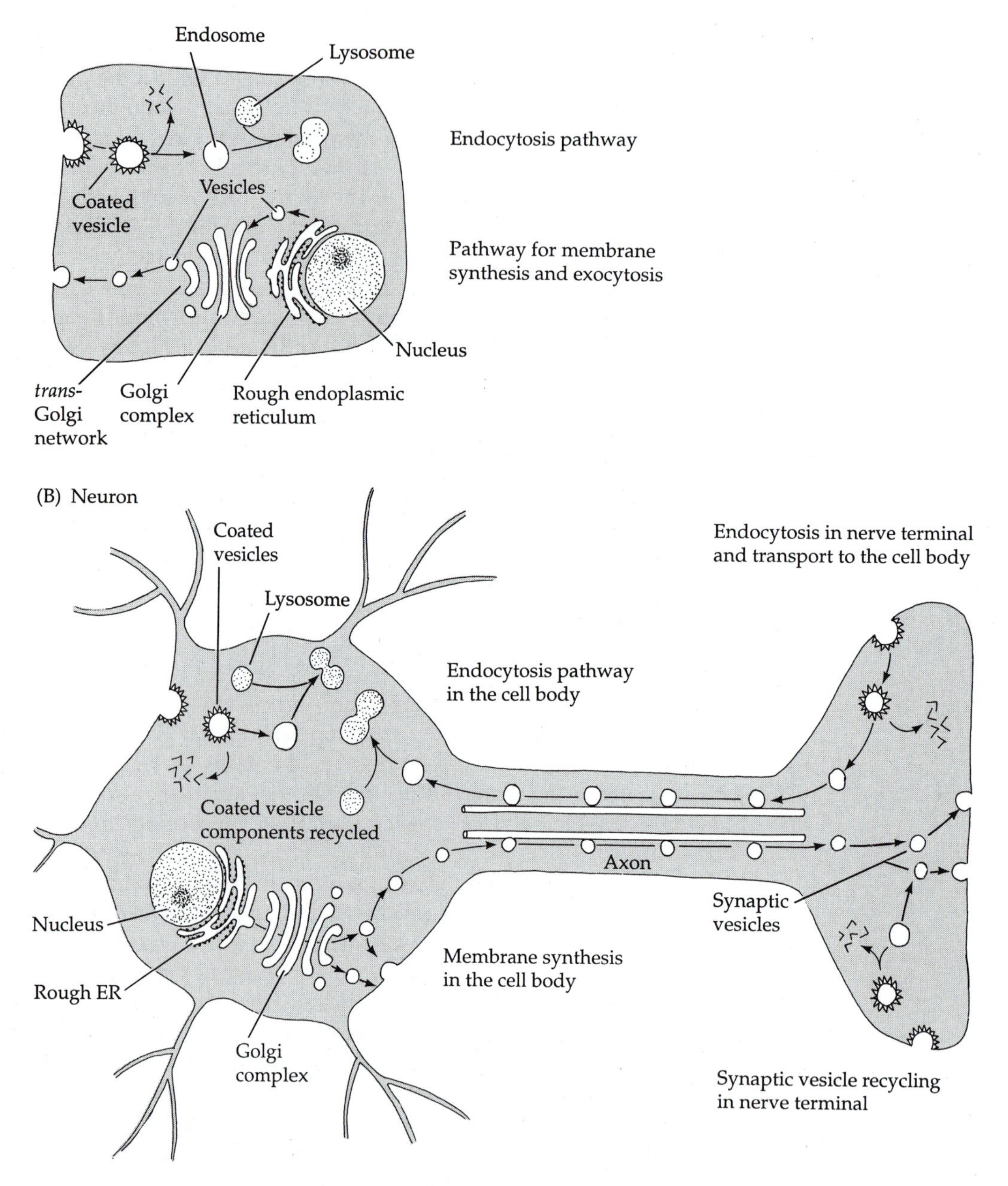

FIGURE 7. Traffic of vesicles to and from the cell surface of a non-neural cell and a neuron. In non-neural cells, the traffic of newly synthesized membrane and membrane proteins goes from the ER to the Golgi complex to the cell surface. Membrane is retrieved from the surface in coated vesicles, then fuses with endosomes, and thence to lysosomes. In neurons the same general pattern is seen, except the terminal is far away from the cell body and nerve terminals have a local pattern of exocytosis and endocytosis.

pH of the endosomes, the iron dissociates and remains in the cell. The transferrin protein, minus the iron but still attached to its receptor, is then returned to the cell surface where it dissociates. Endocytosis is also used for the internalization of protein signaling molecules that are bound to their receptor. Protein growth factors that bind to their receptors on the surface can be removed by endocytosis, taken to endosomes and thence to lysosomes, where they are degraded.

In neurons these same processes occur, except that much of the relevant cell surface is far away from the cell body, which contains the TGN and the lysosomes. Vesicles must be thus be transported over great distances within neurons. Membrane traffic in axons and dendrites goes in both directions. Proteins destined for secretion and for the surface membranes and organelles must be transported out from the cell body to the processes; proteins and other macromolecules taken in at the tips of the processes must be brought to the cell body to be degraded. In addition, axons must also transport to nerve terminals the membrane proteins used for the synthesis and secretion of transmitters.

In the terminals, transmitters are stored in synaptic vesicles. Action potentials that invade the terminal cause the influx of calcium, a signal that causes the vesicles to fuse with the surface membrane and release their contents into the synaptic cleft and surrounding medium. The membrane proteins added to the surface during exocytosis are retrieved by endocytosis and used to make new vesicles that are filled with neurotransmitter that is synthesized locally. Thus, an active local cycle of exocytosis and endocytosis in the terminal is superimposed on the flow of membrane traffic to and from the terminal.

The cytoskeleton provides a highway for transport and a framework for the cell

A system of interconnected macromolecular filaments guides membrane vesicles to their destinations within axons and dendrites and provides neuronal processes with mechanical strength. Three polymeric structures form the basis of this **cytoskeleton: actin filaments, microtubules,** and **intermediate filaments** (Figure 8). Microtubules and intermediate filaments are oriented longitudinally in axons and dendrites, and are cross-linked to each other by accessory proteins. Actin filaments, also called microfilaments, form a meshwork underneath the membrane of the entire neuron and in axons and dendrites are connected to the microtubules and intermediate filaments. Vesicles move through axons and dendrites along individual microtubules. The vesicles are attached to the microtubules by proteins that act as motors, consuming energy and propelling the vesicles along the linear tracks formed by the microtubules. This process, called **axonal transport**, is also responsible for the movement of mitochondria, which are distributed throughout the cell.

Cell surface adhesion molecules influence the behavior of neurons

In addition to transporters, ion channels, and receptors for synaptic messengers, neurons have on their surfaces a group of molecules that regulate

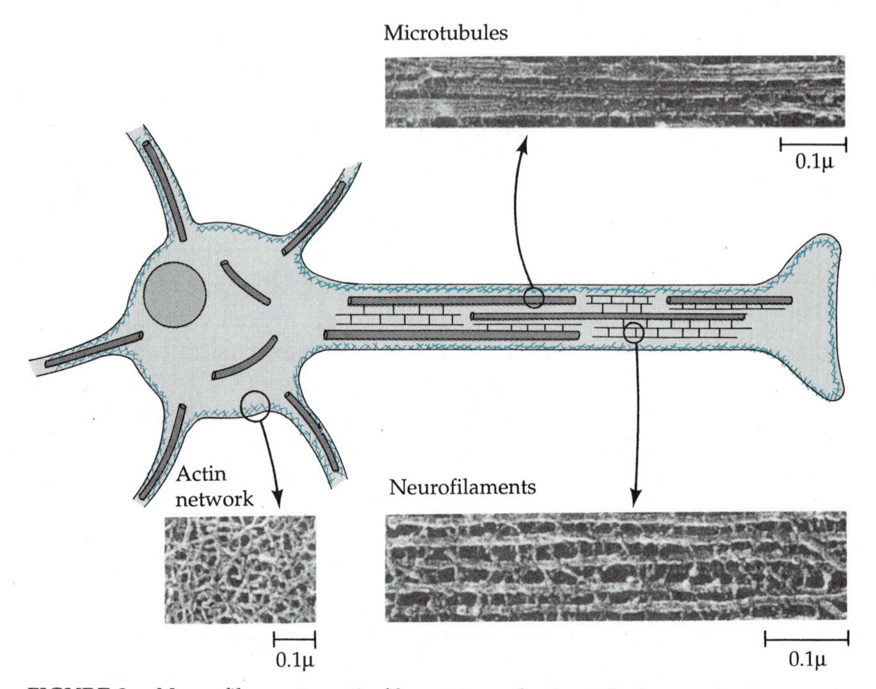

FIGURE 8. Neurofilaments, actin filaments, and microtubules are the three major components of the neuronal cytoskeleton. Microtubules and neurofilaments are concentrated in axons and dendrites where they are oriented longitudinally and are connected by cross-links. Actin filaments are organized in a meshlike network that underlies the surface membrane. Representatives of each of these components are shown in electron micrographs of neural tissue that has been rapidly frozen, fractured, deep-etched, and rotary-shadowed. The microtubules and neurofilaments are from frog axons (N. Hirokawa, 1982. J. Cell Biol. 94: 129–142), and the actin network is from chick hair cells (N. Hirokawa and L. G. Tilney, 1982. J. Cell Biol. 95: 249–261).

the interactions of neurons with each other, with other cell types, and with the extracellular environment. These molecules are particularly important in early development, where they influence pathways of cell migration and neurite outgrowth, but they also play important roles in adult tissues, where they are responsible for the adherence of nerve terminals to their targets and for the interactions between neurons and other cell types. For two families of these molecules, the **CAMs** (cell adhesion molecules) and the **cadherins,** identical molecules on different cells bind to each other, forming the basis of cell-specific adhesion. In other cases, receptors in the cell membrane such as the **integrins** bind molecules in the extracellular matrix, and cell surface receptors bind specific carbohydrates. The binding interactions mediated at the cell surface by these various classes of molecules alter neuronal behavior by changing the cytoskeleton to which they are attached, or by influencing the intracellular network of second messengers.

Glia

Glia are the support cells of the nervous system. They fill the interstices between neuronal cell bodies, ensheathe axons, surround the blood vessels of the brain, and line its internal and external surfaces. In the mammalian brain, they outnumber neurons by at least 10 to 1. The two principal types in the central nervous system, **astrocytes** and **oligodendroglia,** are differentiated by their appearance and function. **Schwann cells** ensheathe axons in the peripheral nervous system. During development and during neurite regeneration (in the peripheral nervous system), glial cells provide an attractive surface for axons to grow on and help guide nerve processes to their proper destinations.

Astrocytes are the major interstitial cells of the brain

To early anatomists astrocytes looked star shaped, with hundreds of processes radiating out from a central nuclear region. In the central nervous system astrocytic processes are intimately intertwined with neurons, surrounding their cell bodies, dendrites, and sometimes their nerve terminals, insulating and isolating them from each other (Figure 9). Astrocytic processes, particularly those in the white matter, are filled with intermediate filaments made from a particular intermediate filament protein, **glial acid-**

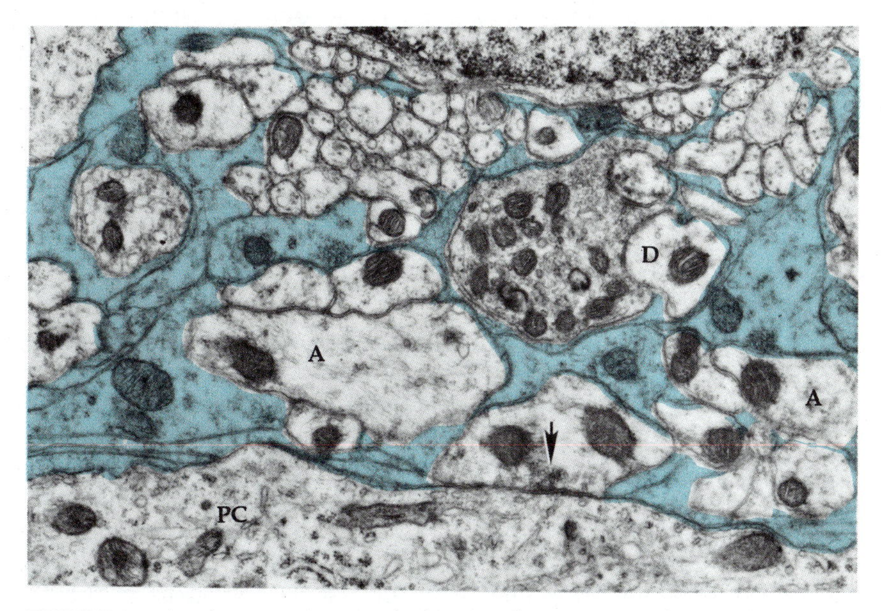

FIGURE 9. An electron micrograph showing the processes of an astrocyte intertwined with neurons and neuronal processes in the rat cerebellum. Astrocytic processes have been colored for easier identification. PC is the cell body of a Purkinje cell; at the top is a portion of a small granule cell with a narrow band of cytoplasm around its nucleus. Some axons (A) and dendrites (D) have been identified; the arrow indicates a synapse. (From S. W. Kuffler and J. G. Nicholls, 1966. Ergeb Physiol. 57: 1–9.)

ic fibrillary protein (GFAP), that is specific for astrocytes. In culture and tissue sections, antibodies to GFAP selectively stain, and thus identify, astrocytes. Their high content of filaments suggests that astrocytes provide structural support for the nervous system.

Astrocytic processes sometimes form swellings, or end feet, at their extremities. In the mammalian brain, end feet surround capillaries and were once thought to represent the principal barrier to diffusion of large molecules from the circulation into the brain (the **blood–brain barrier**). That barrier is now believed to result from the tight junctions that brain capillary endothelial cells form with each other. These very tight junctions, which are found only in the brain, prevent the diffusion of proteins and other charged molecules into the intercellular space. Only lipid-soluble molecules, and molecules such as glucose, for which the brain has specific transport mechanisms, are able to gain efficient access to the brain from the circulation.

Astrocytes are not identical in all parts of the brain but vary in both functional and morphological characteristics. The optic nerve contains two types, which can be distinguished by specific surface molecules. The two have different lineages: type-2 astrocytes are derived from a common precursor with oligodendroglia; type-1 astrocytes arise from a different precursor (Chapter 11). Whether or not this classification extends to other areas of the brain is unclear.

All astrocytes are highly permeable to potassium ions. Because the extracellular space in the brain is very small, intense electrical activity in nerves can cause changes in extracellular K^+. The increased extracellular K^+ concentration causes astrocytes to depolarize. In some cases glial depolarizations account for very slow potential changes that are recorded from the surface of nervous tissue. Some astrocytes also have high affinity uptake systems for γ-aminobutyric acid (GABA) or serotonin, and presumably help remove these neurotransmitters following their release from nerve terminals.

Oligodendroglia and Schwann cells form myelin sheaths

The rapid signaling of neurons in vertebrate nervous systems is made possible by **myelin,** a sheath of stacked membranes that enwraps segments of axons and speeds conduction of electrical signals. In the central nervous system oligodendrocytes produce this sheath by wrapping their enfolded surface membrane around axons (see Chapter 9); in the peripheral nervous system Schwann cells perform this function. Each Schwann cell forms myelin around only a single axon; oligodendrocytes can myelinate up to 50 axons. Smaller axons, which are unmyelinated, are enclosed in oligodendrocytes and Schwann cells. In both cases, a single glial cell can envelop many unmyelinated axons. Schwann cells also cap the nerve terminal at the neuromuscular junction.

Oligodendrocytes and Schwann cells can aid cell migration and axonal growth

One of the most important functions of oligodendrocytes and Schwann cells is the guidance of axons during development and regeneration. In several instances during development, the oligodendrocytes form a surface to guide axonal growth. In the developing cerebellum, for example, a special class of oligodendrocytes called **Bergmann glia** extend themselves in an alignment that is perpendicular to the cerebellar surface and form a strutwork along which granule cell neurons migrate to find their proper location within the cerebellum.

After transection of an axon in the peripheral nervous system, the distal axon degenerates. The Schwann cells remain, however, forming a cellular tube through which the regenerating axons regrow. The Schwann cells provide not only mechanical guidance but an attractive surface for neurite outgrowth. Growing neurites adhere closely to them, and they induce receptors for growth factors in regenerating axons. In the central nervous system, severed or damaged axons do not regenerate, in part because oligodendrocytes do not support regeneration; in fact, central myelin contains surface molecules that actively inhibit neurite outgrowth. Some severed central nervous system neurons are competent to regenerate axons, as shown by their growth into denervated Schwann cell sheaths that are transplanted from the periphery.

The Organization of Neurons within the Nervous System

In both vertebrates and invertebrates, neurons and glia are highly organized into elaborate, three-dimensional structures to form the nervous system. Their organization reflects both the evolutionary history of the organism and the demands of function. In all nervous systems, ranging from the simplest to the most complex, **sensory neurons** detect changes in the internal and external environments of the organism, then transmit this information to a network of **interneurons,** which organizes, processes and stores it, and finally gives instructions via **motor neurons** to the muscle and gland cells that are the effectors of the nervous system. A simple organisms like the nematode *Caenorhabditis elegans* uses 305 cells to perform these functions; the mammalian nervous system uses more than a thousand billion neurons. Mammals have more sensory and motor neurons than do nematodes, but it is the interneurons that makes their brains so large and so remarkable.

The organization of nerves in invertebrates

The most primitive organisms that have neurons are coelenterates (e.g., jellyfish or hydra). In these animals, neurons lie in a single layer between the endoderm and ectoderm, where they form synapses with each other in a two-dimensional **nerve net.** Sensory nerves innervate specialized receptors for light, touch and gravity, and the output of the nerve net regulates swimming and reflex movements of the organism.

In higher invertebrates, neuronal cell bodies are grouped into **ganglia** (Figure 10). In molluscs, such as snails, clams, octopuses, or the marine gastropod *Aplysia,* the ganglia are located in close vicinity to the internal organs or viscera. Segmented animals such as nematodes and arthropods (including insects and crustaceans) have, in addition, a **nerve cord,** a

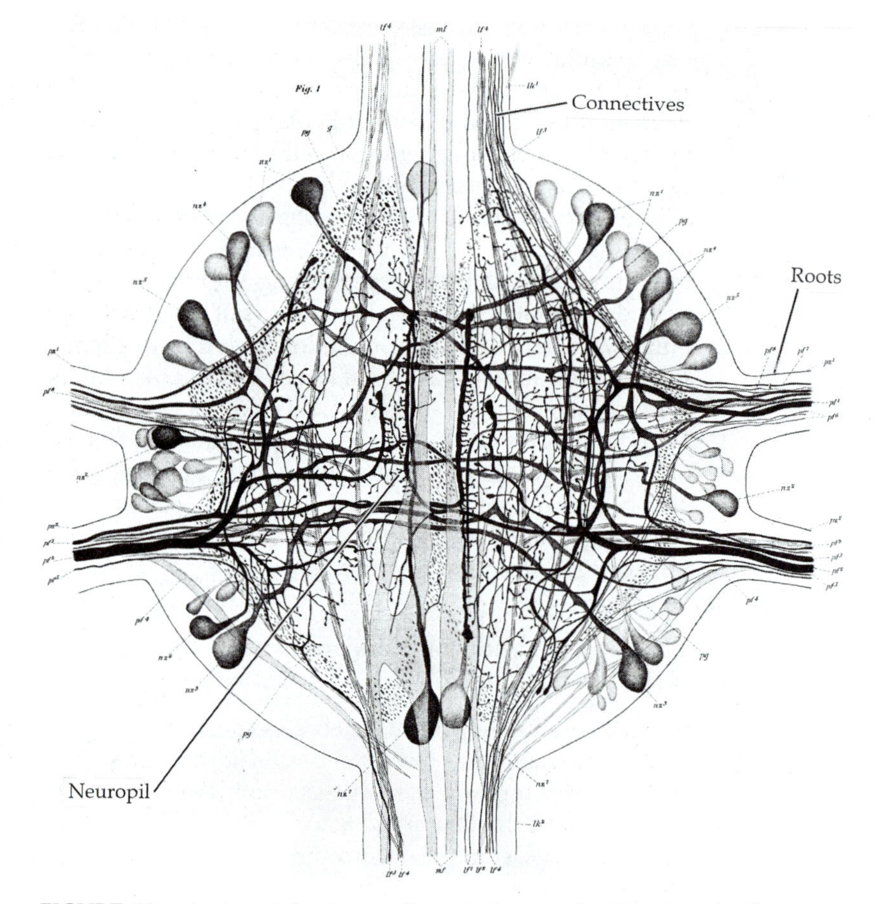

FIGURE 10. An invertebrate ganglion. A drawing by Retzius of cells in a segmental ganglion of the leech stained with methylene blue and viewed from the ventral side. Note that the cell bodies of the neurons are at the periphery of the ganglion and that synaptic connections occur in a central neuropil. In the figure the connectives are at the top and bottom of the ganglion and the roots of the ganglion leading to the periphery are at each side. (From G. Retzius, 1891. Zur Kenntniss des Centralen Nervensystems der Würmer. Biologische Untersuchungen, Neue Folge II, 1–28. Samson and Wallin, Stockholm.)

segmented chain of stereotyped ganglia that runs along the longitudinal axis of the animal. A nerve or bundle of axons goes laterally from each ganglion to innervate the muscles of the body wall and to receive sensory information from it. Some sensory neurons, such as crustacean stretch receptors, have their cell bodies in the periphery. The cell bodies of other sensory neurons are in the ganglia, however, along with those of motor neurons and interneurons.The cell bodies in the ganglion, which themselves receive no synapses, are located at its circumference and they send a single process into the internal **neuropil,** where axons and dendrites branch from it, and where synaptic connections are made. Neurons in the ganglia send axons to other segments via longitudinal **connectives.** In some cases the ganglia are bilateral and are connected at each segment by transverse bundles of axons called **commissures.** At the anterior end of the nervous system several ganglia are fused to form a head ganglion, or primitive brain, that has specialized sensory cells and is larger and more complex than the other ganglia. In higher invertebrates such as insects, the brain can be quite large and complex.

In invertebrate ganglia, each cell occupies a relatively fixed position. Because the number of cells is small, particular neurons can often be reliably identified from preparation to preparation solely by their position (Figure 11). This predictability enormously aids investigation of their properties and function. Each cell in the developing nervous system of *C. elegans,* for example, can be identified by viewing the living organism under the microscope with Normarski optics, allowing the lineage of each cell to be established (see Chapter 11). In other organisms, such as the lobster or *Aplysia,* the cell bodies are large, facilitating electrophysiological recording and direct injection of substances into the cells. The large axons of the squid have been invaluable preparations for investigating mechanisms of the action potential and of synaptic transmission. Most recently, the power of genetics has made *Drosophila* an attractive organism for investigating fundamental problems in neural function and development.

The organization of vertebrate nervous systems

Vertebrates have extremely complex nervous systems that are larger and have many more cells than those of invertebrates. Their most prominent

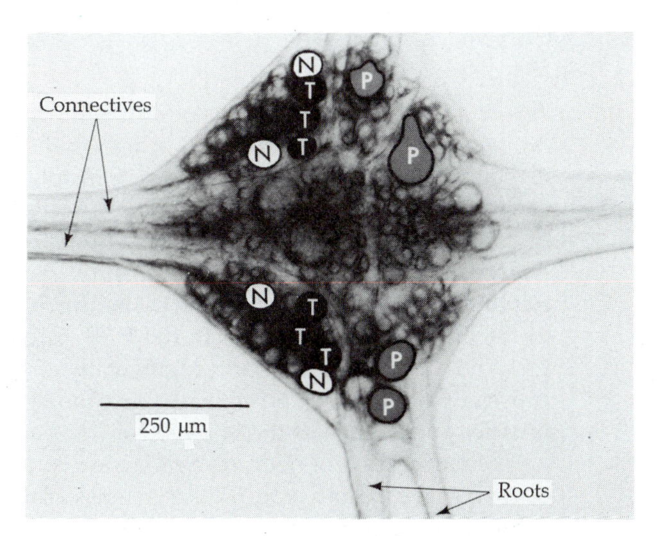

FIGURE 11. Identified cells in a segmental ganglion of the leech viewed from its ventral side. T, P, and N refer to identified sensory neurons that respond to touch, pressure, and noxious stimuli, respectively. AE refers to motor neurons that innervate muscles that raise the annuli on the leech surface into ridges. (From J. G. Nicholls and D. A. Baylor, 1968. J. Neurophysiol. 31: 740–756.)

feature is the very large brain in the head of the animal, which in adult humans reaches a weight of about 1300 g (3 lbs). In spite of their much greater size, the brains of vertebrates retain some familiar features. One is the segmental character of the nervous system. Although the **central nervous system (CNS)** of vertebrates (the brain and spinal cord), forms one large structure, nerves leave the spinal cord and brain stem to innervate the organs and muscle in a segmental fashion. Its organization into segments appears to be an important feature of the developing vertebrate nervous system (Chapter 11).

The vertebrate CNS has several parts: the **spinal cord,** which receives sensory information from, and sends motor information to, the trunk, the limbs, and many of the body organs; the **medulla** and **pons,** which contain motor and sensory neurons for the head and neck and for special senses (hearing, taste, and head movement) as well as neurons that serve as a relay station between the motor cortex and the cerebellum; the **midbrain,** where visual and auditory input are coordinated; the **cerebellum,** which coordinates motor activities; the **diencephalon,** which controls the autonomic nervous system and the pituitary and is also a relay station for sensory information on its way from the periphery to the cortex; and, finally, the **cerebral hemispheres,** which are responsible for the higher activities of the brain, including perception, most kinds of memory, higher motor functions such as speech, and cognitive function (Figure 12). The cerebral hemispheres consist of the **cerebral cortex,** a convoluted plate of cells that forms most of the surface of the brain, and the **basal ganglia** and **hippocampus,** which lie underneath the cortex. Another important part of the CNS is the **reticular formation,** a diffuse network of neurons in the **brain stem** (medulla, pons and midbrain) that are concerned with states of attention and motivation and with regulation of the autonomic nervous system.

As the description of functions of different regions implies, neurons in the brain are organized in functional groups in the brain. In the cerebellar and cerebral cortices, cells are arranged in histologically recognizable layers, each of which contains particular types of cells that make specific patterns of connections and serve specific functional roles. Elsewhere in the CNS neuronal cell bodies are often in groups called **nuclei.** The **peripheral nervous system (PNS)** consists of all neurons and nerve processes outside of the brain and spinal cord. In the PNS, cell bodies of similar types are grouped in **ganglia** that are segmentally organized. The neurons and processes of the PNS go to and from the spinal cord to innervate the muscles and organs of the body.

The segregation of neurons by function can be seen in the organization of the spinal cord and of the PNS that is associated with it. A view of a cross section of the spinal cord (Figure 13) shows it to be divided into **gray matter,** which contains the cell bodies and synapses of neurons in the spinal cord, and **white matter,** which contains the myelinated axons whose high lipid content gives the white matter its appearance. These axons

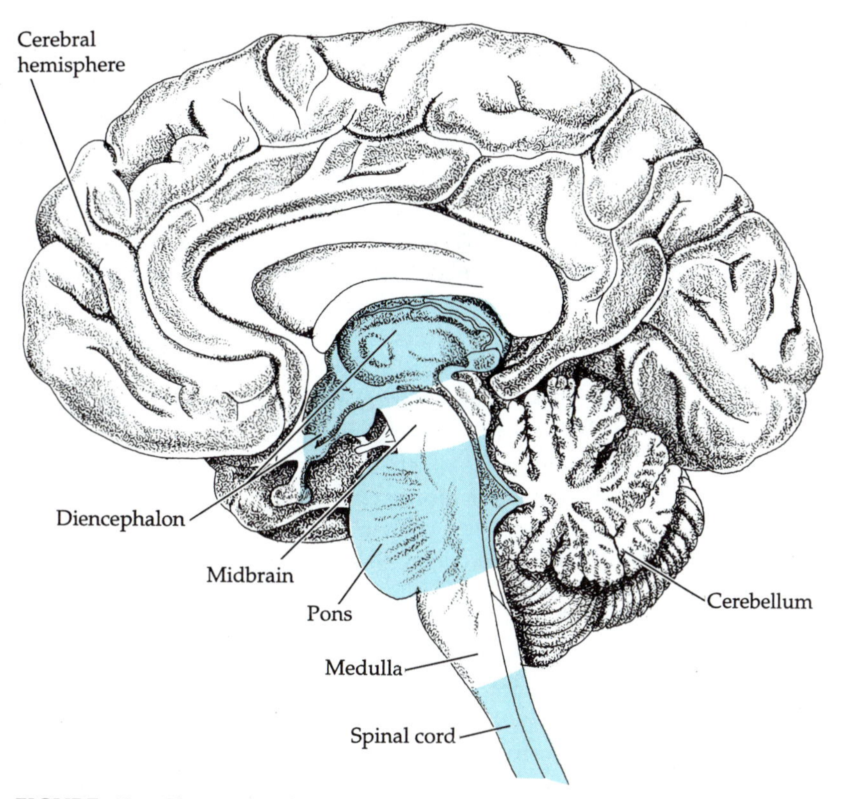

FIGURE 12. The major divisions of the vertebrate central nervous system. A midsagittal plane through the human brain.

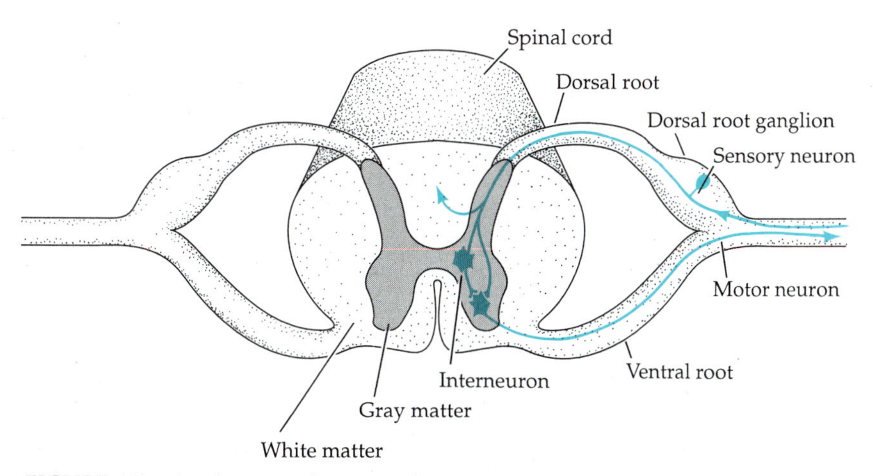

FIGURE 13. A schematic diagram of a cross section through a vertebrate spinal cord. Sensory neurons whose cell bodies are in the dorsal root ganglion send axons through the dorsal root into the spinal cord to synapse with motor neurons in the ventral horn, and with interneurons. Motor neurons send their axons out the ventral root.

ascend and descend the cord carrying information to and from the brain, and to and from other levels of the cord. Sensory axons enter, and motor axons leave the cord by largely separate routes. Motor neurons, which have their cell bodies in the **ventral** (toward the abdomen) half of the spinal cord gray matter, send their axons out to the muscles that they innervate via nerves called **ventral roots** that exit bilaterally on the ventral side of the cord. Most of the axons of sensory neurons, in contrast, enter the cord on the **dorsal** (toward the back) side by nerves called **dorsal roots.** Their cell bodies lie in **dorsal root ganglia** that are attached to the root. Ventral and dorsal roots emerge from the cord at each segment. The two roots then join to form mixed motor and sensory nerves that go to and from the skin and muscles. Some of the sensory axons that enter the spinal cord, such as those from stretch-sensitive receptors in the muscle, make excitatory synaptic contacts directly with motor neurons. These connections form the basis for **monosynaptic** reflexes. Most sensory axons that enter the cord, however, synapse with interneurons that either regulate motor neurons by more complex local circuits or ascend to the brain.

Neurons are also organized topographically

A major principle of the organization of neurons within vertebrate nervous systems is their topographical continuity. This form of organization is seen most clearly within sensory systems. In the retina, for example, the photoreceptors form a two-dimensional array that contains a coherent map of the visual world upon it. At each synaptic relay between the photoreceptors and their central connections, the topography of this two-dimensional map is preserved. Thus, stimulation of adjacent photoreceptors in the retina activates neurons that are near each other in the visual cortex. One can thereby map the visual world on a sheet of cells in the visual cortex. Equal areas of the retina do not receive the same cortical representation. Those near the center of the eye, where vision is most acute, occupy a larger area of cortex than those at the periphery. Nevertheless, the topographical relationships are maintained. Although it may be distorted, a coherent map of the visual field is sustained from one synaptic level to the next. In the cortex, the visual world is represented in over 20 separate, distorted, but coherent maps. Superimposed on the topographical maps in the cortex are other modes of organization based on the response properties of cortical cells. Most prominent are cortical columns, in which cells in different layers with different inputs and outputs nevertheless share certain response properties. In the visual cortex, for instance, **ocular dominance columns,** containing groups of cells that respond preferentially to one eye or the other, form alternating stripes or patches (Figure 14).

The auditory and somatosensory systems are also organized topographically and in columns. In the part of the cerebral cortex that receives input from the body surface (the **somatosensory system**), the body is represented in several distorted, but largely continuous, two-dimensional maps. In the part of the cortex that organizes motor behavior, neighboring neurons direct muscles that are near one another in the body (Figure 15). Although the overall map may not always be so clear as in these examples,

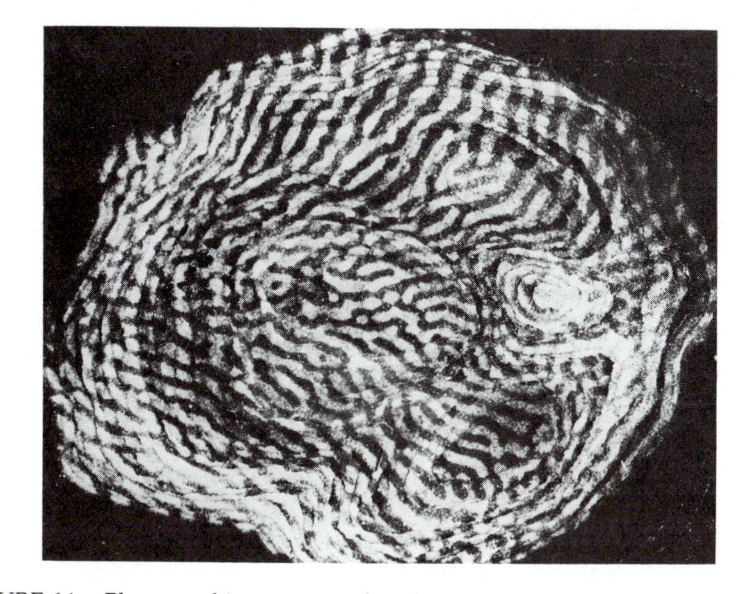

FIGURE 14. Photographic montage of ocular dominance columns in the visual cortex of the monkey. Axons and nerve terminals in the visual pathway from one eye were labeled by injecting radioactive amino acids into the eye. The label was then transported transsynaptically to the cortex. After autoradiography of the cortex, the terminals appear white. The dark bands represent unlabeled axons and terminals from the other eye. (From S. LeVay et al., 1985. J. Neurosci. 5: 486–501.)

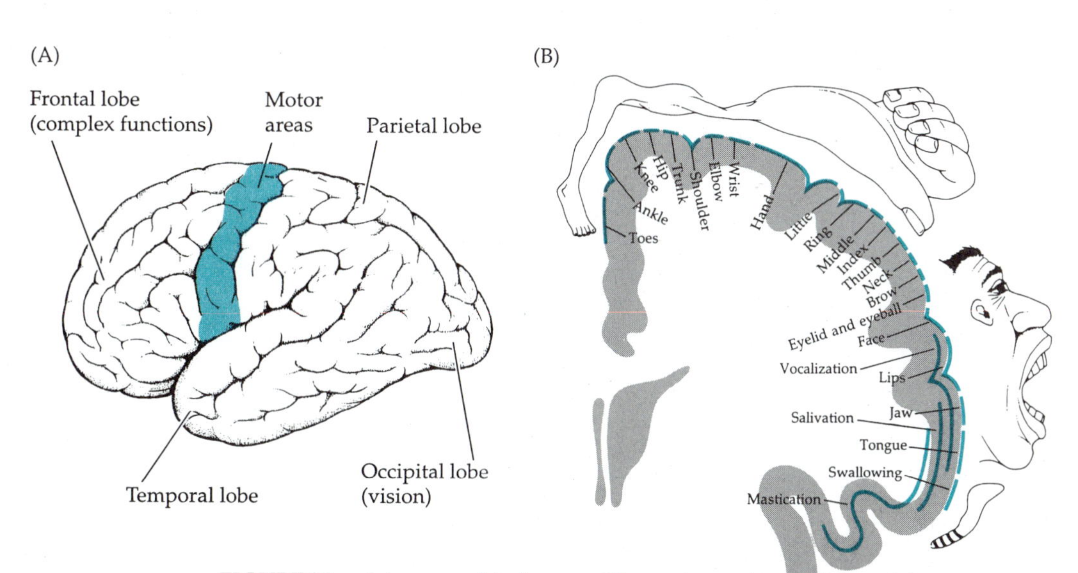

FIGURE 15. A topographical map of the motor cortex. (A) Map of the representation of the body on the human cortex. (B) Body representation in the human motor cortex. (Adapted from W. Penfield and T. Rasmussen, 1950. *The Cerebral Cortex of Man: A Clinical Study of Localization of Function.* Macmillan, New York.)

neurons in many parts of the nervous system can maintain topographical relationships both within axon bundles (or **tracts** as they are sometimes called in the CNS) and within the field of targets that they innervate.

Secretory glands, heart muscle, and smooth muscle throughout the body (in the gastrointestinal, vascular, reproductive, and respiratory systems) are all regulated by the brain via a group of nerves called the **autonomic nervous system** (Figure 16). This system has two parts, the **sympathetic** and **parasympathetic** systems, that have their origins in different parts of the spinal cord and brain stem and often innervate the same targets, where they usually have antagonistic effects. The sympathetic system is geared to stimulate the body for action ("fight or flight"), and the parasympathetic system promotes digestive and restorative functions. Thus sympathetic nerve stimulation increases heart rate and blood pressure, inhibits secretion of saliva, and decreases contraction of smooth muscle in the gastrointestinal system. Parasympathetic stimulation, in contrast, slows heart rate, relaxes vascular smooth muscle, stimulates secretion of saliva, and increases the contraction of smooth muscle in the gut.

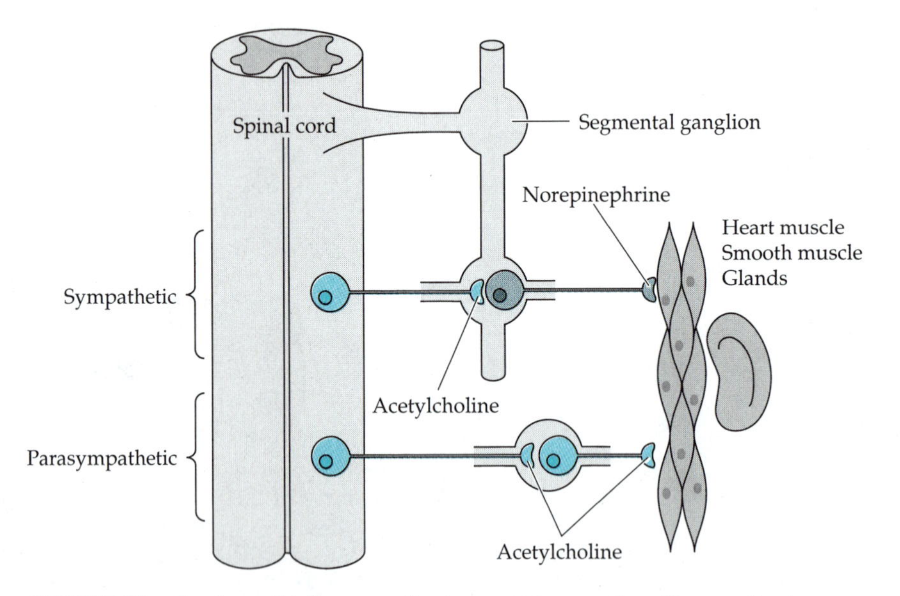

FIGURE 16. A schematic diagram of autonomic innervation. Preganglionic neurons, which arise in the spinal cord, synapse onto postganglionic neurons in the periphery. The postganglionic neurons innervate the glands and organs of the body. The synapses between sympathetic pre- and postganglionic neurons occur in a chain of ganglia near the spinal cord. The synapses between parasympathetic pre- and postganglionic neurons occur in ganglia near the innervated tissues. Acetylcholine is the neurotransmitter released by preganglionic and by parasympathetic postganglionic neurons; norepinephrine is released by sympathetic postganglionic neurons.

In both the sympathetic and parasympathetic systems a chain of two neurons links the spinal cord or brain stem to the target. In each case, **preganglionic** neurons in the spinal cord or brain stem send axons out to peripheral ganglia, where they innervate **postganglionic** neurons that synapse on the target glands or organs. Preganglionic neurons of sympathetic nerves arise from the spinal cord in the chest and abdomen and send their axons out the ventral root to terminate in a chain of segmental ganglia near the spinal cord. In contrast, parasympathetic preganglionic neurons, which arise from the brain stem and the pelvic portions of the spinal cord, synapse with postganglionic cells in ganglia that are near the target tissue. With some exceptions, each system uses a characteristic set of transmitters. Postganglionic parasympathetic cells secrete acetylcholine, along with various peptides, whereas most postganglionic sympathetic cells secrete norepinephrine. The preganglionic axons in both cases secrete a mixture of acetylcholine and peptides.

Development of the Nervous System

The diversity of cells in the nervous system and their intricate and precise interconnections present a developmental problem of daunting proportions. Tens of thousands of cell types must be generated at the right time and place, must extend their axons and dendrites, and must choose among millions of partners to make appropriate synaptic connections. Unraveling the mechanisms by which these events occur in an orderly and timely fashion is the ultimate puzzle in developmental biology.

In vertebrates the nervous system begins at the gastrula stage, when the embryo consists of the three layers: ectoderm, mesoderm, and endoderm. Triggered by chemical signals from the mesoderm, a region of the ectoderm begins to thicken and indent, forming a neural groove on the surface of the embryo along what will become its **rostral–caudal** (head-to-tail) axis. The lips of the neural groove continue to swell, then meet and fuse to form a **neural tube** (Figure 17). This neural tube is the future brain and spinal cord, and its cavity becomes the ventricular system of the brain. The epithelial cells that line the cavity of the neural tube are the precursors of the neurons and glia in the adult brain and spinal cord. Almost all cell division of these precursors, or **neuroblasts,** takes place in the **germinal zone** bordering the ventricular cavity. As the neuroblasts continue to divide, their postmitotic progeny migrate out from the ventricular zone to assume their positions around the ventricles. Cell proliferation is nonuniform along the rostral–caudal axis, giving shape to several swellings that will bend and fold during further development to give the complex structure of the adult brain. In some parts of the brain, such as the cortex, layers are built up by successive migrations of cells from the ventricular zone. In the cortex, those cells that are born first occupy the innermost layers, whereas those that are born last occupy the most superficial layers. Cells differentiate after they have attained their final locations.

Neurons and Schwann cells in the peripheral nervous system arise from the neural crest

During the formation of the neural tube, a group of cells at its dorsal margin separates from it on either side to form bands called the **neural crest** (Figure 17). All of the neurons whose cell bodies lie in the peripheral nervous system, including sensory and postganglionic autonomic neurons, arise from these cells. In addition the neural tube cells are the precursors for Schwann cells, the chromaffin cells of the adrenal medulla, and pigment-containing cells called melanocytes. Precursor cells migrate out from the neural crest to their many locations in the embryo. After they have reached their destination, they continue to divide. Their postmitotic progeny then differentiate appropriately.

Axonal growth in development and regeneration: cells extend axons to contact their targets

One of the earliest steps of differentiation is the extension and growth of axons seeking their targets. Axons often must traverse long distances though a varied terrain to reach their final destinations. **Growth cones** at their growing tips appear to guide them by responding to environmental cues on other cells and in the extracellular matrix. Some of these cues stimulate axonal growth; others inhibit it. Such molecular signals com-

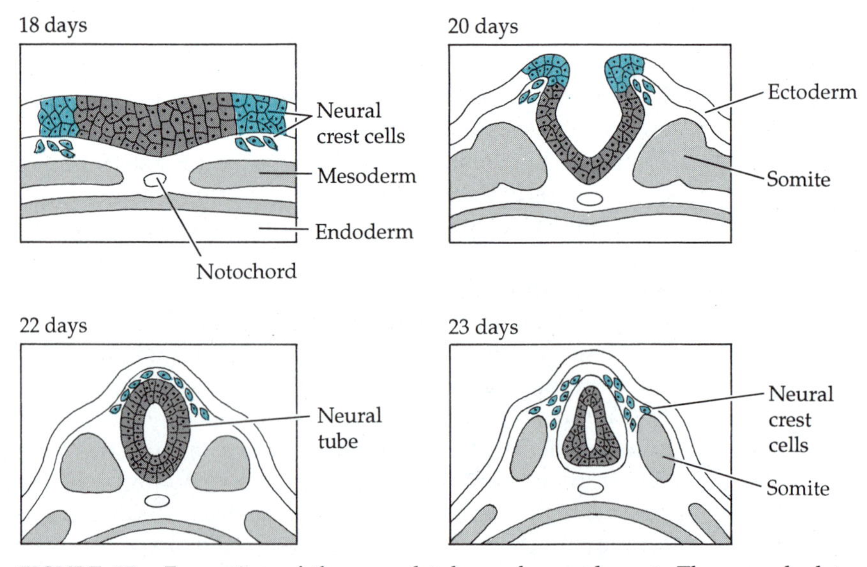

FIGURE 17. Formation of the neural tube and neural crest. The neural plate forms on the dorsal side of the embryo and folds to form a groove that becomes a tube. The neural crest cells (colored) pinch off from the tube, migrate to their final destinations, and develop into sensory ganglion cells, postganglionic autonomic neurons and peripheral glial cells.

municate the positional information necessary for construction of the wiring pattern of the nervous system.

When peripheral nerves are damaged, they regenerate, regrowing their axons and following environmental cues to reestablish connections with their targets. Regeneration is most accurate when damage to the surrounding tissue is minimal and the distance over which the axons must regrow is small. In the CNS, damaged axons do not regenerate, and physically disrupted connections are not reestablished.

Neurons depend upon one another for factors affecting survival, differentiation, and efficiency of signaling

The final structure of the nervous system is shaped not only by the generation of neurons and the establishment of appropriate connections between them, but also by neuronal death. More neurons arise and send out axons than finally appear in the adult nervous system. Neuronal death in part results from a competition between cells for growth, or **trophic,** factors secreted by their targets. Such factors are required for the survival, growth, and differentiation of neurons. These factors are but one example of the many signals that are exchanged between neurons and their targets during synapse formation and in the adult animal. In some cases, the exchange of signals depends upon the activity of the postsynaptic cell. At many synapses, the strength and survival of the synaptic connection depends upon its relative ability to cause an action potential. Target cells can control the chemical and morphological phenotype of the neurons that innervate them and perhaps can even choose which synapses remain on their surfaces; in turn, presynaptic neurons regulate the expression of postsynaptic specializations, the number and type of receptors for neurotransmitters, and their sites of localization in their targets.

Such exchanges are not limited to embryonic life but continue throughout adulthood. Indeed, the influence that neurons have on their synaptic partners forms the basis of the plasticity of the nervous system and of our ability to learn and remember. By studying the molecules that mediate these interactions and discovering how they exert their effects, we may make the first steps toward understanding those properties of the nervous system that seem most remarkable and are of most fascination to us.

Patterns of activity influence the organization of the nervous system

Many of the important elements that determine how the nervous system develops and functions cannot be understood by considering single neurons or pairs of neurons but depend upon the properties of populations of neurons. Thus, in many contexts, the level of activity of a single neuron within a population of apparently identical cells is only meaningful if considered within the context of the spatial and temporal pattern of activity of the entire population. These patterns of activity shape the connections and the functional properties of the nervous system in both the developing and adult animal. The refinement of topographical maps to their adult state of precision, for example, requires patterned neural activity.

In the visual system, the organization of the cortex into ocular dominance columns during development arises because activity in different retinal neurons in one eye is more highly correlated than activity in neurons in the two eyes. The interactions are competitive in that blocking vision in one eye during a developmentally critical period leads to expansion of the representation of the opposite eye in the cortex. Thus, axons in the visual pathway from the two eyes compete to innervate cortical cells. The final arrangement of the cortex into alternating columns of cells dominated by input from one eye or the other, however, cannot be predicted from the behavior of individual cells, but depends instead upon the relative timing and retinal position of activity within the two retinal populations. Competitive interactions are not limited to the developing nervous system. The representation of the body surface on the somatosensory cortex of adult mammals, for instance, is highly dynamic; throughout our lives, it is continually being reshaped in response to the relative activity of sensory nerves from different parts of the body surface.

The nervous system thus operates at molecular, cellular, and integrative levels. Understanding the behavior of individual neurons in terms of their molecules is fascinating in its own right but is also an important piece in the larger puzzle of understanding the behavior of the nervous system. In the end, we hope that the human aspiration of understanding our own nervous systems will give us insight into our behavior both as individuals and as members of the society of human beings.

Summary

Nervous systems are extraordinary for the number of different kinds of cells that they contain and for the precision of their wiring. The two principal cell types are neurons and glia. Neurons receive, process, and transmit information, communicating with each other and with the external and internal environments. They are anatomically specialized with highly branched dendrites to receive signals and process them, and with long axons ending in nerve terminals that transmit signals to other cells. The organelles, cell surface, and cytoskeletal elements of neurons are similar to those of other cells but are adapted to accommodate and produce the extreme dimensions and regional specialization of neurons. Glia are the support cells of the nervous system, ensheathing the different parts of neurons and lining the surfaces of the brain.

Neurons generate both electrical and chemical signals. The electrical signals integrate information in the cell and convey it over long distances. The currents from small local potentials in the dendrites and cell bodies combine in space and time to give integration; large, self-propagating action potentials rapidly transmit signals along the axon to the nerve terminal. Both types of potentials are mediated by combinations of highly regulated and specific ion channels that control the passive movement of ions across the cell membrane. Neurons communicate with each other mostly at chemical synapses, where the nerve terminal of one cell ends on the dendrites or cell body of another. Incoming action potentials trigger the secretion of synaptic messengers that bind specific receptors in the postsynaptic membrane. The interaction of the messengers with their receptors initiates electrical signals in the postsynaptic neuron and, via second messengers, changes metabolic pathways, neuronal structure, and gene expression.

In vertebrates, the nervous system has an elaborate three-dimensional structure in which neurons are grouped according to function and their targets of innervation. Within these groups, neurons are often arranged according to an orderly topographical map that is propagated from synaptic relay to synaptic relay and ultimately represents a sensory field, or the positions of muscles reflected onto the body surface.

The nervous system develops by proliferation of neuronal precursors, differentiation, migration, and, in some cases, neuronal death. Neurons extend axonal processes with growth cones at their tips to make synapses with other cells. The correct formation of synaptic connections occurs through guidance of the growing neurite along cell surface and extracellular pathways, specific recognition between presynaptic and postsynaptic cells, and selective stabilization of synapses by spatial and temporal patterns of activity.

Neurons do not exist in isolation but as part of a larger community of interacting cells whose properties transcend those of the individual. Understanding how individual cells function at the molecular level is the first step to understanding the behavior of this community.

References

Alberts, B., Bray, D., Lewis, J., Raff, M., Roberts, K. and Watson, J. D. 1989. *Molecular Biology of the Cell*, 2nd Ed. Garland, New York.

Hammond, C. and Tritsch, D. 1990. *Neurobiologie Cellulaire.* Doin Editeurs, Paris.

Kandel, E. R., Schwartz, J. H. and Jessell, T. M. 1991. *Principles of Neural Science*, 3rd Ed. Elsevier, New York.

Nicholls, J. G., Martin, A. R. and Wallace, B. G. 1992. *From Neuron to Brain*, 3rd Ed. Sinauer Associates, Sunderland, MA.

Purves, D. and Lichtman, J. W. 1985. *Principles of Neural Development.* Sinauer Associates, Sunderland, MA.

Part One

Signaling in the Nervous System

2

Electrical Signaling

Peter B. Sargent

NEURONS GENERATE a rich variety of electrical signals: large (100 millivolts) and small (100 microvolts); long (hours) and short (milliseconds); depolarizing and hyperpolarizing; with both simple and complex wave forms (Figure 1). Some cells generate periodic electrical signals autonomously; others are silent unless stimulated. All of these signals arise from the action of various combinations of **ion channels,** proteins that form aqueous pores through which ions traverse the membrane (Chapter 3). The opening and closing of the channels is highly regulated; different channels have different ionic specificities, different modes of regulation, and different kinetics of opening and closing.

Ion concentration gradients across the membrane provide energy for electrical signals. When ion channels are open, ions move through the channels down their electrochemical gradients. Their net movement across the membrane constitutes a current that changes the membrane potential and generates an electrical signal. The density, specificity, mode of regulation, and kinetics of ion channels in the membrane determine its electrical properties and the electrical signals that it generates.

In this chapter, we shall first discuss how ion channels influence membrane potential. We shall then see how channels are regulated by voltage and by neurotransmitters, and how different combinations of channels generate the basic electrical signals of the nervous system.

Ion Permeability and the Membrane Potential

The membrane potential of a cell is determined by the relative permeabilities of its membrane to specific ions and by the transmembrane concentration gradients of those ions. To understand precisely how these factors influence membrane potential, we consider the behavior of an imaginary cell whose ionic concentration gradients we specify and whose channels we open and close at will.

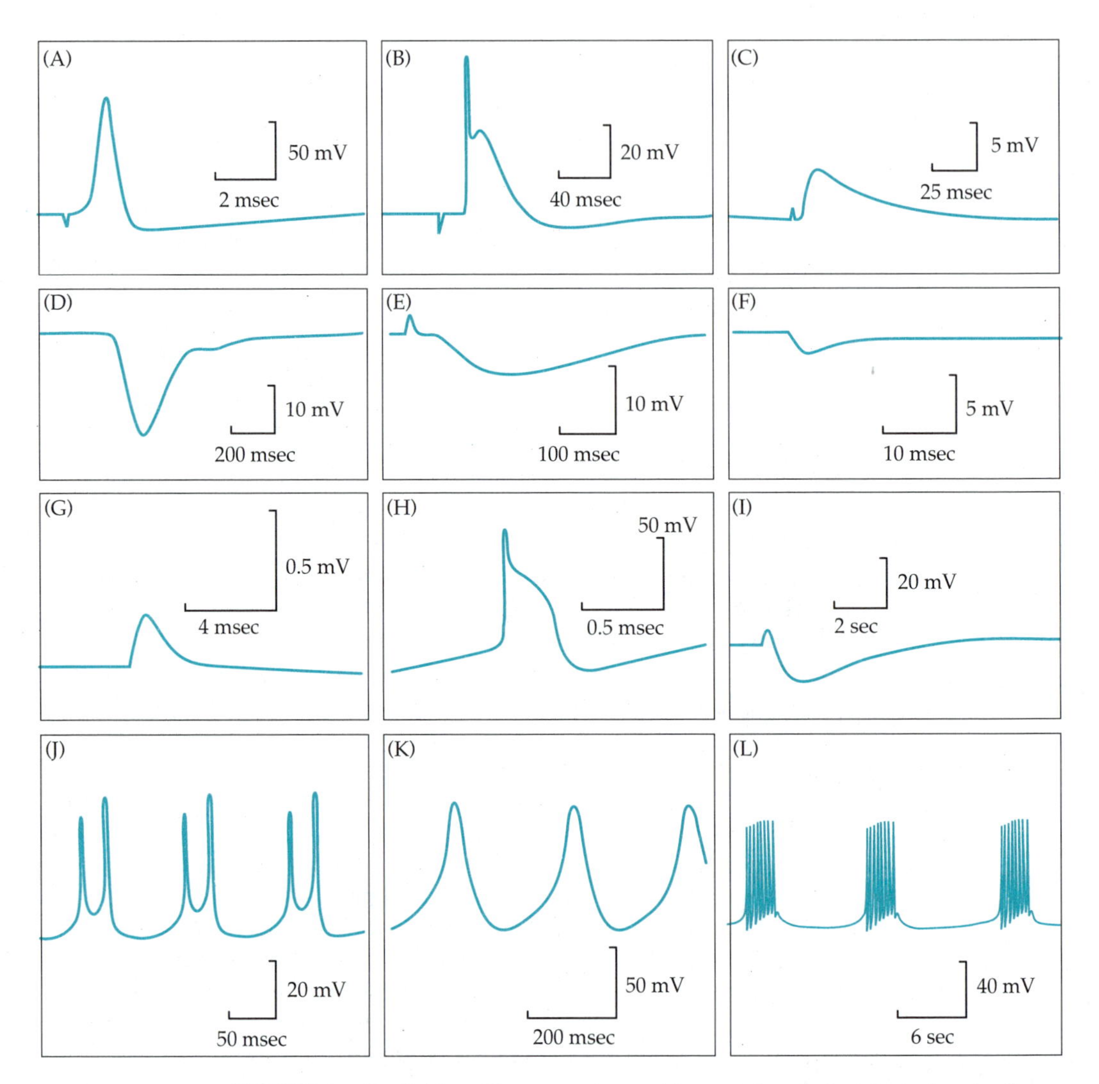

FIGURE 1. Excitable cells generate a wide variety of electrical signals. Some signals are large (e.g., A, B, H), while others are small (e.g., G). Some are depolarizing (C), while others are hyperpolarizing (D). Some last seconds (I), while others last only milliseconds (A). Some have complex waveforms (E, H), while others are relatively simple (F). Finally, some signals are periodic (J, K, L).

The concentration gradient of an ion determines its equilibrium potential

We begin with an imaginary cell that has 100 m*M* KCl on both sides of its surface membrane, is impermeable to both K^+ and Cl^-, and has no potential difference across its membrane (Figure 2A). We now insert into the membrane a few channels that are selectively permeable to K^+. K^+ flows randomly back and forth across the membrane through the channels, but because there is no electrical or chemical gradient, the rate of flow in both directions is the same, and no net flow of K^+ ensues. The system is thus in equilibrium.

If we now change the concentration of KCl outside the cell to 10 m*M*, ignoring the osmotic consequences, a chemical diffusion force on K^+ results, causing a net outward movement of K^+ through the open channels (Figure 2B). Because the channels are not permeable to Cl^-, K^+ moves out alone, thus transferring electrical charge from the inside of the cell to the outside. The resulting separation of charge creates an electrical potential across the membrane, outside positive with respect to the inside, which opposes the outward movement of K^+. (By convention, all membrane potentials are referred to the extracellular fluid as zero potential. Thus, if the inside of the cell is 10 mV more negative than the outside, the membrane potential is said to be –10 mV).

There are now two forces acting on K^+—a chemical diffusion force driving it out of the cell and an electrical force driving it in. As more and more K^+ moves out of the cell, the electrical force increases until, at some value of membrane potential, the electrical force exactly counterbalances the chemical force (Figure 2B). At that time, the inward flow of K^+ equals the outward flow, and there is no net movement of K^+ across the membrane. The membrane potential at which this occurs, called the **equilibrium potential,** depends on the concentration gradient of K^+ across the membrane, and is given by the **Nernst equation:**

$$E_K = \left[\frac{RT}{zF}\right] \ln \left[\frac{[K^+]_o}{[K^+]_i}\right] \tag{1}$$

where E_K is the equilibrium or Nernst potential for K^+, R is the gas constant [8.31 joules/(mole. K)], T is the temperature in degrees Kelvin, z is the valence for K^+ (+1), and F is the faraday (96,500 coulombs/mole). At room temperature (297 °K) this expression simplifies to

$$E_K = (58 \text{ mV}) \log_{10}\left[\frac{[K^+]_o}{[K^+]_i}\right] \tag{2}$$

If the concentration gradient is 10 to 1, inside to outside, as in our example, the equilibrium potential for K^+ is –58 mV.

Notice, in our hypothetical example, that the equilibrium potential for K^+ is unaffected by the presence of other ions. If we add 100 m*M* NaCl to the outside of the imaginary cell and 10 m*M* NaCl to the inside (Figure 2C), there will be a concentration gradient for Na^+. However, this gradient has no effect on the potassium equilibrium potential, and, because the membrane is not permeable to Na^+, it has no effect upon membrane potential.

A cell that is permeable to K^+ resists displacement of the membrane potential from E_K. If we add a small amount of Na^+ to the inside of the cell (Figure 3), thus depolarizing it, K^+ leaves the cell until the membrane is restored to E_K. K^+ leaves the cell because the addition of Na^+ reduces the membrane potential and the electrical force on K^+, making it smaller than the chemical force. The electrical and chemical forces acting on K^+ are no

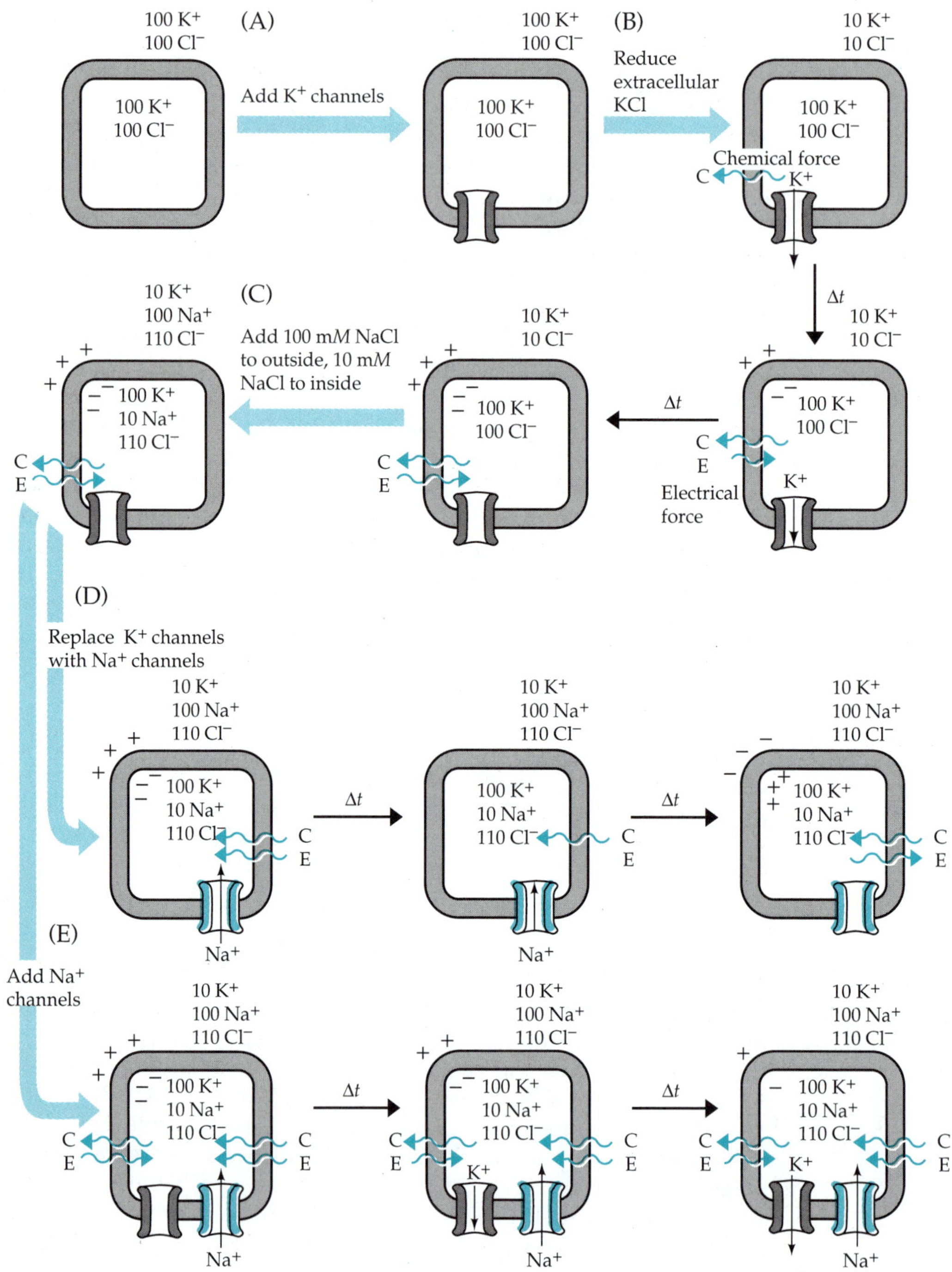

(A)
100 K+
100 Cl−
Add K+ channels
(B)
Reduce extracellular KCl
10 K+
10 Cl−
Chemical force
C
K+
Δt
Electrical force
C
E
(C)
Add 100 mM NaCl to outside, 10 mM NaCl to inside
10 K+
100 Na+
110 Cl−
100 K+
10 Na+
110 Cl−
(D)
Replace K+ channels with Na+ channels
Na+
(E)
Add Na+ channels

FIGURE 2. The electrical potential across the plasma membrane is determined by its permeability and by the concentration gradients of the permeant ions. In this series of schematic drawings of an imaginary cell, both permeability and concentration gradients are changed in order to note their effect upon transmembrane current flow and membrane potential. Large arrows indicate a change in conditions, while small arrows marked by "Δt" indicate time-dependent changes induced by a change in conditions. Osmotic considerations are ignored.

Both intracellular and extracellular ionic constituents are given for each cell. The polarity and magnitude of the membrane potential is indicated in the upper left corner of the cell, and the chemical (C) and electrical (E) gradients are indicated as vectors in the lower half of the cell (K^+ to the left and Na^+ to the right). Potassium- and sodium-selective channels, when present, are shown along the bottom of the cell, and the net current flowing through them is shown in color.

(A) Addition of potassium channels to the membrane of a cell having 100 m*M* KCl on each side has no effect, since there are neither chemical nor electrical gradients acting on K^+.

(B) Reducing the extracellular KCl concentration from 100 m*M* to 10 m*M* creates a concentration gradient of K^+, which leaves the cell until an electrical potential is established across the membrane that matches the chemical gradient. The existence of a chemical gradient of Cl^- has no effect, since the membrane is not permeable to Cl^-.

(C) Addition of 100 m*M* NaCl to the outside and 10 m*M* NaCl to the inside of the cell creates a chemical gradient for Na^+, but Na^+ does not enter the cell because the membrane is not permeable to Na^+.

(D) Replacing the potassium channels with sodium channels results in the inward movement of Na^+, driven both by chemical and electrical gradients. Na^+ continues to enter the cell until the electrical gradient is reversed and becomes equal in magnitude to the chemical gradient, at which point the membrane potential is at equilibrium for Na^+.

(E) Adding sodium channels to a membrane having potassium channels leads to the inward movement of Na^+. As Na^+ enters and depolarizes the cell, a driving force is created for K^+, causing K^+ to leave the cell. Eventually, the inward sodium current is matched by an outward potassium current, and the membrane is at steady state. The membrane potential at steady state lies between E_K and E_{Na} as determined by the ratio of potassium and sodium conductances (Equation 8).

longer equal, and there is a net driving force (or, more correctly, driving potential) on K^+. The outward movement of K^+ continues until the two forces are balanced, and the cell is again at E_K. Whenever the membrane potential is not at E_K, K^+ will always move in the direction to bring it there.

How would the cell in Figure 2C behave if it were selectively permeable to Na^+ instead of K^+? Na^+ is more concentrated outside the cell than inside. Na^+ thus flows into the cell through the open channels, making the inside positive with respect to the outside (Figure 2D). The inward sodium current continues until the membrane potential reaches the equilibrium potential for sodium, E_{Na}, given by the Nernst equation:

$$E_{Na} = (58 \text{ mV}) \log_{10} \left[\frac{[Na^+]_o}{[Na^+]_i} \right] \tag{3}$$

If the gradient is 10-fold, the sodium equilibrium potential, and hence the membrane potential of the cell, will equal +58 mV.

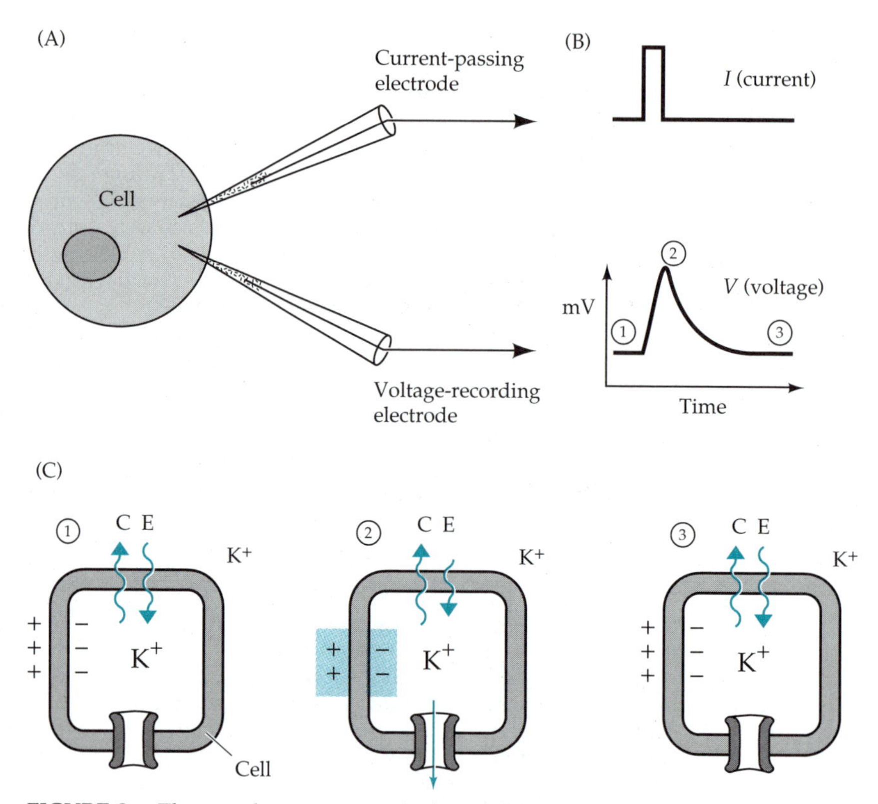

FIGURE 3. The membrane potential of a model cell whose membrane is permeable only to K⁺ returns to E_K when displaced. (A) Two micropipettes are inserted into a cell, one to pass current and one to record the membrane potential. (B) The membrane potential depolarizes during injection of current but returns exponentially to rest following cessation of the current. (C) Before displacement (left), the cell is at equilibrium, with the chemical (C) and electrical (E) gradients balanced (see arrows). Immediately after displacement (center), the electrical gradient is reduced, and there is a net driving force on K^+, which leaves the cell and hyperpolarizes it. K^+ continues to leave until the electrical gradient is once again as large as the chemical gradient, at which point the cell is at equilibrium.

The transfer of very few ions across the membrane changes membrane potential

In our examples ions move back and forth across the membrane until an electric potential is established that balances the diffusion force generated by the ionic concentration gradient across the membrane. Do these movements change the ionic concentration gradients? The relationship between the amount of charge separated by the membrane and the electrical potential across it is given by the electrical **capacitance** of the membrane. The membrane, which consists mainly of a thin, poorly conducting lipid bilayer separating two conducting solutions, behaves like an electrical capacitor. The capacitance of all biological membranes arises mostly from the lipid bilayer structure and has a value of approximately 1 $\mu F/cm^2$. Using the relation

$$Q = CV \quad (4)$$

where Q is the charge in coulombs, C is the capacitance in farads (a constant), and V is the electrical potential in volts, we can calculate how many ions must move across the membrane of a spherical cell 50 μm in diameter to change the membrane potential by 100 mV. The surface area of the membrane is 8×10^{-5} cm^2, and its capacitance is thus 8×10^{-11} F; therefore, 8×10^{-12} coulombs of charge must move across the membrane to establish a potential difference of 0.1 V. Using the value of 96,500 coulombs of charge per mole of univalent ions, one can calculate that this amount of charge is equivalent to approximately 8×10^{-17} moles of potassium ions. If the cell has an internal concentration of 100 m*M* KCl, as in our example, it contains 7×10^{-12} moles of K^+, or about 100,000 times the amount needed to establish a potential difference of 100 mV. Thus, the number of ions of K^+ required for the membrane potential to reach the E_K is much too small to reduce measurably the ion concentration gradient. The same is true in real cells. Electrical signals in the nervous system are created by the movement of very few ions. Over the short term, the effects of these movements on the transmembrane ion concentration gradients are negligible.

In cells that are permeable to two ions the membrane potential lies between the two equilibrium potentials

What are the consequences of making our imaginary cell permeable to both Na^+ and K^+? Suppose we start with a cell having equal and opposite concentration gradients for Na^+ and K^+ as illustrated in Figure 2C. When only potassium channels are present, the membrane potential of the cell will be at E_K, which in this case is –58 mV. If we now insert sodium channels into the membrane, Na^+ enters the cell (Figure 2E). If the membrane potential is at E_K, Na^+ is far from equilibrium, and both chemical and electrical forces drive Na^+ into the cell. As Na^+ enters the cell, potential across the membrane is reduced, making the inside less positive. The decrease in membrane potential means that K^+ is no longer at equilibrium, and K^+ will begin to leave the cell. Thus there are now two opposing currents—Na^+ current flowing into the cell and potassium current flowing out. As long as the sodium current moving into the cell is greater than the potassium current moving out, the membrane potential will continue to move toward zero. Note that the potassium current will continue to increase as the membrane potential moves farther from E_K and that the sodium current will continue to decrease as the membrane potential approaches E_{Na}. At some value of membrane potential, the sodium and potassium currents will be equal and opposite, at which point the system is at steady state. Neither ion is at equilibrium, but the two currents balance each other. The membrane potential is stable: any change in potential (caused by injecting ions into the cell, for instance) puts the sodium and potassium currents out of balance. The resulting net current will restore the membrane potential to its original value.

What determines the potential at which the system comes to rest? To address this question, electrical terms are useful. We express the currents carried by each ion in terms of Ohm's Law, which states that the current across a resistor is equal to the voltage across it divided by its resistance, i.e., $I = V/R$. In the case of membranes we approximate the permeability of the membrane to a given ion by its **conductance,** which is the reciprocal of resistance ($G = 1/R$). The electrochemical **driving force** or driving potential on an ion (i.e., how far it is from equilibrium) is expressed in terms of voltage as the difference between the membrane potential, V_m, and the equilibrium potential for the ion. Thus for K^+, Ohm's law is expressed as:

$$i_K = g_K(V_m - E_K) \tag{5}$$

and for sodium current,

$$i_{Na} = g_{Na}(V_m - E_{Na}) \tag{6}$$

The application of Ohm's law to channels implies that there is a linear relationship between current and driving force for an open channel. This is often not strictly true but is a good approximation for many channels. At steady-state, when the membrane potential is constant, the sodium and potassium currents are equal and opposite. Thus,

$$i_{Na} + i_K = 0 \tag{7}$$

Substituting Equations 5 and 6 into Equation 7 and solving for the membrane potential yields

$$V_m = E_{Na}\left[\frac{g_{Na}}{g_{Na} + g_K}\right] + E_K\left[\frac{g_K}{g_{Na} + g_K}\right] \tag{8}$$

Thus the steady state potential depends upon the relative conductances of the cell to Na^+ and K^+, and on the equilibrium potentials, E_{Na} and E_K, which are determined by the respective concentration gradients of Na^+ and K^+. Since the concentration gradients do not change (in real cells they are maintained by pumps), the membrane potential can be set anywhere between E_K and E_{Na} simply by changing the ratio of sodium to potassium conductances. If the sodium conductance is much greater than the potassium conductance, the membrane potential is close to E_{Na}. Conversely, if the potassium conductance is much greater than the sodium conductance, the membrane potential is close to E_K. If the sodium and potassium conductances are equal, the membrane potential is equidistant between the equilibrium potentials for Na^+ and K^+.

The main points of our discussion so far are: (1) that currents flow through ion-selective channels to move the membrane potential toward the equilibrium potential for that ion; and (2) when several selective ion channels are open, the summed currents through these channels drive the membrane potential to a value that is determined by the relative permeabilities or conductances of the membranes to the ions and by their equilibrium potentials. These principles are useful even when ion channels open so briefly that the membrane potential does not have time to stabilize at a new value. For example, if a cell is permeable only to K^+, and sodium channels are opened briefly, the current that flows through the sodium channels will cause the membrane potential to move transiently toward E_{Na}.

As we shall see, these principles can also be applied to channels that are permeable to more than one ion. Consider a cell whose ionic concentration gradients are the same as those in Figure 2C and that at rest is permeable only to K^+. The membrane potential is –58 mV, the equilibrium potential for K^+. If we now open a population of channels that are equally permeable to Na^+ and K^+, we calculate from Equation 8 that membrane potential is driven toward a value that is halfway between E_K and E_{Na}. In other words, opening these channels is equivalent to opening equal numbers of separate sodium and potassium channels whose conductances are equal. If the relative permeabilities of the channels to Na^+ and K^+ is 4:1, the membrane potential is driven toward a potential (given by Equation 8) that is closer to E_{Na} than to E_K.

Finally, in situations in which several channels are open, the calculations become more complex, but the principles remain the same. The magnitude of the current through each type of channel is equal to the driving force on that ion multiplied by its conductance (the product of single-channel conductance and number of channels). The net effect on membrane potential is determined by the summed currents. When the inward current is greater than the outward current, the membrane becomes more **depolarized** (less inside-negative), and when the outward current is greater than the inward current, the membrane is **hyperpolarized** (more inside-negative).

Resting Potential

In most cells, including glia, there is a stable difference in electrical potential across the cell membrane. In neurons, electrical signals are superimposed on this **resting potential**. Like the signals, the resting potential arises from the selective permeability of the surface membrane and the ionic concentration gradients across it.

None of the major ionic species have the same concentration on the two sides of the membrane. Table 1 gives experimentally determined values for

TABLE 1. Approximate Concentrations of the Major Ionic Constituents Inside and Outside Cells (in m*M*)

	Frog muscle fiber			Squid giant axon		
Ion	**Intracellular**	**Extracellular**	**E_{ion}**	**Intracellular**	**Extracellular**	**E_{ion}**
K^+	140	2.5	−102 mV	400	10	−93 mV
Na^+	10	120	+63 mV	50	460	+56 mV
Cl^-	1.5	77.5	−99 mV	100	540	−42 mV
Organic anions	86	40	—	360	—	—
Membrane potential			−90 mV			−70 mV

Sources: A. Hodgkin, 1965. *The Conduction of the Nervous Impulse,* Liverpool University Press, Liverpool; and B. Katz, 1966. *Nerve, Muscle, and Synapse,* McGraw-Hill, New York.

the major ions in the cytosol and the extracellular fluid of two large, excitable cells in which these measurements are easily made, frog muscle fibers and the squid giant axon. Because squids are marine animals and frogs are fresh water species, the absolute values of extracellular and intracellular concentrations of the ions are different for the two species, but the ratios are approximately the same. In both cases, K^+ is more concentrated inside than outside, and Na^+ is more concentrated outside than in. The chloride concentration is also higher in the extracellular fluid than in the cytosol. The high concentration of organic anions in the cytosol are counter ions for intracellular K^+, satisfying the requirement that the concentration of positive and negative ions on each side of the membrane be the same. (Remember that on the scale of biochemical concentrations, the charge imbalance that underlies the membrane potential is insignificantly small.) In mammals, the intracellular anions arise largely from acidic metabolic intermediates, some of which are phosphorylated. Negatively charged proteins also contribute. The membrane is impermeable to the organic anions, which in turn influence the distribution of the other ions, as described below.

The asymmetric distribution of Na^+ and K^+ across the membrane arises from two sources: the action of the Na^+/K^+ transporter, an ion pump that uses energy derived from the hydrolysis of ATP to pump Na^+ out of the cell and K^+ in; and the presence of impermeant anions inside the cell. The chloride gradient results largely from its passive distribution across the membrane; in some cells, Cl^- is actively transported into or out of the cell, thus reducing or increasing the gradient that results from its passive distribution.

Resting neurons are more permeable to K^+ than to Na^+

As we saw in the previous section, the membrane potential is determined by the relative permeabilities of the surface membrane to different ions, and is closest to the equilibrium potential for those ions to which the membrane is most permeable. Both frog muscle fibers and squid giant axons have resting potentials that are close to E_K (Table 1). The finding that

Box A Intracellular Recording

The development of new techniques has always been a major impetus for progress in biology. Electrophysiology offers a striking example of this truism; over the last fifty years, three major technical innovations have propelled advances in the field: intracellular recording, voltage clamp techniques, and patch clamp recording. This box and Boxes C and D describe these techniques.

During the last century and the early part of this century, researchers used extracellular electrodes to investigate the electrical properties of excitable cells. Their recordings, made on whole nerves or muscles rather than on single fibers, gave estimates of membrane potential that were smaller than the true values due to the shunting effect of the extracellular medium. The rediscovery of the squid giant axon by J. Z. Young in 1936 soon led to the first intracellular recording, accomplished almost simultaneously by Kenneth Cole and Howard Curtis at the Woods Hole Marine Biological Laboratory in Massachusetts and by Alan Hodgkin and Andrew Huxley at the Plymouth Marine Station in England. Both groups threaded a fine wire electrode down the length of an isolated segment of squid axon and measured the potential between this electrode and an electrode in the extracellular fluid both at rest and during an action potential.

Their results yielded an immediate surprise. They found that during the peak of the action potential the membrane potential did not simply go to zero, as had been postulated by Julius Bernstein in 1902, but that it actually reversed to become briefly inside positive. An overshooting action potential cannot be explained by the breakdown of selective permeability, as Bernstein had proposed, and the reversal of the membrane potential led to the hypothesis that the membrane becomes selectively and transiently permeable to Na^+ during the action potential. Whether other cells also had overshooting action potentials, however, could not be determined until new techniques allowed intracellular recording from smaller cells.

In the late 1940s, Gilbert Ling, Judith Graham, and Ralph Gerard developed a technique of recording transmembrane potentials with glass micropipettes. They heated fine glass tubing and pulled it out to produce a sharp capillary electrode with an open tip of less than 0.5 μm. With such a small tip, the electrode could be inserted into a cell without producing irreversible damage. The capillary pipette was filled with a salt solution to give electrical continuity with the cytosol, and connected via a wire to an amplifier. The development of these glass microelectrodes had immediate and profound effects on the field of electrophysiology; within a few years, glass micropipettes had been employed to study the ionic basis of resting and action potentials and of synaptic transmission in a variety of cell types.

muscle cells have a high internal potassium concentration (and hence a Nernst potential that is inside-negative) first suggested to the physical chemist Julius Bernstein in 1902 that the membrane of resting muscle cells is selectively permeable to K^+.

Confirmation of Bernstein's idea comes from experiments in which the extracellular concentration of K^+ is changed. Intracellular recording (Box A) shows that altering the potassium concentration gradient in this way changes resting potential in the direction predicted by the Nernst equation. The magnitude of the change is less than predicted, however, since the membrane is not permeable solely to K^+. In the squid giant axon the potassium concentration gradient can be changed by manipulating the intracellular potassium concentration as well as the extracellular one. When the interior of the axon is perfused with a solution containing the same concentration of K^+ as is present externally, the resting potential is near zero. When the interior of the squid axon is perfused with a solution whose K^+ concentration is similar to that of the normal extracellular fluid, and the axon is suspended in a solution whose K^+ concentration resembles that in squid axoplasm, the polarity of the resting potential is reversed: the inside of the membrane is positive with respect to the outside. The actual membrane potential under a variety of conditions is determined largely by the potassium concentration gradient; the deviation from the value predicted by the Nernst equation is explained by finite permeability to Cl^- and to Na^+. These findings demonstrate that the membrane itself has no inherent polarity and that the resting potential measured across it is deter-

Box B The Donnan Equilibrium

The impermeant anions inside nerve cells alter the distribution of other ions to which the cells are permeable. This effect is most easily seen by considering a cell that is permeable to both K^+ and Cl^- and that contains 120 m*M* potassium salt, KA, where A^- represents an impermeant, monovalent anion. The cell is bathed in an equal volume of 120 m*M* KCl. Cl^- flows into the cell down its concentration gradient. The resulting membrane potential, inside-negative, causes K^+ to flow in as well. Ignoring osmotic forces for the moment, Cl^- and K^+ continue to flow across the membrane, increasing their internal concentrations and reducing their external concentrations, until both ions have equal and opposite concentration gradients across the membrane. At this point, both ions are in electrochemical equilibrium and there is a negative membrane potential such that $V_m = E_K = E_{Cl}$ (Figure A).

At equilibrium,

$$\left[\frac{RT}{F}\right] \ln \left[\frac{[K^+]_o}{[K^+]_i}\right] = -\left[\frac{RT}{F}\right] \ln \left[\frac{[Cl^-]_o}{[Cl^-]_i}\right] \quad \text{(A1)}$$

which can be simplified to

$$[K^+]_o \times [Cl^-]_o = [K^+]_i \times [Cl^-]_i \quad \text{(A2)}$$

which is called the Donnan rule. The Donnan rule states that the products of the concentrations of two ions of opposite sign on each side of the membrane are the same at electrochemical equilibrium.

If, in the example above, we substitute Na^+ for half the K^+ on both sides of the membrane and make the cell permeable to Na^+, then Na^+ distributes itself exactly as K^+ does, and the cell comes to equilibrium at the same membrane potential, which is now equal to E_K, E_{Cl}, and E_{Na} (Figure A, bottom).

The distribution of ions in this example differs from that found in real cells. The most obvious difference is in the sodium gradient, which in real cells is reversed from that seen in the example. In addition, E_K is usually not at resting potential, but at a more hyperpolarized potential, and E_{Cl} may be different as well. These differences arise because of the action of ion pumps. The Na^+/K^+ pump transports Na^+ out of the cell, and because of the low passive permeability of the cell to Na^+, is actually able to reverse the gradient. In contrast, the action of the Na^+/K^+ pump increases the gradient for K^+. In some cells, a chloride pump redistributes Cl^-.

The pump also performs another important function. In the examples given above, the cell is osmotically imbalanced at electrochemical equilibrium. As both K^+ and Cl^-

mined solely by its permeability and by the concentration gradients of permeant ions.

In skeletal muscle, changing the external concentrations of ions reveals that the muscle cell membrane has a high permeability to both K^+ and Cl^- and a low permeability to Na^+. If the contribution of either K^+ or Cl^- to the membrane potential is negated, then the membrane potential is determined largely by the concentration gradient of the other ion, as predicted by the Nernst equation. For example, if the contribution of Cl^- is reduced by replacing extracellular Cl^- with an impermeant anion, the membrane potential in muscle fibers varies according to the Nernst equation over a wide range of external potassium concentrations, deviating from theoretical predictions only at the lowest extracellular concentrations as a result of the finite permeability to Na^+ (Figure 4). This behavior is thus similar to that of the imaginary cell discussed earlier. The resting potential is close to E_K because the membrane is much more permeable to K^+ than to Na^+.

have entered the cell and no ions have left, the osmolarity inside the cell exceeds that of the external solution. The entry of water to correct the imbalance would cause the cell to swell and burst. This problem is solved by the Na^+/K^+ pump and the low membrane permeability to sodium. The pump transports Na^+ from the cell interior to the extracellular fluid, thus restoring osmotic balance.

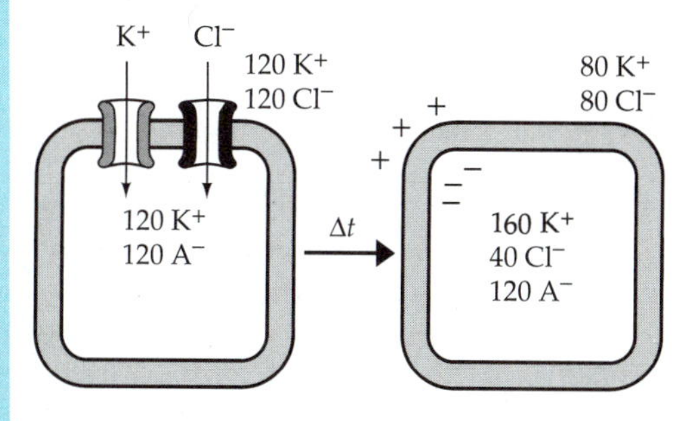

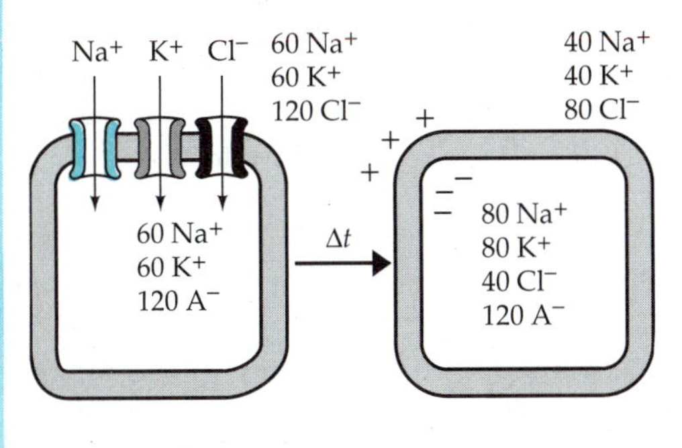

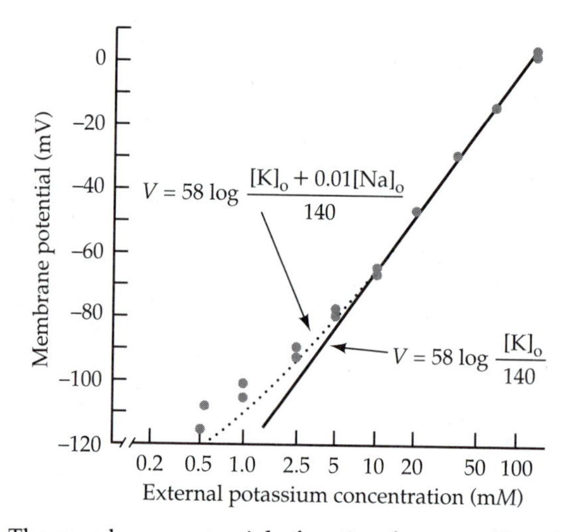

FIGURE 4. The membrane potential of resting frog myofibers is accurately predicted by the Goldman–Hodgkin–Katz equation. The membrane potential (ordinate) was measured with a micropipette after altering the extracellular potassium concentration (abscissa). The contribution of Cl^- was largely negated by replacing external Cl^- with impermeant sulphate anions. The solid line is drawn according to the Goldman–Hodgkin–Katz equation, assuming that the membrane is permeable only to K^+, while the dotted line is drawn assuming that the sodium permeability is 1% of that for K^+. Changes in the external Cl^- concentration also alter the membrane potential; in skeletal muscle chloride permeability is comparable to potassium permeability. (After A. L. Hodgkin and P. Horowicz, 1959. J. Physiol. 148: 133.)

The Na^+/K^+ pump is electrogenic

Although the resting permeability of the membrane to Na^+ is low, the driving force is high. In fact, at rest the influx of Na^+ is approximately equal to the efflux of K^+, whose permeability is high but whose driving force is low because V_m is so close to E_K. Thus, at rest, there is a steady flow of K^+ out of the cell, and a steady flow of Na^+ into the cell. Although these currents are small and have no immediate effect on the concentration gradients, over a period of hours or days the concentration gradients would run down in the absence of a pump, and V_m would approach zero. The concentration gradients are maintained by the Na^+/K^+ pump, which transports Na^+ out and K^+ in. Active transport of Na^+ and K^+ by the pump is not balanced, however. For every molecule of ATP hydrolyzed, three Na^+ are usually transported out, and two K^+ are usually transported in.

The unequal stoichiometry of the Na^+/K^+ pump has three important consequences. First, because the action of the pump causes a net transfer of charge across the membrane, it is **electrogenic.** The current produced by the pump hyperpolarizes the membrane and contributes to the resting potential. The contribution of the pump to the membrane potential depends upon the electrical resistance of the membrane, and in most cells is about 5–10 mV.

The second consequence of the unequal stoichiometry of the pump is that the passive movements of Na^+ and K^+ are unequal. For concentration gradients to be maintained, the sodium influx through ion channels must equal the sodium efflux generated by the pump; likewise, the rate of potassium efflux must match the potassium influx produced by the pump. Thus, at rest the passive inward sodium current is about 1.5 times the outward potassium current.

The third consequence of the unequal stoichiometry of the pump concerns a factor that we have so far ignored, the osmotic balance of the cell. Although the permeant ions, Na^+, K^+, and Cl^-, move passively across the membrane in response to electrical and chemical forces, large internal anions do not. In a cell with no pumps, the ions would distribute themselves so that the number of osmotically active molecules inside the cell would be higher inside than outside (see Box B); water would then move down its concentration gradient into the cell. In animal cells, the Na^+/K^+ transporter effectively prevents the influx of water that would otherwise occur. Because more Na^+ is pumped out than K^+ is pumped in, the pump decreases the osmolarity of the intracellular fluid and balances the effect of impermeant anions in the cytosol.

The impermeant, intracellular anions contribute to ionic gradients

The concentration gradients of Na^+, K^+, and Cl^- arise from two sources—the action of the Na^+/K^+ pump and the effects of the impermeant, intracellular anions. Although the action of the pump is easily understood, it clearly is not the only factor affecting the distribution of ions across the surface membrane. If it were, then one would expect because of the asymmetric action of the pump that the sodium gradient would be steeper than the potassium gradient, but in fact it is less steep.

To understand the influence of the impermeant ions, it is easiest to consider a cell permeable to Na^+, K^+, and Cl^- but having no pumps. At equilibrium, concentration gradients for K^+ and Cl^- would be established that are in the same direction as those found in real cells (Box B). A concentration gradient for Na^+ would also arise, but in the opposite direction from that found in real cells (i.e., the intracellular concentration would be higher than the extracellular concentration). Eventually, all permeant ions would come to equilibrium, and the membrane potential would reach E_K, E_{Na}, and E_{Cl}, all of which would be identical (see Box B).

In real cells, the Na^+/K^+ pump alters this distribution. The pump augments the gradient for K^+ and reverses the gradient for Na^+. The presence of the impermeant anions, however, is reflected in the fact that the sodium gradient is smaller than the potassium gradient. The final distribution of the ions is thus influenced by both pumps and impermeant anions and represents a compromise between the distributions that either alone would achieve.

The equilibrium potential for Cl^- is near resting potential

In many cells, E_{Cl} is very close to the resting potential. In cells lacking active chloride transport, such as skeletal muscle fibers, Cl^- distributes itself passively across the membrane so that it comes into equilibrium, i.e., so that $E_{Cl} = V_m$. If the membrane potential is displaced so that Cl^- is no longer at equilibrium, then Cl^- moves across the membrane so that its concentration gradient balances the electrical gradient. In such instances, the membrane potential determines the concentration gradient of Cl^-, and not the other way around.

In some cells, Cl^- is not passively distributed but is actively pumped. In the squid giant axon, Cl^- is pumped in, yielding an E_{Cl} that is more depolarized than V_m; in other cells Cl^- is pumped out, yielding an E_{Cl} more negative than V_m. Because E_{Cl} is so close to resting potential and chloride permeability is often low, its contribution to resting potential in most neurons is minor. As we shall see, however, the chloride permeability is important in determining how easily a cell is depolarized.

Cells at rest must meet three conditions to maintain chemical, electrical and osmotic stability: (1) the total number of anions must equal the total number of cations on each side of the membrane; (2) the number of osmotically active particles on the two sides of the membrane must be equal; and (3) the net flux across the membrane for each ion must be zero. A consequence of the third condition is that the total current across the membrane is zero. This is used to calculate an approximation to the resting potential. If one ignores the direct contribution of the electrogenic sodium pump to the resting potential, then, at steady state,

$$i_{Na} + i_K + i_{Cl} = 0 \tag{9}$$

Substituting for the current terms and solving for V_m (as in Equations 5–8) yields

$$V_m = E_{Na}\left[\frac{g_{Na}}{\Sigma g}\right] + E_K\left[\frac{g_K}{\Sigma g}\right] + E_{Cl}\left[\frac{g_{Cl}}{\Sigma g}\right] \tag{10}$$

where $\Sigma g = g_{Na} + g_K + g_{Cl}$.

This method is useful for electrophysiological experiments because the conductances can be determined by electrical recording. A more exact relation is given by the Goldman–Hodgkin–Katz (GHK) equation, using permeabilities rather than conductances.

$$V_m = 58 \text{ mV} \log_{10}\left[\frac{p_K[K^+]_o + p_{Na}[Na^+]_o + p_{Cl}[Cl^-]_i}{p_K[K^+]_i + p_{Na}[Na^+]_i + p_{Cl}[Cl^-]_o}\right] \tag{11}$$

The GHK equation differs from Equation 10 in that it is derived theoretically from diffusion equations, using ionic permeabilities (which are measured in radioactive transport experiments) rather than ionic conductances. Although the two are equivalent for most purposes, they are not identical, since conductance depends on ion concentration whereas permeability does not.

Although the conductance and permeability equations are slightly different, the principle in both cases is the same as the one we have enunciated earlier. Membrane potential depends on the relative permeabilities of ions and on their concentration gradients, as expressed in the Nernst equation. As we will see, this holds not only for resting potentials, but for action potentials and synaptic potentials as well.

Action Potentials

Neurons use a single type of signal, the action potential, to transmit information over long distances. Action potentials are large, brief, invariant signals that propagate themselves along axons without decrement. Because action potentials, once initiated, are stereotyped and independent of the stimulus that produced them, they are "all-or-none." The action potential is produced by two classes of ion channels, a sodium channel and a potassium channel, acting in concert. The opening and closing of the two channels is precisely timed to give a transient reversal of membrane potential, which travels along axons at speeds up to 120 meters per second. In this section we will examine the properties of these two channels and how they work together to produce action potentials.

The action potential has a characteristic shape. Depolarization of the membrane past a threshold value initiates a stereotyped series of changes: the membrane rapidly depolarizes to become briefly inside-positive and then rapidly repolarizes to a potential near rest. In many cells the repolarization continues beyond the original resting potential, so that the membrane becomes transiently hyperpolarized, and then slowly returns to the original resting potential. The last phase is called the **after-hyperpolarization**.

The rapid depolarization of the membrane is due to inward sodium current, caused by a sudden, large increase in sodium conductance (Figure 5). During the rising phase of the action potential, the sodium conductance exceeds the potassium conductance, and the net inward current drives the membrane potential toward E_{Na}. As the action potential reaches its peak, the sodium conductance declines, and the potassium conductance increases. The outward potassium current soon exceeds the inward sodium current, and the membrane potential begins to repolarize and move toward E_K. Thus, during the action potential the membrane progresses from one that is permeable mostly to K^+ to one that is permeable mostly to Na^+, and then returns to being mostly permeable to K^+. Chloride permeability does not change during the action potential.

The conductance changes that occur during the action potential are explained by the molecular properties of the sodium and potassium chan-

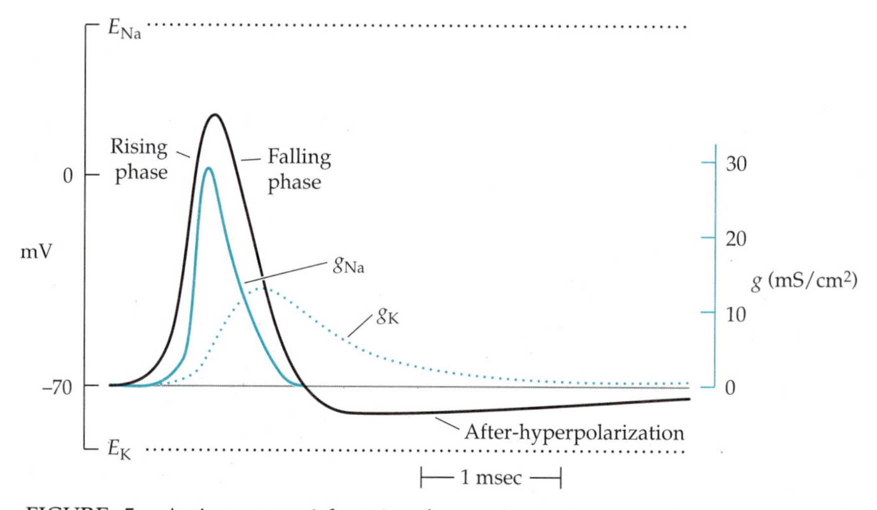

FIGURE 5. Action potentials arise from phasic increases in g_{Na} and, subsequently, in g_K. The action potential in the squid giant axon (in color) consists of three phases: a rising phase, a falling phase, and an after-hyperpolarization. g_{Na} (in color) rises rapidly and then falls and is responsible for the rising phase of the action potential, while g_K rises more slowly and is responsible for the falling phase of the action potential. The decrease of g_{Na} is due to sodium channel inactivation, while that of g_K is due to repolarization. (After A. L. Hodgkin and A. F. Huxley, 1952. J. Physiol. 117: 530.)

nels, which are the keys to understanding the action potential. The sodium channel is a voltage-regulated channel whose equilibrium between open and closed states is influenced by membrane potential (Figure 6). Each channel moves randomly back and forth between open and closed states. At rest, the probability that the channel will open is small, and most channels are in the closed state. Depolarization of the membrane increases the probability that a channel will open. The depolarization that initiates the action potential causes a few sodium channels to open, and Na^+ enters the cell, depolarizing it. This depolarization causes more channels to open, which results in more sodium entry, which depolarizes the membrane further. The original depolarization thus sets in motion a positive feedback system that causes an explosive, regenerative event. Because of this positive-feedback loop, the depolarizing phase of the action potential is self-sustaining and, once started, requires no additional stimulus to run its course.

The decrease in sodium permeability that begins as the action potential approaches its peak results from a second property of the sodium channel, its **inactivation.** Channels in the open state soon become inactivated, so that they no longer respond to depolarization, regardless of its magnitude. The inactivation itself is not voltage-dependent, but by increasing the probability that a channel will open, depolarization also increases the probability that it will become inactivated (Figure 6). After repolarization of the membrane, inactivation is slowly reversed. Depolarization thus has

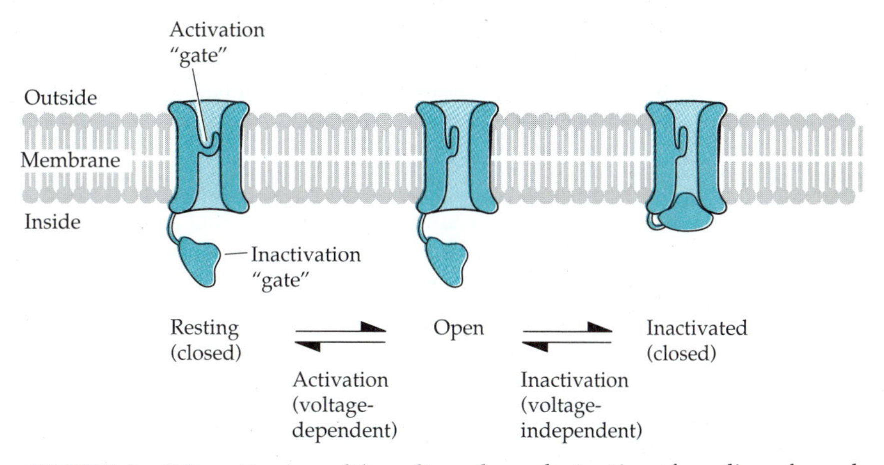

FIGURE 6. Schematic view of the voltage-dependent gating of a sodium channel. Depolarization leads to the opening of the activation "gate." Once the channel is activated, it can close by the action of an inactivation "gate," or it can return to the closed state.

two effects on sodium conductance: it causes a rapid increase, due to activation or opening of the sodium channels, and then a decrease, due to inactivation.

Potassium conductance is increased slowly during the action potential

Potassium channels are also voltage-regulated channels whose probability of opening is increased by depolarization. In contrast to sodium channels, however, potassium channels respond more slowly to voltage changes. During the action potential the potassium conductance does not begin to increase until the action potential is near its peak, and it remains high during the falling phase (Figure 5). Repolarization causes the potassium conductance to decrease, but also with a delay. The delayed activation of potassium channels, in combination with sodium channel inactivation, is responsible for the falling phase of action potential and the after-hyperpolarization. The net outward current through the potassium channels brings the membrane potential back toward resting values. During the after-hyperpolarization the potassium conductance is higher than normal, the sodium conductance is lower than normal, and the membrane potential is driven even closer to the equilibrium potential for K^+ than it is at rest (see Equation 10).

Voltage-gated potassium channels are not essential to the action potential mechanism. If there were no voltage-dependent potassium channels, sodium channel inactivation would still lead to repolarization of the membrane due to outward current flowing though chloride channels and through potassium channels that are open even at rest. Repolarization is faster, however, when voltage-dependent potassium channels are present. Voltage-dependent potassium channels thus shorten the duration of action potentials and shape it; they permit neurons to generate action potentials at higher frequencies than would be possible in their absence, and by their kinetics they determine its duration. Some voltage-dependent potassium channels are open even at rest and contribute to the resting potential.

Immediately following an action potential there is a **refractory period**, when the threshold for initiation of another action potential is higher than normal. The causes of the refractory period are the same as those of the after-hyperpolarization: residual inactivation of sodium channels and re-

Box C Voltage Clamp

By 1950, there was good reason to believe that action potentials result from a transient increase in sodium permeability followed by a transient increase in potassium permeability. These changes appeared to be caused by depolarization, but the precise effect of depolarization on ionic permeabilities could not be determined without a method of holding membrane potential constant. Then Alan Hodgkin and Andrew Huxley developed the voltage clamp technique, based on the earlier work of Kenneth Cole and George Marmont.

During voltage clamping, a feedback circuit containing a differential amplifier is used to hold the membrane potential constant by injecting current across the membrane. One input to the amplifier is the cell's membrane potential; the other is a "command" potential set by the experimenter. A difference in voltage between the two inputs results in a flow of current through the output that acts to eliminate the difference. The amplifier can be set to hold or clamp the cell's membrane potential at any desired level. The output of the amplifier is a measure of the membrane current, except that its sign is reversed.

Voltage clamping has two advantages. First, transmembrane currents accurately reflect changes in membrane conductance caused by the opening and closing of ion channels in the membrane. In a voltage clamped cell, the current records are directly proportional to changes in conductance. In contrast, the potential changes produced by these currents, measured by conventional intracellular recording, depend not only on changes in conductance, but also on the passive electrical properties of the membrane. Second, as noted above, voltage clamping is invaluable in investigating the behavior of voltage-regulated channels. Conventional recordings, which do not control membrane potential, cannot be used to reveal how voltage regulates conductance. Because the opening and closing of channels itself changes membrane potential, an external source of current is necessary to maintain a constant voltage during an experiment.

Hodgkin and Huxley used voltage clamping to investigate the properties of the voltage-gated sodium and potassium channels in the squid giant axon (Figure A). They inserted two wire electrodes down the interior of the axon: one monitored membrane potential and was connected to one input of the differential amplifier; the other, which was used to pass current, was connected to the output of the amplifier. In a voltage clamped axon in which the command potential is set to match the resting membrane potential, no current flows through the feedback circuit. If, however, the command potential is suddenly set to a different voltage, such as 0 mV, the amplifier responds by injecting current into the axon to depolarize it to 0 mV. This depolarization opens voltage-gated sodium channels; normally the inward current through these would depolarize the axon further, and a full-

sidual activation of potassium channels. Sodium channels that remain inactivated after the falling phase of the action potential cannot be opened by depolarization. With fewer sodium channels available to open, a smaller number will actually open in response to a depolarization, and less inward current will flow. A larger depolarization is thus required to elicit an action potential. The removal of inactivation depends upon time and voltage and may take several milliseconds at repolarized values of membrane potential. During the refractory period, not only is a larger depolarization required to initiate an action potential, but more inward current is required to achieve that depolarization because outward current

scale action potential would ensue. In the voltage clamped axon, this inward current through the membrane is opposed by outward current generated by the feedback amplifier in the voltage clamp circuit, thus holding the membrane potential at 0 mV. At 0 mV, the sodium channels soon become inactivated, and voltage-regulated potassium channels begin to open, producing an outward current through the membrane that would normally repolarize the cell. The amplifier now supplies inward current to match this outward current and prevent the repolarization. The current output of the amplifier thus provides a direct measure of the membrane current. From Ohm's law, the current is directly proportional to conductance (Equations 5, 6), and thus the amplifier output can be used to measure changes in membrane conductance.

Hodgkin and Huxley analyzed the time- and voltage-dependence of the Na^+ and K^+ currents in squid giant axons. They developed empirical expressions for the changes in conductance that produced them and showed that these changes in conductance could account fully for the action potential. Their analysis was published in a remarkable series of five papers that appeared in the *Journal of Physiology* in 1952 (the first paper was co-authored with Bernard Katz). For this work, they were awarded the Nobel Prize in 1963.

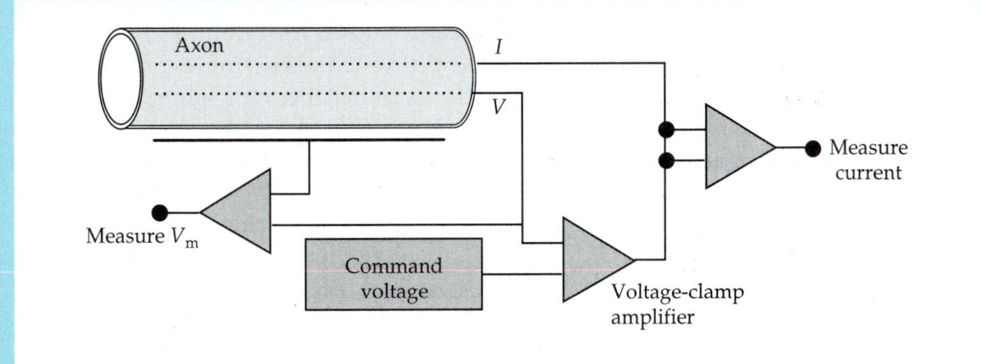

through residually activated potassium channels more effectively opposes inward current. The refractory period limits the frequency of action potentials that an excitable cell can sustain. For some cells the upper limit may be as low as 100 per second, while for other cells it may be as high as 1000 per second.

Changes in sodium and potassium conductance during the action potential are measured by voltage clamp experiments

In 1952 Alan Hodgkin and Andrew Huxley demonstrated in a landmark series of experiments that the action potential was produced by voltage-dependent sodium and potassium conductances. To measure sodium and potassium conductances and to establish their dependence on membrane potential and time, Hodgkin and Huxley voltage clamped squid giant axons (Box C).

Hodgkin and Huxley's experiments were designed to measure the currents that flow across the membrane as a result of a step change in membrane potential from its resting level. After a depolarizing step, they observed a transient inward current followed by a sustained outward current (Figure 7). Hodgkin and Huxley demonstrated that the transient inward current was carried by Na^+ by stepping the membrane potential to E_{Na}, at which no sodium current flows since there is no driving force on Na^+. At this potential, only a sustained outward current was obtained. The outward current remaining when the membrane potential was stepped to E_{Na} was not affected by changes in E_{Cl} and was presumably carried by K^+.

When Hodgkin and Huxley replaced some or all of the external Na^+ with the impermeant cation choline, they noted a reduction in the transient inward current in response to step depolarizations. This would be expected if Na^+ carried the inward current, since the driving force on Na^+ is reduced by lowering its external concentration. The difference between the membrane current in normal sea water and zero-sodium sea water could be taken as a measure of the sodium current, thus permitting Hodgkin and Huxley to generate a family of sodium current curves in response to step depolarizations of different magnitude. The difference between the sodium current and the total membrane current was assumed to be the potassium current. Thus, both sodium and potassium currents could be expressed as a function of time at different membrane potentials.

The current records obtained during voltage clamp were used to generate conductance curves for Na^+ and K^+. Since the voltage during the step of depolarization is constant, the current is simply divided by the driving force to obtain a record of conductance with time (from Ohm's law; see Equations 5 and 6). Hodgkin and Huxley then used the family of conductance curves to write empirical equations that expressed each conductance as a function of membrane potential and time. These equations were then used to predict the behavior of a nerve fiber following a small depolarization as well as those that exceeded threshold. The calculations, which were carried out in an iterative fashion (by hand!—computers were not then available), showed close agreement between the predicted responses and those actually recorded. These experiments demonstrated that the properties of an excitable membrane could be entirely accounted

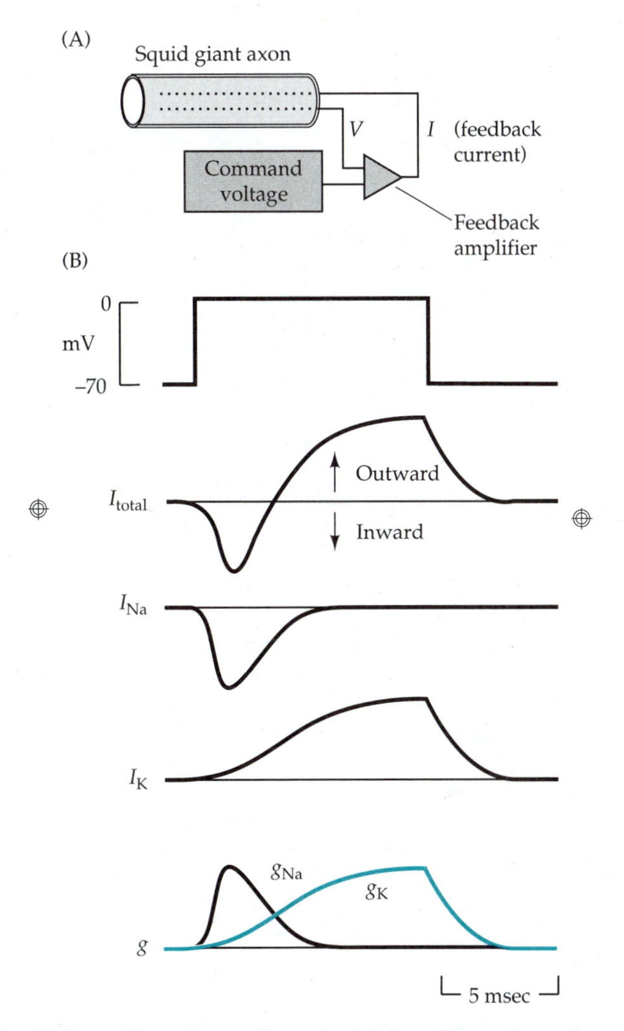

FIGURE 7. Voltage clamp recordings reveal the effects of membrane potential upon sodium and potassium conductance. (A) A length of the squid giant axon is voltage clamped to levels set by the experimenter by means of a feedback amplifier. (B) A depolarizing step in voltage leads to a transient inward current followed by a sustained outward current, which decays after the potential is returned to rest. By comparing current traces obtained in low external sodium and traces obtained in normal sea water, Hodgkin and Huxley were able to separate total membrane current into sodium current and potassium current. The sodium and potassium conductances (bottom trace) are directly related to the currents by the driving force (Equations 5 and 6), which is constant during the step.

for by the time-dependent effects of voltage upon sodium and potassium channels in the membrane.

Patch clamping reveals the properties of a single channel

The properties of the sodium and potassium channels that are inferred from Hodgkin and Huxley's experiments are seen directly in **patch clamp** experiments (see Box D) in which the currents flowing through a single

Box D Patch Clamping

The methods of recording discussed so far measure the electrical behavior of membranes containing large numbers of ion channels. The properties of individual channels can only be inferred from the behavior of the entire population. One means of making such inferences, called noise analysis or fluctuation analysis, was introduced in the 1970s. This method analyzes the increase in electrical noise produced by the opening and closing of single ion channels to deduce the amplitude and time course of single-channel events. Although useful, this method still gives only indirect information about the behavior of individual channels.

Erwin Neher and Bert Sakmann first directly observed the current flowing through single channels in 1976, using a new method of recording called patch clamping. In this method, the fire-polished tip of a glass micropipette having a tip diameter of 1–5 μm is brought into close contact with the surface of a cell that has been treated with proteases to remove the extracellular matrix and adherent debris. The contact between the tip of the pipette and the membrane electrically isolates the small patch of membrane within the tip from which recordings are made. In an important later refinement introduced by Neher, gentle suction to the pipette induces an extremely tight seal (a gigaohm seal) between the pipette orifice and the lipid bilayer. The tightness of the seal has two important consequences. First, the resistance of the current pathway between the inside of the pipette and the extracellular fluid is so high—up to 100 gigaohms (10^{11} ohms)—that the small changes in resistance caused by the opening and closing of single channels in the membrane can be detected. Second, the seal between the pipette and the membrane is mechanically very strong.

The mechanical strength of the gigaohm seal can be exploited to produce several different recording configurations, each with particular advantages (Figure A). In a cell-attached patch, the electrode is used to isolate electrically the patch of membrane within the pipette orifice. Single-channel activity can then be observed directly. Alternatively, additional suction can be used to break the membrane within the tip, creating a low resistance pathway into the cell interior. This mode, called whole-cell recording, generally causes less damage than penetration with a conventional microelectrode and provides a low resistance pathway into the cell interior, reducing the noise of the recording. Also, the pipette can be used to introduce substances into the cell by diffusion, creating both experimental difficulties and opportunities. Channels whose activation is mediated by G proteins, for example, (see Chapter 6), are inactive unless GTP is supplied through the pipette.

Because of the mechanical strength of the gigaohm seal, the pipette can be used to pull the patch of membrane at its tip away from the surrounding membrane. When the pipette is withdrawn from the cell after generation of a cell-attached patch, the membrane forms a vesicle at the end of the pipette whose external surface can be broken by briefly lifting the pipette into the air. This leaves attached to the pipette an inside-out membrane patch whose cytoplasmic surface now faces the external solution. Alternatively, if one begins from the whole-cell configuration and withdraws the pipette, an outside-out patch results: the external surface of the membrane faces the bulk solution. These techniques give the experimenter easy access to either side of the membrane for application of small molecules, proteins, or other reagents. In the usual

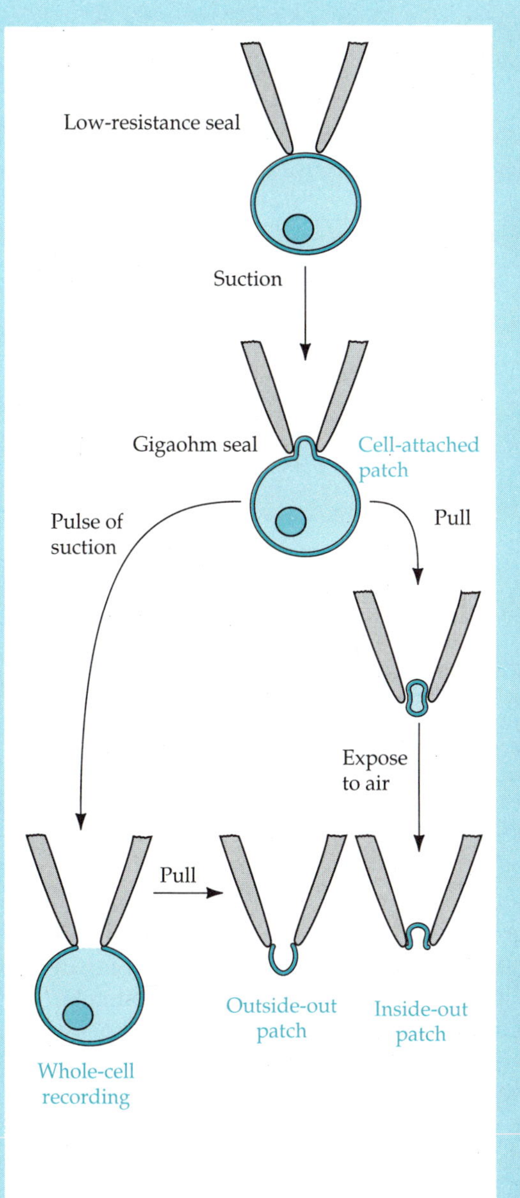

mode of recording, a voltage clamp circuit gives control of membrane potential as well.

The introduction of the gigaohm seal has created a modern revolution in electrophysiology, not only allowing the study of individual channels in unprecedented detail, but also allowing a variety of other applications. Whole-cell recordings can be made from cells and organelles whose small size results in irreversible damage after penetration by conventional microelectrodes, and from lipid vesicles containing purified membrane proteins. In another variation, whole-cell recordings can also used with appropriate electronic circuitry to measure cell capacitance. Because exocytosis and endocytosis change the area of the surface membrane, and hence the membrane capacitance, these physiological processes can be studied in a new way. The use of isolated patches allows the interactions of purified proteins with ion channels to be investigated. Finally, isolated patches containing channels that are sensitive to small amounts of biologically active molecules can be used as biological "sniffers." Outside-out patches of myofiber membrane containing ACh receptors have been used in this way to detect the ACh released from single growth cones.

The various techniques of patch clamp and whole-cell recording have now been applied to many types of cells and ion channels. Single-channel analysis of *Xenopus* oocytes and of transfected cells expressing mutationally altered ion channels has been particularly important for the exploration of the molecular basis of ion channel function (see Chapter 3). Although these techniques of recording are used most easily with cultured cells, recent advances allow their application to thin slices of nervous tissue from the vertebrate CNS. We may thus expect a new wave of progress in our understanding of the complex interactions of synapses in the CNS.

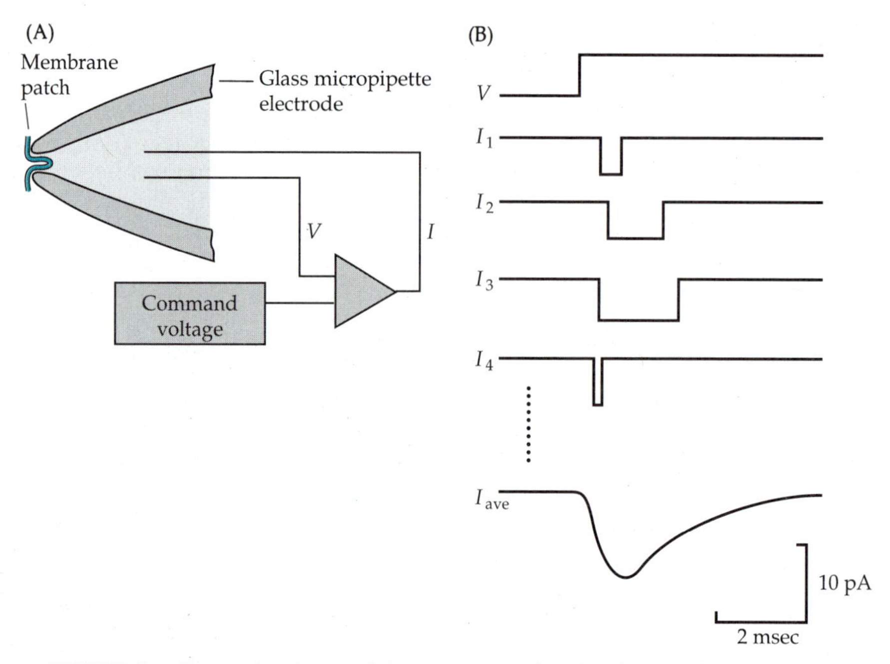

FIGURE 8. The activation and inactivation of individual sodium channels in response to depolarization are observed in patch clamp experiments. (A) An excised patch of membrane is sealed to the orifice of a glass pipette and subjected to voltage clamp. (B) Step depolarizations lead to abrupt changes in membrane current caused by the opening and closing of individual channels. Each current trace ($I_1, \ldots I_4$) represents a separate trial. The average response (I_{ave}) has the same wave form as the sodium current curve obtained from voltage clamping the squid giant axon (Figure 7) and is a measure of the probability that a channel will be open as a function of time after depolarization.

channel are observed (Figure 8). When a voltage clamped patch of membrane containing sodium channels is depolarized, fluctuations in current are seen that represent the opening and closing of single channels. The **single-channel conductance** can be calculated from the amplitudes of the current by using Ohm's law (Equation 6). Analysis of the records indicates that the channels close or inactivate after opening, even though the depolarization is sustained. By repetition of many trials, the summed **ensemble response** (Figure 8, I_{ave}) is obtained. This response has the same appearance as the sodium current curves measured by Hodgkin and Huxley (Figure 7, I_{Na}), as expected, since the macroscopic sodium current (Figure 7) should be equal to the number of open channels multiplied by the current flowing through a single channel (Figure 8). As the probability of opening increases, the number of channels in the open state increases and the current increases. The form of the current curve vs. time (or of the conductance curve vs. time) gives the probability of opening for a single channel. Single-channel recordings show that the probability of opening is increased at more depolarized membrane potentials.

At threshold inward sodium current becomes greater than outward potassium current

So far we have referred to threshold simply as the potential beyond which a depolarization becomes regenerative and leads to an action potential. To understand what happens at threshold, imagine injecting a brief, small pulse of positive current into a resting cell whose membrane contains voltage-gated sodium and potassium channels (Figure 9). After a small

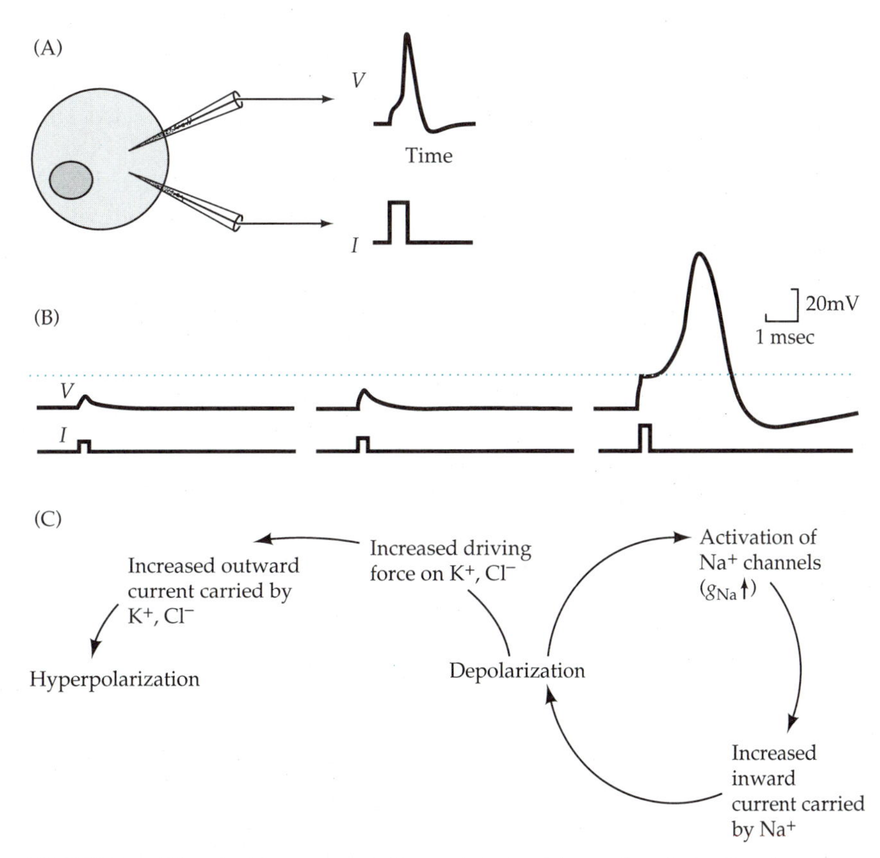

FIGURE 9. Threshold represents the point at which a depolarization becomes regenerative and leads to an action potential. (A) Two micropipettes are inserted into a neuron, one for passing brief pulses of current and one for recording membrane potential. (B) An action potential occurs when the membrane is depolarized to threshold (dotted line). (C) Depolarization has two effects. First, it increases the driving force on K^+ and Cl^-, and the increased outward current carried by these ions tends to repolarize the cell (in this example it is assumed that chloride ions are passively distributed). Second, depolarization opens voltage-dependent sodium channels, and the increased inward current resulting from increased g_{Na} tends to depolarize the cell. If the depolarization is large enough, inward current exceeds outward current and the result is a regenerative action potential.

depolarization (that does not activate sodium channels), the driving force on both K^+ and Cl^- increases, and the potassium and chloride currents increase as well. K^+ leaves the cell and Cl^- enters it, producing an outward current that repolarizes the membrane (note that inward movement of Cl^- corresponds to outward current). If the original depolarization is larger, some sodium channels are activated, and an inward sodium current results. As long as the outward current elicited by a depolarization is larger than the inward current, the membrane repolarizes (Figure 9). Larger depolarizations open correspondingly more sodium channels, however, and a sufficiently large depolarization elicits an inward current that exceeds the outward current. After such a depolarization, the net inward current depolarizes the cell further; this depolarization opens more sodium channels, and an action potential results. Threshold, then, is the value of membrane potential at which the inward current exceeds the outward current.

The concepts of threshold and of "all-or-none" go hand in hand. The existence of threshold means that a depolarization either elicits an action potential or it does not—there is no middle ground. Threshold does not represent an absolute value of membrane potential. Its value changes as the result of the cell's immediate past activity. Thus, threshold is higher than normal during the refractory period that follows an action potential. The value of threshold also depends on the time course of the stimulus. Because of sodium channel inactivation, a rapid depolarization is more effective in eliciting an action potential than a slow depolarization of the same magnitude. The value of threshold depends upon the density of sodium channels in the membrane. Threshold is lowest at the neuron's initial segment (see Figure 2 in Chapter 1), where the density of sodium channels is higher than in the surrounding membrane; thus, action potentials are initiated there.

Thus far we have discussed the action potential as if it occurred everywhere at once. To understand how action potentials propagate along axons, we must consider another class of electrical signals, called **local potentials,** which are subthreshold changes in membrane potential. Local potentials do not propagate, but decay with distance from their origin; if sufficiently large, they initiate action potentials, however, and are responsible for the propagation of action potentials. In addition, they are important for integration of electrical signals in neurons.

To understand the behavior of local potentials, we must first discuss the electrical resistance and capacitance of neurons, which are often called their **passive electrical properties**. These are usually measured under conditions in which the number of open channels in the membrane does not change. The passive electrical properties of a neuron depend on the density of open channels, the capacitance of the membrane, and the geometry of the cell.

Lipid bilayer membranes have both electrical resistance and capacitance. Their resistance depends upon their ability to carry ionic current when there is a driving force—just as the resistance of a copper wire depends upon its ability to carry a current of electrons when there is a voltage difference between its two ends. Ions traverse biological membranes through channels; the higher the density of channels, the lower the electrical resistance and the higher the conductance.

The electrical capacity of the membrane arises from the hydrophobic interior of the lipid bilayer, which is a poor conductor. When charge is separated across a thin layer of a nonconducting medium (a capacitor), positive and negative charges, attracted to each other, accumulate on either side of the bilayer. The separated charge produces a voltage difference across the bilayer whose size depends upon its properties. Because the charge that produces the voltage is supplied at a finite rate, the capacitance of the membrane affects the time course with which voltage changes occur across it.

A patch of membrane may be represented electrically as a parallel RC circuit: a resistor and a capacitor in parallel (Figure 10). When a current is applied across the membrane, part of the current (the capacitative current) is used to change the transmembrane voltage, while the other part (the resistive current) flows across the membrane through the open channels. Immediately following an instantaneous change in applied voltage, most of the current is capacitative, but as the membrane potential reaches a steady value, the capacitative current declines, and the current becomes entirely resistive.

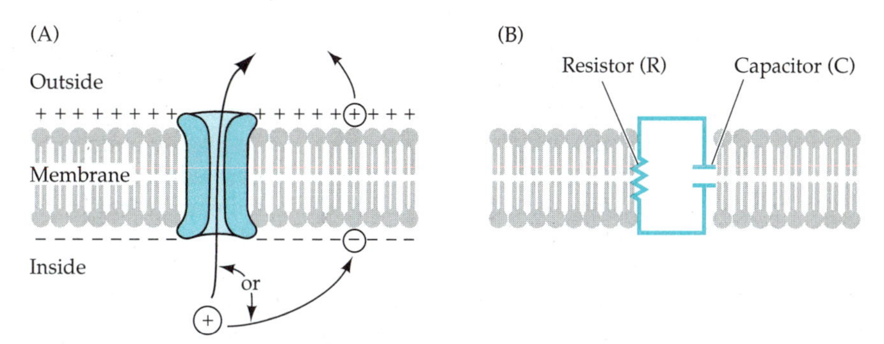

FIGURE 10. Transmembrane current is either resistive (ionic) or capacitative. (A) Resistive current occurs when an ion flows across the membrane through a channel. Capacitative current occurs when one ion approaches one surface of the membrane and another is expelled from the other surface. (B) A parallel RC circuit represents the passive electrical behavior of a patch of membrane.

The geometry of the cell influences its passive electrical properties

The contribution of the resistance and capacitance of the membrane to the electrical properties of neurons is seen by considering a small spherical cell in which the resistance to ion flow within the cell is negligible compared to the transmembrane resistance, making the interior of the cell isopotential. The individual RC elements, each symbolizing a patch of membrane, may be represented by a single RC circuit, whose resistor represents the equivalent resistance of the cell membrane, called its **input resistance,** and whose capacitor represents the equivalent capacitance of the membrane (Figure 11).

A glass micropipette inserted into the cell is used to inject current; a second micropipette in the cell measures the change in membrane potential produced by the injected current (Figure 11A). When a rectangular pulse of current (carried by ions in the micropipette) is injected into the cell, the voltage across the membrane rises exponentially to a steady value during the pulse and falls exponentially at the end (Figure 11B). If the experiment is repeated with both depolarizing and hyperpolarizing current pulses of different amplitudes, a family of voltage curves is generated. If the current pulses are small, so that no voltage-dependent channels are affected by the change in membrane potential, the steady-state value of potential is proportional to the current injected, i.e., the membrane obeys Ohm's law, and its input resistance is calculated from the slope of the current–voltage curve (Figure 11C).

The delay in achieving the steady-state voltage at the beginning of the current pulse reflects the capacitance of the cell, which must be charged by the current delivered through the pipette, and which discharges after the current pulse has ceased. As in an RC circuit, the rate at which the voltage moves towards its final value depends on both the capacitance of the cell and its input resistance. The exponential time course of the voltage change at the beginning and end of the pulse is characterized by the time constant, τ, which is experimentally measured as the time required for the voltage to fall to 37% ($1/e$) of its steady state value after cessation of the current pulse (Figure 11D). Tau equals the product of the input resistance and the equivalent capacitance of the cell: thus, $\tau = RC$. Changes in either R or C change the time constant and thus change the time required for the voltage to change in response to a change in applied current.

Although a neuronal cell body may be spherical in shape, axons and dendrites are more analogous to cylinders. Because the internal axial resistance in an axon is not negligible compared to the transmembrane resistance, the interior of an axon is not isopotential; axons thus cannot be reduced electrically to a single equivalent RC circuit. Each small annulus of membrane is equivalent to an RC circuit, however, so that the equivalent circuit of a cylindrical axon is represented as an array of parallel RC circuits, as illustrated in Figure 12A.

When a pulse of current is injected into an axon, some of the injected charge leaves the membrane at the site of injection (through the RC circuit), depolarizing the membrane there. Some of it, however, flows longitudi-

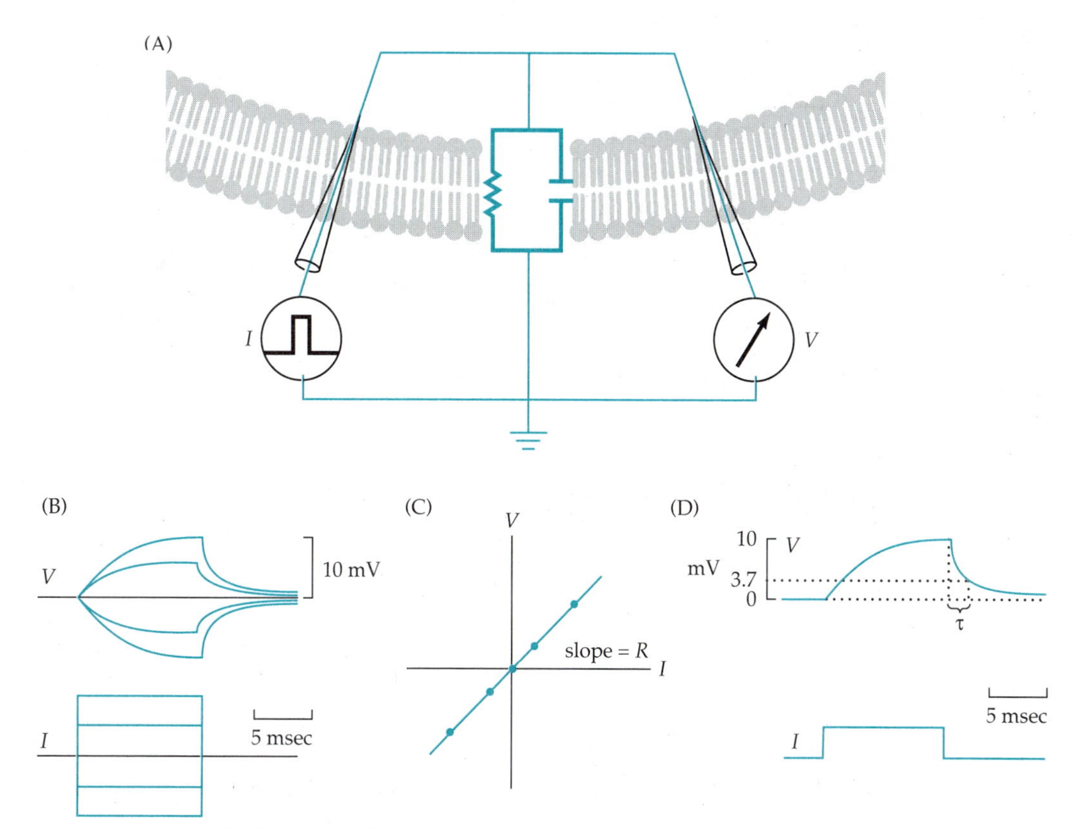

FIGURE 11. The equivalent resistance and capacitance of a cell membrane can be measured by noting its response to rectangular pulses of current. (A) Two glass micropipettes are inserted into a spherical cell, one for passing current and one for recording the resultant changes in membrane potential. Current flows through the current electrode into the interior of the cell and then across the membrane to the extracellular fluid. Some of the current charges the capacitor and some of the current passes across the resistor. (B) The current flowing across the membrane produces a voltage change that increases exponentially to a steady state value and falls exponentially to its initial value when the current is turned off. (C) The steady state voltage response of the cell membrane is linearly related to the current used to produce it as long as the voltage deflections are small and do not activate voltage-dependent channels. When the voltage has reached a steady state value, all the current is passing through the resistor, and the slope of the voltage–current relation for small deflections in voltage is equal to the equivalent resistance of the cell. (D) The time constant, τ, of the cell is the time required for the voltage to fall to 37% (1/e) of its steady state value after the current has been shut off. The time constant of the cell is the product of its equivalent resistance and equivalent capacitance. The equivalent capacitance can be calculated from the measured values of input resistance (R) and τ(RC).

nally along the axon to exit in subsequent segments, depolarizing them. (Note that most current within the axon is carried by K^+, as it is the major ionic species with a high mobility). At each point along the axon, as charge flows out across the membrane, correspondingly less charge is available to flow farther along the axon. The steady state change in potential produced

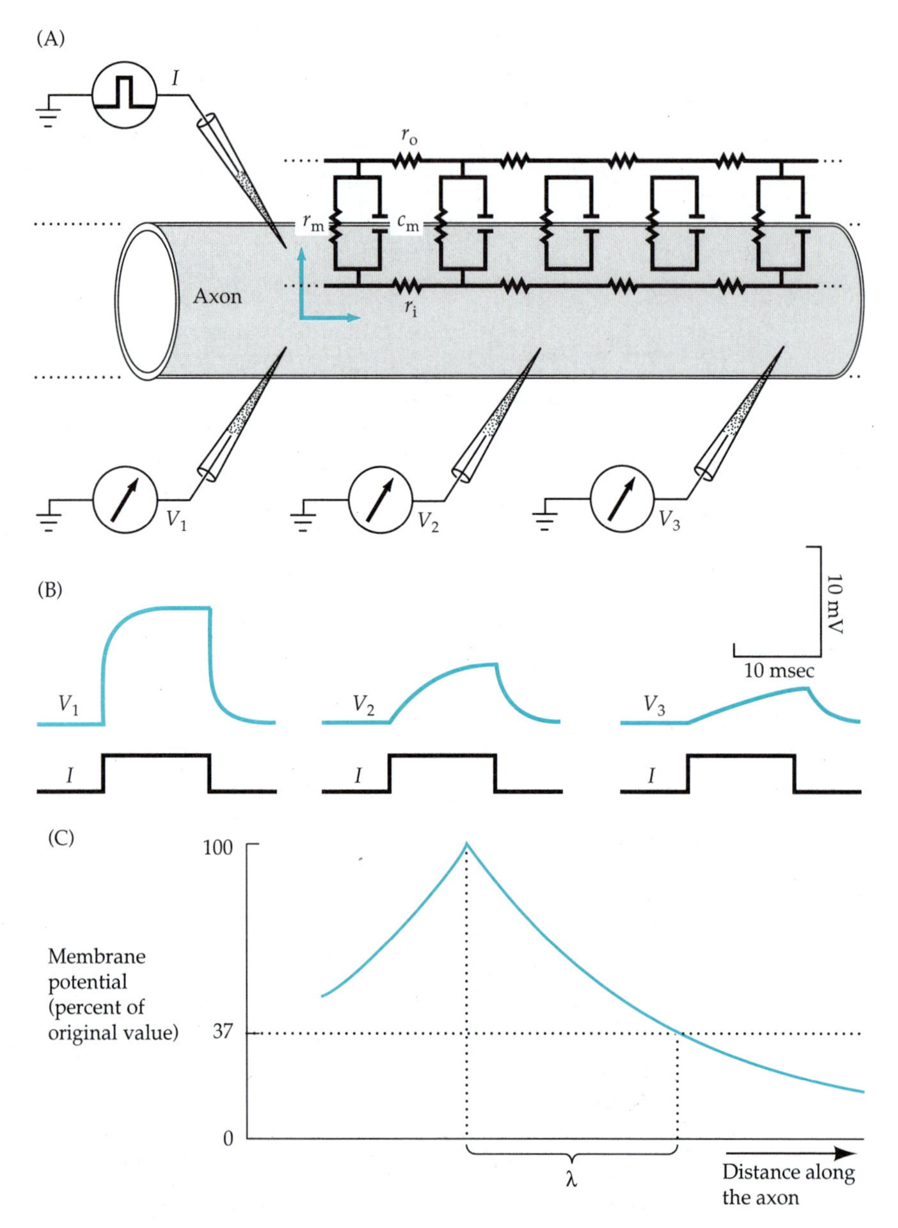

FIGURE 12. Local potentials in cylindrical structures (cables) decay as a function of distance from the site of origin. (A) A portion of a long cable is shown with a micropipette for passing current and three micropipettes for recording changes in voltage. Charge injected into the cell may leave the cylinder immediately or may pass some distance down its interior before exiting (the first choice point is marked by colored arrows. (B) Voltage responses at three different sites along the cable (V_1, V_2, and V_3) are shown in response to the injection of a rectangular pulse of current at site #1. Both the steady state voltage response and the speed with which the response reaches its steady state value decline with distance from the site of current injection. (C) The steady state voltage response decays exponentially as a function of distance from the site of current injection. The length constant, λ, is the distance over which the steady state response declines to 37% (1/e) of its original value.

by a current pulse thus declines monotonically with distance from the site of current injection (Figure 12B).

The relative values of the internal resistance and the membrane resistance across the axon determine the rate of decay of potential with distance along the axon. The lower the internal resistance, compared to the membrane resistance, the more current flows along the axon and the farther away the membrane is depolarized. Conversely, if the transmembrane resistance is low compared to the internal resistance, most of the current flows across the membrane at the site of injection, and very little remains to depolarize distant axonal segments.

The exponential profile of decay of the steady state potential along the axon is characterized by a length constant (or space constant), λ, which represents the distance over which the steady state value of voltage falls to 1/e (37%) of its initial value (Figure 12C). For most cylinders,

$$\lambda = \sqrt{\frac{r_m}{r_i + r_o}} \qquad (12)$$

where r_m is the membrane resistance of each annular axonal segment, r_i is its internal axial resistance, and r_o is the axial resistance of the external medium (r_o is usually ignored since it is negligible compared to r_i). If r_m is large relative to r_i, the length constant is large, and the potential decays slowly with length; alternatively, if r_m is small compared to r_i, the length constant is small, and the decay of voltage with distance is steep.

Both r_m and r_i are altered by changing the diameter of the cylinder. Doubling the diameter results in a twofold decrease in r_m, since there is twice as much membrane and twice as many channels per unit length (assuming that channel density remains constant). Doubling the diameter causes a fourfold decrease in r_i, however, because the cross-sectional area of the axon is increased by a factor of four. Therefore, doubling the diameter increases λ. For skeletal muscle fibers, the length constant is as large as two millimeters, and, for small, unmyelinated axons, as small as 100 μm.

Because of membrane capacitance, the potential changes recorded in an axon at a distance from the site of current injection are not only smaller than those recorded close to the site, but they are slower (Figure 12B). As the current flows down the axon, the capacitance of each successive segment of membrane must be charged, thus slowing the rate at which the axial current reaches more distal segments. Thus, local potentials produced by brief episodes of transmembrane current are both smaller and slower the farther from their site of origin they are recorded.

Now that we understand how local potentials decay with time and distance, we can consider how action potentials propagate and what determines their speed of propagation. As an action potential travels down an axon, part of the inward sodium current that produces the rising phase flows down the interior of the axon to produce a local potential in advance of the action potential. The local potential depolarizes the membrane past threshold, and the action potential advances to the next segment of axon. The speed with which the action potential advances depends upon how far ahead the local potential brings the membrane potential to threshold. This distance is determined by the passive electrical properties of the axon.

Consider an axon in which we freeze in time an action potential that is propagating from right to left (Figure 13). Each portion of the axon is thus undergoing a different phase of the action potential, so that a plot of membrane potential versus distance resembles a conventional figure of the action potential plotted as a function of time. At the segment where the membrane potential is rising, sodium current is entering the axon. Just as with the current delivered by a micropipette, some of the charge delivered by the current flows out across the membrane and some of it flows longitudinally down the axon. The charge that flows in advance of the axon depolarizes the membrane, activating voltage-gated sodium channels. When enough sodium channels are activated to reach threshold, there is a rapid and explosive increase in inward current that further depolarizes the membrane. This inward current then acts as the source for a local change in potential farther downstream. This process, repeated over and over again, moves the action potential along the axon.

The inward current during the rising phase of the action potential not only flows downstream, but also flows back along the axon (Figure 13). While this current may depolarize the membrane, it does not initiate an action potential because it encounters an axon segment in its refractory period, during which potassium conductance is high and sodium channels are inactivated. Thus action potentials propagate in only one direction.

Two strategies are used to increase axonal conduction velocity

The speed of conduction is determined by the resistance and capacitance of the membrane, and by the internal resistance of the axon. The longer the length constant, the further down the axon the local potential extends, and

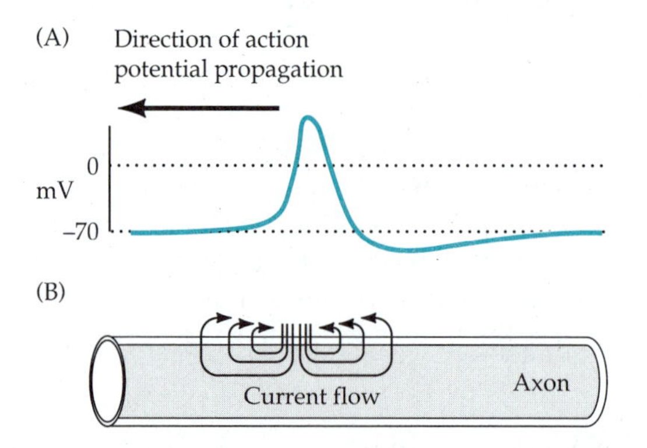

FIGURE 13. The entry of Na^+ into the active segment of a cylinder (axon, myofiber) during action potential propagation leads to local potentials that depolarize the next segment to threshold. (A) The membrane potential at an instant in time is shown vs. distance as it propagates from right to left. (B) The current entering the active part of the membrane flows in loops, just as it does when injected with a micropipette. This current depolarizes regions of the membrane in advance of the active segment and is responsible for propagation of the action potential.

the faster the action potential propagates. The smaller the capacitance, the more quickly the membrane potential at distant sites is brought to threshold. Thus, rapid action potential propagation is favored by a high transmembrane resistance, low internal resistance, and low membrane capacitance.

One strategy for increasing the rate of action potential propagation is to increase the diameter of the axon. As we have seen, increasing diameter has a greater effect on r_i than on r_m and thus increases the length constant. Some axons in invertebrates, such as the squid giant axon, are large and have high conduction velocities. With large numbers of neurons, this strategy is less successful, and another mechanism, myelination, is more often used. Glial cells wrap themselves many times around segments of axons to produce multiple layers of membranes (**myelin**) that ensheathe the axons (see Chapter 9). The myelin is periodically interrupted at the nodes of Ranvier by a short gap of unmyelinated axon. Myelin segments are typically about 1 mm in length, and the nodes of Ranvier about 10 μm.

Myelination has two effects on the passive electrical properties of axons: transmembrane resistance is increased, and membrane capacitance is decreased. Both effects arise from the stacked membranes, which are represented electrically as RC elements in series (Figure 14). The addition of membrane resistances in series increases r_m roughly by a factor equal to the number of membranes present, which may be as high as 200, thus increasing the length constant. The addition of membrane capacitances in series reduces the effective capacitance of the membrane. Thus less capacitative current is required to change membrane voltage and more is available to flow down the axon to depolarize downstream segments. Both effects, increased resistance and decreased capacitance, increase conduction velocity.

In myelinated axons an action potential in one node produces a local potential in the next node that depolarizes the membrane past threshold. In these axons the action potential jumps from node to node, a method of propagation known as **saltatory conduction.** Sodium channels are not needed in the internodal membrane and are confined to the node of Ranvier. Myelination not only increases the speed of axonal conduction but also decreases its metabolic cost. By restricting large ionic currents to the nodes fewer ions enter and leave the axons and fewer must be returned by active transport.

Myelination is very effective in increasing conduction velocity. Squid axons, which have a diameter of 500–1000 μm, conduct action potentials at about 20 meters per second at 20° C. By contrast, the axons of small myelinated sensory neurons, which have a diameter of only 1–5 μm, conduct action potentials at about the same rate. Myelin thus represents a means of increasing conduction velocity without requiring large increases in the volume of the nervous system.

Synaptic Potentials

Most communication between neurons occurs at **chemical synapses,** where a chemical transmitter released from the presynaptic nerve terminal

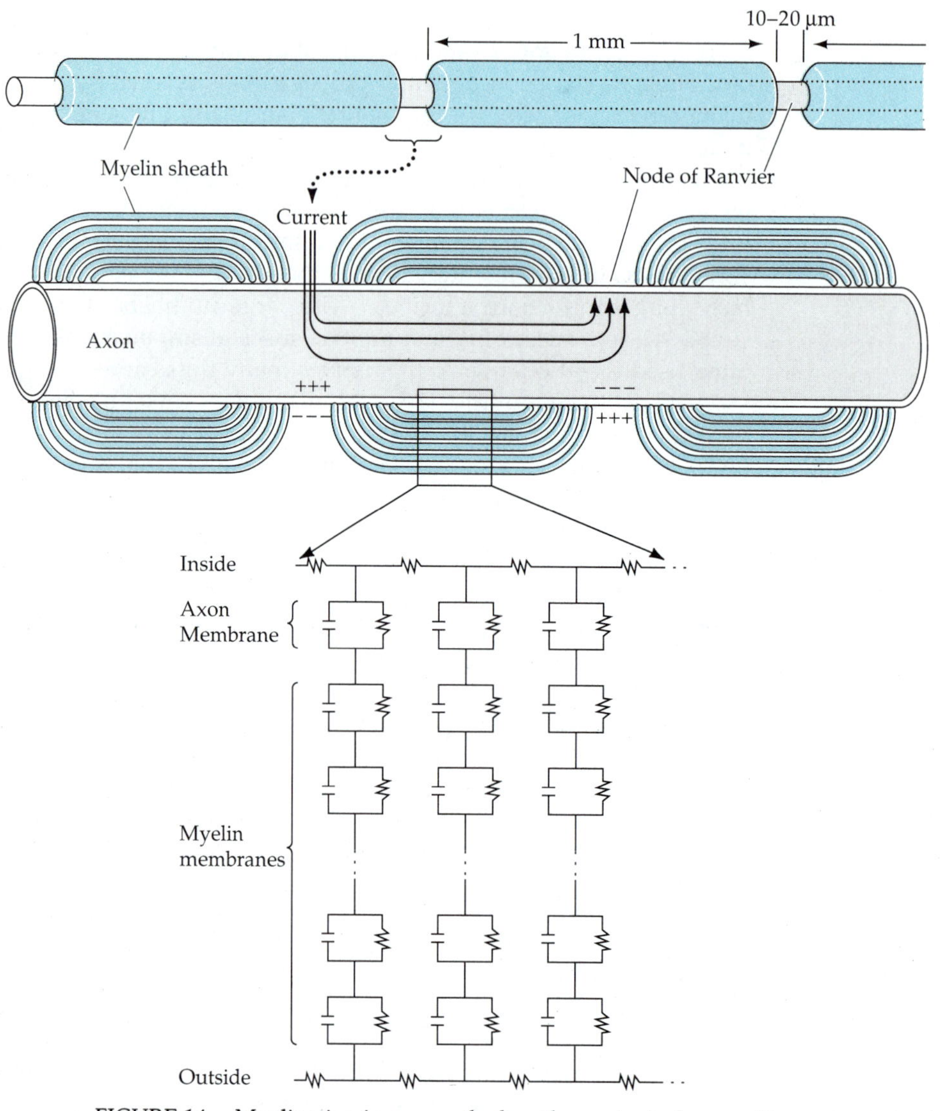

FIGURE 14. Myelination increases the length constant of axons. Current entering a myelinated axon at a node exits at the next node, unlike the continuous exit of current that occurs in unmyelinated axons (see Figure 12). The effects of myelin result from the high resistance and low capacitance afforded by the many concentric glial membranes that ensheathe the axon.

binds to receptors in the postsynaptic membrane and causes a permeability change that generates a local **synaptic potential**. Some neurons communicate, however, via **electrical synapses**, at which there is direct electric coupling between the cells.

Neurons are normally electrically isolated from each other. Even at chemical synapses in which cells are closely apposed, the 20– to 50–nm cleft between the pre- and postsynaptic cells acts as a low-resistance shunt so that very little of the current that flows across the presynaptic membrane during the action potential enters the postsynaptic cell (Figure 15A). Some neurons, however, are joined by **gap junctions** in which protein channels span the two membranes, allowing current to flow directly from one cell to the other (Figure 15B; also see Box A in Chapter 3).

The electrical coupling at gap junctions is seen by injecting current into one cell and then simultaneously recording from both it and its neighbor with intracellular microelectrodes. A current pulse that produces a depolarization in cell A also produces a smaller depolarization in cell B. Some of the injected current flows out across the membrane of cell A, and some of it passes through the channels of the gap junction to depolarize cell B. The ratio of the potential change in B to that in A, called the **coupling ratio,** depends on the relative values of the transmembrane resistance in cell A and the resistance for current that flows through the gap junction channels and across the membrane of cell B.

The gap junction channels, which are formed by proteins called **connexins,** are quite large; they are freely permeable to all molecules up to a molecular weight of about 1500 (see Box A in Chapter 3). Electric synapses transmit signals very rapidly (within 100 μsec) and, in invertebrates, are often found in reflex pathways involving escape or defense, where speed is at a premium. Electrical synapses are limited by the requirement that the pre- and postsynaptic cells be matched electrically. Thus the current generated by the presynaptic cell must be large enough to depolarize the postsynaptic cell. At most chemical synapses the presynaptic terminal is much smaller than the postsynaptic cell, and the current produced during an action potential in the terminal is insufficient to depolarize the postsynaptic cell, even if they were electrically coupled.

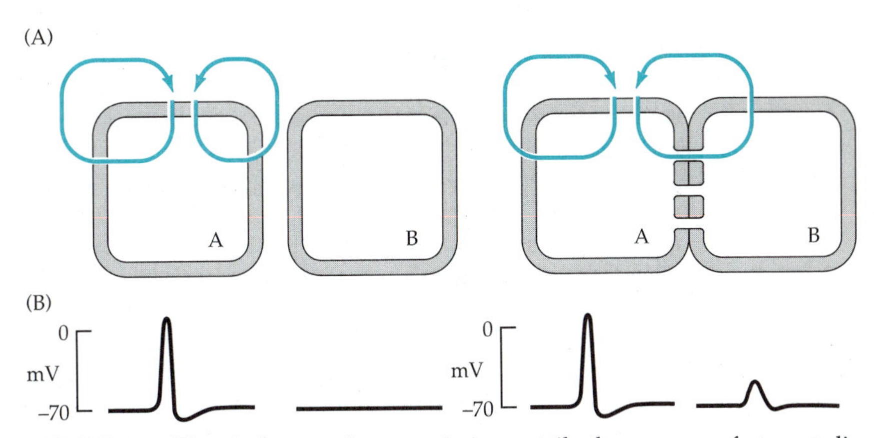

FIGURE 15. Electrical synaptic transmission entails the passage of current directly from the presynaptic to the postsynaptic cell. (A) The 20–50 nm gap separating most cells from their synaptic partners serves as a shunt for current. Very little of the current entering the presynaptic cell A during the rising phase of an action potential enters the postsynaptic cell B. (B) When excitable cells are electrically coupled, some of the current entering one cell passes directly into the other and depolarizes it.

The endplate potential at the neuromuscular junction is a local potential that reaches threshold

Chemical synaptic transmission, although slower than electrical transmission, allows amplification and flexibility of signal transfer between cells. The intensive study of a few simple and accessible synapses has provided much of our understanding about the principles of chemical transmission. One of these is the neuromuscular junction, the synapse between motor neurons and skeletal muscle fibers. Muscle fibers are long, cylindrical cells innervated near their midpoint by the nerve terminals of a motor neuron, which release acetylcholine (ACh) as the neurotransmitter. The released ACh binds the ACh receptors that are concentrated in the postsynaptic membrane, thus producing a synaptic potential, which at the neuromuscular junction is called the **endplate potential (epp)**. The endplate potential is unusual in that it always depolarizes the muscle cell past threshold, eliciting an action potential that is mediated by voltage-regulated sodium channels in the muscle membrane. Thus transmission is "one-to-one," as every action potential in the nerve invariably leads to an action potential in the muscle fiber. The effectiveness of neuromuscular transmission results both because the epp is large and because the threshold is low since there is also a high density of voltage-dependent sodium channels at the endplate membrane (Chapter 8). The action potential, once initiated, propagates from the endplate in both directions and induces muscle contraction.

The time course of the epp is normally obscured by the action potential that it elicits, but if the ACh receptors are partially blocked by a competitive antagonist such as curare, the epp is reduced to subthreshold levels and is observed without contamination. After partial blockade with curare, the epp recorded with an intracellular micropipette consists of a rapid depolarization, occurring in about 1 or 2 msec, followed by a gradual repolarization to the resting potential that occurs over several tens of milliseconds. The epp, like other local potentials, decays with distance. If the recording electrode is placed farther from the endplate, the epp is smaller and slower, exactly as would be predicted from the passive electrical properties of the muscle fiber. The epp behaves as though it were produced by a brief pulse of inward current at the endplate.

The current that produces the epp may be measured by voltage clamping the endplate to resting potential (Figure 16). Stimulation of the nerve then produces an inward **endplate current (epc)** lasting about 2–4 msec. At an unclamped endplate, a brief surge of inward current through the ACh receptor channel thus depolarizes the endplate membrane. After cessation of the current, the membrane repolarizes as K^+ and Cl^- flow down their electrochemical gradients. The rate at which membrane potential returns to the resting level is determined by the time constant of the membrane, i.e., by its resistance and capacitance.

Cation-selective channels generate the endplate current

The binding of ACh to ACh receptors, whose channels are normally closed, increases the probability of opening of ACh receptors, just as depolarization increases the probability that voltage-gated sodium channels open. The massive release of ACh at the endplate causes many ACh receptor channels to open simultaneously. ACh then dissociates from the receptors, and as it does, the ACh receptor channels close. The ACh is rapidly removed from the synaptic cleft by acetylcholinesterase (Chapter 5) and does not rebind receptors. The epc produced by nerve stimulation thus consists of a rapid rising phase, followed by an exponential falling phase that is the result of channels closing as ACh dissociates from its receptor.

Which ions carry the epc? Because the current is inward at normal resting potential, a plausible hypothesis is that the ACh receptor channel is selective for sodium ions. This hypothesis can be tested by voltage clamping the endplate membrane to the equilibrium potential for Na^+. If the channel were specific for sodium ions, stimulation of the nerve at this potential would produce no change in current across the membrane, because there would be no driving force on Na^+. A different and surprising result is obtained, however. Stimulation of the nerve produces a net *outward* current through the receptor channels (Figure 17). As the clamped potential is reduced from E_{Na}, the outward current becomes smaller and smaller. At about –15 mV stimulation produces no current, and clamping the potential at more and more hyperpolarized potentials results in an inward current of increasing size. The current thus reverses at a potential, the **reversal potential,** that is approximately halfway between E_K and E_{Na}. Recall from the earlier discussion of the imaginary cell that this behavior is precisely what is expected of a channel that is approximately equally permeable to Na^+ and K^+ (see Equation 8). At the reversal potential, inward sodium current equals outward potassium current, so that the net flow of current through the channel is zero. In a normal, unclamped muscle fiber at a resting potential of about –90 mV, stimulation of the nerve thus produces a net inward current that drives the membrane toward –15 mV. Endplate channels are also permeable to Ca^{2+} and Mg^{2+},

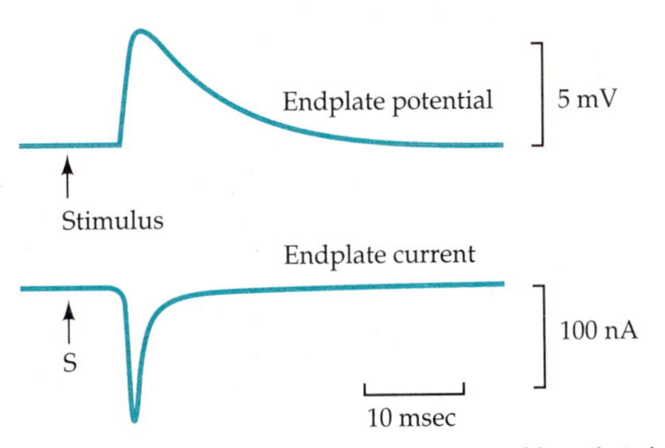

FIGURE 16. The endplate potential (epp) is generated by a brief surge of current, the endplate current. The currents that underlie the epp, which can be recorded by voltage clamp (lower trace), are considerably briefer than the potential change they produce (upper trace). Much of the falling phase of the epp occurs after the endplate current has returned to baseline. The time course of this part of the epp is determined by the passive electrical properties of the myofiber.

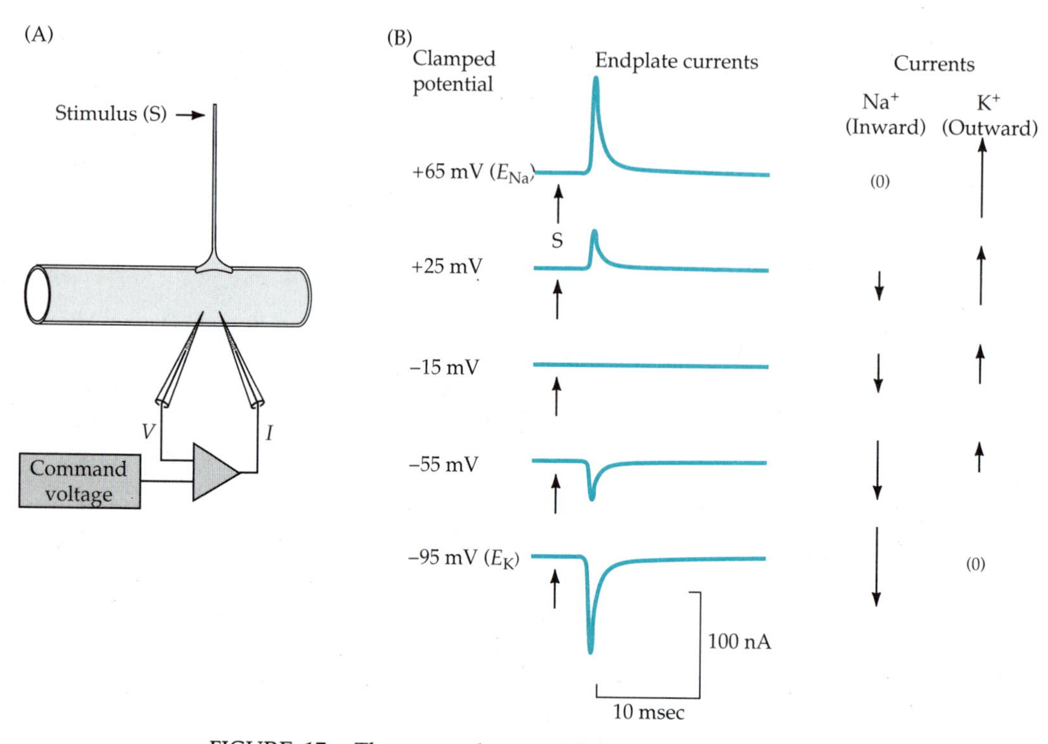

FIGURE 17. The reversal potential for acetylcholine-activated channels at the skeletal neuromuscular synapse is about −15 mV. (A) The membrane potential in the endplate area is clamped using a pair of micropipettes (one for recording membrane potential and one for passing current). (B) The membrane potential is stepped to different voltages, and the motor axon is then stimulated to evoke transmitter release (stimulus applied at "S"). The endplate current is inward and carried by Na^+ when the potential is held at E_K, and it is outward and carried by K^+ when the potential is held at E_{Na}. The polarity of the endplate current is reversed about midway between E_K and E_{Na}, at about −15 mV.

but these ions carry only a negligible proportion of the total current because of their relatively low concentration. The ACh receptor is thus a relatively nonselective cation channel. This conclusion, originally reached on the basis of electrophysiological studies of the neuromuscular junction, has been confirmed by study of ACh receptor channels in patch-clamp recordings (see Chapter 3).

In summary, ACh released by nerve terminals at the neuromuscular junction binds to ACh receptors in the postsynaptic membrane and thereby opens many cation-specific channels in the postsynaptic membrane. The flow of cations through the channels produces a large inward current that declines exponentially as the channels close. The inward current depolarizes the membrane past threshold, setting off an action potential that propagates from the endplate over the muscle fiber in both directions and causes the muscle to contract.

The same general mechanisms that produce the epp at the neuromuscular junction also operate at other excitatory synapses. The transmitter is usually not ACh, but is commonly glutamate, which is the major excitatory transmitter in the vertebrate CNS and a transmitter at the neuromuscular junction of many invertebrates. In both instances, glutamate opens a cation-selective channel having a reversal potential that is near 0 mV. One important difference between **excitatory postsynaptic potentials (epsp's)** and epp's is that the former are almost always below threshold and are often a millivolt or less in size. Small epsp's are important in the CNS because they sum with other potentials and allow integration to occur.

The synaptic potentials produced at inhibitory synapses **(inhibitory postsynaptic potentials,** or **ipsp's)** are usually hyperpolarizing instead of depolarizing, but result from the same mechanisms that are responsible for epsp's. Inhibitory neurotransmitters, commonly GABA or glycine, bind to receptors that open channels for K^+ or Cl^-. Because the equilibrium potentials for these ions are close to resting potential, increasing the conductance of the membrane to them stabilizes membrane potential near resting potential and makes depolarization of the cell to threshold more difficult. Hence, their effect is inhibitory.

Neurons in the vertebrate CNS receive hundreds or thousands of synaptic inputs that are integrated into an output signal of action potentials that travels down the axon toward the nerve terminal. Release of transmitter from each synapse on the dendrites or cell body generates an epsp or an ipsp, so that at any moment the cell may have hundreds of local synaptic potentials. Because the density of voltage-gated sodium channels on dendrites and cell bodies is usually low, action potentials do not ordinarily arise there. Instead, the local synaptic potentials are integrated by **spatial** and **temporal summation**. In spatial summation, the effects of synaptic currents arising from different sites on the cell are integrated according to the passive electrical properties of the neuron. Temporal summation refers to addition of successive synaptic potentials at a single synapse. From moment to moment, the currents produced at excitatory and inhibitory synapses sum to give a membrane potential that fluctuates in time and space. The fluctuating potential is "read out" at the initial segment, where there is a high density of voltage-dependent sodium channels and where threshold is correspondingly low. If the membrane potential at the initial segment exceeds threshold at any given moment, then an action potential is sent down the axon.

The Release of Neurotransmitters

Transmitter is released at nerve terminals by exocytosis at the active zone, a region of the terminal that directly apposes the postsynaptic membrane.

Electrophysiological experiments in which the postsynaptic membrane is used to detect transmitter release have given insight into the intermediate steps between invasion of the nerve terminal by an action potential and the release of transmitter.

Calcium entry couples action potentials to transmitter release

The arrival of an action potential at the nerve terminal increases the probability that neurotransmitter will be released. The first suggestion that Ca^{2+} linked the two events was the observation that omission of Ca^{2+} from the bathing medium blocked neuromuscular transmission. Removal of Ca^{2+} does not affect the presynaptic action potential or the action of ACh on the postsynaptic membrane, suggesting that it plays a role in transmitter release. Elevated extracellular Mg^{2+} also reduces transmitter release, apparently by antagonizing the effect of Ca^{2+}.

Experiments at the squid giant synapse, where the presynaptic nerve terminal is large enough to permit the insertion of microelectrodes, show that an increase in intracellular Ca^{2+} in the absence of depolarization stimulates transmitter release. Thus, Ca^{2+} is both necessary and sufficient for secretion. If a calcium-sensitive dye (see Figure 9 in Chapter 7) is injected into the presynaptic terminal, the time course of the increase in intracellular Ca^{2+} is seen to lag only slightly behind the action potential, consistent with its entry through voltage-gated channels. In voltage clamped squid nerve terminals, calcium currents display activation in response to depolarizing step changes in membrane potential but do not rapidly inactivate as do voltage-dependent sodium channels. Calcium channels activate slowly compared to sodium channels, accounting for much of the 0.5 msec delay that intervenes between depolarization and release. The relationship between calcium current and transmitter release is highly nonlinear, suggesting that several Ca^{2+} ions must bind simultaneously to a site within the terminal to trigger release.

Transmitter is released spontaneously

Paul Fatt and Bernard Katz discovered an important clue to the mechanism of transmitter release in 1951 when they observed small (approximately 0.5 mV), spontaneous depolarizations in unstimulated muscle cells (Figure 18B). Because the depolarizations had the same shape as the epp, they called them **miniature endplate potentials (mepp's)**. Mepp's are found only at endplates and are abolished by curare or by denervation, suggesting that they originate from the release of ACh by nerve terminals. At the neuromuscular junction, mepp's occur at random intervals and at an average frequency of about 1 per second.

Do mepp's represent the action of single ACh molecules? Fatt and Katz found that the application of ACh directly to the endplate did not increase the frequency of mepp's, and further, that the size of the mepp's could be reduced by curare. They therefore concluded that mepp's arise from the release of multimolecular packets, or **quanta,** of ACh. More recent experiments indicate that two molecules of ACh are required to open a single receptor channel and that opening a single channel at the frog neuromuscular junction causes a depolarization of about 0.2 μV. Hence, each mepp thus arises from the action of about 5000 molecules of ACh on the postsynaptic membrane.

Evoked transmitter release is quantal

Jose del Castillo and Bernard Katz proposed that quanta were the units of evoked release as well as of spontaneous release of neurotransmitter. Their

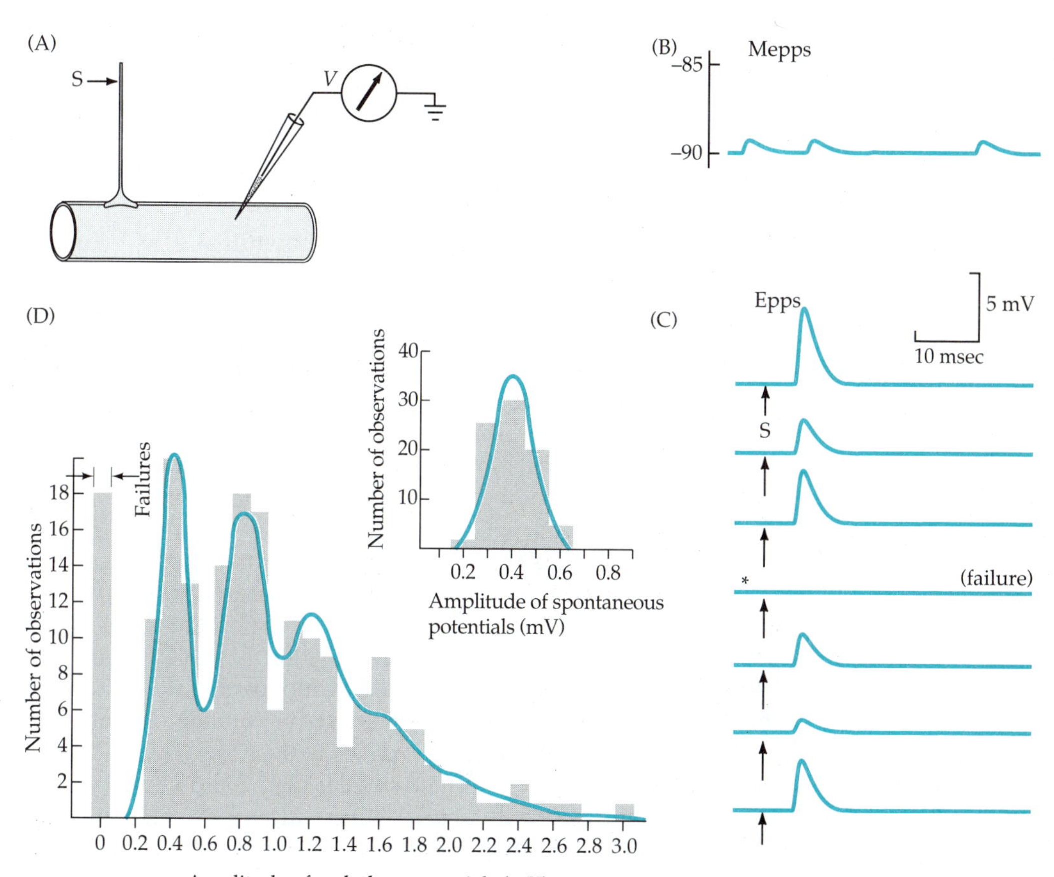

FIGURE 18. Transmitter is released in multimolecular packets called quanta. An action potential evokes the simultaneous release of many quanta. (A) The membrane potential at the endplate regions of a skeletal myofiber is monitored with a micropipette, and a stimulating electrode is placed on the motor axon. (B) Randomly occurring miniature endplate potentials (mepp's) of about 0.5 mV can be recorded from the endplate in the absence of stimulation. (C) Endplate potentials (epp's) fluctuate from trial to trial in response to a stimulus applied to the motor axon (indicated by an "S") when the muscle is bathed in a solution containing low Ca^{2+} and high Mg^{2+}. In one instance, marked by an asterisk, stimulation led to a "failure." (D) The amplitude distribution of epp's consists of a series of peaks that are integrals of the smallest response, which is identical in size to spontaneously occurring mepp's. The distribution of epp amplitudes is accurately predicted by the Poisson equation (solid line). (D from I. A. Boyd and A. R. Martin, 1956. J. Physiol. 132: 83.)

proposal was based on a careful analysis of the statistics of evoked neurotransmitter release. They found that as the concentration of Ca^{2+} was progressively decreased at the neuromuscular junction, the size of the epp at first decreased smoothly, but at low concentrations of Ca^{2+} and high concentrations of Mg^{2+} obvious fluctuations in epp size were apparent. In fact, some stimuli delivered to the motor nerve failed to elicit any response whatsoever (Figure 18C). The smallest epp's that were produced were approximately equal in size to the mepp. On the basis of their experiments, del Castillo and Katz proposed that the normal epp is composed of the same quanta that generate mepp's. Resting terminals release quanta randomly and at low probability; nerve stimulation sharply increases the probability of release. During an action potential in the motor nerve terminal, calcium entry raises the frequency of release by about five orders of magnitude, from one quantum per second to a few hundred quanta per millisecond.

A variety of subsequent experiments have shown that each quantum corresponds to the ACh contained in a single synaptic vesicle. The spontaneous release of ACh thus results from the fusion of individual vesicles with the presynaptic membrane. After an action potential, the simultaneous fusion of many vesicles results in a large potential change whose quantal components are obscured. Statistical analysis of the variation in endplate potentials at the neuromuscular junction gives an estimate of 200–300 for the number of quanta in a single epp, or its **quantal content.** At central synapses, the quantal content of postsynaptic potentials is smaller (see below).

The quantal content, m, is equal to the product of two variables: n, the number of quanta available for release, and p, the probability of release. At many synapses, the number of quanta released by stimulating the nerve over a number of trials is accurately predicted by the binomial distribution, which implies that each quantum is released independently. When the extracellular calcium concentration is lowered, the quantal content is reduced, and the binomial distribution simplifies to the Poisson distribution, which is applicable whenever p is much smaller than 1 and n is very large (Figure 18D). Reducing extracellular Ca^{2+} thus reduces p, the probability of release, rather than n, the number of quanta available for release.

The exact significance of n is uncertain. n is considerably less than the total number of vesicles in the terminal, and thus not all vesicles are available for release. One hypothesis is that n represents a subpopulation of vesicles that have undergone an initial step of exocytosis and are "cocked," awaiting the Ca^{2+} trigger (see Chapter 5). Another hypothesis is that n represents the number of release sites.

At the neuromuscular junction of crustacea and in the vertebrate CNS, inhibitory axons synapse directly on excitatory nerve terminals. As at conventional inhibitory synapses, the transmitter at these synapses is usually GABA or glycine. Stimulation of the inhibitory axon reduces the amplitude of the action potential that propagates into the excitatory terminal, reduces the amount of Ca^{2+} that enters, and thus decreases the amount of transmitter released. The ionic mechanisms that underlie presynaptic inhibition appear to be the same as those responsible for postsynaptic inhibition. Increasing the conductance of the nerve terminal to Cl^- or to K^+ (ions whose equilibrium potentials are close to the resting potential) reduces the depolarization caused by the action potential, since there is more outward current to oppose the inward current carried by sodium ions. The entry of Ca^{2+} through voltage-gated channels is reduced, thus decreasing transmitter release. In other cases, synapses on nerve terminals increase transmitter release by increasing the amount of Ca^{2+} that enters during the action potential (also see Chapter 14).

At many synapses, the response to the second of two paired stimuli is larger than the first response (Figure 19). The application of paired pulses of transmitter from a micropipette onto the postsynaptic neuron does not reproduce this effect; moreover, quantal analysis reveals that more transmitter quanta are released in response to the second stimulus than to the first. The **facilitation** of release in response to the second action potential appears to be caused by residual Ca^{2+} in the terminal that is incompletely removed following the first action potential. Following its entry into the terminal, the calcium concentration is reduced rapidly by a number of sequestration mechanisms (see Chapter 5). If a second presynaptic action potential invades the terminal before the internal calcium concentration is reduced to its basal levels, the total free calcium concentration in the terminal is greater than immediately after the first action potential, and an increased rate of release results.

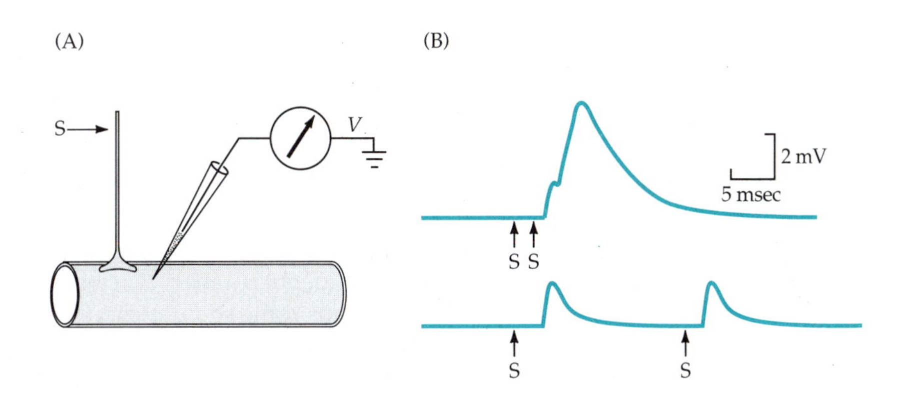

FIGURE 19. Previous electrical activity facilitates transmitter release. (A) A micropipette is inserted into a myofiber near the endplate, and a stimulating electrode is placed on the axon. (B) Facilitation occurs when the response of the target cell to the second of two stimuli delivered to a synaptic input is greater than that to the first (upper trace). Facilitation is not observed when the stimuli are widely spaced (lower trace). Quantal analysis shows that facilitation is explained by an increase in quantal content.

Facilitation represents an instance where the response of a target cell to stimulation of a presynaptic nerve terminal is dependent upon the past history of activity in that terminal. Other such examples are post-tetanic potentiation and long-term potentiation. As the names suggest, these phenomena describe prolonged changes in the response of the target cell resulting from intense but brief trains of stimuli delivered to a presynaptic neuron. Post-tetanic potentiation may last several minutes, while long-term potentiation can last hours or days.

Summary

Information is transmitted within the nervous system by electrical signals, which are produced when ions flow across the plasma membrane through ion channels. Ion concentration gradients provide the energy for electrical signaling, and each cell's complement of ion channels determines the electrical signals it generates.

The membrane potential of a cell at any time is determined by its relative permeability to different ions and by the concentration gradients of those ions across its membrane. Each ion moves across the membrane to bring the cell toward its equilibrium potential, which is determined by its concentration gradient as specified by the Nernst equation. The balance of the different ion currents then determines membrane potential. The more permeable a cell is to a particular ion, the closer its membrane potential will be to the equilibrium potential for that ion.

Ions are distributed asymmetrically across the cell membrane. The cytosol is rich in K^+ and in impermeant organic anions, while the extracellular fluid is rich in Na^+ and Cl^-. The asymmetric distribution of permeant ions results both from the presence of impermeant anions and from the Na^+/K^+ pump, which transports Na^+ out of the cell and K^+ in. In resting cells the permeability of the membrane to K^+ is greater than that to Na^+, and the membrane potential therefore lies close to the equilibrium potential for K^+, which is inside-negative. At rest neither K^+ nor Na^+ are at equilibrium, and each flows down its electrochemical gradient. The pump maintains the gradients for these ions, and the gradient provides stored energy for electrical signaling.

Action potentials are large, brief, positive-going signals that propagate long distances without decrement. The rising phase of the action potential results from inward sodium current, while the falling phase results from outward potassium current. A depolarization that brings the membrane potential beyond threshold opens some sodium channels, and the inward current through these channels opens still more channels, producing an explosive, regenerative signal. The action potential is terminated by the inactivation of sodium channels that have opened and by the time-delayed activation of potassium channels. The current entering the axon during the rising phase of the action potential spreads down the axon to depolarize neighboring regions to threshold. Myelin represents a specialization that facilitates the axial flow of current and that increases conduction velocity.

At chemical synapses the binding of a neurotransmitter to its receptor opens channels and permits ions to flow down their electrochemical gradients. At the neuromuscular junction, the resultant current produces a large endplate potential (epp) that elicits an action potential. Epp's are produced by the opening of cation-selective channels through which both Na^+ and K^+ flow. The net current through these channels moves the membrane toward a potential near –15 mV. A similar ionic mechanism occurs at central excitatory synapses. Inhibitory postsynaptic potentials are produced when transmitter opens channels permeable to Cl^- or K^+, whose equilibrium potentials are close to resting potential. Increasing the permeability to these ions makes depolarization to threshold by excitatory inputs more difficult.

Neurotransmitter is released in multimolecular packets, or quanta. Quanta are released randomly and at low frequency in the absence of stimulation. The arrival of an action potential at the terminal activates voltage-dependent calcium channels and leads to calcium entry, which evokes the simultaneous release of many quanta.

References

General References

Aidley, D. J. 1971. The *Physiology of Excitable Cells.* Cambridge University Press, Cambridge, Chapters 3–7.

Hodgkin, A. 1965. *The Conduction of the Nervous Impulse.* Liverpool University Press, Liverpool.

Kandel, E. R., Schwartz, J. H. and Jessell, T. M. 1991. *Principles of Neural Science,* 3rd Ed. Elsevier, New York, Chapters 6–11, 13.

Katz, B. 1966. *Nerve, Muscle and Synapse.* McGraw-Hill, New York.

Nicholls, J. G., Martin, A. R. and Wallace, B. G. 1992. *From Neuron to Brain,* 3rd Ed. Sinauer Associates, Sunderland, MA, Chapters 1–11.

Resting Potential

*Baker, P. F., Hodgkin, A. L. and Shaw, T. I. 1962. The effects of changes in internal ionic concentrations on the electrical properties of perfused giant axons. J. Physiol. 164: 355–374.

Bernstein, J. 1902. Untersuchungen zur Thermodynamik der bioelektrischen Strome. Pflügers Arch. 82: 521–562.

Boyle, P. J. and Conway, E. J. 1941. Potassium accumulation in muscle and associated changes. J. Physiol. 100: 1–63.

Hodgkin, A. L. and Horowicz, P. 1959. The influence of potassium and chloride ions on the membrane potential of single muscle fibres. J. Physiol. 148: 127–160.

Hodgkin, A. L. and Keynes, R. D. 1955. Active transport of cations in giant axons from *Sepia* and *Loligo.* J. Physiol. 128: 28–60.

Action Potentials

*Hamill, O. P., Marty, A., Neher, E., Sakmann, B. and Sigworth, F. J. 1981. Improved patch-clamp techniques for high-resolution current recording from cells and cell-free membrane patches. Pflügers Arch. 391: 85–100.

Hodgkin, A. L. and Huxley, A. F. 1952. The components of membrane conductance in the giant axon of *Loligo.* J. Physiol. 116: 473–496.

Hodgkin, A. L. and Huxley, A. F. 1952. Currents carried by sodium and potassium ions through the membrane of the giant axon of *Loligo.* J. Physiol. 116: 449–472.

Hodgkin, A. L. and Huxley, A. F. 1952. The dual effect of membrane potential on sodium conductance in the giant axon of *Loligo.* J. Physiol. 116: 497–506.

*Hodgkin, A. L. and Huxley, A. F. 1952. A quantitative description of membrane current and its application to conduction and excitation in nerve. J. Physiol. 117: 500–544.

Hodgkin, A. L., Huxley, A. F. and Katz, B. 1952. Measurement of current-voltage relations in the membrane of the giant axon of *Loligo.* J. Physiol. 116: 424–448.

*Hodgkin, A. L. and Katz, B. 1949. The effect of sodium ions on the electrical activity of the giant axon of the squid. J. Physiol. 108: 37–77.

Hodgkin, A. L. and Rushton, W. A. H. 1946. The electrical constants of a crustacean nerve fibre. Proc. R. Soc. Lond. B 133: 444–479.

Huxley, A. F. and Stämpfli, R. 1949. Evidence for saltatory conduction in peripheral myelinated nerve fibres. J. Physiol. 108: 315–339.

Synaptic Potentials

Coombs, J. S., Eccles, J. C. and Fatt, P. 1955. The specific ionic conductances and the ionic movements across the motoneuronal membrane that produce the inhibitory post-synaptic potential. J. Physiol. 130: 326–373.

*Fatt, P. and Katz B. 1951. An analysis of the end-plate potential recorded with an intracellular electrode. J. Physiol. 115: 320–370.

Furshpan, E. J. and Potter, D. D. 1959. Transmission at the giant motor synapses of the crayfish. J. Physiol. 145: 289–325.

Takeuchi, A. and Takeuchi, N. 1960. On the permeability of end-plate membrane during the action of transmitter. J. Physiol. 154: 52–67.

Augustine, G. J., Charlton, M. P. and Smith, S. J. 1985. Calcium entry and transmitter release at voltage-clamped nerve terminals of squid. J. Physiol. 367: 163–181.

Boyd, I. A. and Martin, A. R. 1956. The endplate potential in mammalian muscle. J. Physiol. 132: 74–91.

*del Castillo, J. and Katz B. 1954. Quantal components of the end-plate potential. J. Physiol. 124: 560–573.

Dudel, J. and Kuffler, S. W. 1961. Mechanisms of facilitation at the crayfish neuromuscular junction. J. Physiol. 155: 530–542.

Fatt, P. and Katz, B. 1952. Spontaneous subthreshold activity at motor nerve endings. J. Physiol. 117: 109–128.

3

Ion Channels

Zach W. Hall

ION CHANNELS ARE the fundamental elements of signaling in the nervous system. They generate the electric signals of neurons; they regulate secretion of neurotransmitters; and they convert into electrical responses both the chemical and mechanical stimuli that act upon the cell from the outside and the chemical signals that are generated within cells. The properties of ion channels are also intimately connected with the long-term changes in signaling that underlie the plasticity of the nervous system. Their utility for signaling arises from two sources: their ionic specificity and their susceptibility to regulation. Ionic specificity is important because it allows different ions to flow across the membrane independently. The independent movements of Na^+ and K^+, for example, make the action potential possible. The regulation of ion channels by a variety of stimuli, often in complex ways, gives flexibility to neural signaling.

During the last twenty years, new techniques of protein biochemistry, patch-clamping, and molecular cloning have revolutionized the study of ion channels. In 1970, only a few different voltage-sensitive ion channels and a handful of neurotransmitter receptors were known. Today we know of at least 75 different channels, and expect to discover as many more in the next few years. The unexpected variety of channels means that signaling in the nervous system is more varied and more intricate than previously believed. Definition of the physiological and pharmacological properties, molecular structures, cellular locations, and physiological roles of ion channels will help us understand the basis of these complex signals and the functional roles that they serve. Amid their diversity, however, the fundamental molecular problems of ion channel function remain the same: how are they constructed; how do they facilitate the movement of ions across the membrane; what is the basis of their specificity; and how are they regulated?

Molecular biology and biophysics give us the tools to address these questions. Although the task seems discouragingly large, we are helped by the emergence of several general principles. One is that many ion channels are built according to a common design in which homologous polypeptides or protein domains surround a central aqueous pore. Moreover, only a few basic patterns govern the molecular structure of the polypeptide chains, so that most ion channels belong to one of several large families. In this chapter, we will first discuss the principles of ion permeation in simple systems and then go on to consider in detail two of these families, the channels that are gated by neurotransmitters, and voltage-gated channels.

Ion Permeability of Membranes

Neuronal signaling is based on the movements of ions across membranes. To understand how protein channels facilitate their movement we first need to examine the behavior of ions in aqueous solution, what impedes their movement across membranes, and how simple molecules aid their transit.

Ions in solution bind water molecules

Although ions diffuse freely in aqueous solution, they rarely cross phospholipid bilayer membranes. They fail to do so because of the different chemical structures of water and the bilayer membrane, and the different interactions of ions with each. Water molecules are small dipoles that interact with each other via hydrogen bonds. The collective strength of these bonds, which at physiological temperature are constantly being made and broken, gives water many of its unusual properties, including its relatively high viscosity, dielectric constant, and surface tension.

Because they are dipoles, water molecules interact with ions via electrostatic forces. A single ion in solution, such as Na^+, is surrounded by water molecules that are partially oriented so that on average their more negative end is closest to the positively charged ion. Because the electrostatic forces are not strong enough to hold them in a fixed position, the water molecules around an ion continually exchange with others at a rate of approximately 10^9 per second. In contrast, water molecules make and break bonds with each other at a rate of about 10^{11} per second. In spite of their transient quality, the bonds between ions and water are, in aggregate, surprisingly strong. The energy required to remove a sodium ion from aqueous solution is about the same as that required to extract it from a crystal of sodium chloride.

The ability of an ion to structure the water molecules around it affects its mobility in solution. When a sodium ion moves in an electric field, for example, the water molecules around it pull against it as it moves and to some extent are moved by it. The ion in solution moves more slowly than is expected from its size in a crystal and thus appears to have a larger radius.

Small ions interact more strongly with water than large ions, as is reflected by their relative **hydration energies** (Table 1), the energy gained by transferring an ion from a vacuum into aqueous solution. According to one explanation, water molecules can get closer to the center of charge of smaller ions, and by Coulomb's Law (electrostatic force is proportional to the inverse square of the distance between charges), the force, and hence

TABLE 1. Ionic Radii and Hydration Energies of the Alkali Metal Cations

Cation	Ionic radius (Å)[a]	Free energy of hydration (kcal/mol)[b]
Li^+	0.60	–122
Na^+	0.95	–98
K^+	1.33	–80
Rb^+	1.48	–75
Cs^+	1.69	–67

Sources: [a]L. Pauling, 1960. *Nature of the Chemical Bond and Structure of Molecules and Crystals,* 3rd Ed. Ithaca, NY: Cornell University Press. [b]J. Edsall and W. McKenzie, 1978. Adv. Biophys. 10: 137–207.

the total energy of interaction, between an ion and its surrounding water molecules will be larger for small ions than for large ones. A large ion such as Cs^+ can shed its associated water molecules more easily and rapidly than a small ion such as Li^+. As we will see later, this difference has important consequences for the ion selectivity of channels.

Membranes pose a barrier to ion movement

The basis of biological membranes is a **phospholipid bilayer** whose interior consists of the long, nonpolar, fatty acid side chains of the phospholipid molecules. The interior of the membrane thus forms a thin, oil-like film around the cell. Because they are nonpolar, the hydrocarbon side chains interact with each other only by weak, short-range forces. In aqueous solution, a hydrocarbon chain structures the water around it and disrupts water–water interactions. To minimize the energetically unfavorable contact of hydrocarbon chains with water molecules, phospholipids in solution form a bilayer.

One can readily see why ions have such difficulty crossing lipid bilayer membranes. They must either drag water molecules with them and thus increase the contact of water with the hydrocarbon side chains, or they must divest themselves of the water molecules with which they strongly interact and come in direct contact with the hydrocarbon side chains with which they interact almost not at all. Either alternative is energetically unattractive: one can calculate that an ion, given a choice, is 10^{72} times more likely to be in aqueous solution than in a hydrocarbon environment.

Pores and carriers

Relatively simple molecules can increase specific ion movements across membranes. They form two groups that operate by different mechanisms: **carriers** and **pores.** The antibiotic valinomycin, for example, is a carrier that specifically increases the potassium permeability of membranes to which it is added. A valinomycin molecule in the membrane forms a specific complex with K^+ at one surface, diffuses across the membrane, and releases the ion on the other side. Gramicidin, in contrast, forms a relatively stable aqueous pore across the membrane through which small cations, but not anions, pass. Although the mechanisms in each case are different, the fundamental principle is the same. In both cases, polar and nonpolar groups are combined in a single molecule that adopts a configuration in the membrane that allows the polar groups to interact with the ion, while the nonpolar groups interact with the **hydrophobic** interior (Figure 1). The interaction of the ion with the carrier or pore offers at least partial compensation for the lost interactions with water and makes partition of the ion into the membrane energetically feasible.

The most important difference between carriers and pores is the maximum speed that they permit ions to flow. Ions move more quickly through pores just as cars cross a river more rapidly through a tunnel than by ferry. The rate of diffusion of the carrier–ion complex limits the maximum rate of carrier-mediated transport to about 10^5 ions per second. Ions go through pores at much faster rates, up to 10^8 ions per second. Although the idea that ions move through pores in biological membranes dates back almost 150 years, estimates of the rates of ion flow through individual ion channels in biological membranes have only been made within the last 20 years. The high rates observed (about 10^7 ions per second) have made it clear that ion channels function as pores and not as carriers.

Pores are like enzymes

Just as enzymes do not determine the direction of chemical reactions, but only speed them up, pores do not determine the direction of ion flow, but

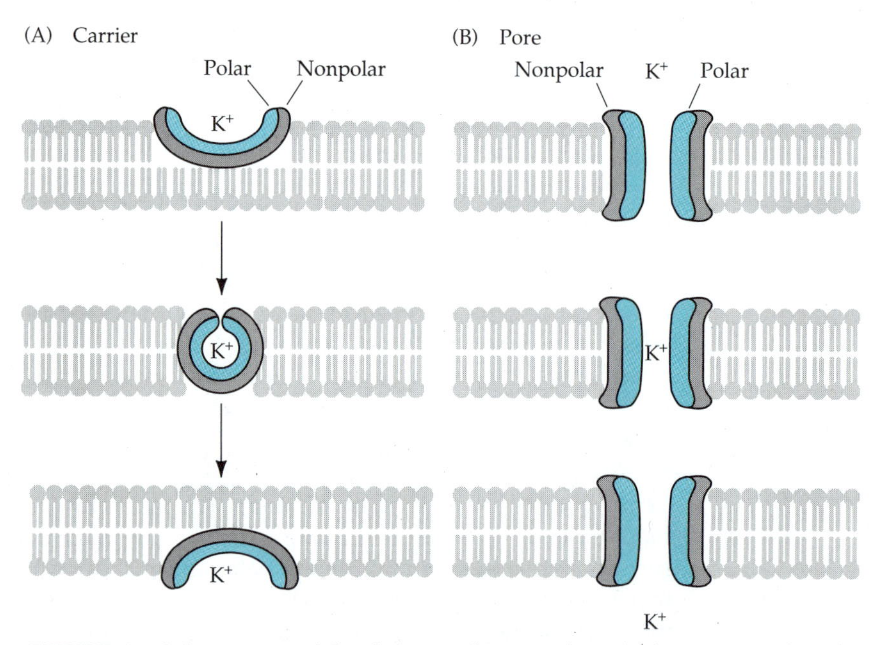

FIGURE 1. Schematic models of the mechanisms by which carrier and pores facilitate ion movement across membranes. Carriers and pores work by the same principle. The pore (or carrier) has both nonpolar (shaded) and polar (open) parts. The nonpolar portion of the molecule interacts with the membrane and the polar part interacts with the ion.

change its rate (Figure 2). The *direction* of ion flow depends on the relative concentrations of ions and the difference in electrical potential on the two sides of the membrane as expressed in the Nernst equation (Chapter 2). The *rate,* on the other hand, depends on the height of the highest energy barrier between the initial and final states. In a chemical reaction the height of this barrier is the activation energy; for ion flow, it corresponds to the energy required to move an ion into the hydrophobic interior of the membrane. Pores reduce the height of the energy barrier, which, as we have seen, is virtually insurmountable in their absence. As we discuss below, most ions must at least partially dehydrate in order to enter a channel. Pores provide chemical groups or sites within the membrane to which the dehydrated ions can transiently bind. Because there is a fixed number of sites, ion flow through a pore "saturates" at high levels of substrate, just as enzyme reactions do. Interaction of the sites with ions is brief; otherwise ions would not move through the pore.

To make flow more rapid, why not have a pore large enough to allow ions to pass through without dehydrating? A large pore, such as gap junctions form (see Box A), permits a rapid flow of ions, but at the price of losing ion specificity. As discussed below, both the small size of the pore and the nature of the ionic interaction with the site appear to be important for determining ion specificity.

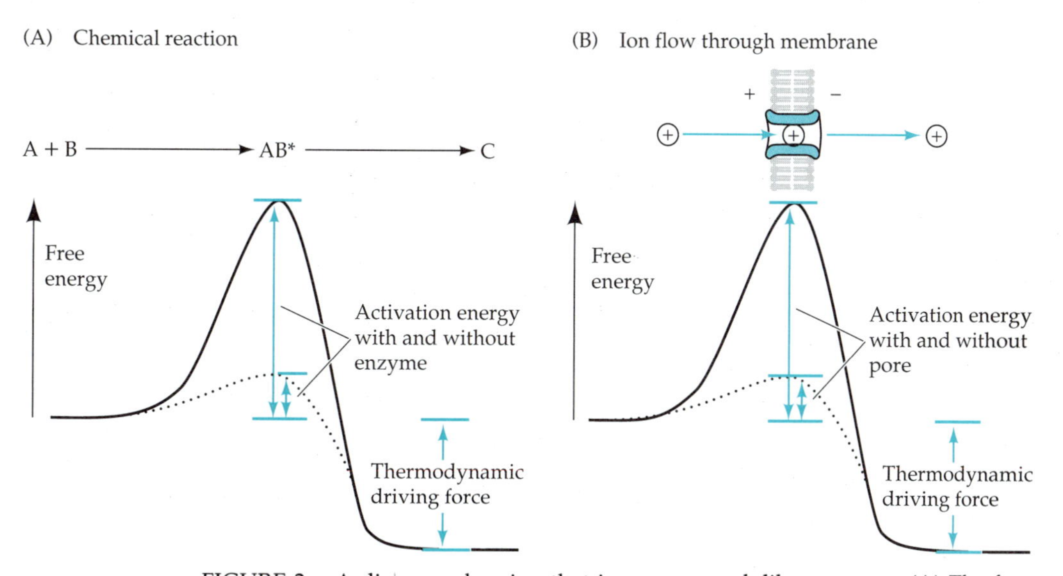

FIGURE 2. A diagram showing that ion pores work like enzymes. (A) The free energy of a chemical reaction plotted against the reaction coordinate (which represents the progress of the reaction). The rate of the reaction in either direction depends on the activation energy; its direction depends on the difference between the energies of the initial and final states. An enzyme lowers the activation energy of the reaction without affecting the energies of the initial and final states. (B) The free energy profile of transport as an ion moves across a lipid bilayer membrane. The rate at which the ion moves depends upon the height of its highest energy state, which is in the middle of the membrane. A pore decreases the height of this energy barrier without affecting the electrochemical gradient across the membrane.

Box A Gap Junctions

Electrical synapses between cells are formed by gap junctions, which allow ions to flow from the interior of one cell directly to the interior of another without passing through the extracellular fluid (see Chapter 2). The channel connecting the two cells is quite large (about 1.6 nm) and allows the passage not only of ions but also of other small molecules such as metabolites and second messengers. Gap junctions between excitable cells allow rapid signaling and coordination of electrical activity between neurons. Invertebrates use them for rapid reflexes; in vertebrates they are not only found in some neurons but also connect the muscle cells of the heart and some smooth muscles. Gap junctions between astrocytes presumably increase the cells' ability to buffer the potassium concentration in the extracellular space. Although they do not open and close rapidly, gap junctions can be regulated by changes in the intracellular concentrations of calcium or of protons.

The structure of gap junctions follows the motif seen in other channel proteins in which protein subunits surround a central pore. In this case, two hexameric half-channels in each cell are joined to give a continuous pore linking the two cell interiors (Figure A). A single protein, **connexin**, forms the gap junction. Several homologous forms of connexin are expressed in different tissues; the two main types have molecular weights of 32 or 43 kD. Each subunit has N- and C-terminal cytoplasmic domains and four transmembrane domains.

Structural analysis of gap junctions, facilitated by their simple composition and regular arrangement in the membrane, has not only given us a clear picture of their architecture but has also suggested a model of how they might open and close. The transmembrane segments of connexin are α helices that are not strictly aligned with the long axis of the pore but are slightly tilted relative to it. Interestingly, at high calcium concentrations, a condition that closes the channel, the degree of tilt is decreased, causing a displacement at one end of the subunits of up to 9 Å, and a narrowing of the pore diameter. The change in tilt of the helices is accompanied by a slight twisting, which would change the position of the amino acid residues lining the pore. Such a twist could easily move bulky residues into the channel, thus blocking it. A similar mechanism might mediate channel gating in other regulated ion channels.

(1)

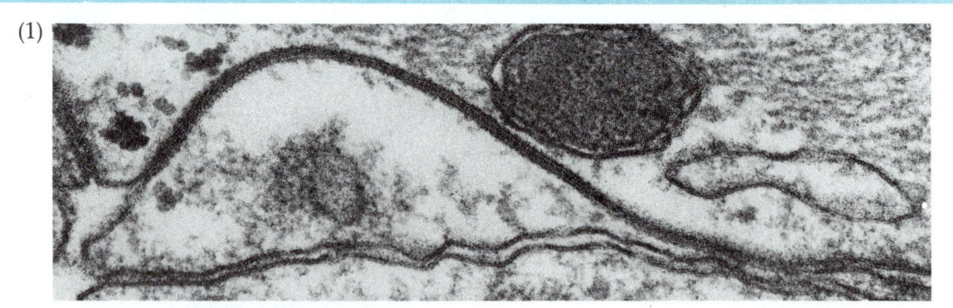

(2)

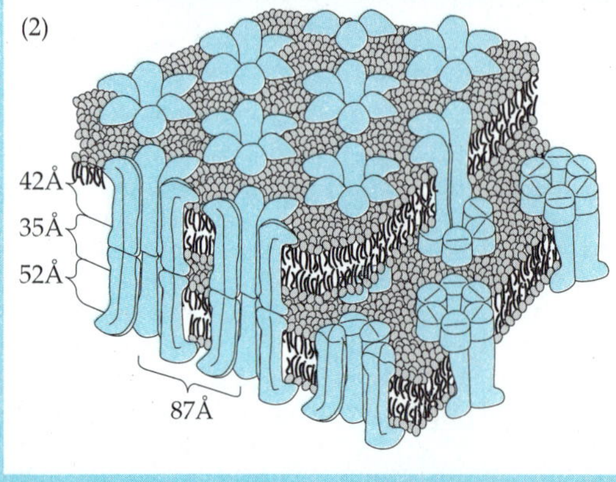

FIGURE A. The structure of the gap junction. (1) A thin-section electron micrograph showing the close apposition of the membranes at a gap junction formed by two astrocytes. (Micrograph courtesy of D. McDonald.) (2) A schematic diagram of the molecular structure of the junction. Six connexon molecules in each cell assemble to form a half-channel that is joined to its partner to make a pore that connects the cytoplasm of the two cells.

Simple peptides can form specific ion pores

The antibiotic **gramicidin** forms a simple channel whose chemical and physical structures are known. Gramicidin is a polypeptide of fifteen hydrophobic amino acids, alternately in D- and L- forms, with both N and C terminals blocked so that the molecule is completely uncharged and is readily soluble in a phospholipid membrane. In a bilayer, the alternating D- and L- amino acids form an unusual helix around a central aqueous pore. The hydrophobic amino acid residues are on the outside of the helix and the relatively polar amide nitrogens and carbonyl oxygens of the peptide backbone line the pore. Two helices align head-to-head in the membrane to give a dimer long enough (30 Å) to span the bilayer.

The pore formed by gramicidin is permeable to water and is much more permeable to small cations than to anions. Among the alkali metal cations, Cs^+, whose ionic radius is largest (1.69 Å), moves through most easily, and Li^+, the smallest (0.60 Å), with most difficulty. (A plausible explanation for this apparent anomaly is discussed below.) Kinetic experiments show that ions traverse the channel in single file, separated by water molecules. Given the ionic radii of the alkali metal cations (Table 1), the size of water molecules (2.8 Å), and the diameter of the channel (4 Å), ions clearly must partially dehydrate in order to fit through the pore. Interactions with the oxygens and amide nitrogens on the pore interior replace the lost interactions with water molecules.

Polypeptides in animal cells contain only L-amino acids, and form α helices without a central pore. α helices are a favored configuration for polypeptide segments that traverse membranes because the amide nitrogens and carbonyl oxygens in the peptide backbone are hydrogen-bonded to each other and are thus unavailable to water molecules. If the amino acid residues in the helix are hydrophobic, there are no exposed or nonbonded polar groups, so that the helix associates with the membrane rather than with water.

Although a single, transmembrane α helix does not itself make a pore, several can associate to form a central pore. A simple ion pore can be made by a 21 amino acid synthetic peptide consisting only of serine and leucine in the ratio 2:1 (Figure 3). The sequence of the peptide is designed to make an α helix with the relatively polar serine residues aligned along one side. When added to a lipid bilayer membrane, six helices spontaneously assemble to form an aqueous pore lined by serine hydroxyl groups. The pore, which appears to have a diameter of 8 Å, is permeable to cations. Remarkably, it has an ion specificity and rate of ion flow that is similar to that of the acetylcholine (ACh) receptor. Clearly, ion channels do not need to be very complicated molecules.

Ion specificity depends on pore size and the strength of the interaction with the pore site

One of the most important and remarkable features of channels in excitable cells is their ion specificity. Some channels, such as the ACh receptor, are highly specific for small cations, but show only slight preferences among

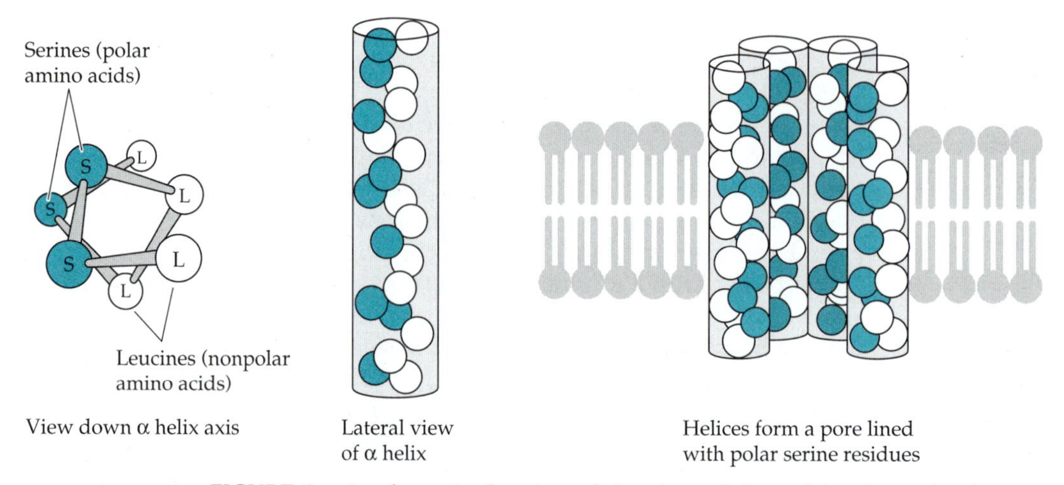

FIGURE 3. A schematic drawing of the channel formed by the serine-leucine peptide. The peptide is designed to form a helix with the relatively polar serine groups on one side and the nonpolar leucine groups on the other. Six α helices are thought to associate in the membrane with their polar sides forming the walls of an aqueous pore.

them. Others, such as potassium channels, are extremely selective, preferring K^+ to Na^+, for example, by ratios as high as 100:1.

How can channels discriminate between different ions? Enzymes recognize their substrates by their distinctive shapes or chemical structures. As Na^+ and K^+ have the same shape and electrical charge, how can they be distinguished? The apparent key is ionic size.

The simplest way to regulate the ions that flow through a channel is to restrict the pore size (Figure 4). When the rates of flow of a series of cations of known size are tested on the channel formed by the serine-leucine peptide, for example, monovalent cations such as glucosamine (10 Å) do not traverse the pore easily, whereas ions with sizes of 8 Å or smaller do. Similar experiments, carried out on the ACh receptor, which is a relatively nonspecific cation pore, indicate a pore diameter of approximately 6.5 Å.

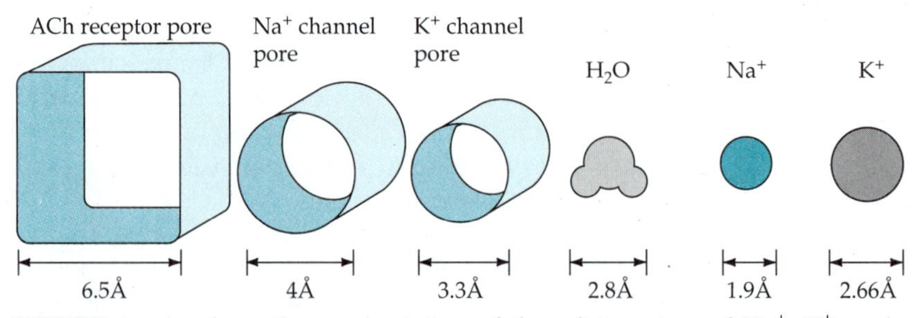

FIGURE 4. A schematic representation of the relative sizes of Na^+, K^+, and a water molecule; the pore sizes of the ACh receptor of frog muscle; and the voltage-sensitive sodium channel and voltage-sensitive potassium channel of frog nerve, as estimated by electrophysiological methods.

The estimated pore diameter of the voltage-regulated sodium channel is 4 Å, and that of the voltage-regulated potassium channel even smaller (3.3 Å).

Consideration of potassium channels, however, indicates that more than size must be involved in ion specificity. We can understand how the channel excludes larger ions such as Cs^+ or Rb^+ on the basis of size. But how do potassium channels exclude Na^+, a smaller ion? A possible clue comes from the preference of the gramicidin pore for Cs^+ rather than for the smaller Na^+. For Na^+ to enter the pore, it must lose its association with water molecules; interactions with the relatively weak polar groups lining the pore offer only partial recompense. Theoretical calculations suggest that in any channel in which the interaction of the permeant ions with the groups lining the pore is relatively weak, the dominant factor that determines which ion is preferred is the ease of dehydration. Since large ions can shed their associated water molecules more easily than small ions (Table 1), a pore with weakly polar sites such as those seen in the gramicidin pore will prefer large ions to small ones.

In channels whose pores have strongly charged sites, the interaction of the dehydrated ion with the site may be more important in determining preference than the ease of dehydration. In those cases, smaller ions have an advantage over larger ions because they can approach the site more closely and interact more strongly with it. Although we do not yet know their molecular structures, we thus anticipate that the pore of a channel specific for Na^+ will be lined by residues that are more polar than a channel specific for K^+. Because at least some channels in biological membrane exhibit more extreme specificities than those of simple peptides, we may also expect other mechanisms to be at play.

Armed with knowledge about how ions flow through simple pores, we can now consider the more complicated channels of excitable cells. Although we know which molecules are involved, we as yet know relatively little about how they work; we will find that they are much larger, with more complex modes of regulation, and that they have additional mechanisms for increasing specificity and rates of transport.

Ligand-Gated Ion Channels

Neurotransmitters regulate the opening and closing of an ion pore by binding to a site on the channel protein. The best-known members of this group respond to the extracellular neurotransmitters ACh, γ-aminobutyric acid (GABA), glycine, and glutamate, and mediate rapid synaptic transmission in the central and peripheral nervous systems. Many of these channels share a common structure whose subunits have homologous protein sequences.

Other channels are regulated by internal ligands. Homologous channels with intracellular binding sites for cGMP and cAMP play important roles in visual and olfactory transduction, respectively (see Box B in Chapter 7). The molecular structure of these channels appears to be related distantly, if at all, to the neurotransmitter channels. The cytoplasmic messengers

Box B Fish Electric Organs: A Neurochemist's Dream

When David Nachmansohn went to the 1937 World's Fair in Paris, where he was working at the time, he saw for the first time an electric fish—a marine ray (*Torpedo marmarota*). Nachmansohn, who was trained as a biochemist, had become interested in the role of ACh as a neurotransmitter and was curious about how it was made and broken down. He had detected acetylcholinesterase in muscle, and found it also to be widely distributed in the nervous system. Because he had read that the electric organs of some fish are related to muscle, he thought it would be interesting to assay acetylcholinesterase in this tissue. When the fair was finished, he requested the *Torpedo* for assay. To his astonishment and delight, acetylcholinesterase activity in the electric organ was prodigious. In one hour, a gram of tissue could hydrolyze several times its own weight in acetylcholine!

This observation initiated a new era in the biochemistry of the nervous system. The *Torpedo* electric organ is a concentrated, abundant, and homogeneous source of cholinergic synapses that has provided several generations of biochemists and molecular biologists with material for study. The first cell-free synthesis of acetylcholine was observed in an electric organ extract in 1944, and ACh receptors, acetylcholinesterase, proteins of the postsynaptic density, cholinergic synaptic vesicles, and synaptic vesicle proteins were all first purified or cloned from the electric organs of *Torpedo* or the fresh water eel, *Electrophorus electricus*.

In both organisms, the electric organ is used to generate signals for navigation and to stun prey. The tissue consists of parallel arrays of long stacks of cells, like rolls of coins packed together (Figure A). Each of the flat cells (the electrocytes or **electroplax**) is innervated on one of its faces by a cholinergic nerve. In the case of *Torpedo,* the electroplax do not conduct action potentials, so the innervated face is literally covered with cholinergic nerve terminals, and the entire innervated face is one large postsynaptic membrane. The stacks are arranged so that extracellular current cannot flow from one side of an electroplax to the other; the extracellular fluid on the two sides are thus electrically isolated. When the cholinergic nerves innervating the organ are simultaneously stimulated, the innervated face of each electroplax is depolarized. The inside of the cell is isopotential; there is thus a potential difference between the extracellular fluid of each electrocyte on its two sides. The transcellular potentials in a stack of cells add, like the potentials across plates in a battery, to create a massive electric discharge, which in the electric eel can be hundreds of volts.

phosphoinositol bisphosphate (IP_3), the metabolites of arachidonic acid, and Ca^{2+} modulate the activity of other channels (Chapter 7). Whether the latter two regulate channels by binding directly or act via an accessory protein is not known.

The ligand-gated channel that we know most about is the nicotinic ACh receptor in the postsynaptic membrane at the vertebrate neuromuscular junction and at similar synapses in the electric organs of the marine ray *Torpedo* and the eel *Electrophorus* (Box B). This channel was the first ligand-regulated channel to be investigated with intracellular and patch clamp recording, the first to be biochemically purified and cloned, and the only

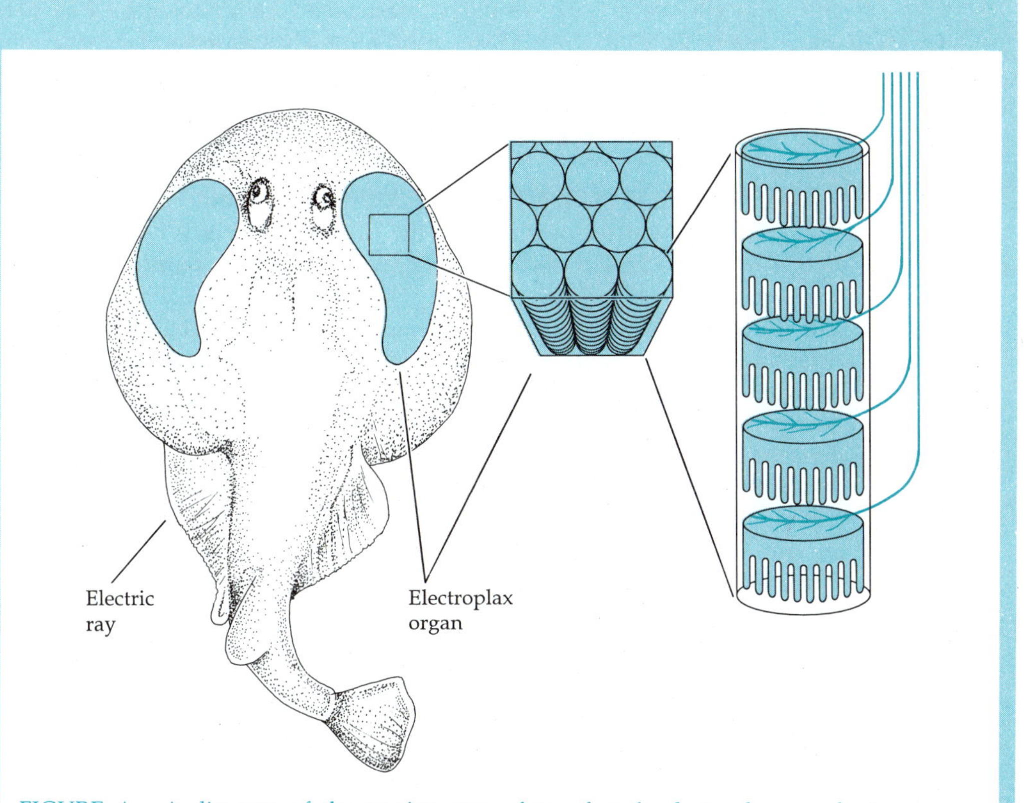

FIGURE A. A diagram of the marine ray *Torpedo*. The electric organ, which fills each wing, consists of vertical stacks of cells (electroplaques) that are innervated by a cholinergic nerve on one surface. Stimulation of the nerve causes depolarization of the innervated face of each electroplax, producing a potential difference between the two sides of the cell. The potentials across the cells in each column add to produce a large electrical discharge.

one about which we have even rudimentary information about three-dimensional structure. The muscle ACh receptor has served as a prototype for the entire group of ligand-regulated channels, particularly those that act as neurotransmitter receptors.

Ligand binding and pore opening of the ACh receptor can be separated by physiological experiments

The ACh receptor responds to a synchronous burst of ACh released in response to a presynaptic action potential. At the neuromuscular junction, the released ACh is rapidly removed through diffusion and the action of the extracellular enzyme acetylcholinesterase (see Chapter 5). The net effect is a pulse of ACh that lasts approximately 1 msec and produces a concentration of about 0.5 m*M* in the synaptic cleft. The muscle cell responds with a brief surge of inward current, mediated by the ACh receptor (see Figure 16 in Chapter 2) that depolarizes the membrane and triggers an action potential.

The response of the receptor to ACh can be separated into two steps. Each receptor, which is normally in the closed state, binds two ACh molecules to form a ligand-receptor complex; this complex then undergoes a conformational change to open the pore, which is permeable to Na^+, K^+ and Ca^{2+} (see Chapter 2). The two steps, binding and channel opening, which were first postulated by del Castillo and Katz almost 40 years ago, are most clearly observed in single-channel recordings (Figure 5). The dissociation constant for ACh is high (approximately 100 μ*M*), and the binding and unbinding steps are relatively slow; transitions to and from the open state of the pore, in contrast, are relatively rapid. Thus channel openings occur in short bursts lasting several milliseconds, which represent the lifetime of the ligand-receptor complex. During the burst, the channel flickers open and shut. **Antagonists,** such as *d*-tubocurarine and α-bungarotoxin bind to the receptor and block ACh binding but do not open the pore. Because they thus block neuromuscular transmission, these toxins are potent paralytic agents when administered to intact animals.

If the receptor is exposed to ACh or other agonists for seconds or minutes, rather than milliseconds, the receptor becomes **desensitized** or unresponsive. Conversion of the ligand-receptor complex to the desensitized state occurs over several seconds at a rate that is influenced by the extracellular Ca^{2+} concentration and the state of phosphorylation of the receptor (see Chapter 7). Upon removal of the agonist, the receptor recovers slowly to its original state. Receptor desensitization is responsible for the paralyzing effect of anti-cholinesterase drugs (insect poisons and nerve gases), which, by inhibiting acetylcholinesterase, prolong the lifetime of ACh in the synaptic cleft.

The ACh receptor is formed of homologous subunits that form a channel at their center

To study the structure and function of ion channels at a molecular level, they must be solubilized and purified. Their purification is complicated by the fact that functional assays for them are not easily made once they are removed from the membrane. Ion channel purification has generally only been possible when specific, high-affinity ligands can be used in binding assays. The discovery of α-bungarotoxin in snake venom proved to be the key for purification of the ACh receptor. This small protein (8000 kD), which binds tightly and specifically to the ACh receptor, can be radioiodinated and used to measure the number of acetylcholine receptors either in the membrane or after solubilization. Because two toxin molecules bind to each receptor, the binding reaction provides a stoichiometric assay. A toxin affinity column can also be used to purify the ACh receptor.

The ACh receptor, originally purified from *Torpedo* electric organ, has four different polypeptide subunits, α, β, γ, and δ. Each subunit is a

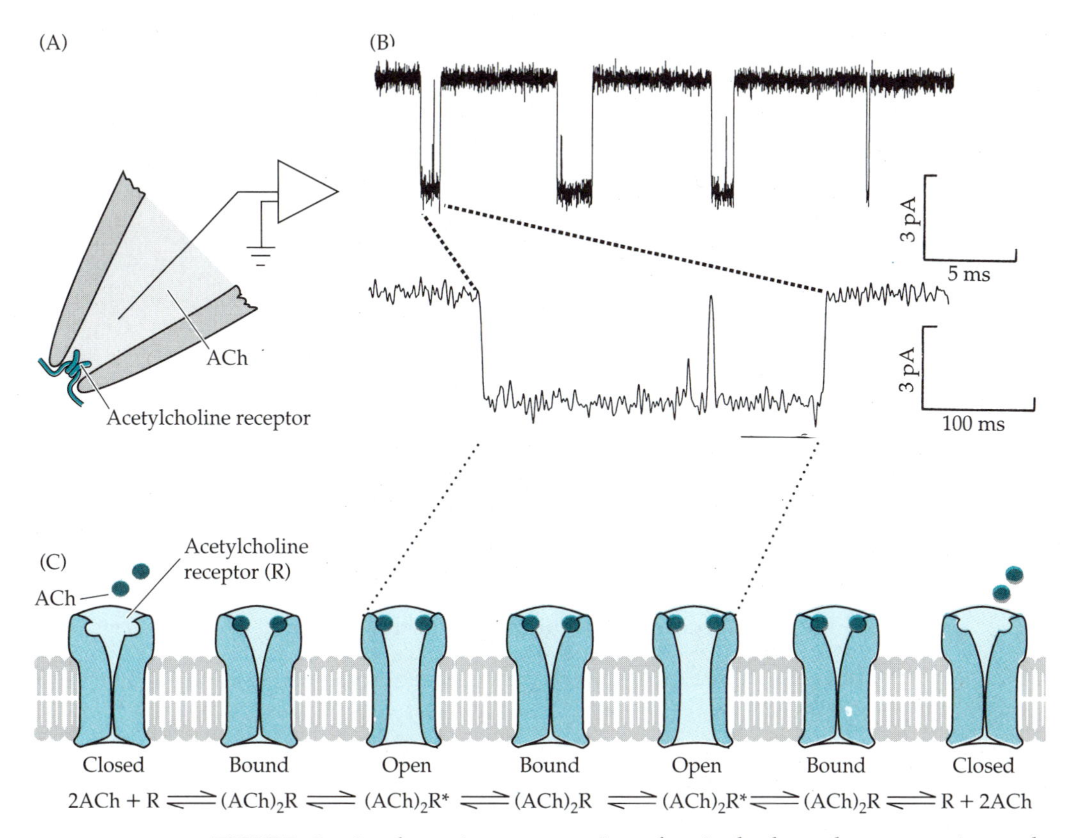

FIGURE 5. A schematic representation of a single-channel pore opening and closing in a patch clamp recording from a membrane containing the ACh receptor. (A) An inside-out patch clamp recording is made from a membrane containing the ACh receptor. A dilute concentration of ACh is in the pipette. Membrane potential is held constant across the membrane and changes in current through the patch of membrane are recorded. The ion concentrations and voltage across the patch are arranged to resemble those normally present in the cell. (B) A record of current across the membrane is shown in two time scales. Increased current from the inside of the pipette to the outside (shown as downward) represents channel opening. The channel pore is either open or closed. Channel openings occur rapidly and in bursts. (C) A diagrammatic model that accounts for the pattern of current. Two molecules of ACh bind to the ACh receptor, which then undergoes a further conformational change to open the pore. The rapid flickering represents rapid transitions between ligand-bound and open states; the beginning and end of the bursts represent the slower association and dissociation of ACh from the receptor (R*). (Record in B courtesy of D. Colquhoun.)

glycoprotein of approximate molecular mass 55 kD that traverses the membrane. The stoichiometric ratio of the subunits is $\alpha_2\beta\gamma\delta$, giving the ACh receptor a total molecular weight of about 275,000. Compared to the simple peptide channels that were described earlier, the size of the ACh receptor is enormous.

X-ray analysis and electron microscopy of paracrystalline arrays of the receptor in the membrane show a structure that looks like a chalice (Figure 6). Much of its mass is on the extracellular side of the membrane with a

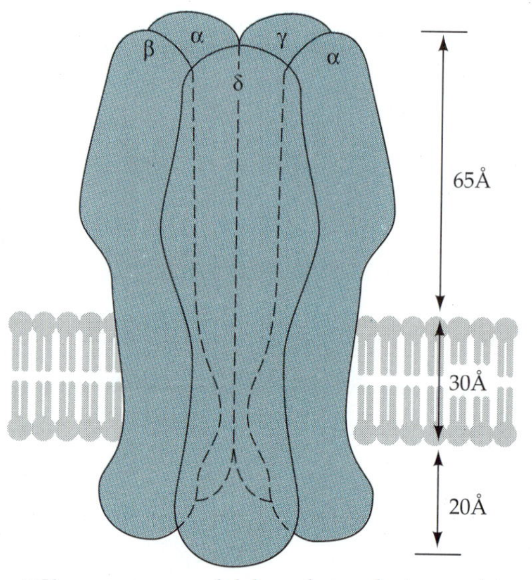

FIGURE 6. An ACh receptor model based on electron microscopy and X-ray diffraction experiments. The order of the subunits is based on experiments suggesting that ACh-binding sites are at the junctions of the α and γ and the α and δ subunits. Other experiments suggest that the positions of the β and γ subunits are reversed.

smaller extension on the cytoplasmic side. The five subunits are arranged symmetrically about a central axis that is perpendicular to the membrane. An aqueous channel enclosed by the protein extends along this axis through the center of the molecule. The openings at either end of the channel are quite large, about 25 Å in diameter, and form roughly cylindrical vestibules of about 65 Å and 20 Å on the extracellular and cytoplasmic sides of the membrane, respectively. The two vestibules are connected by a constricted region about 30 Å long that traverses the hydrophobic core of the membrane. This region, the pore, is the only part of the long channel enclosed by the protein that provides a substantial obstacle to ion flow. Physiological experiments suggest that transit of an even smaller length of the pore (5–10 Å) may be rate limiting.

Biochemical purification of the ACh receptor led to the molecular cloning and DNA sequencing of each of the four subunits. Based on comparison of the deduced amino acid sequences, the different receptor subunits are highly homologous. Comparison of any two (α vs. β, for example) shows that about 20% of the amino acids are identical and another 20% represent conservative substitutions. Their homology and common structure suggests that all four subunits evolved from a common ancestor that presumably formed a homo-oligomeric channel.

The deduced amino acid sequences of ACh receptor subunits reveal their structure in the membrane

The amino acid sequence of the ACh receptor subunits, deduced from the cDNA sequences, gave unexpected insight into their structures. Unlike most soluble proteins, whose amino acid sequence offers little information about protein conformation, the amino acid sequence of membrane proteins reveals hydrophobic segments that could plausibly span the membrane. As membrane-spanning regions are usually α helical, about 20 amino acids are required to form a helix long enough (30 Å) to traverse the hydrophobic portion of the lipid bilayer. Hydrophobic segments that are potential membrane-spanning regions can be identified by plotting the average hydrophobicity of the amino acid residues of a protein along its sequence (Figure 7A). Possible membrane-spanning regions are often immediately apparent from such a profile.

The hydrophobicity plot for the sequence of the α subunit of the ACh receptor allows one to readily see that it has a long, hydrophilic N-terminal segment followed by three hydrophobic regions, a stretch of hydrophilic residues, and a fourth hydrophobic segment near the C-terminus. This plot, along with experimental data on the location of the N-terminal segment, immediately suggests a model for the transmembrane orientation of the α subunit (Fig 7B). An N-terminal signal sequence that is cleaved, along with experiments using antibodies and receptor ligands that bind to the N-terminal sequence (see below), indicate that the long, N-terminal hydrophilic segment of the α subunit is extracellular. Then follow three transmembrane segments (M1–M3), a long cytoplasmic loop, and M4, with the C-terminal also on the extracellular side. Although not completely verified, this model is presently accepted by almost all investigators in the field.

Cloning of the other subunits of the ACh receptor shows that each of them has a similar overall structure. The most highly conserved segments of the subunits are the four transmembrane regions; the cytoplasmic loop is the least conserved. If the intact protein is roughly symmetrical around a central axis, the long N-terminal segments form most of the cup or vestibule that is on the extracellular side of the channel, and the cytoplasmic loops between M3 and M4 of each subunit form the intracellular vestibule of the receptor. The pore itself is formed by α helical membrane-spanning segments from each subunit, much like the postulated structure of the serine-leucine peptide channel.

Acetylcholine binds near a disulfide bond in the N-terminal segment of the α subunit

Although the exact location of the ACh binding site is not known, several kinds of experiments locate it in the N-terminal domain of the α subunit. Experiments on the ACh receptor in situ show that a disulfide bond near the ACh binding site is particularly susceptible to reduction and that the resulting cysteine groups can be labeled by an alkylating agent. A radioactive alkylating agent identified the reactive residues as cysteines 192 and 193 of the α subunit (Figure 7). The unusual juxtaposition of two cysteines, which form a disulfide in the native ACh receptor, occurs only in the α subunit of the ACh receptor, and not in the other three. As the receptor has two α subunits, it also has two ACh binding sites, consistent with the physiological evidence that two ACh molecules must bind to the receptor

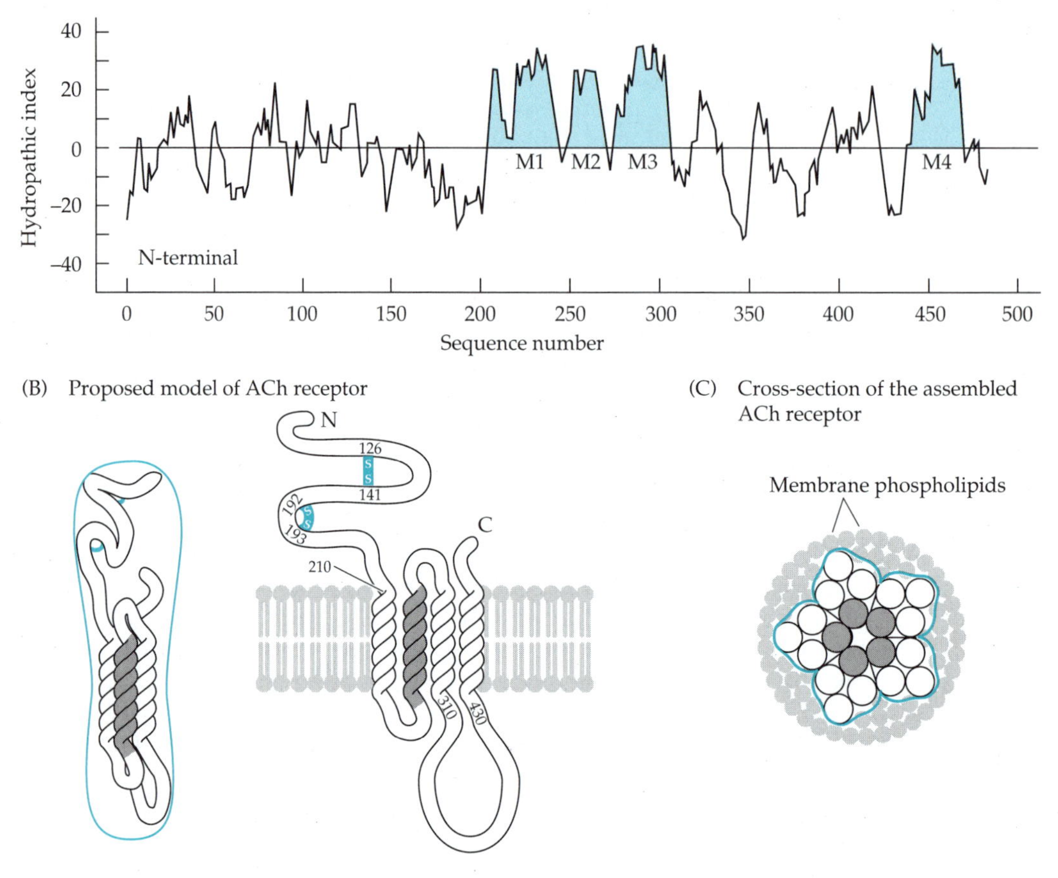

FIGURE 7. The topographical orientation of the ACh receptor subunits in the membrane. (A) A hydrophobicity plot of the sequence of the γ subunit of the *Torpedo* ACh receptor. To construct such a plot each amino acid is given a value for hydrophobicity based on its solubility in ethanol, ranging from +4 (the most hydrophobic) to −4 (the most hydrophilic). A running average is then made for the entire sequence; each amino acid position is assigned the average value of hydrophobicity for the seven amino acids centered on that position. Regions of hydrophobicity in the sequence that extend over twenty amino acids are potential membrane-spanning regions. (B) A model for the transmembrane orientation of the α subunit based on the hydrophobicity plot in (A). The numbered positions represent cysteine residues in the N-terminal hydrophilic domain. (C) A schematic diagram of the assembled ACh receptor. (A from T. Claudio et al., 1983. Proc. Natl. Acad. Sci. USA 80: 1111–1115.)

to fully open the channel. Each of the two binding sites has slightly different pharmacological properties that appear to arise from the different neighbors of each α subunit.

Experiments with α-bungarotoxin, whose binding is competitively blocked by ACh, also indicate that part of the binding site for ACh is near cysteine 192. α-bungarotoxin binds to the isolated α subunit with relatively high affinity, and with somewhat lower affinity to a 30 amino acid synthetic peptide whose sequence is identical to residues 172–201 of the α subunit. No other subunit binds α-bungarotoxin.

Other ligand-gated channels have a structure that is similar to the muscle ACh receptor

Several other ligand-gated ion channels have been purified from brain and their subunits cloned. The glycine receptor, the $GABA_A$ receptor, and the neuronal ACh receptor all mediate rapid postsynaptic responses and appear to be generally modeled after the nicotinic muscle ACh receptor, with the important difference that the GABA and glycine receptors have ionic specificities that are different from the ACh receptor. GABA and glycine are the two major inhibitory neurotransmitters in the mammalian CNS, and their receptors have channels that are permeable to Cl^-, rather than to cations. Nicotinic ACh receptors also occur at some excitatory synapses in the CNS. Although their ion channels appear to be generally similar to those of the muscle ACh receptor, there are clear differences both with respect to pharmacological specificity and channel properties.

Structurally, all members of this group appear to be pentameric proteins, each with two to four different kinds of subunits of size 50–60 kD. Biochemical experiments show that in each case, one of the subunit types binds the ligand (Table 2). The purified glycine receptor, for example, has two polypeptide chains. One of them binds glycine and the high affinity antagonist strychnine, and the other does not. The purified neuronal ACh receptor also has two subunits: α, which bears an acetylcholine binding site, and β, which does not.

The $GABA_A$ receptor appears to be more complicated. The purified receptor has two polypeptide chains that have been identified: β, which binds GABA, and α, which binds the benzodiazepines, a class of drugs that are tranquilizers and that act to increase the affinity of the receptor for GABA. Expression of α and β subunits in *Xenopus* oocytes, however, shows that each can form a functional channel and thus must have a binding site for GABA.

Cloning of the subunits of the neuronal ACh receptor and the $GABA_A$ and glycine receptors revealed a surprising degree of similarity among them. Each of the subunits has the same overall hydrophobicity pattern and has significant homology in amino acid sequence to other subunits in

TABLE 2. Subunit Structure of Members of the AChR Family

Receptor	Subunit	Binding site	Variants
Neuronal AChR	α	ACh	α_2–α_7
	β	—	β_2–β_4
$GABA_A$	α	Benzodiazopines and GABA	α_1–α_7
	β	GABA	β_1–β_3
	γ	(Required for high affinity benzodiazopine binding)	
	δ	—	
Glycine receptor	α	Glycine and strychnine	α_1–α_3
	β	—	

the group and to the subunits of the muscle ACh receptor. All have a long extracellular N-terminal hydrophilic sequence and four transmembrane segments with a lengthy cytoplasmic loop connecting segments 3 and 4. In general, the hydrophobic sequences are the most highly conserved, but several other features appear in all subunits. These include two cysteine residues corresponding to residues 128 and 142 in the α subunit of the muscle ACh receptor, and a proline residue in the middle of M1. The neighboring cysteine residues at 192 and 193 that are associated with the ACh binding site in the muscle α subunit occur in the ACh binding subunit (α) of the neuronal ACh receptor, and also in the glycine-binding subunit (β) of the glycine receptor.

The high degree of homology between subunits means that cDNAs for new subunits can be sought by screening cDNA libraries from brain with probes from conserved regions. Such a strategy has led, for instance, to the identification of cDNAs for several new subunits of the GABA receptor that had not been identified in the purified protein. The only one to which a tentative functional role can be assigned is the γ subunit. The γ subunit itself does not bind GABA or benzodiazepines, but, when paired with αand β subunits in the oocyte expression system, it confers high-affinity binding of benzodiazepine on the oligomeric receptor. The discovery of new subunits, and the heterogeneity of those already known (see below), suggest that the subunit composition of ligand-gated channels may be more complex than previously expected.

Ligand-gated receptor subunits show great diversity

An unexpected dividend of screening libraries with probes for the subunits of the neuronal ACh receptor and the GABA and glycine receptors was the discovery that each subunit has several closely related subtypes. Although pharmacological experiments had indicated that not all GABA receptors, for example, were identical, the degree of diversity has surpassed all expectation. Thus five different variants of the α subunit of the neuronal ACh receptor have been cloned from brain, each coded by a separate gene. All have the characteristic neighboring disulfide at positions corresponding to α 192–193 in the muscle ACh receptor, and most form functional channels in a *Xenopus* oocyte expression system, when paired with a β subunit. The β subunit itself comes in four varieties. The channels formed by various combinations of α and β are all activated by acetylcholine and have the same ion specificity but differ subtly from each other in kinetic or pharmacological properties. GABA and glycine receptor subunits have a comparable degree of diversity.

Each of the subunit types thus appears to consist of a family of closely related polypeptides that in some cases can substitute for each other to form a variety of receptors with functionally distinct properties. The variation in number of subunits along with the possibility of combining them in different ways gives the nervous system an impressive array from which to choose for a particular task. In situ hybridization of brain tissue shows that each member of a family has a characteristic pattern of expression in different regions of the brain, leading to the idea that different cells in the nervous system, or even different parts of the same cell, may express different variants of a particular receptor type. Because of similarities between different families, subunits of the glycine and $GABA_A$ receptors, along with those of the neuronal and muscle ACh receptors, may be said to form a superfamily related by a common structure that reflects a common evolutionary origin.

Glutamate receptors are a separate or more distantly related family

Glutamate receptors are of special interest for two reasons. First, glutamate is the most common excitatory transmitter in the mammalian CNS. Virtually all CNS neurons respond to glutamate iontophoresed onto their surfaces. Second, glutamate receptors are thought to be intimately involved in mechanisms of synaptic plasticity and of neuronal cell death (Chapters 14 and 15, respectively). In spite of their great interest we know little of their biochemistry. Very recent progress in cloning of the glutamate receptors, however, should lead to rapid progress in the next few years.

Pharmacological and physiological experiments have defined three types of rapid glutamate responses, presumed to be mediated by different receptors. The receptors are given the names of their specific agonists and are called **kainate, quisqualate/AMPA** (α-amino-3-hydroxy-5-methyl-isoxazole-4-propionic acid), and N-methyl-D-aspartate **(NMDA)** receptors. These three types differ not only in their pharmacologies, but also in the kinetics of the responses that they mediate (the NMDA response is slower than the others), and in their ion specificities (all are permeable to the monovalent cations, but only the NMDA channel is significantly permeable to calcium). The NMDA receptor also has an unusual form of voltage-regulation in which the open channel is occluded at normal resting potential by magnesium ion in the extracellular fluid. Depolarization drives the magnesium ion out of the channel, allowing other ions to pass. Thus, glutamate is most effective in opening the NMDA channel when the cell is depolarized. Two other glutamate receptors, one sensitive to quisqualate and one to ABP (L-2-amino-4-phosphonobutyrate), are G protein-linked receptors that mediate slow responses (see Chapter 6).

Recent cloning experiments have identified six different cDNAs that encode presumptive glutamate receptor subunits. In addition, four of the cloned channels have a short segment in which either of two versions can be inserted by alternate splicing ("flip and flop"). The predicted polypeptide sequences of the glutamate receptors indicate that each encodes a protein of about 100 kD. Although the precise identification of the transmembrane domains is uncertain, all can be fit into a model with four transmembrane domains arranged in a pattern similar to that of the nicotinic ACh receptor. The main difference between the glutamate receptor subunits and members of the ACh receptor family is that the N-terminal domain is much longer—450–500 amino acids instead of 200. Although the general pattern of transmembrane domains is similar to that of the ACh receptor subunits, there is little overall homology, indicating that the glutamate receptors form a separate family that is only distantly related to the nicotinic receptors. Among themselves, four of the six are closely related (70% homology); the other two are also related to each other, but less closely (40% homology). When expressed in *Xenopus* oocytes or in mammalian cells, the first four form channels that respond with high affinity to AMPA; one of the others forms a channel with high affinity for kainate. None respond to NMDA. Each receptor, including the flip and flop variants, also shows a distinctive pattern of expression in the mammalian CNS.

M2 lines the channel pore

The common structure of the ligand-gated receptors suggests that they share a common mechanism of ion permeation. Experiments on the muscle-type ACh receptor subunits have identified transmembrane sequences that form the walls of the pore. If the pore is 6.6 Å, as inferred from physiological experiments, only a single α helix from each subunit is needed to surround it. Several lines of evidence suggest that the M2 domain from each subunit lines the pore. At first glance, the sequences of this domain are unremarkable and give no hint that they might line an aqueous pore. In each case the M2 sequence contains several serine and threonine residues that are especially concentrated at the cytoplasmic end, with the remainder being mostly hydrophobic residues (Figure 8A).

(A) The M2 sequence from the mouse muscle AChR subunits

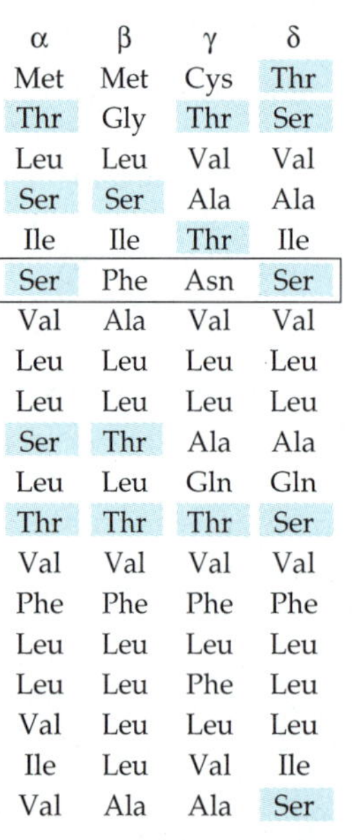

α	β	γ	δ
Met	Met	Cys	Thr
Thr	Gly	Thr	Ser
Leu	Leu	Val	Val
Ser	Ser	Ala	Ala
Ile	Ile	Thr	Ile
Ser	Phe	Asn	Ser
Val	Ala	Val	Val
Leu	Leu	Leu	Leu
Leu	Leu	Leu	Leu
Ser	Thr	Ala	Ala
Leu	Leu	Gln	Gln
Thr	Thr	Thr	Ser
Val	Val	Val	Val
Phe	Phe	Phe	Phe
Leu	Leu	Leu	Leu
Leu	Leu	Phe	Leu
Val	Leu	Leu	Leu
Ile	Leu	Val	Ile
Val	Ala	Ala	Ser

(B) Amino acids near the M2 segment of the ACh and GABA$_A$ receptors

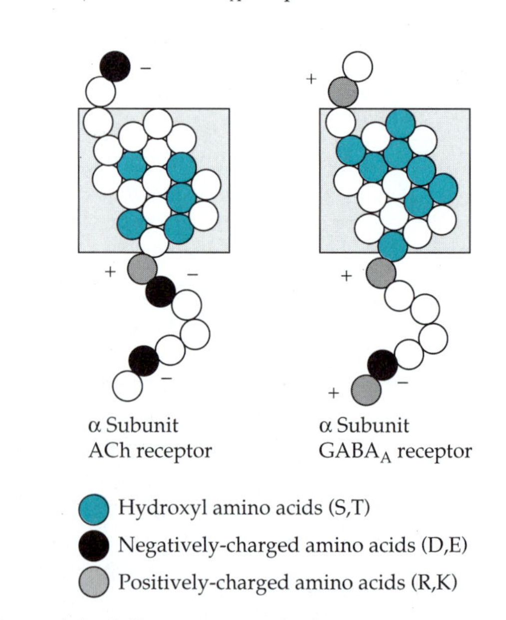

FIGURE 8. The amino acid sequence of the M2 segment and adjoining regions. (A) The amino acids in the M2 segment of the four subunits of the bovine muscle AChR. The boxed amino acid residues are at the position labeled by open-pore blockers. Most amino acid residues in the M2 segments are hydrophobic; the hydroxyl amino acids serine and threonine are noted in color. (B) Amino acids near the M2 segments of the alpha$_1$ subunits of the bovine ACh and GABA$_A$ receptors. The amino acids at the ends of M2 have a net negative charge in the case of the AChR, which is a cation channel, and a net positive charge in the case of the GABA receptor, which is an anion channel.

Box C Xenopus *Oocytes: The Synapse in an Egg*

Xenopus oocytes unite two powerful technologies for studying ion channels: recombinant DNA techniques and biophysics. These large cells, measuring several millimeters in diameter, were first used as a research tool in molecular and developmental biology by John Gurdon and his colleagues, who found that exogenous mRNA, injected into them, is faithfully and efficiently translated. Eric Barnard, Ricardo Miledi, and Katumi Sumikawa introduced the preparation to neurobiologists in 1982 by showing that oocytes express new ion channels in the surface membrane after injection of mammalian brain mRNA. *Xenopus* oocytes are now part of the armamentarium of any self-respecting channelogist.

Oocytes have been especially valuable in two ways: for the biophysical study of ion channels and for their cloning. The ion channels expressed after injection of mRNA, isolated from tissues or synthesized from cloned cDNA, can be easily characterized using voltage-clamp or patch-clamp electrophysiological recording. The channels seen after injection of tissue mRNA in almost all cases faithfully duplicate the properties observed in vivo. Oocytes have been most powerfully used with mRNAs made from cloned cDNAs, however. Channel properties can be studied after the injection of various combinations of channel subunit mRNAs, as with ligand-gated channels, or after in vitro mutagenesis to alter the amino acid sequence of channel proteins. These techniques have now become so standardized that the range of experiments is limited only by the investigator's imagination.

Oocytes have also been useful as an expression system for the cloning of new ion channel cDNAs. cDNA libraries can be made using vectors for transcription, and mRNA produced from the library assayed by injection in oocytes. If a positive response is obtained, the library can be subdivided further and further, using the oocytes as an assay system, until one or more positive clones have been identified.

Oocytes can be used for the cloning of other receptors or transport proteins as well. Oocytes contain chloride channels that can be activated by calcium through a G-protein linked phosphatidylinositol bisphosphate pathway, and muscarinic receptors for ACh, 5-hydroxytryptamine receptors, and substance P receptors from mammalian brain all activate this pathway when expressed in oocytes. Finally, the recent demonstration that nicotinic ACh receptor clustering can be produced in *Xenopus* oocytes by coexpression of the ACh receptor with the 43-kD protein (see Chapter 8) shows that other properties of receptors can be studied in the oocytes as well. The description of oocytes as a "synapse in an egg," a term coined by zealous investigators, thus seems only slightly fanciful.

Physiological experiments with the ACh receptor have shown that several agents bind reversibly to a site in the open pore and block it. Radioactively labeled **open-pore blockers** can be induced by irradiation to covalently label amino acids near the binding site. Residues in three of the four subunits are labeled, and in all cases the residues are in the M2 segment, indicating that these residues must be accessible from inside the pore.

The amino acids that bind open-pore blockers have been more completely specified by in vitro mutagenesis experiments using *Xenopus* oocytes to express the altered proteins (Box C). Binding of open-pore blockers to the ACh receptor can be *decreased* by substitution of alanine for serine residues near the cytoplasmic end of M2 and *increased* by substitution of alanine residues for serine near the extracellular end of the presumptive M2 helix. Each of the residues that are important for binding lie on adjacent turns of one side of an α helix. These results are consistent with a model in which the open-pore blockers, which are charged at one end and hydrophobic at the other, bind to sites on the M2 helices and occlude the channel.

Amino acids around the pore opening increase the rate of specific ion movement through the pore

If the pores in all of the related ligand-gated channels are lined by the M2 transmembrane α helices, examination of the amino acid residues lining the pores might be expected to give clues about how ions are recognized by the channel. In particular, a comparison of sequences from the ACh receptor with those of the GABA (or glycine) receptor should prove instructive, since both are based on the same common plan, and yet the two have very different specificities—i.e., for cations and anions, respectively. Curiously, the important differences appear to be not in the transmembrane region of the pore, but on either side of it (Figure 8B). In both receptors, the M2 helix contains mostly hydrophobic residues, with a strip of serine residues along one side, presumably the side facing the aqueous pore. At the ends of M2, both receptors have charged residues—positively charged residues in the case of GABA subunits, and negatively charged for the ACh receptor subunits.

In the case of the muscle ACh receptor, in vitro mutagenesis experiments have shown that the identity of the amino acids just outside of the pore influences the rate of ion movement through it. Surprisingly, the serines lining the pore appear not to have a specific role. Changing as many as three of the serines in M2 (on different subunits) to alanine has a relatively small effect on the movement of cations through the pore. A larger effect results from changing amino acid residues near the mouth of the pore on either end. In all four subunits of the ACh receptor, negatively charged glutamic and aspartic acid residues are concentrated on either side of the M2 sequence. When uncharged or positively charged amino acids are substituted for glutamate or aspartate on the extracellular side of the pore (the C-terminal side of M2, according to the model in Figure 7), the rate of inward ion flow is reduced with little effect on outward ion flow. Conversely, decreasing the negative charge on the cytoplasmic end of M2 (N-terminal of M2) selectively diminishes the outward current. These experiments lead to the conclusion that the charges near the mouth of the channel on either side increase the efficiency of ion movement through the pore. Negative charges near the mouth of the pore increase the local cation concentration, thus making it easier for ions to enter the pore (Figure 9A). Charges at the opposite end of the pore would have the same effect on ions moving in the opposite direction. Serines lining the pore may have a relatively nonspecific role; their hydroxyls may collectively provide a water-like environment for ions and water molecules in the channel. Thus, one can imagine ions moving through a pore, making and breaking bonds with the hydroxyls that line the pore much as they make and break

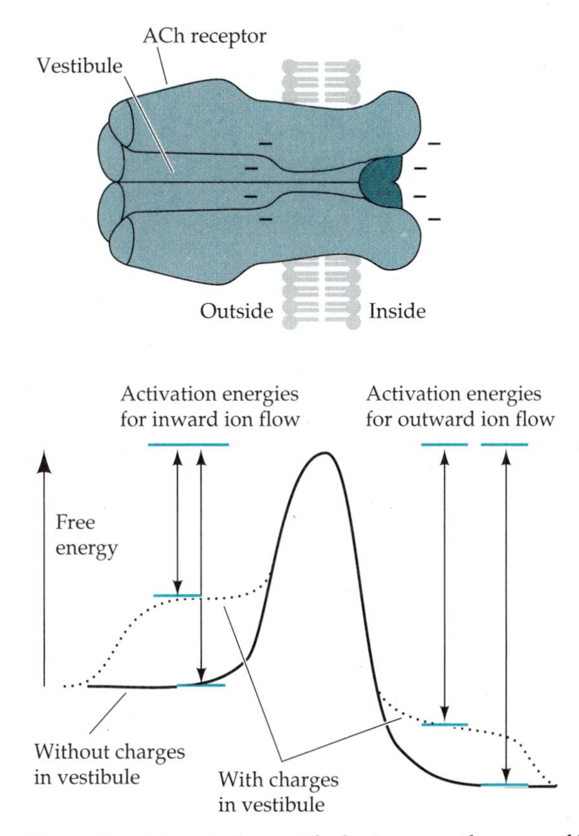

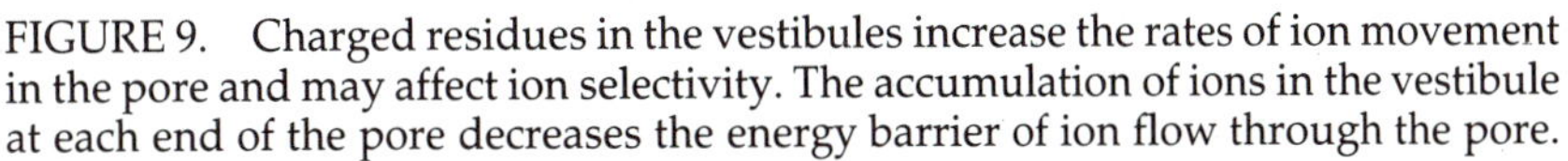

FIGURE 9. Charged residues in the vestibules increase the rates of ion movement in the pore and may affect ion selectivity. The accumulation of ions in the vestibule at each end of the pore decreases the energy barrier of ion flow through the pore.

bonds with water molecules as they move through aqueous solution.

The charges at the mouth of the pore may also serve as a filter to screen out ions with the wrong charge. This idea is supported by a comparison of the charges just outside M2 in the ACh and GABA receptors. A preponderance of positive charges surrounds the opening of the GABA pore, which is permeable to Cl^-, and a preponderance of negative charges surrounds the ACh receptor pore, which is permeable to cations. (Figure 8B). What is surprising is that the M2 regions themselves are quite similar; both, in fact, resemble the serine-leucine peptide that forms a channel that is mostly cation permeable. We can postulate that the pore alone, although preferring cations, is also permeable to anions. When flanked by positive charges, however, the cations are screened out, and the pore becomes anion-selective. These and other ideas of ion selectivity will come under intense experimental scrutiny during the next few years.

Ligand-gated channels are modulated by other ligands and by phosphorylation

Ligand-gated channels are not only controlled by the neurotransmitters that they bind but can also be regulated by other agents. Some receptors have sites for other ligands. As we have seen, benzodiazepines bind to GABA receptors and increase their affinity for GABA. The normal function of the benzodiazepine site is not known, but its presence suggests that there may be a class of endogenous ligands that have the same action. Also, micromolar concentrations of glycine increase the response of the NMDA receptor to glutamate, apparently by binding directly to it. How glycine acts, and how this effect is used to modulate channel function, are unknown.

Other agents modulate channels indirectly. Substance P, a peptide contained in the nerve terminals of preganglionic sympathetic nerve fibers, increases the rate of desensitization of the neuronal ACh receptor in the postsynaptic membrane of the ganglion cell. In cell-attached patch recordings, substance P applied to the cell outside of the patch increases the rate of desensitization of ACh receptors in the patch, indicating that substance P does not act directly on the ACh receptor, but through a diffusible second messenger. Because specific inhibitors of protein kinase C block the response, phosphorylation of the receptor or an associated protein may be responsible for the effect.

Phosphorylation of the muscle ACh receptor by cAMP-dependent protein kinase either in vitro or in vivo increases its rate of desensitization by up to 10-fold. The kinase phosphorylates a serine residue in the large cytoplasmic loop of the δ subunit (see Chapter 7). The physiological role of the change in desensitization is not known, but calcitonin gene-related peptide (CGRP), a peptide in the nerve terminal (Chapter 4), can increase the levels of cAMP in the muscle and activate the kinase.

Subunits of the neuronal ACh receptor, and of the GABA and glycine receptors, also carry potential sites for phosphorylation in the large cytoplasmic loop. In addition, phosphorylation of voltage-sensitive channels can change their properties, suggesting that phosphorylation is a general mechanism for regulating ion channels.

Subunits of the ACh receptor change during development and after denervation of the neuromuscular junction

The diversity of subunits among ligand-gated receptors suggests that the nervous system requires many different types of receptors with slightly different properties and that some of these are generated by combinatorial variation. Receptors with different properties may be expressed in different cells (e.g., the ACh receptor in muscle cells and in sympathetic neurons); in different parts of the same cell; or at different times in development (flip and flop variants of the kainate receptor). The most completely understood example of a change in subunits occurs during the development of the neuromuscular junction. When the synapse between nerve and muscle first forms in rat or mouse embryos, the ACh receptor at the endplate has the four subunits, $\alpha_2\beta\gamma\delta$. Shortly after birth, synthesis of γ subunit mRNA ceases, and a new subunit, ε, is made. All ACh receptors subsequently inserted into the membrane have the ε subunit rather than the γ (see Chapter 13). Electrophysiological recordings show that although their reversal potentials (and hence their ion specificities) are the same, γ ACh receptors have a smaller single-channel conductance and a longer burst duration that do ε ACh receptors. A computer model in which the two were compared shows that the channel properties of the γ ACh receptor are important because they make it more effective than the ε ACh receptor in depolarizing embryonic muscle fibers that have smaller diameters. As we begin to determine the properties and distribution of receptors formed with various combinations of subunits in the CNS, a more precise and clearly developed rationale for the abundance of subunit variants will undoubtedly emerge.

Voltage-Regulated Channels

Voltage-regulated ion channels are the hallmark of excitable cells. Although they are expressed in many cell types, their fundamental roles in signaling have particularly identified them with excitable cells. Because of their importance in propagation of the action potential, much work has focused on voltage-regulated sodium channels. As the attention of neurobiologists has turned to the more complex processes of signal integration and plasticity, however, appreciation of the versatility and importance of potassium and calcium channels has grown. Potassium channels control excitability and shape electric signals; calcium channels regulate levels of intracellular Ca^{2+}, which not only triggers transmitter release, but modulates many other cellular activities.

Electrophysiological experiments have identified several types of each of these three voltage-regulated channels. As with ligand-gated channels, molecular cloning of voltage-regulated channels suggests an even greater diversity than previously suspected. All that have so far been described, however, appear to be evolutionarily related and to work by common principles. The voltage-regulated channels thus form another large family of ion channel proteins, distinct from the ligand-gated channels.

Voltage-regulated channels have both activated and inactivated states

Alan Hodgkin and Andrew Huxley, in their analysis of the action potential, experimentally separated Na^+ and K^+ currents and derived empirical equations that described the activation of the channel for each ion as a function of membrane potential and of time (see Chapter 2). Following a sustained depolarization of the squid giant axon, they found that sodium current first increased to reach a peak (activation) and then decreased to baseline values (inactivation).

The processes of activation and inactivation can be seen in single-channel recordings of voltage-regulated sodium channels in which the probability of channel opening and the duration of the open state depend on both membrane potential and time. This pattern of channel opening can be understood in terms of a simple (and simplified) scheme in which the sodium channel has three states: closed, open (activated), and inactivated, roughly analogous to the open, closed, and desensitized states of the acetylcholine receptor (see Figure 14). Channels in the closed state are opened by membrane depolarization; transitions to the inactivated state occur from the open state. The rates of transitions between closed and open states, and thus the relative proportions of the channel molecules in each state, depend on voltage.

Hodgkin and Huxley found in the squid axon that the potassium current, in contrast to the sodium current, responds to depolarization by slowly increasing to a constant plateau. The potassium channel that carries this current (sometimes called the delayed rectifier) can thus be described as having only two states, open and closed. Other potassium channels, such as the potassium A channel that is the product of the *Shaker* gene in *Drosophila melanogaster* (see below), show inactivation similar to that exhibited by the sodium channel.

Voltage-regulated channels have subtypes with different physiological properties

Channels differ, not only in their ion specificities, but also in the time and voltage dependence of activation and inactivation, their single-channel conductances (which express how easily ions move through the pore), and their susceptibilities to particular pharmacological agents. One class of calcium channels, found in neurons and in heart muscle, for example, has a small single-channel conductance, is rapidly activated by small depolarizations, and rapidly inactivates. These channels, called "transient," or **T channels,** are thought to generate pacemaker currents. Another class of channels, found in heart muscle and also in nerve terminals, has a larger single-channel conductance, activates more slowly and at larger depolarizations, and inactivates slowly or not at all. These channels, called "long," or **L channels,** carry current during the plateau phase of the cardiac action potential and may be involved in excitation-secretion coupling: they can be either opened or blocked by members of a class of compounds called dihydropyridines. Yet other calcium channels, some of which occur in neurons and are sometimes called **N channels**, have properties that are intermediate between those of T and L channels.

The kinetic properties of voltage-regulated channels, that is, the rates of transition between states, are critical to their physiological roles. The rapid activation and slow inactivation of the sodium channel and the slow activation of the potassium channel following a depolarization give the action potential its characteristic shape. Channel kinetics is a major distinguishing feature of different varieties of ion channels, suggesting that this variable is an important one in generating the variety of signals that the nervous system uses.

Sodium and calcium channels consist of one large channel protein associated with accessory proteins

As with ligand-gated channels, an important approach to understanding how voltage-regulated channels work is to purify them, find out how many subunits they have, how the subunits are arranged, and, by modifying different parts of the protein, determine what roles they play. Because voltage-regulated channels generally occur at low density, purification has been possible only for sodium and calcium channels; in each case, success

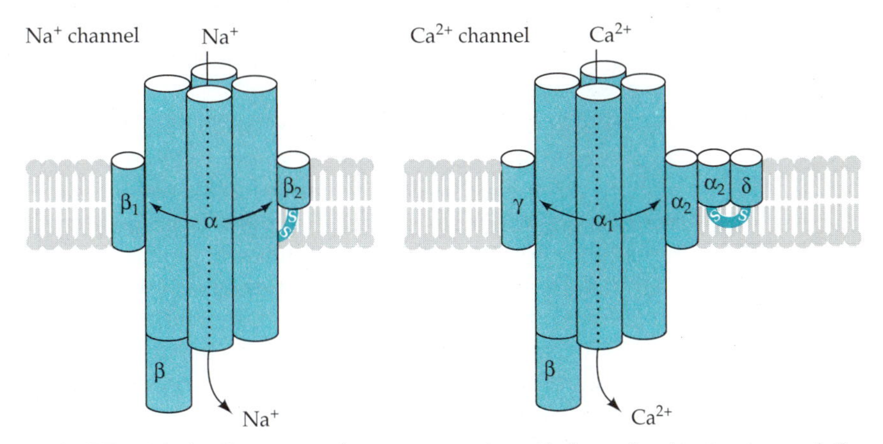

FIGURE 10. A diagrammatic representation of the subunit structure of the purified voltage-gated sodium and calcium channels.

has depended upon the specificity and high-affinity binding of a toxin or ligand that could be used for assay and for affinity purification.

The most useful toxins that disrupt sodium channel function have been tetrodotoxin and saxitoxin, which act from the extracellular side of the membrane to prevent the transition of the channel to the open state. Using these toxins sodium channels have been purified from brain, muscle, and the eel electric organ. The purified sodium channel from mammalian tissues is an oligomer consisting of a single large polypeptide (α) of approximately 260 kD, associated with two smaller subunits (β_1 and β_2) of about 35 kD. Experiments in intact membranes with antibodies raised against the purified subunits and with impermeable toxins show that the α subunit and β_1 extend across the membrane, whereas β_2, which is linked to α by disulfide bonds, is apparently only exposed on the extracellular surface (Figure 10). The purified oligomer can be reconstituted into phospholipid membranes where it mediates a voltage-dependent sodium permeability that is tetrodotoxin-sensitive, thus proving that the purified protein is indeed the channel.

Which of the subunits form the channel? Expression of the cloned α subunit protein in *Xenopus* oocytes shows that it alone forms a voltage-regulated sodium channel. The function of the smaller subunits is not well understood, but these subunits may alter the kinetics of opening and closing of the channel or stabilize it in an active conformation. The sodium channel thus appears to differ from ligand-gated channels in that a single protein, rather than several subunits, form the channel.

The calcium channel has been purified from the transverse tubules of muscle, where it plays a role in excitation-contraction coupling. The purified channel has one large subunit α_1, tightly associated with two smaller subunits, β and γ, and more loosely associated with α_2 and δ subunits. As with the sodium channel, however, only the largest subunit, α_1 (about 170 kD), is required to form a channel.

The large subunit of sodium and calcium channels has repeating domains

Amino acid sequence information from the purified proteins led directly to cloning of the cDNA of the α subunit cDNA from both sodium and calcium channels. The deduced amino acid sequences for the two channel proteins are strikingly similar. Each is a large polypeptide with four internally homologous domains of 300–400 amino acids (I–IV) (Figure 11). Within each domain, there are six regions (S1–6) that appear to be membrane-spanning. Four of these are highly hydrophobic; a fifth (S4) has an interesting amphipathic structure in which every third amino acid residue is an arginine or lysine. All are long enough to form a membrane-spanning α helix. In addition to these membrane-spanning regions, recent evidence indicates that the region betweeen S5 and S6 may form a nonhelical hairpin structure that lines the pore (see below).

The internal homology between the four domains of the α and α_1 subunits suggests that these domains may form the ion channel in the same way as do the five subunits of the ACh receptor. Viewed in this way, the sodium or calcium channel is formed by four homologous "subunits" that are not separate proteins as in the ACh receptor, but are tied together in one long polypeptide chain. Note that if the domains are to have homologous structures, each must have an even number of membrane-spanning segments, or else the polypeptide chain at the start of each domain will be on opposite sides of the membrane. The idea that the pore of the sodium and calcium channels is formed by four instead of five subunits is consistent with estimates of pore size, which indicate that the aqueous pore formed by both sodium and calcium channels is smaller than that of the ACh receptor.

Potassium channels are made from individual subunits

Much of what we know about potassium channels comes from the study of *Shaker,* a curious mutant of *Drosophila* that was originally isolated on the basis of a bizarre (and still unexplained) phenotype in which the mutant flies jerk their legs convulsively when exposed to ether. Electrophysiological experiments on the neuromuscular junction of larval flies originally suggested that the mutants have a defect in potassium channels that leads to a prolonged action potential in the nerve terminal. Later work showed the defect to be in a specific current, called the A current, that is rapidly activated and inactivated by depolarization. Genomic clones were isolated from a region of the *Drosophila* genome in the vicinity of the *Shaker* gene, and a chromosomal "walk" using overlapping clones led to the *Shaker* gene. Note that in this approach, the protein was not directly isolated and in fact was not identified until after the gene had been cloned.

Two aspects of the *Shaker* gene were surprising when first found. First, the gene encodes a protein that is about 70 kD, much smaller than that of other voltage-regulated channels. Second, by alternate splicing mechanisms, the gene gives rise to at least five related, but different, gene products.

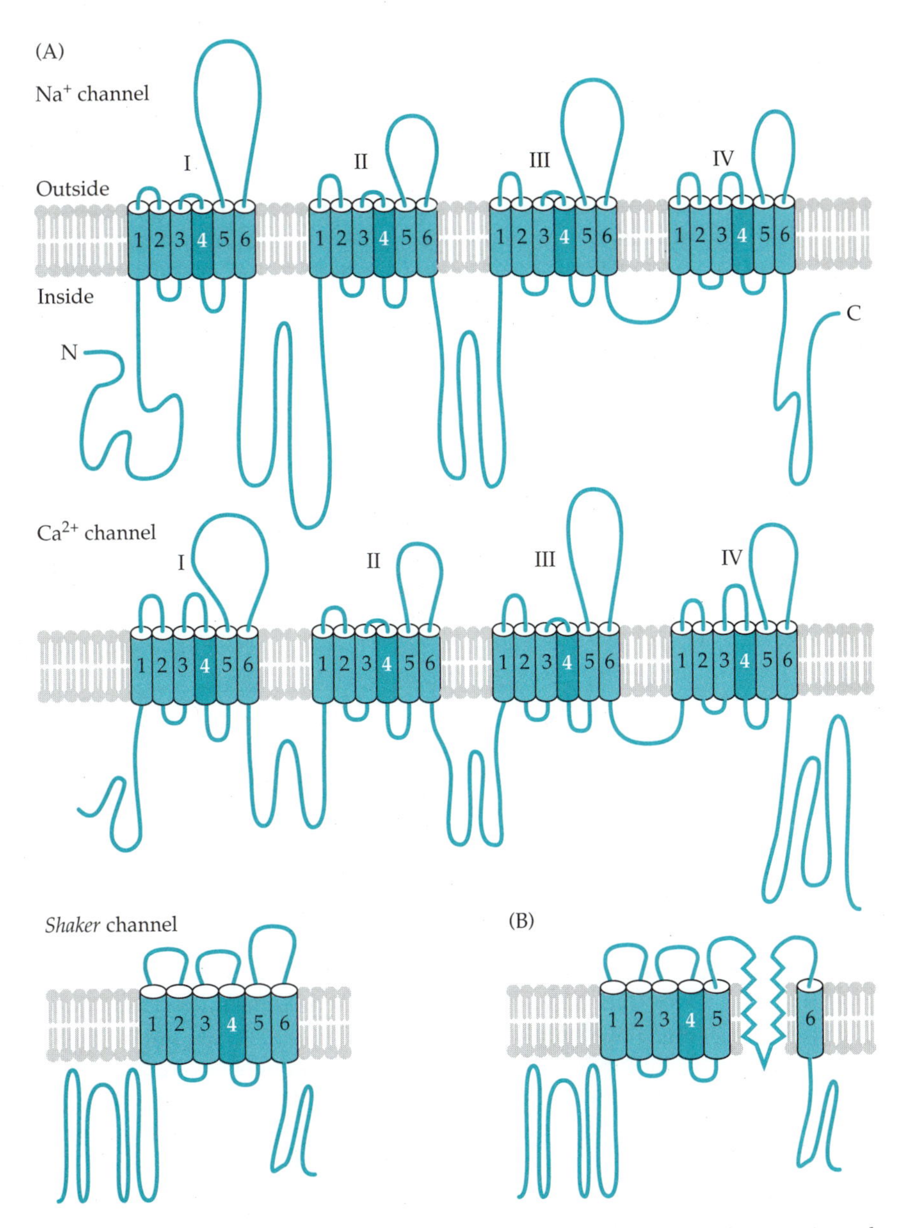

FIGURE 11. Models of the transmembrane orientation of the voltage-activated sodium, calcium, and Shaker potassium channels. (A) Models based on hydrophobicity profiles of the amino acid sequence. (B) Recent experiments on the Shaker protein indicate that the segment between S5 and S6 forms a non-helical hairpin structure across the membrane that lines the aqueous pore (see Figure 15). Similar structures between S5 and S6 probably occur in the sodium and calcium channels as well. (A After W. A. Catterall, 1988. Science 242: 50–61.)

Although the size of the protein(s) encoded by the *Shaker* gene seems small when compared to the sodium or calcium channel proteins, its structure is familiar. The deduced protein sequence has six relatively hydrophobic segments that could form a membrane-spanning α helix; moreover, the fourth segment has a sequence in which every third amino acid is a lysine or arginine like the S4 segment of the sodium channel. In other words, the gene for the Shaker channel encodes a protein whose overall hydrophobic pattern is strikingly similar to that of one of the domains of the sodium or calcium channels (Figure 11). When examined in detail, the amino acid sequence of the Shaker protein has little homology to the sodium or calcium channels, however, except for the region around the S4-like sequence.

The appearance of a new potassium A-like channel in *Xenopus* oocytes after injection of mRNA produced from Shaker cDNA demonstrates that the Shaker gene encodes a potassium channel. Since a single Shaker mRNA gives rise to a potassium channel, no other protein is needed for channel formation, suggesting that the Shaker protein could form a homo-oligomeric ion channel in vivo.

Multiple transcripts generate diversity in potassium channels via homo-oligomeric and hetero-oligomeric channels

Close comparison of the channels made in oocytes when each of the five Shaker gene products is separately injected shows that all activate and inactivate in response to a depolarization but that the rates at which these transitions occur are significantly different in each case. All five Shaker proteins share a common internal sequence that extends across most of the membrane-spanning segments of the protein (Figure 12), but they differ in their N- and C-terminal domains. Two or more variations are possible at each position, allowing several distinct Shaker sequences to be generated combinatorially. Because only the terminal segments differ between the five transcripts, they must be important in determining rates of activation and inactivation.

If mRNAs for two different sequences are injected, a new channel appears with properties that are intermediate to those of the homo-oligo-

FIGURE 12. Four of the variant protein products of the *Shaker* gene in *Drosophila* showing variation at the N and C terminals. (From C. Miller, 1988. Trends Neurosci. 11: 185–186.)

meric channels formed by either mRNA species alone. Thus different subunits can be combined into a single hetero-oligomer to give further variation in the kinetic parameters of potassium channels.

Three other genes (*Shab, Shaw, Shal*) identified in *Drosophila* have products that resemble the Shaker protein and appear to encode potassium channels. Potassium channel cDNAs resembling those of each of the *Shaker* variants and of the *Shaker*-related genes are found in mammalian brain. The mammalian Shaker variants, like those of *Drosophila,* form hetero-oligomers in vitro. Potassium channels in the brain show an enormous variability in physiological properties, perhaps more so than any other channel type. This variability presumably reflects their importance in generating the variety of electrical signals used by the nervous system. To meet the need for a large number of channel types, several evolutionary strategies have been adopted. First, several different genes encode potassium channels. Second, multiple splicing from a single gene gives rise to further variants. Finally, by having an oligomeric channel further diversification can be generated by different combinations of subunits.

The S4 segment of voltage-regulated channels may be a voltage sensor

Although the sequences of voltage-regulated channels do not immediately suggest which parts of the protein form the pore, they suggest a compelling hypothesis about how channel gating might be regulated by membrane potential. The strong conservation of sequence in the S4 segment of the voltage-regulated channels indicates its importance and suggests that it may play a role in the voltage-sensitivity of the protein.

For a protein to respond to a change in the electric field, some charged part of it (the **voltage sensor**) must move. This idea was originally suggested by Hodgkin and Huxley, who postulated in 1952 that gating of the channel was accompanied by movement of charge within the membrane. The steepness of the curve for the voltage-dependence of channel activation allows one to estimate the equivalent number of charges that must be moved to activate it. For the voltage-activated sodium channel this number is approximately 6. Such a movement of charge within the protein itself constitutes a small current (the **gating current**) that can be measured at the beginning and end of a brief depolarization.

One puzzling aspect of the idea that several charges move across the membrane each time the sodium channel opens is the energetics. The energy required to move a single ion into a lipid bilayer is enormous; it is even more difficult if one imagines moving a charged segment of a protein across the membrane. A neat solution to this difficulty has been recently proposed. In the S4 region, the basic amino acids that occur at every third residue form a spiral of positive charge on the outside of an α helical segment (an α helix makes a complete turn for each 3.4 amino acids of sequence) (Figure 13). In a transmembrane segment, each of these charges must be paired with negative charges from neighboring helices. At normal resting potential, the S4 helix is in a stable configuration, held in the membrane in part by the positive potential on the extracellular side of the membrane. If this positive charge is reduced, the force holding the helix in the membrane is diminished, and the helix rotates. The result is to displace

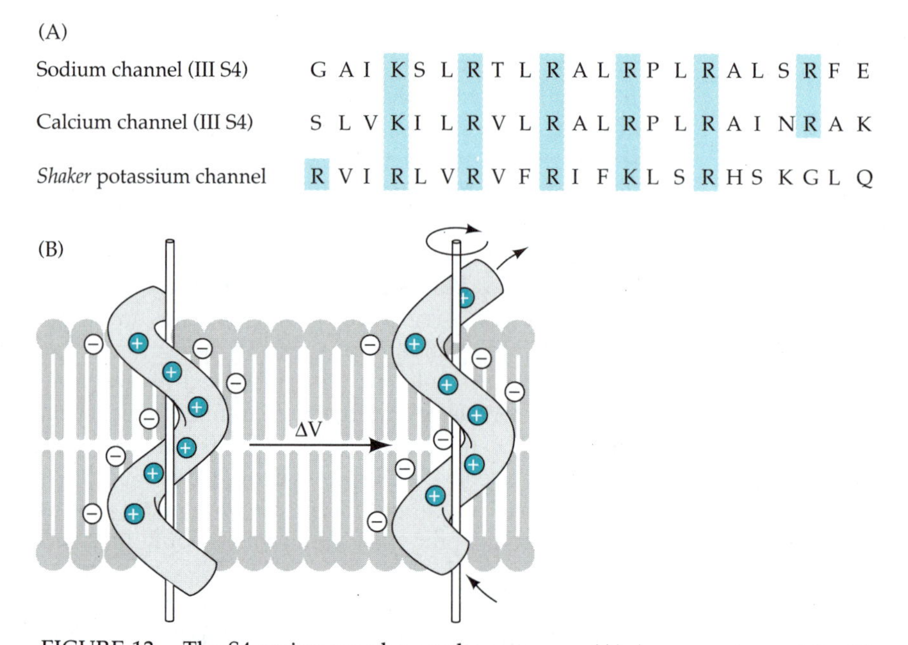

FIGURE 13. The S4 region may be a voltage sensor. (A) A comparison of the S4 sequences from domain III of the rat brain sodium channel, domain III of the rabbit muscle calcium channel, and the Shaker potassium channel. Basic amino acids, either lysine (K) or arginine (R) occur at every third position. (B) A schematic model of the S4 region of voltage-gated ion channels indicating how it could act as a voltage sensor. The charged groups on the outside of the S4 helix are presumed to interact with fixed negative charges in the membrane. When the membrane is depolarized, the charges move toward the extracellular side of the membrane by rotation of the helix.

each charge on the outside of the helix to the position of its next charged neighbor where it will be stabilized; the net effect of rotation of the entire S4 helix will be to remove one positive charge on the inside of the membrane and to add a positive charge on the outside of the membrane. A charge will thus have been transferred from one side of the membrane to the other without requiring the movement across the membrane of any one segment of protein. Rotation of the helix by 60° would result in a net translational movement of only about 5 Å.

The calculation that the movement of approximately 6 charges accompanies channel gating for sodium and potassium channels is in approximate agreement with the idea that each of the four domains or subunits lining the channel has an S4 segment and that movement of all four is required to open the channel. Although in practice the movement of the protein is unlikely to be so simple, this model provides an attractive mechanism that could explain how the change in membrane potential is translated into a change in protein conformation. How this change could open a channel remains unclear.

The idea that S4 is the voltage sensor has been put to experimental test by in vitro mutagenesis experiments with both the sodium channel and

Shaker proteins. In the case of the sodium channel, one, two, or three of the positively charged amino acids in the S4 segment of the first domain were changed to uncharged or negatively charged amino acids. The result was to decrease the slope of the curve relating channel activation to membrane voltage. In general, the larger the change in charge, the less steep the curve becomes. Although the quantitative aspects of this result cannot be easily interpreted in terms of a specific molecular mechanism, it gives strong evidence that S4 is the voltage-sensor for activation of the voltage channel.

A cytoplasmic domain of the Shaker channel mediates inactivation

Experiments performed almost 20 years ago showed that perfusion of isolated squid axons with the protease trypsin had a remarkably specific effect on sodium channels in the squid axon membrane. Protease treatment did not alter the permeability properties and voltage-dependent activation of the channels but completely abolished inactivation. This result not only validated the Hodgkin–Huxley idea that activation and inactivation of sodium channels were two separable processes, but also suggested that inactivation is mediated by a polypeptide on the cytoplasmic surface of the membrane that is perhaps part of channel itself. An explicit "ball-and-chain" model (Figure 14) proposed that inactivation resulted from the movement into the open channel of a positively charged, cytoplasmic segment of the channel protein (the ball) attached to a hydrophilic segment (the chain), thus occluding it.

A series of in vitro mutagenesis experiments made with the Shaker potassium channel has shown that the N-terminus of the protein plays an important role in inactivation and has validated this simple model in remarkable detail. The first 19 amino acids of the Shaker protein appear to constitute the ball; it consists of 11 hydrophobic residues, followed by 8 hydrophilic amino acids, of which 4 are positively charged. Deletions within the ball region abolish inactivation, and mutation of pairs of the 4 positively charged amino acids to neutral amino acids slows it. Mutation of the hydrophobic amino acids also alter inactivation; they appear to be necessary to maintain the structure of the ball. The 60 amino acids following the ball region appear to constitute the chain; deletions within this region (which shorten the chain) speed inactivation, while insertions (which lengthen the chain) slow its rate. If the ball is entirely removed, so that inactivation is lost, a synthetic peptide whose sequence is that of the first 20 amino acids of the Shaker protein, restores inactivation. This mechanism may be a general one. In vitro mutagenesis experiments indicate that a positively charged cytoplasmic region between domains III and IV plays a similar role in inactivation of the sodium channel.

The pore of voltage-regulated potassium channels is formed by a non-helical region between S5 and S6

Three kinds of experiments have recently identified a sequence in the Shaker protein and in other voltage-regulated potassium channels that lines the aqueous pore. The sequence is unusual for two reasons. First, most models of the protein based on its hydrophobic plot (e.g., Figure 11)

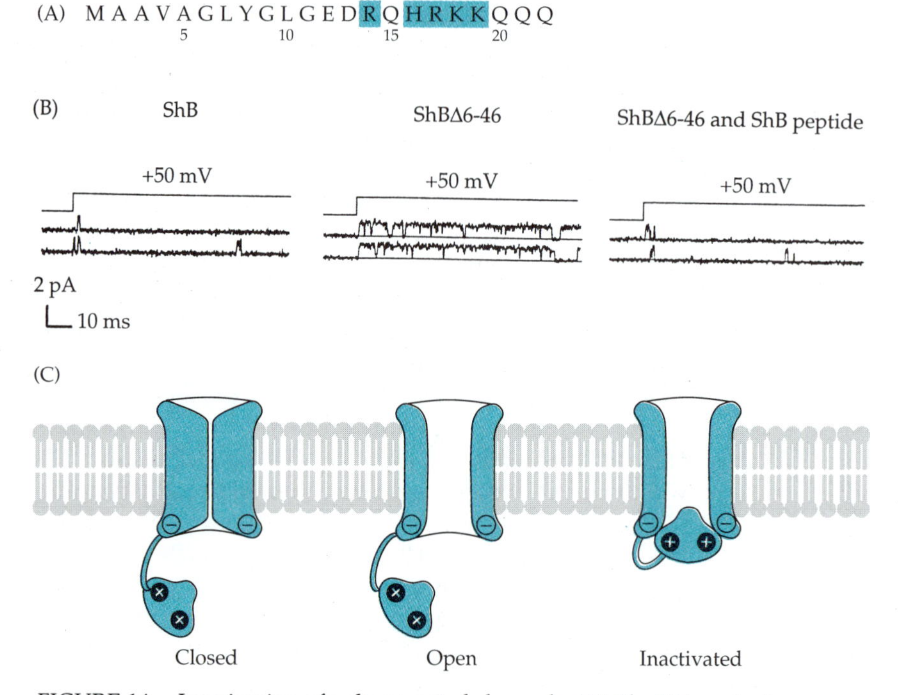

FIGURE 14. Inactivation of voltage-gated channels. (A) The N-terminal sequence of the Shaker B protein. These amino acids are thought to form a "ball" that physically occludes the pore and produces inactivation. Positively charged amino acids are in color. If the two lysines at positions 18 and 19 or the two arginines at positions 14 and 17 are neutralized by in vitro mutagenesis, inactivation is slowed. (B) Single channel openings for the wild-type Shaker B protein, for a mutant protein in which residues 6-46 are deleted, and for the mutant protein in the presence of a 20 residue peptide with the sequence shown in (A). The channels were induced to open by a voltage step from –80 to +50 mV. Channel openings are represented by an upward deflection of the current record. (From C. Miller, 1991. Science 252: 1092–1096.) (C) A schematic diagram of the ball-and-chain model of inactivation of voltage-gated ion channels. Depolarization opens the channel, exposing a binding site for a positively charged segment on the cytoplasmic side of the channel. Binding of the positively charged segment blocks the channel and inactivates it. The transition from the closed to the open state is voltage-sensitive, but the transition from the open to the inactivated state is not.

had placed this sequence extracellularly rather than in the membrane. Second, the sequence does not appear to form an α helix, but a more extended β structure.

The first series of experiments arise from physiological observations on the mechanism of block of potassium channels by the large cation, triethylammonium (TEA^+). TEA^+ blocks potassium channels by binding to sites within the open pore and occluding it. In the case of the Shaker channel, TEA^+ can block the pore from either side of the membrane by binding to separate sites at each end of the pore. Specific amino acids corresponding to each of the sites have now been identified by in vitro mutagenesis of the region of polypeptide (sometimes called H5) between

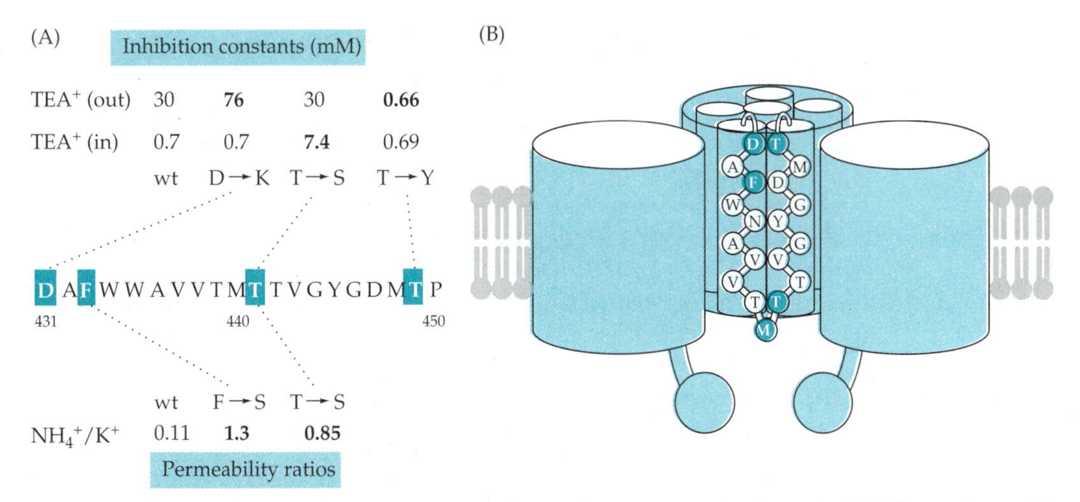

FIGURE 15. The pore of the Shaker potassium channel. (A) The amino acid sequence lining the pore in the Shaker potassium channel and the effects of mutations in the sequence. Mutations that affect the inhibition of ion movement through the pore by TEA^+ in the external or internal solutions are indicated above the sequence, and mutations that affect the relative permeabilities of NH_4^+ and K^+ through the pore are indicated below. (From G. Yellen et al., 1991. Science 251: 939–942; and A. J. Yool and T. L. Schwarz, 1991. Nature 349: 700–704.) (B) A hypothetical model of the sequence lining the transmembrane pore. The residues indicated in color represent those shown in color in (A).

segments S5 and S6. Mutations at two sites, an aspartic acid at position 431 and a threonine at 449, selectively reduce inhibition by externally applied TEA^+, whereas mutation of a threonine residue at 441 specifically diminishes the sensitivity of the channel to internally applied TEA^+ (Figure 15). These experiments strongly suggest that the sequence from 431 to 449 traverses the membrane twice in a hairpin-like structure. Because an α helix of this length (17 amino acids) is too short to span the membrane twice, the polypeptide sequence is postulated to form an anti-parallel β sheet in which the amino acids are in a more extended structure.

Two other types of experiments support the identification of H5 as the region lining the pore. The first makes use of two cloned potassium channel proteins from mammalian brain that have differences in their single-channel conductances and in their susceptibilities to TEA^+. Exchange of a segment containing 21 amino acids in the H5 region between the two proteins conferred upon each of them the channel properties of the other.

Finally, alteration of several of the residues in the H5 region of the Shaker channel change its ionic specificity. Three mutations (Figure 15), F433S (i.e., a phenylalanine at position 433 is changed to serine), T441S and T442S alter the permeability of the channel for Rb^+ and NH_4^+. The Shaker channel is ordinarily poorly permeable to these ions, which are larger than K^+, as well as to Na^+, which is smaller than K^+ (Table 1). Each of the mutations increases the permeability of the channel to $Rb^+$and NH_4^+ relative to K^+, without altering the permeability of the channel to Na^+. These experiments not only tell us that amino acids in the H5 region of the Shaker channel are responsible for the ion selectivity of the pore, but suggest that at least two mechanisms are responsible for selectivity, one for larger ions and one for smaller ions. As each of the mutations that increases the permeability to large ions represents the substitution of a smaller amino acid for a larger one, steric factors may be most important in reducing the permeability of the channel to ions larger than K^+.

Because of the similarities in overall structure between the Shaker channel and voltage-regulated sodium and calcium channels, we may expect that the region between S5 and S6 of these channels may also form a pore. In vitro mutagenesis experiments will quickly determine which amino acids in these sequences are responsible for their ion selectivities.

In summary, the remarkable properties of voltage- and ligand-gated ion channels that have seemed for so long to be mysterious in molecular terms are now yielding to the powerful methods of molecular genetics and biophysical analysis. The next large step will depend on the further application of structural analysis to give concrete form to our ideas of channel function.

Summary

Ion channels facilitate ion movement across phospholipid membranes by forming an aqueous pore through the membrane that is lined by polar groups. The pore does not determine the direction of ion flow but only its rate. Because of the size of the pore, ions go through in single file and must partially dehydrate. Interactions of the ion with the polar groups in the pore replace the energy lost by dehydration. Pores can be made by very simple molecules, including short polypeptides that associate in the membrane to form an aqueous pore at their center.

Cellular ion channels form pores by the association of similar subunits or domains into oligomeric structures that surround a central aqueous pathway. The nervous system has a very large number of channels (>75) distinguished by their ion specificities, the time courses of their activation and inactivation, and the stimuli that regulate them. These are grouped into several large families, of which two have been clearly identified, neurotransmitter-gated channels and voltage-gated channels.

Neurotransmitter-gated channels are exemplified by the nicotinic ACh receptor, a pentamer of four highly homologous subunits in the stoichiometric ratio $\alpha_2\beta\gamma\delta$. All of the subunits have a common structure in which a long N-terminal extracellular sequence is followed by three transmembrane domains (M1–3), a long cytoplasmic loop and a fourth transmembrane domain (M4). ACh binds to a region in the N-terminal domain of the α subunit near the first transmembrane segment. The walls of the pore are formed by the M2 helix from each subunit. The M2 sequences have mostly hydrophobic amino acids and serine or threonine residues; negatively charged amino acids at either end of the pore facilitate ion movement through it.

The subunits of the $GABA_A$ receptor, the glycine receptor, and the ACh receptor from brain have a similar pattern of hydrophobic residues and share homology with the muscle ACh receptor subunits, indicating a common evolutionary origin. All are oligomeric receptors containing two to four subunits that apparently form channels in the same way as the muscle ACh receptor. Each of the variants has multiple variants, allowing the formation of many different oligomeric receptors with slightly different functional or pharmacological characteristics.

The neural ACh receptor, like the muscle ACh receptor, is a cation channel, but the $GABA_A$ and glycine receptors are chloride channels. The chloride channels have more positive charges than negative charges around the pore at each end, suggesting that they determine the anionic or cationic specificity of the channel.

Voltage-gated channels for Na^+ and Ca^{2+} have a large subunit with a molecular weight of about 250,000 that may be associated with other, smaller subunits but can itself form a voltage-sensitive channel. The large subunits of both sodium and calcium channels have four internally homologous domains, each with six segments that are possible transmembrane regions. In each of the four domains, the S4 segment has a distinctive structure in which every third amino acid is a lysine or arginine residue. The *Shaker* gene in *Drosophila* encodes a potassium channel subunit of 70 kD that has a structure similar to a single domain of the sodium or calcium channel. The most highly conserved region among the three proteins is the S4 segment. This segment may be the voltage sensor that moves upon depolarization of the membrane, an idea supported by in vitro mutagenesis experiments. Such experiments also support a ball-and-chain model of inactivation in which a positively charged part of the protein moves into the open channel to occlude it. Recently, a region between S5 and S6 of voltage-regulated potassium channels has been identified as forming a nonhelical hairpin structure that lines the ion pore.

Several splice variant products of the *Shaker* gene form homo-oligomeric channels with differing kinetics of activation and inactivation. Combinations of subunits also form channels with distinctive properties. Gene-splicing and hetero-oligomeric combinations thus make available to the nervous system a wide range of potassium channels with different kinetic properties.

References

General References

Hille, B. 1992. *Ionic Channels of Excitable Membranes*, 2nd edition. Sinauer Associates, Sunderland, MA. (See especially Chapters 10 and 11 and pp. 355–361.)

Miller, C. 1989. Genetic manipulation of ion channels: A new approach to structure and mechanism. Neuron 2: 1195–1205.

Unwin, N. 1989. The structure of ion channels in membranes of excitable cells. Neuron 3: 665–676.

Peptide Ion Channels

Lear, J. D., Wasserman, Z. R. and DeGrado, W. F. 1988. Synthetic amphiphilic peptide models for protein ion channels. Science 240: 1170–1181.

Ligand-Gated Ion Channels

Betz, H. 1990. Ligand-gated ion channels in the brain: The amino acid receptor superfamily. Neuron 5: 383–392.

Charnet, P., Labarca, C., Leonard, R. J., Vogelaar, N. J., Czyzyk, L., Gouin, A., Davidson, N. and Lester, H. 1990. An open-channel blocker interacts with adjacent turns of α-helices in the nicotinic acetylcholine receptor. Neuron 2: 87–95.

Claudio, T. 1989. Molecular genetics of acetylcholine receptor-channels. In *Frontiers in Molecular Biology: Molecular Neurobiology*. D. M. Glover and B. D. Hames (eds.). IRL Press, Oxford. pp. 63–142.

*Hollmann, M., O'Shea-Greenfield, A., Rogers, S. W. and Heinemann, S. 1989. Cloning by functional expression of a member of the glutamate receptor family. Nature 342: 643–648.

*Imoto, K., Busch, C., Sakmann, B., Mishina, M., Konno, T., Nakai, J., Bujo, H., Mori, Y., Fukuda, K. and Numa, S. 1988. Rings of negatively charged amino acids determine the acetylcholine receptor channel conductance. Nature 335: 645–648.

Keinanen, K., Wisden, W., Sommer, B., Werner, P., Herb, A., Verdoorn, T. A., Sakmann B. and Seeburg, P. H. 1990. A family of AMPA-selective glutamate receptors. Science 556–560.

Mishina, M., Tobimatsu, T., Imoto, K., Tanaka, K., Fujita, Y., Fukuda, K., Kurasaki, M., Takahashi, H., Morimoto, Y., Hirose, T., Inayama, S., Takahashi, T., Kuno, M. and Numa, S. 1985. Location of functional regions of acetylcholine receptor α-subunit by site-directed mutagenesis. Nature 313: 364–369.

Pritchett, D. B., Sontheimer, H., Shivers, B. D., Ymer, S., Kettenmann, H., Schofield, P. R. and Seeburg, P. H. 1989. Importance of a novel $GABA_A$ receptor subunit for benzodiazepine pharmacology. Nature 338: 582–585.

Toyoshima, C. and Unwin, N. 1988. Ion channel of acetylcholine receptor reconstructed from images of postsynaptic membranes. Nature 336: 247–250.

Voltage-Regulated Channels

Catterall, W.A. 1989. Structure and function of voltage-sensitive ion channels. Science 242: 50–61.

Isacoff, E. Y., Jan, Y. N. and L. Y. Jan. 1990. Evidence for the formation of heteromultimeric potassium channels in *Xenopus* oocytes. Nature 345: 530–534.

*Hoshi, T., Zagotta, W. N. and Aldrich, R.W. 1990. Biophysical and molecular mechanisms of *Shaker* potassium channel inactivation. Science 250: 533–538.

Jan, L. Y. and Jan, Y. N. 1989. Voltage-sensitive ion channels. Cell. 56: 13–25.

Noda, M., Shimizu, S., Tanabe, T., Takai, T., Kayano, T., Ikeda, T., Takahashi, H., Nakayama, H., Kanaoka, Y., Minamino, N., Kangawa, K., Matsuo, H., Raftery, M. A., Hirose, T., Inayama, S., Hayashida, H., Miyata, T. and Numa, S. 1984. Primary structure of *Electrophorus electricus* sodium channel deduced from cDNA sequence. Nature 312: 121–127.

Schwarz, T. L., Tempel, B. L., Papazian, D. M., Jan Y. N. and Jan, L.Y. 1988. Multiple potassium-channel components are produced by alternative splicing at the *Shaker* locus in *Drosophila.* Nature 331: 137–142.

Stühmer, W., Conti, F., Suzuki, H., Wang, X. D., Noda, M., Yahagi, N., Kubo, H. and Numa, S. 1989. Structural parts involved in activation and inactivation of the sodium channel. Nature 339: 597–603.

Tanabe, T., Takeshima, H., Mikami, A., Flockerzi, V., Takahashi, H., Kangawa, K., Kojima, M., Matsuo, H., Hirose, T. and Numa, S. 1987. Primary structure of the receptor for calcium channel blockers from skeletal muscle. Nature 328: 313–318.

*Yellen, G., Jurman, M. E., Abramson, T. and R. MacKinnon. 1991. Mutations affecting internal TEA blockade identify the probable pore-forming region of a K^+ channel. Science 251: 939–942.

*Yool, A. J. and Schwartz, T. L. 1991. Alteration of ionic selectivity of a K^+ channel by mutation of the H5 region. Nature 349: 700–704.

4

Chemical Messengers at Synapses

Richard H. Scheller and Zach W. Hall

CELLS IN THE NERVOUS SYSTEM communicate with each other at synapses, mostly via chemical messengers. Thirty years ago only a few synaptic messengers were known and their actions were thought to be relatively simple. Each opened an ion channel that either hyperpolarized or depolarized the cell. Today dozens of synaptic messengers are known, many of whose actions are much more subtle and varied than simply opening and closing ion channels. Synaptic messengers, often acting through intracellular second messengers, can affect the numbers of active ion channels, their kinetics, and their voltage sensitivities, resulting in both short- and long-term changes in cellular excitability. Synaptic messengers can also change cellular metabolism, the structure of the cytoskeleton, and gene expression. The function and structure of the synapse itself can be altered to change the efficiency of both release and reception mechanisms.

These many effects are mediated by two pathways of synaptic communication that exist side by side at many synapses. One is responsible for rapid communication; the other, an evolutionarily much older system, operates more slowly and mediates many of the long-term effects of synaptic communication. In both pathways membrane vesicles play a pivotal role. They are the means by which the secretory products are selected, conveyed to the site of release, and secreted by exocytosis. In some cases, they also play important roles in the synthesis or processing of synaptic messengers. At each synapse multiple messengers are released. Some are peptides, associated exclusively with the slow synaptic system; others are small molecules, which are mostly associated with the fast synaptic system but can also mediate slow responses. Because substances released by nerve cells can have many effects on target cells, we refer to them generically as **chemical messengers.** The term **neurotransmitter** is reserved for those chemical messengers that directly or indirectly affect the opening or closing of ion channels. Various aspects of these two signaling pathways will be considered in the next few chapters. We focus here on the identification of neurotransmitters and other chemical messengers, and on the synthesis, storage, and physiological actions of one important class of messengers, the peptides.

Fast and Slow Chemical Transmission

The most familiar form of synaptic transmission is exemplified by the action of acetylcholine at the neuromuscular junction. The neurotransmitter, stored in synaptic vesicles, is secreted by exocytosis at a specialized region of the nerve terminal. The released neurotransmitter binds to an ion channel in the postsynaptic membrane, resulting in a brief transmembrane current. The entire process takes only a few milliseconds (Chapter 2). This pathway of rapid communication appears to have evolved specifically in neurons. It forms the basis of synaptic signaling in vertebrate and invertebrate nervous systems and is used at most synapses. The common neurotransmitters that mediate signaling of this type are **glutamate** and **acetylcholine (ACh)**, which are usually excitatory neurotransmitters, and **glycine** and **γ-aminobutyric acid (GABA)**, which are usually inhibitory neurotransmitters (Figure 1).

Many synapses have a second signaling pathway that resembles rapid chemical transmission in some ways: the messengers are packaged into vesicles, are released by exocytosis in response to electrical signals, and interact with a membrane-bound receptor in the postsynaptic cell. In con-

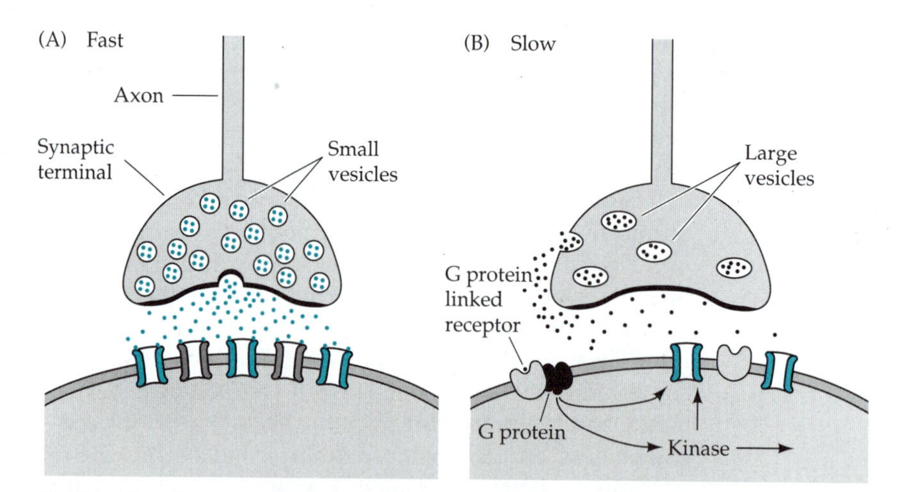

FIGURE 1. Presynaptic and postsynaptic pathways of fast and slow synaptic transmission. (A) Small-molecule neurotransmitters are rapidly released from small, clear vesicles at the active zone, act directly on postsynaptic ion channels of one or more types, and are rapidly removed. (B) Other chemical messengers, including peptides, are stored in large, dense-cored vesicles, are released from non-specialized sites, and act on G protein-linked receptors, which influence channels and other processes indirectly. At many synapses, the two pathways coexist, or fast and slow elements from each pathway are mixed. Amines are stored in both small and large vesicles, but act mostly on G protein-linked receptors. Most rapidly acting neurotransmitters act on G protein-linked receptors as well as on ion channel receptors.

trast to rapid transmission, however, this pathway often uses **neuropeptides** as messengers and is much slower, both in onset of the response (hundreds of milliseconds) and in duration (seconds, minutes, or hours). The receptors in the postsynaptic membrane that bind messengers in the slow pathway are not directly connected to ion channels but affect them or alter the levels of intracellular second messengers through intermediary **G proteins**. Finally, the two systems, slow and fast, use different types of vesicles, whose synthesis and release occur via distinct pathways. Peptides are synthesized and packaged into large vesicles in the cell body; fast transmitters are synthesized in the nerve terminals and packaged into smaller vesicles that arise by local endocytosis. In general, the small vesicles are secreted from sites that are specialized for rapid release and are directly apposed to the postsynaptic cell; large vesicles are secreted from many sites (Chapter 5).

Slow synaptic transmission is closely related to forms of communication in other cells

Many non-neural cells use peptides as chemical messengers. The essential features of the synthesis, storage, release, and reception of peptide messengers in these cells is often remarkably similar to that seen in neurons. One of the simplest eukaryotes, baker's yeast (*Saccharomyces cerevisiae*), secretes peptides as part of its mating behavior. Mating between the two haploid cell types, α and a, which fuse to become diploid, begins with the reciprocal exchange of peptide mating factors. The signaling pathway used by α, which is a 13 amino acid peptide, has many of the elements of signaling pathways in higher organisms. The α mating factor is synthesized as part of a larger precursor molecule, is packaged in membrane-bound vesicles, and is released by exocytosis. The factor then diffuses to the target a cells and binds to a G protein-linked receptor in the membrane to initiate biochemical changes required for mating.

In higher organisms, local, or **paracrine,** communication between cells in the same tissue occurs by mechanisms similar to those seen in yeast (Figure 2). Cells secrete messengers locally into the extracellular fluid which bind to receptors on neighboring cells. A further specialization in vertebrates is **endocrine** communication, in which messengers are secreted from specialized islands of cells into the bloodstream to reach distant targets. Adrenocorticotropic hormone (ACTH), for example, is stored in large vesicles in anterior pituitary cells, is released by exocytosis into the circulation, and ultimately interacts with its target cells in the adrenal medulla through a G protein-linked receptor that increases the levels of the intracellular second messenger cAMP.

Neurosecretory cells have characteristics of both neurons and endocrine cells. Neurons in the mammalian hypothalamus extend axonal processes into the posterior pituitary, where the terminals end near small capillaries. Some neurons secrete the peptide hormone oxytocin, which stimulates uterine smooth muscle; others secrete vasopressin, which acts on kidney and bladder cells to increase water transport. Each is synthesized in the cell body, packaged into secretory granules, transported into the terminals and released into the bloodstream as a hormone.

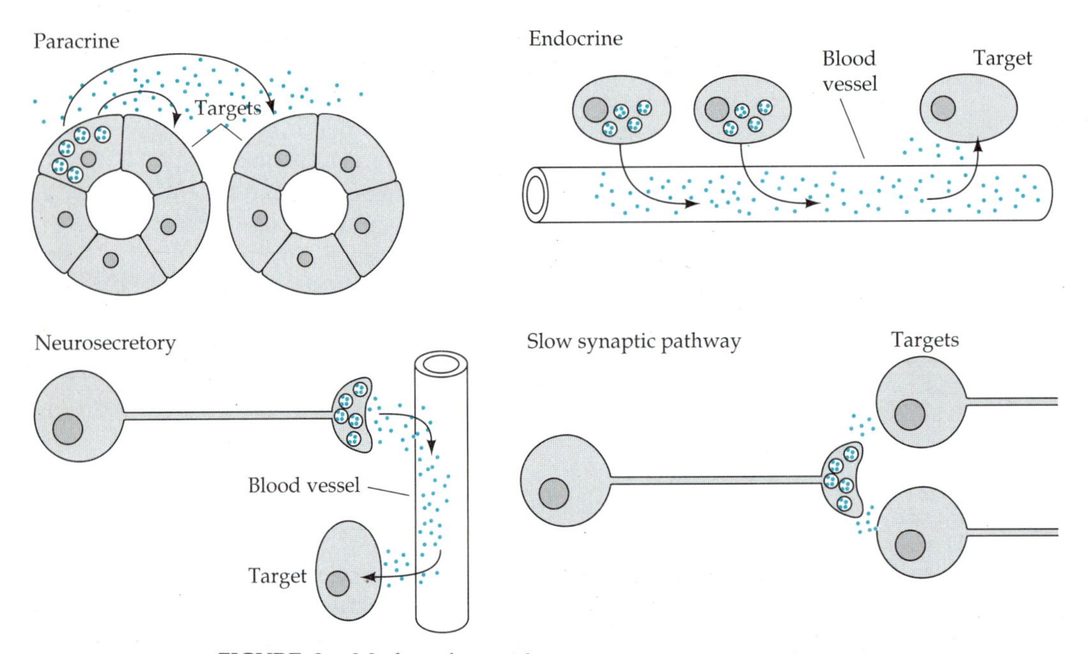

FIGURE 2. Modes of peptide communication. Peptides are released into the extracellular fluid to act on targets that are nearby or far away. In paracrine communication, neighboring cells are the target. In endocrine communication, the peptides or hormones are released into the circulation and can affect targets all over the body. In neurosecretion, neurons release hormones or chemical messengers into the bloodstream. (After D. Krieger, 1983. Clin. Res. 31: 342–354.)

Neurosecretory cells play important roles in invertebrates. The bag cell neurons in *Aplysia,* for instance, release a set of peptides that control egg-laying behavior. The bag cells, which are electrically coupled, fire a synchronous burst of action potentials that continues for up to 30 minutes. The peptides released by these action potentials act on central neurons and peripheral targets to coordinate egg deposition with appropriate respiratory and muscular activities.

Peptides released from neurons and neurosecretory cells in vertebrates can also produce widespread and dramatic behavioral changes by acting on networks of neurons that control behavioral circuits. As an example, injection of angiotensin II into the lateral ventricle of the mammalian brain elicits drinking. Other peptides, such as substance P, counteract this effect. Yet another peptide, cholecystokinin (CCK), may regulate eating behavior.

The widespread use of peptides as messengers and the similarities in the mechanisms by which they are used in different biological systems is useful to investigators of peptidergic synaptic transmission in the nervous system. Thus insights gained from simpler systems can be applied to neuronal problems. Because of the ease of genetically manipulating them, yeast are an especially important experimental system; studies on mutant strains of yeast that are defective in mechanisms of peptide processing and secretion aid the understanding of these processes in neurons.

Signaling at many synapses has characteristics of both slow and fast pathways

In contrast to the rapid synaptic transmission mediated by neurotransmitters such as ACh and glutamate and the slower communication mediated by neuropeptides, synaptic communication mediated by many amines has aspects of both fast and slow transmission. The **biogenic amines**—a group that includes **serotonin (5-hydroxytryptamine), histamine,** and the catecholamines **norepinephrine, epinephrine,** and **dopamine**—are messengers that are used by endocrine and other cells as well as by neurons.

Norepinephrine and epinephrine are secreted by chromaffin cells of the adrenal medulla, which are closely related to the postganglionic sympathetic neurons that secrete norepinephrine (Chapter 1). Chromaffin cells store catecholamines in large granules that fuse with the surface membrane to release catecholamines into the bloodstream. The granules also contain and release peptides, such as the chromogranins and enkephalin. The released catecholamines act on their target cells in the heart, smooth muscle, and secretory glands via G protein-linked receptors.

Sympathetic nerves innervate, and release catecholamines onto, these same cells. In the sympathetic nerve terminals the catecholamines are released from specialized sites along the axon and in the terminal and are stored in small vesicles made near the sites of release. The catecholamines released by the sympathetic nerves, however, like those released into the circulation from the adrenal medulla, act on diffusely distributed G protein-linked receptors. Thus in sympathetic nerves, synthesis of the neurotransmitter norepinephrine is local, the messenger is stored in vesicles whose size resembles those used by other small molecules, and the transmitter is released from specialized sites. The catecholamines act, however, on a G protein-linked receptor whose effects are mediated by intracellular second messengers. This system is an effective one for cardiac and smooth muscle because both tissues are intrinsically excitable; that is, they generate repetitive electrical signals in the absence of innervation, and the amines act to modify or **modulate** the excitability of the tissue over an extended period of time.

Neurotransmitters that are usually associated with rapid synaptic communication systems may also act on G protein-linked receptors. ACh released by the parasympathetic nerves that innervate cardiac and smooth muscle acts on muscarinic ACh receptors, which are not ion channels but are G protein-linked receptors. The neurotransmitters GABA and glutamate, which usually mediate rapid responses, also act on G protein-linked receptors at some synapses. Indeed, both fast and slow receptors may mediate the effects of the same transmitter at a single synapse. Sympathetic postganglionic cells have both fast, nicotinic and slow, muscarinic ACh receptors that mediate a rapid postsynaptic potential (a few milliseconds) and a slow excitatory potential (tens of milliseconds), respectively (see Figure 12, below). Neurotransmitters may also act on more than one rapidly acting receptor at a single synapse. Some synapses in the vertebrate CNS, for example, have two types of glutamate receptors, an NMDA type and a non-NMDA type (see Chapter 3). The NMDA response is different from the non-NMDA response in that it is slower, it only occurs at depolarized potentials, and it is accompanied by calcium entry into the cell.

Synaptic communication thus offers many possibilities for variation. In addition to transmitters that mediate rapid synaptic communication, many synapses release one or more messengers that elicit slower responses. Some synapses have messengers for both fast and slow responses, and at others a single transmitter may elicit both types of response. Before we examine how these responses can work together, we will first discuss the identification and localization of transmitters, how peptide messengers are synthesized and stored, and the mechanisms that neurons use to generate different combinations of messengers.

Identification and Localization of Neurotransmitters

The idea that chemicals might mediate the actions of nerves arose in the early years of this century, and was discussed but never critically tested until Otto Loewi demonstrated in 1921 that stimulation of the vagus nerve innervating the frog heart (which slows the heart rate) causes a factor to be released into the medium that slows the beating of a second, nonstimulated heart. This factor was subsequently shown to be ACh. Later experiments in the 1930s by Sir Henry Dale and his colleagues demonstrated that ACh is also the transmitter at ganglionic synapses in the autonomic nervous system and at the vertebrate neuromuscular junction. In spite of the rigor of these experiments, the idea that a chemical messenger could mediate an event as rapid as neuromuscular transmission met stiff resistance, and a period of intense controversy ensued until the opponents of chemical transmission conceded in the late 1940s.

Transmitters are most rigorously identified by collection and application experiments

During the period when the concept of chemical transmission was still new, a series of criteria were developed to test whether or not a substance could qualify as a transmitter. Two of the criteria are critical: 1) the demonstration that a substance is released when the nerve is stimulated; and 2) the demonstration that application of the substance to the postsynaptic cell produces the same effect as nerve stimulation. Ideally, the two should be quantitatively matched, i.e., the amount of substance released by the nerve per stimulus should be enough to account for the postsynaptic response that a stimulus evokes. In practice, this correspondence is hard to establish because of the difficulty of experimentally mimicking the efficient use of transmitter achieved by the very close apposition of pre- and postsynaptic membranes at the synapse. Only in the case of the neuromuscular junction has sustained and exacting experimentation over a period of years brought the two figures close. There, less than 10,000 molecules of ACh have been shown to cause a potential change equivalent to that caused by the release of a single vesicle, which is estimated from other experiments to deliver about 8000 molecules of ACh into the extracellular fluid at the endplate.

These experiments constitute a rigorous and complete proof that ACh is the neurotransmitter at the neuromuscular junction.

Similar experiments have established in a less quantitative way that norepinephrine is the transmitter secreted from mammalian sympathetic nerves and that GABA is the transmitter at the inhibitory synapse of crustacean muscles. Over the last 25 years, a score of other transmitters have been similarly identified, mostly at invertebrate or mammalian peripheral synapses. These include the amino acids glutamate and glycine; the amines epinephrine, serotonin, histamine, and octopamine; and the purines adenine and ATP.

Immunocytochemical methods can identify potential transmitters

The use of the classical criteria to identify transmitters in the mammalian CNS is much harder than in the PNS. The anatomical complexity of the brain and the inaccessibility of most synapses make unambiguous transmitter collection and application experiments difficult; thus, other methods of identifying prospective transmitters have been adopted. One strategy is based on the high-affinity uptake of transmitters by the nerve terminals that secrete them (Chapter 5). Incubation of spinal cord slices in radioactive glycine, for example, selectively labels a class of terminals on motor neurons that are inhibitory.

By far the most powerful strategy for detecting potential neurotransmitters, however, is the use of antibodies directed against either the neurotransmitter or against an enzyme in its biosynthetic pathway. Antibodies either to the transmitters themselves (e.g., serotonin, GABA, or glutamate) or to specific biosynthetic enzymes can be used to selectively stain and identify these neurons. Neurons that secrete GABA, ACh, or catecholamines are specifically labeled by immunostaining with antibodies to glutamic decarboxylase (Figure 3), choline acetyltransferase, or tyrosine hydroxylase, which are the respective biosynthetic enzymes for these neurotransmitters (Chapter 5). In the spinal cord, for example, antibodies to choline acetyltransferase stain motor neurons, and in the cerebellum, antibodies to glutamic decarboxylase identify the cell bodies and processes of Purkinje cells.

Peptides are selectively concentrated in many neurons

Although useful for the localization of neurons secreting known transmitters, antibodies have been most useful in the identification of new synaptic messengers. Their use in immunocytochemical experiments has shown that neurons in the mammalian central nervous system contain a bewildering variety of peptides, each with a distinctive distribution among the varied cell types of the brain. Each peptide that is found in a limited group of neurons is presumed, like the small-molecule neurotransmitters, to be released by nerves. In some cases, release of the peptide has been demonstrated; in others, immunocytochemical experiments with the electron microscope have shown that the peptide is contained in secretory vesicles. In yet other cases, the release of the peptide is inferred only from its selective concentration in certain neurons.

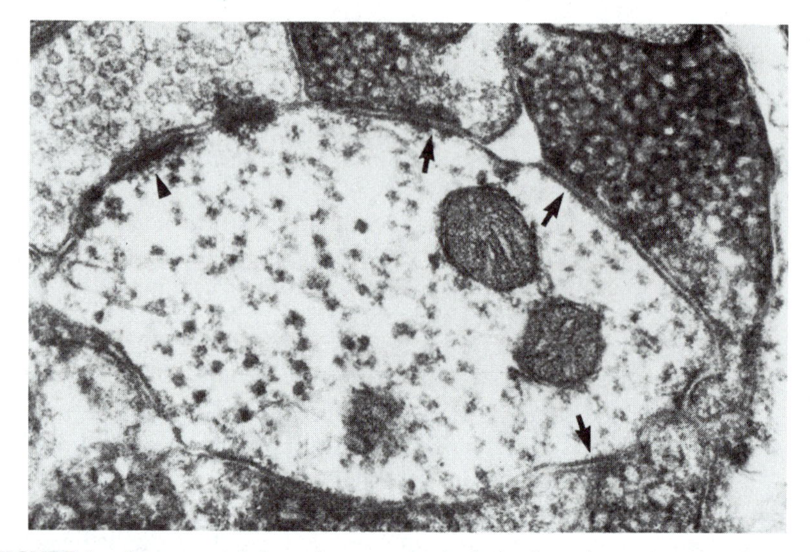

FIGURE 3. Immunostaining of nerve terminals in the substantia nigra that synthesize GABA. Sections of rat brain were fixed, embedded, and then incubated with a rabbit antiserum to purified glutamic decarboxylase, the enzyme that synthesizes GABA (Chapter 5). Antibody binding was detected by an indirect method using the enzyme peroxidase, which generates an electron-dense reaction product. The picture shows several nerve terminals ending on a dendrite. The center and right-hand terminals at the top of the picture are positive for glutamic decarboxylase (arrows), whereas the terminal on the left (arrowhead) is not. (From C. E. Ribak et al., 1976. Brain Res. 116: 287–298.)

At present, over 40 **neuropeptides** have been identified in CNS neurons (Table 1). Many of these had been previously identified as chemical messengers in other tissues. Some, such as glucagon or calcitonin, are hormones secreted by endocrine glands into the bloodstream. Others, such as thyrotropin-releasing hormone (TRH), corticotropin-releasing factor (CRF), or luteinizing-hormone–releasing hormone (LHRH) were originally identified as peptides that regulate the release of hormones from cells in the anterior pituitary (see Box A). A third class are peptides originally isolated from the gut as inhibitors or activators of glandular secretion or of smooth muscle contraction. These include vasoactive intestinal peptide (VIP), CCK, and substance P. One important group of peptides, the opioid peptides, were first identified because of their ability to bind opiate receptors in the CNS (Box B).

A single cell can contain several chemical messengers

Immunocytochemical studies of the distribution of peptides in the CNS led to a surprising conclusion: many peptides are selectively concentrated in neurons and nerve terminals that also contain small-molecule transmitters. This overlap is most clearly seen in double- or triple-label immunocytochemistry experiments, in which the distribution of two or three compounds can be determined in the same tissue section (Figure 4). The **co-localization** of chemical messengers in neurons is now known to be a

TABLE 1 Peptides in the mammalian brain[a]

Hypothalamic-releasing hormones	*Gut–brain peptides*
Thyrotropin-releasing hormone (TRH)	Vasoactive intestinal peptide (VIP)
Gonadotropin-releasing hormone	Cholecystokinin (CCK)
Somatostatin	Gastrin
Corticotropin-releasing factor (CRF)	Motilin
Growth-hormone–releasing hormone (GHRH)	Pancreatic polypeptide
Luteinizing-hormone–releasing hormone (LHRH)	Secretin
	Substance P
	Substance K
Pituitary peptides	Bombesin
Adrenocorticotropic hormone (ACTH)	Neurotensin
Growth hormone (GH), somatotropin	Gastrin-releasing peptide (GRP)
Lipotropin	
Alpha-melanocyte–stimulating hormone (α-MSH)	*Opioid peptides*
	Dynorphin
Prolactin	Beta endorphin
Luteinizing hormone	Met-enkephalin
Thyrotropin	Leu-enkephalin
Neurohypophyseal hormones	*Others*
Vasopressin	Bradykinin
Oxytocin	Carnosine
Neurophysin(s)	Neuropeptide Y
	Epidermal growth factor (EGF)
Circulating hormones	Atrial natriuretic factor (ANF)
Angiotensin	Calcitonin gene-related peptide (CGRP)
Calcitonin	Calcitonin
Glucagon	Neuromedin K
	Galanin

[a]These are some of the peptides that are likely to be used as chemical messengers in the brain. Many of the compounds were initially discovered in non-neuronal tissues, such as gut, and were later suggested to function in the nervous system.

frequent, rather than an exceptional, finding. Numerous examples are known in which particular peptides are co-localized with each of the classical transmitters, and in which two or more peptides are co-localized in the same neuron (Table 2). A few cases are also known in which two classical transmitters (serotonin and GABA, for example) appear to be co-localized in the same neuron.

Immunostaining methods that demonstrate the distribution of peptides have further emphasized the biochemical heterogeneity of the nervous system and revealed differences that were previously unsuspected. For instance, virtually all neurons in the sympathetic ganglia that innervate the guinea pig gut contain norepinephrine. Immunostaining with peptides divides this apparently homogeneous population into three groups, according to whether they also stain for somatostatin, for neuropeptide Y, or for neither. The subsets of cells defined in this way occupy distinct regions within the ganglion and project to different target organs, indicating that the distinction is a functional and anatomical one, as well as a biochemical one.

Box A Hypothalamic Releasing Factors

In the mammalian brain, the secretion of hormones by anterior pituitary cells is regulated by neurosecretory cells in the hypothalamus. The axons of these cells terminate in the median eminence, where they secrete releasing factors or hormones into a special system of capillaries that are connected by a portal system to capillaries in the anterior pituitary (Figure A). Because of this direct connection, the releasing factors reach the pituitary before being diluted into the general circulation. The factors both stimulate and inhibit release of specific anterior pituitary hormones. Because they are released close to their site of action, only small amounts of the factors are secreted, and their isolation and identification required a heroic effort.

Two laboratories, one headed by Roger Guillemin and the other by Andrew Schalley, undertook the daunting task. Over 10 years of work and over 50 tons of hypothalamic tissue culminated in the purification of the first factor, thyrotropin-releasing factor (TRF), in 1969. Extensive analysis and peptide chemistry yielded an unconventional structure, a tripeptide of Glu-His-Pro in which the N-terminal glutamic acid is cyclized to the pyro form and the C-terminal is amidated. Once the first peptide was found, other peptide factors quickly followed: gonadotropin-releasing hormone (GnRH, also called luteinizing-hormone–releasing hormone or LHRH), corticotropin-releasing hormone (CRH), and growth-hormone–releasing hormone (GHRH), all stimulate release of pituitary hormones. The peptide factor somatostatin inhibits growth hormone release, and dopamine inhibits prolactin secretion. Each factor acts on a specific G protein-linked receptor to affect hormone release from pituitary cells.

Once the purified releasing factors and antisera to them were available, they were soon found elsewhere in the peripheral and central nervous systems. Somatostatin, for example, is released by nerves that innervate the pancreas, and inhibits secretion of insulin and glucagon by pancreatic cells, and LHRH mediates a slow potential in sympathetic ganglia (see text). The function of these peptides in regions of the brain outside the hypothalamus remains an intriguing mystery.

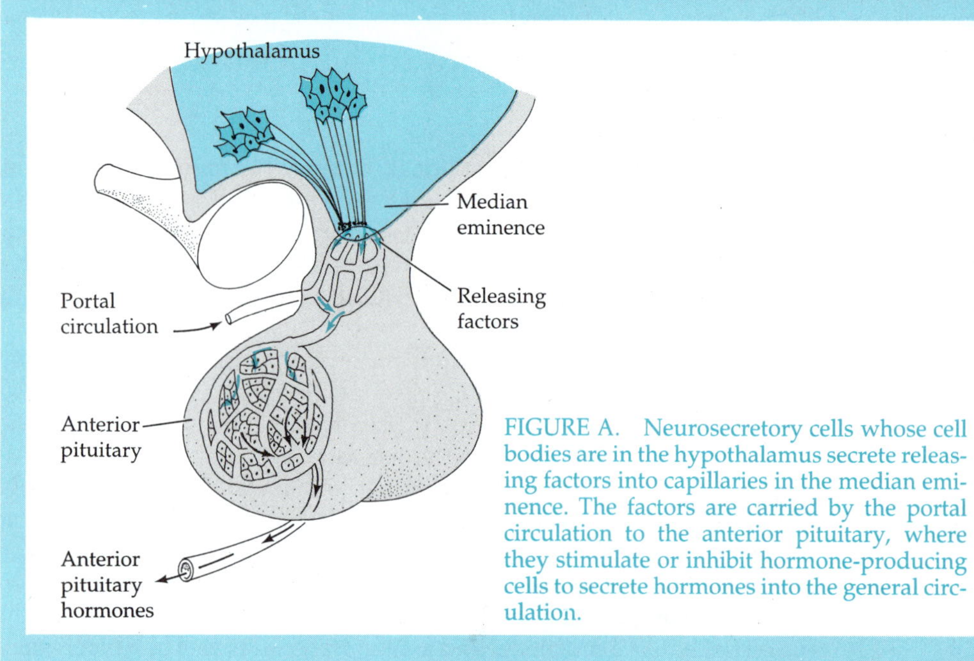

FIGURE A. Neurosecretory cells whose cell bodies are in the hypothalamus secrete releasing factors into capillaries in the median eminence. The factors are carried by the portal circulation to the anterior pituitary, where they stimulate or inhibit hormone-producing cells to secrete hormones into the general circulation.

Box B Opiates and Their Receptors

The discovery of the opioid peptides reversed the usual paradigm: the receptors were found first and their natural ligands later. Because of their importance for analgesia and addiction, pharmacologists had long sought the brain receptors for morphine and its derivatives. In 1973, three independent groups, those of Solomon Snyder and Candace Pert, Lars Terenius, and Eric Simon, identified a binding site in brain membranes for radiolabeled naloxone, a potent opiate antagonist. Following an earlier idea of Avram Goldstein's, they used the known difference in biological activity of stereoisomers to demonstrate the specificity of the site and to identify it as a biologically important site. In Pert and Snyder's experiments, levorphanol, an opiate agonist, blocked naloxone binding, but dextrorphan, its inactive optical isomer, did not. Opiate antagonists also inhibited binding, and their relative potencies in the binding assay paralleled their biological activities. These experiments were important for two reasons. By identifying an opiate receptor and a method for its study, they opened the door to a new generation of opiate research in the brain. They also introduced a new technology to neuroscience (popularly called "grind and bind") that has been widely applied to the identification and pharmacological study of many other brain receptors.

The discovery of the opioid receptors also raised a new question: Why are they there? Thus began the search for a natural ligand. Because morphine is known to inhibit smooth muscle contractility, John Hughes and Hans Kosterlitz looked for a similar activity in brain extracts. Using this assay they purified two biologically active pentapeptides, **Met-** and **Leu-enkephalin** (Tyr-Gly-Gly-Phe-Met and Tyr-Gly-Gly-Phe-Leu, respectively), whose actions on smooth muscle were blocked by naloxone. Other biologically active peptides, including the endorphins and the dynorphins, were identified in the pituitary and other tissues. Currently, at least 18 endogenous peptides with opiate-like activity are known. All begin at the N terminus with the sequence of either Leu- or Met-enkephalin.

More refined binding studies have shown that opiate receptors have at least three forms: delta, kappa, and mu. All appear to be G protein-linked receptors. At several synapses, mu receptors, which have the highest affinity for morphine-like opiates, mediate presynaptic inhibition; the cellular function of the others is not known. All three receptors are found in pathways mediating pain in the brain and spinal cord. At present, none have been cloned.

Individual nerve terminals contain two types of vesicles with different messengers

Electron microscopy of nerve terminals shows that most contain two types of vesicles: small vesicles, of about 50 nm in diameter, similar to those seen at the active zones of the neuromuscular junction; and larger vesicles, 90–250 nm in diameter, whose interior, or core, appears dense in the electron beam. The small vesicles, which are numerous, are usually clustered near the face of the terminal that opposes the postsynaptic cell; the larger vesicles are fewer, and tend to be scattered more randomly in the terminal (Figure 5A). We refer to the large, dense-cored vesicles that occur both in nerves and in other tissues as **secretory vesicles,** and to the small vesicles seen in nerve terminals as **synaptic vesicles.** The dense appearance of secretory vesicles comes from their content of peptides. Except for

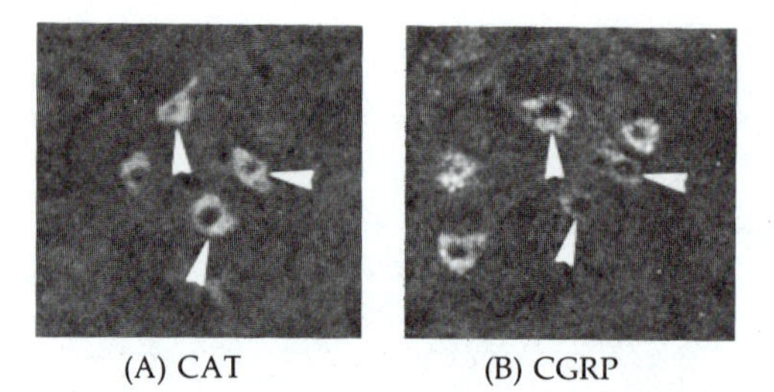

(A) CAT (B) CGRP

FIGURE 4. Neurons of the facial nucleus of the rat contain both choline acetyltransferase and calcitonin gene-related peptide (CGRP). Alternate sections of rat brain were stained with an antibody to choline acetyltransferase (A), the enzyme that synthesizes ACh, and with antibodies to CGRP (B). The sections were then incubated with a second antibody conjugated to a fluorescent chromophore and viewed by fluorescence microscopy. Arrowheads mark the cell bodies that contain immunoreactivity for both primary antibodies. (From K. Takami et al., 1985. Brain Res. 328: 386–389.)

amine-containing vesicles (see below), synaptic vesicles have a low density in the electron beam and are clear in electron micrographs.

When nerves containing peptides are examined by immunogold methods, antibodies to the peptides bind exclusively to the secretory vesicles, whereas antibodies to the neurotransmitters GABA and ACh bind to the synaptic vesicles. Double-label experiments show that the same terminal can contain a peptide transmitter in secretory vesicles and a small molecule transmitter in synaptic vesicles (Figure 5B). Large vesicles often contain more than one peptide. Peptide combinations that are unrelated and are derived from different precursors, such as somatostatin and tachykinin, or substance P and enkephalin, are shown by double-labeling experiments to be contained in the same vesicles.

Amines occur in both large and small vesicles

Sympathetic nerve terminals, which secrete norepinephrine as the neurotransmitter, contain two sizes of vesicles, with mean diameters of about 50 nm and 80 nm. Both contain norepinephrine. The smaller vesicles, which are the predominant type, can be distinguished from the small vesicles in other terminals because osmium, used in fixation of tissues for electron microscopy, reacts with the catecholamine to yield a compact, electron-dense core. These small, dense-cored synaptic vesicles are seen after osmium fixation in all nerve terminals that secrete catecholamines, both in the central and peripheral nervous systems, and identify them as containing amines.

In many terminals, the secretory vesicles contain both biogenic amines and peptides. The large vesicles in **adrenergic** terminals (those that secrete norepinephrine) contain both norepinephrine and a group of peptides called chromogranins (see below). Some nerve terminals contain amines as well as amino acid transmitters. Several different terminals that contain GABA, for example, also contain either serotonin, dopamine, or histamine. The existence of amines and peptides in the same secretory vesicle not only occurs in neurons, but is also common in the endocrine system.

TABLE 2 A partial list of small-molecule neurotransmitters and peptides that are co-localized in neurons[a]

Small molecule	Peptide
ACh	Enkephalin
	VIP
	CGRP
	Substance P
	Somatostatin and enkephalin
	LHRH
	Neurotensin
	Galanin
Dopamine	CCK
	Enkephalin
	Neurotensin
Ephinephrine	Enkephalin
	Neuropeptide Y
	Neurotensin
	Substance P
GABA	CCK
	Enkephalin
	Somatostatin
	Neuropeptide Y
	Substance P
	VIP
Glutamate	Substance P
Glycine	Neurotensin
Norepinephrine	Enkephalin
	Neuropeptide Y
	Neurotensin
	Somatostatin
	Vasopressin
Serotonin	CCK
	Enkephalin
	Substance P and TRH
	TRH

[a]Most of these combinations are based on immunohistochemistry. The structure of the immunoreactive material has not been determined in most cases. Abbreviations: CCK, cholecystokinin; VIP, vasoactive intestinal peptide; TRH, thyrotropin-releasing hormone; CGRP, calcitonin gene-related peptide; LHRH, luteinizing-hormone–releasing hormone.

Isolation of synaptic vesicles and secretory vesicles reveals their contents

Isolation of synaptic vesicles and analysis of their contents demonstrates that they contain several components. The purest preparations come from the nerve terminals in the electric organ of *Torpedo,* a rich and homogenous source of cholinergic terminals (see Box B, Chapter 3). The purified vesicles contain a high concentration of ACh (0.5 *M*), ATP, a small amount of GTP, and a negatively charged proteoglycan. The ATP, which is present at about one-third the concentration of the ACh, is released along with the ACh. The function of the proteoglycan is not known, but it may serve to neutralize charge inside the vesicle.

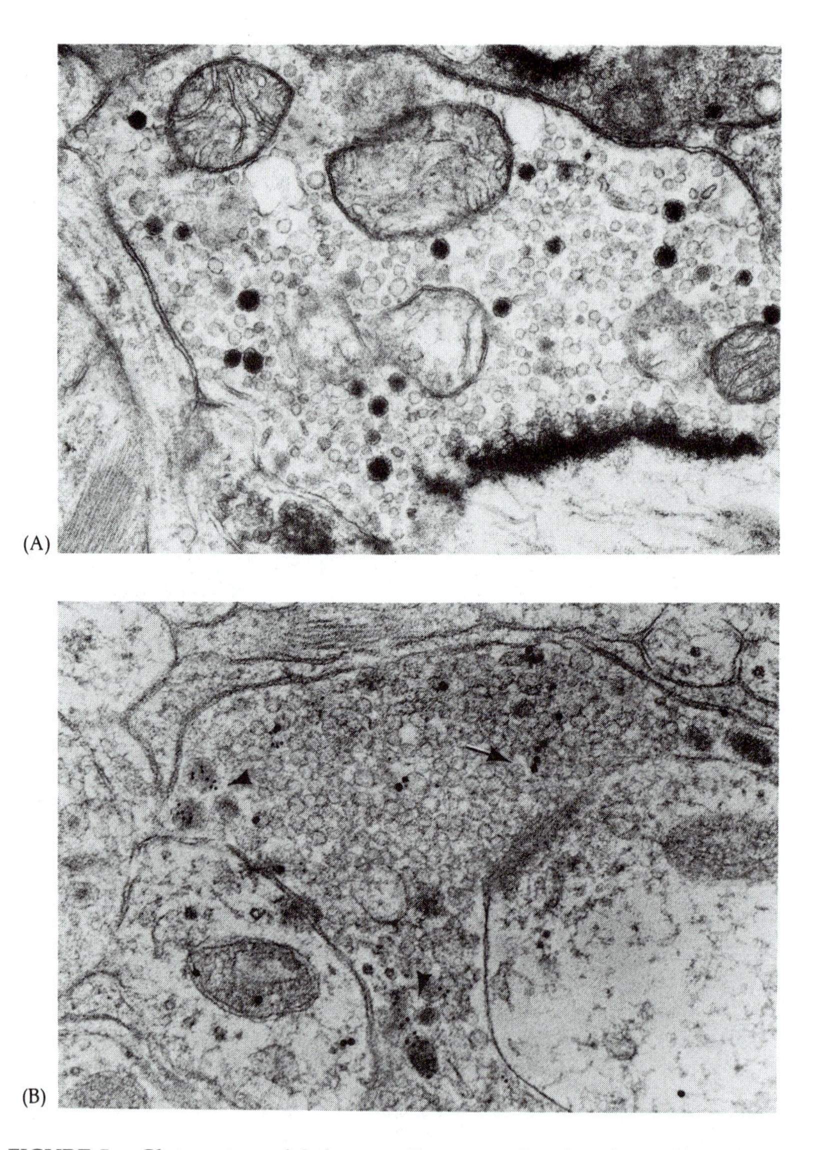

FIGURE 5. Glutamate and Substance P are associated with small and large vesicles in nerve terminals in the spinal cord. (A) An electron micrograph of a nerve terminal in the spinal cord containing both small synaptic vesicles and large, dense-cored secretory vesicles. (B) Localization of glutamate and Substance P in a nerve terminal. Fixed and embedded rat spinal cord sections were stained sequentially with rabbit antibodies to substance P and to glutamate. After treatment with anti–substance P antibodies, the sections were incubated with secondary antibodies conjugated to 10-nm gold particles; the sections were then exposed to paraformaldehyde vapors to destroy the immunoglobulin binding sites and incubated with antibodies to substance P, followed by a second set of secondary antibodies conjugated to 20-nm gold particles. The large gold particles, which represent glutamate immunoreactivity, are associated with small, clear vesicles (arrow); the small gold particles, which represent substance P immunoreactivity, are associated with large, dense-cored vesicles (arrowheads). (A courtesy of D. Ralston and H. R. Ralston. B from S. De Biasi and A. Rustioni, 1988. Proc. Natl. Acad. Sci. USA 85: 7820–7824.)

Synaptic vesicles can also be purified from brain, either by differential centrifugation, or more recently, by immunoadsorption. Brain synaptic vesicles are highly homogenous in size, but are heterogenous in their contents, as they are derived from many types of nerve terminals. Amino acid analysis shows that the mixed population of vesicles contains GABA, glutamate, and glycine.

Because of their heterogeneity and small number, purified secretory vesicles have not been isolated from neurons. Homogeneous preparations of secretory vesicles (often called granules) can be easily isolated, however, from non-neural secretory cells such as chromaffin cells of the adrenal medulla or mast cells. Those from chromaffin cells are of particular interest because they resemble the secretory vesicles in sympathetic nerves. Chromaffin granules contain a high concentration of catecholamines (0.5 *M*), ATP (0.125 *M*), ascorbic acid, and a variety of peptides. These include a family of related proteins of unknown function called chromogranins, the opioid peptide enkephalin, and a soluble form of dopamine beta-hydroxylase, an enzyme in the synthetic pathway for norepinephrine and epinephrine (see Chapter 5).

Several membrane proteins are common to all synaptic vesicles

Examination of purified synaptic vesicles from brain and from the electric organ of marine rays shows them to contain several abundant integral membrane proteins. Many of these proteins have been purified and cloned, and specific antibodies made to them. Immunostaining and immunoprecipitation experiments indicate that four of the proteins—synaptophysin, SV2, p65 (synaptotagmin), and VAMP (synaptobrevin)—are common to all synaptic vesicles (Figure 6). Some of these proteins, most notably synaptophysin, are found at low levels, if at all, in the secretory vesicles of endocrine or exocrine cells, or in the secretory vesicles of nerve terminals.

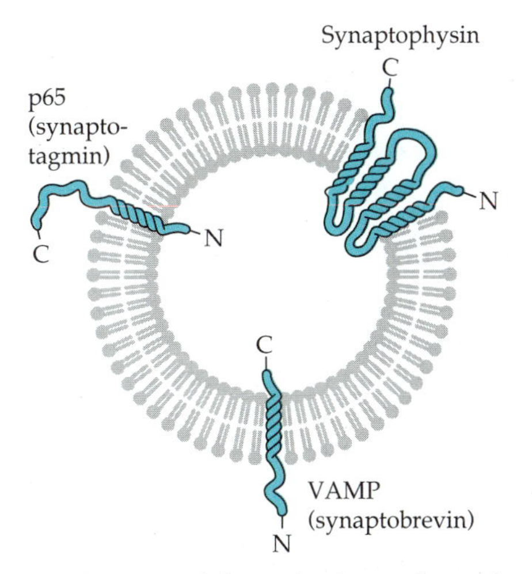

FIGURE 6. Schematic diagram of the major integral membrane proteins of synaptic vesicles. These proteins are common to all synaptic vesicles.

The widespread distribution of these proteins suggests that they are essential to synaptic vesicle function, but in no case has a specific role been assigned to them. Synaptophysin, a 39-kD protein, may function in exocytosis, and p65 could play a role in excitation-secretion coupling (see Chapter 5). There are few clues to the functions of VAMP or SV2. In addition to their possible roles in exocytosis, these proteins may be important for the local production of synaptic vesicles in nerve terminals. Unlike secretory vesicles, which arise from the *trans*-Golgi network in the cell body (see below), synaptic vesicles are generated by endocytosis in the nerve terminal.

Synthesis and Storage of Peptide Transmitters

The central organelle in the synthesis, storage, and secretion of neurotransmitters is the synaptic vesicle. As we have seen, neurons have two types, small vesicles that contain transmitters associated with rapid synaptic transmission, and large vesicles that can contain peptides as well as small molecules. The large vesicles are formed in the Golgi apparatus of the cell body and transported to the terminals where their contents are released. Because the pathways by which the large, peptide-containing vesicles are formed are similar to those of secretory vesicles in many other cell types, much is known about them. Much less is known about the formation of synaptic vesicles, which occurs mostly in nerve terminals. We consider here the synthesis and processing of peptide transmitters and their sequestration in large granules. The synthesis of small molecule neurotransmitters and the formation of small vesicles is considered in the chapter on nerve terminals (Chapter 5).

Synthesis of peptide transmitters follows the secretory pathway

Experiments on non-neural cells that secrete large amounts of protein, such as the acinar cells of the pancreas, have defined a pathway, from synthesis to release, that the secreted proteins follow through the compartments of the cell (Figure 7). The proteins, which are usually made as part of a longer precursor, are synthesized on ribosomes, transported into the lumen of the rough endoplasmic reticulum (ER), and then to the Golgi complex, where they are sorted into secretory vesicles. Specific proteolytic cleavages and modifications occur in each compartment. Upon stimulation by an appropriate signal, the proteins are released into the extracellular fluid by fusion of the secretory vesicles with the surface membrane. This **regulated secretory pathway**, which is common to almost all secretory cells, is distinct from the pathway responsible for replenishing the surface membrane and extracellular matrix. In the latter pathway, called the **constitutive pathway,** vesicles fuse continuously with the surface membrane in the absence of any signal.

Neurons are secretory cells par excellence, and the peptides that they secrete follow the same pathway as that seen in other secretory cells. They are synthesized in the ER, modified and sorted into vesicles at the Golgi, and transported to their principal sites of release in the terminals. The

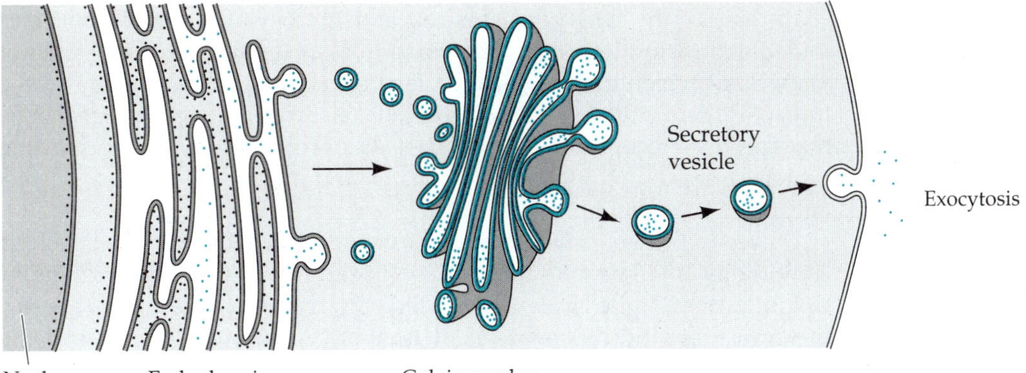

FIGURE 7. The secretory pathway. Secreted proteins in neurons and other cells follow a defined pathway through the cell from synthesis to secretion. The secreted protein or its precursor is synthesized on ribosomes and translocated into the lumen of the endoplasmic reticulum, where modifications of the protein begin. The protein is then transported via small vesicles to the *cis* aspect of the Golgi complex, and traverses the compartments of the Golgi to the *trans* aspect. Further modifications of the protein occur in each compartment. In the *trans*-Golgi network, the proteins are sorted into secretory vesicles and are then transported to their sites of release. A change in a cytosolic second messenger then triggers release by exocytosis.

major difference between the secretion of peptides in neurons and in other cells is the very long route of transport that is interposed in neurons between the sites of synthesis and secretion.

Peptide precursors are made in the ER and processed in the Golgi complex

Precursor proteins containing peptides that are to be secreted are directed to the ER by a relatively hydrophobic sequence of 20–30 amino acids, the **signal sequence,** which is usually at the N terminus of the protein. During synthesis of the protein on the ribosome, the signal sequence in the nascent protein is recognized and bound by a complex of protein and RNA, the **signal recognition particle (SRP)**. This particle, which is recognized in turn by an SRP receptor on the ER, guides the nascent protein and attached ribosome to the ER, where the N terminal signal sequence inserts into the membrane and the growing polypeptide is transported into the lumen. As the nascent protein chain enters the lumen, the signal sequence is cleaved and rapidly degraded. Translation of the precursor and its transport across the membrane continue until the entire protein is in the lumen of the ER.

Most secreted proteins are glycosylated. The initial steps in the pathway of N-linked glycosylation take place in the ER. A preformed oligosaccharide enriched in mannose residues is attached to the amide nitrogen of asparagine residues of the growing polypeptide. The asparagines that are glycosylated are part of a glycosylation consensus sequence, either Asn-X-Ser or Asn-X-Thr, where X can be any amino acid except proline.

In the ER, the newly synthesized and glycosylated protein becomes folded and disulfide bonds are formed in a process whose details are not known but which is thought to be facilitated by proteins resident in the lumen. Protein disulfide isomerase catalyzes disulfide exchange reactions that speed correct pairing of cysteines, and proteins such as BiP (binding protein) bind to unfolded peptide chains and may act to promote correct folding of the peptide.

Correctly folded proteins are transported from the ER to the Golgi complex by small vesicles that pinch off from the ER and fuse with the *cis* Golgi. In the Golgi complex, the *N*-linked oligosaccharide chains are further processed by the removal of terminal mannose residues and the addition of galactose, *N*-acetylglucosamine, and sialic acid residues. For many peptides, further proteolytic processing also takes place in the Golgi complex.

Multiple peptides arise from a single precursor

For some secreted proteins, such as the hormone prolactin, cleavage of the signal sequence in the ER is the only proteolytic modification made before secretion. In the case of many small peptides, however, the physiologically active peptide is embedded in a longer protein, the **prohormone,** from which it must be released by further proteolytic cleavage (Figure 8). In the case of insulin, only a small peptide is released from the prohormone to generate the active polypeptide. In the case of substance P, on the other hand, the prohormone is many times larger than the active peptide from which it is released. There are two important variations on the prohormone theme: one in which a single precursor encodes more than one physiologically active peptide, and one in which multiple copies of the same peptide are encoded in a single precursor. The large precursors, encoding multiple active peptides, are sometimes referred to as **polyproteins.** A precursor protein secreted by bag cells in the sea hare, *Aplysia,* for example, encodes four different physiologically active peptides, which are released as part of a complex program related to egg-laying behavior.

Several precursors contain multiple copies of a single physiologically active peptide that are cleaved and secreted. The precursor for alpha mating factor from yeast contains four copies of the peptide, whereas that for the enkephalins contains six copies of Met-enkephalin and one copy of Leu-enkephalin. Perhaps the most spectacular example, however, is the 28 copies of the FMRFamide peptide (Phe-Met-Arg-Phe-amide) contained in a precursor in *Aplysia.* Multiple copies of peptides in prohormones most likely evolved through duplication within the gene as the result of unequal crossover events. In some precursors the duplicated units have diverged, producing peptides of related, but unique, sequence. The stoichiometry of the peptides, which may act on different receptors and have different functions, is thus controlled by the precursor structure. In this way the evolution of polyprotein content shapes the physiological response.

The generation of peptides from larger prohormones depends on the specificity of the proteases responsible for cleavage of the precursor. The most common cleavage recognition sites are the dibasic sequences Lys-Arg

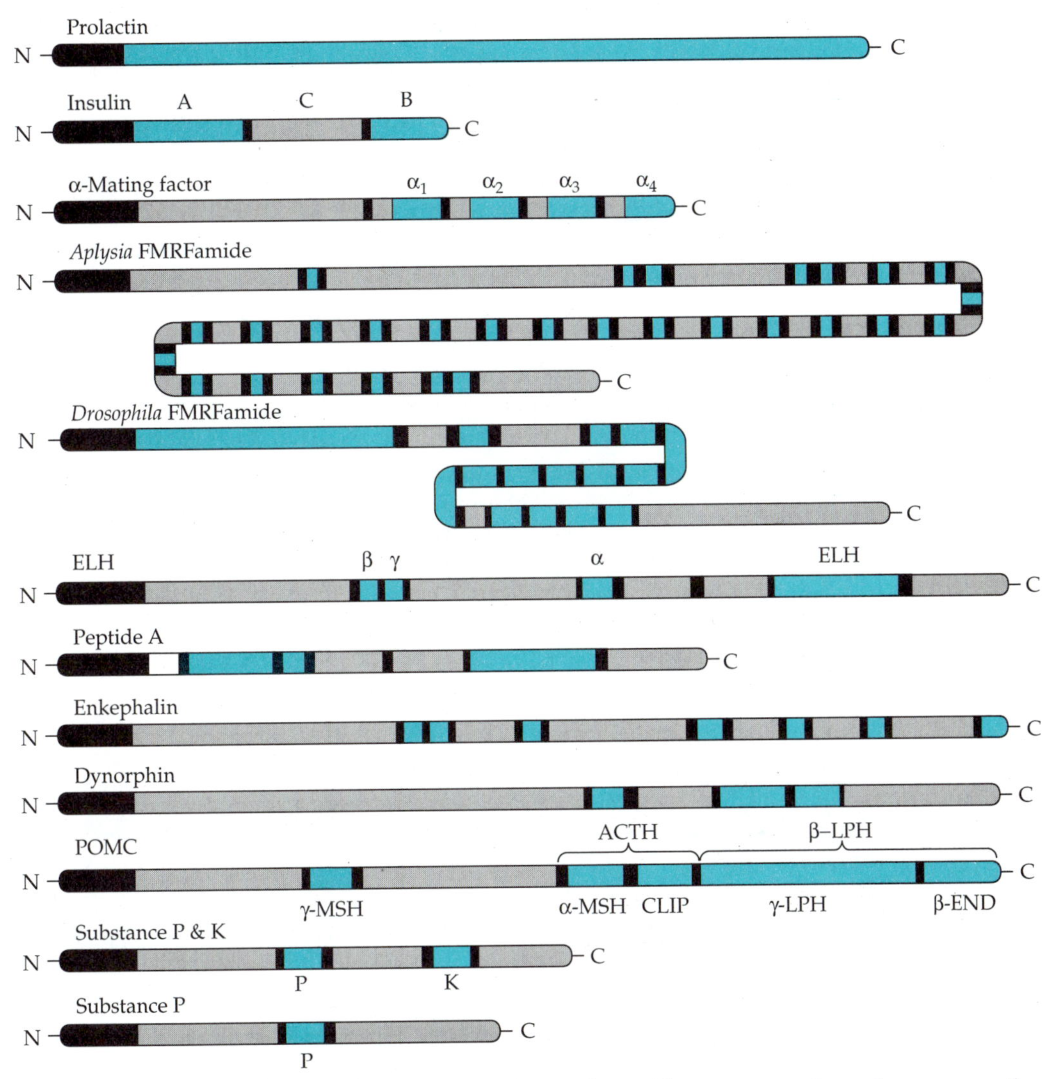

FIGURE 8. The structures of several peptide precursors. In each precursor, the signal sequence is indicated by the dark area at the left and the final products by colored areas. Vertical lines indicate sites of endoproteolytic cleavage. Dark areas between vertical lines represent trimming by diaminopeptidyl peptidase. The names of some of the products are given. (After W. Sossin et al., 1989. Neuron 2: 1407–1417.)

or Arg-Arg, which are hydrolyzed at their C termini. Cleavage sometimes occurs at a single basic residue, so that the structure of the precursor must also determine cleavage sites. After cleavage, the basic residues are removed by carboxypeptidase E. If the new C terminus is glycine, an elimination reaction yields a C-terminal amide.

The endoprotease(s) responsible for cleavage at the dibasic residues has not been definitely identified in mammals but may be related to a calcium-dependent serine protease encoded by the KEX2 locus that carries out this function in yeast. Several mammalian genes related to KEX2 have been cloned recently; thus the long search for these elusive proteins may be over. In a polyprotein with many cleavage sites, the cleavages are not random, but occur in a fixed sequence. The cleavage mechanisms in mammals are probably common to many secretory cells since prohormones expressed in foreign secretory cells are processed to the correct products. Thus, proinsulin and proenkephalin are not processed to generate active products in transfected L cells, a fibroblastic cell line, but are processed correctly when their cDNAs are transfected into AtT-20 cells, cells derived from the anterior pituitary that do not ordinarily synthesize these peptides.

A single precursor may be processed differently in different cells. The pro-opiomelanocortin precursor in the anterior lobe of the pituitary is processed to adrenocorticotropin (ACTH) and two other products, β-lipotropin (β-LPH) and an N-terminal peptide, whose functions are unknown (Figure 9). In the intermediate lobe of the pituitary, ACTH and β-LPH are further processed. ACTH is cleaved to yield α-melanocyte–stimulating hormone (α-MSH) plus another peptide (called CLIP), and β-LPH is cleaved to yield endorphin and γ-lipotropin. Thus, although the same

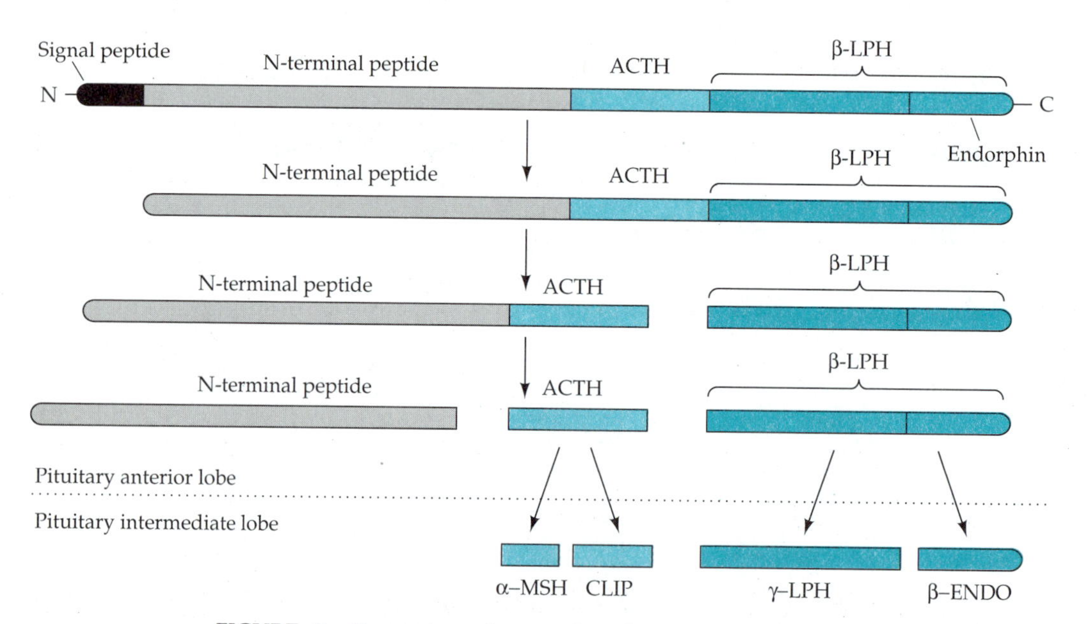

FIGURE 9. Processing of pro-opiomelanocortin (POMC) in the anterior and posterior lobes of the rat pituitary. The cleavage steps above the line occur in both anterior and intermediate lobes. β-LPH is β-lipotropin; ACTH is adrenocorticotropin; α-MSH is α-melanocyte–stimulating hormone; β-endo is β-endorphin; and CLIP is corticotropin-like intermediate lobe peptide. Those below the line occur only in the intermediate lobe. (After E. Herbert et al., 1984. Cold Spring Harbor Symp. Quant. Biol. 48: 375–384.)

precursor protein is expressed in cells in both lobes of the pituitary, different sets of peptides are released.

Peptide intermediates and products are sorted into secretory granules in the trans-*Golgi network*

Secretory vesicles form by budding from the *trans*-Golgi complex. Their formation involves a process of sorting, by which the membrane proteins of the vesicle and its protein contents are separated from other proteins in the Golgi that are destined for other membrane compartments of the cell such as the lysosomal or surface membrane. The sorting occurs in a system of membranes associated with the *trans*-Golgi known as the ***trans*-Golgi network (TGN).** The mechanisms by which proteins destined for the secretory vesicle are distinguished from other proteins in the Golgi are not known. No common peptide sequence or pattern of glycosylation has been detected that could be responsible for targeting the proteins to secretory granules.

The timing and localization of the proteolytic cleavage of each prohormone is relevant to its packaging. Cleavage in the Golgi, before sorting into vesicles, allows the possibility of independent packaging of the products into separate vesicles. Cleavage after packaging, however, assures that the multiple products of a precursor are secreted simultaneously from the same vesicle. In the posterior pituitary, vesicles can be obtained at different stages of transport from the cell bodies to the terminals. In vesicles (granules) from proximal axons the precursor for the hormones oxytocin and vasopressin is largely intact, while those taken from more distal axons have mostly peptide products. Clearly, in this case, cleavage of the prohormone occurs in the vesicles during transport down the axon.

A different mechanism is seen in *Aplysia,* in which the initial cleavage of the prohormone containing the egg-laying hormone (ELH) and other peptides occurs in the TGN. The two intermediates are sorted into separate vesicles, and each intermediate is cleaved to its final products in the vesicles. Each of the vesicle types is then targeted to different locations within the neuron. The set of vesicles containing peptides that act locally to modulate electrical activity is transported to processes that release them in a restricted region of the ganglion. The other vesicles contain ELH, which acts on distant as well as nearby neurons, thereby controlling the physiology of peripheral tissues. These vesicles are transported into processes that efficiently deliver the product to the circulatory system.

The finding that different processes contain and presumably secrete different peptides is a potential exception to an early and often useful generalization made by Sir Henry Dale in the 1930s of the metabolic unity of the neuron. Stated in modern terms, Dale hypothesized that a neuron releases the same set of chemical messengers from all of its processes. This hypothesis has rarely been put to a critical test, particularly with respect to the combination of transmitters released at each terminal. As morphological techniques for examining the distribution of transmitters within neurons become more sophisticated, other exceptions to Dale's hypothesis may become apparent.

Box C CGRP: A Novel Neuropeptide Discovered by Cloning

The identification of most physiologically active peptides begins with a crude tissue extract that has a biological activity such as increasing smooth muscle contraction or stimulating secretion. Then follow the long and often arduous tasks of purifying the active substance using the bioassay and establishing its chemical identity (see Box B). The discovery of calcitonin gene-related peptide (CGRP) proceeded in an entirely different manner: the peptide was discovered first, through cDNA cloning, and its localization and biological activities later.

In 1982 the laboratories of Ronald Evans and Michael G. Rosenfeld isolated cDNA clones for calcitonin, a peptide hormone secreted by the thyroid that regulates blood levels of Ca^{2+}. Analysis of mRNAs showed that there are two species, only one of which encodes calcitonin. The two mRNAs arise from a single gene by alternate RNA processing of a single primary transcript. One mRNA directs the synthesis of a precursor protein that is cleaved to yield calcitonin, plus a peptide encoded by a region common to both mRNA products. The other mRNA directs the synthesis of a precursor protein that is processed to the common-region peptide plus CGRP (Figure A). In the thyroid gland the primary transcript is processed to the mRNA encoding calcitonin, whereas in the brain, the primary transcript is processed to the mRNA encoding CGRP.

Then began the search for the peptide and its biological activity. A specific antiserum made to a synthetic polypeptide from the CGRP sequence showed that a CGRP-like peptide was present in many neurons in the CNS and PNS, including motor neurons (Figure 5) and some dorsal root ganglion cells; chemical analysis of extracts identified the brain peptide as authentic CGRP. Later experiments showed CGRP in large, dense-cored vesicles in motor nerve terminals. Although the precise physiological role of CGRP is not known, it increases ACh receptor synthesis in cultured muscle cells (see Chapter 13) and also appears to inhibit motor nerve sprouting. In addition to motor nerves, CGRP is present in sensory ganglia and in fibers innervating blood vessels. As CGRP is a potent vasodilator, it may also play a role in control of vascular function.

Diverse combinations of peptides are generated by several mechanisms

Single neurons usually secrete a combination of chemical messengers, often including several peptides. As we have seen, alternate proteolytic processing allows different, but related, peptides to be expressed in different tissues. Selective packaging allows different combinations of peptides to be released from different terminals of the same cell. Other mechanisms for diversity operate at the level of DNA or RNA.

Gene families represent a mechanism for introducing variety at the genetic level. For ELH in *Aplysia* and the opioid peptides in mammals, small families of genes encode precursors for different sets of related peptides. Differential expression of these genes in particular groups of neurons allows different combinations of peptides to be expressed. In *Aplysia,* four homologous genes encode a set of related peptides that govern egg-laying behavior. Differential expression in nervous, endocrine, and exocrine cells results in the expression of a variety of secretory products. Interestingly, amino acid substitutions between the different members of the gene family are concentrated in regions of the prohormones that alter the proteolytic processing pathways, thus producing a different set of peptides from the related precursors.

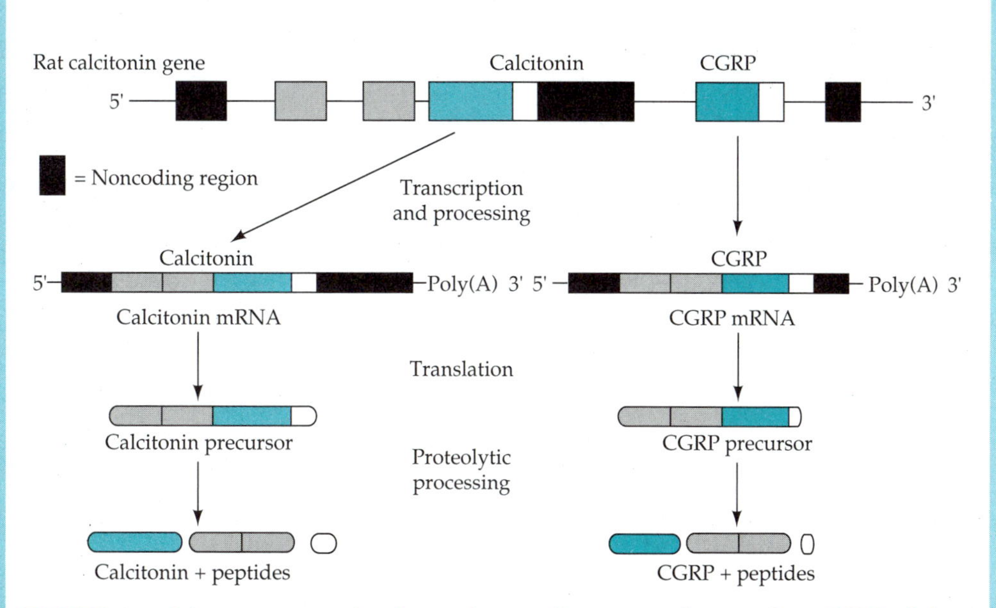

FIGURE A. Calcitonin gene–related peptide (CGRP) arises from alternative RNA and protein products of the calcitonin gene. Two mRNAs arise from the primary transcription product of the calcitonin gene. The precursor protein encoded by one of them is proteolytically processed to produce CGRP. Calcitonin is derived from a different precursor protein made from the other mRNA (see also Figure 17 in Chapter 10). (After M. G. Rosenfeld et al., 1983. Nature 304: 129–135.)

In mammals at least 18 different peptides have the biological activity of opioids. All of the known peptides derive from three genes that encode the polyprotein precursors, pro-opiomelanocortin, proenkephalin, and prodynorphin (Box B). While the diversity of receptors for these molecules is just beginning to be understood, their differential expression in specific brain regions underlies the varied cell–cell interactions that they mediate.

Another important mechanism for generating diversity of peptide messengers from a single gene is by alternate RNA processing (Chapter 10). The calcitonin/CGRP (Box C) and the protachykinin-A genes are the best examples. The protachykinin-A gene comprises seven exons, one encoding substance P and another encoding substance K, with the remainder encoding the rest of the transcription unit. In the nervous system, precursors that give rise to either substance P, or to substance P and substance K, are expressed. In peripheral tissues, however, only the precursor expressing both peptides is found. Regulatory events that determine which exons are included in the final mRNA are a very specific means of regulating peptide content of different tissues.

Physiological Actions of Peptides

If peptides are released from synaptic terminals, what are their physiological effects, and how do these effects differ from those of transmitters that mediate rapid synaptic transmission? Although we are beginning to understand some of the actions of peptides, our ignorance is vast. The information that we do have about peptides mostly comes from the peripheral nervous system, where experimental manipulation of synapses is easier and functional relations more clearly defined. Their prevalence in the central nervous system, however, suggests that peptides play important and varied roles at many synapses. One could say that they give color and variety to a basic black-and-white system of excitatory and inhibitory potentials mediated by the small-molecule neurotransmitters, subtly changing the size, shape, and time course of these responses. Slow synaptic transmission mediated by peptides is thus not just a system reserved for special situations, such as stress, pain, or egg-laying, but is part of the warp and woof of synaptic communication throughout the nervous system.

LHRH mediates a slow excitatory potential in the frog sympathetic ganglion

In some cases, peptides act as conventional neurotransmitters, producing potential changes in the postsynaptic cell that resemble those produced by other neurotransmitters. In the frog sympathetic ganglion, for example, stimulation of the presynaptic nerve produces four different potentials in the postsynaptic cell (Figure 10). The most rapid of these, the fast excitatory postsynaptic potential (epsp), is caused by the action of ACh on a nicotinic ACh receptor. The fast epsp, which can be elicited by a single stimulus, lasts a few tens of milliseconds. The slowest, called the late, slow epsp, arises through the action of a peptide that is closely related to the mammalian pituitary releasing hormone LHRH. The late, slow epsp, which is seen only after a train of repetitive stimuli, does not begin until approximately 100 msec after the stimulus to the nerve and lasts for up to 40 minutes.

The long duration of the late, slow epsp is due to the persistence of the peptide in the extracellular fluid. Thus, addition of an antagonist of LHRH that binds to the receptor but does not activate it can truncate the postsynaptic potential even if added after the response has begun (Figure 10C). This finding suggests that repeated activation of the receptor is required to elicit the normally long response. The slow removal of the peptide transmitter stands in sharp contrast to the rapid removal of ACh, the transmitter for the fast epsp, at the same synapse. ACh is hydrolyzed by extracellular acetylcholinesterase within a few milliseconds after its release, thus allowing only a single round of receptor activation.

LHRH activates both postsynaptic and neighboring cells

A further difference between the two transmitters is revealed by their immunocytochemical localization. The frog sympathetic ganglion contains two types of cells: B cells, which receive innervation from one set of spinal

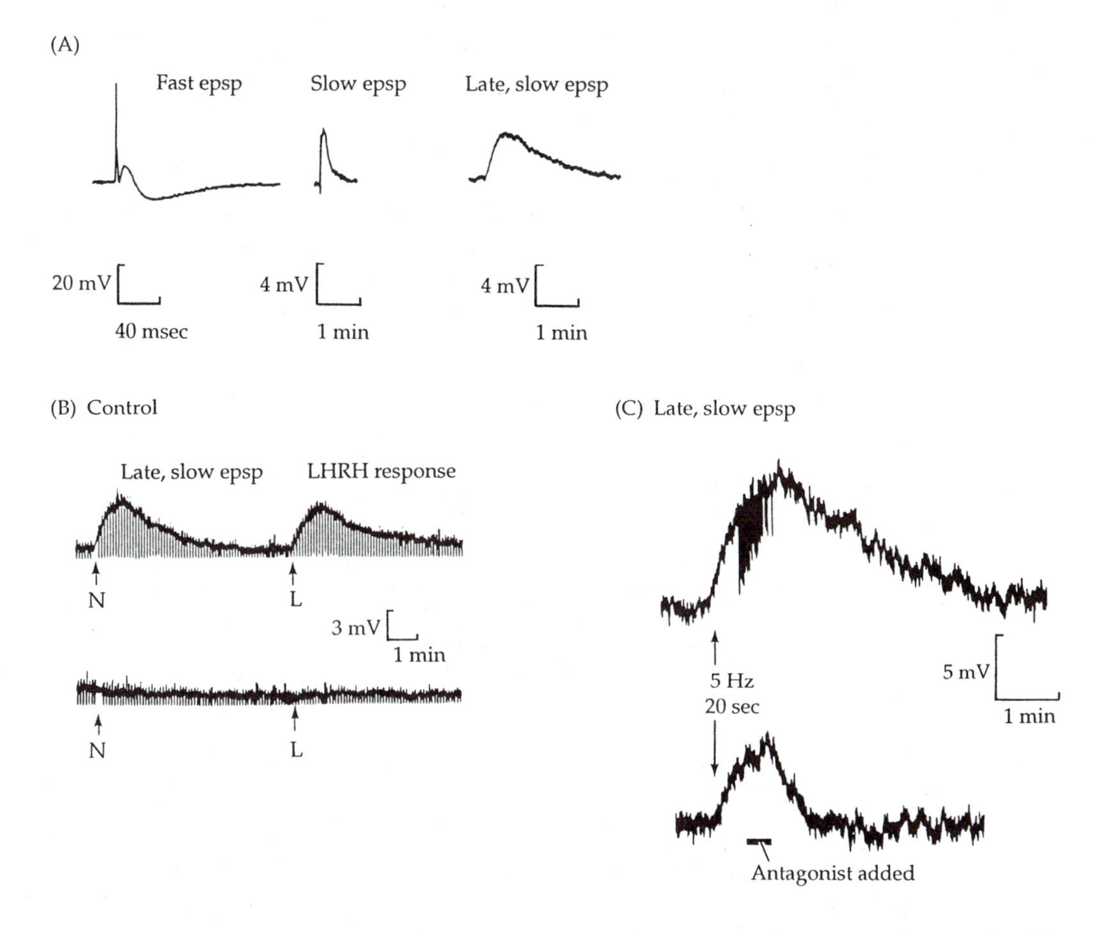

FIGURE 10. Complex synaptic communication in frog autonomic ganglia. (A) Three types of synaptic responses are elicited by two chemical messengers acting on three types of receptors. The fast epsp (30–50 milliseconds) results from the action of ACh on nicotinic receptors. A second response, the slow epsp (30–60 seconds) is mediated by the action of ACh on muscarinic receptors. A third response, the late, slow epsp, is mediated by a peptide messenger resembling mammalian luteinizing-hormone–releasing hormone (LHRH), and lasts many minutes. (B) The late, slow epsp has the same onset, magnitude, and duration as the LHRH-induced response. (C) Addition of an LHRH antagonist prematurely terminates the response. (From L. Y. Jan and Y. N. Jan, 1982. J. Physiol. 327: 219–246.)

nerves; and C cells, which are innervated by a different set. If terminals on both cell types are examined for immunoreactivity with antibodies to LHRH, only those on C cells are shown to contain the peptide; both, however, are positive for choline acetyltransferase, the enzyme responsible for ACh synthesis. Stimulation of the nerves to C cells causes a fast epsp only in C cells, but a late, slow potential in both cell types. Stimulation of the nerves to B cells, in contrast, produces only a fast epsp in B cells, and no response in C cells.

These observations have led to the proposed scheme shown in Figure 11 in which ACh, released from both sets of terminals, acts only on the cells with which the terminals make synaptic contact. LHRH, although released only from terminals on C cells, not only binds to receptors on C cells, but also diffuses into the medium, and acts on neighboring B cells. The ability of LHRH to activate neighboring cells is possible only because of its long persistence in the extracellular medium. The action of LHRH in the frog sympathetic ganglion resembles the paracrine communication, discussed earlier (Figure 2), that is seen in epithelial and gland cells, in which a messenger released from one cell diffuses locally to activate neighboring cells.

ACh and VIP are released from the same terminal at different frequencies of stimulation

The way in which two chemical messengers work together can be seen in experiments on the submandibular salivary gland of the cat, in which cells are arranged around ducts into which they secrete saliva. Both sympathetic and parasympathetic nerves innervate the gland, ending on or near both secretory cells and smooth muscle cells (Figure 12). In the terminals,

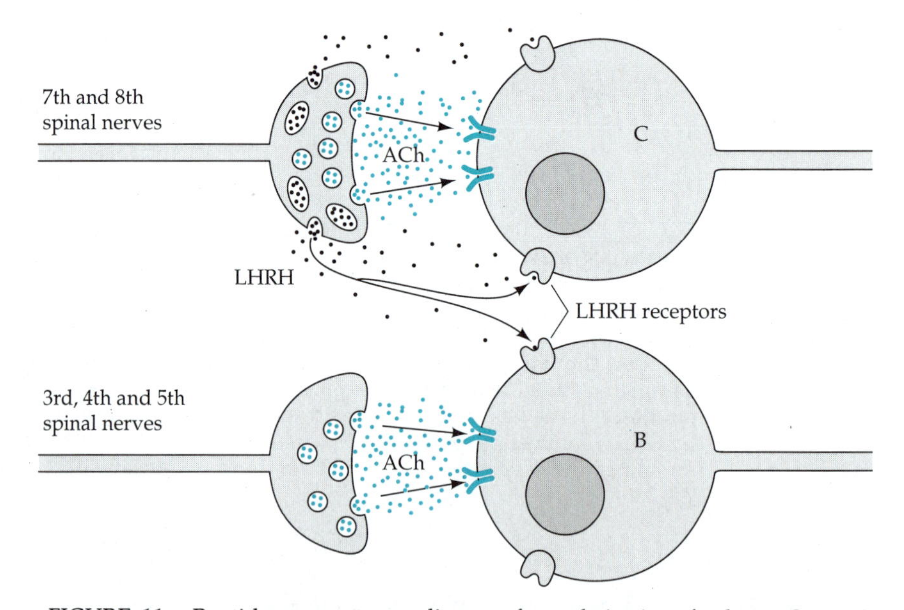

FIGURE 11. Peptides can act at a distance from their site of release. Synaptic communication in the ninth and tenth sympathetic ganglia of the frog. Cholinergic neurons from the seventh and eighth spinal nerves innervate C cells, while those from the third, fourth, and fifth nerves innervate B cells. Only the C cells have terminals on them that are immunoreactive for LHRH. Stimulation of the seventh and eighth spinal nerves produces a late, slow epsp in both B and C cells, suggesting that the peptide diffuses from its site of release at the surface of C cells to activate receptors on B cells. (After L. Y. Jan and Y. N. Jan, 1982. J. Physiol. 327: 219–246.)

ACh is contained in small synaptic vesicles, and VIP in large secretory vesicles. Stimulation of the parasympathetic nerves, which secrete ACh and VIP, causes increased secretion from the gland, and also increased blood flow through neighboring capillaries (Figure 12). At low frequencies of stimulation (2 Hz), both the increased secretion and increased blood

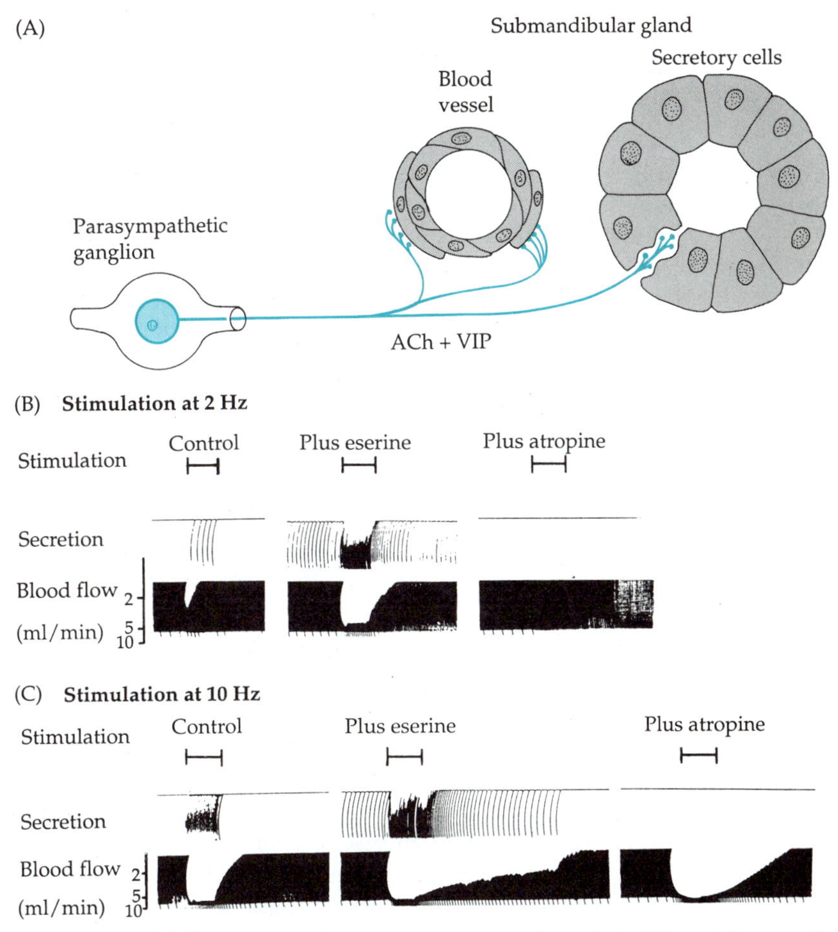

FIGURE 12. Different synaptic messengers are released at different frequencies of stimulation of presynaptic terminals. (A) Parasympathetic neurons innervate the blood vessels and the acinar or secretory cells of the cat submandibular gland. The nerve cells secrete ACh and vasoactive intestinal polypeptide (VIP) as chemical messengers. The messengers are stored in separate vesicles and released by separate mechanisms. (B) Low-frequency stimulation (2 Hz) of the parasympathetic nerves causes a low level of secretion from the salivary gland and a small increase in blood flow (a). Both effects are enhanced by eserine (b), which blocks the degradation of ACh by acetylcholinesterase, and inhibited by atropine (c), which blocks muscarinic ACh receptors. (C) High-frequency stimulation (10 Hz) causes a higher level of secretion and a larger increase in blood flow (a). Both effects are enhanced by eserine (b). Atropine blocks the effect of stimulation on salivation but not on blood flow. (From J. M. Lundberg et al., 1982. Acta Physiol. Scand. 115:525–528.)

flow are enhanced by eserine, an inhibitor of acetylcholinesterase, and inhibited by atropine, an antagonist of the action of ACh at muscarinic (G protein-linked) receptors. At higher frequencies of stimulation (10 Hz), both effects are enhanced by eserine, but, in this case, only salivation and not blood flow is blocked by atropine. Thus VIP seems to be selectively released at high frequencies of stimulation. In addition to its action on blood vessels, VIP enhances the effect of ACh on the secretory cells by increasing the affinity of the receptors for it.

Experiments like these at several synapses (e.g., LHRH at frog sympathetic ganglia) have suggested that peptides are not released by a single stimulus but only by trains of high-frequency stimuli. The different frequency response for release of different transmitters not only increases the physiological flexibility of the synapse but also indicates that the two release mechanisms may be different. One possibility is that the peptide-containing secretory vesicles may be released from different sites in the presynaptic terminal than are the small vesicles, which contain other transmitters. Such a mechanism would be consistent with what we know about exocytosis of secretory vesicles in non-neural cells, and, as we will discuss in Chapter 5, may also provide a plausible explanation for the different frequency response of large and small vesicles.

Summary

Nervous systems are communication networks that rely on the secretion of chemical messengers for intercellular signaling. Small-molecule neurotransmitters, such as glutamate, ACh, and GABA, are synthesized and packaged into small vesicles in nerve terminals. They act on postsynaptic receptors that are directly linked to ion channels and mediate responses in the millisecond time scale; they may also act on G protein-linked receptors to alter concentrations of second messengers. At a single synapse a transmitter may act on multiple receptors. Their actions on postsynaptic cells are terminated quickly.

Peptides, which form a second class of chemical messengers, are synthesized and packaged into larger vesicles in the cell body and are then transported to the nerve terminals, where they are released via a separate secretory pathway. Peptides act exclusively on G protein-linked receptors to mediate slow postsynaptic responses. The biogenic amines, which are synthesized in nerve terminals, are packaged into both large and small vesicles and act almost exclusively on G protein-linked receptors. Virtually all nerve terminals contain small vesicles; many also contain large vesicles. The secretion of multiple messengers and the variety of postsynaptic responses that are possible give rich diversity to synaptic communication in the nervous system.

Neuropeptides are generally synthesized as part of larger precursors that are proteolytically processed to active products. In many cases more than one physiologically active peptide can be cleaved from a single precursor. The precursors are synthesized in the cell body on the rough ER, travel through the Golgi apparatus, and in the TGN are sorted from other proteins and packaged into vesicles. Proteolytic processing of the precursor can take place both before packaging and within the vesicle.

At some synapses, neuropeptides, which appear to be secreted only at high frequencies of stimulation, are not rapidly degraded, but can diffuse to act on several postsynaptic cells. Their persistence in the extracellular fluid allows them to act over a much longer period of time (seconds or minutes) than the small molecule transmitters. Neurons thus appear to interact via two superimposed and overlapping systems of communication, allowing many responses with widely varying time courses and yielding different postsynaptic responses.

References

General references

Dale, H. 1935. Pharmacology and nerve endings. Proc R. Soc. Med. 28: 319–332.

*Hokfelt, T., Millhorn, D., Seroogy, K., Tsuruo, Y., Ceccatelli, S., Lindh, B., Meister, B., Melander, T., Schalling, M., Bartfai, T. and Terenius, L. 1989. Coexistence of peptides with classical neurotransmitters. Experientia Suppl. 56: 154–179.

Kelly, R. 1985. Pathways of protein secretion in eukaryotes. Science 230: 25–32.

Sossin, W. S., Fisher, J. M. and Scheller, R. H. 1989. Cellular and molecular biology of neuropeptide processing and packaging. Neuron 2: 1407–1417.

Identification and localization of neurotransmitters

Dale, H. H., Feldberg, W. and Vogt, M. 1936. Release of acetylcholine at voluntary motor nerve endings. J. Physiol. 86: 353–380.

Guillemin, R. 1978. Peptides in the brain: The new endocrinology of the neuron. Science 202: 390–402.

*Hughes, J., Smith, T. W., Kosterlitz, H. W., Fothergill, L. A., Morgan, B. A. and Morris, H. R. 1975. Identification of two related pentapeptides from the brain with potent agonist activity. Nature 258: 577–579.

*Kuffler, S. W. and Yoshikami, D. 1975. The number of transmitter molecules in a quantum: An estimate from iontophoretic application of acetylcholine at the neuromuscular junction. J. Physiol. 251: 465–482.

McLaughlin, B. J., Barber, R., Saito, K., Roberts, E., and Wu, J. Y. 1975. Immunocytochemical localization of glutamate decarboxylase in rat spinal cord. J. Comp. Neurol. 164: 305–322.

Synaptic vesicles and secretory granules

De Biasi, S. and Rustioni, A. 1988. Glutamate and substance P coexist in primary afferent terminals in the superficial laminae of spinal cord. 85: 7820–7824.

Sudhoff, T. C. and Jahn, R. 1991. Proteins of synaptic vesicles involved in exocytosis and membrane recycling. Neuron 6: 1–20.

Trimble, W., Linial, M. and Scheller, R. 1991. Cellular and molecular biology of the presynaptic nerve terminal. Annu. Rev. Neurosci. 14: 93–122.

Wagner, J. A., Carlson, S. S. and Kelly, R. B. 1978. Chemical and physical characterization of cholinergic synaptic vesicles. Biochemistry 17: 1199–1206.

Peptide synthesis and processing

Jung, L. J. and Scheller, R. H. 1991. Peptide processing and targeting in the neuronal secretory pathway. Science 251: 1330–1335.

Mains, R. E., Eipper, B. A. and Ling, N. 1977. Common precursor to corticotropins and endorphins. Proc. Natl. Acad. Sci. USA 74: 3014–3018.

Nakanishi, S., Inoue, A., Kita, T., Nakamura, M., Chang, A., Cohen, S. and Numa, S. 1979. Nucleotide sequence of cloned cDNA for bovine corticotropin-B-lipotropin precursor. Nature 278: 423–427.

Nawa, H., Kotani, H. and Nakanishi, S. 1984. Tissue-specific generation of two preprotachykinin mRNAs from one gene by alternative RNA splicing. Nature 312: 729–734.

*Roberts, J. L. and Herbert, E. 1977. Characterization of a common precursor to corticotropin and β-lipotropin: Cell-free synthesis of the precursor and identification of corticotropin peptides in the molecule. Proc. Natl. Acad. Sci. USA 74: 4826–4830.

Rosenfeld, M., Mermod, J., Amara, S., Swanson, L., Sawchenko, P., Rivier, J., Vale, W. and Evans, R. 1983. Production of a novel neuropeptide encoded by the calcitonin gene via tissue-specific processing. Nature 304: 129–135.

Physiological action of peptides

*Jan, L. Y. and Jan, Y. N. 1982. Peptidergic transmission in sympathetic ganglia of the frog. J. Physiol. 327: 219–246.

Jan, L. Y., Jan, Y. N. and Brownfield, M. S. 1980. Peptidergic transmitters in synaptic boutons of sympathetic ganglia. Nature 288: 380–382.

Lundberg, J. M. and Hokfelt, T. 1983. Coexistence of peptides and classical neurotransmitters. Trends Neurosci. 6: 325–333.

5

The Nerve Terminal

Zach W. Hall

THE NERVE TERMINAL is the most distant outpost of a neuron. Although it receives electrical signals from the cell body within milliseconds, metabolic communication requires hours or days. Over the short term, the terminal must be self-sufficient, generating the energy and materials needed to carry out its task, which, in the adult, is the secretion of neurotransmitter. Nerve terminals are thus filled with mitochondria and have an elaborate array of enzymes and transporters for the synthesis of specific neurotransmitters. Nerve terminals also assemble the small synaptic vesicles that contain and release neurotransmitter, using proteins that are transported from the cell body. To assure rapid and efficient release of transmitters onto postsynaptic cells, the terminals have a specialized release apparatus that is not found in other cells. As we have seen earlier, nerve terminals not only secrete small molecules as neurotransmitters, but also release peptides and amines from large vesicles using a separate secretory pathway. Finally, terminals have surface membrane receptors that receive external signals, and a network of proteins and small molecules that translate these signals via second messengers into responses that modify or modulate transmitter synthesis and secretion. Placed at the end of a long, relatively inert axon, the terminal is thus a veritable beehive of activity.

When viewed by conventional electron microscopy, two features of nerve terminals quickly stand out (Figure 1). First, they are filled with small **synaptic vesicles** about 40 nm in diameter. These are not randomly distributed within the terminal but are clustered near the second distinguishing feature of the terminal, the **active zone,** one or more areas of the surface membrane that are somewhat electron dense and are usually in close apposition to the postsynaptic cell. Nerve terminals characteristically contain many mitochondria to fuel their intense biosynthetic and membrane recycling activities. In addition, a few large vesicles with dense cores are often seen. These are not clustered at the active zone, but are often near the surface membrane elsewhere in the terminal (see, for example, Figure 5 in Chapter 4).

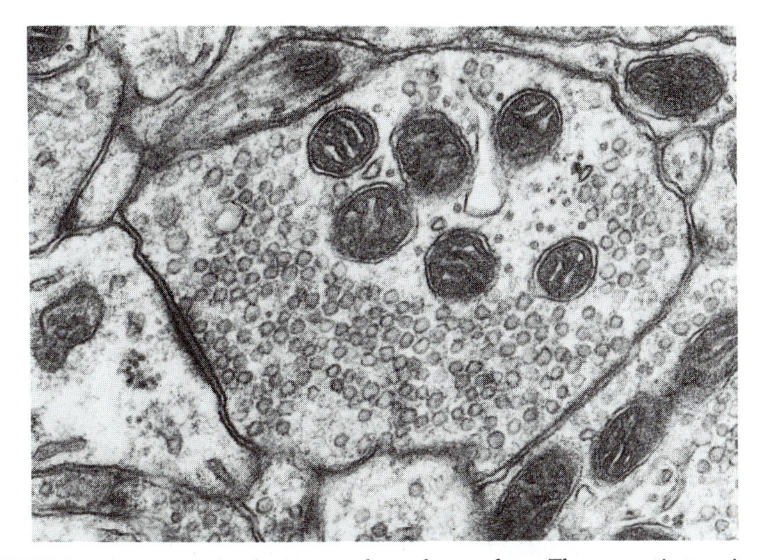

FIGURE 1. A synapse in the interpeduncular nucleus. The synaptic terminal on the left is filled with synaptic vesicles that are concentrated near an electron-dense portion of the presynaptic membrane, the active zone. The active zone is coextensive with the postsynaptic membrane, which is lined with electron-dense "fuzz." The synaptic cleft between the two is slightly widened, indicating the presence of synaptic cleft material. (Micrograph courtesy of D. McDonald.)

The release of neurotransmitters and other messengers is the final response of a nerve to the excitatory and inhibitory inputs that converge upon it. Although we commonly think of signal integration as occurring at the cell body and dendrites of cells, the nerve terminal is also a site of integration as the signals that it receives can modify the secretory response to an action potential. Because it is the site of final signal output, the nerve terminal is an especially sensitive and critical point of control for the neuron. As we shall see (Chapter 14), signals that modify neuronal secretion are thought to play important roles in synaptic plasticity. To understand how these modifications occur, we need to know as much as possible about how nerve terminals work.

Preparation: Transmitter Synthesis And Storage

Rapid, high-frequency signaling between neurons depends upon a ready supply of small synaptic vesicles filled with neurotransmitter. To meet this need, nerve terminals retrieve vesicle components from the surface membrane after exocytosis and use them in the formation of new vesicles. Neurotransmitter that is made locally is then transported into the newly formed vesicles. Because nerve terminals do not contain the biochemical apparatus for protein synthesis, the transmitters that fill small vesicles must be small molecules, such as acetylcholine (ACh), the amino acid neurotransmitters, or one of the catecholamines. The enzymes that synthesize these transmitters are made in the cell body and transported to the terminals. Although some synthesis and storage of neurotransmitter may occur in the cell body or in the axons during transport, most transmitter synthesis takes place in the terminals. Because they control transmitter levels in the terminals, the pathways of neurotransmitter synthesis and storage are the targets of many drugs that alter neuronal function and are thus of interest from a pharmacological, as well as a physiological, point of view.

The transmitters ACh, GABA, dopamine, norepinephrine, and serotonin (5-hydroxytryptamine) are synthesized by specialized biosynthetic pathways that are present only in the neurons that secrete them. Precursors for synthesis are either provided by mitochondrial metabolism or by uptake from the surrounding medium. Because the pathways are specialized, their presence in a cell is often used to identify the transmitter secreted by the cell. The amino acid transmitters glutamate and glycine, in contrast, are generated by enzymes that are present in all cells using metabolic intermediates produced by mitochondria in the terminals.

Enzymes that metabolize transmitters are also present in terminals. They act on the small pool of cytosolic neurotransmitter that is in equilibrium with transmitter stored in vesicles. The concentration of neurotransmitter in the cytosol thus represents a steady-state balance between synthesis and degradation. Drugs that block the degradative enzymes cause an increased cytosolic concentration of neurotransmitter. Under normal conditions, the cytosolic transmitter represents a relatively small fraction of the amount stored in vesicles. This small cytosolic pool can be important, however, in controlling the rate of transmitter synthesis. For several of the neurotransmitters, the rates of synthesis are tightly coupled to rates of release, so that when large amounts of transmitter are secreted, more is quickly made to replace it.

ACh and GABA are synthesized in the cytosol

In cholinergic nerve terminals, ACh is made in the cytoplasm by a single enzyme, **choline acetyltransferase (CAT),** which transfers an acetyl group from acetyl CoA to choline (Figure 2). **Acetylcholinesterase** hydrolyzes the ester bond of ACh to yield choline and acetate. CAT is a soluble enzyme; acetylcholinesterase in the terminal is probably membrane-bound. The choline used for the synthesis of ACh comes from the extracellular fluid. Although many cells take up choline with low affinity, nerve terminals are specialized in having a high-affinity uptake mechanism that co-transports choline and Na^+. Choline transport into terminals is probably the rate-limiting step in the synthesis of ACh. ACh that is released is hydrolyzed by extracellular acetylcholinesterase, and about half of the resulting choline is taken up and used in the synthesis of new transmitter.

GABA is synthesized by a single step in terminals that secrete it. The precursor glutamate is decarboxylated to yield GABA in a virtually irreversible reaction catalyzed by **glutamic acid decarboxylase (GAD),** a soluble cytosolic enzyme. GABA is metabolized by two successive reactions of transamination and oxidation, respectively, to yield succinic acid (Fig-

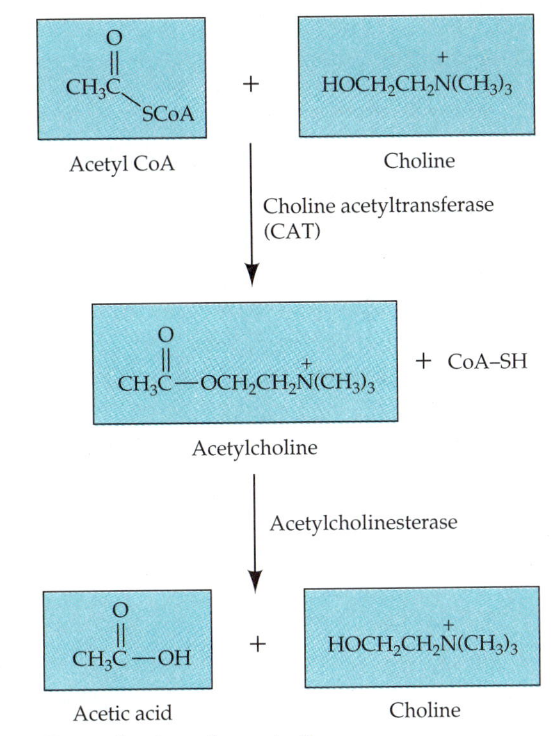

FIGURE 2. ACh synthesis and metabolism.

ure 3). These three reactions, decarboxylation, transamination, and oxidation, form a cycle in GABA neurons that regenerates the precursor glutamate, with the net conversion of alpha-ketoglutarate to succinate. Because the GABA pathway represents an alternative to the usual oxidative decarboxylation of alpha-ketoglutarate to succinate in the tricarboxylic acid cycle, it is sometimes called the GABA shunt. In nerve terminals, the most important product of the pathway is GABA, which is stored in vesicles.

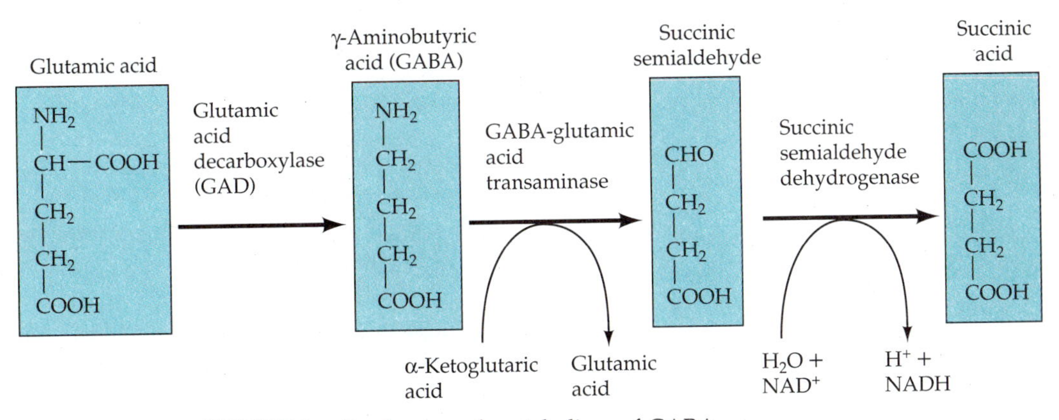

FIGURE 3. Synthesis and metabolism of GABA.

FIGURE 4. Catecholamine synthesis and metabolism. (A) Pathway of catecholamine synthesis. BH_4 and BH_2 represent reduced and oxidized forms, respectively, of tetrahydrobiopterin. (B) Pathway of catecholamine degradation. The pathway is illustrated for norepinephrine, but is the same for epinephrine and dopamine. Note that monoamine oxidase and catechol-*O*-methyltransferase can act in either order to produce the final products (in color).

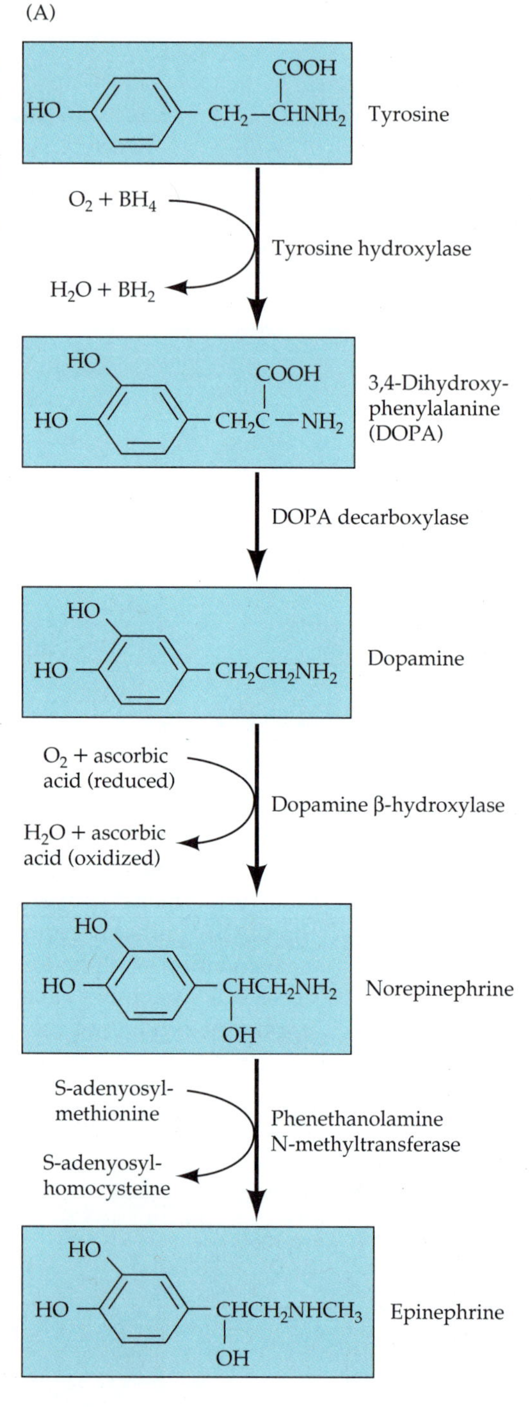

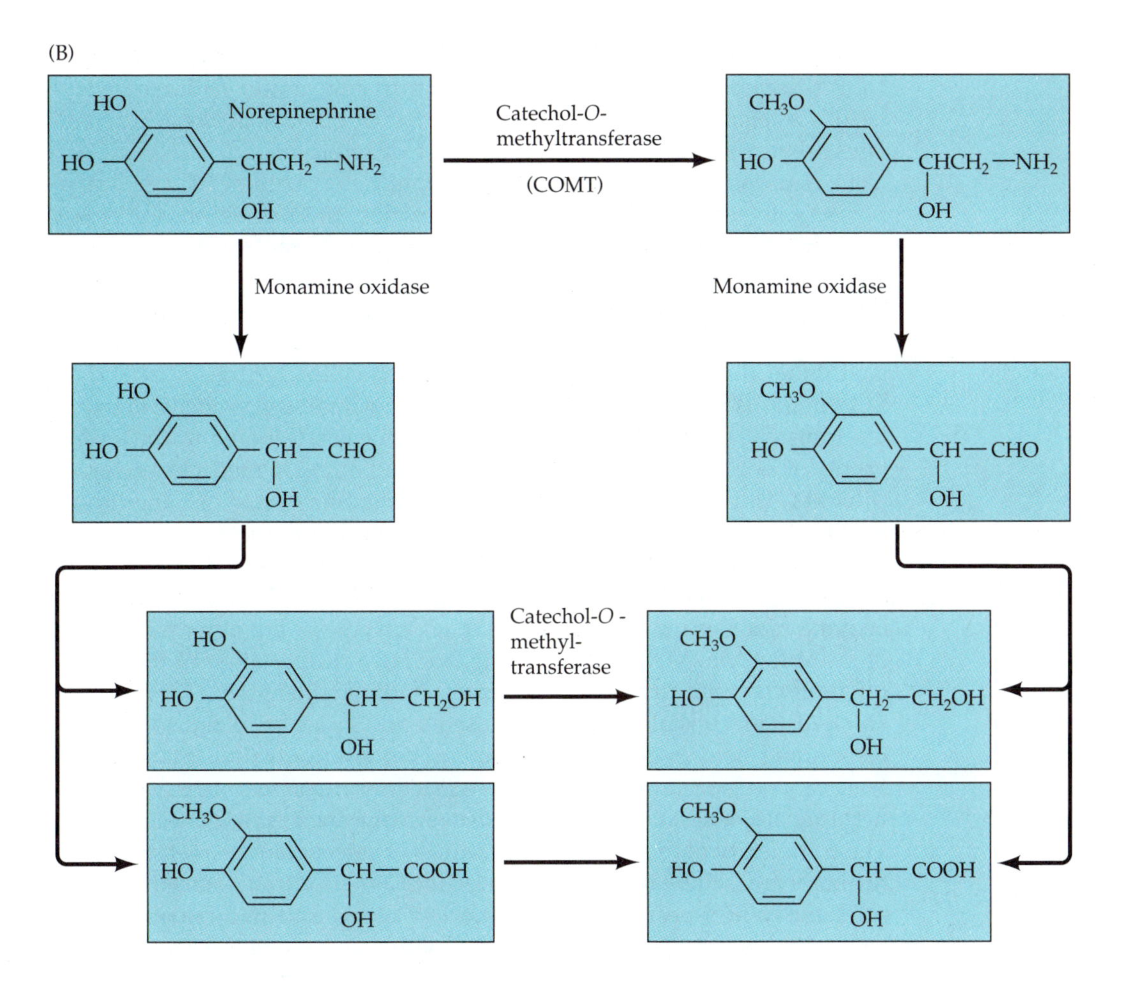

Dopamine, norepinephrine, and epinephrine are synthesized by a pathway whose enzymes are in both the cytosol and synaptic vesicles

A single biochemical pathway is responsible for the synthesis of the catecholamines dopamine, norepinephrine, and epinephrine, all of which are used as transmitters by neurons in vertebrate and invertebrate nervous systems (Figure 4A). The starting point of the pathway is the amino acid tyrosine, which is synthesized within neurons by the hydroxylation of phenylalanine and is also taken up from the extracellular fluid into nerve terminals by a sodium-dependent transporter.

The first two steps in the catecholamine biosynthetic pathway occur in the cytosol. First, tyrosine is converted to L-dihydroxyphenylalanine (L-DOPA) by **tyrosine hydroxylase,** in a reaction that requires molecular oxygen and a cofactor, tetrahydrobiopterin (BH_4), which is the source of hydrogen and electrons. L-DOPA is then converted to dopamine by **DOPA-decarboxylase.** This is the terminal step in the pathway leading to dopamine, which is then taken up into vesicles for storage and secretion. Because the decarboxylase has a low K_m and is present in excess, little

L-DOPA accumulates in nerve terminals that secrete catecholamines. DOPA-decarboxylase (sometimes called aromatic amino acid decarboxylase) is a relatively nonspecific enzyme that in other neurons decarboxylates 5-hydroxytryptophan, the precursor of serotonin.

In neurons that synthesize norepinephrine, dopamine is also transported into vesicles, but, within the vesicles, undergoes an additional reaction to yield norepinephrine. **Dopamine beta-hydroxylase,** most of which is attached to the lumenal surface of the vesicles, catalyzes the hydroxylation of the ethylamine sidechain of dopamine to form norepinephrine. This enzyme also requires molecular oxygen, but uses ascorbic acid rather than a pteridine cofactor as a source of electrons.

In neurons that synthesize epinephrine, the final step in the pathway is the transfer of a methyl group from S-adenosylmethionine to the amino group of norepinephrine by **phenylethanolamine N-methyltransferase (PNMT).** Curiously, PNMT is a cytoplasmic enzyme. Norepinephrine, produced within the vesicle, must return to the cytoplasm to be converted to epinephrine, and is then transported back into the vesicle to be secreted.

All of the catecholamines are metabolized by a relatively nonspecific enzyme **monoamine oxidase (MAO),** an enzyme of the outer membrane of mitochondria. MAO catalyzes the oxidative deamination of the amine (or methylamine) group to the corresponding aldehyde. The resulting aldehydes are unstable and are subsequently either oxidized to the corresponding acid or reduced to the corresponding alcohol by separate, relatively nonspecific enzymes. These reactions result in a variety of products that are released into the blood stream and then excreted (Figure 4B). The complexity of the mixture of products of catecholamine metabolism is further increased by the action of **catechol-*O*-methyltransferase (COMT),** a relatively nonspecific enzyme found in blood and in sympathetically innervated target tissues. The enzyme methylates one of the ring hydroxyls of secreted catecholamines and of their alcohol and aldehyde products. Metabolites of catecholamines thus include a variety of methylated and nonmethlyated amines, aldehydes and alcohols that are ultimately excreted. An altered pattern of catecholamine metabolites in the urine of humans can be a clue to disorders of enzymes of amine metabolism.

The synthesis of serotonin in neurons proceeds by biochemical steps that are analogous to those of the dopamine biosynthetic pathway. The starting material is tryptophan, which is hydroxylated by **tryptophan hydroxylase** to 5-hydroxytryptophan, which is then decarboxylated by dopamine decarboxylase to serotonin. Like catecholamines, serotonin is degraded by monoamine oxidase, and its product is then either oxidized or reduced.

The first step in catecholamine biosynthesis is rate limiting and is regulated

Even when stimulated at high frequency, nerve terminals maintain a constant store of transmitter. At high rates of stimulation, cholinergic nerve terminals in sympathetic ganglia may secrete half of their transmitter in 5 minutes. This rate of loss is 70-fold higher than the rate of transmitter metabolism at rest, and yet active nerve terminals suffer no apparent diminution of transmitter level. To maintain the steady state during periods of rest and activity, rates of transmitter synthesis must be capable of wide variation and must be closely regulated. The regulation of transmitter synthesis has been most thoroughly studied in adrenergic nerve terminals, where a complex mixture of both short-term and long-term mechanisms regulate the biosynthetic pathway and adjust the supply of transmitter to its demand.

The control point for regulation of catecholamine biosynthesis is the hydroxylation of tyrosine. Both enzymatic assay and kinetic studies of flux through the pathway show that this step is rate limiting for the production of norepinephrine. The rate of tyrosine hydroxylation increases following nerve stimulation. The activity of tyrosine hydroxylase is inhibited by the end product of the pathway, norepinephrine, which competes with the required tetrahydropteridine cofactor. Feedback inhibition is thus one mechanism by which catecholamine synthesis is controlled, and is probably most important at rest.

A second mechanism of regulation of tyrosine hydroxylase is phosphorylation. The enzyme is a substrate for at least three different kinases: cAMP-dependent protein kinase, type II calcium/calmodulin-dependent protein kinase, and protein kinase C. These kinases phosphorylate a partially overlapping group of sites on tyrosine hydroxylase (see Chapter 7 for a more complete discussion). The phosphorylated or "activated" enzyme has a lower K_m for the pteridine cofactor and, in some cases, also has a higher K_i for norepinephrine. Thus the phosphorylated enzyme is less susceptible to end product inhibition. Although the concentration of pteridine cofactor in terminals is not known, its concentration in chromaffin cells in the adrenal medulla appears to be much lower (10 μM) than its K_m (2 mM), suggesting that cofactor concentrations in situ limit enzyme activity.

During high-frequency stimulation, calcium entry into the terminal stimulates type II calcium/calmodulin-dependent protein kinase, which phosphorylates tyrosine hydroxylase, increasing its activity. The enzyme is also activated by phosphorylation when peptides, transmitters, or growth factors increase the levels of cAMP or those of the second messengers of the phosphoinositide pathway (see Chapter 7).

A chronic increase in electrical activity, over many hours rather than minutes, stimulates the synthesis of mRNA for both tyrosine hydroxylase and dopamine beta-hydroxylase, resulting, after a lag of several days, in an increase in the amounts of the proteins in the nerve terminal. The mechanism by which synthesis of these mRNAs is specifically stimulated is unknown.

Specific transporters concentrate neurotransmitters in synaptic vesicles by a proton-dependent mechanism

After their synthesis, neurotransmitters are transported into synaptic vesicles for storage and release. The concentrations of the transmitters inside the vesicles are 10- to 1000-fold higher than their concentrations in the cytosol. The mechanisms by which vesicles concentrate neurotransmitters have been investigated in purified preparations of cholinergic synaptic vesicles from the electric organ of the marine ray, *Torpedo,* in purified synaptic vesicles from brain, which are heterogenous with regard to neurotransmitter, and in secretory granules from chromaffin cells in the adrenal medulla, which contain epinephrine and norepinephrine. Adrenal secretory granules are plentiful and because of their high density can be easily obtained in pure form. Catecholamine uptake has also been investigated in partially purified vesicles from sympathetic nerves. Transport in all of these preparations is measured by the uptake of radiolabeled neurotransmitters or other compounds.

In each type of vesicle, transport is mediated by a carrier protein that is highly specific for the neurotransmitter. In adrenal medullary granules and in sympathetic nerve vesicles, a single carrier is responsible for the uptake of dopamine, norepinephrine, and epinephrine. Comparison of the K_m's of the transporter in different types of vesicles shows that they are matched to the cytosolic concentration of the neurotransmitter. Thus the transporters for glutamate and GABA, whose concentration in the cytosol is millimolar, have K_m's in the millimolar range, whereas the catecholamine transporter has a K_m of a few μM. The catecholamine carrier can be specifically and irreversibly blocked by the drug reserpine, which causes depletion of norepinephrine from vesicles in nerve terminals.

Because transport generates a concentration gradient of neurotransmitter, energy must be expended. In all synaptic vesicles (and secretory granules) that have been investigated so far the energy for transport comes from an electrochemical proton gradient generated by a pump, or ATPase, that is associated with the vesicles.

The proton pump in chromaffin cell granules, which has been purified, is highly homologous to the protein responsible for generating proton gradients in endosomes and lysosomes, perhaps indicating a common evolutionary origin for synaptic vesicles and secretory granules. The proton pump couples hydrolysis of cytosolic ATP to the transport of protons into the vesicle. Depending on the permeability properties of the vesicle, proton transport generates either a proton gradient, a membrane potential, or both. If the vesicle is freely permeable to an anion such as Cl^-, the transport of protons across the membrane causes anions to move also, so that a proton gradient is produced without a change in membrane potential. If anions cannot move across the membrane at the same rate as the protons, the transport of a small number of unaccompanied protons (see the discussion of capacitance, Chapter 2) sets up a countervailing membrane potential that will oppose further transport. In either case, the proton pump ultimately provides the driving force for neurotransmitter uptake. Uptake of some transmitters is driven best by a potential difference across the membrane; for others, uptake is driven preferentially by the proton gradient.

The most thoroughly investigated transport mechanism is that of catecholamines into the granules from the adrenal medulla. In this case, the action of the proton pump produces both a proton gradient (internal pH

5.5) and a membrane potential (+60 mV) in which the lumen is positive relative to the cytosol. The catecholamine transporter apparently binds uncharged catecholamine in the cytosol, which is then translocated into the granule interior where it becomes protonated at the low internal pH of the granule, and thus becomes impermeable (Figure 5). This scheme satisfactorily explains much experimental data on catecholamine transport in adrenal medullary granules in vitro. It provides no explanation, however, for the return of norepinephrine to the cytosol to be methylated in those cells that synthesize and secrete epinephrine. In addition to the catecholamine transporter, adrenal medullary granules also contain a nucleotide transporter that is also coupled to the electrochemical proton gradient and is responsible for the accumulation of adenine nucleotides in the granules (Chapter 4).

The internal concentration of other ions in vesicles, and how they affect uptake, is unknown. Empty vesicles in nerve terminals, which are derived by endocytosis from the surface membrane, may contain high concentrations of Na^+ and Cl^-. The downhill movement of these ions out of the vesicles could provide energy for the transport of transmitter or nucleotides into vesicles.

Storage of neurotransmitters

The high concentration of neurotransmitters in synaptic vesicles originally suggested that they must be present as part of an insoluble complex. This may be true for chromaffin granules, in which the concentration of catecholamines (0.5 *M*), is higher than iso-osmotic concentration. The estimated concentrations of neurotransmitters in synaptic vesicles, however, appear to be within the normal range of osmolarity (50–150 m*M* for mammalian vesicles; 0.4 *M* for the cholinergic vesicles from *Torpedo*), and the neurotransmitter within these vesicles appears to be osmotically active. Electroneutrality must be maintained within the vesicle for those neuro-

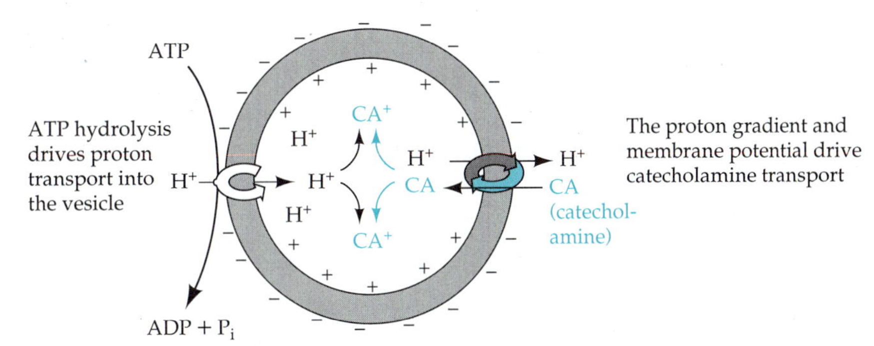

FIGURE 5. Schematic diagram of the postulated mechanism of catecholamine (CA) uptake into secretory granules of adrenal chromaffin cells. The proton pump hydrolyzes ATP and produces a proton gradient and membrane potential across the granule membrane that drive the transport of the amine. The amine transporter appears to bind the uncharged amine, which, after transport, becomes protonated at the low pH of the granule interior.

transmitters that have a net charge (at physiological or acidic pH, ACh, serotonin, and catecholamines are cations; glutamate is an anion at physiological pH and may or may not be charged in the vesicle interior, depending on pH). ATP is present in both cholinergic and adrenergic vesicles, but its concentration is too low to neutralize all of the charge. Other, unidentified anions must thus be present.

Transmitter Release and Its Control

Exocytosis is a fundamental mechanism used by eukaryotic cells for addition of protein and lipid to the surface membrane, for secretion of proteins and small molecules to the cell exterior, and, in concert with endocytosis, for traffic between intracellular membrane compartments. Because the fundamental molecular mechanisms are likely to be similar in all cases, we may look for clues to neurotransmitter secretion in other systems of exocytosis. Transmitter release has several distinctive features, however, that separate it from other forms of exocytosis. First, neurotransmitters, like the secretory products of exocrine and endocrine cells, are released by a form of **regulated secretion,** in which the products are accumulated in vesicles until exocytosis is stimulated by an intracellular signal. In contrast, proteins and membranes are added to the surface of cells by a separate **constitutive pathway of secretion** in which vesicles are transported to the surface and fuse with the surface membrane as they are made.

Secondly, neurotransmitter secretion differs from most other forms of regulated secretion by its incredible speed. Secretion from a pancreatic acinar cell or mast cell typically occurs seconds or tens of seconds after the initial stimulus. In contrast, only several milliseconds elapse between an action potential and the beginning of the postsynaptic response. As we shall see, the extraordinary rapidity of the secretory response in neurons is achieved in several ways: by having Ca^{2+} rather than a metabolic intermediate as the primary regulatory agent; by placing calcium channels very close to the sites of release to minimize the time required for diffusion; and by having the exocytotic event itself occur quickly. Opening of voltage-gated calcium channels and the influx of Ca^{2+} is responsible for much of the synaptic delay; exocytosis itself is estimated to require less than 200 μsec. To achieve such rapid exocytosis, vesicles that are about to undergo secretion appear to be already docked at the surface membrane, as part of a preformed complex on the verge of membrane fusion. Ca^{2+} then acts as the trigger.

To understand transmitter release from nerve terminals, we thus need to consider what is known about the molecular mechanism of exocytosis; the transport and docking of vesicles at the active zone; how Ca^{2+} triggers exocytosis; and, finally, how neurotransmitter release is regulated by other synaptic messengers.

Quantal release of transmitter at nerve terminals reflects single exocytotic events

The initial idea that cellular products are secreted by exocytosis arose from the study of neurons. In 1952 Paul Fatt and Bernard Katz discovered that the release of ACh from motor nerve terminals is quantized, i.e., that a roughly constant number of neurotransmitter molecules are released together—a number that we currently estimate at about 10,000 (see Chapters 2 and 4). Shortly thereafter, electron microscopy of motor nerve terminals showed them to be filled with vesicles which, when later isolated, were found to contain ACh. These findings suggested that the number of transmitter molecules released in each quantum corresponds to the number of molecules stored in each vesicle. This hypothesis not only provided a satisfying link between the two discoveries, but suggested that synaptic vesicles are the agents of release, secreting their contents into the extracellular fluid by fusing with the external membrane.

The idea that secretion of cellular products occurs via exocytosis was rapidly extended to endocrine and exocrine cells, which contain large secretory vesicles or granules. Two decisive experimental observations shortly established exocytosis as the mechanism of secretion in these cell types. First, because secretory vesicles often contain multiple substances, one can test whether or not the substances are released together. The secretory granules of chromaffin cells in the adrenal medulla, for example, not only contain norepinephrine, but also ATP and several soluble proteins, including the chromogranins (see Chapter 4). Upon stimulation, each of these is secreted in exact proportion to their ratio in the vesicles. In contrast, lactic dehydrogenase, a cytosolic protein, is not released. These results give strong evidence that the granule contents are released directly into the medium.

The most important experimental evidence supporting exocytosis, however, arises from electron microscopy. Stimulation of secretion in endocrine or mast cells induces readily visible characteristic "omega" figures in which partially open vesicles are continuous with the surface membrane. Because the contents of some vesicles are electron-dense, they can be seen as they are extruded from the vesicle interior (Figure 6A).

Because transmitter release from nerve terminals occurs so quickly after a stimulus and is of such short duration, vesicles can be caught in the act of exocytosis only rarely (Figure 6B). A decisive series of experiments, however, has firmly linked exocytosis to transmitter release. These experiments used several technical innovations that allowed exocytosis at the frog neuromuscular junction to be visualized and quantitated. First, the drug 4-aminopyridine was used to block potassium channels in the terminal and prolong the action potential, thus increasing Ca^{2+} entry and the number of quanta released in response to a single nerve stimulus. Second, nerve terminals were quickly frozen at a precise interval after a stimulus by slamming them into a block of metal chilled by liquid helium. The frozen tissue was then fractured, and a metal replica of the surface examined in the electron microscope. Unlike thin-section microscopy, freeze-fracture exposes large surfaces of membrane, making it easier to detect sites at which vesicles are fusing with the surface membrane. Under the conditions used in these experiments, the number of exocytotic events obtained at different concentrations of 4-aminopyridine was directly proportional to the number of quanta released (Figure 7).

Several other kinds of experiments also support the idea that trans-

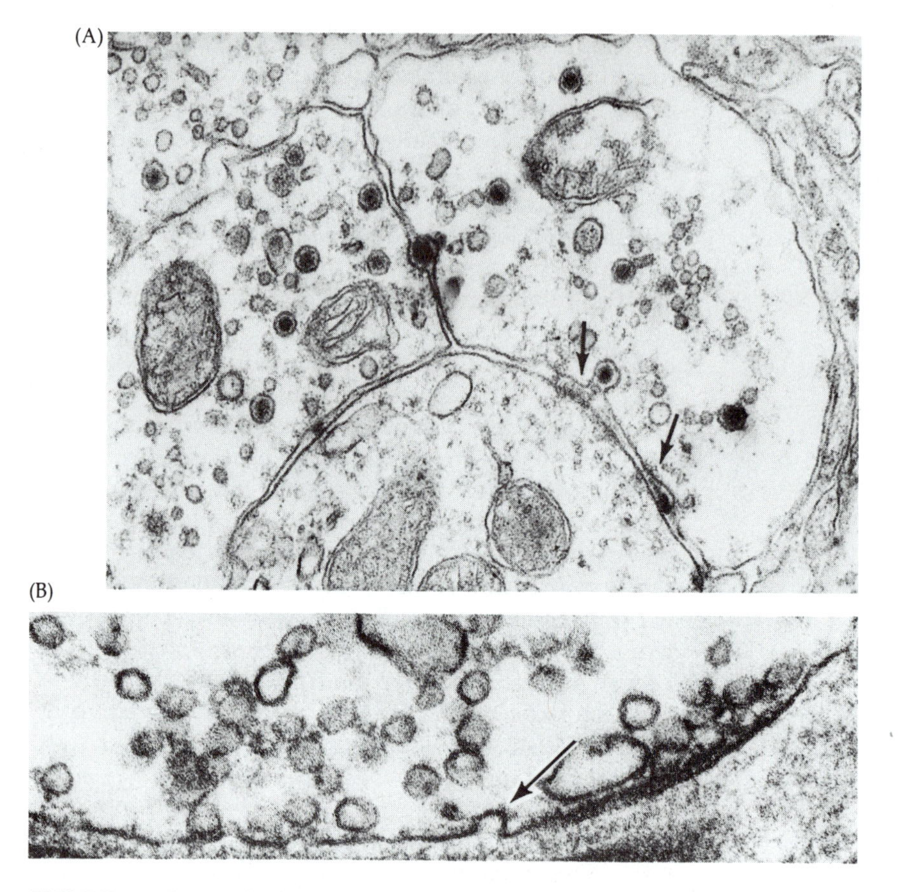

FIGURE 6. Omega figures illustrating secretion by exocytosis. (A) Exocytosis of secretory vesicles containing dense cores from a synaptic bouton in the rat hypothalamus. Note that exocytosis (arrows) occurs from several sites in the terminal. (B) A nerve terminal at the frog neuromuscular junction. An omega figure is seen (arrow) at the left edge of the presynaptic dense material that marks the active zone. (A from J. F. Morris and D. V. Pow, 1988. J. Exp. Biol. 139: 81–103. B from J. Heuser et al., 1984. J. Neurocytol. 3: 109–131.)

mitter is released at nerve terminals by exocytosis. Alpha-latrotoxin, a protein in black widow spider venom that causes massive release of transmitter from motor nerve terminals, drastically decreases the number of vesicles in the terminal and dramatically increases surface membrane area. Additional confirmation comes from immunological experiments showing that epitopes of synaptic vesicle proteins on the internal surface of vesicles become exposed to the extracellular fluid during transmitter release. Thus, antibodies against these epitopes do not normally stain the surface membrane of nerve terminals, but do so after stimulation of transmitter release.

Calcium entry into nerve terminals links the action potential and exocytosis

Electrophysiological experiments by Bernard Katz and his collaborators on the frog neuromuscular junction and on the squid giant synapse first

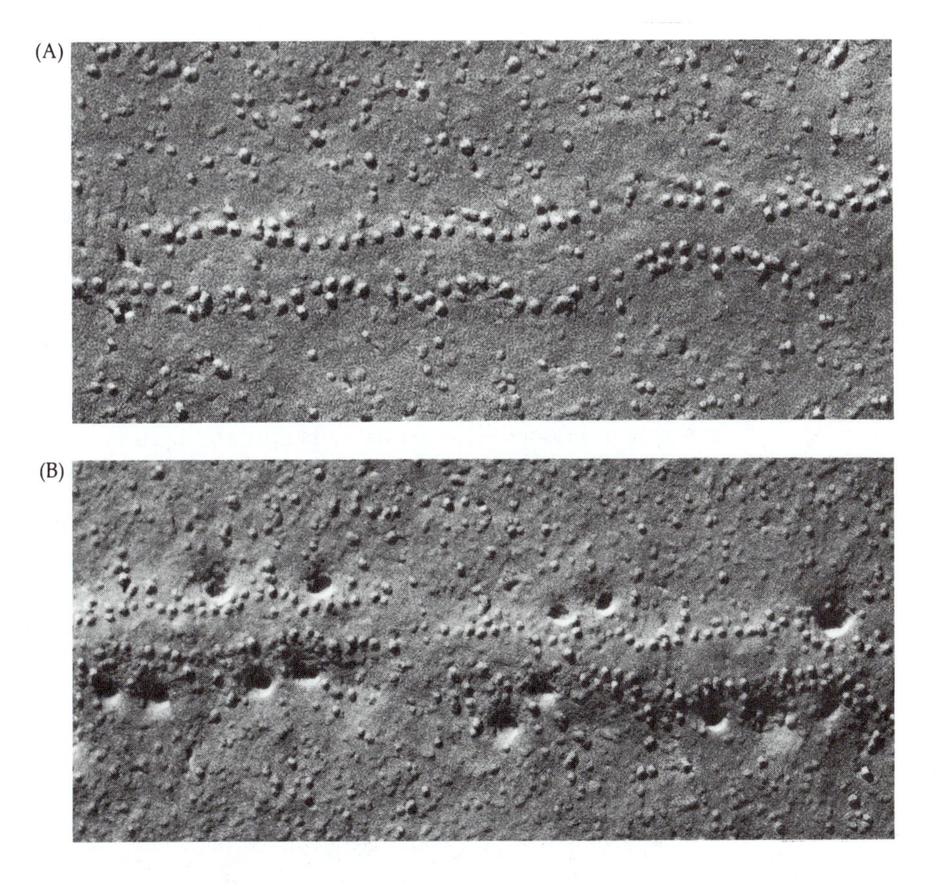

FIGURE 7. Increased exocytosis following an action potential at the frog neuromuscular junction. After a nerve stimulus, the tissue was rapidly frozen and a freeze-fracture replica made of the cytoplasmic face of the nerve terminal membrane at an active zone. The large intramembranous particles probably represent calcium channels. To increase exocytosis following the action potential, the nerve was incubated in 4-aminopyridine, which blocks potassium channels and prolongs the action potential. (A) Immediately after the stimulus to the nerve, before exocytosis has begun. (B) Approximately 4 msec after a stimulus to the nerve. Numerous pits at the edge of the active zone indicate the sites at which vesicles are joined to the membrane by exocytosis. (From J. Heuser and T. Reese, 1981. J. Cell Biol. 88: 564–580.)

demonstrated the essential role of Ca^{2+} in coupling the release of neurotransmitter to an action potential in the nerve terminal membrane (see Chapter 2). By withholding Ca^{2+} from the extracellular medium, they blocked transmitter release, then restored it by supplying Ca^{2+} with a pipette positioned near the nerve terminal. Their finding that Ca^{2+} was required during or just after the action potential suggested that Ca^{2+} enters the nerve terminal through voltage-sensitive channels and that the increased concentration of intracellular Ca^{2+} triggers transmitter release.

The precise characteristics of the voltage-sensitive calcium channels in the presynaptic membrane of most vertebrate neurons are not known. In most of the few cell types that have been investigated, these channels do not correspond to L- or T-type channels (see Chapter 3) but appear to resemble calcium channels seen in other tissues with properties that are intermediate between the two. At the squid synapse, which has been voltage-clamped, the calcium channels inactivate slowly and only in response to large depolarizations. Because the channels inactivate slowly, the duration of the action potential determines the amount of Ca^{2+} that enters.

Calcium channels are concentrated in the active zone

The calcium channels in presynaptic nerve terminals are not uniformly distributed over the surface membrane, as is thought to be the case in exocrine or endocrine secretory cells, but are highly concentrated in the nerve terminal membrane directly opposing the postsynaptic membrane. The localization of calcium channels to the active zones in nerve terminals is based on several observations. At the frog neuromuscular junction, presynaptic calcium channels are identified by the binding of a highly specific toxin, ω-conotoxin, that blocks them and inhibits transmitter release. Fluorescence methods show that the toxin is bound only on the side of the terminal facing the muscle cell, where it is confined to narrow strips that directly oppose the bands of ACh receptor in the postsynaptic membrane (Figure 8).

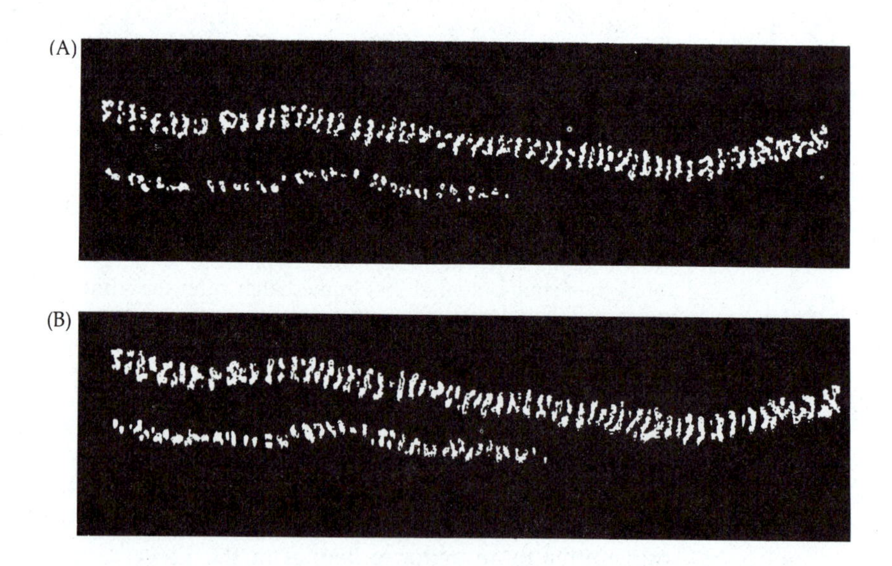

FIGURE 8. Calcium channels in the presynaptic nerve terminal are concentrated at the active zone. A frog muscle was incubated with ω-conotoxin labeled with one fluorescent dye (Texas Red) and α-bungarotoxin conjugated to another (rhodamine). The distribution of the labeled molecules was identified by immunofluorescence, using filters to visualize the two dyes separately. (A) Distribution of ω-conotoxin. (B) Distribution of α-bungarotoxin. Omega-conotoxin, which labels certain types of calcium channels, is distributed in bands on the face of the nerve facing the muscle that correspond exactly to the bands of ACh receptors in the postsynaptic membrane. (From R. Robitaille et al., 1990. Neuron 5: 773–339.)

Freeze-fracture images of the presynaptic membrane show that a transmembrane protein, seen as a 10-nm particle, is concentrated and aligned in two narrow bands along the active zone (Figure 7). In hair cells of the frog, which form synapses with afferent nerves, and at the squid synapse, where particles of similar size are seen, careful counts have shown that their number is equivalent to the estimated number of voltage-sensitive calcium channels. Thus the 10-nm particle is assumed to represent the calcium channels that regulate transmitter release. In hair cells, current flow through calcium-sensitive potassium channels that are also located near the active zone permit the local concentration of free Ca^{2+} following a depolarization to be estimated. These experiments suggest that the local concentration of internal Ca^{2+} at active zones may rise as high as 1 m*M*.

Concentrating calcium channels near the site of exocytosis speeds transmitter release in three ways. First, because release sites are nearby, only the Ca^{2+} concentration near the active zone needs to be increased and not the concentration throughout the terminal. Having the channels close to the release sites reduces the time required to reach a critical concentration and uses the Ca^{2+} that enters more efficiently. Second, the same number of channels are more effective in achieving a rapid rise in local concentration when concentrated than when scattered over the entire terminal. Third, release can be rapidly terminated as the local concentration of Ca^{2+} under the membrane is quickly reduced by diffusion. These factors permit rapid, high-frequency release of transmitter in response to action potentials.

Release shows a steep dependence on calcium ion concentration

To understand the dynamics of transmitter release, we need to know quantitatively how it depends on Ca^{2+} concentration. What Ca^{2+} concentration must be reached to trigger transmitter release, and how are the number of quanta that are released related to Ca^{2+} concentration? Unfortunately, the means of controlling and measuring Ca^{2+} concentrations at sites of release in nerve terminals are not yet adequate to answer these questions. They can be addressed, however, by experiments in other types of secretory cells that can be permeabilized to allow the entry and exit of small molecules. When cells in suspension are exposed to high-voltage electric fields, the surface membrane undergoes dielectric breakdown, resulting in small holes in the cell that are porous to small molecules but not to proteins. Remarkably, when chromaffin cells are treated in this way and incubated in a solution containing Ca^{2+} and Mg-ATP, the permeabilized chromaffin cells undergo calcium-dependent exocytosis. In the permeabilized cells, in which the internal and external concentrations of Ca^{2+} are equal, half-maximal secretion is observed at a Ca^{2+} concentrations of 1 μ*M*. When measured at different concentrations of Ca^{2+}, secretion varies according to the square of the Ca^{2+} concentration. In intact anterior pituitary cells in which intracellular Ca^{2+} concentration is measured with a calcium-sensitive dye (see Chapter 7), secretion depends on the fourth or fifth power of the cytosolic Ca^{2+} concentration.

Indirect estimates at the frog neuromuscular junction suggest that transmitter release depends on the third or fourth power of the internal Ca^{2+} concentration. The steep dependence of release on Ca^{2+} concentration is crucial for understanding the regulation of release, because it means that secretion can be effectively turned off and on by relatively modest changes in the intracellular calcium ion concentration. Moreover, in the steep part of the curve for Ca^{2+} dependence, very small differences in the amount of Ca^{2+} that enter during the action potential can cause large changes in the number of quanta released (see below and Chapter 14).

Calcium is removed quickly

If the secretory response of a nerve terminal is to faithfully mimic the high-frequency electrical signals that produce it, the Ca^{2+} concentration at the site of release must not only increase rapidly, but must also rapidly diminish. The quick decline in transmitter release after a single action potential suggests that the increased Ca^{2+} at release sites disappears within less than a millisecond (Figure 9). Ca^{2+} that enters the terminal during electrical activity is ultimately returned to the extracellular fluid by calcium-ATPase and the sodium–calcium exchanger, as described in Chapter 7. Conventional transporters are too slow to account for the rapid rate of removal, however; the initial decline in Ca^{2+} concentration at release sites probably occurs by diffusion from the localized sites of entry into the remainder of the terminal and by binding to cytosolic proteins in the terminal that act as Ca^{2+} buffers. In addition to transport across the surface membrane, some Ca^{2+} is also removed by transport into intracellular membrane vesicles that store Ca^{2+} (see Chapter 7). These membranes obviously

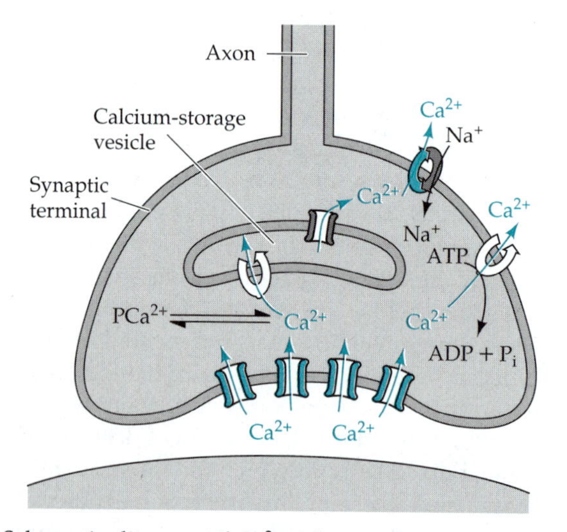

FIGURE 9. Schematic diagram of Ca^{2+} influx and removal from the cytosol following an action potential in the presynaptic nerve terminal. Ca^{2+} enters through voltage-gated calcium channels at the active zone and diffuses into the terminal, where it binds proteins (P), is taken up into a Ca^{2+} storage compartment, and is actively transported out of the cell. As the cytosolic concentration falls, the storage compartment slowly releases Ca^{2+} to be transported out of the cell by a Ca^{2+} ATPase and an Na^+/Ca^{2+} exchange pump (see Chapter 7). At high concentrations of cytosolic Ca^{2+} (0. 5 μM), mitochondria also take up Ca^{2+}.

cannot continue to take up Ca^{2+} indefinitely, but are thought to sequester the excess Ca^{2+} temporarily. As the Ca^{2+} concentration in the cytoplasm falls, they slowly release Ca^{2+} into the cytosol, from which it is then transported back to the extracellular fluid. The entire process of reequilibration of Ca^{2+} after a single action potential is estimated to require 0.1–1 sec. During this slow process of removal the cytoplasmic concentration of Ca^{2+} remains slightly elevated. This "residual" Ca^{2+} adds to any Ca^{2+} that enters during subsequent action potentials and is thought to be one basis for the short-term **facilitation,** or increased release of transmitter, that occurs when one action potential immediately follows another.

Exocytosis of synaptic vesicles requires a protein catalyst

What are the molecular mechanisms of exocytosis? They are presently unknown, but several possible clues arise both from physiological studies of secretion and from investigation of membrane fusion in other systems. Membrane fusion during exocytosis almost certainly requires proteins. Under physiological conditions, phospholipid bilayers are very stable and do not easily fuse; their stability is the basis of the orderly arrangement of cellular membranes and of the traffic between them. The energy barrier that prevents spontaneous membrane fusion is presumably reduced by specific proteins that join the hydrophobic interiors of surface and vesicle membranes and facilitate their union.

Viruses that fuse with surface or internal membranes offer one model of how such proteins work. Influenza virus particles are taken into cells by endocytosis; the endocytotic vesicles then fuse with lysosomes to discharge the intact virus into their lumen. The low internal pH of the lysosomes induces a conformational change in a protein on the surface of the virus (the hemagglutinin or HA protein). The conformational change exposes a hydrophobic domain of the HA protein that then inserts into the lysosomal membrane, causing it to fuse with the viral membrane. The nucleic acid within the virus is thus delivered directly into the cytoplasm of the host cell. One could imagine that synaptic vesicles had a comparable protein on their surfaces whose conformation changed in response to Ca^{2+}, rather than hydrogen ions, and that would promote the fusion of the vesicle and surface membranes. Compared to exocytosis at a nerve terminal, however, viral fusion is very slow, perhaps reflecting the difficulty of inserting a protein into a bilayer. Studies of membrane fusion in other physiological contexts may offer alternative models (Box A).

The first event of exocytosis is the formation of a fusion pore

Experiments on secretion in a mutant mouse suggest one way in which a transmembrane protein might facilitate membrane fusion. In experiments on nerve terminals, single release events are monitored by recording from the postsynaptic cell and using the postsynaptic membrane as a detector. Recent advances have made it possible to record single exocytotic events directly in secretory cells by measuring changes in the electrical capacitance of the cell. Because the cellular capacitance depends on surface membrane area, the addition of membrane to the surface caused by vesicle

Box A Mechanisms of Fusion in Non-Neural Systems

All cells have a pathway of constitutive secretion by which they transport membrane proteins to the cell surface. Two current investigations of membrane fusion events in this pathway may offer particularly important clues to the molecular mechanisms of neurotransmitter release: the analysis of mutants of baker's yeast (*Saccharomyces cerevisiae*) that are defective in constitutive secretion, and the development of in vitro membrane fusion systems related to transport between membrane compartments in mammalian cells. As is increasingly the case in modern biology, results from two biological systems complement and illuminate each other.

In 1980 Peter Novick and Randy Schekman carried out a massive screen for temperature-sensitive mutants defective in constitutive secretion of the protein invertase, a component of the yeast cell wall. When changed to the restrictive temperature (37° C), cell surface growth in the secretion mutants was blocked but protein synthesis continued. As a result the mutants became more dense, allowing their separation from other yeast cells. The mutants not only failed to secrete invertase but also other components of the membrane and cell wall.

The identified secretory mutants represent defects in 23 different genes (the *SEC* genes), and comprise three classes, each blocked in a different step of the secretory pathway: from the endoplasmic reticulum to the Golgi complex; from the Golgi complex to the formation of secretory vesicles; or from secretory vesicles to the surface membrane. Mutants in the latter class accumulate secretory vesicles at the restrictive temperature, presumably because they are unable to fuse them with the surface membrane. Although mutants in this class appear to be the most relevant to neurotransmitter secretion, all of the mutants are defective in some aspect of endocytosis, exocytosis, or membrane targeting, and are therefore of potential interest. Many of the *SEC* genes have been cloned, so that the sequences of their protein products are known. One of these, the *SEC4* gene product, is a GTPase that is similar to the ras proteins; it belongs to a subclass of the ras family called rab. The rab3 protein that is attached to synaptic vesicles (see text) is another member of this family.

The second experimental system is one developed by James Rothman and his collaborators in which the vesicular transport of proteins from one Golgi compartment to another is reconstituted in vitro. Golgi membranes from a mutant mammalian cell defective in glycosylation are mixed with those from a normal cell. The transfer of labeled viral protein from the defective Golgi to normal Golgi can be detected by its glycosylation. A late step in the process is sensitive to the sulfhydryl reagent, *N*-ethylmaleimide (NEM). NEM causes the accumulation of vesicles that are bound to the target Golgi membrane but that do not fuse with it. A protein has been purified that restores the fusion activity, and several accessory proteins that bind this protein to the membranes have been identified. Interestingly, this protein and one of the accessory proteins each have amino acid sequences that are related to those of *SEC* gene products in yeast.

The future study of these systems may identify proteins that are related to, or models for, proteins that mediate neurotransmitter secretion. Most importantly, they indicate that targeted membrane fusion is unlikely to be a simple event mediated by a single protein but in each case may require a battery of proteins that form a fusion machine, tailored to particular physiological needs.

fusion increases it. Individual synaptic vesicles are too small to cause a detectable increment in capacitance, but such events can be detected in cells with large secretory vesicles, such as chromaffin cells or mast cells. The most detailed measurements have come from a mutant strain of mice (beige mice) whose mast cells have giant granules. Instead of several thousand vesicles, each about 0.8 μ in diameter, mast cells in beige mice have only 10–40 giant vesicles with diameters up to 5 μ.

After stimulation of secretion in beige mast cells, discrete increases in membrane capacitance are observed, corresponding to the addition of single vesicles to the cell surface. In general, the addition of membrane to the surface is rapid and irreversible. Membrane retrieval occurs by a slower, later process of endocytosis.

In such recordings, the first step in exocytosis, observed before the increase in membrane capacitance, is the formation of a narrow fusion pore whose dimensions appear to be only slightly larger than those of an ion channel. At rest, the vesicle membrane has an electrical potential between the lumen and the cytoplasm whose size is determined by the proton pump of the vesicle (see above); in mast cells this potential ranges from 0 to 160 mV, inside positive. When the potential across the plasma membrane is held at 0 mV with a voltage clamp, there is a large potential difference between the inside of the vesicle and the extracellular medium, causing current to flow between the two as soon as a pathway is available. By determining the potential that drives this brief flow of current, the conductance of the initial opening can be calculated. The conductance is surprisingly small (200–300 pS), about the size of a single gap junction channel, which corresponds to an opening of about 2 nm. Within a few milliseconds the pore rapidly expands. One model that could explain how a channel across two membranes could form so quickly postulates that two half-channels, one in the vesicle and the other in the surface membrane, are aligned prior to fusion to form a channel that is complete, but closed. A Ca^{2+} trigger then opens the channel. Once open, the channel could expand by the addition of lipid, thus leading to fusion of the two membranes (Figure 10) and transmitter release.

Several synaptic vesicle proteins may play a role in exocytosis and its regulation

Specific proteins that facilitate membrane fusion during transmitter release have not been identified. Several proteins associated with synaptic vesicles, however, have intriguing properties that mark them as potential candidates for a role in the exocytotic event. Synaptophysin, a 38-kD protein that is found in all synaptic vesicles, has been reported to form hexameric ion channels in lipid bilayers whose conductance (150 pS) resembles that of gap junctions. These similarities raise the interesting possibility that synaptophysin, like the gap junction protein, could be part of a channel spanning two membranes, thus acting as a fusion pore.

Calmodulin, the calcium-binding regulatory protein (see Chapter 7), shows calcium-dependent binding to a 65-kD protein (p65) in synaptic vesicles, suggesting that it might be important in mediating the effects of

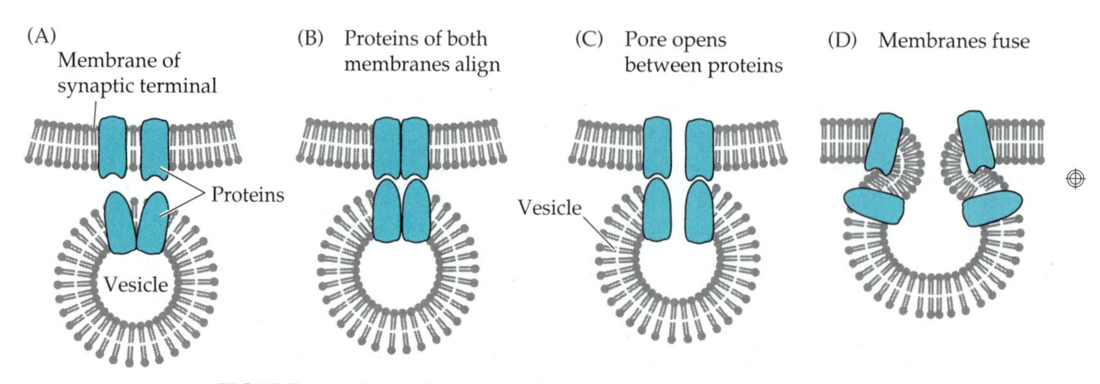

FIGURE 10. Hypothetical scheme of the formation of a fusion pore as an early step in exocytosis. (A) The vesicle first docks, aligning the proteins in the vesicle and surface membranes that will form the pore (A, B); the pore then opens (C); phospholipids from the membrane then insinuate themselves between the subunits of the pore to enlarge it and fuse the two membranes (D). (After W. Almers, 1990. Annu. Rev. Physiol. 52: 607–624.)

Ca^{2+} on neurotransmitter release. Antibodies against calmodulin block exocytosis in several other secretory cells. The most dramatic example is the massive exocytosis of cortical vesicles seen in sea urchin eggs following fertilization. Vesicle secretion, which is triggered by Ca^{2+}, can be observed in an in vitro preparation in which the external face of the surface membrane is attached to a glass surface, with vesicles adhering to the exposed cytoplasmic face. Fusion of the large vesicles with the membrane, monitored visually, is almost completely blocked by antibodies to calmodulin. Calmodulin's role as a regulatory protein for protein kinases and other enzymes is well established; the rapidity of neurotransmitter secretion, however, makes it unlikely that it triggers secretion in nerve terminals via activation of a kinase. It may be important, however, in regulating secretion or in the transport of vesicles to the release site (see below).

GTPase attaches and detaches from synaptic vesicles during the release cycle

Another protein associated with purified synaptic vesicles is rab3, a small (25 kD) member of the GTPase family that is related to ras p21 (see Chapter 6). GTP is required for secretion in mast cells, and mutations in genes encoding ras p21–like proteins block the constitutive pathway of secretion in yeast (see Box A). These observations have suggested that GTP-binding proteins play a role in secretion, perhaps by guiding the attachment of vesicles to the target membrane prior to exocytosis, or by acting as the trigger for membrane fusion. The relation of rab3 binding to exocytosis has been investigated in **synaptosomes,** which are pinched-off nerve terminals produced by homogenizing brain tissue in isotonic sucrose to prevent osmotic lysis. Much more rab3 is associated with vesicles purified from resting synaptosomes than from those stimulated to secrete. How rab3 associates and dissociates from the vesicle membrane and how this is related to GDP- and GTP-bound forms of the protein is not known. A protein has been isolated from brain, however, that stabilizes the GDP form and prevents its association with vesicles, suggesting that association may be regulated. At present, the observations relating GTP and rab3 to vesicle secretion are too fragmentary to allow more than speculation about their roles, but it seems likely that this protein will be closely involved in transmitter secretion.

Vesicle transport and docking

We have assumed so far that exocytosis at the active zone of nerve terminals arises from a subset of vesicles that are in a preformed complex at the presynaptic membrane. Although this assumption is made to help explain the speed of exocytosis, several physiological and anatomical observations indirectly support it. First, nerve terminals often show vesicles that are docked, or in very close contact with, the membrane at the active zone. Second, analysis of the statistics of quantal release (see Chapter 2) at many synapses suggests that only a small fraction of the total number of vesicles are immediately available for release. An attractive hypothesis is that these correspond to vesicles already docked at the membrane. Because the calculated number of vesicles available for immediate release is much smaller than the number in close contact with the membrane, only a subset of the docked vesicles may be primed for release at any one time.

As vesicles fuse with the surface membrane during exocytosis, nearby vesicles must be able to rapidly replace them. Electron microscopy of nerve terminals after rapid freezing and etching shows that many of the vesicles near the presynaptic membrane are attached by short connecting links to long filaments (Figure 11). In some cases, the vesicles are attached to actin filaments; in others, to long unidentified filaments, perhaps containing fodrin, that extend from the presynaptic membrane at the active zone into the cytoplasm. In both instances, the connecting link appears to be synapsin I, a phosphorylated protein found exclusively in nerve terminals. Synapsin I is about 35 nm long, with a globular head and a long, collagen-like tail. The molecule has binding sites for both actin and synaptic vesicles. The protein also binds tubulin and fodrin. By attaching synaptic vesicles to elements of the cytoskeleton that are localized near the presynaptic membrane, synapsin I concentrates them near the sites of release.

The binding of synapsin I to synaptic vesicles and to actin is regulated by phosphorylation. Synapsin I is a substrate for type I and type II Ca^{2+}/calmodulin-dependent protein kinases and for cAMP-dependent protein kinase. Phosphorylation by type II Ca^{2+} calmodulin-dependent protein kinase decreases the affinity of binding of synapsin I to synaptic vesicles and to actin. Phosphorylation by the other kinases also decreases the number of actin binding sites on the protein (see Chapter 7). The role of synapsin I in transmitter release was investigated by injecting it into the presynaptic terminal of a synapse in the squid (the giant synapse) whose terminal is large enough to permit penetration by a microelectrode. Injection of the dephosphorylated form of the protein inhibited transmitter release, whereas injection of the phosphorylated form had no effect. In-

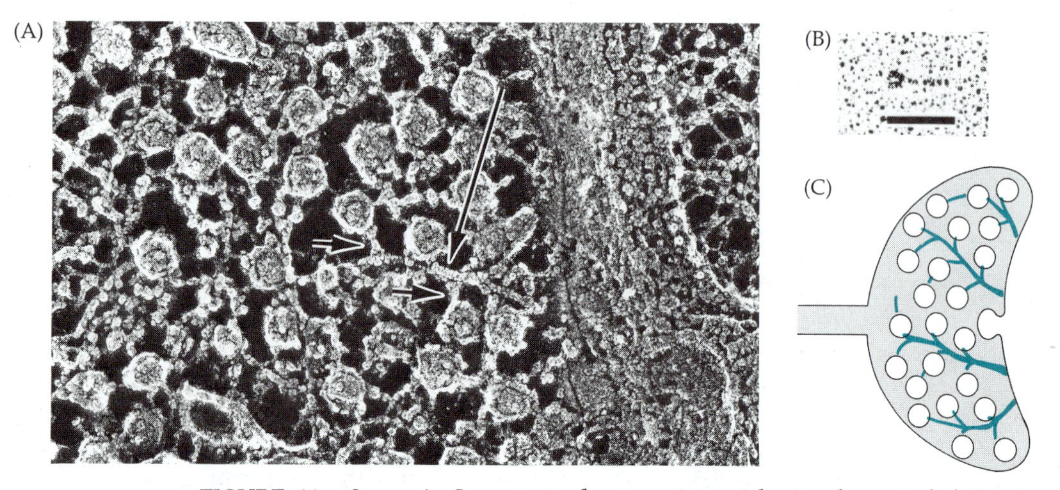

FIGURE 11. Synapsin I may attach synaptic vesicles to the cytoskeleton in presynaptic terminals. (A) Linkage of synaptic vesicles to the cytoskeleton at the frog neuromuscular junction. The tissue was rapidly frozen, freeze-fractured, and deep-etched. Short strands (20–30 nm; short arrows) link the synaptic vesicles to each other and to long actin filaments (long arrow) near the surface membrane. The postsynaptic membrane is shown to the right, with its attached cytoskeleton. (B) Rotary shadowed image of purified synapsin I. The bar is 50 nm. (C) Schematic diagram of the presynaptic terminal. (A and B from N. Hirokawa et al., 1989. J. Cell Biol. 108: 111–126.)

jection of Ca^{2+} / calmodulin-dependent kinase II also enhanced the release of transmitter. One interpretation of these results is that the vesicles that are not docked are held near the active zone by the cytoskeleton; synapsin phosphorylation then releases them to allow docking. The phosphorylation of synapsin I is stimulated by increased intracellular Ca^{2+} and cAMP; thus it can be regulated both by electrical activity in the terminal and by extracellular signals that influence adenylyl cyclase. By stimulating phosphorylation of synapsin I, both types of stimuli mobilize vesicles for exocytosis.

Transmitter secretion is regulated by extracellular signals

At many synapses the average number of quanta that are released in response to a single action potential can be altered by extracellular messengers that bind to receptors in the nerve terminal membrane. At some synapses, the released neurotransmitter not only binds to postsynaptic receptors but also acts on presynaptic receptors in the nerve terminal membrane to affect its own release; at other synapses, synaptic messengers secreted by other nerves alter transmitter release. In each case, the messengers alter transmitter release by changing the amount of Ca^{2+} that enters during the action potential, or by changing the sensitivity of the release process to Ca^{2+}. The ability of presynaptic receptors to change transmitter release rates gives enormous flexibility to synaptic transmission and is one source of the synaptic plasticity that underlies behavioral changes in the nervous system (Chapter 14).

Presynaptic inhibition (Chapter 2) is one of the simplest examples in which one neuron changes transmitter release by another. At the crayfish neuromuscular junction inhibitory nerve terminals end directly on excitatory ones. Stimulation of the inhibitory axon reduces the number of quanta of excitatory transmitter (glutamate in this case) released by a stimulus to the excitatory axon. The inhibitory transmitter GABA binds to its receptor, a chloride channel, in the nerve terminal. The increased chloride conductance produced by GABA reduces the depolarization produced by action potentials that invade the terminal, thus decreasing the amount of Ca^{2+} that enters during a single action potential (Figure 12). Presynaptic inhibition also occurs in the mammalian CNS and is particularly prominent in the spinal cord.

Other synaptic messengers act on G protein–linked receptors to reduce Ca^{2+} current during an action potential. As discussed in detail in Chapter 14, serotonin facilitates (increases) transmitter release in the sensory neurons of *Aplysia* by decreasing the potassium current during an action potential, thus prolonging the action potential and allowing more Ca^{2+} to enter. Acting through a G protein, serotonin increases cAMP and activates a cAMP-dependent protein kinase that reduces the number of functional potassium channels via phosphorylation.

At many synapses, the secreted transmitter acts on receptors in the terminal membrane to decrease its own secretion, a process called **autoinhibition.** In sympathetic nerves, for example, binding of norepinephrine to α2 receptors in the nerve terminal activates a G protein that reduces

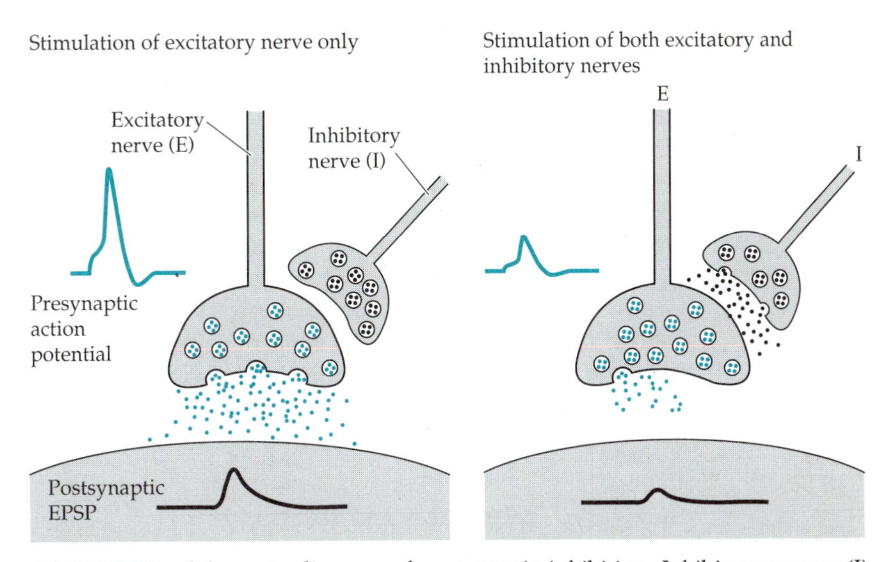

FIGURE 12. Schematic diagram of presynaptic inhibition. Inhibitory nerves (I) terminate on excitatory nerve (E) terminals. The inhibitory transmitter opens chloride or potassium channels in the presynaptic membrane of the excitatory nerve, reducing the size of the nerve terminal depolarization and the amount of calcium that enters the terminal. Transmitter release by the excitatory nerve and the postsynaptic excitatory postsynaptic potential (epsp) are thus decreased by inhibitory nerve stimulation.

Ca^{2+} current through the channels that mediate transmitter release. The G protein does not act through a second messenger but appears to affect the channel directly, changing its kinetics to reduce the open time of the channel. The precise location of the receptor and G protein within the terminal are not known, but their efficiency would obviously be highest near the release sites in the active zone.

Large peptide-containing vesicles and small synaptic vesicles are secreted by different pathways

In addition to the small synaptic vesicles filled with small molecule neurotransmitters (including catecholamines), nerve terminals also have larger vesicles that contain peptides and amines (Chapter 4). In most nerve terminals, these are far less numerous than the small vesicles; profiles of terminals in electron micrographs may have one, or several, or sometimes none. Also, unlike small vesicles, they are not concentrated at the active zone but occur throughout the terminal, often near the periphery. As noted previously (Chapter 4), the secretion of large vesicles may be much more closely related to secretion in endocrine and exocrine cells, which occurs without apparent morphological specialization. Their location throughout nerve terminals fits with the idea that large vesicles can release their contents by exocytosis outside the active zone. Electron micrographs show profiles of large vesicles undergoing exocytosis, not only at active zones, but also at sites in the nerve terminal that appear undifferentiated (see, for example, Figure 6). In invertebrate neurons, exocytosis of large vesicles has even been seen in cell bodies.

Exocytosis of large vesicles away from the active zone offers a potential explanation for the observation that peptides, which are released from large vesicles, require a higher frequency of stimulation for their release than do small-molecule neurotransmitters (Chapter 4). Following an action potential, Ca^{2+} concentration rises much more quickly at the active zone, where calcium channels are concentrated, than elsewhere in the terminal. Because of efficient removal of Ca^{2+}, a single action potential may cause little change in the bulk Ca^{2+} concentration, and a train of action potentials may be required to increase Ca^{2+} throughout the terminal. Recent studies in which ionophores were used to increase the Ca^{2+} concentration uniformly throughout synaptosomes suggest that release of peptide may in fact be more sensitive to Ca^{2+} concentration than the release of small-molecule transmitters. The high density of Ca^{2+} channels near the release sites for small vesicles ensures that they are exposed to large changes in Ca^{2+} following an action potential (Figure 13).

Recovery: Transmitter Inactivation and Vesicle Recycling

Once transmitter is released, two actions are necessary to prepare for the next round of transmitter synthesis and secretion: removal of the released transmitter, and reformation of the vesicles following retrieval of the synaptic vesicle proteins from the surface.

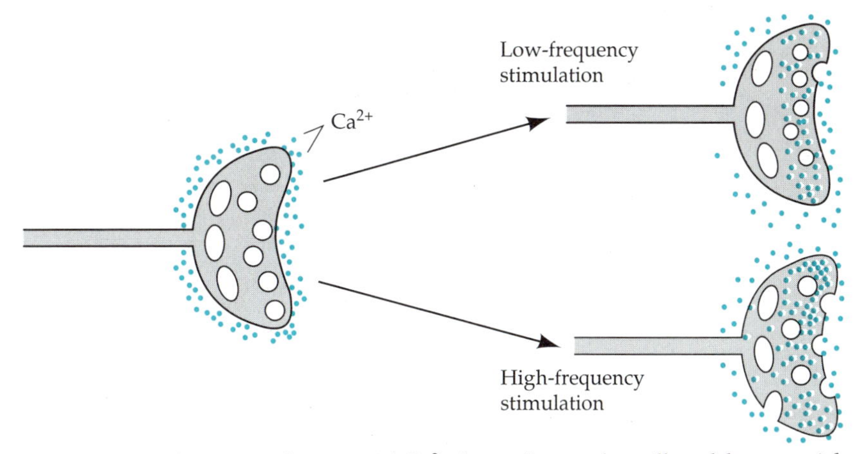

FIGURE 13. Schematic diagram of Ca^{2+} dependence of small and large vesicle secretion. At low frequencies of stimulation, Ca^{2+} concentration is increased in the vicinity of the active zone, where small vesicles are secreted, but not in the bulk cytosol. Large vesicle secretion is therefore little affected. At a higher frequency of stimulation, Ca^{2+} is increased both at the active zone and in the bulk cytosol.

Efficient removal terminates the action of transmitters on the postsynaptic membrane

After their release from nerve terminals, neurotransmitters are rapidly removed from the synaptic cleft either by enzymatic action or by uptake either back into the terminal or into glial cells. Reuptake into the terminal is the most common mechanism (Figure 14). The catecholamines, serotonin, glutamate, glycine, and GABA are all taken up into nerve terminals by high-affinity transporters that are specific for the neuron secreting that particular transmitter. In addition, some glial cells have high-affinity transport systems for GABA and serotonin. In each case, the transmitters are co-transported with sodium, whose concentration gradient across the membrane provides the energy for uptake. In several cases, the transporters can be blocked by specific drugs. Cocaine, for example, blocks the reuptake of dopamine into nerve terminals after secretion, an action that is the basis of some of the central nervous system effects of this drug. The transporter for GABA has been recently cloned using *Xenopus* oocytes as an expression system. The protein, which appears to have 12 transmembrane segments, has no significant sequence similarity to other cloned transport proteins, such for those for glucose or proline.

After transport into the terminal, the transmitters are accumulated in synaptic vesicles and released again. Uptake is highly efficient: in sympathetic terminals, as much as 50% of the released transmitter is taken back up into the nerve endings. If uptake of the transmitter into sympathetic nerve terminals is blocked, the effects of nerve stimulation in the target tissue are increased, indicating that uptake limits the physiological action of the transmitter.

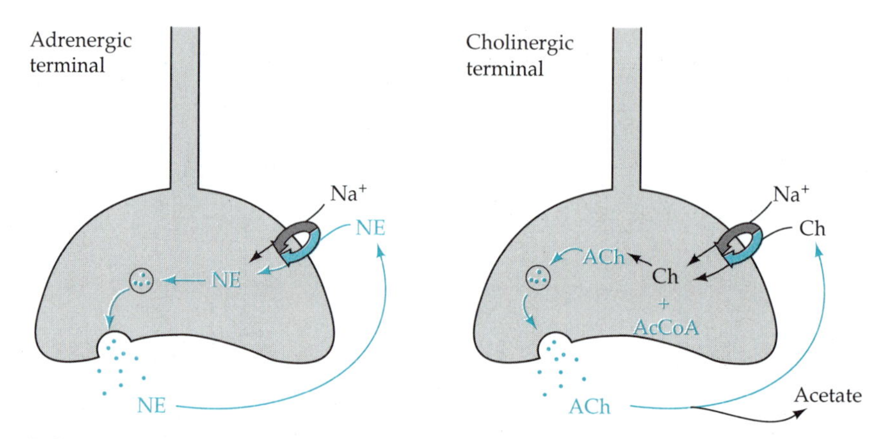

FIGURE 14. Schematic diagram of transmitter reuptake cycles in adrenergic and cholinergic nerve terminals. In adrenergic terminals, secreted norepinephrine is taken up by a sodium-dependent transporter in the nerve terminal membrane, transported into vesicles, and secreted again. In cholinergic terminals, the released ACh is hydrolyzed to choline and acetate by acetylcholinesterase. The choline is then taken up and used in the synthesis of new ACh. NE is norepinephrine, Ch is choline.

The ACh released from cholinergic terminals is not taken up but is hydrolyzed by the enzyme acetylcholinesterase to the inactive products choline and acetate (Figure 2). Choline is then taken back up into the terminals to be used in the synthesis of new ACh, as described above. As with uptake at adrenergic terminals, inhibition of acetylcholinesterase at cholinergic synapses prolongs the postsynaptic action of the neurotransmitter. Because the ACh receptor at the neuromuscular junction desensitizes upon prolonged exposure to the transmitter (Chapter 3), cholinesterase inhibition in vertebrates results in paralysis of the neuromuscular junction and ultimately death from respiratory failure. The most common nerve gases (Sarin and Tabun) are acetylcholinesterase inhibitors.

Most of the acetylcholinesterase at the neuromuscular junction is associated with the basal lamina in the synaptic cleft. The catalytic subunits of the enzyme are bound by disulfide linkage to a long, collagen-like tail that attaches the enzyme to heparan sulfate proteoglycan, a component of the basal lamina (Figure 15). The high density of acetylcholinesterase in the basal lamina at the endplate is the basis for a classic histochemical stain, marking the neuromuscular synapse. The synaptic enzyme appears to be made by the muscle and deposited in the extracellular matrix.

Synaptic vesicles are assembled in the nerve terminal

Just as some of the secreted transmitter or its products are retrieved from the extracellular fluid to be reused in a new cycle of release, so are the protein components of the vesicle membranes retrieved from the surface by endocytosis and used to make new synaptic vesicles. In this respect, small synaptic vesicles differ from large vesicles, which must be made from the Golgi apparatus and transported to the terminals. When motor

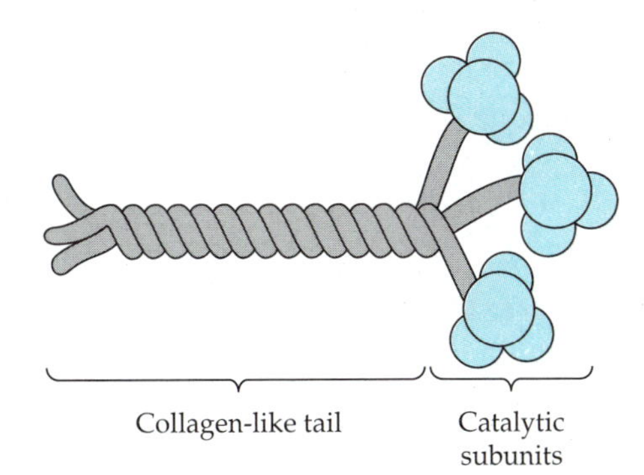

FIGURE 15. Structure of acetylcholinesterase in the basal lamina at the neuromuscular junction. Three tetramers formed of the catalytic subunit of the acetylcholinesterase are attached by disulfide bonds to a long, collagenous tail. The tail is thought to bind to a heparan sulfate proteoglycan in the basal lamina.

nerves are stimulated over a period of several hours, transmitter release continues, although the number of quanta released per impulse declines slightly. The number of quanta released during this period is several times the original number of vesicles in the nerve terminal. (A similar observation was made above with respect to the transmitter content of terminals.) Clearly, during the experiment new vesicles have been formed and refilled with transmitter. In such a short time, new vesicles cannot have been supplied from the cell body but must have been made locally. Experiments such as this one dramatically illustrate the capacity of the nerve terminal to maintain high rates of transmitter release over long periods of time.

The new vesicles arise through endocytosis from the nerve terminal membrane. The cycle of exocytosis and endocytosis in nerve terminals may be followed by using an impermeable extracellular tracer, such as horseradish peroxidase (HRP), an enzyme whose insoluble reaction product can be easily detected in the electron microscope. The tracer is taken up by endocytosis and can be followed through the membrane compartments of the terminal.

Only small amounts of tracer are taken up if resting terminals are incubated in HRP. If the nerve is stimulated, however, HRP is taken up and appears in synaptic vesicles. These results show not only that synaptic vesicles are derived from membrane pinched off from the cell surface but also that endocytosis is induced following nerve stimulation. This idea makes sense: exocytosis adds membrane to the terminal that must subsequently be recovered by endocytosis.

The particular pathway followed during the cycle of exocytosis and endocytosis can be seen by incubating the nerve briefly with HRP during stimulation, removing the extracellular HRP, and following the subsequent fate of HRP taken up into the terminal (Figure 16). The HRP first appears in coated vesicles, then in membranous cisterns, and finally in synaptic vesicles. In cholinergic nerves, the vesicles containing the marker can be identified as synaptic vesicles by isolating them from the terminals, separating them from other vesicles, and showing that they contain ACh

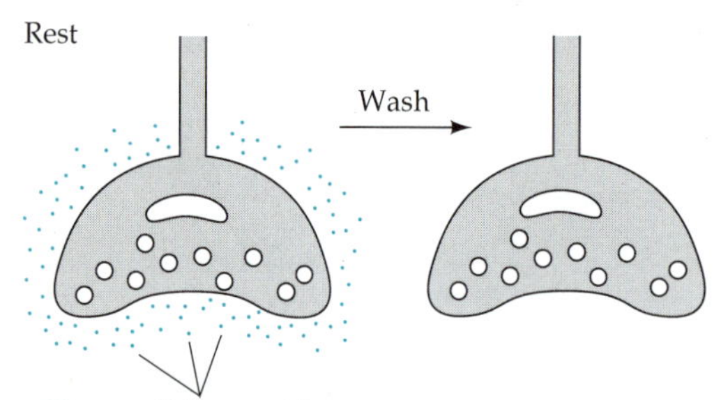

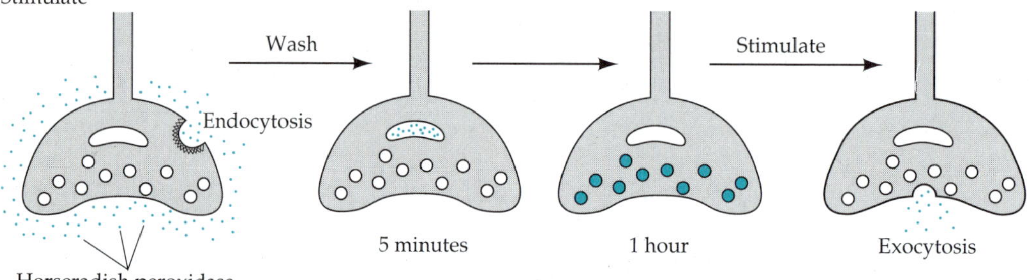

FIGURE 16. Diagram of vesicle recycling pathway as revealed by uptake of horseradish peroxidase. Nerve terminals bathed in horseradish peroxidase do not take up the enzyme unless they are stimulated. Five minutes after a short train of stimuli, enzyme activity is found in cisternae in the terminals; within one hour enzyme activity is seen in synaptic vesicles. After a second train of stimuli, no vesicles with enzyme activity are seen. These experiments show that synaptic vesicles formed by endocytosis are reused in further rounds of exocytosis. (After J. E. Heuser and T. Reese, 1973. J. Cell Biol. 57: 315–344.)

and ATP. Furthermore, following stimulation, vesicles that contain the marker disappear from the nerve terminals. The marker thus reveals the complete round of the cycle of exocytosis and endocytosis that synaptic vesicles follow during and after nerve stimulation in the terminals.

The timing and site of endocytosis has been determined at the frog neuromuscular junction using the rapid freeze and fracture technique discussed above. Nerves were stimulated and rapidly frozen at timed intervals after the stimulus. Endocytosis was first seen several hundred milliseconds after stimulation and occurred over a broad area of membrane several hundred nanometers from the active zone.

The fusion of vesicles with the surface membrane during exocytosis poses a potential problem. How are the synaptic vesicle proteins, which are different from surface membrane proteins, recognized and selectively retrieved? Although the answer to this question is not known, some experiments indicate that the vesicle proteins may not disperse but remain as a coherent cluster in the presynaptic membrane, where they can be conveniently and efficiently recovered as a group.

The endocytosis of surface membranes following stimulation occurs through a clathrin-dependent mechanism. Once internalized, the vesicles shed their clathrin coats and fuse with a larger system of internal membranes whose exact nature is ill-defined but which may be related to the early endosome compartment that is the target in many cells for vesicles that have been endocytosed. New vesicles are presumably then formed by pinching off from these membranes. This pathway is currently thought to be the origin of virtually all the small synaptic vesicles seen in nerve terminals.

Although vesicle membrane proteins are originally made in the cell body and probably arise from the Golgi, very little is known about the form in which they are transported along the axon. In sympathetic axons, vesicles in axons contain transmitter, but have a somewhat larger diameter than the small synaptic vesicles in terminals, suggesting that they may be a precursor form of the synaptic vesicles in the terminal. In the cholinergic axons leading to the *Torpedo* electric organ, on the other hand, synaptic vesicle membrane proteins appear to be associated with vesicles that do not contain transmitter. Thus, the vesicle components may be assembled in the cell body into a special form for transport, and the true synaptic vesicles may be assembled only in the nerve terminals.

Summary

Nerve terminals synthesize vesicles by endocytosis from the surface membrane, fill them with neurotransmitter, and store them until an action potential triggers the release of their contents by exocytosis into the synaptic cleft. Although they generate their own source of energy from mitochondria, and either synthesize biosynthetic precursors or accumulate them from the extracellular fluid, they are dependent on the cell body for the proteins and new membrane needed to accomplish these tasks.

Except for the amino acid transmitters glycine and glutamate, which are made in all cells, neurotransmitters are synthesized by specialized pathways found only in the neurons that secrete them. The transmitters are taken up into synaptic vesicles by specific transporters that are driven by the proton gradient and/or the membrane potential generated by a proton pump in the vesicle.

Vesicles are concentrated near the sites of release, or active zone, by attachment to the cytoskeleton, to which they are linked by synapsin I. Phosphorylation of synapsin I then releases them for docking at specialized release sites on the membrane. Transmitter release is triggered by the entry of Ca^{2+} into the terminal through voltage-gated channels concentrated near the active zone. The high Ca^{2+} concentration at the active zone is quickly diminished by diffusion, and the Ca^{2+} is ultimately returned to the extracellular fluid by active transport. The steep dependence of release on Ca^{2+} results in an increase in release rate of over 10^5 after an action potential. The molecular mechanism of exocytosis is unknown, but the first detectable step appears to be the formation of a fusion pore that extends across both vesicle and surface membranes, connecting the vesicle interior with the extracellular space. After exocytosis, vesicle membrane proteins are retrieved from the membrane by clathrin-dependent endocytosis, and new vesicles are formed. The released transmitter is rapidly metabolized by extracellular enzymes or is taken up in the terminal by high-affinity transport to be used again.

References

General References

*Almers, W. and Tse, F. W. 1990. Transmitter release from synapses: Does a preassembled fusion pore initiate exocytosis? Neuron 4: 813–818.

Katz, B. 1969. *The Release of Neural Transmitter Substances.* Liverpool University Press, Liverpool.

Kelly, R. B. 1988. The cell biology of the nerve terminal. Neuron 1: 431–438.

Landis, D. M. D., Hall, A. K., Weinstein, L. A. and Reese, T. S. 1988. The organization of cytoplasm at the presynaptic active zone of a central nervous system synapse. Neuron 1: 201–209.

Smith, S. J. and Augustine, G. J. 1988. Calcium ions, active zones and synaptic transmitter release. Trends Neurosci. 11: 458–464.

Trimble, W. S., Linial, M. and Scheller, R. H. 1991. Cellular and molecular biology of the presynaptic nerve terminal. Annu. Rev. Neurosci. 14: 93–122.

Transmitter synthesis and storage

Birks, R. I. and MacIntosh, F. C. 1961. Acetylcholine metabolism of a sympathetic ganglion. Can. J. Biochem. Physiol. 39: 787–827.

Maycox, P. R., Hell, J. W. and Jahn, R. 1990. Amino acid neurotransmission: Spotlight on synaptic vesicles. Trends Neurosci. 13: 83–87.

Weiner, N. and Rabadjija, M. 1968. The effect of nerve stimulation on the synthesis and metabolism of norepinephrine in the isolated guinea-pig hypogastric nerve-vas deferens preparation. J. Pharmacol. Exp. Ther. 160: 61–71.

Zigmond, R. E., Schwarzschild, M. A. and Rittenhouse, A. R. 1989. Acute regulation of tyrosine hydroxylase by nerve activity and by neurotransmitters via phosphorylation. Annu. Rev. Neurosci. 12: 415–461.

Transmitter release

*Augustine, G. J., Charlton, M. P. and Smith, S. J. 1985. Calcium entry and transmitter release at voltage-clamped nerve terminals of squid. J. Physiol. 367: 163–181.

Blaustein, M. P. 1988. Calcium transport and buffering in neurons. Trends Neurosci. 11: 438–443.

Breckenridge, L. J. and Almers, W. 1987. Currents through the fusion pore that forms during exocytosis of a secretory vesicle. Nature 328: 814–817.

Fischer von Mollard, G., Sudhof, T. C. and Jahn, R. 1991. A small GTP-binding protein (rab3A) dissociates from synaptic vesicles during exocytosis. Nature 349: 79–82.

*Heuser, J. E., Reese, T. S., Dennis, M. J., Jan, Y., Jan, L. and Evans, L. 1979. Synaptic vesicle exocytosis captured by quick freezing and correlated with quantal transmitter release. J. Cell Biol. 81: 275–300.

Knight, D. E. and Baker, P. F. 1982. Calcium-dependence of catecholamine release from bovine adrenal medullary cells after exposure to intense electric fields. J. Membr. Biol. 68: 107–140.

*Llinás, R., McGuinness, T. L., Leonard, C. S., Sugimori, M. and Greengard, P. 1985. Intraterminal injection of synapsin I or calcium/calmodulin-dependent protein kinase II alters neurotransmitter release at the squid giant synapse. Proc. Natl. Acad. Sci. USA 82: 3035–3039.

Neher, E. and Marty, A. 1982. Discrete changes of cell membrane capacitance observed under conditions of enhanced secretion in bovine adrenal chromaffin cells. Proc. Natl. Acad. Sci. USA 79: 6712–6716.

Roberts, W. M., Jacobs, R. A. and Hudspeth, A. J. 1990. Colocalization of ion channels involved in frequency selectivity and synaptic transmission at presynaptic active zones of hair cells. J. Neurosci. 10: 3664–3684.

Verhage, M., McMahon, H. T., Ghijsen, W. E. J. M., Boomsma, F., Wiegant, V. M. and D. G. Nicholls. 1991. Differential release of amino acids, neuropeptides and catecholamines from isolated nerve terminals. Neuron 4: 577–524.

Transmitter removal and vesicle recycling

*Heuser, J. E. and Reese, T. S. 1973. Evidence for recycling of synaptic vesicle membrane during transmitter release at the frog neuromuscular junction. J. Cell Biol. 57: 315–344.

Taylor, P. 1991. The cholinesterases. J. Biol. Chem. 266: 4025–4028.

Part Two

The Cell Biology of Neurons and Glia

6

G Proteins and Receptors in Neuronal Signaling

Elliott M. Ross

A LARGE FAMILY of receptors mediates the effects of neurotransmitters, light, odorants, hormones and other extracellular messengers by interaction with a group of GTP-binding, regulatory proteins known as G proteins. In comparison to the rapid responses mediated by ligand-gated ion channels, G protein-coupled receptors initiate responses that are slower and more diverse. They regulate ion channels, but also cause changes in growth, metabolism, cytoskeletal structure, and gene expression. G proteins do not produce most of these effects directly, but act by changing the concentration of intracellular signaling molecules known as **second messengers.**

The basic elements of G protein-mediated signaling pathways are shown in Figure 1. The binding of a specific extracellular messenger to a receptor causes the receptor to activate a G protein, which then activates one or more effector proteins. Effectors are either ion channels or enzymes that alter the concentrations of second messengers. Each of these elements comes in many forms. Over 100 different receptors, including those for amines, peptides, eicosanoids, light, and odorants, communicate through various G proteins. The G proteins themselves form a family of at least 20 distinct proteins, and there is probably a similar number of G protein-regulated effectors.

Each neuron has distinctive subsets of receptors, G proteins, and effectors that form part of a complex signaling network in the cell. The spectrum of neuronal responses that an extracellular signal evokes depends upon which particular members of these sets are expressed. The action of G protein-coupled receptors is both much more complex and slower than that of ligand-gated ion channels. The reward for this increased complexity is a signaling system that has increased sensitivity, diversity of response, and, most importantly, flexibility.

This chapter focuses on the earliest steps in the pathways that involve G proteins, receptors, and effectors. The next chapter discusses second messengers and how they produce their effects.

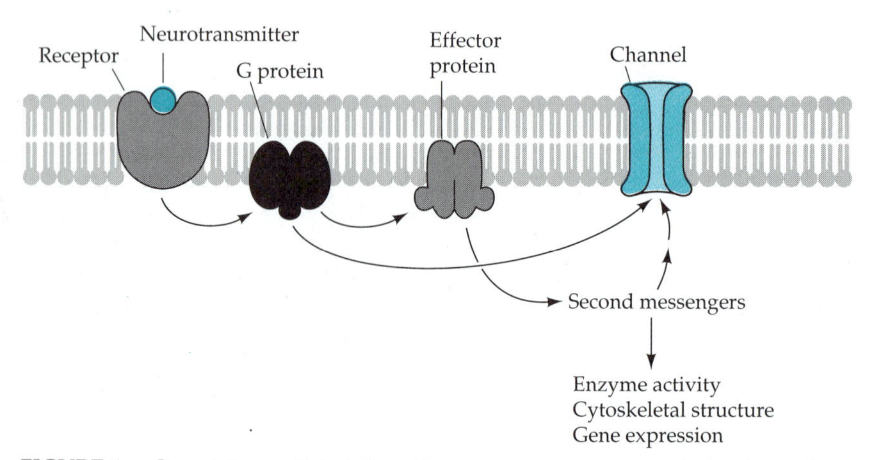

FIGURE 1. G protein-mediated signaling systems are composed of at least three separate proteins. Receptors span the plasma membrane; they bind neurotransmitters on the extracellular face and catalyze G protein activation on the cytoplasmic face. Activated G proteins regulate the activities of multiple effector proteins, which can be ion channels or enzymes that control the concentrations of intracellular second messengers.

Receptors Catalyze the Activation of G Proteins by GTP

G proteins play a pivotal role in signaling. They amplify the responses of receptors and influence the timing and magnitude of signals. They also sort signals by coupling receptors to their appropriate effectors.

Receptors activate G proteins by promoting GTP binding

G proteins exist in two conformations: an active form bound to GTP and an inactive form bound to GDP. The interconversion of these two forms constitutes a GTPase cycle whose control is the key to regulating G protein activity (Figure 2). In the cycle, the exchange of GTP for GDP activates the G protein. Activation is transient, however, because the G protein is a *GTPase* and hydrolyzes bound GTP to GDP. The concentrations of the two forms and, consequently, the relative activity of the G protein, are determined by the rates of the two basic steps in the cycle, the hydrolysis of GTP, and the exchange of GTP for GDP. In unstimulated cells, GTP/GDP exchange is the slowest step in the cycle. The GDP-bound, inactive form of the G protein predominates (~99%). In fact, GDP binding is so tight that purified G proteins retain bound GDP after weeks of storage. Agonist-liganded receptors increase the rate of GTP/GDP exchange and thus increase the amount of GTP-bound, active G protein. Under maximal stimulation by receptors, as much as 60% of the G protein can be maintained in the active form.

The GTPase activity of a G protein acts as a molecular timer, limiting the lifetime of the activated state. Measured in vitro, G proteins are slow catalysts, yielding typical lifetimes for bound GTP of 3–15 sec. Although altering the rate of GTP hydrolysis is a potential mode of regulating G

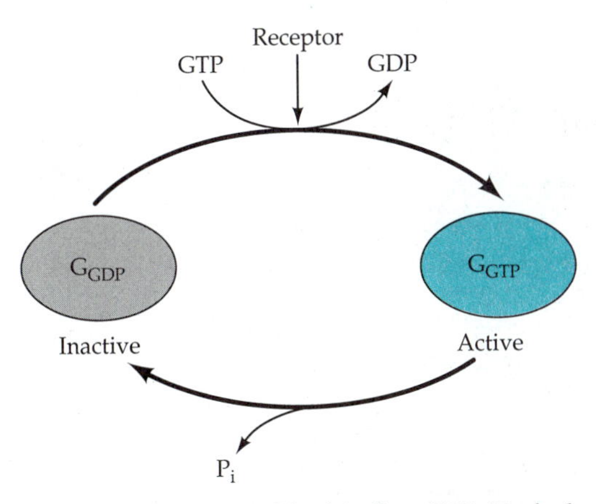

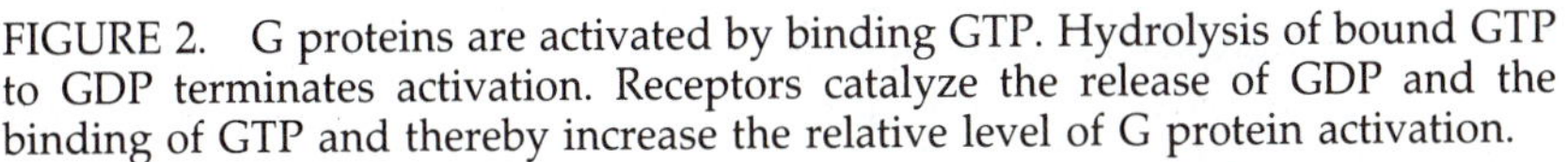
FIGURE 2. G proteins are activated by binding GTP. Hydrolysis of bound GTP to GDP terminates activation. Receptors catalyze the release of GDP and the binding of GTP and thereby increase the relative level of G protein activation.

protein activation, cellular regulators of GTP hydrolysis have not been found. Regulatory proteins promote deactivation of other members of the GTPase superfamily, however, and analogous proteins might influence the extent of G protein-mediated responses in neurons.

Each receptor molecule can activate many G protein molecules

The relatively long lifetime of the activated state of a G protein and the rapidity with which a receptor can stimulate GTP/GDP exchange means that each agonist-liganded receptor acts catalytically by activating many G protein molecules (see Box A). Such amplification can be striking. In the dark-adapted frog retina, a single molecule of bleached rhodopsin (formed by the action of one photon) can activate about 37,000 molecules of G protein. Even in less specialized cells, a single receptor can activate 10–20 G protein molecules. Because effector proteins are themselves catalysts, a significant cellular response requires only a few molecules of neurotransmitter.

Signaling depends upon the relative number of receptors and G proteins and upon the catalytic activity of receptors

Because a single receptor molecule can activate many G protein molecules, maximal activation of effectors can frequently be achieved when only a fraction of the total number of receptors is bound to agonist. The disparity between the total number of receptors and the number of agonist-bound receptors needed to give a maximal response causes the concentration-response curve for an agonist to differ from its equilibrium binding curve (Figure 3). As shown, the EC_{50} for an agonist (the concentration at which it exerts half-maximal effect) can be far below its K_d. This discrepancy is often said to result from "spare receptors" because it appears that more receptors are present than are needed. For G protein-coupled systems,

Box A Receptor-Catalyzed Nucleotide Exchange: The Beauty of Noncovalent Catalysis

All receptors are catalysts. Receptors increase the rates of key regulatory reactions that are thermodynamically favored but which are, for one reason or another, slow. A receptor is thus like a switch, thrown by an agonist, that turns on an otherwise favored process. The thermodynamic driving force for that process provides the energy that amplifies the signal. For example, a channel allows the flow of ions down a gradient established at the expense of considerable metabolic energy, but only minimal energy is needed for the conformational change that causes the channel to open. Similarly, G protein-coupled receptors catalyze GTP/GDP exchange, an otherwise slow reaction that is thermodynamically driven by the high intracellular [GTP]/[GDP] ratio.

The mechanism whereby the G protein-coupled receptors catalyze nucleotide exchange is a form of allosteric interaction in which two ligands bind to a protein at separate sites and the binding of one ligand decreases the affinity of binding of the other. This mechanism drives GTP/GDP exchange by elongation factor EF-Tu (see Box B), allows the activation of the cyclic AMP (cAMP)-dependent protein kinase by cAMP (Chapter 7), and activates the ATPase activities of the molecular motors kinesin and dynein (Chapter 8).

Why establish a system in which two ligands negatively influence each other's binding? *Speed!* Allosteric interactions are commonly considered in terms of changes in affinity, but the way in which these interactions change the speed of the association and dissociation reactions that determine affinity is the key to understanding their power. Regulatory ligands usually occur at low concentrations and their detection requires receptors to bind them with high affinity. However, high affinity and speed impose conflicting demands.

For the binding of a ligand A to a protein P, the equilibrium dissociation constant, K_d, equals the ratio of the association and dissociation rate constants, k_{-1}/k_1. Because the association rate constant k_1 is limited by diffusion to about $10^7\ M^{-1}\cdot sec^{-1}$, high affinity binding can only be achieved with a low dissociation rate constant, k_{-1}. For a diffusion-limited binding reaction whose K_d is 10^{-1} M, for example, k_{-1} would be 10 sec^{-1}.

$$P + A \underset{k_{-1}}{\overset{k_1}{\rightleftharpoons}} P\cdot A \qquad K_d = \frac{k_{-1}}{k_1}$$

Negative allosteric regulation uses the binding of one ligand to increase the rate of dissociation of another ligand that would otherwise dissociate slowly. The key to generating a rapid response while maintaining high overall affinity is the transient formation of the doubly liganded protein, shown for the two ligands A and B in the diagram below as P·A·B. P·A·B can be formed from either P·A or from P·B. Because both dissociation rate constants for P·A·B (k_{-4} and k_{-2}) are high, it will

however, the appearance of spare receptors has a kinetic basis; G proteins are usually far in excess of receptors.

The utility of spare receptors is that they allow the EC_{50} for a physiological response to be determined both by receptor number and by the catalytic efficiency of the coupling process. If the catalytic action of a receptor is enhanced, sensitivity to agonist will be increased. If the total number of receptors on a cell is increased, the fraction of agonist-occupied receptors needed to produce a maximal response will decrease and the cell's sensitivity to agonist will also be correspondingly augmented. In other words, increasing the total number of receptors shifts the dose-

rapidly dissociate to yield one or the other monoliganded species.

$$\begin{array}{ccc} P + A & \underset{k_{-1}}{\overset{k_1}{\rightleftharpoons}} & P\cdot A \\ {\scriptstyle +B}\; k_3 \Big\downarrow\!\Big\uparrow k_{-3} & & {\scriptstyle +B}\; k_4 \Big\downarrow\!\Big\uparrow k_{-4} \\ P\cdot B & \underset{k_{-2}}{\overset{k_2}{\rightleftharpoons}} & P\cdot A\cdot B \end{array}$$

Note that the presence of the intermediate P·A·B allows rapid interchange among the two monoliganded proteins P·A and P·B without ever passing through the unliganded state, P. Indeed, because k_{-1} and k_{-3} are both small, unliganded P is almost never formed.

G proteins use this mechanism to allow the exchange of guanine nucleotides. The agonist-liganded receptor (R·H) and guanine nucleotide (Nuc—either GTP or GDP) each promote the dissociation of the other. Thus, even though the dissociation of nucleotide directly from G·Nuc is slow (k_{-1}), R·H facilitates dissociation (via the k_{-2} step) by transiently forming the unstable G·Nuc·R·H complex.

$$\begin{array}{ccc} G & \underset{k_{-1}}{\overset{k_1}{\rightleftharpoons}} & G^{-Nuc} \\ k_3 \Big\downarrow\!\Big\uparrow k_{-3} & & k_4 \Big\downarrow\!\Big\uparrow k_{-4} \\ G_{-R\cdot H} & \underset{k_{-2}}{\overset{k_2}{\rightleftharpoons}} & G^{-Nuc}_{-R\cdot H} \end{array}$$

If agonist-liganded receptor decreases the affinity of G protein for nucleotide, then thermodynamic considerations demand that nucleotide reciprocally decrease the affinity of G protein for the agonist-liganded receptor. In a cell, where there is far more GTP than GDP, these reactions yield a cycle in which receptor displaces GDP and is in turn displaced by GTP, allowing it to promote the activation of other G protein molecules (Figure A).

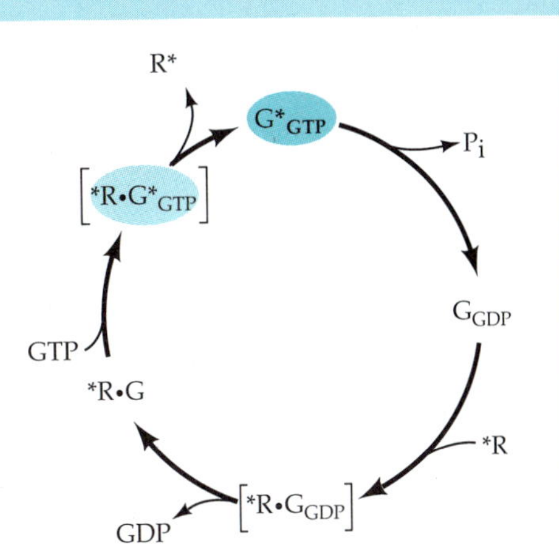

FIGURE A. Receptor catalyzes nucleotide exchange; nucleotide catalyzes receptor recycling. The receptor-catalyzed GTPase cycle depends on the formation of two transient intermediates, shown in brackets. GDP is stably bound to G protein until agonist-liganded receptor R* binds to form the first unstable intermediate, R*·G·GDP, from which GDP dissociates. GTP binds to form the second intermediate, R*·G·GTP, from which R* dissociates. Without the receptor, G protein would be trapped in the GDP-liganded, inactive state. Without cytosolic GTP, the receptor would be trapped in the G protein complex, R*·G, and unable to service other G proteins.

response curve to the left (Figure 3). If the number of receptors is decreased, the dose-response curve is shifted to the right. As the number of receptors becomes smaller still, the dose-response curve and binding curves will become coincident. Further loss of receptors will then diminish the maximal response. By controlling the number of receptors on its surface, the cell can thus regulate its sensitivity to messengers over a wide range of concentrations.

Alterations in the number of G proteins and effectors can also alter both the size of the maximal response and the position and shape of the response curve. Quantitative variations in responses among different cells

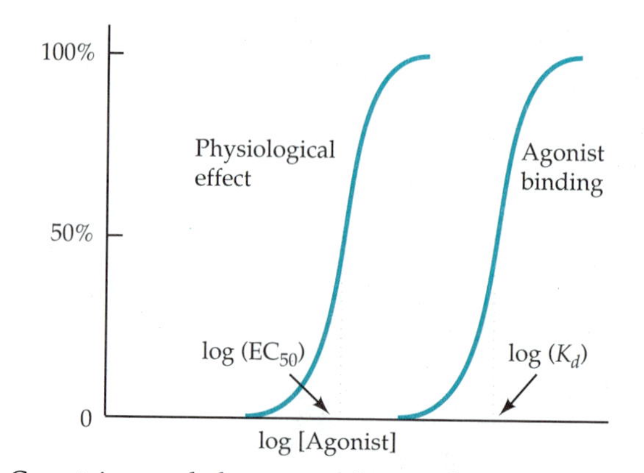

FIGURE 3. G protein-coupled receptors frequently act at concentrations of agonist that are far lower than what would be predicted based only on the affinity of agonist binding. Such behavior might arise if there were excess or "spare" receptors, such that only a small fraction of receptors would have to bind agonist in order to saturate the pool of effector proteins. However, there are usually fewer G protein-coupled receptors than there are G proteins or effectors. The catalytic activity of receptors allows the receptors both to stimulate maximally the pool of G proteins and, frequently, to do so at subsaturating concentrations of agonist. Maximal signaling occurs when only a fraction of the receptors bind agonists, and the activation isotherm therefore lies to the left of the binding isotherm. The EC_{50} for agonist, the concentration needed to produce half-maximal effect, is less than the K_d, the equilibrium binding constant.

no doubt reflect such alterations. The use of recombinant DNA techniques to express receptors, G proteins, and effectors in recipient cells will be a powerful tool for defining further the quantitative relationships between biological responses and the expression of signaling proteins.

G Proteins Are Heterotrimers Whose α, β, and γ Subunits Have Both Independent and Interactive Functions

The ability of G proteins to integrate and direct information within the signaling network is dependent on the regulatory activities of their subunits. Each G protein is a heterotrimer of a GTP-binding α subunit and a stable complex of regulatory β and γ subunits.

The α subunits bind GTP and convey information from receptor to effector

The α subunit determines a G protein's identity. The α subunit binds GTP and, when activated, selectively activates appropriate effector proteins in the absence of βγ. Thus, both G proteins and their α subunits are identified by mnemonic subscripts that refer to their originally recognized regulatory activity. For instance, G_s stimulates adenylyl cyclase, G_i inhibits it, and G_p activates phospholipase C (see Table 1).

The α subunits are a family of homologous proteins about 40 kD in size. Because they share similar structures, they are difficult to separate by

TABLE 1. G Proteins and Their α Subunits

α Subunit	**Effectors**[a]	**Receptors**[b]
α_s	Adenylyl cyclase (↑) Calcium channel (↑)	Epinephrine (β-adrenergic), histamine, LH, TSH, glucagon, dopamine, prostaglandin, prostacyclin
α_{olf}	Adenylyl cyclase (↑)	Odorant receptors
α_t	cGMP phosphodiesterase (↑)	Rhodopsin
α_i	Adenylyl cyclase (↓) Potassium channel (↑) Calcium channel (↓)	Epinephrine (α_2-adrenergic), opiate, peptide, prostaglandin, serotonin, angiotensin, dopamine, acetylcholine (M2, muscarinic)
α_o	Potassium channel(↑) Calcium channel (↓)	Acetylcholine (M2, muscarinic) in vitro; probably others
α_p[c]	Phospholipase C (↑)	Acetylcholine (M1, muscarinic); probably many others
α_z	Not identified	Acetylcholine (M2, muscarinic) in vitro; not well defined

[a]The arrows refer to the effect of G proteins in either increasing (↑) or decreasing (↓) the activity of the effectors.
[b]Several of the receptors listed have isoforms, each of which is coupled to a different G protein.
[c]Recent cloning experiments have identified two species of α_p, α_{11} and α_q.

traditional biochemical techniques. Recombinant DNA technology has revealed the number of α subunits (Table 1) and defined subfamilies of α subunits with similar structures and functions. Thus, the three α_i's are quite closely related and can substitute for each other in many cases. α_o is closely related to the α_i's. The two α_p's, which activate phospholipase C, form a separate group, as do α_s and α_{olf} or the two α_t's.

The conservation of amino acid sequence among the α subunits and the similarity of their biochemical properties suggest that they have similar overall structures and share definable functional domains. Three regions of highly conserved sequence in the α subunits are responsible for GTP binding and hydrolysis. These same regions are also found in two distantly related GTP-binding proteins, EF-Tu and $p21^{ras}$, whose three-dimensional structures have been solved by X-ray crystallography (Box B). Each α_s subunit must also have distinct domains that selectively recognize receptors and effectors. These recognition domains are presumably conserved with respect to overall structure and regulatory function but must be sufficiently distinct to provide selectivity. Genetic engineering of chimeric α subunits that combine homologous sequences of different α subunits is now helping to identify the domains that confer such selectivity.

The βγ subunits modulate activation of α subunits and help anchor α subunits to the plasma membrane

In contrast to the functional individuality of the α subunits, the β and γ subunits of different G proteins are more or less interchangeable. The β and γ subunits are purified as a complex and have not been separated under nondenaturing conditions. Genetic studies of βγ in yeast suggest

Box B A Superfamily of GTP-Binding Proteins Performs Diverse Cellular Functions

The G protein α subunits form a distinct family within a much larger superfamily of homologous GTP-binding proteins. Another prominent group in the superfamily are the 20–25 kD monomeric GTP-binding proteins. These soluble and peripheral membrane proteins are involved in the control of cell growth and development and the routing and fusion of subcellular organelles and secretion. The best-known member of the group is $p21^{ras}$, the product of the *ras* proto-oncogene, whose activation by point mutation is the cause of many cancers. Members of another group of proteins, called rabs, are associated with synaptic vesicles (Chapter 5).

A third important group of GTP-binding proteins consists of the initiation and elongation factors required for the synthesis of proteins on ribosomes. Research by Yoshito Kaziro and others on the bacterial elongation factor EF-Tu has provided a model for understanding the nucleotide exchange reactions of G proteins. GTP bound to EF-Tu is hydrolyzed at each round of translational elongation. A different elongation factor, EF-Ts, is required to catalyze GTP/GDP exchange to reactivate EF-Tu, a role analogous to that of a G protein-coupled receptor.

The three-dimensional structures of both EF-Tu and $p21^{ras}$, determined by X-ray crystallography, are strikingly similar in the regions that surround the nucleotide binding site. The amino acid sequences of the two proteins are highly conserved in the segments that comprise this domain, and these sequences are also the most conserved in the G protein α subunits. The phenotypes of α subunits with amino acid replacements in these domains are consistent with predictions based on the EF-Tu/ras consensus structure.

that they act as a dimer and do not have independent functions. Although there are at least four different β subunit genes and probably more than five γ subunits (and hence at least 20 possible βγ dimers), significant regulatory differences among preparations of βγ have not been observed. This finding is consistent with the extraordinary conservation of sequence among the β subunits (>90%), but leaves open the question of why so many different β and γ subunits are needed.

The βγ subunits have several mechanistically related activities. Most importantly, they are required for the receptor-catalyzed activation of α subunits. Although activated free α subunits can efficiently regulate effector proteins, receptors cannot stimulate GTP/GDP exchange on α subunits in the absence of βγ. The βγ subunits modulate GTP/GDP exchange directly, but the role of βγ in promoting receptor–α-subunit coupling is probably more closely related to the second major function of the βγ subunits, which is to anchor α subunits to the plasma membrane. Receptor–G protein binding is of relatively low affinity. By anchoring α subunits at the plasma membrane, βγ subunits increase the local concentration of α subunits to permit efficient coupling. Anchorage is also important for targeting α subunits to the correct subcellular organelle. The βγ subunits inhibit activation of the α subunit, both directly and by stabilizing the binding of GDP at the nucleotide-binding site (see Box C). They thus maintain a low rate of G protein activation in the absence of stimulation by receptors.

G Protein-Coupled Receptors: Transmitting Information across the Plasma Membrane

G protein-coupled receptors are a family of membrane proteins that use a single molecular architecture to convert the stimulus of ligand binding at the outer face of the plasma membrane to the activation of G protein on the

Box C A Mechanism for Receptor-Mediated Inhibition of G Protein Activation

When a G protein α subunit is activated, its affinity for $\beta\gamma$ decreases. A G protein that has been persistently activated by a nonhydrolyzable guanine nucleotide such as GTPγS dissociates into free α and $\beta\gamma$ subunits. As required by thermodynamics, the $\beta\gamma$ subunits also allosterically inhibit the binding of GTP and consequent activation. In contrast to GTP, GDP and the $\beta\gamma$ subunits each increase the affinity of the α subunit for the other. The $\beta\gamma$ subunits thus inhibit activation, both by stabilizing the binding of GDP and by inhibiting the binding of GTP.

Regulation of nucleotide binding by $\beta\gamma$ subunits has two important effects on G protein-mediated signaling. First, the stabilization of the inactive, GDP-liganded G protein by $\beta\gamma$ subunits maintains a low basal level of activity in the absence of stimulation by receptors. Second, regulation by $\beta\gamma$ subunits provides a mechanism for the receptor-mediated inhibition of G protein signaling. When a G protein is activated, its dissociated $\beta\gamma$ subunits can interact with other α's (Figure A). Thus, activation of any G protein will inhibit the activation of other G proteins in the membrane. The magnitude of such inhibition is related to the size of the pool of the G protein that releases $\beta\gamma$ subunits and to the affinity of $\beta\gamma$ subunits for G proteins. These considerations suggest some reasons why the inhibition of adenylyl cyclase is the only $\beta\gamma$-mediated inhibition that is frequently observed. The affinity of α_s for $\beta\gamma$ is high relative to that of other α subunits and the concentration of α_s in most cells is low relative to that of G_i. Thus, a major mechanism by which G_i can inhibit adenylyl cyclase is by releasing its $\beta\gamma$ subunits and thereby driving the deactivation of G_s.

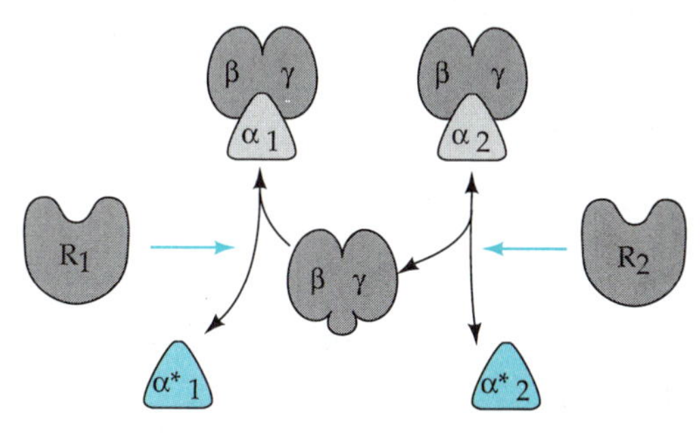

FIGURE A. $\beta\gamma$-mediated inhibition of G protein signaling. The $\beta\gamma$ subunits inhibit activation of the α subunit by GTP. Thus, if G protein 1 is activated, it will release its $\beta\gamma$, which will suppress the activation of G protein 2.

inner face. The receptors are globular glycoproteins whose structure is based on a bundle of seven membrane-spanning hydrophobic helices connected by hydrophilic loops. The glycosylated amino terminus is on the extracellular face of the membrane and the carboxyl terminus is on the cytoplasmic side (Figure 4). For the dozens of receptors that fit this general plan, the most highly conserved regions are the membrane-spanning helical domains. The N- and C-terminal regions and the cytoplasmic loop that links spans 5 and 6 display the most variation. A combination of site-directed mutagenesis and biochemical studies have helped to define distinct domains in this structure and show which parts of a receptor serve which functions.

Two modes of binding accommodate large and small agonists

The basic structure of the G protein-coupled receptors accommodates the binding of a remarkable range of ligands: amines, large and small peptides, eicosanoids, and the visual chromophore retinal. Despite the chem-

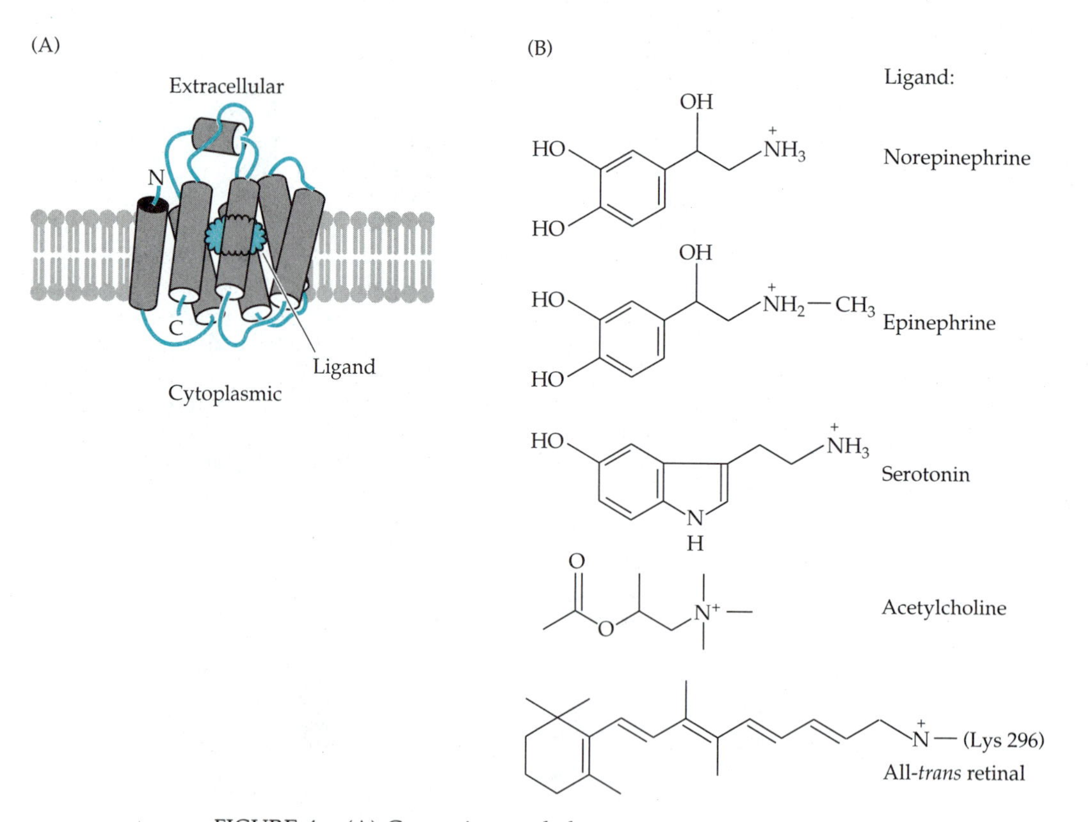

FIGURE 4. (A) G protein-coupled receptors are based on a core of seven membrane-spanning helices arranged as shown. The positioning of the more hydrophilic domains is not known. (B) Small, cationic and relatively hydrophobic ligands bind to a site within the bundle of helices sketched in (A). A large, extracellular amino terminal domain is required to bind protein ligands such as thyrotropin or luteotropin.

ical diversity of the ligands, their binding sites fall into two general groups. For those receptors that bind hydrophobic cations, including adrenergic and muscarinic cholinergic receptors and the opsins, the agonist-binding site lies deep within the bundle of membrane-spanning helices and orients the agonist more or less parallel to the plane of the plasma membrane (Figure 4). It is easy to see how the binding of an agonist at this site could twist, lever, or otherwise jiggle the packing of the helices to alter the conformation of a G protein-binding domain on the cytoplasmic face.

Such a small binding site cannot accommodate large glycoprotein agonists such as luteinizing hormone or thyrotropin. Instead, their receptors have long amino terminal extensions that are required for ligand binding. By analogy to the catecholamine receptors or rhodopsin, the amino terminal ligand-binding domain may orient a small part of the agonist in a cleft within the bundle of helices. The binding of small peptide agonists is less well understood, even though several of these receptors have been sequenced.

G proteins are regulated by cationic structures on the cytoplasmic face of the receptors

A receptor binds G proteins at a site composed of several cationic regions on the cytoplasmic face. These regions are the short second loop that connects spans 3 and 4; 5–20 amino acids in both stalks of the large loop that connects spans 5 and 6; and, probably, cytoplasmic regions near the end of span 7. Each of these regions is required for G protein activation. The remainder of the large third cytoplasmic loop and the C terminal domain are not required for regulation of G proteins. Most eukaryotic membrane-spanning proteins have several positive charges near the cytoplasmic ends of membrane-spanning sequences, but the concentration of positive charge in these key regions of the receptors is also important in regulating GTP/GDP exchange on G proteins. Consistent with this idea, short, amphipathic cationic peptides can also catalyze nucleotide exchange.

The G protein-binding domains determine the receptor's selectivity among homologous G proteins. Chimeric receptors in which short sequences from the second and third cytoplasmic loops of a donor receptor replace the homologous sequences in a recipient receptor are not only active, but they selectively activate the G protein targets of the donor receptor rather than the targets of the recipient. Thus, these two regions of cytoplasmic sequence determine which G proteins a receptor regulates. They presumably form the part of the G protein binding site that is responsible for selectivity. The independent function of these cytoplasmic domains and their ability to be regulated by the agonist-binding core of a heterologous receptor testifies to a common mechanism whereby receptors relay information through the membrane-spanning helices to the cytoplasmic surface.

Each G protein-coupled receptor has several forms

Pharmacological studies over the last fifty years have demonstrated multiple forms of receptors for a single agonist (Box D). Molecular cloning has

Box D Classification of Receptors: Types and Subtypes, K_d's and Clones

J. N. Langley, in about 1880, first proposed that hormones and neurotransmitters are recognized by **receptors**. The selectivity of receptors for biological signaling molecules has formed the basis for their classification ever since. Conceptually, this now seems trivial: cholinergic receptors respond to acetylcholine (ACh) and serotonergic receptors respond to serotonin. A more detailed examination of structure-activity relationships yields much finer distinctions, however, and these investigations have provided us with the pharmacological taxonomy of receptors that we use today.

Henry Dale first differentiated receptor subtypes in 1914 when he noted that the action of ACh at the neuromuscular junction could be selectively mimicked by **nicotine** but that its actions in the autonomic nervous system were mimicked by **muscarine**, another plant alkaloid. He postulated that the different actions of muscarine and nicotine were caused by two distinct cholinergic receptors with different specificities. The observation that each receptor also responded selectively to different competitive antagonists supported this distinction. Curare selectively blocked nicotinic receptors and atropine blocked muscarinic receptors. During the 1940s, Raymond Ahlquist differentiated α- and β-adrenergic receptors by comparison of the effects of epinephrine and norepinephrine in different tissues; this distinction was soon confirmed with the use of selective synthetic agonists and antagonists.

The ingenuity of toxic plants and pharmaceutical chemists has produced even more selective ligands that allow further subdivisions of these receptors. One class of muscarinic receptors in the brain (M1) is blocked selectively by pirenzepine. Others, in the heart (M2), are far less sensitive to pirenzepine and quite sensitive to himbacine, and a third type, in some glandular tissues (M3), fits in neither of the preceding classes. Likewise, β-adrenergic receptors have been divided into three subtypes (β_1, β_2, and β_3), α-adrenergic receptors into at least two (α_1 and α_2), and there are hints that sub-subtypes exist. Such pharmacologic classification of receptor types and subtypes has proved remarkably effective in defining structurally and functionally distinct receptors. The differential tissue distribution of different subtypes supports the physiological importance of the classification of specific receptor subtypes according to their pharmacological specificity.

The recent molecular cloning of numerous receptors has generally supported the pharmacological classification of receptors, even showing that the pharmacological studies had underestimated the diversity of receptor subtypes. Expression cloning, low stringency probing of tissue-specific cDNA libraries with known receptor probes, and the recent introduction of the polymerase chain reaction have shown that each neurotransmitter receptor is chosen from several possible subtypes. Understanding the physiological significance of a cell's use of a specific receptor sub-type is the subject of intense basic and clinical research.

demonstrated even greater diversity. Like the channels discussed in Chapter 3, each G protein-coupled receptor has multiple isoforms that are expressed in different tissues and at different stages of differentiation. The existence of multiple isoforms has forced us to consider what their distinct functions may be and has prompted a complete evaluation of their taxonomy. Why have so many receptors? Surely it is not for the convenience of the clinical pharmacologist who wants to develop specific drugs. Selectivity among natural agonists also seems inadequate to explain the differences among receptor types and subtypes, even in the case of some of the adrenergic receptors that can differentiate between epinephrine and norepinephrine.

One clear advantage of receptor subtypes is that, in different cells or tissues, a single agonist can stimulate different G proteins to initiate distinct second messenger responses. Thus, the M1, M3, and M5 muscarinic cholinergic receptors activate phospholipase C by direct action on the two G_p's. The M2 and M4 receptors, in contrast, inhibit adenylyl cyclase and stimulate a potassium channel via the G_i's and also cause weak activation of phospholipase C through these or other G proteins. The α_1 and α_2 adrenergic receptors, D1 and D2 dopaminergic receptors, H1 and H2 histaminergic receptors, and the multiple serotonergic receptors all display distinct selectivities for G proteins. In these cases, their differential responses to subtype-selective drugs probably results from fortuitous divergent evolution near the ligand binding site.

We can imagine other reasons for the existence of multiple receptor subtypes. For example, β_1, β_2, and β_3 adrenergic receptors all catalyze the activation of G_s, leading to the activation of adenylyl cyclase and calcium channels. Variations in the ability to activate G proteins would provide different gradations of response in different tissues. Last, receptor subtypes probably vary in the rates or extents to which they are desensitized in response to chronic exposure to agonists. This would allow some cells to be more acutely responsive to transient increases in ligand concentration while allowing others to maintain stable responses under relatively constant levels of stimulation.

Effector Proteins: Enzymes, Channels, or Transporters that Regulate the Metabolism of Second Messengers

When G_s was identified in 1977, the only proteins known to be regulated by GTP were adenylyl cyclase and the retinal cyclic GMP phosphodiesterase. The number of recognized G protein-regulated effectors is now about ten and growing quickly. In contrast to G proteins and the G protein-coupled receptors, the effectors are a remarkably diverse group. They include enzymes that synthesize or degrade second messengers, ion channels that alter membrane potential or allow the influx of the regulatory ion Ca^{2+}, and membrane transport proteins. Structurally, effectors may be integral membrane proteins or peripheral proteins that interact only weakly with the membrane surface. Moreover, G proteins regulate effectors by diverse mechanisms, and new forms of G protein-effector coupling continue to emerge.

Adenylyl cyclase is both stimulated and inhibited by G proteins

Adenylyl cyclase is responsible for producing the ubiquitous cytoplasmic second messenger **cyclic AMP (cAMP)**, the molecule for which the term "second messenger" was coined by Earl Sutherland, its discoverer. Adenylyl cyclase is found in almost every animal cell. It is a relatively large (110 kD) integral membrane glycoprotein composed of two bundles of six membrane-spanning sequences and two large cytoplasmic domains arranged as a pseudodimer. This structure resembles several membrane transport proteins or part of an ion channel (Chapter 3). Why an enzyme whose substrate and product are both soluble should have such a structure is unknown, but it suggests that adenylyl cyclase evolved from a transport protein or that it retains some transport function in addition to its known enzymatic activity.

G proteins both stimulate and inhibit adenylyl cyclase and its regulation is an example of the complexity of G protein pathways (Figure 5). G_s stimulates adenylyl cyclase in response to a large number of receptors: β-adrenergic, D1 dopaminergic, E-type prostaglandins, VIP, and glucagon, among others. Activated G_s forms a cyclase-G_s-GTP complex whose activity is 20-fold greater than that of the cyclase alone. There are at least two forms of α_s—alternative products of RNA splicing—that are differentially expressed in the nervous system. These G_s isoforms differ slightly in kinetic properties and in responsiveness to receptors. A variant of G_s known as G_{olf} is the product of a separate gene that is expressed only in the olfactory epithelium, where adenylyl cyclase mediates the effect of odorant receptors.

Another large group of receptors inhibits adenylyl cyclase via G_i. These include M2 and M4 muscarinic cholinergic, α_2-adrenergic, 5HT1B serotonergic, and enkephalinergic receptors. The mechanism of inhibition by G_i is still somewhat controversial. One well-documented mode of inhibition is simply the chelation of activated α_s by the βγ subunits that are freed when G_i is activated and the activated α_i dissociates (see Box C). Thus, inhibition in most cases is simply a blockade of activation. However, α_i also mediates inhibition of adenylyl cyclase activity by a distinct mechanism that does not depend on α_s. This mechanism is not understood and may be indirect.

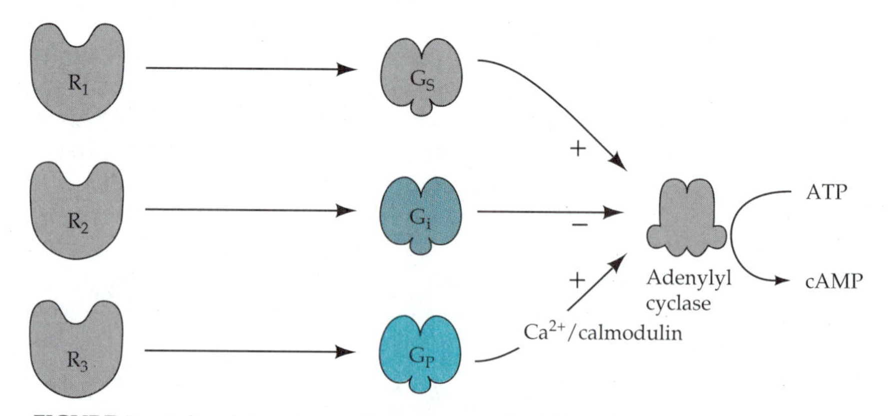

FIGURE 5. Adenylyl cyclase activity is controlled directly by at least three signaling pathways and indirectly by others. Adenylyl cyclase is activated by G_s and, in neurons and some other cells, by Ca^{2+}-calmodulin. This allows the cyclase to be activated by signals that either open Ca^{2+} channels or promote inositol-trisphosphate release. The cyclase is inhibited by G_i, which inhibits activation of G_s and also inhibits the cyclase by some other mechanism.

In neuronal tissue, adenylyl cyclase is also activated by **calmodulin**, a soluble Ca^{2+}-binding regulatory protein. In *Drosophila*, the genetic loss of the neuronal calmodulin-sensitive cyclase in the *rutabaga* mutant causes a learning deficit. Stimulation of neuronal adenylyl cyclase by calmodulin potentiates stimulation by G_s. This potentiation provides a mechanism whereby receptors that stimulate adenylyl cyclase via G_s and receptors that elevate cytosolic Ca^{2+} (by any of several mechanisms) can reinforce each other. Such potentiation of cAMP synthesis could be a short-term mechanism for learning.

A cyclic GMP phosphodiesterase mediates phototransduction in vertebrate photoreceptor cells

In the vertebrate retina, a photosensitive G protein-coupled receptor, rhodopsin, initiates a signaling pathway that ultimately leads to the closing of cation channels in the plasma membrane and consequent hyperpolarization of photoreceptor cells (Figure 6). Rhodopsin is covalently bound to an endogenous antagonist, the chromophore 11-*cis*-retinal. Light isomerizes retinal to the all-*trans* form, which is an agonist. Activated rhodopsin activates G_t, which in turn activates a cyclic GMP (cGMP) phosphodiesterase. Phosphodiesterase activity decreases the concentration of cGMP, which activates a cell surface cation channel (see Box A in Chapter 7) and, in the dark, holds membrane potential at a relatively depolarized level.

The cGMP phosphodiesterase of the photoreceptor cell is a peripheral protein bound to the photoreceptor membrane. It is an $\alpha\beta\gamma_2$ tetramer whose catalytic α and β subunits are inhibited by the γ subunits. In contrast

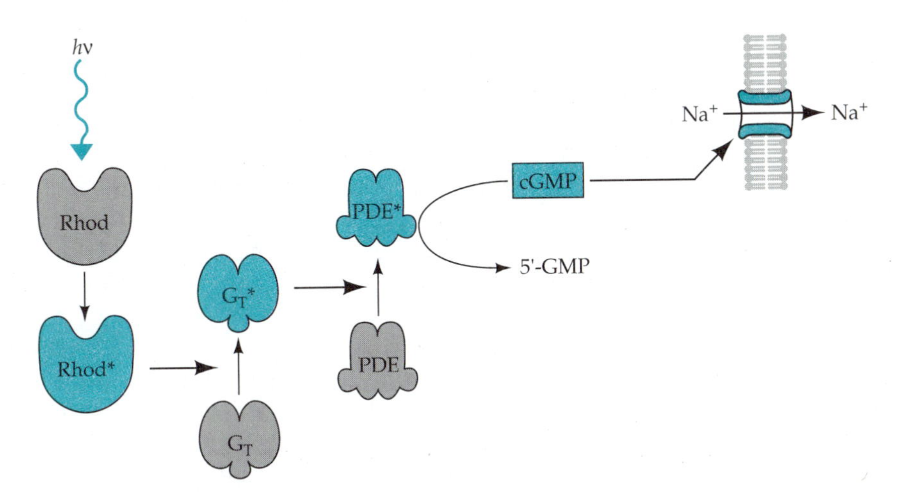

FIGURE 6. Rhodopsin is activated when a photon (hν) converts bound 11-*cis*-retinal to all-*trans*-retinal. Rhodopsin catalyzes the activation of G_t at the surface of the photoreceptor membrane (the disc membrane in vertebrate rods), a specialized structure in which rhodopsin and G_t are highly concentrated. Active G_t activates a cyclic GMP (cGMP) phosphodiesterase, which increases the hydrolysis of cGMP. The decreased concentration of cGMP decreases the current through a cGMP-activated Na^+ channel in the plasma membrane. Consequent hyperpolarization causes a decrease in the rate of neurotransmitter release.

to the mechanism discussed for adenylyl cyclase, activated G_t does not activate phosphodiesterase directly, but instead binds to and removes the inhibitory γ subunits. Deactivation of G_t by GTP hydrolysis releases the phosphodiesterase γ subunits, which then bind and inhibit the enzyme.

The visual phototransduction system is, along with the adenylyl cyclase system, a principal prototype for G protein-mediated signaling. The high concentration of the signaling proteins in photoreceptor cells has allowed experiments that have provided many of our insights into the mechanism of G protein-mediated signaling. The phosphodiesterase system is specialized to provide rapid detection, rapid return to ground state, and the huge dynamic range that is required for vision both in dim and bright light. Key features in the system are (1) the ability of a single rhodopsin molecule to catalyze the activation of over 30,000 molecules of G_t, (2) a complex mechanism for desensitization of rhodopsin through covalent modification, and (3) a neuronal network in the retina itself that maximizes acuity.

The visual system uses two types of photoreceptor cells. Rods mediate sensitive, monochromatic vision, and cones mediate the less sensitive perception of color. Rods and cones contain separate isoforms of α_t and may contain different phosphodiesterases. Cone cells contain one of three different rhodopsin-like molecules. Each color-specific opsin binds the same chromophore, 11-*cis*-retinal, but alters its absorption spectrum by providing a distinctively different environment within the protein.

Phospholipase C produces two second messengers: Inositol trisphosphate and diacylglycerol

Phosphatidylinositol bisphosphate (PIP_2), a phospholipid found primarily on the inner face of the plasma membrane, is hydrolyzed by **phospholipase C** to form two important second messengers. **Inositol trisphosphate (IP_3)** activates a calcium channel in a specialized portion of the endoplasmic reticulum to cause the release of Ca^{2+} into the cytosol. The other product of the phospholipase, **diacylglycerol**, activates **protein kinase C** (see Chapter 7). There are at least three families of PIP_2-specific phospholipase C's (PLCs) of which one, the PLC-β's, are regulated by G proteins. The PLC-β's are peripheral membrane proteins of molecular mass 120 kD.

Two classes of G proteins regulate PLC through separate pathways, which have been defined by the sensitivity of the G proteins to pertussis toxin. Thrombin, M2 and M4 muscarinic cholinergic receptors, and several other receptors act through G proteins that are sensitive to pertussis toxin, probably the G_i's. M1 and M3 muscarinic receptors, α_1-adrenergic receptors and neurokinin receptors stimulate PLC activity through the action of pertussis toxin–insensitive G proteins known as G_p's. Stimulation of phospholipase C through the G_p pathway is usually more robust than through G_i. Two G_p's have recently been identified as products of separate genes. The purified G_p's respond to muscarinic receptors and activate phospholipase C in vitro. Other G_p's may respond to different receptors.

G proteins regulate ion channels independently of second messengers

G proteins regulate ion channels by two distinct mechanisms, both processes that are particularly important for neuronal function. By altering the production of second messengers that regulate protein kinases, G proteins influence channel function through phosphorylation reactions (Chapters 3, 7, and 14). They also regulate channels independently of second messengers. The first evidence for this independent pathway came from electrophysiological studies on the potassium channels in heart muscle that are regulated by acetylcholine (ACh).

ACh, released from vagus nerves, acts on muscarinic receptors in the heart to activate these channels and slow the heart rate (the system used by Loewi to identify Vagusstoff as ACh; see Chapter 4). When these channels were studied by cell-attached patch clamp recording, ACh applied outside the patch was unable to activate them, whereas they could by activated by ACh in the pipette (Figure 7). In a whole-cell patch clamp, they could be activated by ACh, but only if GTP was included in the pipette. Moreover, the response to ACh was blocked by pertussis toxin. Taken together, these results suggested that a G protein mediates the response to ACh but does so without the participation of a diffusible, cytoplasmic second messenger. This idea was substantiated by the demonstration that purified α subunit of the G protein, preactivated by GTPγS, could activate potassium channels from heart muscle when added directly to the cytoplasmic surface of an isolated membrane patch. Whether the G protein binds directly to the channel or to a protein associated with the channel is not known.

A number of channels are now known to be regulated by this direct mechanism, including a variety of potassium, calcium, and possibly sodium channels. The calcium channels that regulate transmitter release in sympathetic nerve terminals, for example, appear to be inhibited by a G protein that is activated by α-adrenergic receptors for norepinephrine (see Chapter 5).

Why have a G protein regulate a channel directly when there are so many second messenger pathways that already regulate channels? One good answer is speed. For example, activation of the calcium channel in cardiac muscle by G_s significantly precedes its activation through cAMP, even though both responses are initiated by β-adrenergic stimulation. Why regulate ion channels through G proteins rather than directly, as is the case for the GABA or ACh receptor? This question is harder to answer, but a plausible explanation is suggested by the fact that G protein-regulated channels respond to G proteins that simultaneously mediate other regulatory signals. For example, calcium channels are activated by G_s, which also activates adenylyl cyclase; potassium channels are activated by α_i's and α_o's that inhibit adenylyl cyclase and stimulate phospholipases C and A_2. Regulation through G proteins thus links channels into a large, highly coordinated signaling network.

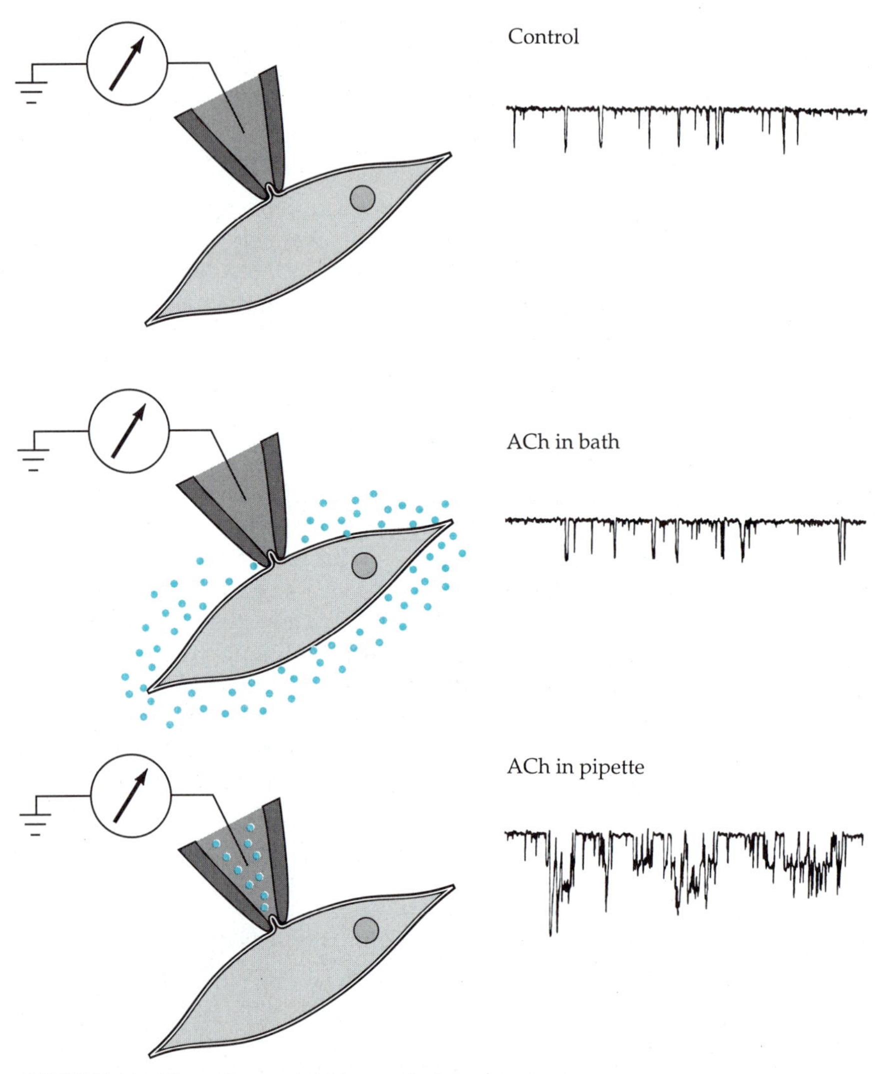

FIGURE 7. The effects of ACh applied via the bath and via the pipette on the potassium channel openings in rabbit atrial muscle cells. Downward deflections represent channel opening. ACh applied in the pipette opens the channel, whereas ACh applied outside the pipette does not. Pertussis toxin treatment of the cell blocks the effect of ACh added to the pipette. (After M. Soejima and A. Noma, 1984. Pflügers Arch. 400: 424–431.)

How many other effectors are there?

G proteins have been implicated in various physiological responses on the basis of the sensitivity of the response to pertussis or cholera toxin, its dependence on guanine nucleotides, or its stimulation by nonhydrolyzable GTP analogs or Al^{3+}/F^- (see Box E). None of these indicators is unambiguous, nor does any of them indicate that a G protein is acting directly upon an effector. Nevertheless, the number of probable G protein targets continues to increase.

The most likely "not quite proven" G protein-regulated effector is the phospholipase A_2 that liberates arachidonic acid, the precursor of prostaglandins, thromboxanes, and leukotrienes. These compounds may be presynaptic mediators both in the neurons in which they are synthesized and in adjacent neurons (see Chapter 7). The evidence that G proteins regulate phospholipase A_2 proteins includes sensitivity to pertussis toxin and activation by Al^{3+}/F^-. Transport proteins are another tantalizing group of "not quite proven" effectors. Indirect evidence suggests that glucose transporters, a Mg^{2+} transporter, and the Na^+/proton exchanger are regulated by G proteins. Regulation may be direct, or novel intermediary signaling molecules may be involved.

G Proteins Are the Basis of a Branched Signal-Processing Network

The many receptors, G proteins, and effectors in a plasma membrane form a complex communications network that balances multiple positive and negative inputs to yield an appropriate array of cellular responses (Figure 8). The network has the capacity to integrate signals from multiple receptors to yield a single response or to transmit the signal from a single receptor to multiple effectors. By differentially expressing G proteins and effectors, different cells can respond to the same array of neurotransmitters and hormones with distinctly different spectra of second messenger responses. Feedback regulation and receptors for growth factors, steroid hormones, and other trophic hormones can further regulate the system. The G protein network thus displays a complexity that almost rivals that of the nervous system itself!

Signals can be integrated by either G proteins or effectors

Convergent regulation of a single G protein by different receptors was first demonstrated in the late 1960s, before the existence of G proteins was recognized. Experiments on adipocytes, which express five different receptors that activate adenylyl cyclase, showed that maximal stimulation of the cyclase by one receptor could not be increased by concurrent stimulation of another. All of the receptors thus act upon a common pool of adenylyl cyclase. When G_s was discovered in the late 1970s, it became clear that these receptors regulate a common pool of G_s and that G_s integrates and conveys their signals to the cyclase. Many G proteins are now known to use this pattern of integration.

Effectors also act as integrators by responding to multiple G proteins. For example, the PLC that hydrolyzes PIP_2 responds to both G_p's and to one or more pertussis toxin–sensitive G proteins. A more complex example is the L-type calcium channel in dorsal root ganglion cells. When this channel is inhibited by bradykinin, G_o, $G_{i,1}$ or $G_{i,2}$ can each act as a transducer with approximately equal effect. When the channel is inhibited by neuropeptide Y, G_o is more efficient than $G_{i,1}$, and $G_{i,2}$ has only slight activity. The different G proteins apparently are differentially sensitive to receptors.

Box E Nucleotide Analogs, Toxins, and Fluoride Ion Are Valuable Probes of G Protein-Mediated Signaling

A wide variety of reagents is available for detecting the participation of G proteins in neuronal signaling and for manipulating the activation and deactivation of G protein pathways. Each of these probes acts at one or more of the steps in the GTPase cycle.

Nonhydrolyzable analogs of GTP, such as GTPγS or Gpp(NH)p (Figure A), activate G proteins essentially irreversibly and cause persistent and extensive activation of effectors. The high affinity of these analogs makes them especially useful as activators of signaling and as radioligands for the measurement of nucleotide binding to G proteins. They can be microinjected into cells, and activation of signaling pathways via GTPγS is frequently the first hint that a G protein is involved.

Cholera toxin has a similar effect. It prevents the deactivation of G_s by catalyzing the ADP-ribosylation of its α subunit near the GTP binding site and thereby inhibits GTP hydrolysis. Cholera toxin thus causes extensive activation even in the absence of stimulation by receptors. Cholera toxin is relatively specific for G_s, and stimulation of a cellular response by treatment with cholera toxin indicates the involvement of cyclic AMP.

Fluoride ion activates G proteins by activating the GDP-liganded protein. Al^{3+} plus F^- binds GDP-liganded G protein at the site on the α subunit that is usually occupied by the γ-phosphoryl group of GTP, forming a GDP-Al^{3+}-F_3^- complex that mimics bound GTP (Figure A). Activation of G proteins by F^- is thus a convenient and general means of stimulating G protein-mediated pathways.

Pertussis toxin is a highly selective probe for many G proteins. It catalyzes ADP-ribosylation at a cysteine residue near the carboxyl terminus of the α subunits of G_i, G_o and G_t. ADP-ribosylation does not substantially alter the G protein's activity but does block the ability of receptors to promote nucleotide exchange. Thus, the ability of pertussis toxin to inhibit signaling is good evidence of the involvement of a G protein. Pertussis toxin is doubly useful. When [^{32}P]NAD is used as a substrate, the toxin transfers [^{32}P]ADP-ribose to those G proteins that are toxin substrates, thereby identifying them. (Cholera toxin can also be used in this way, but its selectivity for G_s has made it less generally useful.)

The large number of natural toxins that attack at each step of G protein-mediated signaling pathways testifies to their importance in cellular regulation. The mastoparans, peptide toxins from wasp venoms, mimic agonist-liganded receptors by catalyzing GTP/GDP exchange on G proteins. Other organisms produce toxic "effectors." Several pathogenic bacteria secrete adenylyl cyclases that are efficiently endocytosed by mammalian cells, and animals in several phyla produce phospholipases. Forskolin, a plant terpene, directly activates adenylate cyclase. These toxins also provide useful reagents for studies of cellular signaling.

Why the complexity? In some cases, the different G proteins that regulate a single effector clearly respond preferentially to different receptors. In other cases it is not so clear. Different G proteins display different patterns of activation kinetics that may be important in determining quantitative aspects of their responses. This may be the key difference among the three G_i's or between the two forms of G_s. The differential expression of closely related G proteins in different cells or during different stages of development suggest that their distinctive properties are indeed important. The complexity of the signaling network, which is so important for

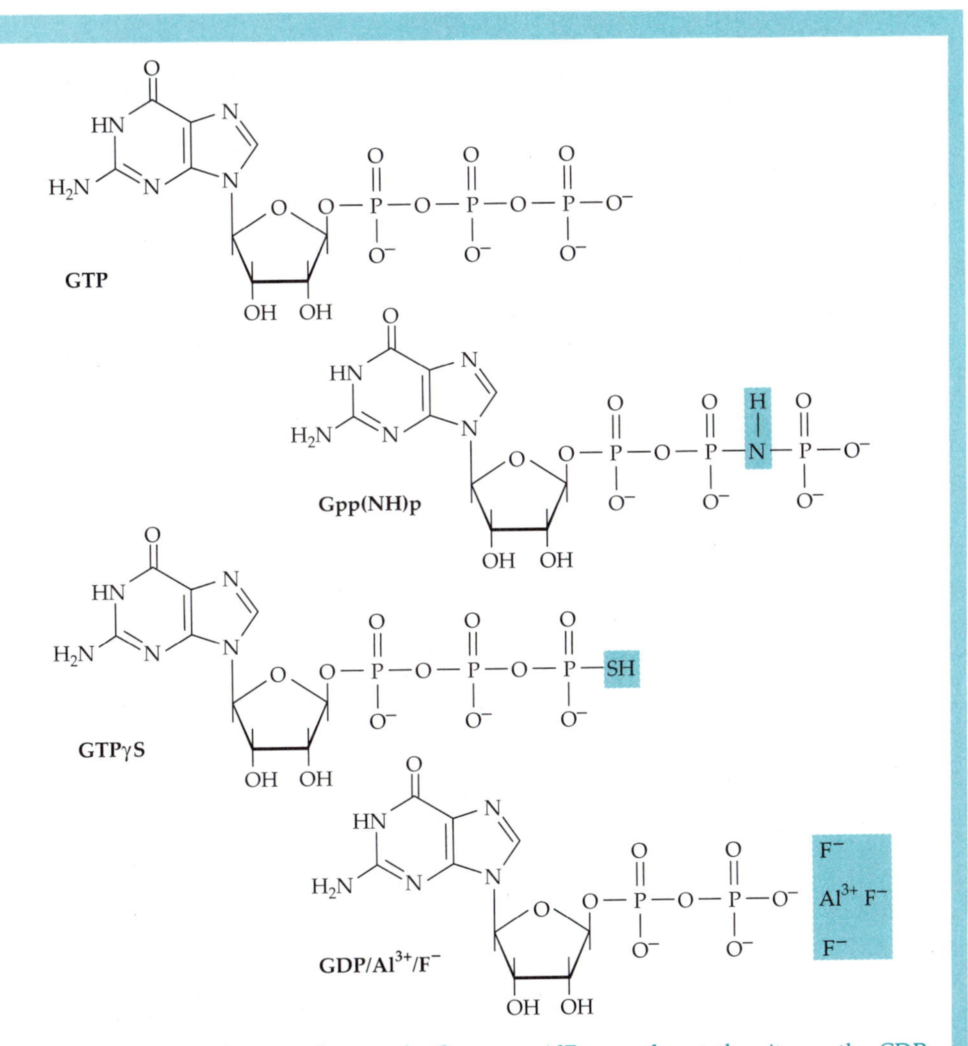

FIGURE A. GTP analogs used to study G proteins. GTPγS and Gpp(NH)p bind to G proteins but are not hydrolyzed, and thus cause persistent activation. Al^{3+} plus F^- bind as an AlF_3 complex at the site on the GDP-liganded G protein where the terminal phosphoryl group of GTP would bind, thereby mimicking the structure of GTP.

cellular regulation, makes both biochemical and genetic approaches to these questions challenging.

Divergent coupling: signals from one receptor can stimulate multiple effectors

A single receptor initiates a characteristic physiological response, but it may do so by generating multiple second messenger outputs within a single cell. To direct an incoming signal simultaneously to several effectors, cells use divergent coupling at both the receptor-G protein and the G

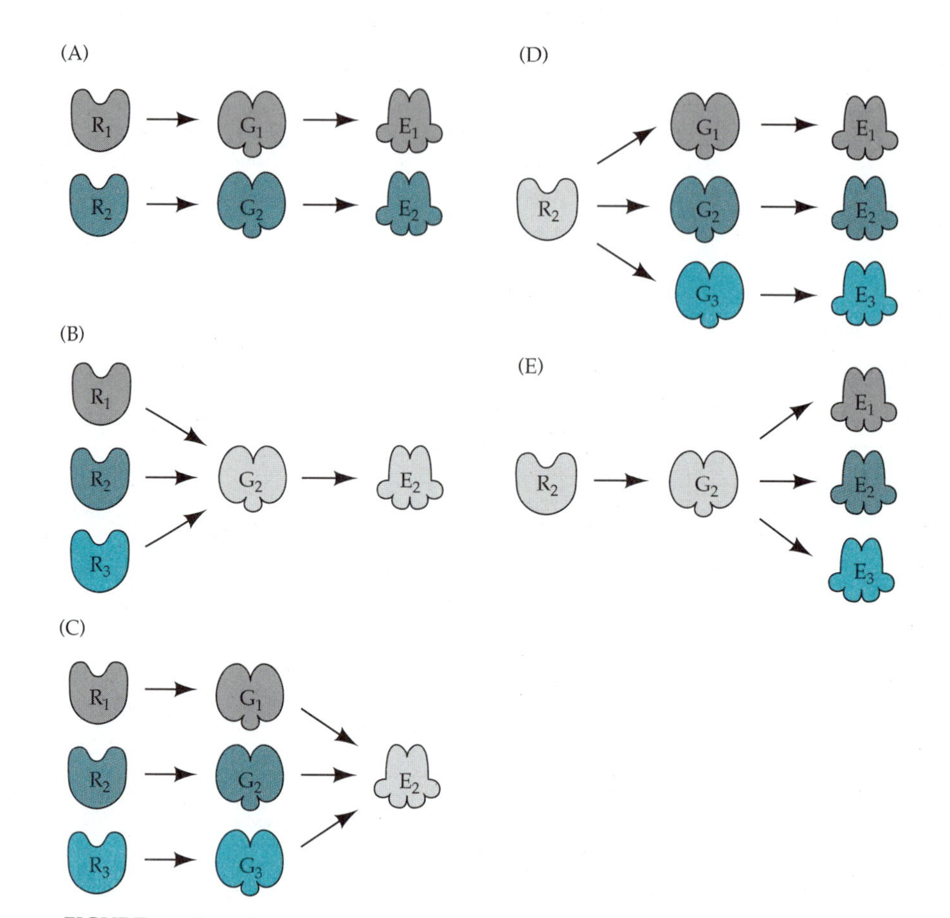

FIGURE 8. Signals can converge or diverge at each level of a G protein signaling network. In a cell, these pathways coordinate multiple second messenger outputs according to multiple receptor inputs.

protein-effector interfaces. The idea that a single receptor can regulate several G proteins arose from experiments in which purified receptors were coreconstituted in phospholipid vesicles with one of several different G proteins. The introduction of recombinant receptors into novel cells where they interact with endogenous G proteins has also demonstrated the spectrum of G protein targets of a single receptor.

The most striking example of divergent coupling at the G protein-effector level is the ability of a single G protein to regulate multiple ion channels as well as the previously recognized effector enzymes. Such channel regulation is of particular importance in neuronal cells. G_s, the G_i's, and G_o can all regulate one or more ion channels depending upon the cell in which each is expressed, providing a richer output of information from a single hormonal signal.

Desensitization: Complex Feedback Regulation of Complex Pathways

Biological sensors modulate their sensitivity in response to the intensity and frequency of stimulation. G protein-coupled receptors are subject to several different mechanisms of feedback regulation that reflect both the stimulation of the receptor itself and the overall stimulation of the downstream signaling pathway. Feedback regulation is usually referred to as desensitization or refractoriness because it is usually observed as a decreased response to continued stimulation (Figure 9). Two criteria help classify the diverse mechanisms of desensitization that are observed in G protein-mediated signaling: (1) Is desensitization unique to a single receptor or does it affect multiple receptors that use the same pathway? (2) Is desensitization achieved through regulation of proteins in a signaling pathway or does it involve regulation of the concentration of these components?

Desensitization that is limited to a single receptor (*homologous* desensitization) most often involves modification of the receptor itself. Phosphorylation of a receptor by a protein kinase that is specific for the agonist-

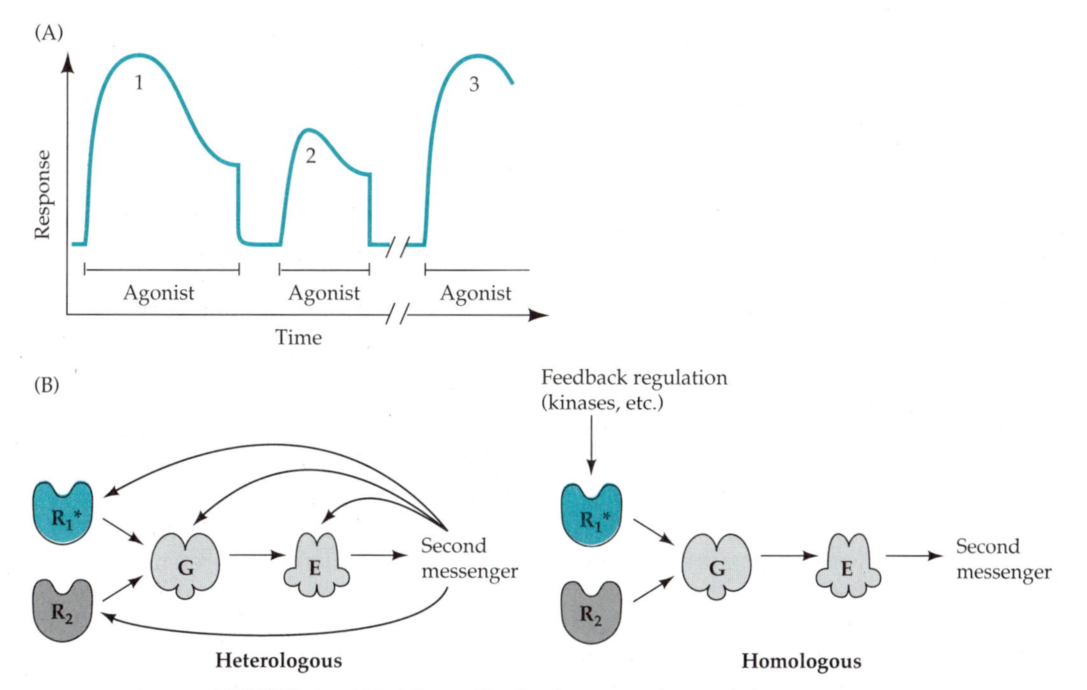

FIGURE 9. (A). Most cells display some form of desensitization in response to neurotransmitter. Upon exposure to agonist, the response peaks and then declines to some tonic level above basal but below the maximum (1). If the agonist is removed, desensitization generally remains, such that a repeat exposure elicits a smaller maximal response (2). In the absence of agonist for an extended time, the cell recovers from desensitization (3). (B). Desensitization can be restricted only to the receptor for the stimulating agonist (homologous), or an entire signaling pathway can be desensitized to input from several receptors (heterologous). Generally, heterologous desensitization is initiated by the second messenger product of the signaling pathway and can be manifested as a decrease of activity of receptors, G protein, or effector. Homologous desensitization usually is some direct modification (covalent or conformational) of the agonist-liganded receptor. Homologous desensitization can be intramolecular, as in the case of the nicotinic receptor, or it can reflect the action of an extrinsic feedback regulator such as a protein kinase.

liganded form may decrease the receptor's regulatory activity. In the case of rhodopsin, phosphorylation both decreases its activity and enhances the binding of an inhibitory protein, arrestin. Similar kinases and an arrestin homologue also regulate other receptors, although their mechanisms of action are less clear. An alternative mode of homologous desensitization is agonist-induced endocytosis, which essentially removes receptors from contact with G proteins and effectors. Receptor phosphorylation is reversible by phosphatases, and endocytosed receptors can be recycled to the plasma membrane. In extreme cases, however, prolonged exposure to agonist leads to receptor "down-regulation," the actual destruction of receptors, which may take place after their endocytosis.

Stimulation of one receptor may also diminish the responsiveness of all the receptors that utilize its signaling pathway, a process referred to as *heterologous* desensitization. For example, stimulation of a β-adrenergic receptor in some cases diminishes the stimulation of adenylyl cyclase by other receptors (β-adrenergic, adenosine, and prostaglandin E). Heterologous desensitization is usually initiated by the second messenger that is the product of the signaling pathway. Thus, heterologous desensitization of adenylyl cyclase systems is generally initiated by cAMP acting upon cAMP-dependent protein kinase (see Chapter 7). Heterologous desensitization is probably a much more general phenomenon than is homologous desensitization and is frequently initiated in response to lower levels of stimulation. It provides homeostatic feedback control of a pathway regardless of the source of the stimulatory input.

Desensitization of G protein-coupled systems can occur over seconds to minutes, but long-term regulation of G protein-coupled pathways by regulation of the synthesis of the components of the pathways is also common. All components of G protein pathways—receptors, G proteins, and effectors—are regulated according to the level of input from agonists. These long-term regulatory changes complement acute desensitization to allow a cell to regulate its responses as stimuli and metabolic needs change over time.

The second messenger product of one pathway can also alter the function of a separate pathway. For example, stimulation of protein kinase C by Ca^{2+} and diacylglycerol can feed back on adenylyl cyclase pathways to either enhance or diminish responses, depending upon the cell in question and its needs. Similar effects of cAMP can cause changes in cytoplasmic Ca^{2+} concentrations.

The neuron's G protein-mediated signaling network is strikingly reminiscent of the nervous system itself, encompassing multiple interactive pathways that receive, sort, and amplify information. As we will see in following chapters, the spectrum of second messengers generated by those pathways controls diverse cellular activities: synaptic reception and transmission, motility, and differentiation. The importance of G protein systems makes them tempting subjects for pharmacologic intervention, and their complexity challenges those who are so tempted.

Summary

Many neuronal receptors use a family of GTP-binding proteins, known as G proteins, to convey their signals to intracellular effector proteins. These receptors promote the binding of GTP to specific G proteins. GTP binding activates the G protein such that it can then regulate the activities of specific effector proteins. Effectors include ion channels, enzymes that synthesize cytoplasmic second messengers, and transport proteins. G proteins are thus involved in regulating the synthesis and release of neurotransmitters, the sensitivity of synaptic receptors, general cellular metabolism, and cellular differentiation and growth.

Over 100 receptors communicate through G proteins. They include receptors for catecholamines and most other biogenic amines, muscarinic cholinergic receptors, and receptors for eicosanoids and for numerous peptide hormones and neuromodulators. The number of distinct G proteins is probably near 20. There probably exists a similar number of G protein-regulated effectors and, hence, a large number of cytoplasmic second messengers. Individual G proteins amplify signals from multiple receptors and direct them to one or more appropriate effector proteins. Signals from different receptors can be integrated through one or more G proteins to stimulate a single second messenger pathway. Stimulatory and inhibitory signals can be balanced at the G protein level to yield a damped messenger output. Information from a single receptor can also be directed to several different effector systems using one or more G proteins as transducers in the pathway. G proteins are thus the basis of a complex information processing network in the neuronal plasma membrane.

References

General References

Bourne, H., Sanders, D. and McCormick, F. 1991. The GTPase superfamily: Conserved structure and molecular mechanism. Nature 349: 117–127.

*Gilman, A. G. 1987. G proteins: Transducers of receptor-generated signals. Annu. Rev. Biochem. 56: 615–649.

Kaziro, Y., Itoh, H., Kozasa, T., Nakafuku, M. and Satoh, T. 1991. Structure and function of signal-transducing GTP-binding proteins. Annu. Rev. Biochem. 60: 349–400.

Ross, E. M. 1989. Signal sorting and amplification through G protein-coupled receptors. Neuron 3: 141–152.

Receptors

Buck, L. and Axel, R. 1991. A novel multi-gene family may encode odorant receptors: A molecular basis for odor recognition. Cell 65: 175–187.

Dohlman, H. G., Thorner, J., Caron, M. G. and Lefkowitz, R. J. 1991. Model systems for the study of seven-transmembrane-segment receptors. Annu. Rev. Biochem. 60: 349–400.

*Hulme, E. C., Birdsall, N. J. M. and Buckley, N. J. 1991. Muscarinic receptor subtypes. Annu. Rev. Pharmacol. Toxicol. 30: 633–673.

Perkins, J. (ed.). 1991. *The β-Adrenergic Receptors.* Humana Press, Clifton, NJ.

Strader, C. D., Sigal, I. S. and Dixon, R. A. F. 1989. Structural basis of β-adrenergic receptor function. FASEB J. 3: 1825–1832.

Vision

Nathans, J., Piantanida, T., Eddy, R., Shows, T. and Hogness, D. 1986. Molecular genetics of inherited variation in human color vision. Science 232: 203–210.

*Stryer, L. 1986. Cyclic GMP cascade of vision. Annu. Rev. Neurosci. 9: 87–119.

Desensitization

*Perkins, J., Hausdorff, W. and Lefkowitz, R. 1991. Mechanisms of ligand-induced desensitization of β-adrenergic receptors. In J. P. Perkins (ed.), *The β-Adrenergic Receptors.* Humana Press, Clifton, NJ, pp. 73–124.

Phospholipase C

Berridge, M. J. and Irvine, R. F. 1989. Inositol phosphates and cell signalling. Nature 341: 197–205.

Maeda, N., Kawasaki, T., Nakade, S., Yokota, N., Taguchi, T., Kasai, M. and Mikoshiba, K. 1991. Structural and functional characterization of inositol 1,4,5-trisphosphate receptor channel from mouse cerebellum. J. Biol. Chem. 266: 1109–1116.

Rhee, S., Suh, P.-G., Ryu, S.-H. and Lee, S. 1989. Studies of inositol phospholipid-specific phospholipase C. Science 244: 546–550.

Smrcka, A., Hepler, J., Brown, K. and Sternweis, P. 1991. Regulation of polyphosphoinositide-specific phospholipase C activity by purified Gq. Science 251: 804–807.

G protein-gated channels

*Brown, A. M. and Birnbaumer, L. 1990. Ionic channels and their regulation by G proteins. Annu. Rev. Physiol. 52: 197–213.

GTPase cycle

Higashijima, T., Ferguson, K. M., Smigel, M. D. and Gilman, A. G. 1987. The effect of GTP and Mg^{2+} on the GTPase activity and the fluorescent properties of G_o. J. Biol. Chem. 262: 757–761.

Higashijima, T., Ferguson, K. M., Sternweis, P. C., Smigel, M. D. and Gilman, A. G. 1987. Effects of Mg^{2+} and the βγ-subunit complex on the interactions of guanine nucleotides with G proteins. J. Biol. Chem. 262: 762–766.

Kaziro, Y. 1978. The role of guanosine 5'-triphosphate in polypeptide chain elongation. Biochim. Biophys. Acta 505: 95–127.

Adenylyl cyclase

Krupinski, J., Coussen, F., Bakalyar, H. A., Tang, W.-J., Feinstein, P. G., Orth, K., Slaughter, C., Reed, R. R. and Gilman, A. G. 1989. Adenylyl cyclase amino acid sequence: Possible channel- or transporter-like structure. Science 244: 1558–1564.

Multiple signaling pathways

Ashkenazi, A., Peralta, E. G., Winslow, J. W., Ramachandran, J. and Capon, D. J. 1989. Functionally distinct G proteins selectively couple different receptors to PI hydrolysis in the same cell. Cell 56: 487–493.

Lechleiter, J., Hellmiss. R., Duerson, K., Ennulat, D., David, N., Clapham, D. and Peralta, E. 1990. Distinct sequence elements control the specificity of G protein activation by muscarinic acetylcholine receptor subtypes. EMBO J. 9: 4381–4390.

Senogles, S. E., Spiegel, A. M., Padrell, E., Iyengar, R. and Caron, M. G. 1990. Specificity of receptor-G protein interactions. Discrimination of G_i subtypes by the dopamine D_2 receptor in a reconstituted system. J. Biol. Chem. 265: 4507–4514.

Wong, S. K.-F., Parker, E. M. and Ross, E. M. 1990. Chimeric muscarinic cholinergic:β-adrenergic receptors that activate G_s in response to muscarinic agonists. J. Biol. Chem. 265: 6219–6224.

Yatani, A., Imoto, Y., Codina, J., Hamilton, S. L., Brown, A. M. and Birnbaumer, L. 1988. The stimulatory G protein of adenylyl cyclase, G_s, also stimulates dihydropyridine-sensitive Ca^{2+} channels. J. Biol. Chem. 263: 9887–9895.

7

Second Messengers and Neuronal Function

Mary B. Kennedy

THE NETWORK of interacting biochemical reactions that enables living cells to detect diverse stimuli, integrate information, and respond to their environment is in many ways like a nervous system in miniature. External stimuli such as growth factors, neurotransmitters, and mechanical stimuli trigger changes in the concentration or activity of many thousands of cell signaling molecules in the cytoplasm. These in turn regulate the metabolism, synthetic activity, and behavior of the cell, working through molecular "circuits" that are the product of over three billion years of biological evolution.

The molecular details of these signaling reactions are only partly understood, but we know already that they are amazingly sophisticated. Receptors, G proteins, second messengers, and protein kinases interact through subtle and highly interconnected biochemical reactions that operate throughout the life of the cell. The number of different types of signaling molecules in the cell is still unknown and research continues to reveal new components. (The number of distinct protein kinases in a eukaryotic cell, for example, may be as many as a thousand.) As we learn more and more we find that even individual molecules can, in a sense, store information by being covalently modified in different parts of their sequence or by adopting distinct conformations. Were we able to view these reactions in real time as they change inside a living cell, it would be like watching the lights of Los Angeles from the air at night with myriad fluctuations in concentration and activity occurring in different regions every second, some organized into logical patterns, others seemingly at random.

Although cell signaling reactions are present in all cells, it is likely that they reach their highest complexity in the cells of the nervous system. The storage and retrieval of information, which is the primary function of the nervous system, has evolved out of the same cell signaling pathways that are present in the cells of the liver or gut. Thus, we surmise that the intracellular signals that enable a migrating cell to steer along specific paths will be present in their most highly developed form in the growth cone, which achieves the most complex tasks of navigation. Similarly, it is reasonable to suppose that the molecular circuitry by which all cells, even bacteria, can record changes and in a sense "remember," will reach its highest level of diversification and subtlety within neuronal synapses.

An overview of major pathways within the interactive network that regulates and coordinates neuronal function is presented in Figure 1. In this chapter we will focus on the four major **second messenger pathways** in neurons: those mediated by **cyclic AMP;** by **cyclic GMP;** by **calcium ion;**

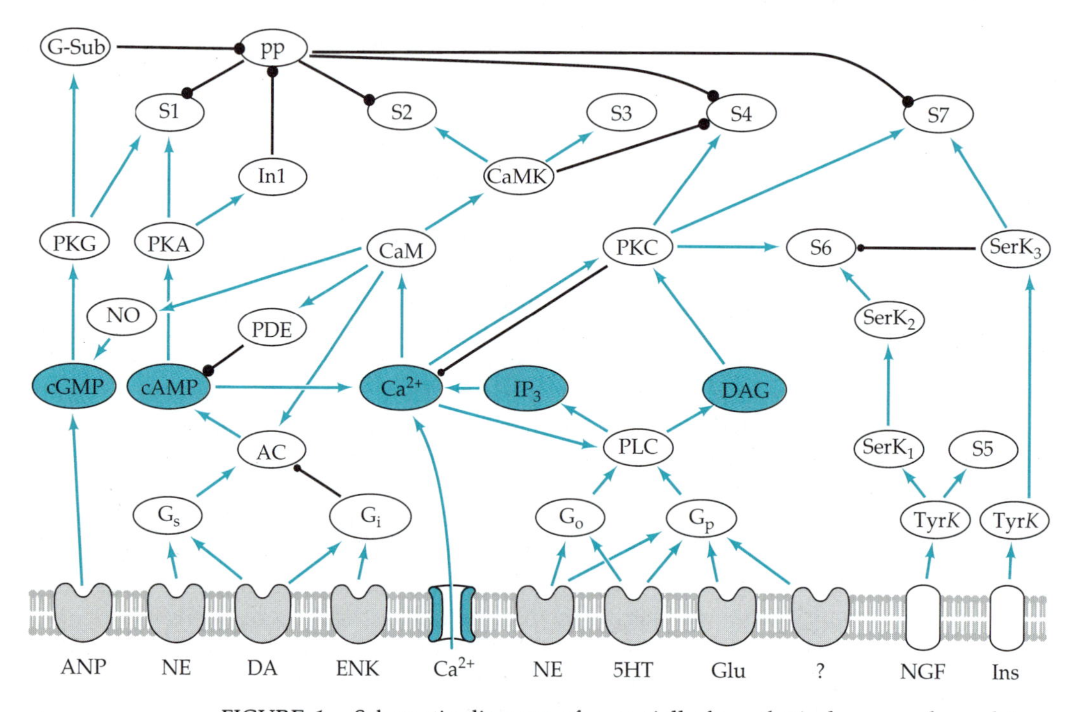

FIGURE 1. Schematic diagram of a partially hypothetical neuronal regulatory network. The diagram illustrates convergence, divergence, and amplification through enzyme cascades. Information can converge from several receptors onto one or a few effector molecules. Conversely, information diverges when one receptor molecule activates two or more distinct regulatory pathways. Enzyme cascades, in which a series of enzymes activate each other in sequence, can produce powerful amplification of a signal because each individual enzyme catalyzes activation of many others at each step in the cascade. ANP, atrial natriuretic peptide; NE, norepinephrine; DA, dopamine; ENK, enkephalin; 5HT, serotonin; Glu, glutamate; NGF, nerve growth factor; INS, insulin; G_s, G_i, G_o, G_p, G proteins (Chapter 6); AC, adenylyl cyclase; PLC, phospholipase C; cGMP, cyclic guanosine monophosphate; cAMP, cyclic adenosine monophosphate; IP_3, inositol-1,4,5 trisphosphate; DAG, diacylglycerol; NO, nitric oxide; CaM, calmodulin; PDE, phosphodiesterase; PKC, C kinase; PKA, cAMP-dependent protein kinase; PKG, cGMP-dependent protein kinase; CaMK, calmodulin-dependent protein kinase; In1, inhibitor-1; G-Sub, G substrate; PP, protein phosphatase; TyrK, tyrosine kinase; SerK1–3, serine kinase cascade; S1–7, substrate proteins regulated by protein phosphorylation. Not included are regulatory pathways associated with arachidonic acid metabolites or specific proteases.

and by the two products of phosphatidylinositol phosphate hydrolysis, **diacylglycerol** and **inositol trisphosphate.** Additional small messenger molecules that can operate both intracellularly and intercellularly include **arachidonic acid** and its metabolites, and the gas **nitric oxide.** For clarity, each pathway will be discussed separately, but in the cell the pathways interact at almost every step.

The Cyclic Nucleotides

Adenosine 3',5'-cyclic monophosphate (cAMP; Figure 2) was discovered in 1957 by Earl Sutherland and T. W. Rall, who coined the term "second messenger" to describe its function. Studies on cAMP, which is widely distributed in the nervous system and plays an important role in many signaling pathways, have shaped many of our ideas about second messenger function, including the important concept that within individual cells cAMP-dependent protein kinase can act as a final common pathway for the action of many different hormones and neurotransmitters. cGMP is less widely distributed, and plays more specialized regulatory roles in excitable tissues. cGMP mediates phototransduction in the vertebrate retina (Chapter 6), is a prominent second messenger in cerebellar Purkinje cells, and regulates muscle tension in smooth muscle.

The concentration of cAMP is controlled by hormone-sensitive adenylyl cyclase

The cytosolic concentration of cAMP is determined by the balance in activities of its synthetic enzyme, hormone-sensitive **adenylyl cyclase,** and its degradative enzyme, **cyclic nucleotide phosphodiesterase.** Because excess cAMP is rapidly hydrolyzed to 5'-AMP, the increase in concentration produced by activation of the cyclase is usually brief, lasting only seconds to minutes. The activities of adenylyl cyclase, a membrane protein, and some forms of the phosphodiesterase, a cytosolic protein, are both regulated: the cyclase by the receptor-linked G proteins, G_s and G_i, and the

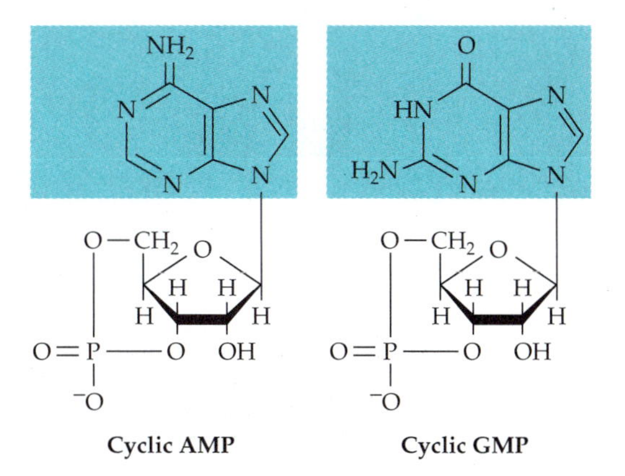

FIGURE 2. Structures of the second messengers cAMP and cGMP.

phosphodiesterase by calcium. At least one isozyme of adenylyl cyclase is regulated by calcium as well; thus the cAMP and calcium second messenger pathways interact at multiple points.

cAMP activates a protein kinase

With the exception of a cAMP-regulated channel in the olfactory epithelium, the only known effector regulated by cAMP in eukaryotes is the **cAMP-dependent protein kinase.** Protein kinases catalyze transfer of the gamma phosphate from ATP to specific serine, threonine, or tyrosine residues on proteins. The large negative charge introduced by the phosphate group changes the folding of the polypeptide chain, altering its function. This mechanism for changing protein conformation is used to regulate receptors, channels, enzymes, and structural proteins. Rapid turnover of protein phosphate groups, maintained by the balance between protein kinase and phosphatase activities, permits rapid adjustments of their concentrations.

The cAMP-dependent protein kinase is a tetramer composed of two heterodimers, each containing a regulatory and a catalytic subunit (Figure 3). The regulatory subunits, which comprise a small family of 50-kD globular proteins, bind to the 42-kD catalytic subunits and inhibit their catalytic activity (Box A). cAMP activates protein kinase activity by binding to the regulatory subunits, altering their conformation and causing them to release the catalytic subunits in an active form.

Although the cAMP-dependent protein kinase can phosphorylate many sites on different proteins, certain sites, occurring on a smaller set of proteins, are preferred. These sites are located in a larger, common **consensus sequence** that identifies them as potential substrates for the cAMP-

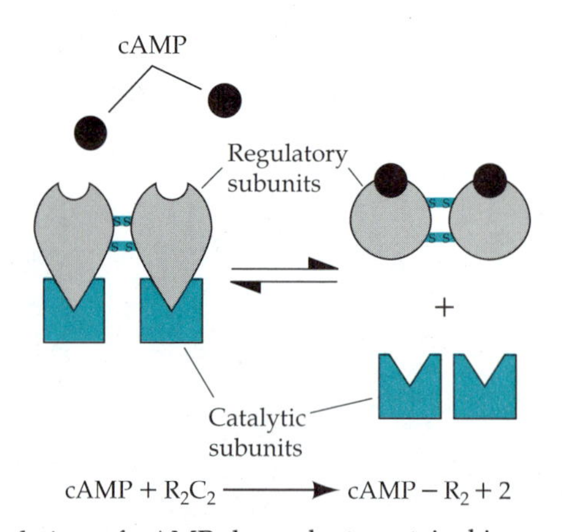

FIGURE 3. Regulation of cAMP-dependent protein kinase. In contrast to the other second messenger regulated protein kinases, the regulatory and catalytic domains of the cAMP-dependent protein kinase are located in separate proteins. Binding of cAMP to the regulatory subunits (RI or RII) induces a conformational change that releases active catalytic subunits (C). (After M. B. Kennedy, 1987. Nature 329: 15–16.)

dependent protein kinase. Other protein kinases have other consensus sequences (Table 1). Because the three-dimensional structures of proteins also influence their binding to kinase active sites, some sites within consensus sequences may not actually be phosphorylated under normal physiological conditions.

The catalytic subunit of the cAMP-dependent protein kinase has only a single functional form, but the regulatory subunit has two: RI and RII. The RI subunit is expressed in most tissues. Two RII subunits, RII-H and RII-B, are similar in sequence, but are encoded by distinct genes and expressed in a tissue-specific manner. RII-B is expressed in brain, and RII-H, which is closely related to RII-B, is expressed in heart and non-neural tissues. Although RI and RII regulatory subunits differ slightly in their affinities for cAMP and for the catalytic subunits, the most important distinction between them is that RII subunits are autophosphorylated by the catalytic subunit whereas RI subunits are not. The autophosphorylation slows the rate of reassociation of RII with the catalytic subunit and may therefore regulate the timing of functional changes produced by activation of the kinase.

By having unique binding sites for subcellular structures, regulatory subunits can determine the cellular location of the kinase. For example, the RII subunits associate strongly with the microtubule-associated protein MAP2. Approximately one third of the cAMP-dependent protein kinase holoenzyme in brain is bound to microtubules through association with MAP2, and is thus concentrated in dendrites (see Chapter 8).

cGMP is synthesized by two distinct guanylyl cyclases

Although cGMP and cAMP are closely related chemically (Figure 2), the structure and regulation of the enzymes that synthesize them are quite different. There are at least two forms of **guanylyl cyclase**, one soluble and the other tightly bound to the membrane; the two are regulated in different ways and have different regulatory roles.

TABLE 1. Sequences of Phosphorylation Sites for Second-Messenger–Regulated Protein Kinases

Protein kinase	Phosphorylation site sequence[a]
cAMP-dependent kinase	R R X S* X
cGMP-dependent kinase	R X X S* R X
Protein kinase C (α, β, γ)	X R X X S* X R X
Type II CaM kinase	X R X X S* X
CaM kinase I	N Y L R R L S* D S N F
CaM kinase III	R A G E T* R F T* D T* R K
Myosin light-chain kinase	X K K R X X R X X S* X

Source: B. E. Kemp and R. B. Pearson, 1990. Trends Biochem. Sci. 15: 342–346.
[a]Phosphorylation sites are usually surrounded by a characteristic pattern of arginine residues that is different for each protein kinase. This pattern is referred to as a substrate consensus sequence. The three-dimensional structure of the protein substrate determines which consensus sequences are actually phosphorylated. Asterisks indicate the phosphorylated residue.

Box A A Regulatory Motif of Protein Kinases

The regulatory domains of most protein kinases contain sequences of 10 to 20 amino acids that are similar to the sequences surrounding sites of phosphorylation in kinase substrates. These autophosphorylation sites, or, in some cases, "pseudosubstrate" sequences, play an important role in regulating kinase activity and also provide a basis for the design of specific peptide inhibitors that are powerful tools for investigating the physiological functions of kinases.

In 1978, Jackie Corbin and co-workers reported that chemical modification of an arginine residue in the cGMP- and cAMP-dependent protein kinases activated the kinases in the absence of cyclic nucleotide activators, and also prevented the autophosphorylation that normally occurs upon activation. This result suggested that, in the absence of cyclic nucleotides, autophosphorylation sites might serve as inhibitory sequences that block the active site and inhibit protein kinase activity. In this model, the cyclic nucleotides activate kinases by causing a conformational change in their regulatory domains that shifts the inhibitory site to permit its autophosphorylation and thus dissociates it from the active site. Determination of the sequences of the cAMP and cGMP-dependent protein kinases revealed that their autophosphorylation sites are in fact surrounded by arginine residues in patterns similar to those around phosphorylation sites on exogenous substrates (Table A). In addition, a sequence similar to the autophosphorylation site in the type II regulatory subunit of the cAMP-dependent protein kinase, but with an alanine replacing the autophosphorylated serine, was discovered in a peptide inhibitor of the cAMP-dependent protein kinase, isolated from muscle by Donal Walsh. This similarity suggested that a **"pseudosubstrate" sequence** in the Walsh inhibitor binds to the catalytic subunit and inhibits it.

This model for kinase regulation has recently been refined and generalized to other protein kinases. Protein kinase C contains distinct regulatory and catalytic domains within a single subunit (see Figure 8 in this chapter). When the regulatory domain is removed by limited digestion with proteases, the C kinase becomes constitutively active, no longer requiring diacylglycerol or Ca^{2+} for activity. In 1987, Bruce Kemp and co-workers, inspired by the earlier work of Corbin and Walsh, searched the sequence of the C kinase and

TABLE A. Inhibitory Pseudosubstrates and Autophosphorylation Sites in Second-Messenger–Regulated Protein Kinases

Protein kinase	**Sequences**[a]
cAMP-dependent Kinase (type I)	G R R R R G A* I
(Substrate) Liver-pyruvate kinase	G Y I R R A S* V
cAMP-dependent kinase (type II)	R F D R R V S* V
(Substrate) Liver-pyruvate kinase	G Y I R R A S* V
cGMP-dependent kinase	D R T T* R A Q G
(Substrate) Histone H2β	R K R S* R K E
C kinase	R K G A* L R Q K
(Substrate) EGF receptor	R K R T* L R R L
Type II CaM kinase	M H R Q E T* V D C
(Substrate) Synapsin I	A T R Q T S* V S G

Source: G. Hardie, 1988. Nature 335: 592–593.

[a]The table lists sequences in the regulatory domains of several protein kinases that resemble the sequences of phosphorylation sites in specific kinase substrates. One example of a protein substrate phosphorylation site is shown for each kinase. Asterisks mark the phoshorylated residues and pseudosubstrate alanine residues.

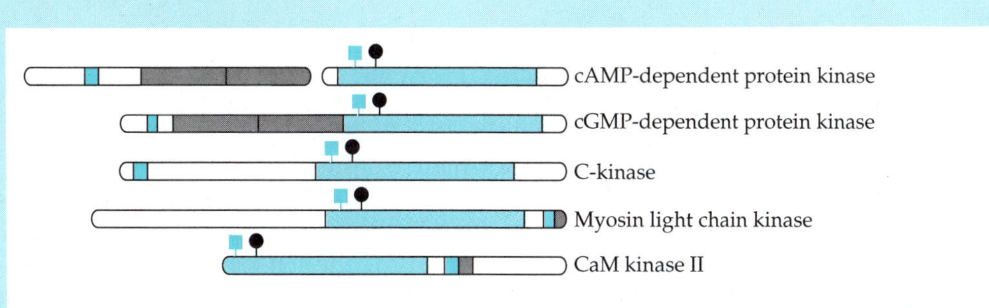

FIGURE A. Pseudosubstrate domains and internal autophosphorylation sites within protein kinases. In the absence of activating secondary messengers that bind to the regulatory domains, these domains interact with the active site to inhibit kinase activity. (After G. Hardie, 1988. Nature 335: 592–593.)

found a region in the regulatory domain, amino acids 19 through 36, that contained a potential pseudosubstrate sequence. They showed that a synthetic peptide with this sequence is a potent inhibitor of the C kinase.

Potential inhibitory autophosphorylation sites or pseudosubstrate sequences have now been identified in at least seven protein kinase regulatory domains (Table A; Figure A). In the case of smooth muscle myosin light chain kinase, a CaM kinase, experiments directly support the model of inhibition by a pseudosubstrate sequence. Removal of the calmodulin binding sequence at the carboxyl terminus by limited proteolysis renders the kinase insensitive to calmodulin activation. Further limited proteolysis of this inactive species removes an additional short segment containing the putative pseudosubstrate sequence and converts the kinase to a constitutively active form.

In addition to providing a structural hypothesis to explain protein kinase regulation, the concept of pseudosubstrate inhibition has guided the synthesis of potent and relatively specific peptide inhibitors of protein kinases for use in biological experiments. The structures of three such inhibitors are listed in Table B. Their use in physiological studies is described in Chapter 14.

TABLE B. Inhibitory Peptides Based on Pseudosubstrate Sequences

Protein kinase	**Inhibitor**[a]
cAMP-dependent protein kinase (Walsh inhibitor 5–24)	T T Y A D F I A S G R T G R R N A* I H D
C kinase (19–36)	R F A R K G A* L R Q K N V H E V K N
Type II CaM kinase (273–302)	H R S T V A S C M H R Q E T* V D C L K K F N A R R K L K G A

Sources: H.-C. Cheng et al., 1989. J. Biol. Chem. 261: 989–992; C. House and B. E. Kemp, 1987. Science 238: 1726–1728; R. Malinow, H. Schulman and R. W. Tsien, 1989. Science 245: 862–866.

[a]These three peptide inhibitors inhibit their respective protein kinases with high affinity and specificity. The peptide sequences are derived from inhibitory sequences in a potent protein inhibitor or in the kinases themselves. Note that the most effective and specific inhibitory peptides are much longer than the short substrate consensus sequences. This apparently reflects the contribution of three-dimensional structure and/or multiple binding domains to protein kinase substrate specificity.

Soluble guanylyl cyclase is activated by nitric oxide

The role of the soluble guanylyl cyclase is best understood in the smooth muscle cells that form blood vessel walls. Acetylcholine (ACh) released by parasympathetic nerves causes dilation of small arterioles in many parts of the body. This control is exerted by a tortuous pathway that begins with activation of muscarinic receptors in the endothelial cells lining the arterioles and ends with relaxation of muscle tension in the neighboring smooth muscle cells. Activation of muscarinic receptors in the endothelial cell membrane increases intracellular inositol trisphosphate (Chapter 6 and below), which releases Ca^{2+} from internal cytoplasmic stores. Increased cytosolic Ca^{2+} concentration activates nitric oxide synthetase, a Ca^{2+}-dependent enzyme that catalyzes the release of the gaseous free radical **nitric oxide (NO)** from the guanidino group of arginine (Figure 4). Nitric oxide has a half life of 3–5 seconds when dissolved in biological fluids containing oxygen. It rapidly diffuses through the membrane of the endothelial cells into the adjacent smooth muscle cells, where it activates soluble guanylyl cyclase by binding to a protoporphyrin heme group on the enzyme. The resulting increase in cGMP concentration stimulates the cGMP-dependent protein kinase (see below), which phosphorylates muscle proteins, leading to muscle relaxation.

The remarkable use of a gas, nitric oxide, as a second messenger apparently occurs in other excitable tissues, including neurons. Several enzymes in the nitric oxide pathway are present in neurons. For example, nitric oxide synthetase is found in high concentration in cerebellar granule cells and guanylyl cyclase in cerebellar Purkinje cells. Activation of glutamate receptors in granule cells may stimulate nitric oxide synthesis (perhaps by entry of Ca^{2+} through NMDA receptors), and NO released from granule cells may diffuse into Purkinje cells and there increases synthesis of cGMP. The cGMP then activates cGMP-dependent protein kinase. Nitric oxide may also carry regulatory information backward from postsynaptic sites to presynaptic terminals during induction of long-term potentiation (Chapter 14).

Membrane-bound guanylyl cyclases are receptors for peptides that regulate fluid and electrolyte homeostasis

Receptors for atrial natriuretic peptides (ANPs) are transmembrane proteins whose cytoplasmic domains exhibit guanylyl cyclase activity. ANPs are a family of circulating hormones synthesized and released into the circulation by cardiac atrial myocytes in response to increased atrial blood pressure. The most important site of action of the peptides is the kidney, where they produce diuresis and enhance the secretion of salt (natriuresis). Thus, this small family of hormones plays a crucial role in control of electrolyte balance and maintenance of cardiovascular homeostasis. The receptors for ANPs are 120–140-kD proteins that span the membrane once (Figure 5). The extracellular ligand binding site is linked directly to a guanylyl cyclase domain via the single transmembrane helix.

A family of ANPs is synthesized in the brain at lower levels than in

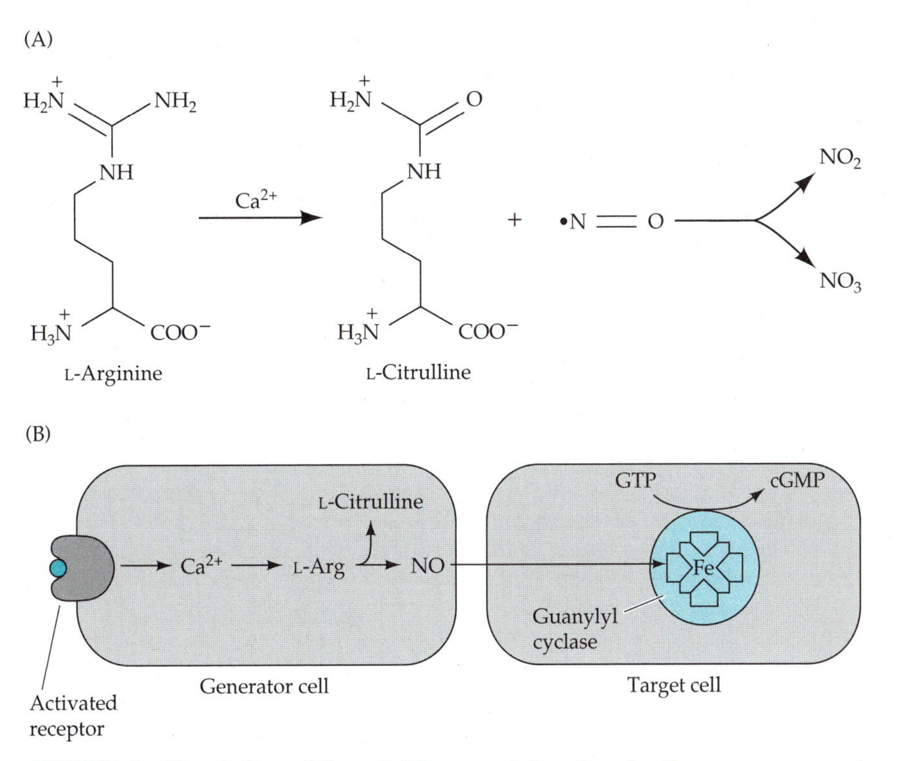

FIGURE 4. Regulation of the soluble guanylyl cyclase by the gaseous second messenger nitric oxide. (A) A Ca^{2+}/calmodulin-dependent enzyme, nitric oxide synthetase, catalyzes the release of the gas nitric oxide (NO) from the amino acid arginine. In the presence of O_2, NO is converted within several seconds to NO_2 and NO_3 by unknown chemical mechanisms. (B) Activation of receptors (R) or ion channels that increase the Ca^{2+} concentration in generator cells activates nitric oxide synthetase. Nitric oxide diffuses through the membrane to neighboring target cells containing soluble guanylyl cyclase, a heme-linked enzyme. Nitric oxide binds tightly to the heme group, thus activating the cyclase. Cyclic GMP produced by the cyclase can activate cGMP-dependent protein kinase. This pathway is most thoroughly understood in arterioles in which the generator cells are endothelial cells and the target cells are smooth muscle cells in the blood vessel walls. Nitric oxide can also activate soluble guanylyl cyclase when it is present within the generator cell. (After J. Collier and P. Vance, 1989. Trends Pharm. Sci. 10: 427–431.)

atrial cells and at least one ANP receptor guanylyl cyclase unique to brain has been described. Thus, the newly discovered receptor-linked guanylyl cyclases will probably turn out to be involved in other forms of regulation than cardiovascular homeostasis.

cGMP activates a protein kinase

In photoreceptor cells cGMP acts directly on an ion channel (Box B); in other cells its principal actions are mediated by a **cGMP-dependent protein kinase**. In contrast to the cAMP-dependent protein kinase, whose regulatory and catalytic functions reside in two separate subunits, a single polypeptide of 76 kD contains both regulatory and catalytic domains of the

Box B cGMP and the Response to Light

In addition to regulation by phosphorylation, membrane ion channels are often regulated by direct binding of second messenger molecules, such as Ca^{2+}, IP_3, or arachidonic acid and its metabolites. Channels that are regulated by cyclic nucleotides play important roles in sensory transduction: a cAMP-regulated channel mediates the response to olfactory stimuli; a cGMP-gated channel mediates photoreception in rod outer segments.

Rods have three main compartments: the outer segment, which contains stacks of membranous disks rich in the light sensing pigment rhodopsin; the inner segment, which contains the nucleus and biosynthetic machinery; and the synaptic terminal, which releases transmitter onto retinal bipolar cells. As in other neurons, a Na^+/K^+ ATPase maintains an ion gradient across the rod membrane. In the dark, the outer segment has a higher permeability to Na^+ and a lower permeability to K^+ than does the inner segment. The high sodium conductance in the outer segment drives the resting potential of the photoreceptor to a partially depolarized potential (~ –30 mV), producing tonic release of transmitter onto bipolar cells in the dark. The absorption of light by rhodopsin in the outer segment leads to the closing of some of the outer segment sodium channels. The resulting hyperpolarization inhibits the tonic release of transmitter, signaling the presence of light. Photoreceptors are exquisitely sensitive; the absorption of a single photon by a dark-adapted rod closes hundreds of sodium channels and produces a hyperpolarization of about 1 mV. The absorption of 30 photons produces half maximal hyperpolarization.

The second messenger cGMP carries information from rhodopsin to the sodium channel. In the dark, the concentration of cGMP, which is high, opens the sodium channel by binding directly to the channel protein. Absorption of light by rhodopsin leads, via a G protein, to activation of a cGMP phosphodiesterase that rapidly hydrolyzes cGMP, transiently reducing its concentration (see Figure 6 in Chapter 6). This reduction allows some of the sodium channels to close, hyperpolarizing the photoreceptor.

Purification of the cGMP-gated sodium channel from bovine photoreceptors led to the isolation of a cDNA encoding a 79-kD channel protein with four to six apparent transmembrane helices (Figure A). Expression of the cDNA in *Xenopus* oocytes revealed a cGMP-gated sodium channel with the properties of the photoreceptor channel. The sequence has no clear similarity to other known ion channels and thus defines a new family of ion channel proteins. The carboxyl terminal portion of the channel protein contains an 80 amino acid segment that is homologous to the cGMP binding domain of the cGMP-dependent protein kinase. Although the channel sequence contains only one cGMP binding site, gating of the sodium channel by cGMP is cooperative, involving two or more cGMP molecules. Thus, the functional channel is likely to be a homo-oligomer composed of several individual subunits, a structure similar to that of the potassium channels (Chapter 3).

cGMP-dependent protein kinase. The two cGMP binding sites are located in the amino-terminal region, and the catalytic domain, which is similar in sequence to the catalytic subunit of the cAMP-dependent protein kinase, is at the carboxyl terminal.

Also unlike cAMP-dependent protein kinase, which is expressed uniformly throughout the brain, cGMP-dependent protein kinase is present at low concentration in most neurons but is highly concentrated in cerebellar Purkinje neurons, which also contain high levels of guanylyl cyclase. The physiological function of the cGMP transduction system in Purkinje neu-

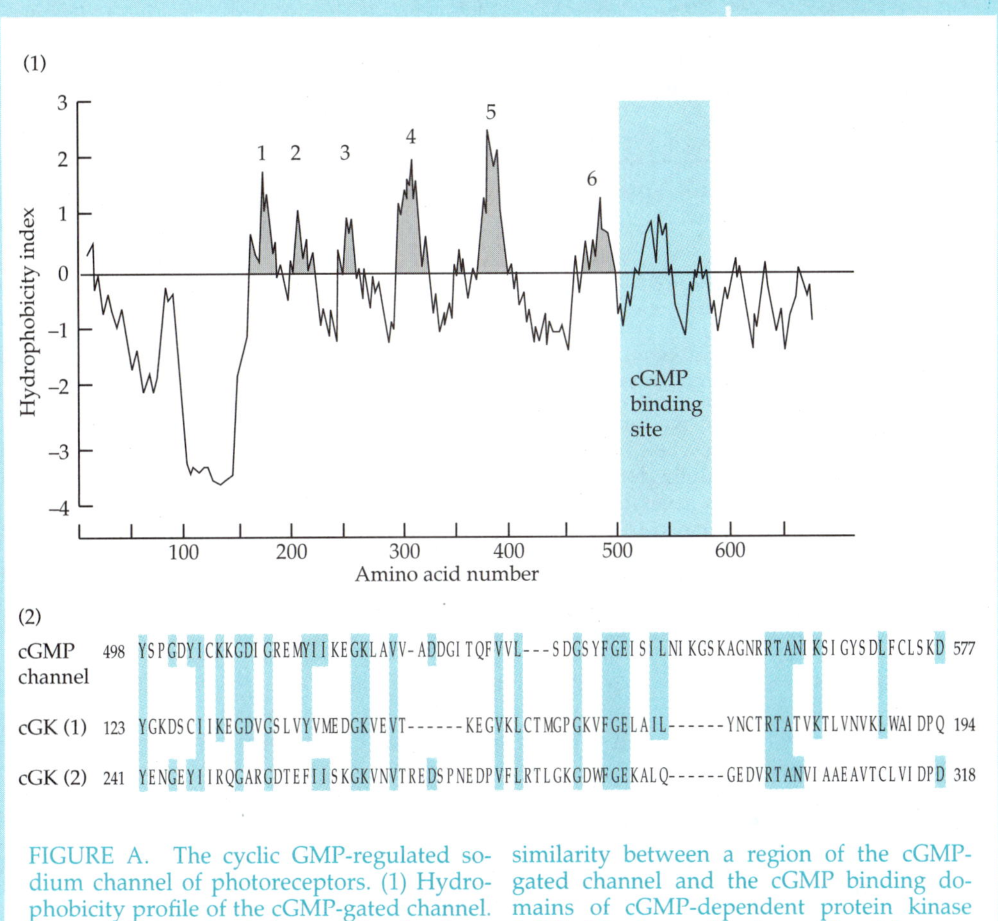

FIGURE A. The cyclic GMP-regulated sodium channel of photoreceptors. (1) Hydrophobicity profile of the cGMP-gated channel. An averaged hydrophobicity index has been plotted against amino acid number. Six potential hydrophobic membrane-spanning domains are highlighted. The putative cGMP binding domain is located on the carboxyl terminal side of the sixth membrane-spanning domain. (2) Amino acid sequence similarity between a region of the cGMP-gated channel and the cGMP binding domains of cGMP-dependent protein kinase (cGK). Numbers of amino acid residues are indicated at both ends. Amino acid identities between the channel and cGK are indicated by colored boxes. The probability that this sequence similarity would occur by chance is 1.2×10^{-11}. (After U. B. Kaupp et al., 1989. Nature 342: 762–766.)

rons is not known; however, a 23-kD protein substrate for the cGMP-dependent protein kinase, which is expressed only in Purkinje cells, offers a clue. This protein, termed "G substrate," has short regions of amino acid sequence that are similar to the sequences of two small protein inhibitors of protein phosphatase-1, called inhibitor-1 and DARPP-32 (see below). These proteins become phosphatase inhibitors only after they are phosphorylated by the cAMP-dependent protein kinase. Thus, one role of the cGMP cascade in the cerebellum may be to inhibit phosphatases through

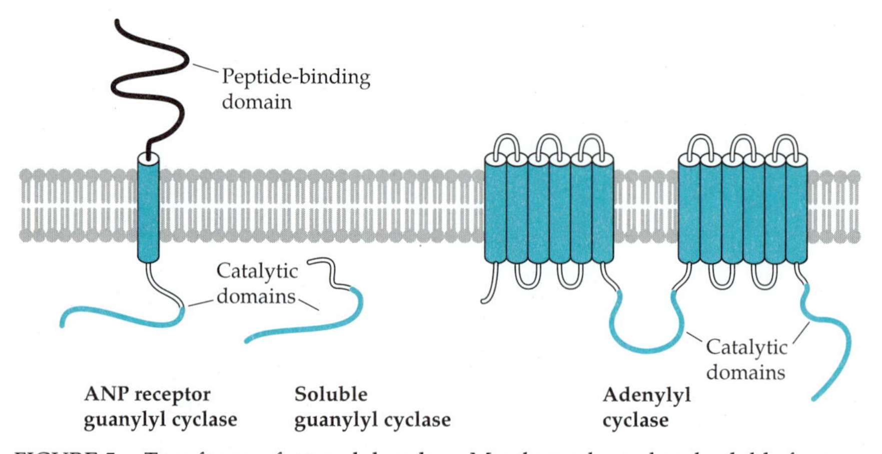

FIGURE 5. Two forms of guanylyl cyclase. Membrane-bound and soluble forms of guanylyl cyclase both contain catalytic domains that are similar in sequence to the two catalytic domains of the membrane-bound adenylyl cyclase. The membrane-bound guanylyl cyclase is a receptor for the peptide hormones atrial natriuretic peptides (ANP). The binding site for the peptides (black) is located in the extracellular space, whereas the guanylyl cyclase domain is located in the cytosol. Colored bars represent regions of sequence homology among membrane-bound and soluble guanylyl cyclases and adenylyl cyclase. (After D. L. Garbers, 1990. New Biologist 2: 499–504.)

the G substrate and thereby prolong the effects of phosphorylation initiated through other signal transduction cascades.

Messengers Produced by Hydrolysis of Phosphatidyl Inositol 4,5-Bisphosphate

The third major group of second messenger molecules are those produced by hydrolysis of the phospholipid **phosphatidyl inositol 4,5-bisphosphate** (**PIP_2**; Figure 6). Many G protein-linked receptors (Chapter 6) stimulate hydrolysis of PIP_2 by activating **phosphatidyl inositol (PI) specific phospholipase C.** The two second messengers produced by this hydrolysis, **diacylglycerol (DAG)** and **1,4,5-inositol trisphosphate (IP_3)**have different fates. IP_3 is water-soluble and diffuses into the interior of the cell, where it releases Ca^{2+} from internal stores. DAG is hydrophobic and remains in the plasma membrane.

Four distinct classes of phospholipase C enzymes specifically hydrolyze phosphatidyl inositol and its phosphorylated derivatives. All of these are cytosolic, Ca^{2+}-stimulated enzymes that can associate with the plasma membrane, perhaps by binding directly to G proteins. Upon purification, the phospholipases have high activity and are not stimulated by G proteins. This curious observation has led to the hypothesis that, within cells, negative repressor proteins inhibit the phospholipases until activated G proteins displace them and relieve the inhibition, a mechanism similar to that for activation of cGMP phosphodiesterase in photoreceptors (Chapter 6). No candidate repressor proteins have yet been identified, however.

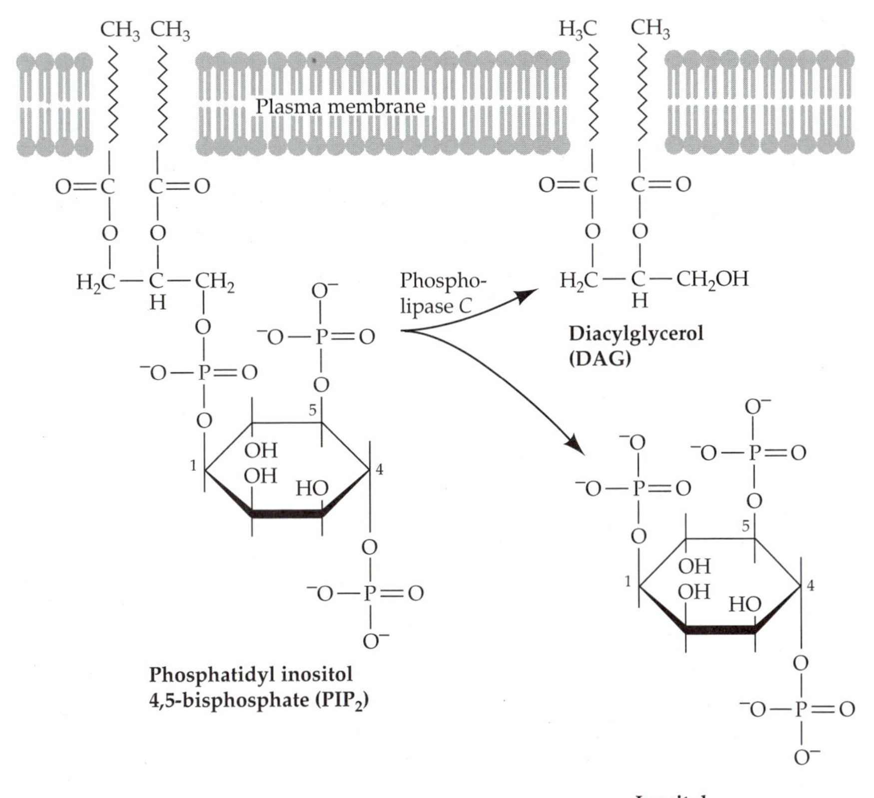

FIGURE 6. Hydrolysis products of phosphatidyl inositol bisphosphate. Phosphatidylinositol 4,5 bisphosphate (PIP$_2$) is a phospholipid composed of a diglyceride backbone and the polar head group 1,4,5-inositol trisphosphate (IP$_3$), a phosphorylated six-carbon ring alcohol. In most neuronal membranes, PIP$_2$ is present at a steady-state level of about 0.4% of total lipid. It is synthesized from phosphatidyl inositol by a series of specific PI-kinases. Receptor-regulated PI-phospholipase C catalyzes hydrolysis of PIP$_2$, producing 1,2-diacylglycerol (DAG) and IP$_3$. Diacylglycerol remains associated with the membrane, while IP$_3$, a hydrophilic molecule, diffuses into the cytosol.

The principal target of diacylglycerol is a family of protein kinases called C kinases

As for the cyclic nucleotides, an important common pathway for the many hormones and neurotransmitters that stimulate hydrolysis of PIP$_2$ is activation of a protein kinase called **protein kinase C,** a ubiquitous enzyme that is activated synergistically by DAG and Ca^{2+} (Figure 7). In the absence of DAG, C kinase activation requires concentrations of Ca^{2+} that are higher than those usually reached under physiological conditions. The presence of DAG lowers the requirement for Ca^{2+} and sets it within the physiological range. Sufficiently high concentrations of DAG can even reduce the Ca^{2+} requirement below resting cellular concentrations (~10^{-7} *M*). Activa-

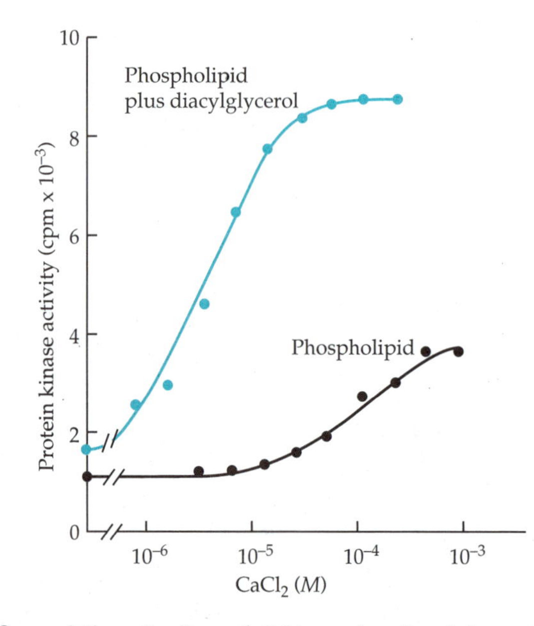

FIGURE 7. Synergistic activation of C kinase by diacylglycerol and Ca^{2+}. The most common isozymes of the C kinase require both Ca^{2+} and phospholipid for activity. The presence of a small amount of diacylglycerol (DAG) potentiates activation by Ca^{2+}. In this experiment, C kinase purified from brain was assayed with different concentrations of Ca^{2+} in the presence of phospholipid alone, or in the presence of phospholipid plus diacylglycerol in a 15:1 ratio (color). (After Kishimoto et al., 1980. J. Biol. Chem. 255: 2273–2276.)

tion, which is accompanied by binding of the kinase to the membrane, involves formation of a complex containing membrane lipids, Ca^{2+}, DAG, and the C kinase protein. Not surprisingly, many proteins that are phosphorylated by the C kinase are membrane proteins. The C kinase participates in regulation of a wide variety of physiological functions, including adaptation of spike frequency in the hippocampus, gating of ion channels, desensitization of receptors, and induction of long-term potentiation in the hippocampus (Chapter 14). Certain hydrophobic esters of the tumor promoter phorbol can diffuse into the membrane from the extracellular fluid and activate the C kinase by mimicking diacylglycerol. Phorbol esters are often used as pharmacological agents to detect the participation of C kinase in physiological events.

The C kinases are a family of at least seven distinct monomeric 77–87-kD proteins (Figure 8). Six of the isozymes—α, β, γ (also referred to as III, II, and I, respectively), δ, ε, and ζ—are products of distinct genes. The transcript of the gene encoding β is alternatively spliced, producing βI and βII isozymes that differ at their carboxyl terminals. Subtle differences in requirements for activation, subcellular locations, and substrate specificities among the C kinase isozymes may be important for fine tuning the signaling properties of different neurons. For example, the γ isozyme is expressed only in the brain and spinal cord; in addition to DAG and Ca^{2+},

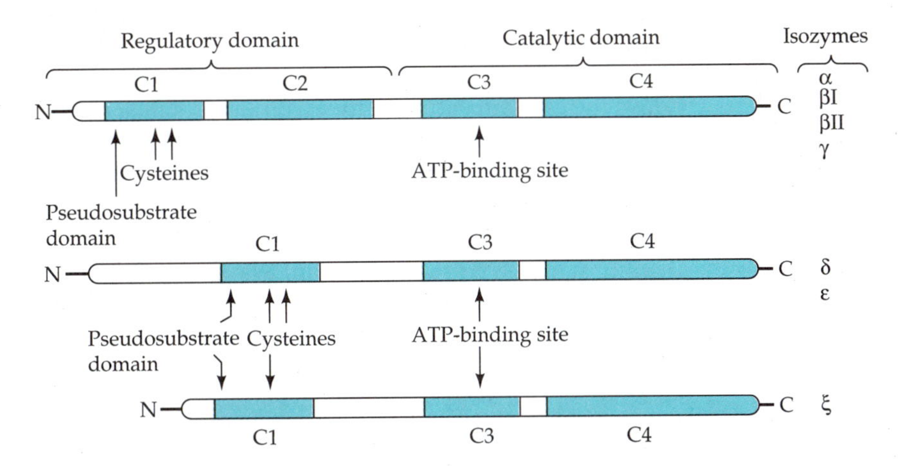

FIGURE 8. Structural domains of the C kinase isozymes. All of the isozymes are divided into an amino terminal regulatory domain and a carboxyl terminal catalytic domain. The regulatory domains of the α, β, and γ isozymes contain two highly conserved regions, C1 and C2. C1 contains the pseudosubstrate sequence that inhibits binding of substrates to the active site in the absence of Ca^{2+} and diacylglycerol (see Box A), and zinc-finger domains (see Chapter 10) with their characteristic pattern of cysteine residues. C2 is absent in δ, ε, and ζ, which are not sensitive to regulation by Ca^{2+}. The catalytic domains contain conserved regions C3 and C4. Small variations in sequence of the isozymes produce differences in regulation, substrate specificity, and subcellular location. (After K.-P. Huang, 1989. Trends Neurosci. 12: 425–432.)

it can be activated by micromolar concentrations of arachidonic acid, which is released from lipids by the enzyme phospholipase A_2 (see below). The δ, ε, and ζ isozymes are activated by DAG but are not regulated by Ca^{2+}.

Calcium Ion

The fourth major second messenger in neurons is Ca^{2+}. Cells in the nervous system, as in other parts of the body, maintain the cytosolic concentration of free Ca^{2+} at 0.1 to 0.2 μ*M*, approximately ten thousand-fold lower than the extracellular concentration of Ca^{2+} (1.8 m*M*). Two proteins in the surface membrane, a **Na^+/Ca^{2+} exchange protein** and a **plasma membrane calcium pump** driven by ATP, sustain this steep concentration gradient. A distinct endoplasmic reticulum ATP-dependent calcium pump also concentrates Ca^{2+} in an internal membrane compartment, where it forms an intracellular storage pool. Rapid increases in the cytosolic Ca^{2+} concentration are used by neurons as regulatory signals. Hormones, neurotransmitters, and electrical activity all control movement of Ca^{2+} across the plasma membrane into the cytosol; the second messenger, IP_3, regulates movement of Ca^{2+} out of the internal storage pool.

Although the basal concentration of free Ca^{2+} in the cytosol is submicromolar, the concentration of total Ca^{2+} is estimated to be 50 to 100 μ*M*.

Most of the Ca^{2+} is bound to proteins, including calmodulin as well as parvalbumin and other high-affinity Ca^{2+} binding proteins. **Calmodulin** relays the calcium signal to many target proteins, as we will discuss. The other proteins appear to serve as Ca^{2+} buffers, limiting the concentration range of free Ca^{2+} within the cytosol. The level of **Ca^{2+} buffering proteins** varies considerably among different neurons. The cerebellum is rich in a vitamin D-dependent calcium binding protein, while GABAergic neurons that fire at a high frequency contain a large amount of parvalbumin. These variations may produce physiologically significant differences in the amplitude and time course of transient changes in Ca^{2+} concentration in different cells.

Fluctuations in intracellular Ca^{2+} concentration can be visualized and measured with Ca^{2+}-sensitive molecules

The accurate measurement of cytosolic Ca^{2+} concentrations poses a challenging problem. The resting internal Ca^{2+} concentration is low and the concentration of potentially interfering ions such as Mg^{2+}, Na^{+}, and K^{+} are 10^4- to 10^6-fold higher, so that a good Ca^{2+} indicator must be both highly sensitive and exquisitely selective. Furthermore, because the intracellular Ca^{2+} concentration often rises to a peak in less than a second and falls again over several seconds, a good indicator must bind and release calcium rapidly enough to accurately reflect the kinetics of Ca^{2+} transients. The first successful indicators were the chemiluminescent protein aequorin and the metallochromic indicator arsenazo III, which were injected into large neurons with microelectrodes. Recently, the introduction of fluorescent Ca^{2+} chelators such as **fura-2,** modeled on EGTA (Figure 9), has expanded the range of cells that can be studied and the sensitivity with which small changes in Ca^{2+} concentration can be detected.

Fura-2 has two advantages over conventional indicators. First, binding of Ca^{2+} to fura-2 (K_d approximately 100 nm) increases the amount of light that the fluorophore emits when excited at 340–350 nm and decreases the light emitted upon excitation at 380–390 nm. Therefore, changes in Ca^{2+} concentration can be detected as changes in the ratio of fluorescence amplitudes resulting from excitation at 350 nm and 385 nm. Measurement of the **fluorescence ratio** cancels out variations in fluorescence intensity produced by variations in dye loading and in cell thickness. A second major advantage of fura-2 is that the charged carboxylate groups that chelate Ca^{2+} can be masked by ester groups, producing lipid-soluble derivatives. These uncharged, lipophilic derivatives diffuse readily across membranes, then are hydrolyzed by cytosolic esterases to the polycarboxylate form. This useful modification permits loading of large populations of cells with dye simply by adding the ester derivative to the medium, thus eliminating the need for microinjection of individual cells.

Video imaging of fluctuations in Ca^{2+} concentration measured with fura-2 reveals that Ca^{2+} concentrations are often controlled differently in dendrites and growth cones than in the soma. For example, ligand and voltage-gated channels are usually the most important mediators of Ca^{2+} increases in dendrites, whereas intracellular stores play a more important role in the soma. With proper instrumentation, local changes in Ca^{2+}

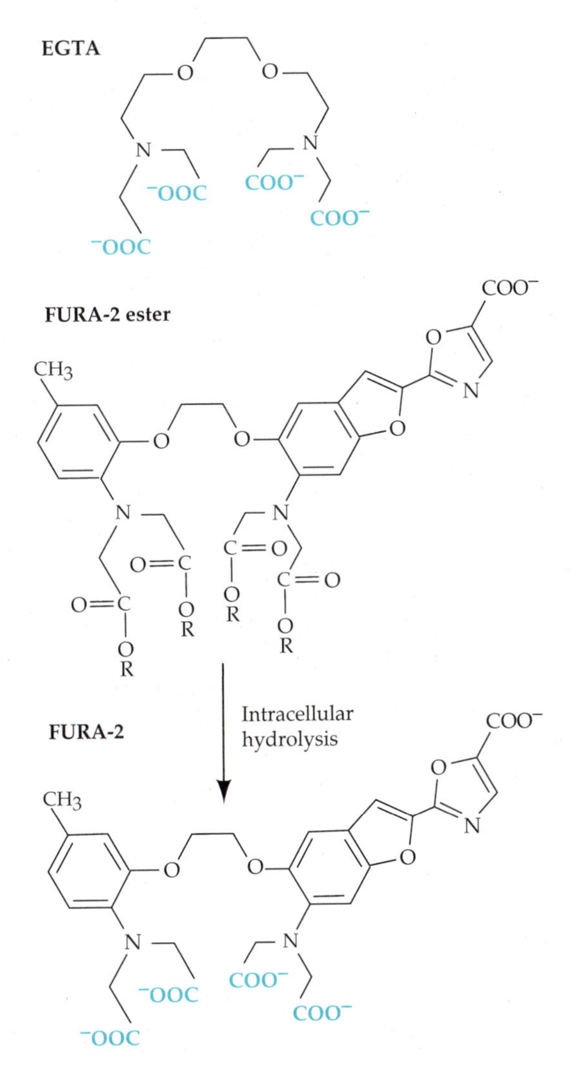

FIGURE 9. Fura-2, a fluorescent Ca^{2+}-sensitive dye. Fura-2 was designed by R. Tsien based upon the structure of EGTA. Its fluorescence emission spectrum is altered by binding to Ca^{2+} in the physiological concentration range. In the absence of Ca^{2+}, Fura-2 fluoresces most strongly at an excitation wavelength of 385 nm; when it binds Ca^{2+}, the most effective excitation wavelength shifts to 345 nm. This property is used to measure local Ca^{2+} concentrations within cells (see text). Cells can be loaded with Fura-2 esters that diffuse across cell membranes and are hydrolyzed to active Fura-2 by cytosolic esterases. The colored carboxyl groups chelate calcium.

concentration can now be correlated with modulation of neuronal functions such as neurite outgrowth (see Chapter 12).

Activation of ligand-gated and voltage-dependent calcium channels initiates flow of Ca^{2+} across the neuronal plasma membrane

Ca^{2+} can flow into the cytosol from two locations, the extracellular fluid and an intracellular storage compartment. Flow of Ca^{2+} across the plasma membrane is initiated by the opening of **Ca^{2+} channels,** which are either **voltage-dependent** or **ligand-gated** (Figure 10A). Voltage-dependent channels that are highly specific for Ca^{2+} (Chapter 3) are found throughout the surface membrane and are especially concentrated in the membrane of presynaptic terminals. The local Ca^{2+} influx caused by depolarization of synaptic terminals triggers fusion of synaptic vesicles with the presynaptic membrane and release of transmitter into the cleft (Chapter 5). At the postsynaptic membrane, some ligand-gated receptors, such as the nicotinic ACh receptor and the N-methyl-D-aspartate (NMDA)-type glutamate receptor, have large pores that permit influx of Ca^{2+} as well as $Na^+$and K^+ (Chapter 3).

To reestablish the several-thousand-fold gradient of Ca^{2+} concentration across the plasma membrane, neurons and other cells pump Ca^{2+} out of the cell (Figure 10A). Neuronal plasma membranes contain a Na^+/Ca^{2+} exchanger that couples movement of Ca^{2+} out of the cell, against its concentration gradient, to movement of Na^+ into the cell, down its concentration gradient. Thus, the exchanger harnesses the energy stored in the Na^+ gradient maintained by the Na^+/K^+ ATPase. Three sodium ions move into the cell for each Ca^{2+} ion pumped out, so that the exchange causes a net movement of charge—it is electrogenic. The affinity of the exchanger for Ca^{2+} and the energetics of transport are such that it can maintain intracellular free Ca^{2+} concentrations as low as 0.3 μM under the usual physiological ionic conditions. To maintain lower concentrations, a Ca^{2+} ATPase is required.

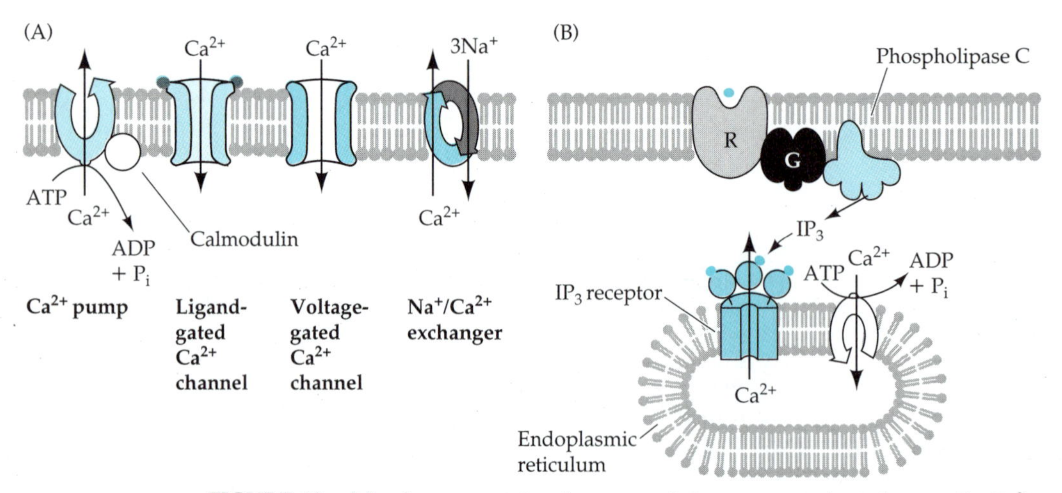

FIGURE 10. Membrane proteins that control the concentration of cytosolic Ca^{2+}. (A) In the plasma membrane, a Ca^{2+}-dependent ATPase and a Na^+/Ca^{2+} exchange protein move Ca^{2+} out of the cytosol. Ca^{2+} enters the cytosol through activated voltage-gated ion channels and ligand-gated ion channels. (B) In the endoplasmic reticulum, a distinct Ca^{2+}-dependent ATPase pumps Ca^{2+} into the internal stores. IP_3 is generated by activation of seven transmembrane-helix receptors. The IP_3 receptor located in the endoplasmic reticulum responds to increased concentration of IP_3 by releasing Ca^{2+} into the cytosol.

The neuronal plasma membrane contains a high-affinity Ca^{2+} transporter or ATPase that couples hydrolysis of ATP to movement of Ca^{2+} out of the cytosol. The plasma membrane ATPase is a 140-kD protein that pumps one molecule of Ca^{2+} for each molecule of ATP hydrolyzed. The rate of pumping is regulated by calmodulin (see below). All of the Ca^{2+} ATPases whose sequences are known contain ten transmembrane domains, four on the amino-terminal end and six at the carboxyl terminal, separated by a central region encoding a large cytoplasmic domain. The exact mechanism by which Ca^{2+} is transported across the membrane by these proteins is unknown. The high affinity of the plasma membrane ATPase for Ca^{2+} (K_M of 0.2 to 0.3 μM) enables it to maintain the cytosolic free Ca^{2+} concentration at basal levels of 0.1 to 0.2 μM.

Inositol 1,4,5-trisphosphate (IP_3) opens a calcium channel in the endoplasmic reticulum, releasing free Ca^{2+} into the cytosol

A second important source of cytosolic Ca^{2+}, in addition to the extracellular medium, is an **intracellular storage compartment.** Neurons and other cells contain internal membrane cisternae, formed from the smooth endoplasmic reticulum, that are analogous to the sarcoplasmic reticulum of muscle (Figure 10B). Their membranes contain two specialized proteins, an ATP-dependent calcium pump that concentrates Ca^{2+} within the lumen, and an **IP_3 receptor protein** that releases Ca^{2+} into the cytosol in response to an increase in cytosolic IP_3. Release of Ca^{2+} from the intracellular stores is triggered by binding of various ligands to G protein-coupled receptors on the cell surface that activate hydrolysis of PIP_2 and generate intracellular IP_3. Increases in cytosolic Ca^{2+} produced by these receptors do not require the presence of Ca^{2+} in the external medium.

The endoplasmic reticulum Ca^{2+} ATPase responsible for concentrating Ca^{2+} in the internal stores is distinct from its plasma membrane counterpart. It is a 105-kD protein that pumps two molecules of Ca^{2+} for each molecule of ATP hydrolyzed and is insensitive to calmodulin. Like the plasma membrane ATPase, it is a high-affinity pump with a K_M of 0.2 to 0.3 μM Ca^{2+}.

The IP_3 receptor, which controls release of Ca^{2+} from internal stores, is a 313-kD protein with 7 potential transmembrane domains clustered at the carboxyl terminal. This region forms a calcium channel that is opened when IP_3 binds to the receptor. The IP_3 binding site at the amino terminal end extends into the cytoplasm. In its active form the receptor is thought to be a tetramer; thus the calcium channel, the IP_3 binding site, or both, may be formed by the intersection of four individual receptor subunits.

The IP_3 receptor is highly homologous to another large membrane protein of similar function called the ryanodine receptor, a calcium channel located in the sarcoplasmic reticulum of skeletal muscle (Figure 11). At low concentrations the drug ryanodine stimulates Ca^{2+} release from isolated sarcoplasmic reticulum, and at high concentrations it inhibits Ca^{2+} release. The amino terminal portion of the ryanodine receptor binds directly to a voltage sensitive protein sensor located in the muscle T-tubule membrane termed the dihydropyridine-sensitive receptor (DHSR). The

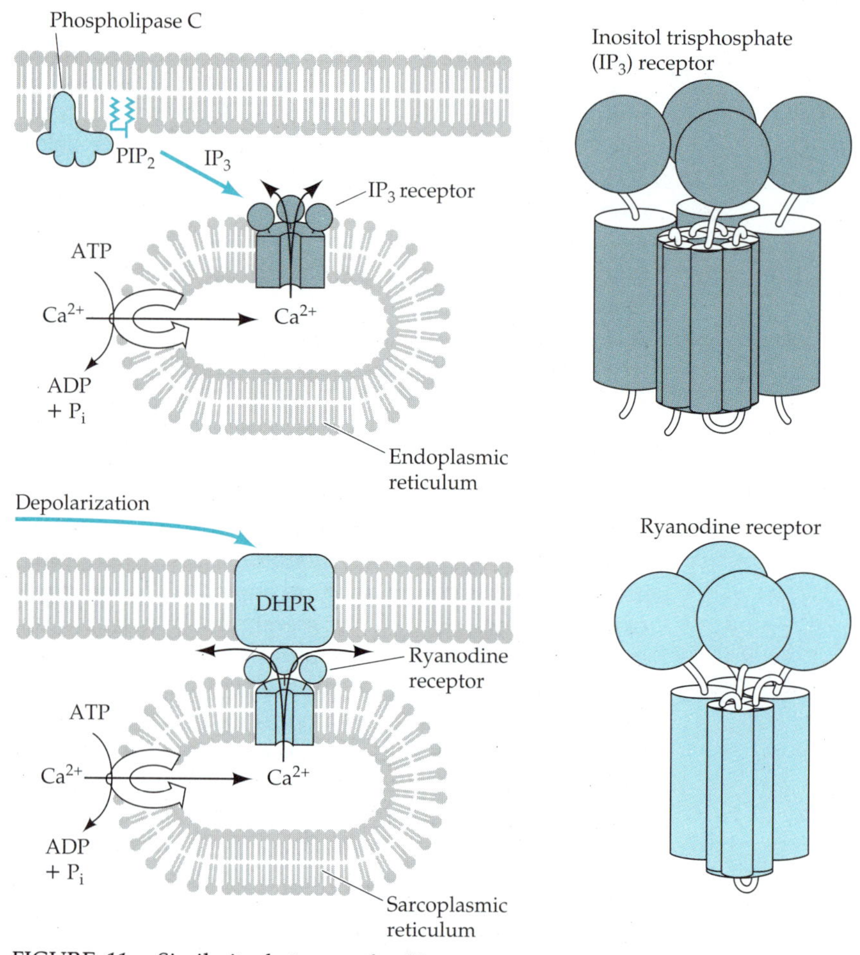

FIGURE 11. Similarity between the IP_3 receptor in nonmuscle tissue and the ryanodine receptor in muscle. The IP_3 receptor is thought to be a tetramer of subunits, each of which consists of seven transmembrane domains and a large cytoplasmic IP_3 binding domain. IP_3 released into the cytosol by the action of phospholipase C binds to the receptor, opening a calcium channel between the lumen of the endoplasmic reticulum and the cytosol. The muscle ryanodine receptor is a tetramer of subunits each containing four transmembrane domains and a large cytoplasmic domain that is in physical contact with the voltage-sensing dihydropyridine receptor (DHPR) in the T-tubule membrane. When the T-tubule membrane is depolarized, the voltage change is commuincated to the ryanodine receptor and a Ca^{2+} channel is opened that releases Ca^{2+} into the muscle cytosol, initiating contraction. The sequences of the IP_3 receptor and the ryanodine receptor are 46% identical in a 134 amino acid segment of their carboxyl terminal domains, suggesting that they have a common evolutionary origin. (After D. L. Gill, 1989. Nature 342: 16–18.)

sequence of the DHSR is highly similar to that of L-type Ca^{2+} channels. The large complex formed by the ryanodine receptor in the sarcoplasmic reticulum and the DHSR in the T-tubule membrane can be seen in the

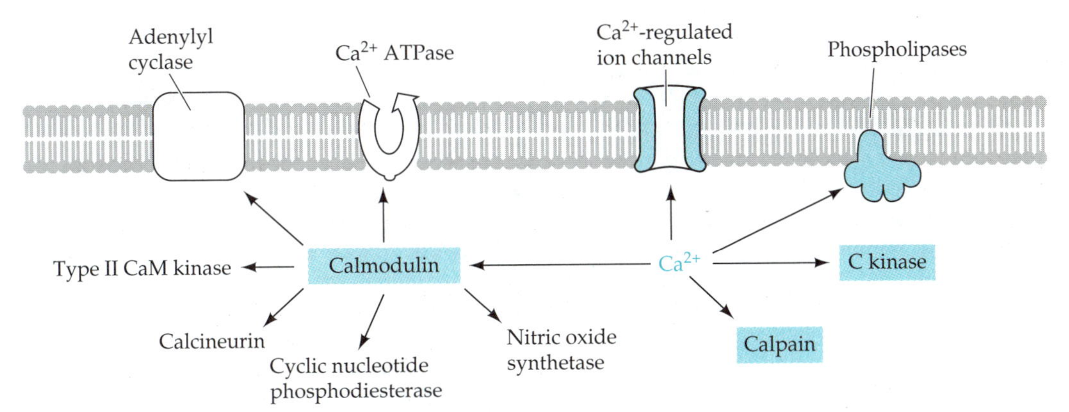

FIGURE 12. Ca^{2+} target proteins. Direct targets of Ca^{2+} are highlighted in color. Many additional proteins are regulated by the Ca^{2+}-bound form of calmodulin.

electron microscope and is called the "SR foot." Depolarization of the T-tubule membrane is communicated via the DHSR directly to the ryanodine receptor, rather than through an intermediate diffusible signal, causing it to open its calcium channel with essentially no delay. Ca^{2+} flows through this channel from the sarcoplasmic reticulum into the cytoplasm to trigger muscle contraction.

In most neurons, the IP_3 receptor is located throughout the endoplasmic reticulum. In cerebellar Purkinje cells, which appear to be specialized to respond to glutamatergic input with a large IP_3-mediated increase in cytosolic Ca^{2+}, the IP_3 receptor is very highly concentrated in the dendritic smooth endoplasmic reticulum.

Ca^{2+}/calmodulin activates a family of protein kinases

The most important Ca^{2+} receptor protein is **calmodulin,** a 15-kD protein that is found in all eukaryotic cells and is present in brain cytosol at a concentration of 30–50 μM (see Box C). Each molecule of calmodulin contains four Ca^{2+} binding domains with dissociation constants of approximately 1 μM. At resting Ca^{2+} concentrations very little Ca^{2+} is bound to calmodulin. As the concentration rises to the μM level, the four binding sites are occupied and calmodulin becomes a multifunctional activator (Figure 12). The Ca^{2+}/calmodulin complex (CaM) regulates a host of other proteins including several protein kinases.

Ca^{2+}/calmodulin-dependent (CaM) protein kinases are more varied in structure and function than either cAMP-dependent protein kinases or the C kinases. Several CaM kinases in brain, including myosin light chain kinase and CaM kinases I and III, have narrow, highly specialized substrate protein specificities. In smooth muscle, myosin light chain kinase phosphorylates myosin and regulates actin-activated ATPase activity. In neurons, the functions of myosin light chain kinase are not known, but it may regulate molecular transport that involves nonmuscle forms of myosin. CaM kinase I phosphorylates the vesicle-associated protein synapsin I at a serine residue that is also phosphorylated by cAMP-dependent

Box C The Structure of Calmodulin

Calmodulin was discovered in 1970 during studies of the cyclic nucleotide metabolizing enzyme cyclic AMP phosphodiesterase. Working independently, Wai Yiu Cheung and Shiro Kakiuchi recognized that the cyclic nucleotide phosphodiesterase catalytic subunit required both Ca^{2+} and a small cytosolic protein for activity. In the next five years, this protein was shown to be a Ca^{2+} binding protein, and other enzymes regulated by it were identified, including brain adenylyl cyclase and the erythrocyte Ca^{2+}-ATPase. It was named the Ca^{2+}-dependent regulator protein (CDR) in 1975; later Cheung, one of its discoverers, proposed the name *calmodulin*. Calmodulin is a small protein that is abundant in most tissues. The sequence is remarkably highly conserved and is nearly identical among all vertebrates.

Calmodulin binds four Ca^{2+} ions with high affinity. In the presence of physiological Mg^{2+} concentrations, binding of Ca^{2+} is cooperative in a pairwise fashion. Two high-affinity sites become occupied first; as the concentration rises, the other two are occupied in rapid succession. The affinity of calmodulin for Ca^{2+} depends strongly upon ionic conditions. In the presence of physiological salt concentration and 1 m*M* Mg^{2+}, reported dissociation constants range from 10^{-6} to 10^{-5} *M*. The four Ca^{2+} binding domains have the **EF hand** structure (Figure A) first described by Robert Kretsinger in the structure of parvalbumin. Bound Ca^{2+} is coordinated with oxygen residues in serine, threonine and acidic amino acids strategically placed in the Ca^{2+} binding loop formed between α helices E and F.

In the absence of Ca^{2+}, the calmodulin chain has a random structure; Ca^{2+} binding confers a fixed and well-defined structure that has been determined by X-ray crystallography. The crystal structure reveals a polypeptide chain whose shape resembles a dumbbell in which each of the two heads contains two bound calcium ions. The heads are connected by a long, central α helix. The nature of the interaction of Ca^{2+}/calmodulin with its target proteins is still the subject of intense study, but Ca^{2+} binding exposes hydrophobic regions within each of the head domains that may be important in these interactions. Many calmodulin binding domains found within calmodulin-regulated proteins are predicted to fold into basic amphiphilic α helices. These helices may form complexes with calmodulin in which the hydrophobic portions of the helix interact with the exposed hydrophobic domains of calmodulin, and the basic residues of the helix form ionic bonds with acidic residues in calmodulin. Different target proteins interact with calmodulin at different, partially overlapping sites, which may account for the wide variation in affinities of target proteins for calmodulin. The most avidly bound target protein, phosphorylase b, binds calmodulin tightly even in the absence of Ca^{2+}. Affinities of other, more conventional target proteins for Ca^{2+}/calmodulin in vitro vary from K_d's of 4 μ*M* to approximately 200 μ*M*. The hierarchy of binding affinities may permit stepwise recruiting of Ca^{2+}-activated proteins as the cytosolic Ca^{2+} concentration rises.

protein kinase. Phosphorylation of synapsin I by cAMP-dependent protein kinase or CaM kinase I modulates its affinity for actin filaments and is thought to regulate the availability of synaptic vesicles for transmitter release (see Box D). No other substrate protein for CaM kinase I is known. CaM kinase III phosphorylates the ribosomal elongation factor, eEF-2. Phosphorylation of eEF-2 inhibits protein synthesis; the physiological role of this inhibition is unknown.

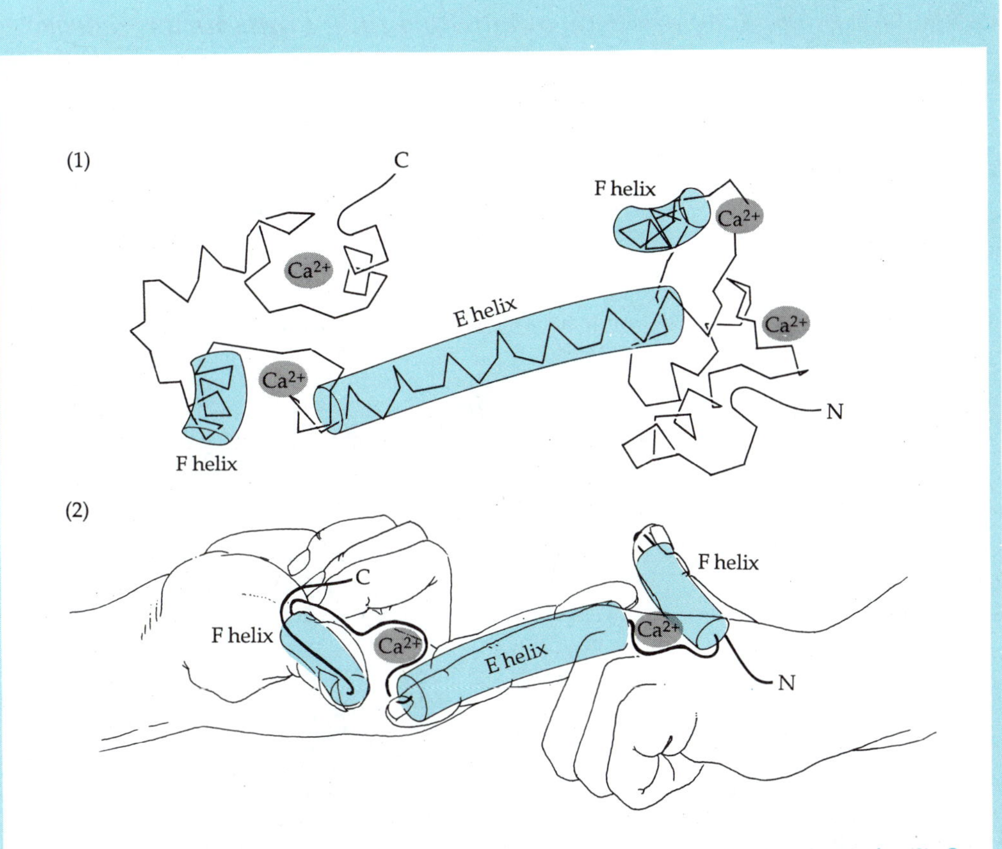

FIGURE A. Schematic diagram of the crystal structure of calmodulin with four bound calcium ions. (1) The crystal structure of calmodulin resembles a dumbbell with two globular domains separated by an extended α helix. The structure of the pockets containing Ca^{2+} is called the EF hand. Each globular domain contains two EF hand motifs. (2) On either side of each Ca^{2+} chelation site are two α helices oriented like a thumb and forefinger extended from a folded hand. This EF hand structure is found in many other high affinity Ca^{2+} binding proteins. (1 After Y. S. Babu et al., 1985. Nature 315: 37–40.)

The most abundant Ca^{2+}/calmodulin-regulated protein kinase in the brain is the type II CaM kinase, a broad-specificity, multifunctional kinase

Type II CaM kinase (or CaM kinase II) is unusually highly concentrated in the brain, particularly in forebrain neurons. For example, it comprises as much as 2% of total hippocampal protein. Within the forebrain, about

Box D Phosphorylation of Synapsin I and Transmitter Release

The amount of transmitter released per impulse at a synaptic terminal is determined by the number of transmitter-containing vesicles that fuse with the presynaptic membrane during each depolarization event. This number can be regulated either by modulating the amount of Ca^{2+} that flows into the terminal during each impulse (Chapter 5) or by regulating the number of vesicles immediately available for Ca^{2+}-triggered fusion at the presynaptic membrane.

One current model holds that in nerve terminals, small synaptic vesicles are bound within a meshwork of actin filaments that associate with an external vesicle protein called **synapsin I** (see Figure 11 in Chapter 5). Synapsin I, in turn, associates with the surface of synaptic vesicles and with intrinsic vesicle membrane proteins (Figure A). Synapsin I was first discovered in 1977 by Tetsufumi Ueda and Paul Greengard as a prominent, brain-specific protein substrate for cAMP-dependent protein kinase. Because synapsin I has no obvious enzymatic activity, its cellular function was not immediately apparent. The presynaptic location of synapsin I in close association with synaptic vesicles prompted the hypothesis that the protein is important for transport of vesicles into nerve terminals or for interactions of vesicles with release zones.

Structural and biochemical studies of synapsin I have refined this hypothesis. Synapsin I is actually a doublet of closely related 83- and 80-kD proteins produced by alternative splicing of a single gene. Each protein has a globular amino-terminal "head" domain of about 50 kD and an extended collagen-like tail domain of about 30 kD. The head domain contains clusters of hydrophobic residues often flanked by charged residues. The tail domain, which is basic, is sensitive to the specific protease collagenase. Synapsin I is not an intrinsic vesicle-membrane protein but associates tightly with the cytosolic surface of vesicles (K_d, 10–100 nM). The head domain binds to phospholipids in the vesicle membrane and appears to partially insert itself into the hydrophobic portion of the bilayer. The tail domain also associates with vesicles, but appears to bind to a specific vesicle membrane protein. In addition to its interaction with vesicles, synapsin I also binds tightly to actin filaments; two high-affinity actin binding sites are located in the head domain.

The associations of synapsin I with vesicles and with actin are modulated by phosphorylation. Synapsin I is phosphorylated at three sites by three different protein kinases (Figure A2). Serine 9 in the head domain (site 1) is phosphorylated by both the cAMP-dependent protein kinase and a very specific Ca^{2+}/calmodulin-dependent protein kinase called CaM kinase I. Serines 566 and 603 in the tail domain (sites 2 and 3) are phosphorylated by type II Ca^{2+}/calmodulin-dependent protein kinase (CaM kinase II). Phosphorylation of sites 2 and 3 reduces the affinity of synapsin I for synaptic vesicles and abolishes its ability to bind actin filaments. Phosphorylation of site 1 has no effect on association of synapsin I with vesicles, but diminishes its ability to bind actin.

Experiments by Paul Greengard, Rodolfo Llinás, and colleagues, in which dephospho- and phosphosynapsin I were injected into the presynaptic terminal of the squid giant synapse, revealed the potential utility of these protein–protein interactions for regulation of transmitter release. Injection of dephosphosynapsin I significantly decreased the amount of neurotransmitter released during each presynaptic impulse, while injection of synapsin I that had been phosphorylated at sites 2 and 3 had no effect. When type II CaM kinase was injected into the terminal, the amount of transmitter released per impulse increased dramatically after the first few impulses. In contrast, injection of heat-inactivated kinase had no effect.

Such experiments are not possible with mammalian presynaptic nerve terminals. However, a related experiment was performed with partially purified **synaptosomes**, pinched off nerve terminals purified

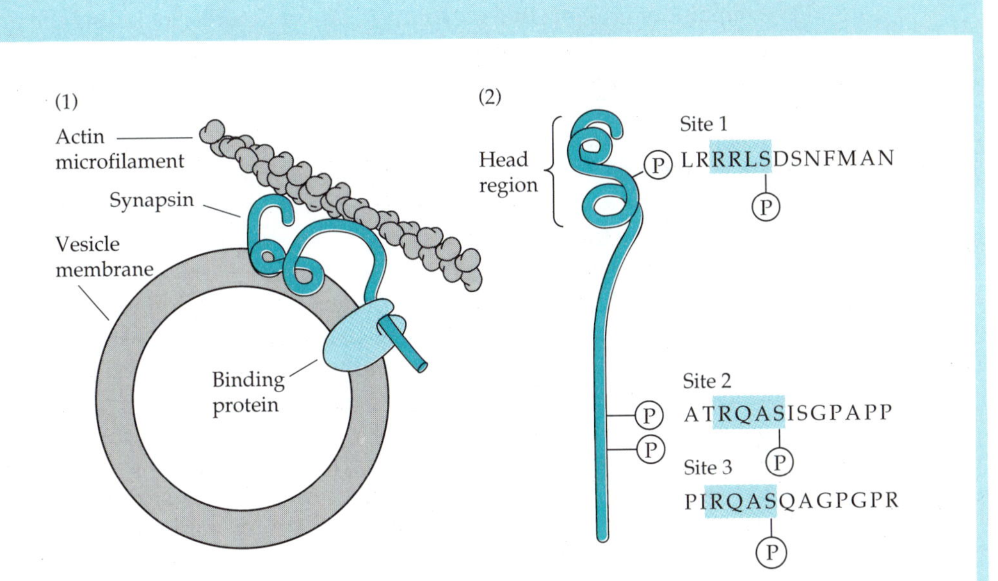

FIGURE A. Association of synapsin with synaptic vesicles is regulated by phosphorylation. (1) The extended, collagen-like tail of synapsin associates with a protein in the vesicle membrane; the globular head group inserts into the vesicle lipid bilayer. Both the head and the boundary between the head and tail contain actin filament binding sites. These binding domains on synapsin are thought to mediate association of synaptic vesicles with actin filaments in synaptic terminals. (2) Binding of synapsin I to synaptic vesicles and to actin filaments is regulated by phosphorylation at two sites. Site 1 is located in the head region and is phosphorylated by the cAMP-dependent protein kinase and by a specialized CaM-dependent protein kinase called CaM kinase I. Phosphorylation of site 1 decreases the affinity of synapsin I for actin filaments. Sites 2 and 3 are located in the collagen-like tail region and are phosphorylated by CaM kinase II. Phosphorylation of sites 2 and 3 decreases the affinity of synapsin I for synaptic vesicles and also reduces its affinity for actin filaments. Phosphorylation of these sites in vivo is thought to regulate transmitter release.

from homogenates of the forebrain. Type II CaM kinase that had been activated by autophosphorylation was introduced into synaptosomes by a rapid freeze/thaw technique. The presence of activated kinase specifically increased the amount of glutamate released from the synaptosomes when they were depolarized in high K^+. Conversely, introduction of a small peptide inhibitor of the kinase decreased depolarization-evoked glutamate release from the synaptosomes.

These experiments support the idea that synaptic vesicles are sequestered in the vicinity of release zones by their association with the actin cytoskeleton (Chapter 8). They suggest that this association is mediated by synapsin I and is regulated by Ca^{2+}/calmodulin-dependent phosphorylation. In this model, addition of extra dephosphosynapsin I removes more vesicles from close proximity to release zones and reduces transmitter release. In contrast, enhanced phosphorylation of synapsin I frees vesicles from association with actin filaments. They then become available for immediate fusion upon depolarization, increasing the amount of transmitter released per impulse.

half of the kinase is distributed throughout the cytosol of the entire neuron. The remainder is associated with particulate structures, including the **postsynaptic density,** a prominent specialization of the submembranous cytoskeleton that is attached to the postsynaptic membrane of CNS synapses (see Chapter 8). The kinase is 20–30% of the total protein in fractions enriched for the postsynaptic density. This association makes it a likely target for Ca^{2+} entering through postsynaptic, ligand-gated ion channels. A particularly intriguing possibility is that CaM kinase II is a major target for Ca^{2+} entering through NMDA receptors during the generation of long-term potentiation (Chapter 14).

CaM kinase II is a family of large, dodecameric oligomers composed of homologous catalytic subunits. There are at least five distinct subunits, α (54 kD), β (60 kD), β' (a 58-kD alternatively spliced version of the β subunit), γ, and δ (each 57 kD). The α and β subunits are expressed only in neurons and are the predominant subunits in brain. The α subunit is present at highest levels in mature forebrain neurons while the β subunit is expressed more uniformly in all neurons. Although the subunits always associate into heteromultimers containing approximately twelve subunits, the proportion of α and β subunits varies in different brain regions. For example, in the forebrain, the ratio of α to β subunits is approximately 3:1 whereas in the cerebellum the ratio is 1:4. The γ and δ subunits are found in many non-neural tissues; in the brain they are relatively minor components.

Neural substrate proteins for CaM kinase II include MAP_2, a microtubule-associated protein; tyrosine hydroxylase; tryptophan hydroxylase; tubulin; and synapsin I, a synaptic vesicle-associated protein. CaM kinase II phosphorylates synapsin at different sites than CaM kinase I and the cAMP-dependent protein kinase. Phosphorylation of these sites regulates the affinity of synapsin for synaptic vesicles, as well as its affinity for actin filaments, and may ultimately regulate transmitter release (see Box D). Many CaM kinase II substrate proteins, including those in the postsynaptic density, remain to be identified.

CaM kinase II may act as a molecular switch

Each CaM kinase subunit is autophosphorylated when the holoenzyme is activated in the presence of Ca^{2+}/calmodulin, changing the properties of the kinase dramatically. This regulatory mechanism may permit the CaM kinase to act as a kind of switch. Nonphosphorylated CaM kinase is catalytically active only in the presence of Ca^{2+}/calmodulin, but when the kinase is autophosphorylated for a few seconds in the presence of Ca^{2+}/calmodulin, a new Ca^{2+}-independent activity appears. The Ca^{2+}-independent activity is the result of autophosphorylation of a single threonine residue located near the calmodulin binding domain (Figures 13 and 14). The location of the site suggests that autophosphorylation partially mimics the effect of calmodulin binding and prevents refolding of the kinase into an inactive conformation after the Ca^{2+}/calmodulin concentration falls. Activation by autophosphorylation is highly cooperative; autophosphorylation of one or two of the subunits in a holoenzyme produces

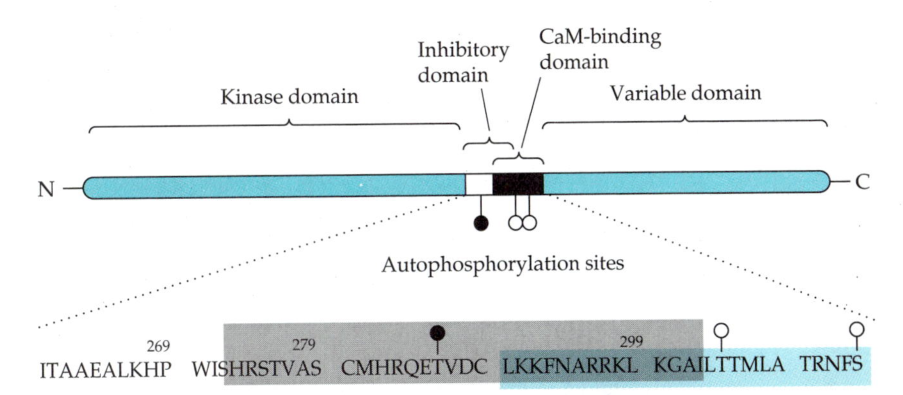

FIGURE 13. Structural domains of type II Ca^{2+}/calmodulin-dependent (CaM) kinase. (A) All of the catalytic subunits of CaM kinase II have a similar functional organization. The catalytic domain is located in the amino terminal portion of the sequence. A 25 amino acid calmodulin binding domain is located in the middle of the sequence. Overlapping this domain on the amino terminal side is an inhibitory domain of about 30 amino acids that inhibits binding of substrates to the active site in the absence of Ca^{2+}/calmodulin. This sequence contains the autophosphorylation site (filled circle) that, when phosphorylated, prevents inhibition by the inhibitory domain after Ca^{2+} is removed. Additional autophosphorylation sites are located within the calmodulin binding domain (open circles; see text). (B) Amino acid sequence of the inhibitory and calmodulin binding domains.

activation of the other subunits, perhaps by allosteric interactions. Autophosphorylation within the holoenzyme also becomes independent of Ca^{2+}; as a result, it can oppose dephosphorylation of the kinase by phosphatases, perhaps slowing the rate of return of the kinase to complete dependence on Ca^{2+}/calmodulin in vivo.

Regulation of CaM kinase II by autophosphorylation can continue through a second stage. When Ca^{2+} is removed after autophosphorylation of the first threonine residue, a second threonine reside, located within the calmodulin binding domain, is exposed and becomes autophosphorylated, causing the kinase to become insensitive to stimulation by Ca^{2+}/calmodulin (Figure 13 and 14). Ca^{2+}-independent activity as well as calmodulin insensitivity are lost after dephosphorylation of the threonines by protein phosphatase.

Experiments with CNS neurons in culture suggest that one function of the autophosphorylation mechanism is to maintain a relatively high level of kinase activity (10 to 30% of maximum) even at basal Ca^{2+} concentrations. The basal Ca^{2+}-independent type II CaM kinase activity appears to be sustained by a dynamic steady state between Ca^{2+}-stimulated (and perhaps also Ca^{2+}-independent) autophosphorylation and dephosphorylation by phosphatases. It has also been widely postulated that the autophosphorylation mechanism may allow CaM kinase II to retain information in situ about prior activating Ca^{2+} signals, thus contributing to memory formation or storage. The retained information would be "read out" as enhanced phosphorylation of functionally significant substrate

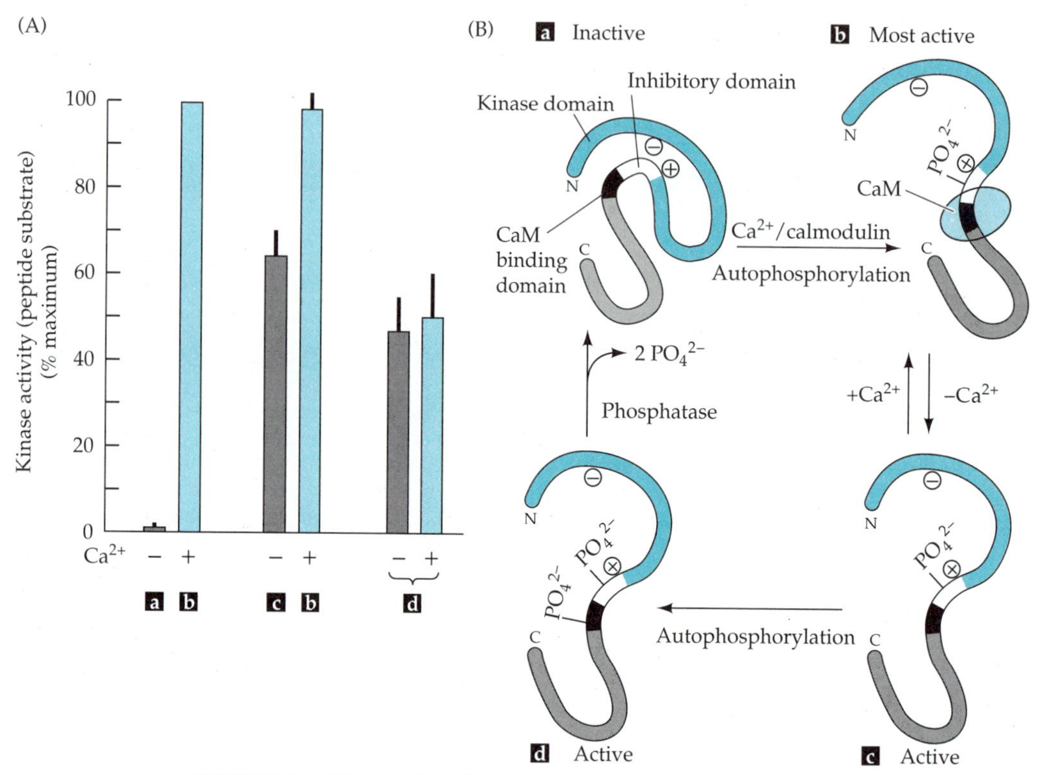

FIGURE 14. Type II CaM kinase can function as a switch. (A) CaM kinase activity in the various autophosphorylated states depicted in (B), measured with a peptide substrate. (B) (a) In its inactive state, the active site of CaM kinase II is inhibited by binding to its own inhibitory domain. (b) After addition of Ca^{2+}, activated calmodulin binds to the calmodulin binding domain, freeing the active site to phosphorylate exogenous substrates. Activation is accompanied by autophosphorylation of threonine-286. (c) When the Ca^{2+} concentration falls, calmodulin diffuses away and the presence of phosphate on threonine-286 prevents binding of the inhibitory domain to the active site. Kinase activity continues at a reduced rate and is now partially independent of Ca^{2+}; the "switch" is on (see A). Readdition of Ca^{2+} causes the kinase to return to state (b), where it displays its maximum catalytic rate. (d) When autophosphorylation continues in the absence of Ca^{2+}, sites within the calmodulin binding domain are autophosphorylated, rendering the kinase insensitive to Ca^{2+}/calmodulin. The kinase is returned to the inactive state (a) by dephosphorylation by protein phosphatases. (After M. B. Kennedy et al., 1990. Cold Spring Harbor Symp. Quant. Biol. 55, 101–110.)

proteins. The length of time that the information would be retained would depend on the balance between the rate of enhanced Ca^{2+}-independent autophosphorylation and the local catalytic rate of cellular phosphatases. This proposed information storage function of autophosphorylation has not yet been demonstrated experimentally.

Ca^{2+}/calmodulin regulates many second messenger–related enzymes

Interactions between regulatory pathways controlled by different second messengers are important for coordinated control of neuronal functions (Figure 1). Many such interactions with the Ca^{2+} pathway are mediated through calmodulin (Figure 12). For example, the brain contains calmodulin-sensitive and -insensitive isozymes of adenylyl cyclase and cAMP phosphodiesterase, and their ratio varies in different neurons and subcellular locations. Thus, a rise in Ca^{2+} concentration at a particular neuronal site may enhance or antagonize production of cyclic AMP depending on the nature of the local cyclases and phosphodiesterases.

Several membrane ion channels are regulated by direct binding of Ca^{2+}

Channels for K^+, Cl^-, and Ca^{2+} itself are activated by direct binding of Ca^{2+} to the channel protein or to proteins closely associated with them. As described later for hippocampal neurons, Ca^{2+}-activated potassium channels are particularly important because they regulate electrical excitability. Ca^{2+} that enters through voltage-gated calcium channels activates these potassium channels, producing a late potassium current that helps to repolarize the membrane.

Ca^{2+} regulates phospholipases and the protease, calpain

Two important families of membrane phospholipases are activated by Ca^{2+} binding. One is phospholipase C, which hydrolyses PIP_2. Ca^{2+} released from internal stores by IP_3 may amplify IP_3 production by prolonging receptor-mediated activation of phospholipase C. The second is **phospholipase-A_2,** which releases fatty acids, including arachidonic acid, from the glycerolipid backbone. Arachidonic acid is a precursor of the paracrine messengers, the eicosanoids. Thus, activation of phospholipase A_2 by Ca^{2+} may stimulate the production of these messengers.

Ca^{2+} binding also activates a group of neutral proteases, the **calpains.** Type I calpain isozymes are activated by micromolar concentrations of Ca^{2+}, while the type II isozymes require millimolar concentrations of Ca^{2+}. Activation of the proteases apparently occurs by an autolytic process initiated by a rise in Ca^{2+} concentration. The autoproteolysis reduces the requirement for Ca^{2+}, thus activating proteolysis of exogenous proteins. More extensive autoproteolysis inactivates the protease. One hypothesis holds that type II calpain is only activated by Ca^{2+} when bound to the membrane and, consequently, it regulates primarily membrane bound or membrane-associated proteins. No specific physiological role for calpain has been demonstrated experimentally in the nervous system. However, because proteins that link cytoskeletal elements are particularly good calpain substrates, it may be that calpain helps to regulate cell shape. Calpain can cleave protein kinase C between its regulatory and catalytic domains, creating an active catalytic fragment referred to as protein kinase M. Similarly, calpain cleaves autophosphorylated type II CaM kinase, releasing an active catalytic fragment. Thus, another physiologically important role of calpain could be the production of constitutively active forms of protein kinase C or CaM kinase II.

Eicosanoids

Metabolites of arachidonic acid can act as both first and second messengers

Activation of norepinephrine, serotonin, NMDA, and FMRFamide receptors stimulate phospholipase A_2, apparently by increasing intracellular Ca^{2+} concentrations. Phospholipase A_2 hydrolyzes fatty acyl esters from membrane phospholipids, releasing free fatty acids, including arachidonic acid (Figure 15). Free arachidonic acid is eventually reincorporated into phospholipid or metabolized to bioactive derivatives, including prostaglandins, leukotrienes, and the HPETEs (hydroperoxyeicosatetraenoic acids), all of which are amphiphilic molecules that diffuse directly across lipid bilayers. The eicosanoids can act as first messengers when they diffuse from their site of synthesis to adjacent cells, or second messengers when they mediate metabolic effects within their cell of origin. As with the gas nitric acid, this property may allow eicosanoids to relay regulatory information from postsynaptic sites back to presynaptic terminals during induction of long-term potentiation in the hippocampus (Chapter 14).

In the nervous system of the mollusc *Aplysia,* metabolites formed from arachidonic acid by 12-lipoxygenase, in particular 12-HPETE (Figure 15), appear to mediate inhibition of transmitter release by the peptide FMRFamide at sensory synapses. When applied to synapses, arachidonic acid and 12-HPETE mimic FMRFamide by increasing the probability of opening of the 5-HT-inactivated potassium channel (I_{KS}, Chapter 14), resulting in inhibition of transmitter release. Furthermore, an inhibitor of lipoxygenase, the enzyme that synthesizes 12-HPETE, blocks activation of the potassium channel by FMRFamide. Eicosanoids can also regulate protein kinase activities; arachidonic acid activates the γ-isozyme of protein kinase C, whereas 12-HPETE potently inhibits type II CaM kinase.

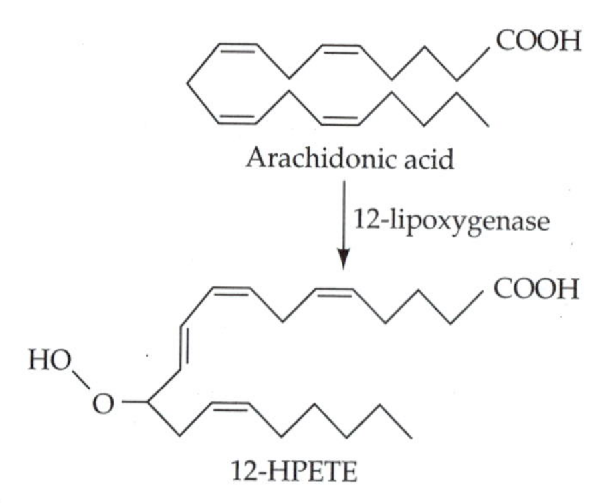

FIGURE 15. Structures of arachidonic acid and 12-HPETE. 12-HPETE ([12s]-hydroperoxyeicosatetraenoic acid) is synthesized from the essential fatty acid arachidonic acid by the enzyme 12-lipoxygenase. Both arachidonic acid and 12-HPETE are thought to have regulatory actions in the nervous system.

Protein Tyrosine Kinases

The protein kinases that are regulated directly by second messengers phosphorylate only serine or threonine residues. A second group of protein kinases that phosphorylate tyrosine residues influence cell signaling during development (Chapter 12) and appear to play regulatory roles in adults as well. There are two distinct classes of protein tyrosine kinases, receptor tyrosine kinases that are located in the cytoplasmic domain of transmembrane receptors, and cytosolic tyrosine kinases that are sometimes associated with the membrane via a fatty acyl group attached to their amino termini. Receptors for many growth factors, including EGF, insulin, and FGF, contain a tyrosine kinase domain that is activated by binding of the growth factor to the receptor. Tyrosine kinase activity mediates the mitogenic effects of the growth factors by initiating a cascade of protein phosphorylation in the cytosol which includes serine/threonine protein kinases that are activated by phosphorylation on tyrosine residues. Like growth factor receptors, cytosolic tyrosine kinases, including c-src and abl, are important for growth control during development; they were discovered when various viral oncogenes were found to be mutant cytosolic tyrosine kinases. However, their mode of regulation and precise role in development is still uncertain. Protein tyrosine phosphorylation can contribute to functional regulation in mature cells, as, for example, by regulation of ACh receptor desensitization. But only a small fraction of total protein phosphate in adult tissues is located on tyrosine residues.

Protein Phosphatases

Steady state phosphorylation of substrate proteins is determined by the balance between rates of phosphorylation and dephosphorylation

The concentration of many phosphoproteins is maintained in situ by continuous phosphorylation and dephosphorylation. The proportion of a protein that is phosphorylated determines the level of its function in the cell, and can be altered by changing its rate of phosphorylation or its rate of dephosphorylation. **Protein phosphatases,** like protein kinases, are tightly regulated. However, in contrast to protein kinases, all but one of the known protein phosphatases are not directly regulated by second messengers. Their structures are less well understood than those of protein kinases, but they can be divided into at least two major classes, those that dephosphorylate phosphoserine or phosphothreonine, and those that dephosphorylate phosphotyrosine. The catalytic subunits of **serine/threonine phosphatases** are small soluble proteins. In contrast, most known **phosphotyrosine phosphatases** are located in the cytoplasmic domains of membrane spanning surface recognition proteins.

There are four major classes of protein serine/threonine phosphatases

Unlike protein kinases, many of the serine/threonine protein phosphatases have broad, overlapping substrate specificities, making them difficult to classify. In one widely used scheme, the phosphatases are divided into types 1 and 2 based on their different specificities for the two subunits of phosphorylase kinase. **Phosphatase-1** is inhibited by two heat and acid stable protein inhibitors, termed inhibitor-1 and inhibitor-2, whereas the type 2 protein phosphatases are insensitive to these inhibitors. Three different type 2 phosphatases can be distinguished by their dependence on divalent cations: **phosphatase-2B** requires Ca^{2+} and is regulated by Ca^{2+}/calmodulin; **phosphatase-2C** requires Mg^{2+}; and **phosphatase-2A** requires neither.

The sequences of catalytic subunits, corresponding to phosphatases-1, 2A, 2B, and 2C have been determined. Together they are thought to account for most, but not all, protein/serine/threonine phosphatase activity. The catalytic subunits of phosphatases-1, 2A, and 2B are about 40% identical, identifying them as members of a gene family. Phosphatase-2C has no sequence homology to the other protein phosphatases and thus defines a second protein serine/threonine phosphatase gene family.

Most phosphatase catalytic subunits associate with regulatory subunits in vivo

Within cells, the catalytic subunit of phosphatase-1 forms complexes with one of several high molecular weight **"targeting" subunits** that bind it to specific subcellular structures and inhibit its activity. The best-studied targeting subunits are those from skeletal muscle where a G subunit binds phosphatase-1 to glycogen particles and an M subunit binds it to myofibrils. Targeting subunits have not yet been characterized in brain. In the cytosol, phosphatase-1 usually exists as an inactive complex with inhibitor-2. Phosphorylation of the targeting subunits or inhibitor-2 by various second messenger-regulated protein kinases releases active phosphatase-1 catalytic units, resulting in indirect activation of phosphatase activity by second messengers. Another cytosolic inhibitor of phosphatase-1 is inhibitor-1, a small protein that, in contrast to inhibitor-2, becomes a phosphatase inhibitor only when phosphorylated by the cAMP-dependent protein kinase. Through this protein, cAMP can inhibit phosphatase activity indirectly by activation of the cAMP-dependent protein kinase.

The catalytic subunit of phosphatase-2A associates with two cytosolic proteins, termed A and B. The functions of these proteins are poorly understood, but they are believed to regulate phosphatase activity. Unlike the targeting subunits of phosphatase-1, they do not appear to bind to subcellular organelles.

Phosphatase-2B, also called calcineurin, is regulated by Ca^{2+}/calmodulin. It is an abundant protein in brain, comprising about 1% of total protein, but it has a relatively narrow substrate specificity. Many of the proteins dephosphorylated by calcineurin are phosphorylated by the cAMP-dependent protein kinase. Among those is inhibitor-1, which inhibits phosphatase-1 only when phosphorylated. Dephosphorylation of inhibitor-1 by calcineurin destroys its ability to inhibit phosphatase-1, forming a regulatory circuit in which activation of calcineurin by Ca^{2+} reduces the functional effects of the cAMP-dependent protein kinase and other protein kinases by indirectly activating phosphatase-1 (Figure 16). Another example of antagonism of the actions of cAMP by calcineurin occurs in

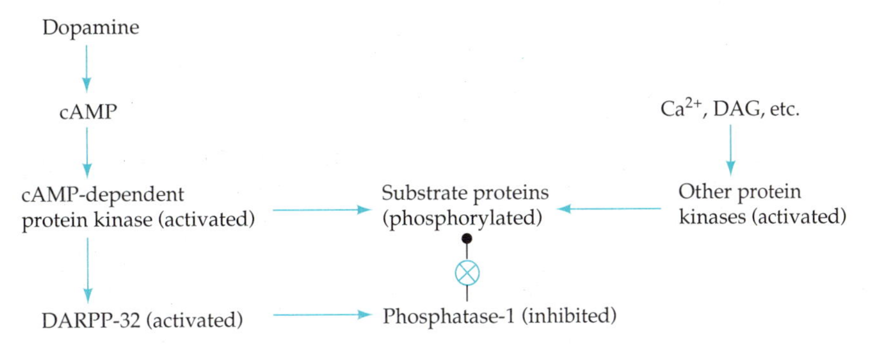

FIGURE 16. Circuit in which dopamine, acting through phosphorylation of DARRP-32 by the cAMP-dependent protein kinase, can potentiate the action of other protein phosphorylation pathways by decreasing phosphatase activity. (After E. Nestler and P. Greengard, 1984. *Protein Phosphorylation in the Nervous System,* Wiley, New York.)

neurons in which phosphorylation of L-type Ca^{2+} channels, or their associated proteins, by the cAMP-dependent protein kinase enhances their activation by depolarization. In these neurons dephosphorylation by a calcium-dependent phosphatase, postulated to be calcineurin, reverses the effect of cAMP.

DARPP-32 is a homologue of inhibitor-1 found only in neurons that express D_1 dopamine receptors

D_1 dopamine receptors, which activate adenylyl cyclase, are abundant in the neostriatum of the forebrain, where they mediate the action of dopamine released by synapses from the substantia nigra, a structure important in control of movement (Chapter 15). Striatal dopaminergic neurons contain high concentrations of a protein, termed **DARPP-32** (*D*opamine and c*A*MP-*r*egulated *p*hospho*p*rotein of apparent molecular weight 32,000), that is both structurally and functionally homologous to phosphatase inhibitor-1. When DARPP-32 is phosphorylated by the cAMP-dependent protein kinase at a specific threonine residue it becomes a potent inhibitor of phosphatase-1. This protein may provide a specialized mechanism through which dopamine can modulate the action of other neurotransmitters by inhibiting a phosphatase that antagonizes the action of second messenger-regulated protein kinases (Figure 16).

Because the substrate specificities of the protein phosphatases are broad and overlapping, there is also overlap in their participation in control of cellular physiology

Many phosphorylated proteins can be dephosphorylated by phosphatases-1, 2A, and 2C, although at different rates. In skeletal muscle, where regulation of the phosphatases has been most carefully studied, a substantial percentage of phosphatase-1 is held within glycogen granules by the targeting protein, G subunit. In contrast, phosphatase-2A is primarily cytosolic. These different cellular locations appear to control access to phosphorylated substrates. Glycogen phosphorylase and glycogen synthase, two major enzymes that control glycogen metabolism and are regulated by phosphorylation, are bound to glycogen particles. Thus, under normal physiological conditions, phosphatase-1 is the principal phosphatase involved in regulation of glycogen metabolism. When glycogen stores are depleted, however, and the phosphorylase and synthase are released into the cytosol, phosphatase-2A may become the more important phosphatase. Thus, intracellular targeting as well as regulation of activity may determine the contribution of each phosphatase to steady state phosphorylation levels of substrate proteins.

Regulation through Multiple Second Messenger Pathways

Few neuronal functions are regulated by only a single second messenger. Instead, signals from different pathways often converge upon critical molecules—for example, by stimulating phosphorylation at the same or different sites within a protein. The different regulatory pathways can either reinforce or antagonize each other. The following three examples illustrate a variety of functional interactions among the major second messenger pathways.

The rate of desensitization of the ACh receptor is regulated by protein phosphorylation

cAMP-dependent protein kinase phosphorylates, and thereby regulates, a number of ion channels. One channel for which both the sites of phosphorylation and its physiological effects are known is the nicotinic ACh receptor. After exposure to ACh for several seconds, the ACh receptor desensitizes, that is, it assumes a conformation in which it is unresponsive to ACh (Chapter 3). At the neuromuscular junction, agents that stimulate synthesis of cAMP, including noradrenaline, calcitonin gene-related peptide (CGRP), and the drug forskolin, a plant alkaloid, enhance the rate of **desensitization** of the ACh receptor as much as tenfold. Thus, modulatory agents acting through cAMP regulate the sensitivity of the motor endplate to acetylcholine.

The cAMP-dependent protein kinase phosphorylates the purified ACh receptor at serine 353 in the γ subunit and serine 361 in the δ subunit. Both sites are contained within the consensus sequence Arg-Arg-X-Ser and are located in a segment of about 20 amino acids in the cytoplasmic loop between the third and fourth transmembrane domains (Figure 17). In purified reconstituted receptors and in living rat myotubes, enhanced desensitization correlates with phosphorylation of serine-361 in the δ subunit and is not related to phosphorylation of the γ subunit.

Other protein kinases, including the C kinase and a protein tyrosine kinase, can also phosphorylate the ACh receptor β, γ, and δ subunits at sites within the same 20 amino acid segments. Phosphorylation of purified receptor by these kinases enhances the rate of desensitization after reconstitution in lipid bilayers. This suggests that other protein kinases, stim-

(A)

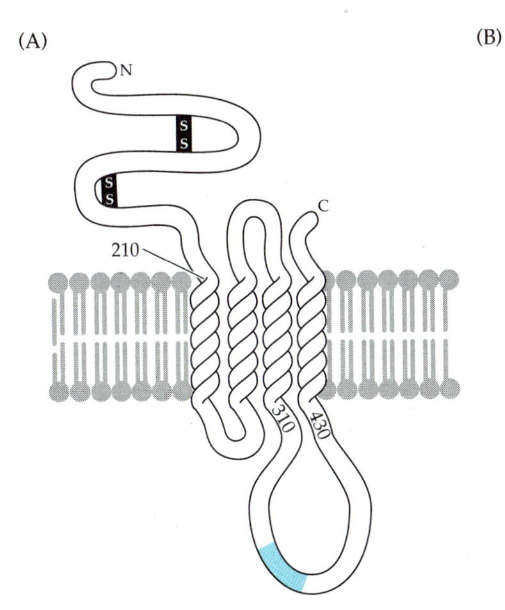

(B)

cAMP-dependent protein kinase

γ (346) Lys Pro Gln Pro Arg Arg Arg Ser Ser Phe Gly Ile (357)

δ (355) Leu Lys Leu Arg Arg Ser Ser Ser Val Gly Tyr (365)

C-kinase

δ (355) Leu Lys Leu Arg Arg Ser Ser Ser Val Gly Tyr (365)
? ? ?

Tyrosine kinase

β (348) Ile Ser Arg Ala Asn Asp Glu Tyr Phe Ile Arg Lys (354)

γ (359) Ile Lys Ala Glu Glu Tyr Ile Leu Lys (367)

δ (368) Lys Ala Gln Glu Tyr Phe Asn Ile Lys Ser Arg (378)

FIGURE 17. Desensitization of the ACh receptor is regulated by phosphorylation. (A) A 30 amino acid segment between the third and fourth transmembrane domains (M3 and M4) of the ACh receptor subunits contains phosphorylation sites for the cAMP-dependent protein kinase, the C kinase, and unidentified tyrosine kinases. Phosphorylation of these sites regulates the rate of desensitization of the receptor. (B) Phosphorylation sites (color) for the cAMP-dependent protein kinase are located in the γ and δ subunits. One for the C kinase is located in the δ subunit. It is not known which of the three serines is phosphorylated by the C kinase. Three tyrosine kinase phosphorylation sites are located in the β, γ, and δ subunits. (After R. Huganir and P. Greengard, 1990. Neuron 5: 555–567.)

ulated by different second messenger pathways, may also regulate ACh receptor desensitization in vivo, providing convergent control of receptor sensitivity. The integration of signaling pathways at the level of a single protein by the selective covalent modification of different amino acids in its sequence is an important theme that recurs throughout neuronal regulatory pathways.

The catalytic rate of tyrosine hydroxylase is regulated by phosphorylation

Stimulation of sympathetic nerves, which use noradrenaline as transmitter, rapidly activates **tyrosine hydroxylase,** enhancing the rate of new transmitter synthesis (Chapter 5). The increase in activity is observed within 1 minute of stimulation and is rapidly reversed when stimulation ends. A similar enhancement is also induced by ACh acting on nicotinic cholinergic receptors, by peptides in the secretin family, and by nerve growth factor. Each of these agents directly or indirectly increases the concentration of cAMP, Ca^{2+}, or diacylglycerol, which then activate different protein kinases in the nerve terminal.

Activation of tyrosine hydroxylase thus provides a second example of

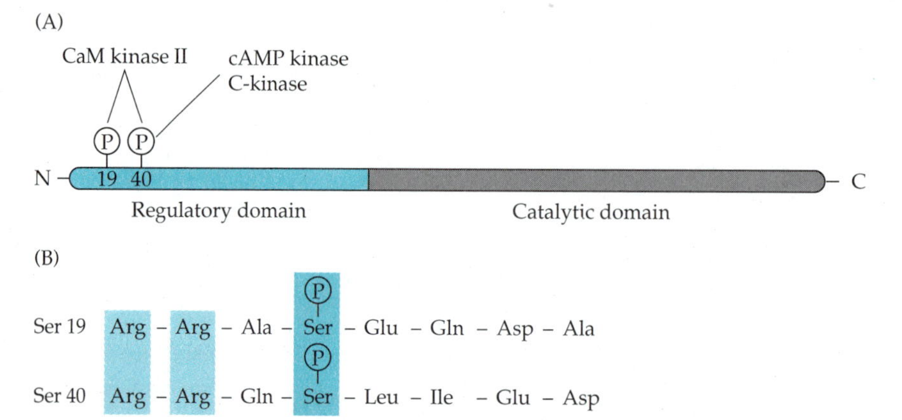

FIGURE 18. The catalytic rate of tyrosine hydroxylase is regulated by phosphorylation. (A) The catalytic domain of tyrosine hydroxylase is located in the carboxyl terminal half and is subject to regulation by residues in the amino-terminal half. Phosphorylation of at least two sites in the regulatory domain, serine 19 and serine 40, influences catalytic rate. These sites are phosphorylated by cAMP-dependent protein kinase (cAMP kinase), type II CaM kinase, and C kinase, as shown. (B) Serine 19 and serine 40 are both contained within consensus sequences for the cAMP-dependent protein kinase and CaM kinase II (see Table 1). Serine 19, however, is phosphorylated most rapidly by CaM kinase II whereas serine 40 is phosphorylated rapidly by the cAMP-dependent protein kinase and the C kinase and only slowly by CaM kinase II. (After D. G. Campbell et al., 1986. J. Biol. Chem. 261: 10,489–10,492.)

integration of signals at the level of protein phosphorylation. The three major protein kinases—cAMP-dependent protein kinase, type II CaM kinase, and the C kinase—phosphorylate tyrosine hydroxylase at two sites within its regulatory domain (Figure 18). cAMP-dependent protein kinase and the C kinase phosphorylate serine 40, which is also phosphorylated at a slower rate by the CaM kinase. The second site, serine 19, is the major site phosphorylated by the CaM kinase. Phosphorylation in each case alters the conformation of tyrosine hydroxylase to increase the affinity of the enzyme for tetrahydrobiopterin, an essential cofactor present in the cytosol at subsaturating concentrations. Phosphorylation also increases the maximum catalytic velocity. The mechanisms for enhancing tyrosine hydroxylase activity are not identical for phosphorylation at different sites. For example, activation of tyrosine hydroxylase after phosphorylation of serine 19 by CaM kinase, but not after phosphorylation by the other kinases, requires the participation of a 70-kD cytosolic protein. In each case, however, the ultimate physiological effect is the same: increased transmitter synthesis. Thus, information from different receptor proteins using different second messenger pathways converges to produce enhanced transmitter synthesis.

Adaptation in the hippocampus is modulated through both the cAMP and PIP_2 signaling pathways

Second messenger pathways often regulate ion channels that control critical membrane properties. In the hippocampus, two distinct second messenger pathways converge to regulate a single important channel, the **slow Ca^{2+}-activated potassium channel,** which controls membrane excitability. This regulation provides an example of synergism between the cAMP and DAG pathways, as well as antagonism between these two pathways and the action of Ca^{2+}.

A burst of action potentials in a hippocampal pyramidal cell is usually followed by a slow hyperpolarizing potential lasting several seconds (Figure 19A). During the action potentials, Ca^{2+} enters the cell through voltage-dependent Ca^{2+} channels and activates Ca^{2+}-dependent K^+ channels, causing the hyperpolarizing potential. The **slow after-hyperpolarization** substantially reduces the rate of action potential discharge, thus serving as a mechanism of **adaptation.** When Ca^{2+} entry is prevented by stimulating the neurons in the presence of blockers of voltage-dependent Ca^{2+} chan-

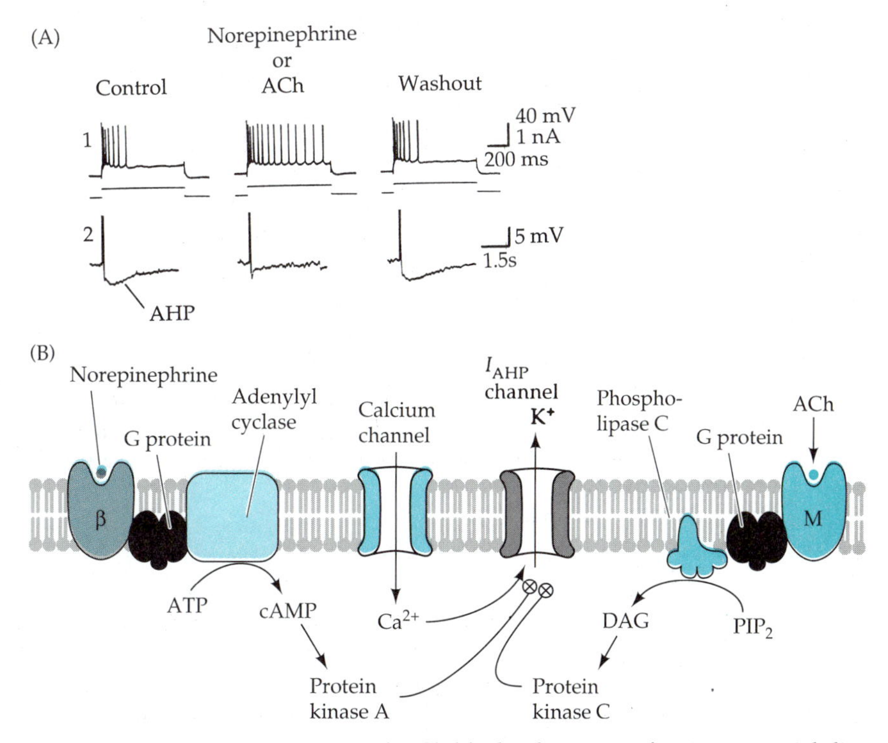

FIGURE 19. Norepinephrine and ACh block adaptation of action potential discharge and the Ca^{2+}-activated K^+ channel in hippocampal pyramidal cells. (A) Control: (1) Action potential discharge in response to a depolarizing current pulse. The current trace is below the voltage trace. (2) Response to a single 100-msec depolarizing current. The single action potential is followed by a hyperpolarizing potential (AHP) generated by the Ca^{2+}-activated K^+ current. Norepinephrine or ACh: (1) Application of norepinephrine or ACh to the bath eliminates adaptation of action potential discharges. (2) The AHP is also eliminated. Washout: After norepinephrine or ACh are washed out of the bath, both adaptation (1) and the AHP (2) return to control levels. (B) Model for inhibition of the Ca^{2+}-activated K^+current by the cAMP-dependent protein kinase, activated by the β-adrenergic receptor, and the C kinase, activated by the muscarinic receptor. (After R. Nicoll, 1988. Science 241: 545–551.)

nels, adaptation is eliminated, so that action potentials continue for the duration of the depolarizing stimulus.

Application of norepinephrine to hippocampal neurons modulates adaptation by reducing the amplitude of the slow after-hyperpolarization (Figure 19A). This action of norepinephrine is mediated by β1-adrenergic receptors, which activate adenylyl cyclase through a G protein (Chapter 5). A membrane permeant analog of cAMP, 8-bromo cAMP, and forskolin, a pharmacological activator of adenylyl cyclase, produce a similar reduction in the after-hyperpolarization. As neither norepinephrine nor cAMP alter the behavior of voltage-sensitive calcium channels, cAMP must act after the entry of Ca^{2+} into the neuron. Phosphorylation and inactivation of the Ca^{2+}-activated potassium channel by cAMP-dependent protein kinase would be the simplest mechanism (Figure 19B); alternatively, channel-associated proteins may be phosphorylated, modulating behavior of the channel.

A second neurotransmitter, ACh, also reduces the slow after-hyperpolarization in hippocampal pyramidal cells. In this case, activation of muscarinic receptors leads to hydrolysis of PIP_2 and the production of DAG. The resulting activation of C kinase then causes inhibition of the slow after-hyperpolarization. When phorbol esters, which enter the neuronal membrane and stimulate C kinase by mimicking DAG, are applied to the hippocampal neurons, they cause inhibition of the slow after-hyperpolarization. Application of analogs of phorbol esters that do not stimulate C kinase have no effect on the slow after-hyperpolarization. These regulatory effects could be explained by phosphorylation and inactivation of the Ca^{2+}-activated potassium channel by the C kinase. The channel has not yet been purified; therefore, its regulation by phosphorylation cannot be confirmed. Nevertheless, it is clear that modulatory transmitters, acting through two distinct second messenger pathways, fine tune the excitability of hippocampal pyramidal neurons by inactivating the Ca^{2+}-regulated K^+ conductance.

Summary

To adapt to a constantly changing environment, neurons have evolved an intricate network of regulatory pathways that permit external stimuli to regulate and coordinate metabolism, membrane excitability, synaptic efficacy, and other critical neuronal functions. Regulation within the network is highly dynamic. Through the network environmental signals continuously modulate the rates of enzymatic reactions and the openings and closings of channels. Transient alterations produced by the network, lasting a few tenths of a second to a few hours, usually result from post-translational modifications of proteins. Modulation of gene expression in the nucleus and of protein synthesis in the cytosol causes slower and more enduring changes, such as changes in cellular structure.

Important intermediaries in the control system are small second messengers that couple surface receptors to an array of cytosolic regulatory effector molecules. The four major second messenger pathways are mediated by cyclic AMP, cyclic GMP, calcium ion, and the hydrolysis products of phosphatidylinositol bisphosphate (diacylglycerol and inositol trisphosphate). Additional small messenger molecules whose importance has been recognized recently include arachidonic acid and its metabolites, and nitric oxide.

Second messengers exert many of their effects through protein kinases that catalyze transfer of phosphate from ATP to specific serine, threonine, or tyrosine residues on proteins. Addition of phosphate to proteins alters their folding and often regulates their function. Rapid turnover of phosphorylated proteins, maintained by the balance between protein kinase and phosphatase activities, permits rapid adjustments in their concentrations. In addition to their action on kinases, second messengers regulate a variety of other enzymes such as Ca^{2+}-activated proteases, membrane transport molecules, and nitric oxide synthetase. They also bind to channel proteins and alter their functions.

A recurrent theme in neuronal regulation by second messengers is the integration of information from several signaling pathways at the level of a single protein through phosphorylation of different amino acids in its sequence. This theme is illustrated in the control of transmitter synthesis and release, in regulation of receptor desensitization, and in the regulation of membrane ion channels that control excitability.

References

Regulatory networks

Bray, D. 1990. Intracellular signalling as a parallel distributed process. J. Theor. Biol. 143: 215–231.

cAMP-regulated effector proteins

Huganir, R. L. and Greengard, P. 1990. Regulation of neurotransmitter receptor desensitization by protein phosphorylation. Neuron 5: 555–567.

Nicoll, R. A. 1988. The coupling of neurotransmitter receptors to ion channels in the brain. Science 241: 545–551.

*Taylor, S. S., Buechler, J. A. and Yonemoto, W. 1990. cAMP-dependent protein kinase: Framework for a diverse family of regulatory enzymes. Annu. Rev. Biochem. 59: 971–1005.

Zigmund, R. E., Schwarzschild, M. A. and Rittenhouse, A. R. 1989. Acute regulation of tyrosine hydroxylase by nerve activity and by neurotransmitters via phosphorylation. Annu. Rev. Neurosci. 12: 415–462.

cGMP-regulated pathways

Collier, J. and Vallance, P. 1989. Second messenger role for NO widens to nervous and immune systems. Trends Pharmacol. Sci. 10: 427–431.

Garbers, D. L. 1990. The guanylyl cyclase receptor family. New Biologist 2: 499–504.

Phosphatidylinositol phosphate-regulated pathways

*Berridge, M. J. and Irvine, R. F. 1989. Inositol phosphates and cell signalling. Nature 341: 197–205.

Huang, K. 1989. The mechanism of protein kinase C activation. Trends Neurosci. 12: 425–432.

Kikkawa, U., Kishimoto, A. and Nishizuka, Y. 1989. The protein kinase C family: Heterogeneity and its implications. Annu. Rev. Biochem. 58: 31–44.

Rhee, S. G., Suh, P.-G., Ryu, S.-H. and Lee, S. Y. 1989. Studies of inositol phospholipid-specific phospholipase C. Science 244: 546–550.

Control of intracellular calcium concentration.

Blaustein, M. 1988. Calcium transport and buffering in neurons. Trends Neurosci. 11: 438–443.

Carafoli, E. 1987. Intracellular calcium homeostasis. Annu. Rev. Biochem. 56: 395–433.

Ferris, C., Huganir, R. L., Supattapone, S. and Snyder, S. H. 1989. Purified inositol 1,4,5-trisphosphate receptor mediates calcium flux in reconstituted lipid vesicles. Nature 342: 87–89.

Furuichi, T., Yoshikawa, S., Miyawaki, A., Wada, K., Maeda, N. and Mikoshiba, K. 1989. Primary structure and functional expression of the inositol 1,4,5-trisphosphate-binding protein P_{400}. Nature 342: 32–38.

Miller, R. 1988. Calcium signalling in neurons. Trends Neurosci. 11: 415–419.

Tsien, R. Y. 1988. Fluorescence measurement and photochemical manipulation of cytosolic free calcium. Trends Neurosci. 11: 419–424.

Tsien, R. W., Lipscombe, D., Madison, D., Bley, K. and Fox, A. 1988. Multiple types of neuronal calcium channels and their selective modulation. Trends Neurosci. 11: 431–438.

*Tsien, R. W. and Tsien, R. Y. 1990. Calcium channels, stores, and oscillations. Annu. Rev. Cell Biol. 6: 715–760.

Calcium-regulated effector proteins

DeCamilli, P., Benfenati, F., Valtorta, F. and Greengard, P. 1990. The synapsins. Annu. Rev. Cell Biol. 6: 433–460.

*Kennedy, M. 1989. Regulation of neuronal function by calcium. Trends Neurosci. 12: 417–420.

Kennedy, M. B., Bennett, M. K., Erondu, N. E. and Miller, S. G. 1987. Calcium/calmodulin-dependent protein kinases. In *Calcium and Cell Function,* Vol. 7, W. Y. Cheung (ed.), Academic Press, New York, pp. 62–107.

Mellgren, R. L. 1987. Calcium-dependent proteases: An enzyme system active at cellular membranes? FASEB J. 1: 110–115.

Melloni, E. and Pontremoli, S. 1989. The calpains. Trends Neurosci. 12: 438–443.

Persechini, A., Moncrief, N. and Kretsinger, R. 1989. The EF-hand family of calcium-modulated proteins. Trends Neurosci. 12: 462–467.

Smith, S. and Augustine, G. 1988. Calcium ions, active zones and synaptic transmitter release. Trends Neurosci. 11: 458–464.

Eicosanoids

Piomelli, D. and Greengard, P. 1990. Lipoxygenase metabolites of arachidonic acid in neuronal transmembrane signalling. Trends Pharmacol. Sci. 11: 367–373.

Protein kinases

Edelman, A., Blumenthal, D. and Krebs, E. 1987. Protein serine/threonine kinases. Annu. Rev. Biochem. 56: 567–613.

Hardie, G. 1988. Pseudosubstrates turn off protein kinases. Nature 335: 592–593.

Hunter, T.1987. A thousand and one protein kinases. Cell 50: 823–829.

Kemp, B. and Pearson, R. 1990. Protein kinase recognition sequence motifs. Trends Biochem. Sci. 15: 342–346.

*Nairn, A. C., Hemmings, H. C. and Greengard, P. 1985. Protein kinases in the brain. Annu. Rev. Biochem. 54: 931–976.

Protein phosphatases

Cohen, P. 1989. The structure and regulation of protein phosphatases. Annu. Rev. Biochem. 58:453–508.

8

The Neuronal Cytoskeleton

Ron D. Vale, Gary Banker,
and Zach W. Hall

FOR NEURONS, perhaps more than for any other cell type, form is linked to function. Neurons have extraordinarily complex and diverse shapes that determine both their signaling properties and the pattern of synaptic connections that they form. Their dimensions are unparalleled. In large animals individual axons may extend for meters with a length-to-diameter ratio that in the extreme approaches 10^7, the equivalent of a pencil lead six miles long. In such a cell, the volume of the axoplasm may be orders of magnitude greater than that of the cell body. This unique architecture clearly poses special problems for cellular function and maintenance. Neurons also exhibit an astonishing variety of axonal and dendritic branching patterns; these are nevertheless so precisely reproduced that a neuron type may be identified at a glance by the shape of its dendritic tree. The shape and dimensions of neurons impose remarkable demands on the neuronal cytoskeleton, which is largely responsible for the generation of axonal and dendritic arbors during development, for their maintenance throughout life, and for their alteration in response to changing function.

The degree of subcellular specialization within each neuron is also unusual and is directly related to function. The differences between dendrites and axons determine the polarity of information flow within each cell. Both axonal and dendritic membranes consist of a patchwork of specialized domains that include the initial segment of the axon, the nodes of Ranvier, and pre- and postsynaptic specializations, each containing its own particular subset of receptors, ion channels, and transporters. By providing the structural substrate for the directed transport of materials into axons and dendrites, and by interacting with membrane proteins to restrict their localization to appropriate domains, the cytoskeleton plays a major role in establishing and maintaining the regional specialization within neurons.

The shapes of neurons are not immutable but change in response to age, experience, electrical activity, denervation, or injury. The cytoskeleton thus not only supports cellular structure but also modifies it. The counterplay between stability and change is a central feature of the cytoskeleton; it is rooted in the fundamental mechanisms that govern assembly and disassembly, and the stabilization and destabilization of individual cytoskeletal polymers within neurons. Although these properties are characteristic of the cytoskeleton in all cells, they are particularly vital for neurons.

Cytoskeletal Components and Their Assembly in Neurons

The cytoskeleton gives neurons their shape

If a cultured neuron is treated with detergent, the lipids and proteins of the membrane are solubilized and the soluble proteins escape, leaving an insoluble cytoskeleton behind. Remarkably, after such treatment, the cytoskeleton retains almost completely the original shape of the neuron. Similarly, the interior of the squid giant axon (the axoplasm) retains its cylindrical shape after being extruded from the plasma membrane. These observations reveal that the cytoskeleton forms a stiff, highly cross-linked gel that fills the entire volume of the cell. This impression is borne out by electron microscopy, which demonstrates that axons and dendrites contain numerous long filaments that are tightly packed and cross-linked to each other (Figure 1). Mitochondria and other membrane-bounded organelles that are embedded in the cytoskeletal matrix are not, however, immobilized by the gel. As we shall see later, these organelles are transported rapidly down the axon, suggesting that the jungle of cytoskeletal cross-links, which appear to pose such an impenetrable barrier, must be dynamically associating and dissociating or "breathing" to allow their passage.

In the electron micrograph shown in Figure 1, two different longitudinal filament types can be discerned: **microtubules,** which are hollow tubes 25 nm in diameter, and **neurofilaments,** whose diameter is approximately 10 nm. Both are cross-linked to each other as well to other filaments of the same type. A third filament type, **actin microfilaments** (8 nm in diameter), form a cortical network just under the surface membrane of axons (Figure 2). Actin microfilaments are primarily at the periphery of axons and are particularly enriched in growth cones. Microtubules and neurofilaments, in contrast, are found throughout the cytoplasm, but are less abundant in the actin-rich cortex.

Each of the filaments is a polymer formed of repeating subunits

The cytoskeleton has the remarkable ability to form both very stable and very dynamic structures and to undergo transitions between these states. In the adult, the cytoskeleton gives mechanical strength to axons and dendrites, and provides, through the microtubules, a track for the transport of materials between the cell body and the nerve terminal. These tasks are executed by stable cytoskeletal structures. In contrast, during development or after transection of peripheral neurons, growth cones can elongate, retract, or rapidly change their shape; all of these events require a highly plastic cytoskeleton.

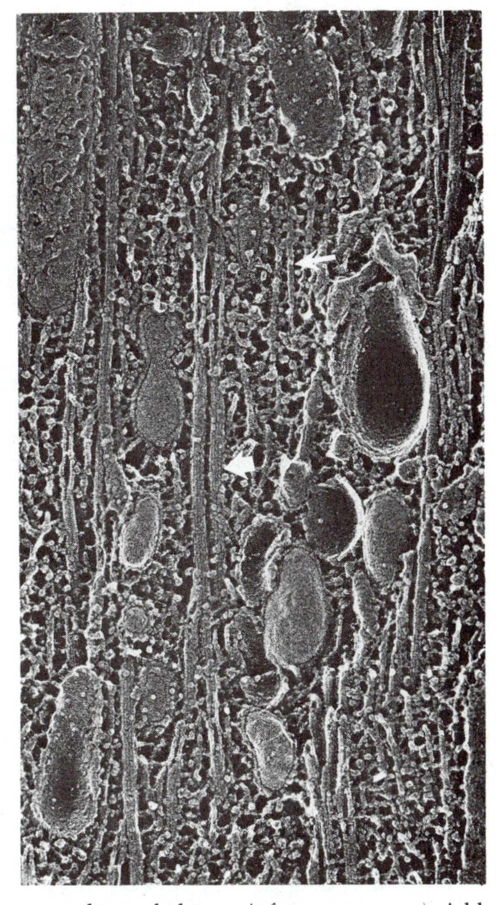

FIGURE 1. The axonal cytoskeleton. A frog axon was quickly frozen, fractured, etched, and then rotary shadowed and viewed in the electron microscope. A cross-linked network of microtubules (thick arrow) and neurofilaments (thin arrow) fills the axon. Vesicles of various sizes and mitochondria are interspersed in the network. (From N. Hirokawa, 1982. J. Cell Biol. 94: 129–142.)

The ability of the cytoskeleton to form both stable and plastic structures lies in the properties of the polymeric filaments, particularly the actin filaments and microtubules. Both of these filaments, as well as the neurofilaments, are composed of noncovalently linked subunits. Microfilaments are formed from **actin**, a 43-kD globular protein that self-assembles into a linear polymer that appears double helical. The monomeric subunit of microtubules is **tubulin**, a heterodimer of two closely related globular proteins of about 50 kD each, called alpha- and beta-tubulin. Microtubules are hollow cylinders whose sides are formed by 13 protofilaments, each of which is a linear array of tubulin heterodimers stacked head to tail. Both actin and tubulin monomers are asymmetric and associate in a specific orientation. As a result, microfilaments and microtubules are inherently *polar* structures whose opposite ends have distinct properties.

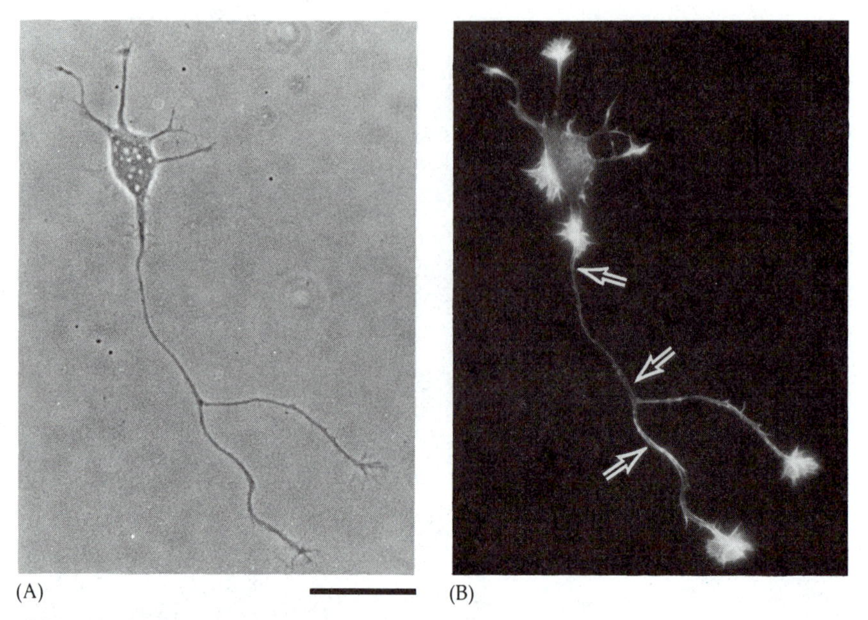

FIGURE 2. The distribution of actin in a cultured hippocampal neuron. (A) A phase-contrast micrograph of the neuron. The bar represents 10 μm. (B) The cell shown in (A) stained with fluorescein-conjugated phalloidin, a small molecule that binds to filamentous actin. Because the growth cones stain most intensely, the cortical staining of the axon (arrows) has been exaggerated during photographic printing. (Photos courtesy of K. Goslin.)

Neurofilaments are composed of three polypeptide subunits: NF-H (high; 112 kD), NF-M (middle; 102 kD) and NF-L (low; 68 kD). In contrast to actin and tubulin, neurofilament proteins are extended molecules with a central α-helical region that forms coiled-coil dimers. The dimers then form tetramers that assemble to form neurofilaments. The neurofilament proteins, which are found only in neurons, belong to a family of related intermediate filament proteins that includes the keratins, vimentin, desmin, glial fibrillary acidic protein (GFAP), and nuclear lamins. Because neurofilaments and GFAP are found only in neurons and glial cells, respectively, their presence is often used to identify these cells by immunofluorescence microscopy. Recently, several new intermediate filament proteins with similar α-helical coiled-coil structures (e.g., peripherin, α-internexin, and nestin) have been found in particular neurons; their functional roles are unknown.

Microtubules and microfilaments are dynamic polymers

Neurons contain both monomeric and polymeric forms of actin and tubulin; the balance between the two is carefully controlled by the cell, which can shift the equilibrium between assembly and disassembly in response to a variety of physiological cues, such as interactions with growth factors or the extracellular matrix. Because of the complexity of the reactions in vivo, the study of filament assembly in a simplified environment in vitro has been extremely useful for understanding their dynamic nature.

Both actin and tubulin can be purified and induced to undergo cycles of polymerization and depolymerization in the test tube. Self-assembly of actin and tubulin requires nucleoside triphosphates and proceeds until the monomer concentration is reduced to a **critical concentration,** which is the concentration at which the rate of subunit addition to the polymer equals the rate of subunit removal. At the critical concentration, assembly and disassembly are balanced. Above this concentration, there is net addition to the polymer, and below it, net removal. Both microfilaments and microtubules are polar structures, and one end of the filament (termed the **plus** end) grows faster during polymerization than the other end (termed the **minus** end). Although neurofilaments also polymerize in vitro, their assembly is not accompanied by nucleotide hydrolysis, and the polymers do not readily disassemble. Because much less is known about their assembly, the remainder of the discussion focuses on actin and tubulin assembly.

Although many self-aggregating molecules such as detergents or viral capsids reach true equilibrium, actin and tubulin are unusual in that both are enzymes that bind and hydrolyze nucleotides during assembly: ATP in the case of actin, and GTP in the case of tubulin. Nucleotide hydrolysis occurs after incorporation of the subunit into the polymer. Surprisingly, nucleotide hydrolysis is not required for polymerization; both actin and tubulin form filaments when nonhydrolyzable analogues of ATP and GTP are used. Why do actin and tubulin consume energy by hydrolyzing nucleoside triphosphates? The answer lies in the increased flexibility of assembly and disassembly that nucleotide hydrolysis provides.

The significance of nucleotide hydrolysis is best seen by considering the polymerization of tubulin in its absence. If tubulin is prevented from hydrolyzing GTP (by using a nonhydrolyzable analogue, for example), the free energies of tubulin dissociating from the plus and minus ends of the microtubule are the same. This is so because the reactants and products are identical and the same bond is broken in the two cases. In other words, the critical concentration (or equilibrium constant) is the same at both ends. Thus, although tubulin associates faster at one end of the polarized polymer than at the other, it also must dissociate faster from that end, so that the equilibrium constants at both ends are balanced. If this were not the case, then net tubulin addition at one end and net tubulin loss at the other could occur under steady state conditions, leading to a net movement of subunits from one end of the polymer to the other at equilibrium. Such **treadmilling** could be used to perform mechanical work, in violation of the first law of thermodynamics. We will see below that treadmilling does occur in actin filaments, but only when energy from nucleotide hydrolysis is supplied.

Nucleotide hydrolysis changes the above scenario dramatically. In the case of tubulin, for example, there are now two types of subunits to consider, GTP-tubulin and GDP-tubulin, and four reactions (association and dissociation of GTP- and GDP-tubulin) that can take place at each end.

Monomeric tubulin is mostly in the GTP form, both because it does not readily hydrolyze GTP, and because it rapidly exchanges bound GDP with GTP in solution. Thus the primary addition reaction of tubulin to microtubules is that of the GTP-form onto the plus end of the microtubule. After it is incorporated into the polymer, tubulin hydrolyzes its bound GTP and retains GDP in its enzymatic site. The hydrolysis of nucleotide apparently also induces a conformational change in the subunit that reduces its affinity for neighboring subunits. Because of its weaker affinity, GDP-tubulin dissociates more readily than GTP-tubulin from the polymer. As discussed below, a slight delay in hydrolysis after polymerization, in conjunction with the differing affinities of GTP- and GDP-tubulin, allows microtubules to exhibit complex behavior.

Nucleotide hydrolysis makes microtubules dynamically unstable and actin filaments treadmill

Tubulin hydrolyzes its bound GTP shortly after it is added to the plus end of the polymer. Because the hydrolysis does not occur instantaneously, the plus end of the polymer is thought to have a "cap" of subunits in the GTP-form. As GTP-tubulin does not easily dissociate from the polymer, the cap at the plus end favors further tubulin addition. The rate of hydrolysis is close enough to the rate of addition, however, that hydrolysis sometimes catches up and exposes GDP-tubulin at the plus end (Figure 3). Whenever this happens, dissociation is favored, and the polymer, now composed almost entirely of GDP-tubulin, rapidly and catastrophically depolymerizes. Thus at steady state (where the total polymer mass is constant) two populations of microtubules coexist: a large population that is slowly polymerizing by addition of GTP-tubulin, and a smaller population that is rapidly depolymerizing by loss of GDP-tubulin. This **dynamic instability** of microtubules is observed both in vitro and in living cells. The short half-life of many microtubules (estimated at 5 min in vivo) allows the microtubule network to rapidly change its configuration unless stabilized. Thus microtubule polymers are inherently plastic; the formation of stable microtubule networks requires additional factors.

The principles governing the assembly of actin are similar to those of microtubules, but kinetic differences in the association and dissociation rate constants as well as in the hydrolysis rate of nucleotide confer on microfilaments different dynamic behaviors than those of microtubules. Polymerized actin subunits hydrolyze their bound ATP relatively slowly. Moreover, although ADP-actin dissociates faster than ATP-actin from polymer ends, the disparities in the dissociation rate constants are not so great as with tubulin. As a result, microfilaments do not exhibit the sudden and drastic changes of length that are observed with microtubules. Nonetheless, like microtubules, ATP-actin is added primarily to the plus end; ADP-actin, in turn, readily dissociates when exposed at the minus end. Because polymerization occurs by one reaction and dissociation by another, the equilibrium constants at the two ends do not have to be the same. Nucleotide hydrolysis thus allows the two ends to have different critical concentrations. At a monomer concentration between the two critical con-

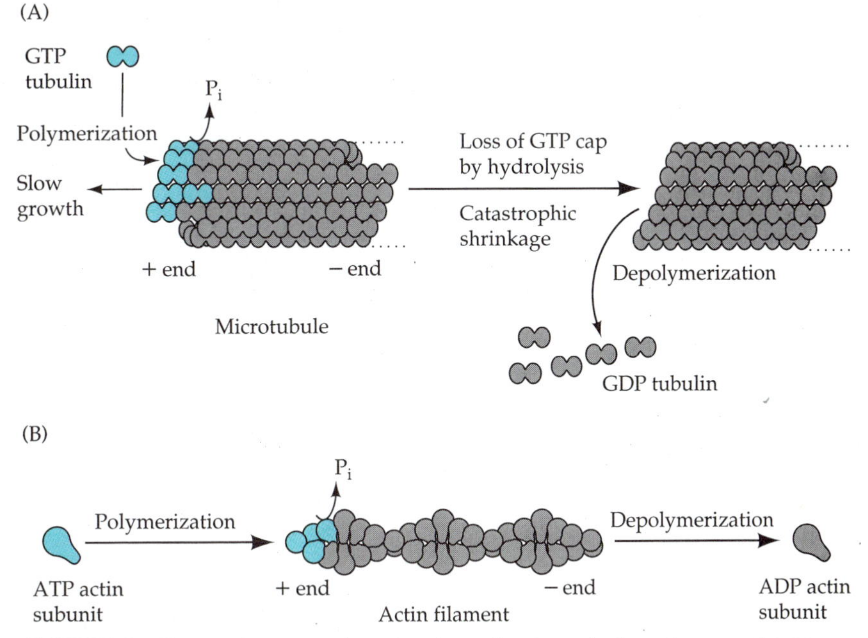

FIGURE 3. Dynamic properties of actin and microtubule polymers. (A) Microtubules, which are anchored by their minus ends to the centrosome in the cell body, polymerize by the addition of GTP-tubulin (color) to their plus ends. The incorporated monomer hydrolyzes its GTP to GDP (gray). Occasionally, hydrolysis catches up to polymerization and exposes GDP-tubulin. Since GDP-tubulin has a very high dissociation rate constant, the microtubule proceeds to depolymerization very rapidly (catastrophe). The microtubule will then disappear completely, unless GTP-tubulin recaps the end (rescue). GDP-tubulin that dissociates from a microtubule rapidly exchanges its nucleotide for GTP. (B) The plus ends of actin filaments, which are anchored by unknown structures at the plasma membrane, incorporate primarily ATP-actin (color). The ATP-actin in the polymer hydrolyzes its bound nucleotide more slowly than do microtubules, and, as a result, actin filaments are thought to have a larger nucleotide triphosphate cap than do microtubules. The minus end of actin has exposed ADP-actin (gray), which readily dissociates from the polymer. Addition of ATP-actin to the minus end occurs but rarely in comparison to the plus end. As a result of these dynamics, actin monomers tend to add onto the filament in the ATP form at the plus end and dissociate from the filament in an ADP state at the minus end. Individual actin subunits thus can move or treadmill through the polymer, a phenomenon that can be used to produce mechanical work in some instances.

centrations, actin subunits are added at one end and lost at the other at the same rate. Thus, at steady state, the polymer treadmills, remaining approximately the same length as subunits move through it (Figure 3). In the cell, treadmilling is a potential source of mechanical energy whose ultimate source is the hydrolysis of nucleotide associated with actin.

Neurons contain both stable and unstable microtubules

Although microtubules can assemble in the absence of other proteins, they grow most readily by elongation from specific nucleating sites within the cell. Cells thus control the organization of microtubules by controlling sites of nucleation. In the neuronal cell body, as in many other cells, microtubules are nucleated from the **centrosome** or **microtubule organizing center,** an amorphous cloud of proteins surrounding the pair of centrioles adjacent to the nucleus. Microtubules are attached to this structure at their minus ends, and they grow out from it by addition to their plus ends. In vivo, dynamic assembly and disassembly thus occurs at the plus end. Microtubule growth in neurons can be observed by fluorescence microscopy after injecting fluorescently labeled tubulin. Some microtubules in vivo show dynamic instability, that is, they grow slowly and collapse suddenly. Others appear to be quite stable, showing no evidence of catastrophic depolymerization over long periods of time.

Some of the stable microtubules that are nucleated from the centrosome in the neuronal cell body extend into the axon. Since the average length of axonal microtubules is only about 200 μm, however, the majority of microtubules in a 1-m axon begin and terminate within the axon. Like the microtubules attached to the centrosome, all of the axonal microtubules are oriented with their plus ends toward the growth cone or nerve terminal. How axonal microtubules are nucleated and maintained in the proper orientation is unknown. Although microtubules in axons are generally stable, they are nonetheless not static structures. When axons are subjected to conditions that cause depolymerization of microtubules (e.g., low temperature or depolymerizing drugs such as nocodazole), part of each microtubule in the axon remains. These stable domains, which are at the minus end, appear to act as nucleating centers for more dynamic growth and disassembly at the plus ends of the microtubules.

Microtubule and actin binding proteins influence filament assembly and structure

A variety of proteins co-purify with microtubules that are isolated from brain. These proteins, called **microtubule-associated proteins (MAPs),** remain with the microtubules through multiple cycles of depolymerization and reassembly. They play important roles in the assembly of microtubules, in cross-linking microtubules to each other and to other filaments, and in transport functions. At least 10 different MAPs have now been identified. A group of high molecular weight MAPs (200–300 kD) and a group of lower molecular weight proteins (the tau MAPs, 60 kD) promote microtubule assembly and stabilize assembled microtubules in vitro. They are thus likely to be very important in controlling microtubule growth, perhaps by altering the transitions between dynamic and stable forms. Consistent with this view, some MAPs are found only in embryonic neurons, during periods of active growth, and others are found only in the adult. Some MAPs appear to link microtubules to each other. Two of them, MAP2 and tau, play special roles in axons and dendrites as discussed later.

Actin filaments in the axons are not longitudinally aligned but form a complex meshwork like steel wool beneath the surface membrane. In this meshwork, the filaments are connected by numerous actin-binding proteins that bundle filaments together and cross-link them to form a gel.

Other proteins control the length of microfilaments by binding to their ends and preventing further polymerization, or by cleaving them into filaments of smaller length. Yet other proteins attach microfilaments to the surface membrane or to cytoskeletal proteins such as fodrin or vinculin (see below). Many of the actin-binding proteins are regulated by second messengers such as calcium, IP_3 or cyclic nucleotides. Actin filaments and the actin network that they form are thus regulated by a complex battery of proteins; these presumably mediate the changes in the actin cytoskeleton that occur during cell movement in response to environmental cues from diffusible factors or the extracellular matrix.

Motor proteins associated with the cytoskeleton hydrolyze ATP and use the derived energy to generate the force necessary to move along a polymer (Table 1). Two types of myosin motors operate on actin filaments. Myosin II resembles muscle myosin in having two globular ATPase heads and a long α-helical stalk that allows the heavy chains to dimerize and to self-assemble into filaments. Myosin I (or "mini-myosin") is a smaller version that has a globular ATPase domain, but does not dimerize as it lacks the α-helical coiled-coil stalk. Both move along actin filaments from the minus end to the plus end. Because it is associated with distal regions of migratory cells, myosin I may play an important role in growth cone motility. The motor proteins associated with microtubules are discussed below in connection with their role in axonal transport.

Neurofilaments are stable polymers that may determine axonal diameter

In contrast to microtubules and actin microfilaments, neurofilaments undergo little turnover. They are assembled in the cell body, transported down the axon, and degraded in the nerve terminals. The three neurofilament subunits, which all have a central α-helical region, assemble together in each neurofilament. The three subunits differ from each other mostly in the length of their C-terminal domains. These domains of NF-M and NF-H protrude from the filaments, allowing them to form cross-bridges with each other, and perhaps between neurofilaments and microtubules, as well.

In large mammalian axons the number of neurofilaments is highly correlated with the cross-sectional area of the fiber. The correlation of

TABLE 1. Motor Proteins in Neurons

Protein	Filament	Preferred direction	Nucleotide	Size
Kinesin	Microtubules	Plus end	ATP	360 kD
Dynein	Microtubules	Minus end	ATP	1200 kD
Dynamin	Microtubules	Not determined	GTP	100 kD
Myosin II	Actin	Plus end	ATP	500 kD
Myosin I[a]	Actin	Plus end	ATP	150 kD

[a]The presence of myosin I has not yet been demonstrated in neurons but it is in a variety of cells that, like the growth cone, move by an actin-based mechanism.

axonal caliber with neurofilament number not only holds for different axons, but also for different parts of the same axon. Many myelinated axons decrease their diameter substantially at the nodes of Ranvier; the number of neurofilaments correspondingly decreases. These observations suggest that axons control their diameter by varying the radial extent of the network of cross-linked neurofilaments that they contain. The caliber of axons is important physiologically because it determines the speed of propagation of electrical signals.

Transport of Macromolecules

Virtually all eukaryotic cells transport protein and lipid to and from the surface membrane, and between intracellular organelles, via small membrane vesicles. This traffic is not random, but is directed along microtubules, which give polarity to the transport. Neurons carry out the same transport processes but face a far more challenging task, as they must transport vesicles over centimeters or even meters. Proteins for transmitter synthesis and secretion are made in the cell body and subsequently transported the length of the axon to the nerve terminal; cytoskeletal proteins must also be supplied to the axon; and growth factors and other proteins taken up by nerve terminals must be conveyed to the cell body. The logistics of transport can be daunting. The volume of cytoplasm in a long axon can be as much as 10,000 times larger than the cell body that supplies it with protein and lipid.

Essential to the transport system in axons are the protein motors that use the chemical energy of ATP hydrolysis to perform mechanical work. In addition to the mechanics of moving objects, the transportation system employs traffic lights and checkpoints to ensure delivery of macromolecules to their proper destinations. Targeting signals, which have not yet been deciphered, must specify whether an organelle undergoes **anterograde** (toward the nerve terminal) or **retrograde** (toward the cell body) **transport**. Since dendrites and axons contain different types of organelles and proteins, sorting mechanisms must separate material bound for these two domains as well. In this section, we shall consider what types of macromolecules are transported in neurons, the motors that drive their movement, and possible mechanisms of controlling the transportation system.

Two systems deliver macromolecules to the nerve terminal at different rates

Axonal transport was first discovered in 1948 by Paul Weiss and his co-workers, who ligated a nerve trunk with a thread and found several days later that axons within the trunk were grossly distended proximal to the ligation, and greatly narrowed distal to it. They interpreted the proximal distension as the damming of axoplasm that was being actively transported from the cell body to the terminals. When the ligature was removed, the material above it advanced down the nerve at a rate of 1–2 mm per day. Weiss pointed out that this was approximately the rate of peripheral nerve regeneration, and suggested that there was a normal flow of material from the cell body that sustained axonal growth.

The **axonal transport** of proteins can also be observed by exposing neuronal cell bodies to a brief pulse of radiolabeled amino acids. The subsequent distribution of the protein within the axon is then followed either by autoradiography or by cutting nerve trunks into segments and measuring the radioactivity in each segment (Figure 4). After the pulse, labeled protein synthesized in the cell body moves down the axon as two discrete peaks of radioactivity: one that travels at 200–400 mm/day, termed **fast axonal anterograde transport;** and a second, **slow transport,** that moves at several mm/day, the rate of axoplasmic movement observed by Weiss. The second peak can be further divided into two components that propagate at 0.2–1 mm/day [slow component A (SC_a)], and 2–8 mm day [slow component B (SC_b)]. Translated into practical terms, these rates correspond to a time of about 1 day for rapid transport and about 1 year for slow transport down the length of an axon that is 35 cm long. A

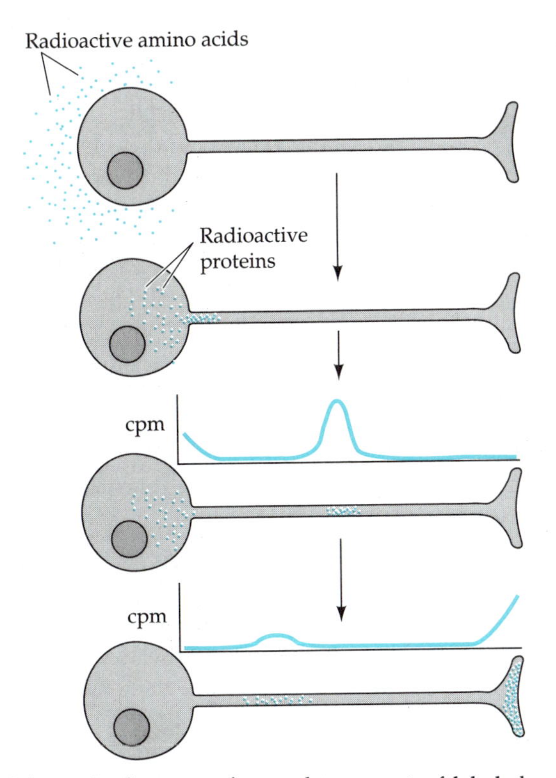

FIGURE 4. Schematic diagram of axonal transport of labeled proteins. Radioactive amino acids are injected into the fluid bathing nerve cell bodies in vivo. The amino acids become incorporated into proteins that are then transported along the axon of the cell. At various times after the initial labeling, the nerve is cut into sections and the radioactivity in each section determined. The profiles of radioactivity along the axon, generated at different times, show two waves of transport down the axon. (After P. J. Hollenbeck, 1989. J. Cell Biol. 108: 223–227.)

retrograde transport system that delivers proteins from the nerve terminal to the cell body was uncovered by introducing labeled lectins, nerve growth factor, or horseradish peroxidase into peripheral target tissues. Neurons internalize these proteins at axon terminals by fluid phase or receptor-mediated endocytosis and then transport them to the cell body at rates similar to those of the rapid anterograde system. Slow retrograde transport has not been demonstrated.

The fast and slow transport systems carry distinct protein species. Constituents of the cytoskeleton—tubulin, actin and the neurofilament proteins—comprise most of the proteins moving at the slow rate. The proteins associated with membrane vesicles, including synaptic vesicles, travel at the fast rates. Mitochondria travel at about 50 mm/day, slightly slower than small vesicles. Electron microscopic studies of ligated nerves show that within hours large numbers of membrane vesicles and mitochondria accumulate on both sides of the ligation. These and other observations led to the idea that membrane organelles are transported bidirectionally at fast rates and that cytoskeletal components move only in the anterograde direction at slow rates.

Fast axonal transport occurs along microtubule tracks

Recent technical advances in microscopy have allowed direct observation of vesicles undergoing fast axonal transport. Using video microscopy (see Box A), membrane vesicles and individual microtubules can be detected, even though they are below the theoretical limit of resolution of the light microscope. When a squid giant axon is viewed by video microscopy, a myriad of small organelles is seen moving bidirectionally at 1–5 μm/sec, the rates expected for fast axonal transport. Remarkably, when the axoplasm is extruded onto a glass slide, organelle transport continues unabated as long as ATP is supplied as an energy source. When the axoplasm is mechanically disrupted, bidirectional transport proceeds along individual filamentous tracks identified by electron microscopy as single microtubules.

Two motors, kinesin and dynein, power fast axonal transport

To identify the force-generating machinery, components of the transport system were isolated and then recombined to reconstitute movement in vitro. Such experiments revealed that organelles isolated from the squid axon could move along microtubules prepared from purified tubulin in the presence of ATP. Since the microtubules were devoid of associated proteins, the motor for fast transport appeared to be attached to the organelles, rather than to the microtubule.

A surprising finding aided the biochemical search for the motor. Factors in a soluble protein extract from the squid giant axon bind to a glass slide and cause microtubules to be transported along the surface (Figure 5). Latex microspheres containing carboxyl groups can also adsorb these factors and be transported along stationary microtubules, or, after injection into squid axons, be transported down the axon. Bead and organelle movements are inhibited by the same pharmacological agents, suggesting

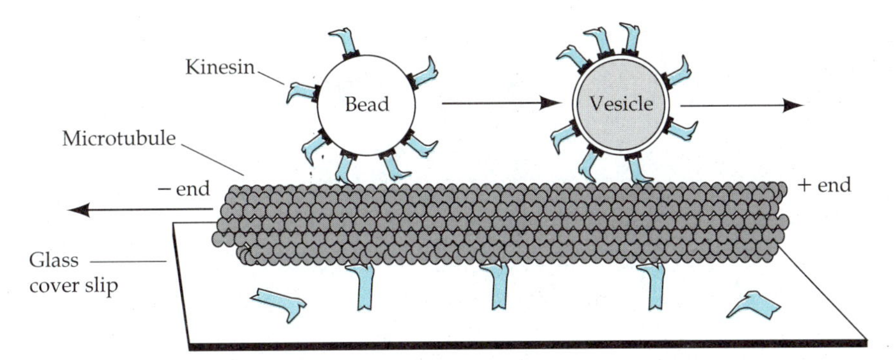

FIGURE 5. A hypothetical scheme showing how the molecular motor kinesin can induce movement of organelles, carboxylated beads, and microtubules. The motor binds the glass randomly, but only those with the proper orientation relative to the microtubule generate force. Kinesin interacts with organelles through a membrane-bound protein, and with the glass bead by nonspecific charge interaction, and causes them to be translocated along microtubules toward the plus end.

that the soluble protein motor that propels the beads along the microtubules, and that causes microtubules to move on glass, is the same one that is responsible for organelle transport in vivo. The movement of microtubules on glass thus provided an assay for the biochemical purification of the motor proteins responsible for organelle movement.

Three force-generating proteins—kinesin, dynein, and dynamin—have now been identified (Table 1). Kinesin and dynein are ATPases whose activity is stimulated by binding to microtubules. **Kinesin** consists of two globular head regions that hydrolyze ATP, an extended α-helical domain, and a small globular domain at the base. The overall molecular structure of kinesin is very similar to that of myosin, although there is no sequence homology between these two proteins. **Cytoplasmic dynein** is related to the family of dynein motors that power the movement of cilia and flagella. Although it has a larger mass than kinesin, it also has two ATP-hydrolyzing heads which are the mechanochemical domains, and which are connected by a stalk to a common globular base. Indirect evidence supports the participation of these two motors in the transport of vesicles. Immunofluorescence experiments have localized kinesin and dynein to vesicles in vivo, and mutations in the kinesin gene of *Drosophila* and a kinesin-like gene (*unc*-104) in *C. elegans* lead to defects in synaptic transmission due to deficient delivery of synaptic vesicles to nerve terminals. Kinesin and dynein occur widely in eukaryotic cells and are even found in single-celled organisms. Neurons have thus adapted a very old mechanism of transport for their specialized needs.

Dynamin appears to be a different type of motor than kinesin or dynein. It uses GTP rather than ATP as an energy source. In vitro it forms a cross-bridge between adjacent microtubules and induces sliding between them. Mutants in the *shibere* gene of *Drosophila*, which has recently been shown to encode dynamin, are blocked in endocytosis and are thus unable

Box A The Power of Video Microscopy

The recent success in isolating the molecules responsible for rapid axonal transport stems from technical advances in light microscopy that allow transport to be observed directly in vivo and in vitro. A light microscope does not give a true image of small organelles such as synaptic vesicles (50 nm) or microtubules (25 nm), which are below its resolving power. (The resolving power of the light microscope, which is limited by the wave length of visible light and the numerical aperture of the objective, is about 200 nm, or the size of a mitochondrion. Two objects separated by less than this distance appear as one.) Small objects such as microtubules generate a diffraction pattern that can produce a detectable image when its contrast is enhanced by differential interference contrast (DIC) microscopy. In Nomarski DIC microscopy, light is passed through a prism that splits it into two overlapping arrays of photons, displaced from one another by a fraction of a micron, and having different planes of polarization. Recombination of the two components of light after they have passed through the specimen gives constructive and destructive interference to produce a strikingly sharp image that appears to be shadowed. The usefulness of DIC optics lies both in the fact that it works well with unstained, living tissue and that it generates images having a very shallow depth of field: the image quality is not markedly degraded by information from out-of-focus planes.

Images of objects as small as microtubules and transport vesicles have extremely low contrast, even when viewed with DIC optics. Our ability to view these images more clearly resulted from the development of video microscopy by Robert Allen, Shinya Inoue, and their colleagues. Video microscopy has many advantages over conventional light microscopy. The response of the video camera is nearly linear with signal intensity, and selected cameras operate at intensity ranges greater or less than that of a human observer. Analog video signals can be manipulated in a variety of ways; background noise can be subtracted from the trace to increase contrast, and the remaining signal can be amplified by increasing the gain of the video amplifier. Still further manipulation is possible with the use of digital processing of the signal, which allows one to average a number of signals to increase the signal-to-noise ratio, to enhance contrast further, to smooth the image in order to reduce unwanted noise, and to subtract information from out-of-focus planes to eliminate problems created by uneven illumination or spatially nonuniform performance by camera and electronics. The remarkable effect of these manipulations is illustrated in Figure A, which shows extruded squid axoplasm before and after video image processing.

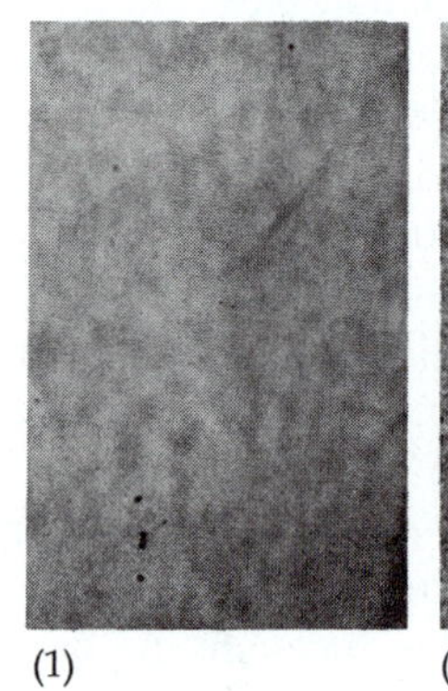
(1)

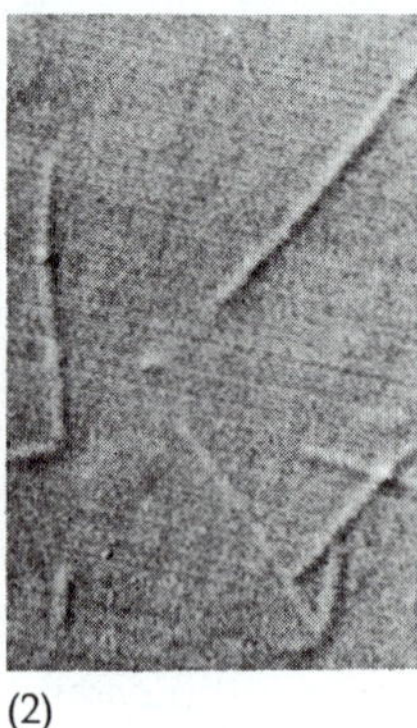
(2)

1 μm

FIGURE A. Microtubules in extruded squid axoplasm viewed by differential interference contrast in the light microscope. (1) The extract seen without image processing. (2) The microtubules (0.025 μm in diameter) become visible after image processing to reduce the background noise and increase the contrast. (From B. Schnapp, 1986. Meth. Enzymol. 134: 561–573.)

to form new synaptic vesicles. This suggests a role for dynamin in vesicle recycling (see Chapter 5) and other endocytic processes.

Kinesin and dynein are anterograde and retrograde motors

Kinesin and dynein are unidirectional motors that move in opposite directions along a microtubule. Kinesin moves toward the microtubule plus end; dynein moves toward the minus end. Having two motors that move in opposite directions along microtubules neatly explains the bidirectional traffic of vesicles within axons. Since more than 90% of the microtubules in axons have their plus ends oriented toward the nerve terminal, microtubule polarity can serve as a compass for an organelle's journey in the axon. According to this hypothesis, organelles with bound kinesin, a plus-end–directed motor, are transported in the anterograde direction toward the terminal. Organelles destined for the cell body, on the other hand, enlist minus-end–directed dynein motors.

For transport to proceed in an orderly fashion, the proper motor be must be attached and activated on the proper vesicle. Indeed, axoplasmic organelles generally proceed unidirectionally along a microtubule, some toward the plus end, and others toward the minus end, as if they had a programmed sense of direction. On the other hand, latex spheres, which bind both kinesin and cytoplasmic dynein, frequently change their direction of movement on a microtubule. Organelles, therefore, must possess something that artificial beads do not, namely proteins that regulate motor attachment or activation. Although little is known about these accessory proteins, they may include soluble docking proteins as well as receptor proteins on the vesicle membrane.

Motors themselves must be transported or recycled in neurons (Figure 6). To transport organelles in the retrograde direction, dynein, for instance,

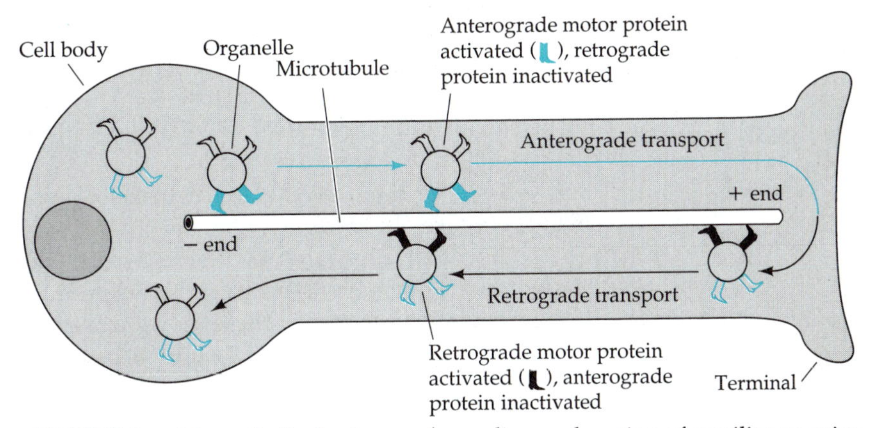

FIGURE 6. A hypothetical scheme of recycling and sorting of motility proteins. Organelles bind kinesin in the cell body, enabling them to be transported to the nerve terminal. The motor may be transported from the terminal back to the cell body in an inactive form. The retrograde motor dynein is transported to the terminal in an inactive form, becomes activated, and then binds organelles, allowing them to be transported in the retrograde direction.

must somehow reach the nerve terminal. Recent work suggests that dynein is delivered to the terminal by organelles moving in an anterograde direction. For this to be true, dynein must travel anterogradely in an inactive form, and must be activated in the terminal for retrograde transport. A similar mechanism could be used to recycle kinesin from the terminal to the cell body. Thus, during transport, both types of motors may be bound to a membrane vesicle, but only one type appears to be active.

Individual cytoskeletal polymers may be transported during slow axonal transport

After proteins in the cell body are pulse-labeled, labeled cytoskeletal proteins move down the axon as a discrete wave (Figure 4), suggesting that they are transported as intact polymers. An entirely different method has recently shown a similar movement of polymerized tubulin in growing axons. Neurons labeled with tubulin conjugated to a photo-activatable dye were cultured and allowed to extend neurites. Then a narrow beam of light was shone across the axon to activate the dye, causing the microtubules in a small segment of the axon to become fluorescent. With time, all of the fluorescence moved as a coherent band down the axon at a rate characteristic of slow axonal transport (Figure 7). When multiple bands along a single unbranched axon were activated at different sites, all moved down the axon at approximately the same rate. The constant intensity of the fluorescence and its resistance to extraction by detergent indicated that it remained associated with polymerized tubulin. These experiments suggest that virtually all the microtubules in a growing axon are continually moving toward the terminal at a constant rate. This movement is thought to supply tubulin to the growing axon and to power its extension (see below). The experimental evidence supporting the idea that all microtubules are moving is not unequivocal, however; some experiments using photobleaching techniques indicate that microtubules may be part of a stationary cytoskeleton. Experiments on neurofilaments indicate that they move down mature axons as polymers and suggest that some of them also become incorporated into a stationary cytoskeletal network. Whether part of the cytoskeleton is stationary and, if so, what the stationary components are, remains a subject of active investigation.

The motive force for the movement of cytoskeletal proteins down the axon remains obscure. Individual filaments might move by a sliding mechanism, analogous to that used by filaments in muscle and cilia that generate movement. Transitory associations between sliding filaments and stationary components could power these movements. Understanding the mechanisms of slow axonal transport is thus a major challenge for the future.

Neurite Extension and Growth Cone Motility

How neurons extend their axons and dendrites is a problem that has occupied neurobiologists for almost one hundred years. This problem is not only fascinating from the perspective of cell motility but is also im-

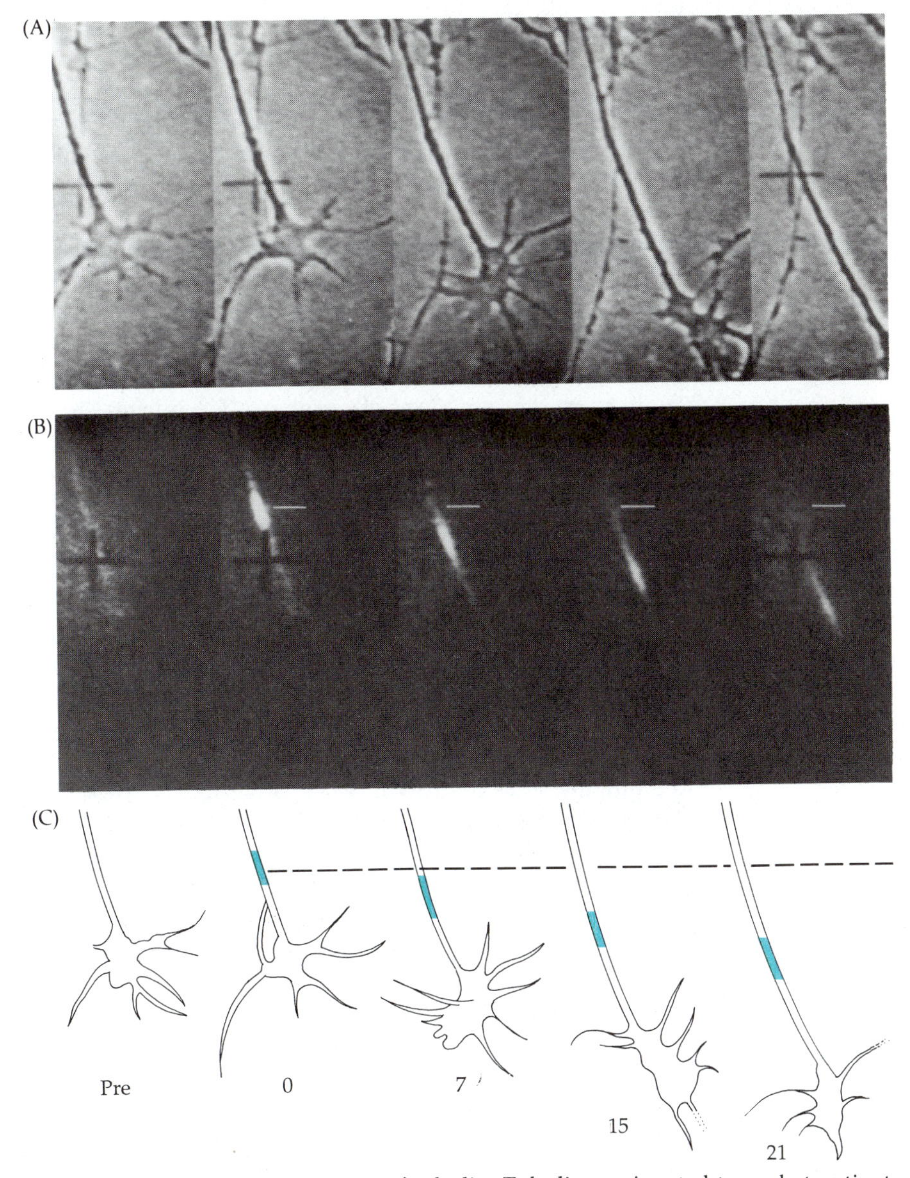

FIGURE 7. Axonal transport of tubulin. Tubulin, conjugated to a photoactivatable dye, was injected into fertilized *Xenopus* oocytes, and neural tube tissue from the embryo was subsequently cultured. The dye was then activated with a band of light across the axon and the axon examined by phase contrast (A) and fluorescence microscopy (B) at various times thereafter. (A) Phase images show an axon from the explanted neural tube. In successive frames the growth cone moves toward the lower right-hand corner. (B) Fluorescence images of the same field. As the axon elongates, the activated band of fluorescence moves distally down the axon. The axon elongated at 90–100 μm/hr, and the fluorescent band moved at about the same speed. Note that the fluorescent band moves with relatively little spreading or loss of fluorescence. (C) Schematic representation of (A) and (B). The numbers indicate the minutes after activation; the dashed line indicates the initial position of the band of fluorescence. (A and B from S. S. Reinsch et al., 1991. J. Cell Biol. 115: 365–379.)

portant for developmental neurobiology, as the precise wiring pattern of the nervous system depends upon the correct growth of axons to their targets. Neurite outgrowth depends upon two separate and interacting cytoskeletal systems: one based on microtubules and the other on actin. In the axon, microtubules are organized into an axially oriented parallel array that is both backbone and supply track for the neurite. Mature axons also contain neurofilaments, but these are added to the axonal cytoskeleton relatively late in development. The extension of the organized system of microtubules is thus what lengthens the neurite. The direction and pattern of neurite growth, however, is determined by the **growth cone,** a highly motile structure at the end of growing axons and dendrites, whose movement is largely actin-based. Although the basic mechanisms of growth are thought to be the same for both axons and dendrites, our discussion will mostly concentrate on axons, whose growth is more extensive and better understood.

Growth cone movement and neurite extension occur by independent, but coordinated, mechanisms

Although growth cones were originally observed by Ramón y Cajal in Golgi-stained tissue (see Chapter 12 for his colorful description of them), their remarkable activity is best appreciated by the continuous observation of cultured neurons. In contrast to the comparatively sedate and linear extension of the neurites to which they are attached, growth cones continuously and exuberantly explore their environment. At the leading edge and along the sides of growth cones, fingerlike processes **(filopodia)** continuously extend and retract. In other areas or in other growth cones, advance is marked by the movement of sheets of membrane **(lamellipodia)** that alternately protrude or move back over the surface of the growth cone as ruffles. The growth cone continually changes its shape, advancing first in one direction and then the other; sometimes it splits, and goes in two directions, thus establishing a branch point. The active motility of the growth cone resembles that of a cultured fibroblast, leading some to call it a "fibroblast on a leash." The basis of its motility, like that of the fibroblast, is a network of actin filaments beneath the plasma membrane.

The activities of the two cytoskeletal systems, based on microtubules and actin filaments, are in some ways independent. When neurons are cultured in the drug cytochalasin, which causes loss of the actin network and collapse of the growth cone, their neurites can grow on adherent substrates at almost normal rates. After treatment with cytochalasin in vivo, however, axons are unable to respond to the environmental cues that normally guide their growth. Growth cones, in turn, do not require intact neurites for their motility; if the neurite is cut, the exploratory activity and movement of the growth cone continue for many hours. There are thus three problems to be considered in understanding neurite outgrowth: how the microtubule network is extended; how growth cones move; and, most importantly, how the two are coupled to allow growth cones to guide neurite extension. A fourth question, how external signals from the environment influence growth cone movement, will be considered in Chapter 12. None of these questions can be answered definitively. But by combining the available data on growth cones with that from other, analogous, motile systems, we can make intelligent speculations and devise plausible models for future experiments.

The organization of actin and microtubules differs in growth cones and in axons

The organized bundle of microtubules that is characteristic of axons ends in the base or neck of the growth cone (Figure 8). Many microtubules extend farther into the growth cone and even penetrate to the membrane at the leading edge and to the base of filopodia. Unlike microtubules in the axon, those in the forward areas of the growth cone are not usually bundled together and are often curved.

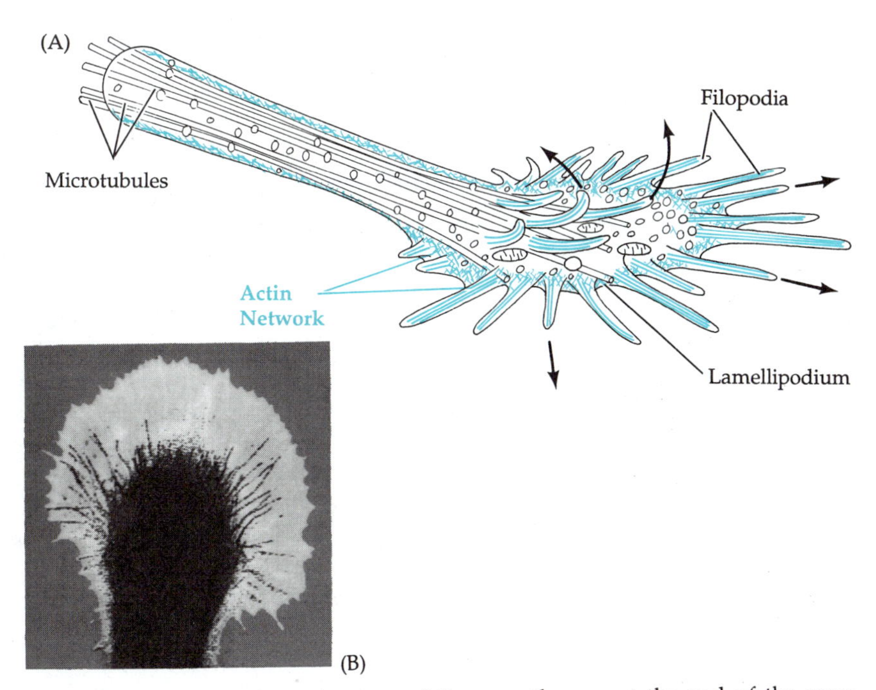

FIGURE 8. (A) A schematic view of the growth cone at the end of the axon. Microtubules in the axon are closely packed and axially aligned, but are many fewer and less organized in the growth cone. An actin network is prominent at the leading edge of the growth cone, and bundles of actin filaments fill the filopodia. The central region of the growth cone is rich in mitochondria and vesicles. The arrows indicate that while the leading edge of the growth cone is advancing in one direction, filopodia and lamellipodia are swept back over the surface of the growth cone in the opposite direction. (B) The distribution of actin filaments and microtubules in the distal neurite and growth cone of an *Aplysia* neuron growing in cell culture. Actin filaments (shown in white) were visualized by staining with rhodamine-conjugated phalloidin. Microtubules (shown in black) were visualized by fluorescein-conjugated antibodies against tubulin. In a more three-dimensional view, actin filaments would be seen above and below the central microtubule domain. (Photograph courtesy S. J. Smith.)

In the axon the actin network is highly stable and forms a thin, cylindrical rind or cortex at the edge of the axoplasm. In the growth cone actin filaments are much more prominent (see Figure 2), occupying the filopodia and leading edge. Some of the actin filaments in the growth cone are organized into tight bundles that extend radially from the center of the growth cone; such bundles, which are rarely seen in axons, form the core of each filopodium. Other actin filaments are assembled into a dense, irregular network that lies just under the membrane of the lamellipodia, and at the base of the filopodia.

Membrane vesicles, delivered to the growth cone by axonal microtubules, rarely penetrate the actin network at the periphery of the growth cone but accumulate in the central region. This region, which is probably the site of membrane addition to the growing neurite, also contains mitochondria, endocytotic vesicles, microtubules, and a loose network of actin filaments.

Extension of the axonal bundle of microtubules occurs by two mechanisms

In the growing axon, microtubules, propelled by an unknown mechanism, steadily move down the axon toward the growth cone, as discussed in the previous section. As they arrive at the base of the growth cone, they splay apart but continue to move forward. The direction in which they move determines the direction of neurite extension. Periodically the disorganized microtubules in the growth cone condense into an organized bundle; this consolidation represents an extension of the axoplasm and growth of the neurite. In areas of the growth cone not invaded by microtubules, filopodia, lamellipodia, and the underlying actin network eventually retract.

The growth cone is also a site of active polymerization of microtubules. As their plus ends are oriented away from the cell body, addition of monomers extends their distal ends and further aids growth of the neurite. Some microtubules in the growth cone are highly labile, undergoing rapid exchange with tubulin by cycles of polymerization and depolymerization. Part of the new addition of microtubules to the growing axon may thus represent stabilization of these labile microtubules.

The growth cone advances through changes in the actin network

Close examination of a moving growth cone reveals a paradox. Although the growth cone is advancing, much of the surface activity is retrograde, with membrane ruffles and attached particles moving back from its leading edge. Filopodia and lamellipodia that do not become incorporated into the growing axon are swept back over the surface of the growth cone and are absorbed into it. Membrane proteins, as well as the underlying actin cytoskeleton, appear to move retrogradely. Thus particles picked up on the growth cone surface near its periphery can be seen to move backward at a rate of several micrometers per second. Clearly two processes are at work in the growth cone, one pushing the cytoskeleton forward, and the other pulling it back. According to current ideas, these two continuous movements, which are spatially separated, could provide a motive force that extends or retracts local regions of the growth cone.

Actin filaments in the filopodia and lamellipodia are in a highly dynamic state. When cytochalasin is added to growth cones, the actin filaments disappear within several minutes (Figure 9). Because cytochalasin acts only to prevent polymerization, the rapid loss of actin filaments means that they normally depolymerize and polymerize at a high rate. After the addition of cytochalasin, actin filaments first disappear from the leading edge. At longer times, the actin network gradually shrinks toward the center of the growth cone and finally disappears. When cytochalasin is subsequently removed, new filaments appear first at the leading edge and gradually extend toward the center. This sequence of events reveals the spatial organization of actin turnover. Active polymerization occurs at the leading edge, which is the position of the plus ends of the filaments. As actin is added, the filaments move toward the center, there to be depolymerized at their minus ends. Actin filaments at the leading edge of the growth cone are thus thought to either treadmill or to move as whole filaments toward the center of the growth cone.

What provides the motive force for moving the actin filaments centripetally? Myosin molecules could play such a role. As actin filaments have their plus ends toward the leading edge, interaction with fixed myosin molecules will generate tension that pulls the filaments toward the center of the growth cone (Figure 10). Immunocytochemical staining shows that myosin is indeed present in the central region of growth cones, near their leading edge, and at the base of filopodia.

The continuous retrograde movement of actin could be the basis for growth cone advance. If filaments at a particular site in the growth cone were to become anchored, thus stopping their retrograde movement, continued polymerization at the plus end would result in protrusion of the actin cytoskeleton and extension of the leading edge (Figure 10). According to this idea, the continual polymerization of actin filaments and their centripetal movement by myosin would correspond to a motor that is idling. Any change in the cytoskeleton that fixed the actin filaments and stopped their retrograde movement would act as a molecular clutch, putting the motor in gear. This idea is an especially attractive one because it provides a plausible mechanism by which the adhesion of growth cones to extracellular substrates, which is known to affect their motility, could be related to growth cone movement. As will be discussed later (Chapter 12), molecules in the surface membrane that mediate attachment to substrates can also attach to actin filaments through intermediary, linking proteins.

Alternatively, anything that increases the rate of polymerization of actin filaments to make them extend at the plus end faster than they are being retracted at the minus end will extend the actin network and advance the leading edge. A clear example of how actin polymerization can cause cytoskeletal movement is seen in the acrosome reaction of sperm. Actin monomers, stored at high concentration in the acrosome body, are released from a binding protein by contact of the sperm with an egg. The actin

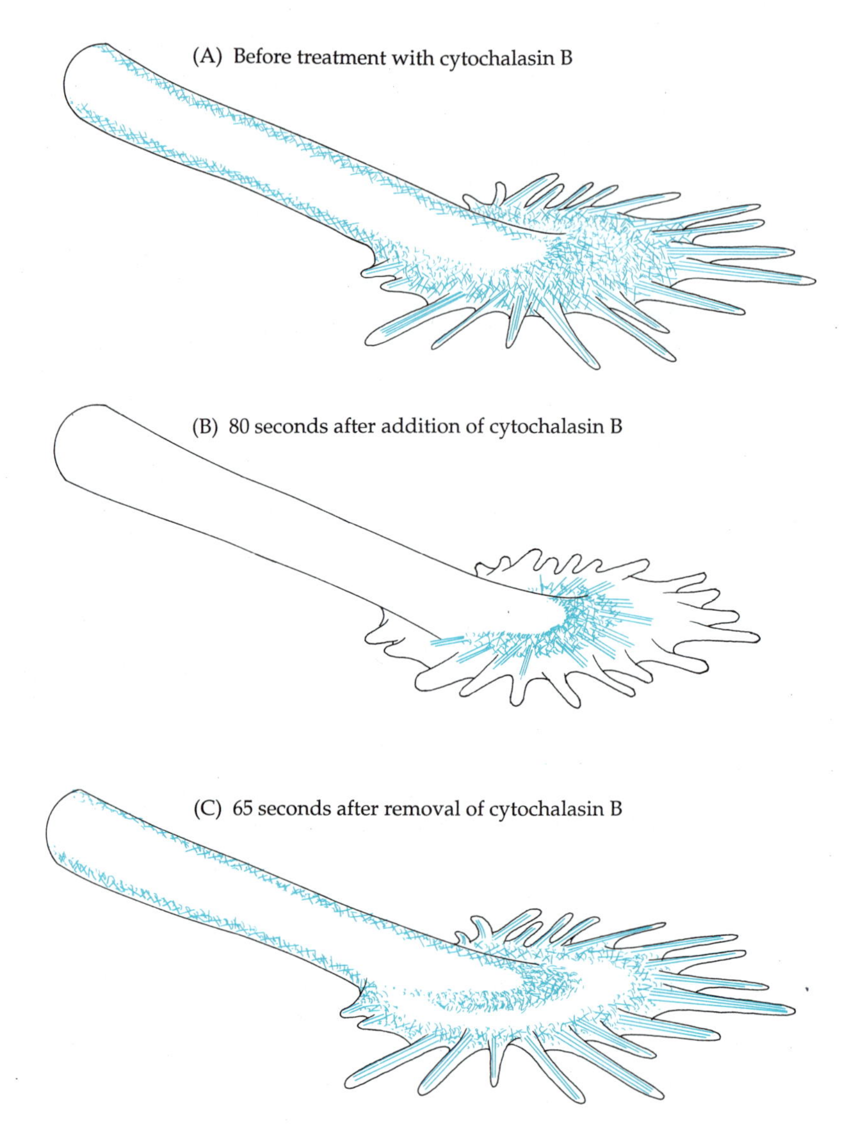

FIGURE 9. The distribution of actin in the growth cone of a cultured *Aplysia* neuron, determined before, during, and after incubation with cytochalasin B, a drug that adds to the plus ends of actin filaments and prevents their polymerization. The drawing indicates both the cellular outlines seen by Nomarski optics and the distribution of filamentous actin as observed in the fluorescence microscope by the binding of phalloidin, a plant alkaloid that binds to actin filaments, conjugated to rhodamine. (A) Originally, actin filaments fill the filopodia, and a dense actin network occupies the region between the leading edge and the central core of the growth cone. (B) Within 80 sec after the addition of cytochalasin, the actin network has shrunk to the central region, and fewer and shorter actin filaments are seen. Filopodia are retracted, and lamellipodial ruffling stops. (C) When cytochalasin is removed, filamentous actin reappears first near the leading edge, indicating that this is the most active site of incorporation of actin into filaments within the growth cone. (After P. Forscher and S. J. Smith, 1988. J. Cell Biol. 107:1505–1516.)

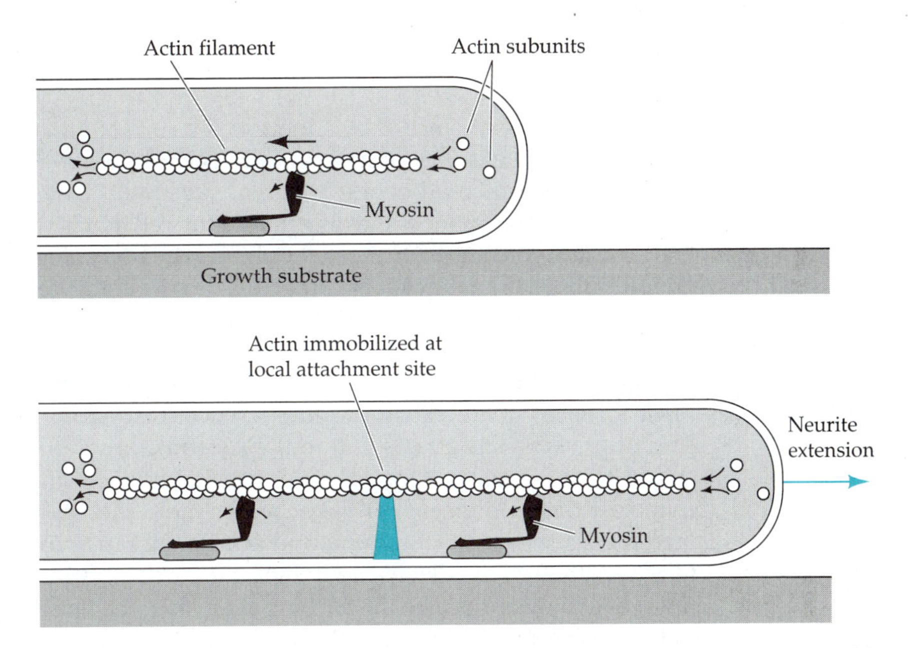

FIGURE 10. A model showing how local attachment to the substrate could convert retrograde movement of the cytoskeleton into neurite extension. Actin filaments polymerize at their plus ends near the front of the growth cone and depolymerize in the central region. At steady state, the filaments are in a constant rearward motion, powered by stationary myosin molecules attached to the substrate or to other parts of the cytoskeleton. If the actin filament is fixed at a substrate attachment site, its continued polymerization at the plus end will advance the leading edge of the growth cone. (After S. J. Smith, 1988. Science 242: 708–715.)

polymers then polymerize onto the plus end of actin filaments, which are attached to the membrane at the head of the sperm, causing the protrusion of a long, spikelike, membrane-bounded process. By analogy, the rapid addition of actin monomers to the distal ends of actin filaments in filopodia could drive their advance. Finally, advance of the leading edge of the growth cone could occur through a local breakdown, or weakening, of the actin network, leading to cytoplasmic protrusion at that point. Such mechanisms are thought to be responsible for ameboid movement in lymphocytes.

Each of these mechanisms invoke local alterations in the organization, tension, or state of polymerization of actin filaments to generate the mechanical force necessary to advance the growth cone. Several actin-binding proteins that are known to regulate these processes have been identified in growth cones. Some of these proteins are controlled by Ca^{2+}, which is known to play an important role in growth cone motility (Chapter 12). Other actin-binding proteins may be activated by other second messengers or by changes in the interaction of surface receptors with the extracellular matrix.

Growth cones adhere to substrates and exert tension on the neurites behind them

As growth cones advance, they adhere to the surface and exert tension on the rest of the cell. This is dramatically seen in cultures of dorsal root ganglion cells grown on poorly adherent substrates. Under these circumstances, the growth cone is the only part of the cell that is strongly attached to the surface, and, as it advances, it pulls the cell body along behind it. The growth cone thus advances faster than the neurite elongates. In cells whose cell bodies are normally adherent, growth cone advancement exerts tension on the neurite that may influence its direction of growth. Tension generated by the actin cytoskeleton may therefore help couple its movement to microtubule extension. Interference microscopy shows that the sites of growth cone attachment to the substrate are in the central region and at the base of the filopodia and lamellipodia. At analogous sites in fibroblasts, actin filaments are attached to molecules in the extracellular matrix via a series of cytoplasmic and transmembrane proteins (see Chapter 12). Although the details of the molecular organization at these sites is unknown, they are clearly the anchors against which tension on the rest of the cell is exerted. Direction of this tension to the plus ends of the advancing and growing microtubules could control the direction of microtubule extension and guide the axon. Such a mechanism could explain the relative preference of growing axons for adherent substrates.

Microtubule extension could also be coupled to advance of the actin network by a mutually inhibitory relationship between the two. A number of observations suggest that the actin network inhibits the extension of microtubules and that, conversely, microtubules exert a restraining effect on expansion of the actin network. Thus, depolymerization of the actin network by cytochalasin leads to the extension of microtubules from the central region to the leading edge; and depolymerization of microtubules by nocodazole (which depolymerizes microtubules) leads to branching along the axonal shaft by extension of the actin network. According to this view, advance of the rear edge of the actin network during growth could open a channel that permits microtubule extension. The microtubule network could expand into this channel and condense there into axoplasm.

Neuronal Polarity and Specialized Domains

Over 125 years ago Otto Deiters examined the morphology of individual neurons that he had laboriously dissected from different regions of the brain and concluded that each of them consisted of two fundamentally different parts: one, the cell body and dendrites; and the other, the axon. Deiters' original distinction, made on anatomical grounds, is a fundamental one. Cell bodies and dendrites are specialized to receive and integrate signals that come from other cells; they have a high density of postsynaptic receptors and associated proteins, and they conduct action potentials poorly—sometimes not at all. Axons, on the other hand, are specialized for the rapid conduction of all-or-none action potentials from the initial segment to the nerve terminals. They also must supply the nerve terminal with the organelles, enzymes, and proteins required for transmitter synthesis, release and regulation. Except in cells that participate in dendro-dendritic synapses, these presynaptic components are not present in dendrites. The dendrites, which arise gradually from the cell body and taper distally, appear to be metabolically continuous with the cell body. Ribosomes, the rough endoplasmic reticulum, and the Golgi complex extend into them for some distance. In contrast, axons, which begin abruptly with a well-defined initial segment, rigidly exclude these organelles.

Neurons are not the only cell types that are polarized. Other cells differentially distribute or sort proteins into different cellular compartments, and, in at least some cases, appear to use mechanisms similar to those employed by neurons. Neurons have a more difficult task than other cell types, however, because they must assemble several specialized surface membrane domains within each compartment. Not surprisingly, the cytoskeleton plays important roles both in sorting cellular constituents and in creating and maintaining specialized surface domains.

Microtubule polarity is different in axons and dendrites

Both axons and dendrites have the same basic cytoskeletal components—microtubules, neurofilaments, and actin filaments. The arrangement and biochemical properties of the components, however, are subtly different in axons and dendrites. Although imperfectly understood, these differences in the cytoskeleton undoubtedly play a role in determining the characteristic shapes of axons and dendrites and their transport properties (Figure 11).

The orientation of microtubules is one of the most provocative of the differences between axonal and dendritic cytoskeletons. All microtubules in the axon are oriented with their plus ends toward the distal tip and their minus ends toward the cell body. In contrast, the polarity of dendritic microtubules is mixed, with roughly equal numbers of microtubules oriented in each direction. The different orientation of microtubules in axons and dendrites could provide a mechanism for selective transport from the cell body into dendrites. Any organelle in the cell body that is specified for transport from the plus to the minus ends of microtubules will be transported into dendrites but not axons.

The proteins associated with microtubules in axons and dendrites are also distinct. MAP2, for example, is only found in the cell body and dendrites, whereas tau is found almost exclusively in axons. Both MAP2 and tau bind to microtubules by homologous domains in the C-terminal half of the protein that include several short, imperfect amino acid repeats. Both also have a nonhomologous projecting arm that, in the case of MAP2, can be seen in the electron microscope. The differences in MAPs in axons and dendrites suggest that microtubular organization in axons and dendrites is different; one tangible difference is the wider spacing of microtubules in dendrites as opposed to axons.

The restriction of MAP2 to dendrites is interesting for another reason. MAP2 not only binds microtubules, but also binds the regulatory subunit of the cAMP-dependent protein kinase; this association is apparently re-

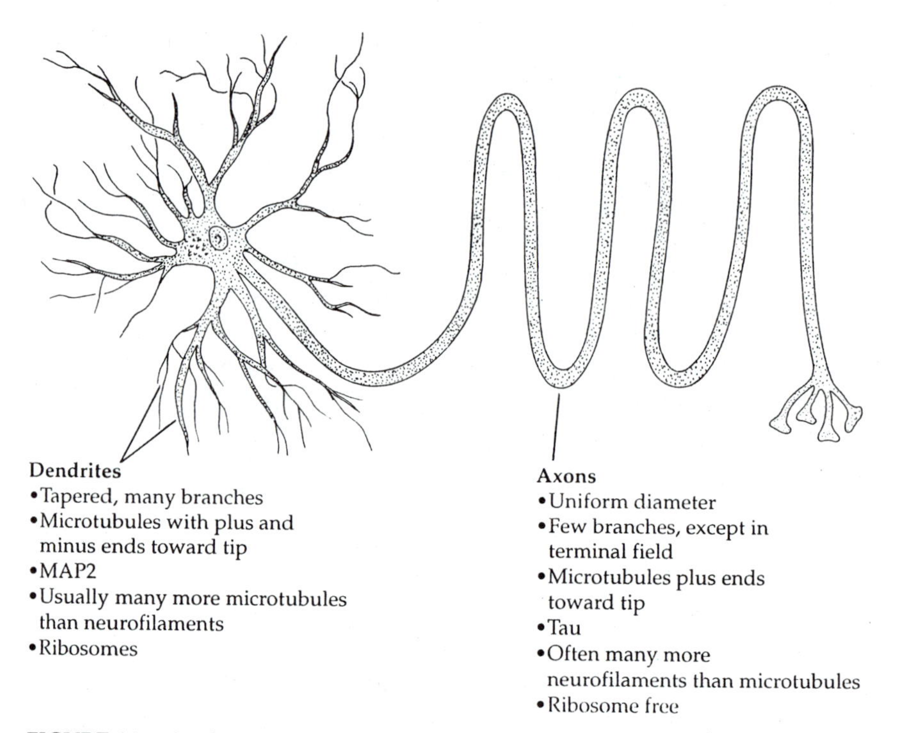

FIGURE 11. A schematic summary of selected differences between dendrites and axons.

sponsible for the concentration of the kinase in cell bodies and dendrites. Axons and dendrites also differ in their relative abundance of microtubules and neurofilaments. Mammalian axons have many more neurofilaments than microtubules, whereas the reverse is true in dendrites. Neurofilament proteins are also more highly phosphorylated in axons than in dendrites.

Axons and dendrites have separate and specific transport systems

How are proteins selectively targeted to axonal and dendritic compartments? Recent clues come from studies of the sorting of viral proteins to apical and basolateral domains in epithelial cells. In epithelial cells, apical and basolateral membrane proteins are both synthesized in the endoplasmic reticulum and pass through the Golgi complex, becoming segregated in the *trans*-Golgi network, where they are sorted into two different vesicle populations, one targeted to the apical membrane, the other to the basolateral membrane. Microtubules play an important role in the subsequent sorting process, particularly in delivering vesicles to the apical domain. Curiously, some viral glycoproteins have basal or apical targeting signals that result in their selective appearance in one or the other of these domains. Newly synthesized G protein of vesicular stomatitis virus, for example, is selectively transported to the basolateral membrane, whereas the hemagglutinin protein of influenza virus appears only in the apical membrane.

When neurons are infected with these viruses, a similar segregation of viral glycoproteins is observed. The G protein is found in dendrites but not axons; conversely, hemagglutinin is found mostly in axons, with little in dendrites. In the case of hemagglutinin, the kinetics of delivery to the terminals suggest that it is transported by fast axonal transport. These results suggest that neurons and epithelial cells use similar sorting mechanisms. Unfortunately, the molecular basis of the recognition signals is known neither in epithelial cells nor in neurons.

Neurons also have a mechanism of transporting mRNA into dendrites. This transport, whose mechanism is unknown, is selective for specific mRNAs. mRNA for MAP2, for example, extends far into dendrites, whereas mRNAs encoding tubulin and the large neurofilament protein are confined to the cell body. Specific transport of mRNA into dendrites and its translation there represents another means by which proteins become selectively distributed within neurons. The presence of particular mRNAs in dendrites may allow the synthesis of specific proteins to be locally controlled, perhaps by activity at nearby synaptic sites.

Isolated neurons develop polarity

During their initial growth neurons extend several processes, only one of which becomes an axon. What determines which neurite becomes the axon? The development of polarity has been most carefully analyzed in cultured neurons, which can be visualized in their entirety and manipulated experimentally. Under appropriate culture conditions, embryonic neurons typically develop one axon and several dendrites, as identified by their shapes and by immunocytochemical staining for selectively distributed proteins such as MAP2 (which is specific for dendrites). The axons and dendrites formed in vitro appear to have almost all of the characteristic properties of these two neurites seen in vivo. The compartmentalization of neurons into distinct axonal and somatodendritic domains thus appears to be a fundamental property of neurons determined by an intrinsic program of development. In contrast, the establishment of at least some specialized membrane domains, such as the active zones or the nodes of Ranvier, appears to require interactions between cells.

When embryonic hippocampal cells are cultured, they initially extend several indistinguishable short processes that contain MAP2 and have microtubules that are oriented with their plus ends toward the growth cone. After several hours, one of the neurites begins to elongate at a more rapid rate than the others (Figure 12). This neurite becomes the axon; the other processes, which continue to grow more slowly, become the dendrites. At about the same time that rapid growth begins, several proteins become selectively segregated into axons. One of the first is **GAP-43 (growth-associated protein),** a protein that is associated with regenerating axons and that is enriched in growth cones. GAP-43 is initially present in all neurites but becomes restricted to the axon at the time polarity is established. After the axon is established, signs of dendritic differentiation appear. MAP2 becomes exclusively localized to dendrites, and microtubules with their minus oriented toward the tip are seen in the proximal segments of the dendrites. Recent experiments suggest that the expression

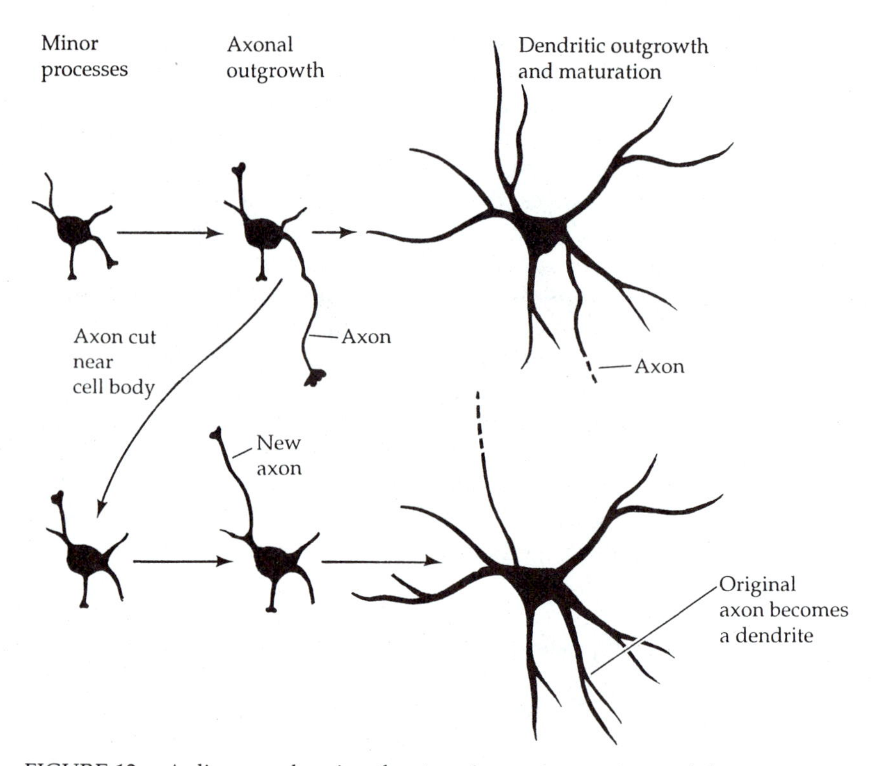

FIGURE 12. A diagram showing the steps in axon extension and the response to axonal transection in a cultured hippocampal neuron. (From C. G. Dotti et al., 1988. J. Neurosci. 8: 1454–1468; and K. Goslin and G. A. Banker, 1989. J. Cell Biol. 108: 1507–1516.)

of tau may also be important for the differentiation of axons. Tau expression is associated with neurite outgrowth; when neurons are cultured with antisense oligonucleotides to suppress the expression of tau, axonal processes fail to extend.

Which of the initial processes ultimately becomes the axon appears to be a matter of chance. If a newly formed axon is transected near the cell body, one of the remaining processes, which otherwise would have become a dendrite, begins to extend rapidly and becomes the new axon. Simultaneously, GAP-43 is rerouted to the new axon. The period of plasticity during which any neurite can become an axon does not last indefinitely. After a week or more in culture, axonal transection near the cell body no longer causes reversal of polarity. Instead, supernumerary axons sprout from the dendrites, a phenomenon also observed following close axotomy of neurons in adult animals. In vivo, the site on the cell body from which nascent axons emerge may not be determined in the apparently random fashion seen in cultured neurons but may be specified by signals in the external environment. Similar mechanisms to those seen in culture, however, probably restrict the delivery of axonal proteins to a single process and thus prevent the development of multiple axons.

Neurons have many specialized surface domains

Although many types of eukaryotic cells have one or more specialized domains on their surfaces, nerve cells are unusual in that their surface consists of a mosaic of specialized domains, each containing a particular subset of the membrane proteins. In axons and terminals, for example, nodes of Ranvier, internodal axonal membranes, nerve terminal membranes, and the active zones all have distinct functional properties and different protein compositions. Since the lipid membrane itself offers no barrier to the lateral diffusion of proteins, the restricted localization of proteins within a domain must depend on their interactions with proteins attached to the cytoskeleton or the extracellular matrix. A specialized membrane domain thus implies a corresponding specialization of the cytoskeleton and/or the extracellular matrix. A major question in neuronal cell biology is to understand how these specialized domains are constructed and organized to give the complex topography of the neuronal surface membrane.

Proteins that immobilize erythrocyte membrane proteins also occur in nerve cells

The interaction of membrane proteins with the underlying cytoskeleton is best understood in the mammalian erythrocyte, which has a simplified cytoskeleton designed to give mechanical strength to the surface membrane as it deforms during passage through capillaries. Although the surface membrane is not divided into domains, the major membrane proteins of erythrocytes are attached to the cytoskeleton and thus have limited mobility.

The principal components of the erythrocyte cytoskeleton are actin and **spectrin,** a large protein with two polypeptide chains, α (240 kD) and β (220 kD) (Figure 13). The two spectrin chains are extended molecules,

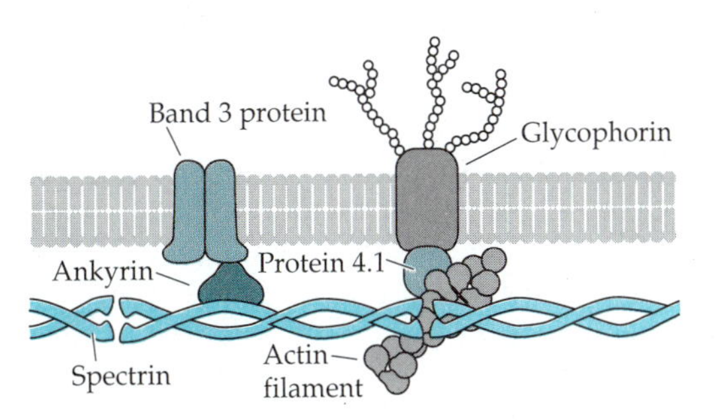

FIGURE 13. A schematic drawing of the arrangement of erythrocyte cytoskeletal proteins. Spectrin dimers that are joined head-to-head form a complex at their tails with other dimers, with actin filaments, and with protein 4.1. These complexes are the nodes of an extensive two-dimensional network that underlies the surface membrane. The network is attached to the membrane through association of protein 4.1 with the membrane protein glycophorin, and through the protein ankyrin, which links the spectrin dimers to the erythrocyte anion channel.

aligned in anti-parallel fashion and wound around each other to form a heterodimer. Two heterodimers then associate head to head to yield long (200 Å), flexible, rodlike tetramers that are the building blocks of the erythrocyte cytoskeleton. Spectrin tetramers are linked at their tails by short actin filaments to form an extensive and continuous network that underlies the entire erythrocyte membrane.

Erythrocyte membrane proteins are bound to the actin-spectrin network by two linking proteins, **ankyrin** and **protein 4.1.** Ankyrin is a large (200 kD), globular protein that binds to a site on the beta chain of spectrin near the head of the spectrin heterodimer. Ankyrin also binds band 3, a chloride/bicarbonate ion channel of the erythrocyte. The link formed by ankyrin immobilizes the band 3 ion channel by attaching it to the actin-spectrin cytoskeleton. The cytoskeleton is also attached via the 4.1 protein, which binds glycophorin, a major erythrocyte membrane protein whose function is unknown. The 4.1 protein is complexed with actin and spectrin at the tail of the spectrin heterodimers.

Homologues of the cytoskeletal and linker proteins of erythrocytes are found in neurons. Spectrins appear to be part of the actin network that underlies the neuronal surface membrane, and several forms of spectrin in brain are differentially distributed within neurons. The cytoskeleton of cell bodies and dendrites contains a form of spectrin like that in erythrocytes, whereas spectrin in axons (also called **fodrin** or **calspectin**) has a different β subunit, and is specific to neurons. During development, the brain-specific form is expressed first; the erythroid form appears at the time of synaptogenesis.

Antibodies to erythrocyte ankyrin and protein 4.1 stain the circumference of neurons, suggesting that homologues of these two proteins are also components of the cytoskeletal network that underlies neuronal membranes. In addition to the diffuse and generalized localization of spectrin and ankyrin at the surface of neurons and their processes, these proteins are concentrated at the node of Ranvier and the postsynaptic membrane, two specialized domains that are discussed below.

Membrane proteins are immobilized at the node of Ranvier by an actin-spectrin cytoskeleton

The node of Ranvier in myelinated axons is specialized to allow ion movement into and out of the cell. The node has a high density of sodium channels, whereas the adjoining myelinated axonal membrane has few, if any. This difference in sodium channel density is the basis of the saltatory conduction of the action potential in myelinated axons (Chapter 2). The node also has a higher density of the Na^+/K^+ ATPase. A layer of dense material on the cytoplasmic side of the membrane is seen in electron micrographs at the node, suggesting the presence of a specialized cytoskeleton. Immunofluorescence and immunogold localization show that the density contains a high concentration of spectrin as well as the erythroid form of ankyrin, which is not seen elsewhere in the axon.

Biochemical evidence from other tissues indicates that both sodium channels and the Na^+/K^+ ATPase bind ankyrin. Sodium channels purified from brain are immunoprecipitated with antibodies to ankyrin, as is the

Na^+/K^+ ATPase from kidney. As spectrin is also immunoprecipitated (in both cases), the erythrocyte model for membrane protein binding appears to be a plausible one to explain the localization of sodium channels and the Na^+/K^+ ATPase at the node of Ranvier.

The postsynaptic membrane at the neuromuscular junction contains two specialized membrane domains

A specialized postsynaptic membrane containing a high density of neurotransmitter receptors is essential for efficient transmission at chemical synapses. The most completely characterized postsynaptic membrane is at the neuromuscular junction. The muscle membrane underneath the motor nerve terminal is highly enfolded, dividing the membrane into two specialized regions: the crests of the folds, which contain acetylcholine (ACh) receptors packed in a near-crystalline array (about 10,000 per μm^2), and the troughs of the folds, which contain sodium channels, but little ACh receptor (< 10 per μm^2) (Figure 14). The organization of the postsynaptic membrane into two interspersed domains allows the endplate potential to efficiently generate an action potential.

Cytoplasmic dense material (the **subsynaptic density**) is associated with the crests of the folds. Antibodies to a protein purified from receptor-rich membranes, the **43-kD** protein, stain the subsynaptic density and show that the protein is localized over exactly the same area of membrane as the ACh receptor. The 43-kD protein appears to play an important role in clustering the ACh receptor because expression of the protein along

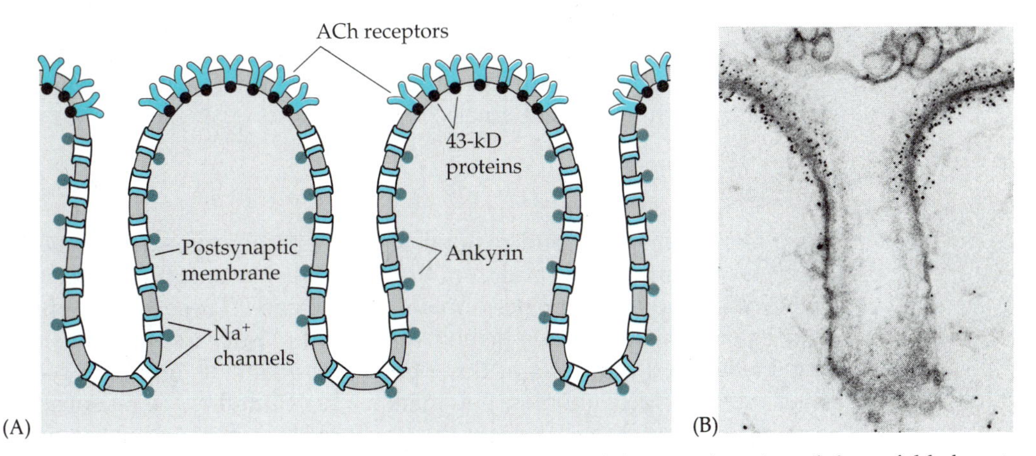

FIGURE 14. (A) Schematic drawing of the two domains of the enfolded postsynaptic membrane at the neuromuscular junction. ACh receptors and the 43-kD protein are confined to the crests and upper sides of the folds, whereas the sodium channel and ankyrin are found only in the depths of the folds. These distributions are based on immunocytochemical localization in the electron microscope. The ACh receptor and 43-kD protein are thought to be in a 1:1 ratio, and their density on the crests approaches a crystalline array. The densities and relative proportions of the sodium channel and ankyrin are unknown. (B) Immunogold localization of the ACh receptor and the 43-kD protein in the postsynaptic membrane of *Torpedo* electroplax. (Photo courtesy of R. Sealock.)

with the ACh receptor in nonmuscle cells causes clusters of receptor to appear in the membrane.

Other cytoskeletal proteins that are concentrated at the neuromuscular junction include actin and a molecule related to β-spectrin. In muscle cells cultured without nerves, clusters of ACh receptors appear spontaneously; these are associated with actin, spectrin, and the 43-kD protein. Careful characterization of the spectrin in these cells shows it to be a novel form of β-spectrin that may form a homotetramer, since no α-spectrin was found. These experiments suggest that in the ACh-receptor–rich domains, the 43-kD protein acts as a linker, tethering the ACh receptor to an actin-spectrin network. The depths of the enfolded postsynaptic membrane form a second specialized domain related to the node of Ranvier in which sodium channels and the erythroid form of ankyrin are concentrated.

Association of membrane proteins with a submembrane actin-spectrin network may be a general mechanism for immobilizing membrane proteins in nerve cells, including regions such as the initial segment (where there is also a cytoplasmic density) and the postsynaptic membrane at which channel proteins are concentrated. In the CNS, postsynaptic membranes are not enfolded but nonetheless have a prominent postsynaptic density beneath the membrane that contains actin and spectrin. The glycine receptor, which is concentrated in the postsynaptic membrane of some inhibitory synapses in the CNS, is associated with a peripheral membrane protein that may play a role similar to that of the 43-kD protein. Other receptors may use forms of ankyrin or of the 4.1 protein as linkers. Another major component of the postsynaptic density in the CNS is type II Ca^{2+}/calmodulin-dependent protein kinase. The presence of this enzyme suggests that the density is not simply an inert anchoring structure for receptors but may play an active role in modifying receptors in the postsynaptic membrane, in maintaining their high density, or in adhesion of pre- and postsynaptic membranes.

Summary

The cytoskeleton forms and maintains the shape of neurons, transports organelles and proteins to and from the extensive neuronal processes, and establishes regional differences on the cell surface. These tasks, which require that the cytoskeleton be both stable and plastic, are accomplished with cross-linked networks of three polymeric filaments and their associated proteins. Microtubules, polymers made of tubulin, are responsible for neurite extension and serve as the tracks for transport within the cells; actin filaments form a cortical framework for the cell that helps localize membrane proteins and is largely responsible for movement of the growth cone; neurofilaments, made of members of the intermediate filament family of proteins, give bulk and strength to neuronal processes.

The polymerization and depolymerization of microtubules and actin filaments within the cell give the cytoskeleton its plastic quality. Both are polar filaments that grow primarily by addition of monomers at one end (the plus end). The rates of polymerization and depolymerization are controlled by accessory proteins and by the nucleotides bound to the monomeric subunits that are hydrolyzed after polymerization. This hydrolysis, which changes the affinity of the subunit for the polymer, gives microtubules a property of dynamic instability and allows actin filaments to treadmill.

The different properties of axons and dendrites are reflected in their cytoskeletons. In axons, microtubules are all aligned with their plus ends pointed toward the neurite tip, whereas in dendrites approximately half are oriented in each direction. Axons and dendrites also have distinct microtubule-associated proteins (MAPs). Neurons establish polarity at an early stage of neurite outgrowth. After several short neurites have been extended, one outgrows the rest and becomes the axon; the others then assume the properties of dendrites.

The microtubules in axons and dendrites serve as tracks along which membrane organelles move, translocated by protein motors that use the energy of ATP hydrolysis to perform mechanical work. The motors kinesin and dynein bind to organelles and translocate them along microtubules to which they transiently attach. Each is a unidirectional motor: kinesin translocates organelles toward the plus end of microtubules, and dynein toward the minus end. In addition to transporting organelles along neurites at a relatively rapid rate (several hundred mm/day), neurons transport cytoskeletal components at a slow rate (several mm/day). Microtubules and neurofilaments appear to move down the axon as polymers; the mechanism of slow transport is unknown.

The movement of microtubules down growing axons extends the microtubule network in the axon and powers neurite outgrowth. The direction of outgrowth, however, is guided by the highly motile growth cone at the end of the neurite, which actively extends and retracts filopodia and lamellipodia. Growth cone movement is based on the dynamic polymerization and depolymerization of the actin network that fills its leading edge. These movements are coupled to the extension of microtubules by an unknown mechanism that may be related to a growth cone's adhesion to its substrate.

In mature neurons, actin and fodrin, a form of spectrin, form a network under the surface membrane. Sodium channels and the Na^+/K^+ ATPase are concentrated and immobilized at the node of Ranvier by attachment to this network via the linking protein ankyrin. Attachment to a submembrane cytoskeleton also immobilizes the acetylcholine receptor at the neuromuscular junction.

References

General

Bray, D. 1991. *Cell Movements.* Garland, New York.

Burgoyne, R. D. 1991. *The Neuronal Cytoskeleton.* Wiley-Liss, New York.

Grafstein, B. and Forman, D. S. 1980. Intracellular transport in neurons. Physiol. Rev. 60: 1167–1283.

Mitchison, T. and Kirschner, M. 1988. Cytoskeletal dynamics and nerve growth. Neuron 1: 761–772.

Smith, S. J. 1988. Neuronal cytomechanics: The actin-based motility of growth cones. Science 242: 708–715.

Steinert, P. M. and Roop, D. R. 1988. Molecular and cell biology of intermediate filaments. Annu. Rev. Biochem. 57: 593–625.

Structure and assembly

Baas, P. W. and Black, M. M. 1990. Individual microtubules in the axon consist of domains that differ in both composition and stability. J. Cell Biol. 111: 495–509.

Hirokawa, N. 1991. Molecular architecture and dynamics of the neuronal cytoskeleton. In *The Neuronal Cytoskeleton.* R. D. Burgoyne (ed.). Wiley-Liss, New York. pp. 5–74.

Hoffman, P. N., Griffin, J. W. and Price, D. L. 1984. Control of axonal caliber by neurofilament transport. J. Cell Biol. 99: 705–714.

*Kirschner, M. and Mitchison, T. 1986. Beyond self-assembly: From microtubules to morphogenesis. Cell 45: 329–342.

Matus, A. 1988. Microtubule-associated proteins: Their potential role in determining neuronal morphology. Annu. Rev. Neurosci. 11: 29–44.

Schnapp, B. J. and Reese, T. S. 1982. Cytoplasmic structure in rapid-frozen axons. J. Cell Biol. 94: 667–679.

Axonal transport

Allen, R. D., Metuzals, J., Tasaki, I., Brady, S. T. and Gilbert, S. P. 1982. Fast axonal transport in squid giant axon. Science 218: 1127–1128.

Davis, L., Banker, G. A. and Steward, O. 1987. Selective dendritic transport of RNA in hippocampal neurons in culture. Nature 330: 477 479.

Hollenbeck, P. J. 1989. The transport and assembly of the axonal cytoskeleton. J. Cell Biol. 108: 223–227.

Nixon, R. A. and Logvinenko, K. B. 1986. Multiple fates of newly synthesized neurofilament proteins: Evidence for a stationary neurofilament network distributed nonuniformly along axons of retinal ganglion cell neurons. J. Cell Biol. 102: 647–659.

*Reinsch, S., Mitchison, T. J. and Kirschner, M. 1991. Microtubule polymer assembly and transport during axonal elongation. J. Cell Biol. 115: 365–379.

*Vale, R. D., Reese, T. S. and Sheetz, M. P. 1985. Identification of a novel force-generating protein, kinesin, involved in microtubule-based motility. Cell 42: 39–50.

Vallee, R. B. and Bloom, G. S. 1991. Mechanisms of fast and slow axonal transport. Annu. Rev. Neurosci. 14: 59–92.

Vallee, R. B., Shpetner, H. S. and Paschal, B. M. 1989. The role of dynein in retrograde axonal transport. Trends Neurosci. 12: 66–70.

Growth cone

*Bentley, D. and Toroian, R. A., 1986. Disoriented pathfinding by pioneer neuron growth cones deprived of filopodia by cytochalasin treatment. Nature 323: 712–715.

*Forscher, P. and Smith, S. J. 1988. Actions of cytochalasins on the organization of actin filaments and microtubules in a neuronal growth cone. J. Cell. Biol. 107: 1505–1516.

Lamoureux, P., Buxbaum, R. E. and Heidemann, S. R. 1989. Direct evidence that growth cones pull. Nature 340: 159–162.

Polarity

Dotti, C. G. and Simons, K. 1990. Polarized sorting of viral glycoproteins to the axon and dendrites of hippocampal neurons in culture. Cell 62: 63–72.

Dotti, C. G., Sullivan, C. A. and Banker, G. A. 1988. The establishment of polarity by hippocampal neurons in culture. J. Neurosci. 8: 1454–1468.

*Goslin, K. and Banker, G. 1989. Experimental observations on the development of polarity by hippocampal neurons in culture. J. Cell Biol. 108: 1507–1516.

Matus, A., Bernhardt, R. and Hugh-Jones, T. 1981. High molecular weight microtubule-associated proteins are preferentially associated with dendritic microtubules in brain. Proc. Natl. Acad. Sci. USA 78: 3010–3014.

Surface domains

Branton, D., Cohen, C. M. and Tyler, J. 1981. Interaction of cytoskeletal proteins on the human erythrocyte membrane. Cell 24: 24–32.

Flucher, B. E. and Daniels, M. P. 1989. Distribution of Na^+ channels and ankyrin in neuromuscular junctions is complementary to that of acetylcholine receptors and the 43 kD protein. Neuron 3: 163–175.

Srinivasan, Y., Elmer, L., Davis, J., Bennett, V. and Angelides, K. 1988. Ankyrin and spectrin associate with voltage-dependent sodium channels in brain. Nature 333: 177–180.

9

Myelin and Myelination

Greg Lemke

MYELIN IS ONE of the signature organelles of the vertebrate nervous system. Familiar to all neurobiologists, this unusual structure is readily recognized, both macroscopically in the white matter tracts of the brain and spinal cord, and microscopically in the spiral rings of membrane that surround all rapidly conducting axons. Compelling evidence that the myelin sheath is an absolute requirement for the integrative functioning of higher nervous systems is provided by those inherited and acquired human diseases in which the sheath never forms, or in which its structural integrity is compromised. Without exception, these diseases, the best known of which is multiple sclerosis, are critically debilitating.

Historically, myelin has proved to be a particularly favorable target for structural studies based on X-ray and neutron diffraction and electron microscopy, largely due to its highly ordered and regularly repeating structure. Similarly, its abundance and exceptionally low buoyant density relative to other biological membranes make it easy to purify and analyze biochemically. As a result, we know a great deal about the general features of myelin structure, of axonal ensheathment and wrapping, and of the physiological basis of impulse conduction in myelinated axons.

More recently, myelin and myelination have become the focus of investigations that exploit the techniques of molecular biology and molecular genetics. For the most part, these studies have evolved from the molecular cloning of myelin-specific genes. Although based on the analysis of only a subset of neural genes and proteins, these investigations have nonetheless provided important general insights into several outstanding problems in molecular neurobiology. They have advanced our understanding of the structure and function of cell adhesion proteins, the genetic perturbation of neural development, the agents that regulate the differentiation of neural cells, and even the mechanisms that underlie cellular autoimmunity in the nervous system. This chapter will consider illustrative examples from each of these lines of investigation.

Myelin Structure and Formation

Myelin is a neural organelle

The myelin sheath is elaborated by two types of specialized glial cells, **oligodendrocytes** in the central nervous system, and myelin-forming **Schwann cells** in the peripheral nervous system. Embryologically, oligodendrocytes derive from multipotential progenitors of the neural tube. Schwann cells, like their neuronal partners in the PNS, derive from the migratory cells of the neural crest (see Chapter 11). It is important to remember that these two cell types are only a subset of peripheral and central glia. The CNS lineage that gives rise to oligodendrocytes, for example, also gives rise to one form of astrocyte, a cell that does not form myelin or express myelin-specific genes (see Chapter 11). Similarly, all peripheral nerves contain both myelinating and nonmyelinating Schwann cells; in many respects, nonmyelinating Schwann cells resemble astrocytes. Just as the gray matter tracts of the CNS are myelin-poor, so some peripheral nerves (e.g., the sympathetic trunk) are made up almost exclusively of nonmyelinated axons that are associated with nonmyelinating Schwann cells.

To a first approximation, the mechanics of myelin deposition by myelinating Schwann cells and oligodendrocytes is the same: each synthesizes a large sheet of plasma membrane that is spirally wrapped and then tightly compacted around target axons. For the largest axons of the PNS, this process may proceed until a sheath of over 100 layers is formed. Morphometric analysis of serial electron micrograph cross sections of the Schwann cells that produce these very elaborate myelin sheaths has demonstrated that their membrane surface area increases several thousand-fold during the course of myelination.

When viewed in the electron microscope in cross section, the final structure of myelin resembles a highly regular array of rings of plasma membrane bilayers (Figure 1). A key feature of the organization of these rings is their compaction. If we imagine that the topology of the myelin sheath is similar to that of a rolled up sleeping bag, then two surface appositions are generated at each turn of sheath, or roll of the bag. One of these corresponds to the bringing together of cytoplasmic membrane surfaces—the inside surface of the sleeping bag. The other corresponds to the apposition of extracellular membrane surfaces—the outside surface of the bag. Structural biologists have referred to these close cytoplasmic and extracellular membrane appositions as the **major dense lines** and **intraperiod lines** of myelin, respectively, based on their appearance in the electron microscope (Figure 1). The distance between one major dense line and the next (or one intraperiod line and the next) defines one turn, or "repeat period" of the myelin sheath. This distance is exceedingly small: electron density profiles generated from high resolution X-ray diffraction of unfixed peripheral and central myelin place the repeat periods at around 170 Å and 150 Å, respectively. Since the width of a single mem-

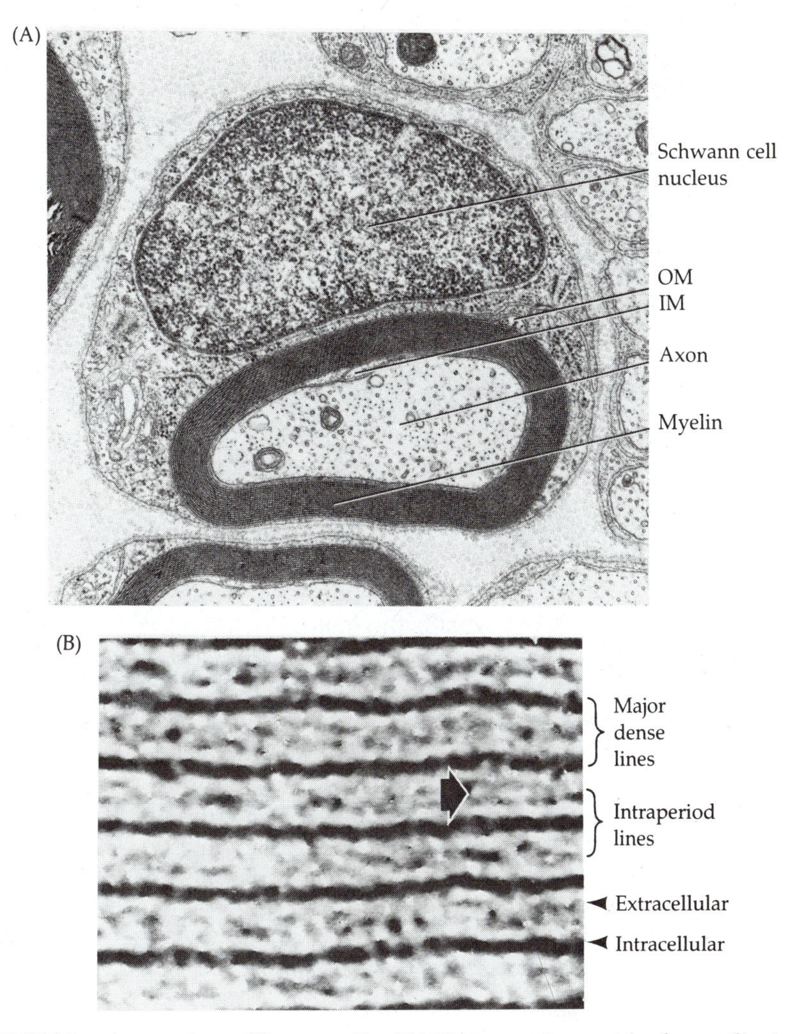

FIGURE 1. A neural-specific organelle. (A) Electron micrograph of a myelinated axon from mouse sciatic nerve in cross section, in the region of the Schwann cell nucleus. This Schwann cell has elaborated a myelin sheath of approximately 15 compacted bilayers. The membrane at the innermost and outermost layers of the sheath (IM and OM, respectively) remains uncompacted. (The large size of the nucleus deforms the normally cylindrical axon.) (B) High resolution electron micrograph enlargement of a small segment of the myelin sheath, illustrating the alternating major dense lines and intraperiod lines. The major dense lines appear as fused cytoplasmic membrane faces. The intraperiod lines correspond to the apposition of extracellular membrane faces and under electron micrograph fixation conditions appear to be less tightly formed. (The large arrow marks a region in which the intraperiod line is clearly resolved into two extracellular membrane leaflets.) (Photograph in A courtesy of Graham Kidd. B from Thomas and Ochoa, 1984. *Peripheral Neuropathy*, 2nd ed., P. J. Dyck, P. K. Thomas, E. H. Lambert, and R. Bunge, eds. W. B. Saunders, Philadelphia.)

brane bilayer is ~50 Å and since there are two bilayers per repeat period, fully compacted regions of the myelin sheath contain very little cytoplasm or extracellular space, just as a well-rolled sleeping bag contains very little air.

Electron microscopic examinations of developing CNS and PNS myelin sheaths have clearly demonstrated that myelin is a specialization of the Schwann cell (oligodendrocyte) plasma membrane, and the innermost layers of the sheath can be traced back and shown to be contiguous with the plasma membrane of the myelin-forming cell. As we shall see below, immunohistochemical and biochemical studies have also demonstrated that even after it is fully formed, the sheath remains in metabolic equilibrium with the remainder of the myelin-forming cell. Like axons and dendrites, myelin should therefore be viewed as a specialized organelle unique to cells of the nervous system.

Is myelin pushed or pulled?

The cellular mechanics of myelination present an interesting problem for cell biologists. How is the myelin sheath actually wrapped around the axon? And what is the motor that drives its assembly? At various times it has been suggested that myelinating cells migrate around the axon, trailing myelin membrane behind them, or that axons in effect wrap themselves by turning within the forming myelin sheath. Studies of early peripheral myelination in vivo, and in myelinating culture systems containing purified Schwann cells and neurons, have demonstrated that neither of these mechanisms applies. Instead, the relative positions of the Schwann cell body and the axon appear, to a first approximation, to be fixed during myelination, with most of the important elaborative movement confined to the growing front (in cross section, the growing tip) of the myelin membrane.

The currently accepted sequence of events for myelination is illustrated in Figure 2. After specific association, Schwann cells (and by extension, oligodendrocytes) curl a sheet of membrane around the axon, a process that appears to be driven by rapid membrane biosynthesis. After completing its first turn, this sheet of membrane must insinuate itself *under* the first wrap of the sheath. In this way, spiral growth of the myelin sheath proceeds through displacement of the previously deposited myelin layer that is in direct association with the axonal surface. In other words, repeated wrapping is achieved through continuous insinuation of the growing inner layer of membrane about the axon, rather than through migration of the Schwann cell body about the axon. This is not to say that cellular migration does not occur, only that it is incidental to the elaboration of myelin. When the Schwann cell body (nucleus) is monitored in myelinating cultures using time-lapse microscopy, its direction of migration is generally found to be the same as that of the moving inner layer, which is exactly the opposite of the relative direction expected if myelin were deposited by cellular migration. Furthermore, the number of times a given Schwann cell circumnavigates an axon during the course of myelination

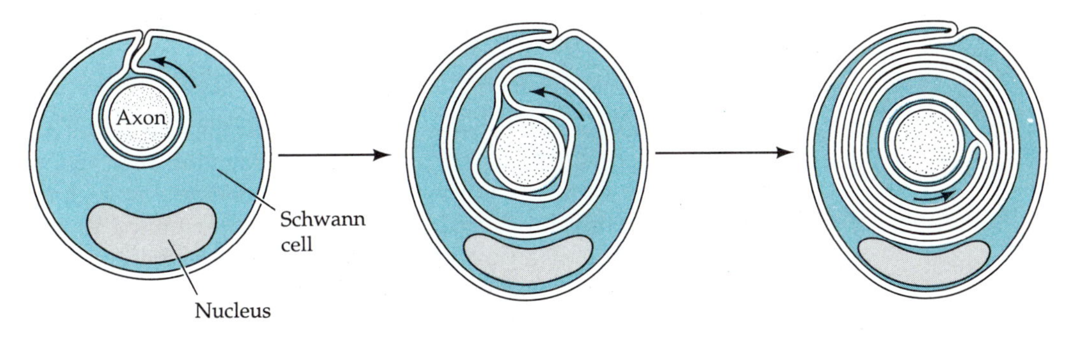

FIGURE 2. The mechanics of myelination. This diagram illustrates the main features of peripheral myelination as viewed in cross section. A Schwann cell first assumes a one-to-one relationship with a peripheral axon. Myelination of the axon is then achieved through repeated "tuck under" wrapping of the innermost loop of the membrane and subsequent compaction of stacked membrane bilayers. Arrows indicate the observed direction of membrane movement during myelination. Nonmyelinating Schwann cells loosely enfold axons in a manner similar to that depicted for the beginning of the myelination sequence, although these nonmyelinating cells are often observed to associate with multiple axons. (After P. Morell and W. T. Norton, 1980. Sci. Am. 242: 88–118.)

(also monitored in myelinating cultures) is much lower than the number of myelin layers deposited. (Occasionally the number of circumnavigations is 0.) As myelin wrapping proceeds, compaction of the developing sheath at the intraperiod line appears to be coincident with the elaboration of new myelin layers, whereas compaction at the major dense line, which effectively corresponds to the removal of cytoplasm, occurs somewhat later. Throughout myelination, vesicles of new myelin membrane are inserted primarily in the vicinity of the cell body, rather than in the interior of the sheath. The great distance of the site of new membrane insertion from the growing front of the inner layer does not present a topological problem, however, since plasma membranes are effectively two-dimensional fluids. In fully formed myelin sheaths, the inner layer of the sheath remains uncompacted.

From the standpoint of final topology, the most important event in this scenario is the initial tucking-under of the myelin sheath at the first turn in its elaboration. If this were not to happen, and the growing sheath were to initially extend around the periphery of the myelin-forming cell, myelination would culminate in a sheath that enclosed both axon and myelin-forming cell. Since this structure is rarely observed in normal nerves or fiber tracts, there must be an effective mechanism that prevents its formation. The initial tucking-under event probably results from the action of a set of strictly localized membrane adhesion molecules that mediate preferential association of the growing myelin sheath either with the axonal surface or with the axon-proximal surface of the first layer of myelin. As we shall see, there are several good candidates for these adhesion molecules, some of which are glial-specific.

Myelin Function and Neural Evolution

Myelinated axons are faster

Why did the elaborate process of myelination evolve in the first place? As discussed in detail earlier (Chapter 2), during the propagation of an action potential few ions move across the axonal plasma membrane in regions that are insulated by myelin. Instead, current flow is restricted to the regularly spaced patches of bare axonal membrane between adjacent myelin sheaths. These patches are called the **nodes of Ranvier,** after the neurologist M. L. Ranvier. Unlike the regions of axon surrounded by myelin, nodes of Ranvier contain an extremely high density of the voltage-sensitive sodium channels required for impulse conduction, and are exposed to a relatively low resistance extracellular environment. This geometry forces current to jump in large loops, in "saltatory" fashion from one node to the next, a mode of conduction that is much faster than conduction along unmyelinated axons.

How much faster? A myelinated axon of average radius conducts nerve impulses at a rate that is approximately ten times that of an unmyelinated axon of the same size. Since the volume occupied by an axon is directly related to the *square* of its radius, a myelinated axon occupies on average only 1/100th the volume of an unmyelinated axon conducting impulses at an equivalent rate. This single consideration has almost certainly been the driving force behind the appearance of myelin and myelin-forming cells, which constitute the last major cellular advance in neural evolution. Of course, the problem of rapid conduction velocities is also faced by invertebrates, which generally lack myelin. These creatures rely on a straightforward solution: they achieve high rates of impulse conduction simply by increasing the radius, and thereby increasing the length constant, of their axons (see Chapter 2). One of the classic preparations of neurophysiology, the squid giant axon, represents exactly this solution. In a big squid, the diameter of this single axon approaches 1 mm! Increasing axon caliber is not a workable solution for the nervous systems of vertebrates, because these nervous systems have a relatively large number of rapidly conducting axons confined to a relatively small space (e.g., our skulls). If we assume that 10% of the volume of the human brain is taken up by myelinated axons (a reasonably conservative estimate), then in the absence of myelin our brains would need to be ten times larger simply to maintain conduction velocity.

Myelination by oligodendrocytes requires less space than myelination by Schwann cells

The pressure to minimize the space occupied by the nervous system may further underlie the evolutionary transition from myelination by Schwann cells to myelination by oligodendrocytes. Phylogenetically, the former is the older process; in the elasmobranch fishes in which true myelin is first observed (e.g., sharks), myelination in both the PNS and CNS is carried out by cells that express a repertoire of Schwann-cell–specific myelin proteins (see below). As one moves forward phylogenetically, this primitive form

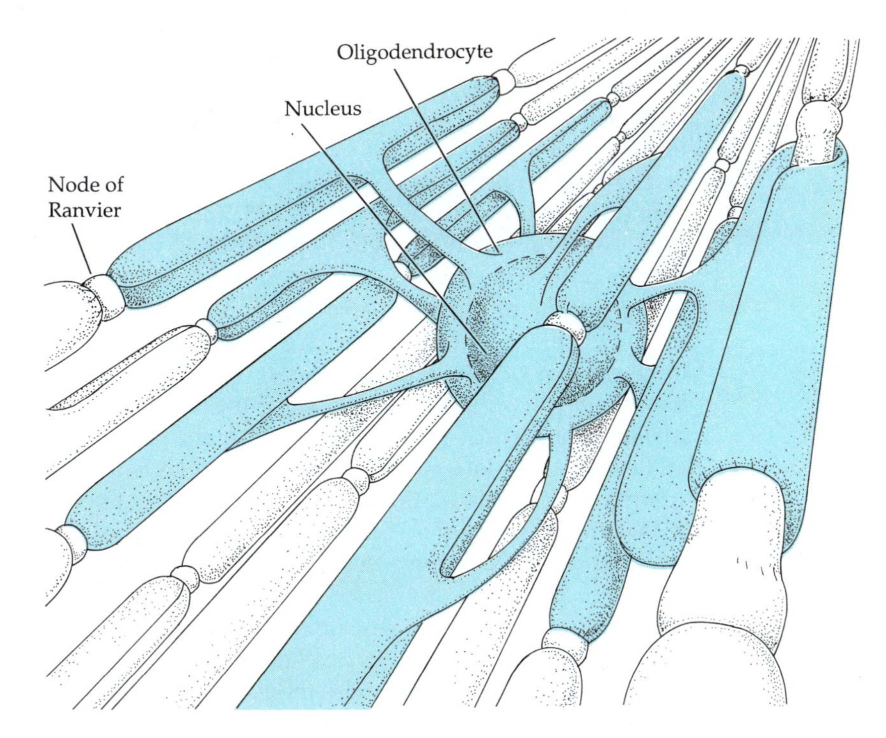

FIGURE 3. Myelination in the central nervous system. Although the mechanics of myelination in the central and peripheral nervous systems are similar, oligodendrocytes have the remarkable ability to extend processes around, and to myelinate, many (>50) different CNS axons. (After P. Morell and W. T. Norton, 1980. Sci. Am. 242: 88–118.)

of myelination and the cells that perform it become segregated to the PNS and replaced in the CNS by oligodendrocyte myelination. One critical difference between these two processes relates to the number of axons that can be myelinated by a single cell. Although one Schwann cell myelinates only a single axon, an oligodendrocyte can elaborate independent myelin sheaths around 50 or more different axons (Figure 3). A 50-fold reduction in the number of cells required for myelination again affords a significant savings in space, an important advantage in the development of a larger, more complex, and more versatile nervous system.

Myelin Genes and Myelin Genetics

Myelinating cells express a set of specialized genes

During their differentiation into myelin-forming cells, Schwann cells and oligodendrocytes activate expression of a group of myelin-specific genes. These genes encode proteins that play roles in the induction of myelination, in the initial deposition of the myelin sheath, and in its wrapping and subsequent compaction about the axon. Grouped into major and minor sets based on the abundance of their expression, the myelin genes have proved to be relatively easy targets for molecular cloning. Thus far, seven of these genes have been successfully cloned, and it is these efforts that have motivated nearly all of the current interest in the molecular genetics of myelination.

The major myelin group includes three genes, those encoding **protein zero (P_0)**, the **proteolipid protein (PLP)**, and **myelin basic protein (MBP).** P_0, an integral membrane glycoprotein of 30 kD, is the major structural protein of PNS myelin, where it accounts for more than 50% of the protein in the peripheral sheath. Its expression is absolutely restricted to myelin-forming Schwann cells; it is not detectably expressed in the CNS or in the large number of Schwann cells that do not form myelin. PLP, an integral membrane proteolipid also of 30 kD, is the major structural protein of CNS myelin, where it also accounts for ~50% of the protein in the central sheath. PLP expression is largely restricted to oligodendrocytes, although extremely low levels of the protein can also be detected in myelinating Schwann cells in higher vertebrates. MBP—actually a family of closely related proteins—is expressed by both oligodendrocytes in the CNS and Schwann cells in the periphery. Unlike PLP and P_0, MBP is not an integral membrane protein; it accounts for ~30% of the protein in CNS myelin and 5–15% of the protein in PNS myelin. The differential (PNS/CNS) expression of the genes encoding these three major myelin proteins is illustrated in the in situ hybridization profiles of Figure 4.

Thus far, four of the minor myelin genes have been molecularly cloned. These genes encode the **myelin-associated glycoprotein (MAG)**, the enzyme **2',3' cyclic nucleotide 3' phosphodiesterase (CNP)**, the fatty acid transport protein **P_2**, and the **oligodendrocyte-myelin glycoprotein (OMgp).** MAG is a heavily glycosylated, myelin-specific integral membrane protein of 100 kD, expressed at low levels by both oligodendrocytes and Schwann cells. MAG is of particular interest because of its location within the myelin sheath and its structure, which is closely related to the neural cell adhesion molecule (N-CAM) described in Chapter 12. Like MAG, OMgp is a large integral membrane glycoprotein exclusively expressed by myelin-forming cells and has a structure that is also similar to those of known adhesion molecules. CNP catalyzes the hydrolysis of several unusual 2',3' cyclic nucleotides. (Note that cAMP and cGMP are cyclized via 3',5' phosphodiester bonds.) Although highly enriched in myelin, the role of CNP in myelination remains a mystery. P_2 is a small (14 kD), myelin-specific, cytoplasmic protein whose amino acid sequence exhibits strong similarity to fatty acid binding and transport proteins expressed by adipocytes, muscle cells, and intestinal epithelia. P_2 is thought to somehow facilitate the exponential increase in plasma membrane biosynthesis associated with peak periods of myelination. The general biochemical properties of these seven myelin-specific proteins are summarized in Table 1.

With the exception of the CNP gene, expression of each of the major and minor myelin genes is specifically activated in oligodendrocytes and Schwann cells only upon their differentiation into myelin-forming cells. As a result, these genes are now routinely used as markers in studies of the regulatory mechanisms underlying glial differentiation. Since their orga-

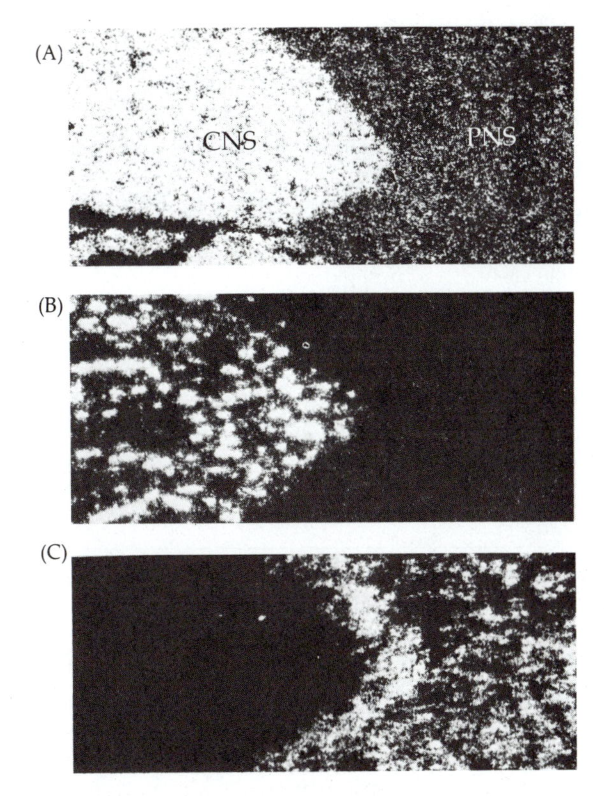

FIGURE 4. Differential expression of the major myelin genes. Oligodendrocytes and Schwann cells express a different repertoire of myelin-specific genes. This figure displays in situ hybridization profiles of the rat trigeminal nerve at the CNS/PNS boundary. As it crosses this boundary, an axon is myelinated first by oligodendrocytes (at left) and then by Schwann cells (at right). Sections A, B, and C have been hybridized with radiolabeled cDNA probes that detect MBP, PLP, and P_0 mRNA, respectively. The MBP gene is expressed both by oligodendrocytes in the CNS and Schwann cells in the PNS, whereas the PLP gene is predominantly expressed by oligodendrocytes, and the P_0 gene is exclusively expressed by Schwann cells. From B. D. Trapp et al., 1987. Proc. Natl. Acad. Sci. USA 84: 7773–7777.)

nization and regulation illustrate principles of general relevance to the molecular biology of neural function and development, we shall first consider the structure, expression, and mutation of several of these genes in detail.

The myelin basic protein family is generated by alternative pre-mRNA splicing

Myelin basic protein plays an essential role in CNS myelin formation. Of the major myelin-specific proteins, MBP is the easiest to purify and has therefore been the most extensively analyzed. It is in fact not one protein, but a family of proteins. Isolation of cDNAs encoding each of the proteins of this family, together with the cloning of the MBP gene, has revealed that

TABLE 1. Myelin Proteins

Myelin protein	CNS	PNS	Molecular weight (kD)[a]	Mouse chromosome, mutation
Protein zero (P_0)	—	>50%	25	1
Proteolipid protein (PLP)	50%	<1%	30 26	X, *jimpy*
Myelin basic protein (MBP)	~30%	~10%	14–21 (multiple forms)	18, *shiverer*
Myelin-associated glycoprotein (MAG)	1%	<1%	60 64	7
P_2 basic protein (P_2)	—	variable	14	ND
2'3'-cyclic nucleotide 3' phosphohydrolase (CNP)	40%	<1%	48 46	ND
Oligodendrocyte-myelin glycoprotein (OMgp)	1%	—	46	17

[a]Molecular weights are based on cDNA sequence and do not reflect posttranslational modifications such as glycosylation (P_0, MAG, OMgp) or phosphorylation (MBP, CNP).

the family is not encoded in multiple genes, but is instead generated by a complex series of alternative pre-mRNA spicing events in which the seven cassette exons of the single MBP gene are variously combined. (See Chapter 10 for the general features of alternative splicing in the nervous system.) Since all of the RNA splice junctions in the MBP gene fall precisely between codons, each of these splicing events (diagrammed in Figure 5) generates an MBP in the same reading frame; family members therefore differ by one or more internal deletions.

Although the functional significance of this complicated pattern of splicing is not yet clear, MBP species lacking the second exon predominate in mature nervous systems, whereas larger forms of MBP that contain sequences encoded in this exon predominate early in postnatal development, during periods of rapid myelination. This may be of functional significance with respect to the subcellular routing of MBP mRNAs and protein isoforms. Just as certain neuronal mRNAs and proteins appear to be specifically transported to, and translated in, dendrites (Chapter 8), MBP mRNA and protein are transported away from Schwann cell and oligodendroglial nuclei and into the myelin sheath. In transfected cells, MBP isoforms containing the amino acids encoded in exon 2 are more efficiently transported away from the nuclei and more diffusely distributed throughout the cytoplasm than are isoforms that lack these exon 2–encoded residues.

The mouse mutation shiverer *implicates MBP in myelin compaction*

How do we know that MBP plays an essential role in the formation of the myelin organelle? Our most direct insight into the function of MBP comes from the study of a recessive neurological mutation that specifically affects myelination in the mouse. It is one of a large number of informative mouse

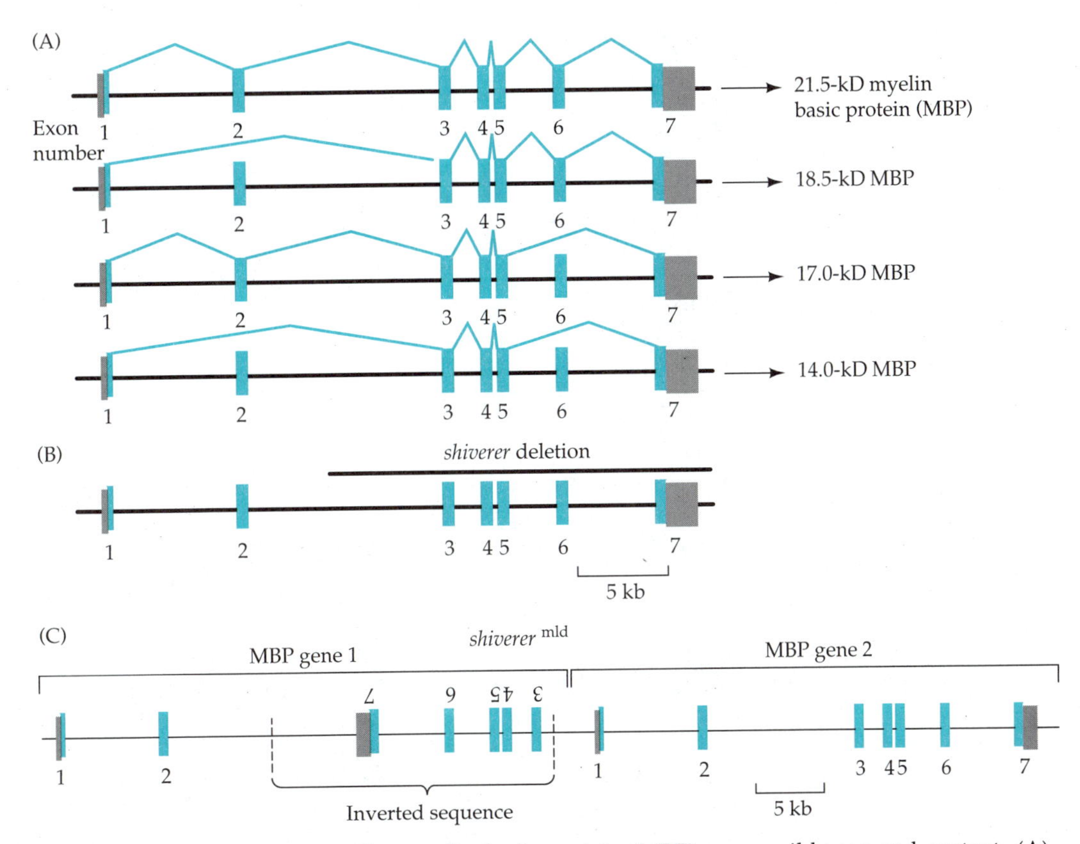

FIGURE 5. The myelin basic protein (MBP) gene: wild-type and mutant. (A) Exon/intron organization and alternative pre-mRNA splicing of the MBP gene. Exons are boxed; colored boxes correspond to coding sequences. Alternative use of the cassette exons of the MBP gene generates MBPs that differ by one or more internal deletions. The major mouse MBP isoforms are depicted. Alternative splicing reactions involving exon 5 have also been described. (B) The mouse *shiverer* mutation is associated with a large deletion within the MBP gene. (C) *Shiverer*^{mld}, an allele of *shiverer*, is associated with an inversion/duplication event that produces two tandem MBP genes, the first of which contains an inversion of exons 3 through 7.

mutations in which various features of neural development are perturbed (see Box A). Called ***shiverer* (*shi*)**, this mutation is located on mouse chromosome 18. When present in two copies, it causes a deficit in CNS myelination that leads first to a generalized intention tremor, then to increasingly frequent convulsions, and finally to premature death between 50 and 100 days after birth. CNS myelin is largely absent from the brains of *shiverer* (*shi/shi*) homozygotes, but where present, it appears as abnormal whorls of membrane, tightly compacted at the intraperiod line but cytoplasm-rich and uncompacted at what would normally be the major dense line. In other words, the major dense line never forms.

Several lines of evidence have shown that the *shiverer* phenotype results

Box A Mutational Analysis of Neural Development in the Mouse

The analysis of neural differentiation in invertebrates has greatly benefitted from the application of classical and molecular genetics. This has been particularly true for easily maintained organisms with short life cycles, such as the fruit fly *D. melanogaster* and the nematode *C. elegans,* whose genomes are well characterized and well saturated with mutations. The notable success of neurodevelopmental studies in these organisms (see, for example, the discussion of the *Drosophila* mutation *sevenless* in Chapter 11) points to the power of combining molecular biology with genetics.

Although the study of neural development in vertebrates has by necessity been based more on direct observation and/or intervention in vivo and on experiments in cell culture, several single-gene mutations in the mouse have provided important insights into how cells of the mammalian nervous system differentiate and function. Mice bearing these mutations have been identified, and are maintained as part of a comprehensive program that has been carried out over the last 40 years at Jackson Laboratories in Bar Harbor, Maine. In general, the "neurological" mutations have arisen spontaneously, affect processes that occur late in neural development, and can be easily identified through aberrant behavioral phenotypes. This latter property is reflected in the names of the mutations, which include *reeler, totterer, staggerer, weaver, spastic, shaker, quivering, shiverer, jimpy,* and *Trembler,* among others. Perhaps the best known of these mutations are those that affect development of the cerebellar cortex. In mice homozygous for the *staggerer* mutation, for example, Purkinje cells fail to develop their mature dendritic trees. Granule cells in *staggerer*s migrate normally and elaborate axons as parallel fibers but subsequently degenerate for want of their normal synaptic targets. In contrast, most of the granule cells in *weaver* mice cerebella degenerate prior to their normal migration. In addition to mutations resulting in movement disorders (most of which have not been molecularly cloned), informative neurological mutations affecting a number of sensory systems have also been described (e.g., *retinal degeneration slow,* for vision).

from the deletion of a large portion of the MBP gene. First, MBP, which is exclusively located at the major dense line in the normally compacted sheath, is specifically deficient in *shi/shi* mice. Second, mRNA isolated from the brains of *shi/shi* mice is devoid of MBP mRNA. Third, the MBP gene maps to the *shiverer* locus on mouse chromosome 18. Fourth, the *shiverer* MBP gene contains an ~20-kb deletion that removes nearly the entirety of this gene from the mutant genome (Figure 5).

Definitive proof of the hypothesis that the *shiverer* phenotype results from the failure to express MBP is provided by a series of experiments in which the cloned MBP gene was reintroduced into the genome of *shi/shi* mice. Wild-type MBP genomic clones were injected into the nucleus of fertilized mouse embryos to produce **transgenic mice.** In some of these mice, the injected DNA was incorporated into the genome of germ cells at a position unlinked to the endogenous wild-type MBP gene, was passed to succeeding generations independently of this gene, and was expressed as

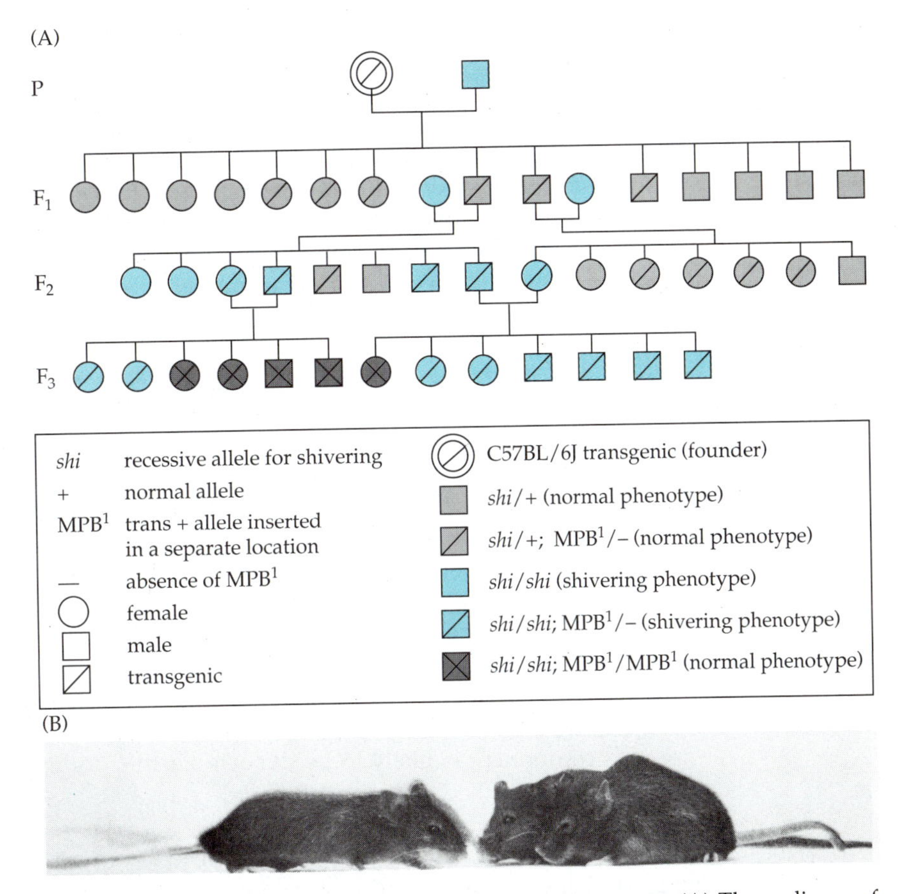

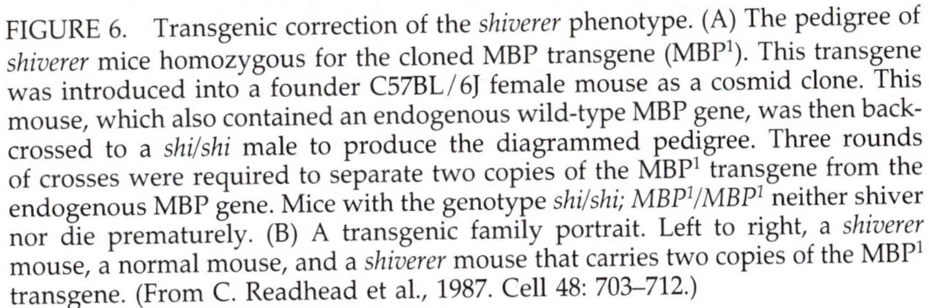

FIGURE 6. Transgenic correction of the *shiverer* phenotype. (A) The pedigree of *shiverer* mice homozygous for the cloned MBP transgene (MBP^1). This transgene was introduced into a founder C57BL/6J female mouse as a cosmid clone. This mouse, which also contained an endogenous wild-type MBP gene, was then backcrossed to a *shi/shi* male to produce the diagrammed pedigree. Three rounds of crosses were required to separate two copies of the MBP^1 transgene from the endogenous MBP gene. Mice with the genotype *shi/shi; MBP^1/MBP^1* neither shiver nor die prematurely. (B) A transgenic family portrait. Left to right, a *shiverer* mouse, a normal mouse, and a *shiverer* mouse that carries two copies of the MBP^1 transgene. (From C. Readhead et al., 1987. Cell 48: 703–712.)

MBP mRNA and protein. Breeding of these transgenic mice with *shiverer*s therefore specifically transferred the cloned MBP transgene into the mutants. By these manipulations, it was found that transfer of two copies of the cloned MBP gene into *shiverer* was sufficient to restore the mutant phenotype to wild-type. The dramatic results of this experiment are illustrated in Figure 6, which shows one normal mouse and two mutant mice, one displaying the behavior typical of *shiverer* and the other apparently normal as a result of transgenic rescue. This was the first transgenic replacement rescue of a mutant phenotype to be reported in vertebrates.

Not surprisingly, a mouse mutation allelic to *shiverer*, designated ***myelin-deficient*** **(*shi^{mld}*)**, also results from mutation of the MBP gene. In this case however, the mutation is not a deletion, but rather a large, inverted duplication. The genome of the *myelin-deficient* mouse contains not one, but two tandemly configured MBP genes, the first of which contains an extensive (~30 kb) inversion at its 3' end (Figure 5). This arrangement of MBP genes results in steady-state oligodendrocyte MBP mRNA levels that are only 2% of wild-type, a reduction that probably results from production of antisense mRNA from the 3' half of the upstream MBP gene. This antisense strand is thought to form a rapidly metabolized duplex with sense mRNA transcribed from the downstream (normally oriented) MBP gene, and to thereby block MBP biosynthesis. *Shiverer* and *shiverer^{mld}* provide two very different examples of mutational inactivation of a nervous system-specific gene.

One curious feature of both the *shiverer* and *myelin-deficient* mutations is that PNS myelin appears to be little affected. Although MBP is essentially absent from both CNS and PNS myelin in *shi* and *shi^{mld}* homozygotes, the characteristically altered appearance of myelin in the mutant—lack of myelin and a missing major dense line—is observed only in the CNS. Peripheral myelin is at most subtly altered and to a first approximation is structurally and functionally normal. These observations suggest that some component specific to peripheral myelin is functionally related to MBP and capable of substituting for the protein in its absence. As we shall see, this component is likely to be the cytoplasmic domain of P_0.

MBP is a potent inducer of cellular autoimmunity

Molecular investigations in one field often lead to important contributions in an initially unrelated field, as is certainly the case for studies of myelin basic protein. MBP has been very widely studied in the context of its ability to induce a T-cell–mediated inflammatory response upon immunization of laboratory animals with purified protein or peptide fragments. This response, referred to as **experimental autoimmune encephalomyelitis (EAE),** is characterized by T lymphocyte invasion of the CNS and PNS and results in demyelination and chronic relapsing paralysis. Historically, EAE has been viewed as an instructive experimental model for multiple sclerosis (MS) in humans, although the latter is clearly a more heterogeneous and complex disease in which MBP autoreactivity is not thought to play a primary causative role.

The disease-inducing cells in EAE and similar autoimmune disorders are T helper (T_H) lymphocytes. These cells recognize antigen after it is processed and presented, usually as denatured peptide fragments, on the surface of antigen-presenting cells. To be efficiently recognized by the T cell antigen receptor (TCR) of T_H cells, antigen must be bound to class II major histocompatibility complex (MHC) proteins on the surface of the antigen-presenting cells. In EAE, the molecular triad of MBP antigen, MHC class II protein, and the TCR cooperates to bring about the disease. These basic features of cellular immunity have been exploited in the design of immunological treatments for EAE, with an eye toward similar treatments for MS.

Two different approaches have met with success. The first of these relies on the fortuitous observation that in EAE the repertoire of MBP-directed TCRs expressed by autoreactive T_H cells is extremely limited. TCRs are made up of separate subunits, α and β, which are each encoded in multiple gene segments, designated V (for variable), D (for diversity), J (for joining), and C (for constant). For both the α and β subunits, the number of individual V, D, and J gene segments present in the genome is very large. These gene segments are combinatorially recombined to generate the diverse repertoire of antigen receptors required for cellular immunity. When different MBP-reactive T_H cell clones were examined in a single strain of mice with EAE, it was observed that among the very large number of V_β gene segments available, only two (V_β 8.2 and V_β13) were actually used. Similarly, only a limited number of available V_α, J_α, and J_β gene segments were used. With this information in hand, monoclonal antibodies specific for particular TCR segments (e.g., V_β8) were used to block TCR recognition of MBP. EAE-associated demyelination is dramatically alleviated by these monoclonal antibody treatments, which have little effect on cellular immunity in general, since the antibodies recognize only a very limited number of TCRs.

The second approach to EAE treatment depends upon perturbation of the interaction between MHC class II proteins and the encephalitogenic MBP peptide. The treatment strategy here is to design peptides that compete with MBP peptides for binding to the MHC protein but do not themselves induce a cellular immune response. As noted above, antigen binding to MHC class II proteins is essential to the T_H-cell–mediated immune response. Using several in vitro procedures, the amino acid residues that are essential for MHC binding can be identified, and peptides that contain these residues but that lack the additional amino acids required for the activation of T_H cells can be synthesized. Injection of such MBP peptide analogs effectively inhibits the induction of EAE in vivo.

P_0 is a member of the immunoglobulin superfamily of cell adhesion molecules

Autoimmune disorders similar to (though generally less robust than) EAE can be induced by immunization with other myelin-specific proteins. In the PNS of higher vertebrates, the most prominent of these proteins is the glycoprotein P_0. In actively myelinating Schwann cells, the P_0 mRNA accounts for over 7% of the entire polyA^+ RNA pool. The structure of P_0, first elucidated through cDNA cloning, is typical of simple, integral membrane glycoproteins. It is synthesized as a precursor containing a cleaved amino-terminal signal sequence, followed by a relatively hydrophobic and glycosylated extracellular domain, a single membrane-spanning domain bounded by charged anchors, and a very basic intracellular domain. This orientation places the P_0 extracellular domain at the intraperiod line of the myelin sheath and the cytoplasmic domain at the major dense line.

The extracellular domain of P_0 has an interesting structure that is related to the reiterated, conserved domains of immunoglobulin (antibody) molecules. This places P_0 in a very large group of immunoglobulin-related proteins, many of which are cell adhesion molecules. Called the **immuno-**

globulin (Ig) superfamily, this group includes the major histocompatibility complex antigens and T cell antigen receptors mentioned above, the CD4 T cell surface antigen that serves as the receptor for the human immunodeficiency virus, the polyimmunoglobulin receptor of epithelial cells, the neural cell adhesion molecule N-CAM, and many other neural adhesion proteins of both vertebrates and invertebrates (see Chapter 12). Although these molecules exhibit structures of varying complexity and function, they are related to the extent that each serves as recognition molecule by binding either homotypically (to itself) or to another immunoglobulin-related structure. Characteristically, binding by members of the Ig superfamily is Ca^{2+}-independent.

P_0 occupies a unique position within this superfamily. It is one of the very few family members to contain a single Ig domain. It is also one of the few members in which the genomic sequences encoding an Ig domain are interrupted by an intron. This is of interest with respect to an old hypothesis that the progenitor Ig domain was originally assembled through the duplication and joining of an ancestral half-domain. This hypothesis was based on the observation that the two halves of many Ig domains show a statistically significant similarity in amino acid sequence and that the solved secondary structures of immunoglobulin and MHC domains exhibit a very clear dyad axis of symmetry. One prediction of the hypothesis is that Ig-related genes that have not undergone extensive evolution might retain remnants of the original gene duplication event encoded as separate exons. The intron that splits sequences encoding the P_0 Ig domain divides the domain into two half-domains of similar length and structure. Its single domain structure and genomic organization are thus consistent with the idea that the P_0 gene has diverged only minimally from the gene encoding the primordial Ig-related recognition molecule.

P_0 acts as a bifunctional adhesion molecule

Like N-CAM, the P_0 extracellular domain appears to function as a homophilic cell adhesion molecule, and in this way promotes the formation of the intraperiod line of PNS myelin. This hypothesis is most directly supported by cell culture experiments in which the cloned P_0 cDNA is stably introduced (by transfection) into tissue culture cells in which the P_0 gene is not normally expressed. These experiments demonstrate that P_0^+ cells aggregate more rapidly with each other than with P_0^- cells, and that they form a tight membrane apposition at which P_0 protein becomes concentrated only when in contact with other P_0^+ cells (see Figure 7).

The participation of the P_0 cytoplasmic domain in formation of the major dense line of peripheral myelin is not established but is indirectly suggested by observations of peripheral myelination in *shiverer*. As noted above, the major dense line in the PNS myelin of *shiverer* homozygotes resembles that of wild-type mice, even though this structure is devoid of MBP and is completely absent from CNS myelin. Like MBP, the P_0 cytoplasmic domain is exceptionally basic, and at the major dense line in the *shiverer* PNS, it accounts for the vast majority of cytoplasmic protein. Together with its Schwann-cell–specific expression, these observations

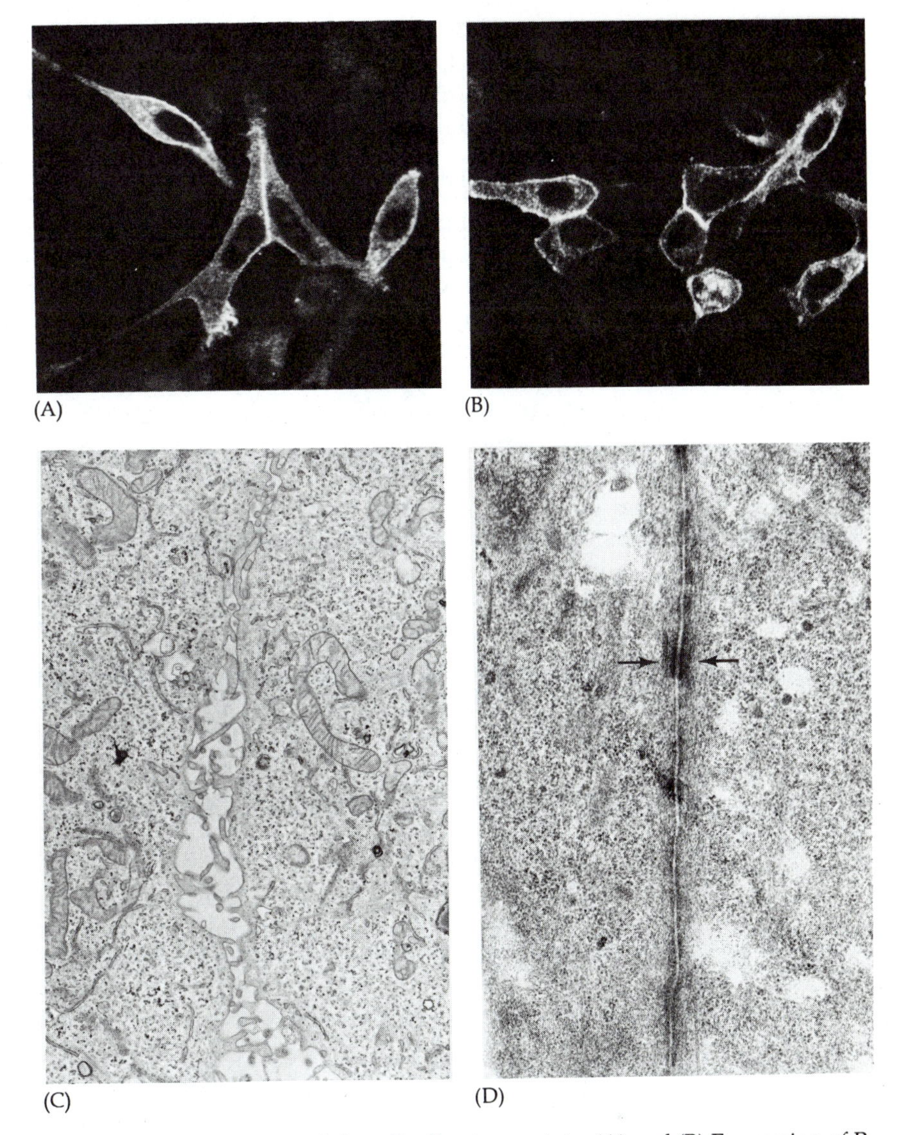

FIGURE 7. P_0 as a homophilic cell adhesion protein. (A) and (B) Expression of P_0 in HeLa cells transiently transfected with a P_0 cDNA expression construct. In these confocal microscopic images, P_0 expression is visualized with a polyclonal anti-P_0 antibody and a fluorescein-conjugated second antibody. Note that intense fluorescence is confined to those regions in which the membranes of two P_0^+ cells are directly apposed. (×300) (C) and (D) Electron micrographs of the ultrastructure of the intercellular border between control cells and stable HeLa P_0 transformants, respectively. Note that this border is made up of interdigitating microvilli in control cells (C), but is characterized by an extended regular apposition in P_0^+ transformants (D). The arrows in (D) point to a prominent desmosome. In (C) and (D), the apical surface of the cells is toward the top. (×30,000) (From D. D'Urso et al., 1990. Neuron 4: 449–460.)

suggest that the P_0 cytoplasmic domain and MBP both act to compact the major dense line of myelin. A model for the functional orientation of P_0 with respect to the major dense and intraperiod lines of the peripheral myelin membrane is illustrated in Figure 8.

The PLP gene is essential for myelination in the CNS

Phylogenetically, oligodendrocyte myelination and PLP appear almost simultaneously, making PLP one of the last neural-specific proteins to arise in vertebrate evolution. The sequence of this protein is not obviously related to that of any other protein currently entered into computer databases, and is remarkably dissimilar to that of P_0, its presumed analog in the PNS. PLP is extremely hydrophobic, but unlike P_0, is not glycosylated. As is the case for many ion channels, the protein crosses the membrane several times, but nonetheless does not contain a cleaved amino-terminal signal peptide. Its sequence exhibits no similarity to those of Ig's, and is not related to that of P_0 in any other way. The sequence is very highly conserved, however: rat and human PLP, each comprised of 276 amino acids, are identical. This very strong conservation, together with the genetic data discussed below, implies that the structure of PLP precisely determines its function in a way that we do not yet understand.

As with MBP, the best clue to the function of PLP comes from work on a mutation that affects CNS myelination in the mouse. This X-linked mutation, designated ***jimpy*** **(*jp*)**, results in axial body tremors, subsequent tonic seizures, and premature death. This phenotype is similar to that of *shiverer*, although *jimpy* mice are more severely incapacitated and tend to die earlier. Curiously, oligodendrocyte cell death is also evident in *jp*/Y animals. Ultrastructurally, the brains of *jimpy* mice exhibit a striking defect in CNS myelin that is distinct from that seen in *shiverer*. When present, this myelin is characterized by an aberrant, improperly spaced, and unstable intraperiod line.

An interesting mutation in the PLP gene is almost certainly responsible for the *jimpy* phenotype. Evidence for this idea first came from the demonstration that the PLP gene maps to the very same position on the mouse X chromosome as the *jimpy* mutation. At about the same time, the PLP mRNA isolated from *jimpy* mice was found to be slightly smaller than the PLP mRNA isolated from wild-type mice. This size difference results from a 74-bp internal deletion in the *jimpy* PLP mRNA, which is in turn the result of a single point mutation in the splice acceptor site preceding the fifth exon of the *jimpy* PLP gene. A transition in this acceptor site—from the normal AG to the mutant GG—forces the pre-mRNA splicing machinery to skip the 74-bp sequence encoded in exon 5, and thereby splice exon 4 directly to exon 6. Unlike the splice junctions of the MBP gene, the junction between PLP exons 4 and 5 does *not* fall between a codon, so the aberrant splicing causes a shift in reading frame and a prematurely terminated PLP protein. This single base mutation almost certainly accounts for the complex pattern of neurological defects displayed by the *jimpy* mouse. A formal proof of this hypothesis, similar to the transgenic rescue of *shiverer* mice discussed above, has not yet been achieved.

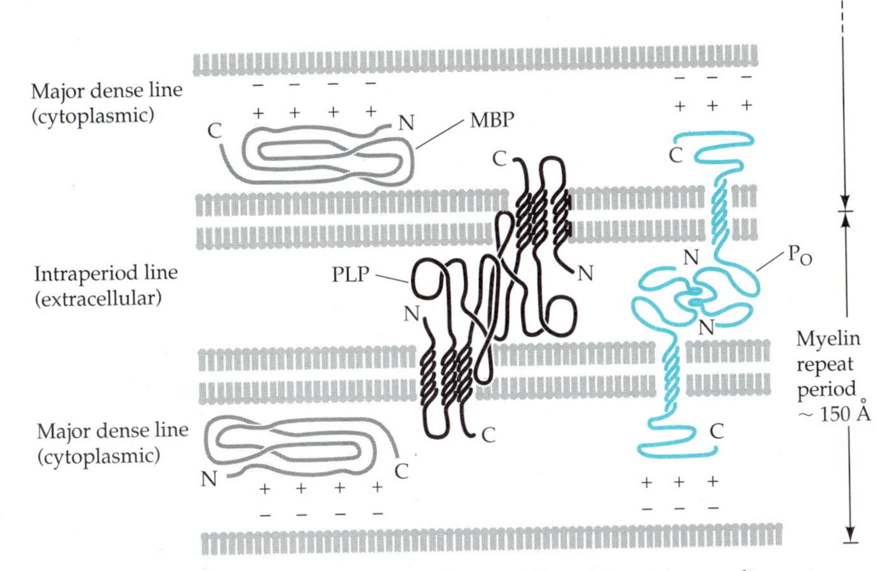

FIGURE 8. Myelin membrane configuration of P_0 and its major myelin partners. This diagram illustrates the spatial segregation and hypothesized functional interaction of the ensemble of major myelin proteins in one-and-a-half 150-Å repeat periods of the compacted myelin sheath. The repeat period shown is a hybrid between CNS and PNS myelin since PLP is found only in the former and P_0 only in the latter. The immunoglobulin-related P_0 extracellular domain functions as a homotypic adhesion molecule to promote compaction at the intraperiod line; PLP may play a similar role in the CNS sheath. MBP and the cytoplasmic domain of P_0, both very basic, promote membrane adhesion at the major dense line by an unknown mechanism. One possible mechanism—an electrostatic interaction between the positively charged proteins and negatively charged lipids of the apposed membrane face—is depicted.

Many mutations in the PLP gene of various species have recently been identified. These include an allele of *jimpy*, *jimpy*^{msd} and a presumed allele *rumpshaker* in mice, *myelin-deficient* (*md*) in rats, and *shaking pup* in dogs. These mutations vary in the severity of neurological phenotype, although all appear to reside in the PLP gene. Remarkably, several involve a single, apparently conservative amino acid substitution. In *jp*^{msd}, for example, an alanine at position 242 is mutated to valine.

Mutations in the PLP gene also account for a set of rare, X-linked, recessive neurological diseases in humans, together named **Pelizaeus-Merzbacher disease (PMD)**. The first symptoms of this disease—nystagmus and poor head control— appear shortly after birth and are followed by ataxia and myoclonic seizures. Most affected boys die within the first three years of life. In each of the three cases of PMD that have been molecularly analyzed, the PLP gene carries a single base mutation that results in a single amino acid substitution in the PLP protein, just as for the animal mutants listed above. The positions of each of the known PLP point mutations, together with the *jimpy* mutation, are indicated in the diagram of Figure 9.

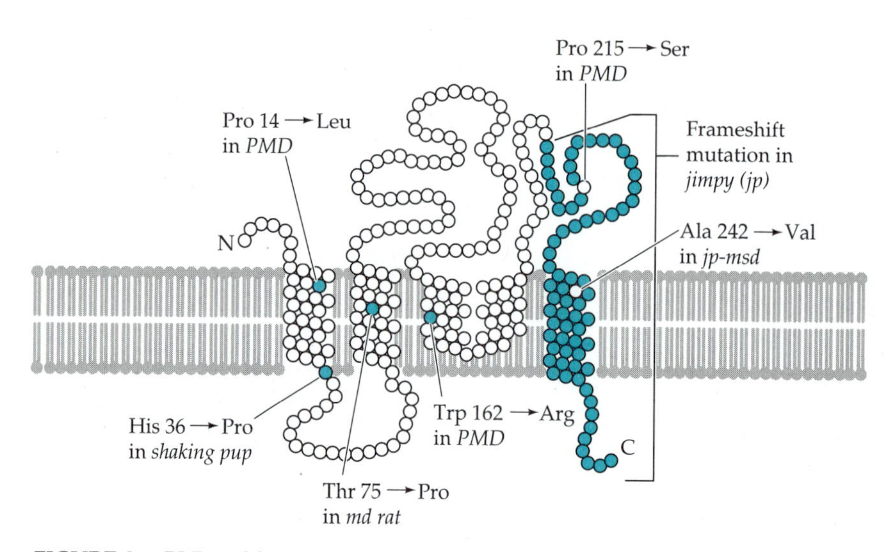

FIGURE 9. PLP: wild type and mutant. In this hypothetical model of proteolipid protein membrane orientation (other models have been proposed), known sites of PLP mutation in mice, rats, dogs, and humans are indicated. With the exception of the splice acceptor mutation in *jimpy* (see text), all of these mutations correspond to single amino acid substitutions. PMD: Pelizaeus-Merzbacher disease, a constellation of lethal, inherited, X-linked CNS myelination disorders in humans. (After a drawing by K. Nave.)

MAG is a cell adhesion molecule closely related to N-CAM

The best-studied of the minor myelin genes encodes the myelin-associated glycoprotein (MAG). In contrast to the major myelin proteins, MAG is restricted to noncompacted regions of the sheath. These regions include the innermost layer of the sheath (immediately adjacent to the axonal membrane), the outermost layer, and a series of cytoplasmic channels that spiral from the cell body through to the innermost layer. These channels are called "Schmidt-Lanterman incisures." In addition to this interesting localization within the sheath, MAG expression can be detected at the very earliest stages of peripheral myelination, prior to the appearance of any of the major myelin proteins. For these reasons, MAG is thought to function as a mediator of the axon-glial adhesion events that precede myelination and to help direct the initial tucking-under of the growing inner layer of myelin. Consistent with this idea, MAG can mediate adhesion between cultured oligodendrocytes and neurons.

Like P_0, MAG has a structure related to immunoglobulins. The protein contains one of two small cytoplasmic domains, generated by alternative pre-mRNA splicing, and a single transmembrane domain. The MAG extracellular domain accounts for most of the protein, and is made up of five homologous Ig domains. The presence of reiterated Ig domains is typical of developmentally regulated neural adhesion molecules such as N-CAM (to which MAG is closely related), L-1, TAG-1, and contactin, and distinguishes this group of proteins from more primitive single domain molecules such as P_0 and Thy-1 (see Figure 10 and Chapter 11). In addition to primary structure, MAG shares a carbohydrate antigen, the HNK-1 epitope, with a large number of neural adhesion proteins, including N-CAM, L-1, and P_0. This conserved moiety, a complex carbohydrate containing glucuronic acid, is thought to potentiate and perhaps thereby confer specificity to MAG binding.

The Trembler *mutation specifically blocks myelination in the PNS*

There are several myelination mutations whose target genes have yet to be identified. Of these, perhaps the most intriguing is ***Trembler***, an autosomal, behaviorally dominant mutation that results in severe hypomyelination by Schwann cells in the PNS but which is without effect on oligodendrocytes in the CNS. Although *Trembler* is dominant in terms of behavioral phenotype—heterozygotes and homozygotes both experience axial body tremors—there are clear gene dosage effects with respect to the extent of myelin loss and the reduction in the steady-state levels of the

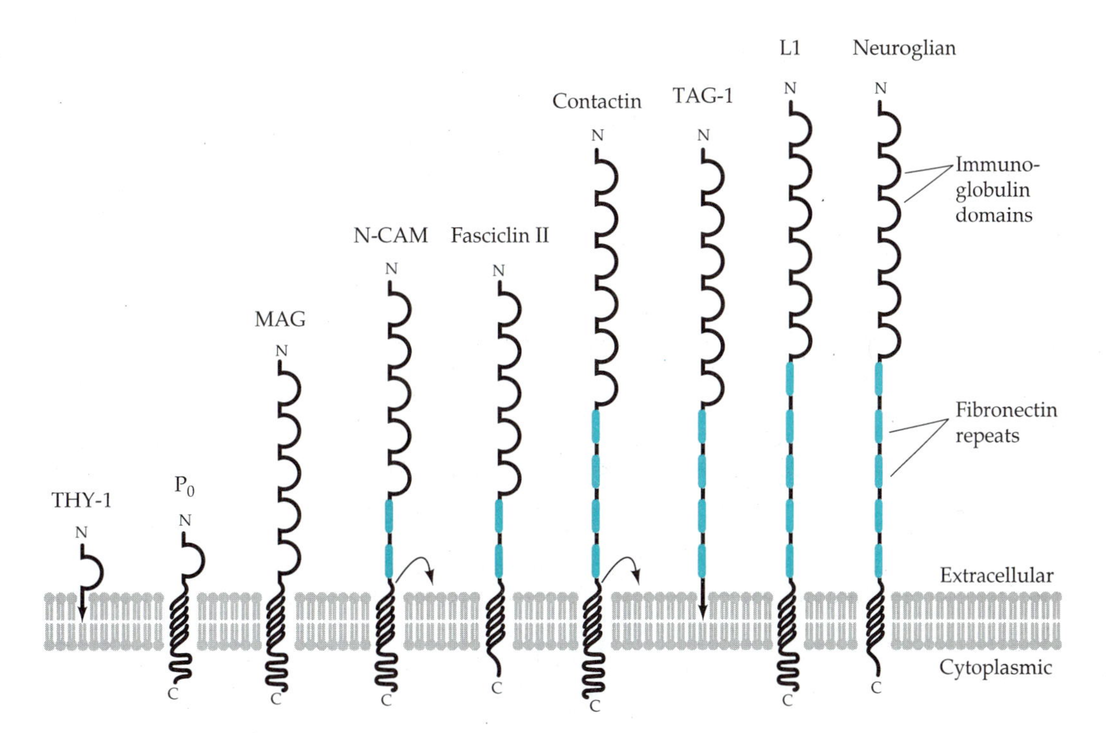

FIGURE 10. MAG and the immunoglobulin (Ig) superfamily of neural adhesion molecules. As described in the text, MAG is a member of a very large family of Ig-related proteins, several of which function as cell surface adhesion molecules in the nervous system. MAG, P_0, TAG-1, contactin, L1, and N-CAM are vertebrate proteins; neuroglian and fasciclin II are *Drosophila* proteins. Ig domains are indicated by half circles, fibronectin type III repeats are indicated by colored boxes, and variant forms of contactin and N-CAM that are attached to the plasma membrane by a phosphoinositol linkage are indicated by arrows.

major myelin proteins. There are two established alleles at the *Trembler* locus, *Tr* and *Tr^j^*. Although *Tr/Tr* homozygotes experience severe tremors and locomote only with great difficulty, if carefully tended they are viable for many months under laboratory conditions. In contrast, Tr^j/Tr^j homozygotes die at about two weeks of age. The basis for this difference in viability is unknown.

Reconstitution experiments performed in myelinating cell cultures in vitro, or using cross-anastomosed nerves in vivo, have strongly suggested that expression of the *Trembler* gene product is Schwann cell autonomous. Thus, wild-type Schwann cells myelinate *Trembler* axons, but *Trembler* Schwann cells do not myelinate wild-type axons. This suggests that the *Trembler* gene product is a Schwann cell-specific protein important for normal differentiation. P_0, an obvious candidate for this protein, has been eliminated from consideration, since the *Trembler* gene maps to mouse chromosome 11 whereas the P_0 gene maps to mouse chromosome 1. In addition to hypomyelination, the peripheral nerves of *Trembler* homozygotes contain many more Schwann cells than do those of wild-type mice. This observation is consistent with a defect early in the myelination program, one that results in the incomplete withdrawal of Schwann cells from the proliferative blast phase that immediately precedes their full differentiation.

The Differentiation of Myelin-Forming Cells

Schwann cell differentiation is driven by neurons

The availability of molecular probes for the major and minor myelin genes has provided a means to directly monitor the differentiation of myelin-forming cells at the level of gene expression. For technical reasons, these studies are more easily carried out in the PNS, where immunoselective procedures can be used to purify and culture large numbers of Schwann cells dissociated from the peripheral nerves of neonatal rats and mice (see Box B), and where in vivo manipulations are more easily performed. In addition, the entire course of Schwann cell differentiation, including full myelination, can be recapitulated in co-cultures of purified neurons and Schwann cells. These experimental advantages have made myelination a useful general model for developmental cell-cell interactions in the nervous system.

Among the most striking features of Schwann cell differentiation is its dependence on axons. Both the acquisition and the maintenance of a myelinating phenotype, as assayed by the expression of myelin-specific genes, require contact with axons. If fully myelinated peripheral nerves are transected, for example, axons distal to the transection site degenerate. In response to this axonal loss, fully differentiated Schwann cells, while remaining viable, undergo **dedifferentiation**—the myelin they have elaborated is phagocytized by macrophages and by the Schwann cells themselves, and expression of the major and minor myelin genes is dramatically decreased. At the same time that high level expression of myelin-specific

Box B Immunoselection and Neuroimmunology

Among the most powerful and widely used techniques in cellular immunology are those in which enrichment of specific cell populations is achieved through the selective maintenance or selective killing of cells in a heterogeneous population, based on the differential expression of cell surface markers. For example, if a cell type expresses a surface marker A that is not expressed by other cell types in a mixed population, A^+ cells can be positively selected with antibodies that specifically bind to the marker A. In a process called "cell panning," the antibodies can be attached to the bottom of a cell culture dish or otherwise immobilized and used to fish out only A^+ cells. Alternatively, the antibodies can be used to fluorescently label the surface of A^+ cells, which can then be sorted from the A^- cells of a mixed population with a fluorescence-activated cell sorter (FACS). This procedure is often used to purify neural cell cultures. An even more powerful variation, involving negative selection by cell killing, can be employed if A^+ cells do *not* express a second surface marker B that is expressed by all other cells in the mixed population. Antibodies directed against marker B can be added to the mixed cell population together with purified complement proteins, which affix to, and are activated by, antibodies bound to the surface of B^+ cells. The activated complement complex specifically kills all B^+ cells via a proteolytic cascade that ultimately results in membrane lysis, leaving a population highly enriched for A^+ cells. Negative selection is routinely used to prepare pure populations of rodent Schwann cells since, unlike all the other cells dissociated from the nerves of neonatal rodents, these cells do not express the cell surface marker Thy-1.

genes is lost, expression of a set of genes that mark premyelinating Schwann cells is reacquired. These genes encode voltage-sensitive sodium channels, N-CAM, and the NGF receptor, among other proteins (see Figure 11). In transected nerves, Schwann cells continue to express these premyelinating genes until they are once again contacted by regenerating axons. When this occurs, normal differentiation is recapitulated; expression of the NGF receptor and N-CAM genes is extinguished and the major and minor myelin genes are reactivated. Neuronal control of Schwann cell phenotype is dependent on direct contact between the surface of axons and Schwann cells. In co-cultures of neurons and Schwann cells, only those Schwann cells that are in actual contact with neurites express the P_0 and MBP genes. When applied to pure Schwann cell cultures, conditioned medium from neuronal cultures does not induce expression of these genes.

The signal on the axonal surface that induces myelination has yet to be identified, but there are clear differences between the inductive capability of axons that are normally myelinated and those that remain unmyelinated. (Remember that many of the peripheral nerves of vertebrates remain unmyelinated. The axons in these nerves are loosely wrapped by nonmyelinating Schwann cells, which, as noted above, do not detectably express myelin-specific genes.) This difference can be demonstrated by cross-anastomosis experiments in which the distal regions of transected myelinated and unmyelinated nerves are swapped prior to the regen-

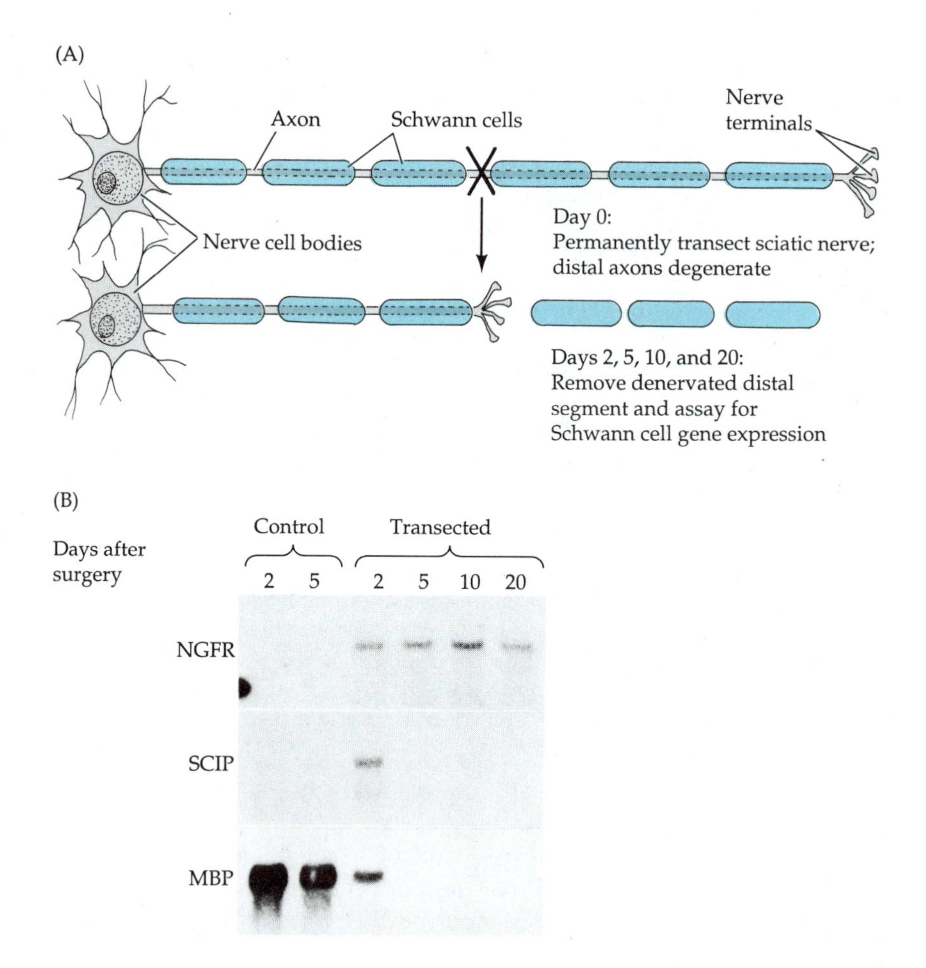

FIGURE 11. Axonal regulation of Schwann cell phenotype. (A) Sciatic nerve transection paradigm. Axons distal to a peripheral nerve transection site degenerate over a period of several days, and Schwann cells in this distal region dedifferentiate in response to axon degeneration. The transection is performed in 30-day-old rats, in which all of the Schwann cells of the sciatic nerve have already assumed a fully differentiated phenotype. (B) Schwann cell gene expression in response to transection. These Northern blots present an analysis of distal Schwann cell expression of three different genes. The two control lanes correspond to RNA isolated 2 and 5 days after a sham operation in which nerves were not transected; the four experimental lanes correspond to RNA isolated from the distal region at 2, 5, 10, and 20 days following a transection. The same blot was successively probed for expression of nerve growth factor receptor (NGFR) mRNA, which marks premyelinating Schwann cells; SCIP mRNA, which encodes a transcription factor that marks proliferative Schwann cells just prior to differentiation; and MBP mRNA, which marks fully differentiated, myelinating Schwann cells. In response to loss of axon contact, Schwann cells stop expressing the MBP gene, and reactivate expression of the NGFR gene. In reverting from a myelinating to a premyelinating phenotype, these cells appear to transiently pass back through the proliferative transition state that is marked by expression of the SCIP gene.

erative regrowth of axons. When such manipulations are carried out, axons from a transected unmyelinated nerve are not myelinated when they regrow into a distal population of Schwann cells that had formerly synthesized myelin, and, similarly, axons from a transected myelinated nerve are myelinated when they regrow into a distal population of Schwann cells that had not previously elaborated a sheath. Whether the difference in the inductive capability of myelinated as opposed to unmyelinated axons is a qualitative or quantitative one is unknown. However, there is a strong correlation between peripheral axon caliber and myelination: small PNS axons ($<1 \mu m$ in diameter) are almost never myelinated, while large axons are always myelinated. Similarly, in vivo manipulations that substantially increase axon caliber, such as experimental enlargement of a peripheral nerve's target field, can convert an axon population from one that is largely unmyelinated to one that is largely myelinated. These observations may be interpreted in several ways, but one possibility is that all axons express a surface inducer whose concentration (per unit area of surface membrane) is fixed, and that Schwann cell exposure to this inducer must reach a threshold level before myelination is triggered. This threshold level could be achieved simply by increasing the size of the axon.

Schwann cell division is also driven by neurons

Proliferation is an additional feature of Schwann cell development that is clearly under axonal control. The number of Schwann cells that initially migrate from the neural crest is too low to accommodate the myelination or loose ensheathment of all of the axons in a peripheral nerve. Accordingly, the developing Schwann cell population is marked by a period of rapid cell division that lasts for several days and immediately precedes full differentiation. In this respect, Schwann cells are similar to most other differentiating cellular systems, such as myoblasts and the developing O-2A glial progenitors of the CNS (see Chapter 11), in which a highly proliferative blast phase precedes the acquisition of a differentiated phenotype. Axonal stimulation of Schwann cell proliferation can be readily demonstrated in cell culture. Purified Schwann cells divide very slowly if grown in a conventional cell culture medium containing 10% fetal bovine serum, but if co-cultured with sensory or sympathetic neurons, they specifically adhere to the surface of the neurites that these neurons extend and subsequently undergo multiple rounds of cell division. Axonal triggering of Schwann cell division also depends on cell-cell contact. Schwann cells do not divide when exposed to conditioned medium from sensory neuron cultures, or even when placed in apposition to neurons but separated from them by a permeable collagen diaphragm. Although the mitogenic activity of the axonal surface can be recovered in membrane fractions prepared from the isolated neurites of cultured neurons or PC12 cells and from axon membrane fractions prepared from brain, it has yet to be purified to homogeneity.

cAMP may be a critical second messenger in neuronal control of Schwann cell phenotype

In cell culture, many of the important features of axonal regulation of Schwann cell phenotype are reproduced by agents that elevate intracellular levels of cAMP. In cultured Schwann cells, this cyclic nucleotide is both a potent trigger of cell division and a partial inducer of major myelin gene expression. Its mitogenic activity probably results from its ability to potentiate the Schwann cell response to polypeptide growth factors such as platelet-derived growth factor (PDGF) and fibroblast growth factor (FGF). As described in Chapter 11, these polypeptide mitogens are also important regulators of the proliferation of O-2A glial progenitors in the CNS. In PNS cultures, agents that raise intracellular concentrations of cAMP synergistically cooperate with PDGF and FGF through a simple biochemical mechanism: they strongly potentiate Schwann cell expression of the genes encoding the PDGF and FGF receptors.

cAMP stimulation of Schwann cell division in vitro occurs in concert with partial activation of major myelin gene expression, even though as noted above, full expression of myelin-specific genes is only observed in actively myelinating cells. Both the P_0 and MBP genes are cAMP-inducible, and the relative sensitivity of these genes to induction by cAMP mirrors their relative level of expression by actively myelinating Schwann cells. However, the *absolute* level of expression achieved in response to cAMP elevation in vitro is much lower than the level of expression observed in actively myelinating Schwann cells in vivo. Thus, cAMP treatment of cultured Schwann cells appears to move them from a premyelinating phenotype, characterized by low-level cell division and high-level expression of pre-myelinating markers such as the NGF receptor gene, to a transitional or blast cell phenotype characterized by rapid cell division, reduced expression of premyelinating markers, strong induction of a transcription factor (SCIP) that marks transitional cells, and partial induction of myelin-specific genes. Many of these cAMP effects are likely to be exercised at the level of mRNA transcription, since when assayed in cultured Schwann cells, the transcriptional activity of the cloned regulatory regions of the major myelin genes has been observed to strongly depend upon elevation of intracellular cAMP. Thus far, the only known method of pushing Schwann cells to a full myelinating phenotype in cell culture is to co-culture them with neurons.

Oligodendrocytes are less developmentally flexible than Schwann cells

Differentiating and mature oligodendrocytes do not exhibit the extreme dependence on axonal contact that is characteristic of Schwann cells. As described in Chapter 11, these cells develop from a multi-potential progenitor—the O-2A cell—that also gives rise to one form of nonmyelinating glial cell, the type-2 astrocyte. Although the proliferation of O-2A progenitors depends on several of the same growth factors that regulate Schwann cell division (e.g., PDGF and FGF), the progenitors adopt and maintain an oligodendroglial phenotype, including the high-level expression of myelin-specific genes, independent of the presence or absence of axons. In neuron-free cultures, O-2A cells follow a time course of differentiation that is very similar to the time course observed in vivo and give rise to oligodendrocytes that express very high levels of myelin-specific genes and that elaborate extensive sheets of myelin-like membrane. The difference in the developmental plasticity of these cells and Schwann cells can be dramatically demonstrated in freshly prepared neuron-free cocultures, in which high-level oligodendroglial expression of the MBP gene is maintained at the same time that Schwann cell expression of the very same gene is extinguished. Once committed to a myelinating phenotype, oligodendrocytes apparently do not have the ability to dedifferentiate.

Summary

Myelin is a cellular organelle unique to and essential for the integrative functioning of higher nervous systems. Like the organelle itself, the cells responsible for its elaboration are relatively late evolutionary inventions. Formed as an enormous sheet of plasma membrane that is repeatedly wrapped and tightly compacted around axons, myelin serves as an electrical insulator to greatly increase the conduction velocity of nerve impulses.

The formation of myelin depends on the activation of a set of cell-specific, differentiation-specific genes that are differentially expressed between oligodendrocytes in the CNS and Schwann cells in the PNS. Of the major myelin genes, Schwann cells express P_0 and myelin basic protein (MBP), while oligodendrocytes express MBP and proteolipid protein (PLP). Each of these proteins is abundantly expressed and plays an essential structural role in the elaboration and subsequent compaction of myelin. Two are associated with multiple neurological mutations in mice and humans.

MBP, located on the cytoplasmic surface of the myelin membrane, is required for compaction at the major dense line of CNS myelin. Mutations in the MBP gene account for the *shiverer* mutation in mice, and for its allele, *shiverer*^{mld}, and can be complemented by transgenic expression of the wild-type MBP gene. Immunization with either the full MBP protein or selected peptides induces experimental autoimmune encephalomyelitis, a T-cell–mediated inflammatory response that can be treated with anti–T-cell receptor antibodies. PLP is an abundant integral membrane proteolipid that plays an important role in extracellular compaction at the intraperiod line of CNS myelin. Mutations in the PLP gene account for the *jimpy* mutation in mice, its allele *jimpy*^{msd}, and a large number of neurological mutations in other species including humans. PLP is among the most highly conserved proteins and is identical in rats and humans. P_0 is an abundant, PNS-specific, integral membrane glycoprotein that functions as an adhesion molecule in the formation of the intraperiod line of peripheral myelin and may also act to promote compaction at the PNS major dense line. Along with the neural cell adhesion molecule N-CAM, it is a member of a large family of immunoglobulin-related adhesion proteins. The structure and genomic organization of P_0 suggest that it is one of the most primitive members of this family. MAG, which resembles N-CAM, is detected very early in the time course of myelination and remains restricted to noncompacted regions of the sheath, including the innermost wrap immediately adjacent to the axon, where it is thought to function as an axon-glial and glial-glial adhesion molecule.

The proliferation and differentiation of myelin-forming Schwann cells are driven by axonal contact. In contrast, the differentiation of oligodendrocytes in the CNS appears to be programmed and does not depend on the presence of neurons. A set of polypeptide growth factors that includes PDGF and FGF regulates the proliferation of the progenitors to both oligodendrocytes and myelin-forming Schwann cells. In developing Schwann cells, the response to these mitogens can be dramatically regulated at the level of expression of the corresponding cell surface receptors. cAMP is probably a critical second messenger in the axonal regulation of Schwann cell proliferation and differentiation, since in cultured Schwann cells, it is a strong inducer of both the PDGF and FGF receptor genes, as well as a partial inducer of the P_0 and MBP genes.

References

General references

*Lemke, G. 1988. Unwrapping the genes of myelin. Neuron 1: 535–543.

Morell, P. (ed.). 1984. *Myelin,* 2nd Edition. Plenum, New York.

Mechanics of myelination/Structure of myelin

*Bunge, R. P., Bunge, M. B. and Bates, M. 1989. Movements of the Schwann cell nucleus implicate progression of the inner (axon-related) Schwann cell process during myelination. J. Cell Biol. 109: 273–284.

D'Urso, D., Brophy, P. J., Staugaitis, S. M., Gillespie, C. S., Frey, A. B., Stempak, J. G. and Colman, D. R. 1990. Protein zero of peripheral nerve myelin: Biosynthesis, membrane insertion, and evidence for homotypic interaction. Neuron 4: 449–460.

Kirschner, D. A. and Ganser, A. L. 1980. Compact myelin exists in the absence of myelin basic protein in the *shiverer* mutant mouse. Nature 283: 207–210.

Morell, P. and Norton, W. T. 1980. Myelin. Sci. Am. 242: 88–118.

Omlin, F. X., Webster, H. deF., Palkovitz, G. G. and Cohen, S. R. 1982. Immunocytochemical localization of basic protein in the major dense line regions of central and peripheral myelin. J. Cell Biol. 95: 242–248.

Myelin genes and genetics

deFerra, F., Engh, H., Hudson, L., Kamholz, J., Puckett, C., Molineaux, S. and Lazzarini, R. A. 1985. Alternative splicing accounts for the four forms of myelin basic protein. Cell 43: 721–727.

Koeppen, A. H., Ronca, N. A., Greenfield, E. A. and Hans, M. B. 1987. Defective biosynthesis of proteolipid protein in Pelizaeus-Merzbacher disease. Ann. Neurol. 21: 159–170.

Lemke, G. and Axel, R. 1985. Isolation and sequence of a cDNA encoding the major structural protein of peripheral myelin. Cell 40: 501–508.

Lemke, G., Lamar, E. and Patterson, J. 1988. Isolation and analysis of the gene encoding peripheral myelin protein zero. Neuron 1: 73–83.

Milner, R. J., Lai, C., Nave, K.-A., Lenoir, D., Ogata, J. and Sutcliffe, J. G. 1985. Nucleotide sequences of two mRNAs for rat brain myelin proteolipid protein. Cell 42: 931–939.

Nave, K.-A., Bloom, F. E. and Milner, R. J. 1987. A single nucleotide difference in the gene for myelin proteolipid protein defines the *jimpy* mutation in the mouse. J. Neurochem. 49: 1873–1877.

*Readhead, C., Popko, B., Takahashi, N., Shine, H. D., Saavedra, R. A., Sidman, R. L. and Hood, L. 1987. Expression of a myelin basic protein gene in transgenic *shiverer* mice: correction of the dysmyelinating phenotype. Cell 48: 703–712.

Roach, A., Boylan, K., Horvath, S., Prusiner, S. B. and Hood, L. E. 1983. Characterization of a cloned cDNA representing rat myelin basic protein: Absence of expression in *shiverer* mutant mice. Cell 34: 799–806.

Trapp, B. D., Moench, T., Pulley, M., Barbosa, E., Tennekoon, G. and Griffin, J. 1987. Spatial segregation of mRNA encoding myelin-specific proteins. Proc. Natl. Acad. Sci. USA 84: 7773–7777.

Experimental autoimmune encephalomyelitis

*Acha-Orbea, H., Mitchell, D. J., Timmermann, L., Wraith, D. C., Tausch, G. S., Waldor, M. K., Zamvil, S. S., McDevitt, H. O. and Steinman, L. 1988. Limited heterogeneity of T cell receptors from lymphocytes mediating autoimmune encephalomyelitis allows specific immune intervention. Cell 54: 263–273.

Wraith, D. C., Smilek, D. E., Mitchell, D. J., Steinman, L. and McDevitt, H. O. 1989. Antigen recognition in autoimmune encephalomyelitis and the potential for peptide-based immunotherapy. Cell 59: 247–255.

Proliferation and differentiation of myelin-forming cells

Monuki, E. S., Kuhn, R., Weinmaster, G., Trapp, B. D. and Lemke, G. 1990. Expression and activity of the POU transcription factor SCIP. Science 249: 1300–1303.

Porter, S., Clark, M. B., Glaser, L. and Bunge, R. P. 1986. Schwann cells stimulated to proliferate in the absence of neurons retain full functional capacity. J. Neurosci. 6: 3070–3078.

*Raff, M. C. 1989. Glial cell diversification in the rat optic nerve. Science 243: 1450–1455.

Weinmaster, G. and Lemke, G. 1990. Cell-specific cyclic AMP-mediated induction of the PDGF receptor. EMBO J. 9: 915–920.

Part Three

Neural Development

10

Gene Regulation in the Nervous System

Greg Lemke

IN THE BACTERIUM *Escherichia coli,* approximately one in ten genes is devoted to encoding proteins that regulate gene expression. Such a substantial genomic investment by this relatively simple prokaryote points to the importance of regulatory proteins in the growth, differentiation, and homeostasis of even a single cell. It is thus not surprising that the development and integrated operation of the vertebrate nervous system appears to require a very large battery of such proteins.

After all, there are hundreds of molecularly distinct neural cell types. This means that individual neurons and glia must *selectively* express only a small subset of the vast number of genes that potentially control their differentiation and define their mature phenotype. They must choose between multiple genes for various neurotransmitter receptors, voltage-sensitive ion channels, vesicle proteins, growth factor receptors, and neuromodulators, and must often make their choices during a narrow window of development. In addition, many of the genes that they express must be regulated in response to extracellular stimuli, such as regular electrical activity at synapses, to which neural cells are uniquely subject. These considerations demand that gene expression in the nervous system be exquisitely regulated.

Although eukaryotic gene expression is controlled at several different levels, two metabolic events are especially sensitive to regulation. The first of these is the rate at which new mRNA transcripts are initiated by RNA polymerase; regulation of this event controls when and where a given gene is transcribed. The elucidation of transcriptional control mechanisms has been one of the great successes of modern molecular biology, and has resulted in the identification of both small *cis*-acting DNA elements that control cell-specific transcription, as well as *trans*-acting proteins (transcription factors) that bind to these elements. The second sensitive event in the regulation of gene expression is the processing of precursor mRNA transcripts (pre-mRNAs) into mature mRNAs. As we shall see, this usually involves the removal of noncoding sequences from the pre-mRNA, a process that is often selective, sometimes highly variable, and sometimes even neural-specific. Both transcriptional control and alternative pre-mRNA processing are means to the same end—the generation of cellular and molecular diversity. The great cellular and molecular diversity of the nervous system therefore makes an understanding of these events of particular importance to neurobiologists.

A Hierarchy of Gene Regulation

Chromatin configuration controls gene expression

Although we shall focus on the events leading to transcript initiation and alternative pre-mRNA splicing, it is important to remember that these events are by no means the whole story of gene regulation (see Figure 1). For example, the accessibility of a given gene to the transcriptional machinery is first and foremost a function of the higher order structure of the DNA molecule in which it resides. This higher order structure includes both the reiterative coiling of the double helix into supercoils and chromosomes, and the stereotyped association of DNA with histones and other nuclear proteins.

These aspects of chromatin configuration are of direct relevance to gene expression. The DNA that we see in the familiar form of metaphase chromosomes, for example, is condensed ~10,000-fold relative to its extended linear length. Such highly coiled and condensed DNA, whether present in chromosomes or in the heterochromatin of interphase nuclei, is almost never transcriptionally active. Molecular biologists refer to the ratio of the extended linear length of a DNA molecule to its actual coiled coil length in cells as the "packing ratio." Whereas the DNA in chromosomes and heterochromatin has a packing ratio in the range of 1000–10,000, the DNA from which mRNAs are transcribed in interphase nuclei has a packing ratio of only 1–10. (Pure DNA in solution has, by definition, a packing ratio of 1.) Although the mechanisms underlying interconversion between highly condensed and transcriptionally active chromatin are not well understood, they operate at the highest level of gene regulation. Perhaps the best demonstration of the regulatory power of chromatin configuration is X chromosome inactivation, in which one of the two X chromosomes in cells that are genetically female is rendered constitutively heterochromatic; hardly any of the genes of inactivated X chromosomes are transcribed.

Much of the condensation of cellular DNA results from the fact that DNA in cells is almost never present as a free duplex of nucleic acid but is instead complexed to a set of DNA binding and scaffolding proteins, the best studied of which are the **histones.** A subset of these proteins, called the core histones, assemble into a 100-kD octameric polyprotein, around the outside of which ~200 bp of duplex DNA is twice looped. This combination of a cylindrical octamer of core histones wrapped by ~200 bp of DNA is called the **nucleosome** and is the fundamental unit of both transcribed and nontranscribed chromatin. Greater than 90% of the DNA in chromatin is wound into nucleosomes, which have a DNA packing ratio of ~6. When relatively extended (i.e., transcriptionally active) regions of chromatin are viewed in the electron microscope, they appear as strings in

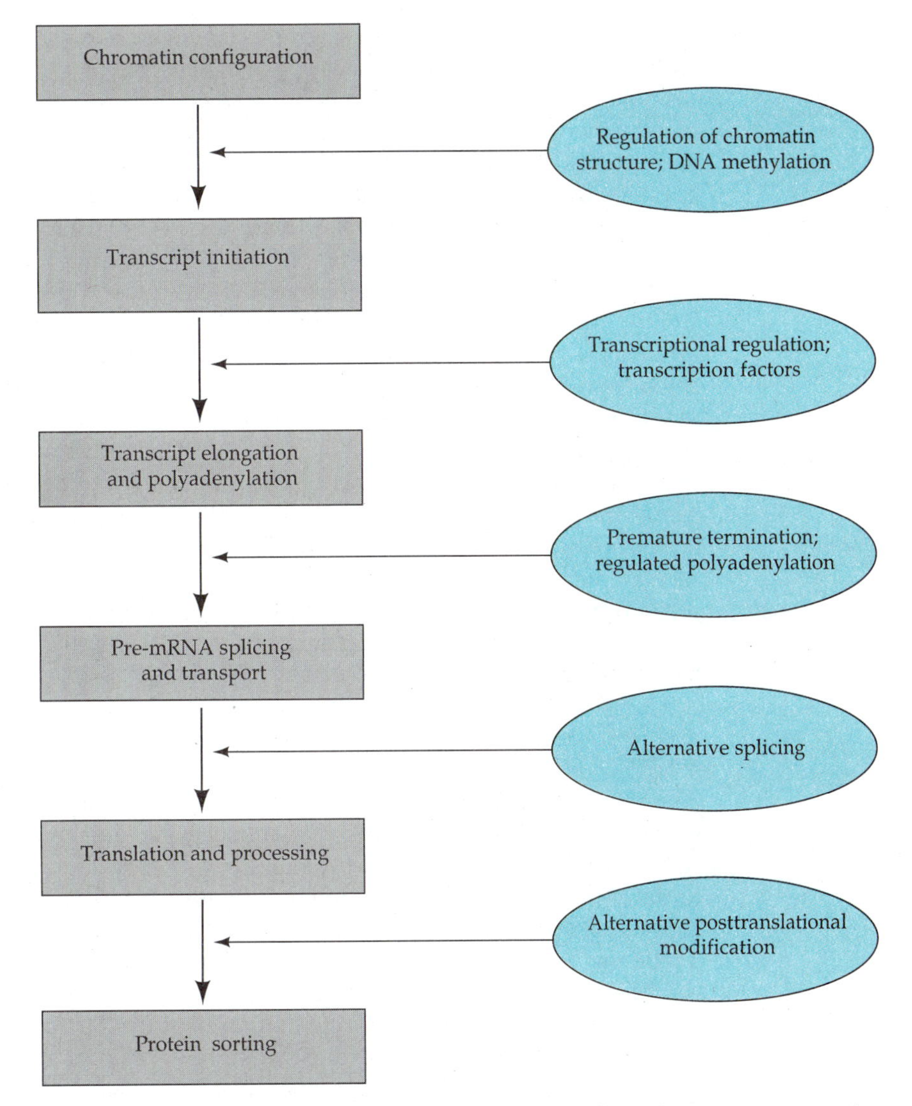

FIGURE 1. A hierarchy of gene regulation. This flow chart summarizes the sequence of metabolic processes through which the expression of genes and proteins is regulated in higher eukaryotes. Boxes correspond to individual events in the expression of genes and proteins; ovals include examples of how these events may be regulated.

which individual nucleosomes resemble regularly and tightly spaced beads. Although generally thought to be important, the extent to which the presence and exact positioning of nucleosomes serve to regulate gene expression is not well understood. Very actively transcribed genes such as those encoding ribosomal RNAs are not organized into nucleosomes, and in certain cases where transcription can be rapidly triggered (e.g., for heat shock genes) nucleosomal association seems to be lost.

In addition to higher order structure, certain features of the primary structure of DNA may be important for gene regulation. The best studied of these is DNA methylation. Approximately 5% of the C residues in animal cell DNA are methylated, and satellite DNA (which is transcriptionally inert) is heavily methylated. (Most methylated C's are found in the doublet 5'CG3', often in so-called "CpG" islands in the genome.) Several lines of circumstantial evidence have tied the methylation state of a given gene to its state of transcriptional readiness. First, many genes that are transcriptionally inactive are characterized by a relative abundance of methylated C residues within their regulatory regions. Second, unmethylated DNA is usually expressed when introduced (transfected) into cultured cells, whereas methylated DNA is not. Third, for some genes, transition from a transcriptionally inactive to an active state is correlated with the disappearance of one or more methylated sites. Fourth, treatment of cultured cells with drugs that result in DNA demethylation activates the expression of certain genes. Finally, the methylation patterns of modified genes are inherited with remarkable fidelity over multiple generations both in cell culture and in vivo. These observations notwithstanding, the case for methylation as a critical *determinant* of transcriptional readiness is compromised by the fact that some heavily methylated genes are nonetheless transcribed, and by the finding that in *Drosophila* and other dipteran insects, DNA methylation does not exist.

There are multiple regulated events between starting a transcript and finishing a protein

Assuming that a given stretch of chromatin is configured to permit transcription, the following (simplified) sequence of events must occur for the cell to synthesize a functional protein. First, RNA polymerase must bind to the DNA and productively initiate an mRNA transcript. Most of this chapter relates to this single event. Second, the polymerase must transcribe a full-length precursor or pre-mRNA that can be polyadenylated—it must not fall off or be bumped off the DNA template prior to completing its task. Although omitted from this discussion, premature transcript termination is a critical and regulated control point in the expression of several genes. Third, most pre-mRNAs must be processed to remove sequences that either do not code for protein, or, alternatively, code for an incorrect protein. Because of its particular relevance to the nervous system, we shall consider the consequences and regulation of alternative pre-mRNA processing in detail. Fourth, the processed mRNA molecule must be transported from the nucleus into the cytoplasm. Fifth, the mature mRNA must reach the ribosome, be translated into a stable protein, and potentially undergo a large number of regulated co-translational and posttranslational modifications (e.g., signal peptide cleavage and glycosylation). Finally, this protein must be routed to its correct location (the nucleus, cytoplasm, cell surface, or outside the cell) and not be shunted to a degradation pathway.

This is a daunting course, to say the least. We should bear in mind that at nearly every stage, the expression of mRNAs and proteins is subject to regulation. This regulation involves both a yes/no decision—will a gene (or protein) be expressed or not?; and more specifically, a decision between appropriate and inappropriate expression—will a gene be expressed in the right cells, at the proper time in development, and in response to the appropriate extracellular stimuli? These decisions, which determine how the early nervous system develops and how the mature nervous system performs, are most effectively made during the initiation of transcription and in the alternative splicing of nascent pre-mRNAs.

Transcription

The regulatory regions of eukaryotic genes are bipartite

The most intensively studied control point in eukaryotic gene expression is the rate at which new mRNA transcripts are initiated by RNA polymerase II (pol II), the large holoenzyme used to synthesize mRNA. The control of transcript initiation has been analyzed both in reconstituted, cell-free systems in vitro, and after transfection-mediated introduction of recombinant reporter constructs into cultured cells. These analyses have consistently demonstrated that the regulatory regions of eukaryotic genes are built up from a set of short (6–20 bp) DNA sequence elements that are the binding sites for regulatory proteins. These sequence elements can be assigned to one of two general regulatory domains.

The first and most basic of these is the **core promoter.** This is the relatively small region of DNA to which pol II and a set of specialized "initiation proteins" bind. To a first approximation, the position and size of the core promoter in any given gene is invariant—it is located immediately upstream of and over the **transcription start site,** the position at which the polymerase begins to transcribe the mRNA(s) that the gene encodes. This start site is often referred to as the "cap site," since the 5' ends of eukaryotic mRNAs are modified ("capped") by the addition of a methylated G residue. Many genes contain more than one core promoter and thus more than one transcription start site. For some genes, these multiple promoters are differentially active in different cell types or at different times in development.

The most important component of the core promoter is the TATA box, a highly conserved A/T-rich sequence located ~25 bp upstream of the transcription start site of most eukaryotic genes. The consensus sequence of the TATA box—GNGTATA(A/T)A(A/T)—is very similar to that of the Pribnow box, a highly conserved RNA polymerase binding site located ~10 bp upstream of the transcription start site of bacterial genes. A series of elegant biochemical and transfection experiments have demonstrated that the principal role of the TATA box and other similarly positioned A/T-rich sequences is to provide a binding site for pol II, to orient this enzyme, and to specify the transcription start site. In other words, this element defines both the 5' → 3' direction in which the polymerase moves on the DNA, as well as the template position at which it starts. As illustrated in Figure 2, the TATA box is specifically recognized and bound by **TFIID,** one of the initiation proteins mentioned above. Following binding of this protein, an elaborate initiation complex is assembled by the sequential addition of TFIIA, TFIIB, pol II, and TFIIE. TFIIB is thought to serve as "bridge" between the TFIID/A complex and pol II, and its size

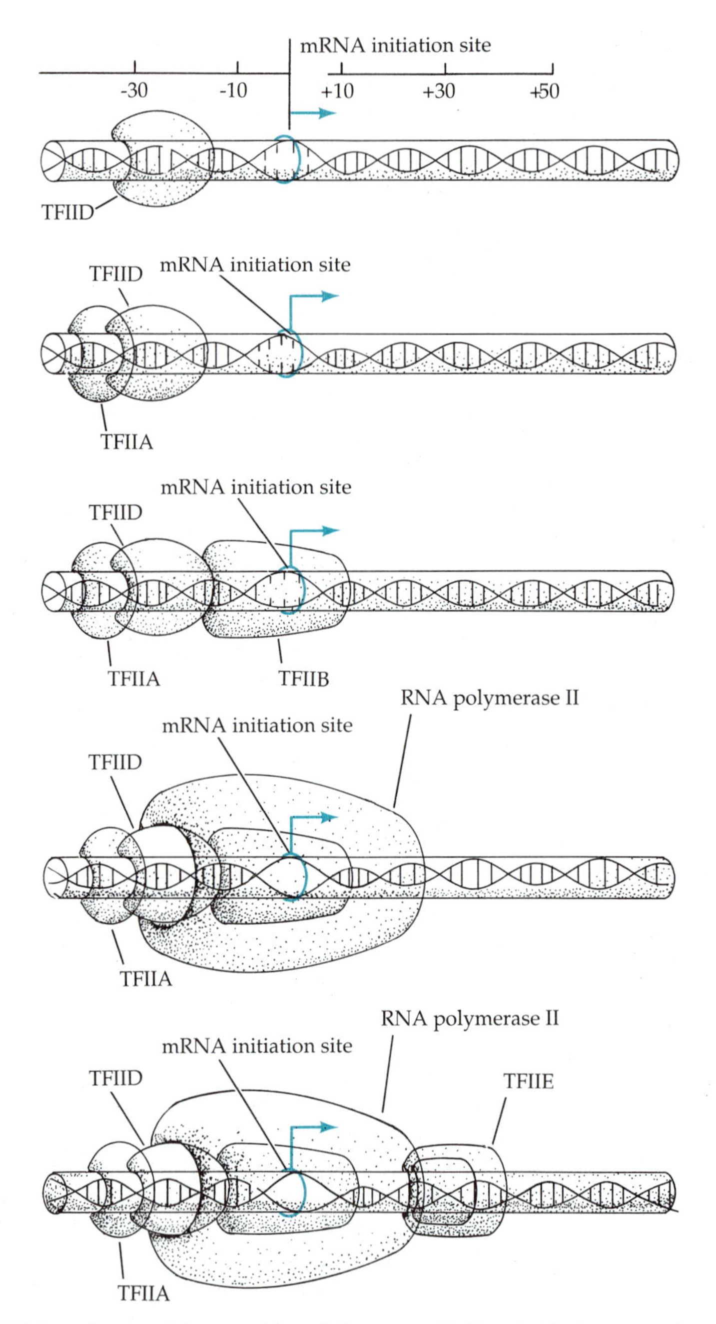

FIGURE 2. Sequential assembly of the transcription initiation complex at the core promoter. The transcription initiation complex is made up of RNA pol II and a set of initiation proteins that sequentially assemble over the TATA box of the core promoter as indicated. Transcript initiation begins at position 0 on the scale, and the direction of transcription is indicated by the arrow. The TATA box is located at –25 to –30 (relative to the start site) and determines the direction in which transcription is initiated. TFIIA, TFIIB, TFIID, TFIIE, and pol II probably represent a complete initiation complex.

and secondary structure may be important in the exact positioning of the catalytic site of the polymerase. Once assembled over the core promoter, this very large complex of proteins (pol II itself is composed of at least ten different polypeptide chains) is poised to locally unwind the DNA template and initiate transcription.

Transcription factors bind to regulatory (enhancer) elements

For most genes in most cells, the TFIIA-E/pol II complex is by itself incapable of initiating transcription at a physiologically significant rate. To attain such a rate, a set of activator proteins must bind to the second major type of sequence element contained within eukaryotic regulatory regions. Elements of this second type have been given more than one name but are most commonly referred to as **enhancer** or **regulatory** or **response elements.** Originally recognized in the control regions of animal viruses such as SV40, these sequences have two distinguishing features. First, unlike the core promoter, they may be located at variable positions relative to the transcription start site. Some enhancer elements are located very close to (<30 bp from) this site, while others are located a great distance (>10 kb) away; for many genes these distances may be experimentally altered without significantly compromising transcriptional efficacy. The positional variability of regulatory elements applies not only to their distance from the transcription start site, but also to their position relative to this site. Although most regulatory sequences have been mapped upstream (to the 5' side) of the core promoter, several have been localized elsewhere. For example, a well-studied transcriptional enhancer of immunoglobulin (Ig) genes—the octamer motif 5'ATGCAAAT3' discussed below—is located within the first introns of these genes. Similarly, an enhancer essential to the erythrocyte-specific expression of the β-globin gene is located at the 3' end of this gene, downstream of the transcription *termination* site. In considering the distance-independence of regulatory elements, it is important to remember that DNA is not an extended (rod-like) duplex in cells—it is looped into nucleosomes that are themselves coiled about one another.

The second defining property of regulatory elements is their orientation independence. Unlike the TATA box, most enhancer sequences work equally well when positioned either in their normal 5' → 3' orientation, or when excised from the DNA and reinserted in an inverted 3' → 5' orientation. This orientation independence is reflected in the fact that the same enhancer element is often found in opposite orientations in different genes. Possible explanations for the orientation independence of regulatory elements are discussed below.

Almost all of the transcriptional regulation that we typically associate with eukaryotic genes, including cell-specific and inducible expression, is achieved by selective expression or activation of the proteins that recognize and bind to regulatory elements. Together, these proteins are referred to as **transcription factors.** As in bacteria, they have been shown to be critical mediators of both transcriptional activation *and* transcriptional repression. Most transcription factors appear to promote or inhibit

Box A Sequence-Specific DNA-Binding Assays: Footprints, Gel Shifts, and the Like

How does one assay for the binding of a transcription factor to a specific nucleotide sequence? How does one determine where transcription factors actually bind within the regulatory region of a cloned gene? Molecular biologists have devised several assays to address these questions. Perhaps the best known of these is the DNase I protection or **footprint assay.** In this procedure, molecules of a cloned DNA fragment are specifically radiolabeled at one end (usually the 5' end, with polynucleotide kinase and γ-^{32}P-ATP), and then incubated in solution with a fraction containing the DNA binding protein or proteins of interest. This fraction may be a relatively crude extract of whole cells that contains many different DNA binding proteins, or a highly purified preparation of a single protein. After allowing sufficient time for DNA binding to occur (about 30 minutes), the DNA–protein complexes are subjected to limited digestion by the endonuclease DNase I. This enzyme is used at a low concentration, so that a small number of cuts (~1) are generated at random positions within each labeled DNA molecule. The assay is then analyzed by running the digestion mixture of DNA fragments on a denaturing polyacrylamide (sequencing) gel, which resolves the mixture into an ordered series of fragments that differ in length by only a single nucleotide. Those regions of individual DNA molecules bound by protein are protected from DNase I digestion. Fragments whose ends are within the protected region are missing in the gel-resolved pattern, thus generating a blank region or "footprint" (Figure A). With this elegant assay one can precisely map the boundaries of DNA binding sites. The assay is relatively insensitive, however, in its ability to detect DNA binding proteins. Because the procedure assays protein binding to a population of DNA molecules, a large fraction of the population must be bound to protein for a footprint to be detected. Strong footprints of the sort illustrated in Figure A, for example, are observed only if nearly all of the labeled DNA molecules are bound to protein during DNase I digestion. As a result, the binding of low affinity or low abundance proteins is difficult to detect by DNase I footprints.

A less frequently used variation on DNase I footprints is the **methylation interference assay,** in which the G residues within segments of DNA bound to protein are protected from modification by the methylating agent dimethyl sulfoxide (DMS). As with DNase I assays, the DMS treatment is performed at limiting concentrations, so that only a small number of G residues within each end-labeled DNA molecule are modified. By the same chemistry as is used in Maxam–Gilbert DNA sequencing, DNA molecules that contain methylated G residues are cleaved at these residues by piperidine while molecules that contain unmodified G residues are not. Since piperidine cleavages do not occur at the unmethylated G positions, a footprint is again obtained when the reaction products are analyzed on a denaturing polyacrylamide gel.

In both the DNase I and methylation interference assays, the experimenter looks for a clustered set of bands to disappear from the gel. Assays that score the appearance of new bands are often easier to do and are usually much more sensitive. In one of these, again a protection assay, the endonuclease DNase I is replaced by the exonuclease **Exo III**. This enzyme binds to the 3' ends of DNA fragments and progressively removes nucleotides as it moves toward the 5' end. In the absence of bound protein, Exo III will eventually de-

the formation of new transcripts through direct protein–protein interactions, either with TFIID, pol II, and other components of the core promoter initiation complex, or with other transcription factors. As we shall see, most transcription factors have multiple domains with distinct functions.

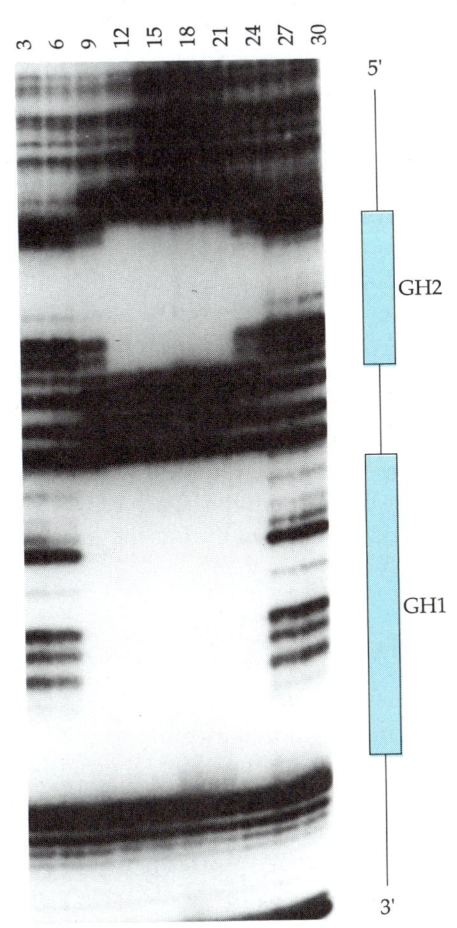

FIGURE A. Pit-1/GHF-1 footprint within the regulatory region of the growth hormone gene. In this example, DNase I footprinting has been used to assay a set of phosphocellulose column fractions for the presence of the transcription factor Pit-1/GHF-1. This transcription factor is detected in fractions 12–21 by virtue of its binding to two sites (GH1 and GH2) in the rat growth hormone promoter. Binding of Pit-1/GHF-1 protects these sites from digestion by DNase I. (Photo courtesy of Harry Mangalam.)

grade the entire fragment. If however, a transcription factor or any other protein is bound to a site within the test DNA, the enzyme is blocked and falls off the DNA, causing a new band, or stop, to appear when the reaction products are resolved on a denaturing polyacrylamide gel. Even if only a very small fraction of 5' end-labeled DNA molecules are bound to protein, their binding will be revealed as a new, binding-dependent "stop site."

The fastest, easiest, and therefore most commonly used DNA binding assay is the gel retardation, or **gel shift** assay, also referred to as the band shift or electromobility shift assay (EMSA). Unlike the assays discussed above, the gel shift is not a protection assay and does not provide information about the boundaries of protein binding sites. Instead, it is used to ask whether a given protein binds at all and, if so, with what affinity, to a given small (10–100 bp) DNA fragment. (These fragments are usually generated on an automated synthesizer as complementary single-stranded oligonucleotides that are then annealed to one another.) The principle behind the assay is that in a native nondenaturing gel, a radiolabeled DNA fragment bound to protein forms a labeled complex that migrates more slowly than the DNA fragment alone. As with the Exo III protection assay, even if only a very small fraction of the labeled DNA population is bound, a new (gel-shifted) band is detected. The great sensitivity of this assay makes it prone to artifact, and appropriate controls must always be performed. These include the demonstration that under conditions in which the concentration of DNA binding protein is limiting, excess unlabeled DNA is capable of specifically competing out the binding of labeled DNA, and that mutant versions of the binding site are not bound by the protein being examined.

One of these is usually a DNA binding domain that recognizes specific sequences within the regulatory regions of genes in much the same way that a cell surface receptor recognizes a specific ligand (see Box A). Very commonly, a transcription factor also carries a physically distinct effector domain that interacts, either positively or negatively, with other transcriptional regulators or with proteins of the pol II complex. Some transcription factors are incapable of binding DNA—they have only effector domains and require interaction with a DNA binding protein in order to reach the DNA and exert their effect. The herpes simplex virus type I VP16 protein, which is one of the strongest transcriptional activators known, is such a protein. An amino terminal region of VP16 allows it to attach to DNA binding proteins (such as the Oct-1 transcription factor discussed below); transcriptional activation is then achieved by the interaction of a separate VP16 activation domain with TFIID. In addition to effector domains, some transcription factors contain one or more regulatory domains that allow their DNA binding and transactivation activities to be modified in response to extracellular stimuli. Proteins of the steroid receptor superfamily, discussed below, fall into this category of DNA binding proteins. The basic modes of action for transcription factors are summarized in Figure 3.

Based on the presence of conserved structural motifs, transcription factors can be segregated into distinct regulatory families. We shall consider the structure and function of five of these families, each of which has been studied in the context of neural development and of the integrative functioning of the nervous system.

Transcription Factors as Signal Transducers

The transcription factors fos and jun are encoded by proto-oncogenes

Many transcription factors were originally studied as proteins that produce an abnormal cellular phenotype when mutated or aberrantly expressed. In vertebrates, the most commonly studied of these abnormal phenotypes has been the "transformed phenotype" defined by neoplastic transformation of cultured cells and by tumor formation in laboratory animals. The genes that confer a transformed phenotype, called **oncogenes,** release cells from normal proliferative restraints; correspondingly, the proteins encoded by oncogenes are called oncoproteins. In many cases, mutant cellular genes behaving as oncogenes have been identified as the resident transforming genes of RNA tumor viruses (retroviruses). The normal cellular counterparts of these retroviral transforming genes are called **proto-oncogenes.**

Several proto-oncogenes have turned out to encode transcription factors. This has generally been established in a step-by-step fashion. Typically, the protein product of a viral oncogene or cellular proto-oncogene is first immunocytochemically localized to the cell nucleus, an observation that immediately makes it a suspect for transcriptional regulation. Second, the purified (and often cloned) protein is demonstrated to bind DNA with high affinity, and sequence-specific binding sites are identified within candidate target genes. Third, identity between the oncoprotein and a protein independently purified on the basis of its ability to bind the target site is recognized. Finally, the protein is demonstrated to directly regulate (enhance or inhibit) transcription of the target gene. (The steps have not

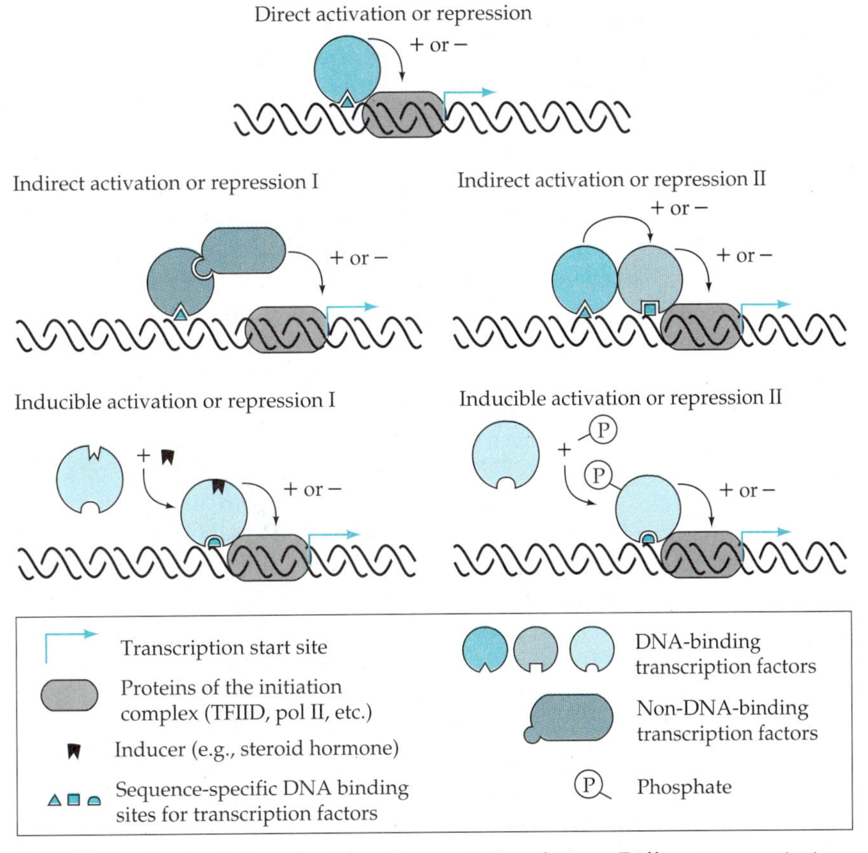

FIGURE 3. Basic modes of action of transcription factors. Different transcription factors exert their effects in different ways. Many bind to sites and then directly interact with (bind to) initiation complex proteins at the core promoter. Some transcription factors are incapable of binding to DNA, however, and must be brought to the transcription template through protein–protein interactions with other transcription factors. (Transactivation by VP16 is an example of this form of transcriptional regulation.) Some transcription factors exert their effect by modulating the activity of other DNA binding proteins. (The yeast homeobox protein α2 acts in this way.) Finally, the dimerization, DNA binding, and transcriptional activities of some transcription factors can be regulated through hormone binding or posttranslational modifications such as phosphorylation. (Transcription factors of the steroid receptor family and the CREB family, respectively, exemplify these forms of regulation.)

always occurred in this order.) Perhaps the best known of the oncogene transcription factors identified in this way are fos and jun. Fos was first identified as viral or v-fos, the resident transforming protein of the FBJ and FBR murine osteosarcoma viruses. Similarly, jun was initially studied as v-jun, the transforming protein of avian sarcoma virus 17 (ASV-17). The FBJ and FBR retroviruses induce bone tumors in infected mice, and the ASV-17 virus induces muscle tumors in chickens. Both viruses confer a neoplastic or transformed phenotype upon infection of cultured fibroblasts. The v-fos and v-jun proteins represent viral variants of normal cellular proteins, which are often referred to as c(for cellular)-fos and c-jun. These proteins are very widely expressed, and function as transcription factors in many different neural and non-neural cells.

Fos and jun form heterodimers that are held together by a zipper

Fos and jun bind to one another to form a heterodimeric protein. This heterodimer recognizes a 7-bp enhancer element, called the TPA response element or **TRE.** (TPA is the phorbol ester 12-O-tetradecanoyl-phorbol-13-acetate, a tumor promoter that strongly activates protein kinase C and rapidly induces expression of genes that contain one or more TREs.) This element is also referred to as the AP-1 site, since it was independently identified as the binding site for a transcription factor complex, called AP-1, purified from nuclear extracts of cultured cells. The consensus sequence of the TRE/AP-1 site is 5'TGACTCA3'; copies of this sequence, or close variants, are found in the upstream regions of many eukaryotic genes. The sequence of the TRE/AP-1 site illustrates an important general principle. Like the consensus binding sites of many other transcription factors, this site is a double stranded palindromic sequence: except for the central nucleotide (C), the same sequence is obtained when the element is read (5' → 3') from the complementary (noncoding) strand of DNA. (Remember that the two strands of DNA run in opposite directions.) The dyad axis of symmetry of palindromic sequences appears to facilitate the binding of dimeric proteins; that is, each monomer binds to one half-site of the recognition element. This is also an important general rule: as we shall see, most transcription factors bind as dimers. The facts that: (a) palindromic DNA sequences are the same in either orientation; (b) most regulatory elements are at least partial palindromes; and (c) most transcription factors bind as dimers probably account for the observed orientation-independence of enhancer activity.

The ability of fos and jun to dimerize depends critically on a conserved structural motif located near the carboxy terminus of each protein, immediately downstream of a basic region that mediates DNA binding. (As will become evident, a stretch of highly basic amino acids is a telltale sign of DNA binding domains.) This conserved motif is the **leucine zipper,** so-named because it consists of a ~28-residue α-helix in which every seventh position is occupied by the amino acid leucine. As illustrated in Figure 4, when these periodic leucine residues are fitted to an idealized α-helix, they align along one face of the helix. Apparently the leucine side chains of the α-helix of one protein (e.g., fos) interact hydrophobically with the leucine side chains of the corresponding helix of its partner (e.g., jun). This interaction results in the formation of an intermolecular coiled coil that stabilizes or "zips up" the heterodimer. Mutations in the leucine zipper domain of either fos or jun destroy the ability of these proteins to dimerize and severely compromise their ability to activate transcription. The leucine zipper is found not only in fos and jun, but in several other transcription factors whose ability to transactivate depends on dimerization (see Figure 4 and below).

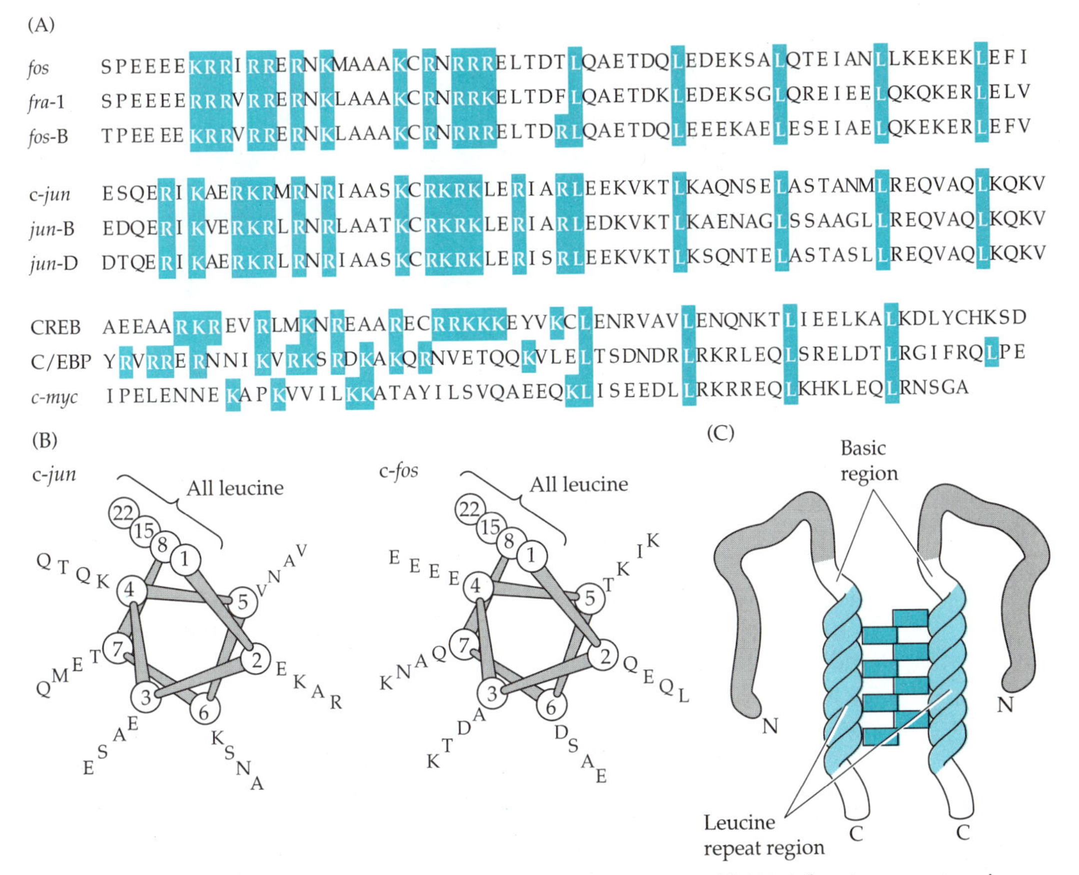

FIGURE 4. Fos, jun, and the leucine zipper motif. (A) A leucine repeat region—the leucine zipper—and an immediately upstream DNA binding domain of basic amino acids are found in fos, jun, and many other transcription factors. Fra-1 and fosB are proteins closely related to fos, and junB and junD are closely related to jun. CREB is the cAMP response element binding protein, C/EBP is the cellular enhancer binding protein in which the leucine zipper was first recognized, and c-myc is a nuclear proto-oncogene like fos and jun. Leucine residues and basic amino acids (lysine and arginine) are indicated by white characters in colored boxes. (B) When mapped to an idealized α helix, all of the leucine residues of the leucine repeat regions of fos and jun align along one face of the helix (positions 1, 8, 15, and 22); that is, seven amino acids make up two turns of the helix. The helical wheels displayed spiral into the plane of the page from position 1. (C) Leucine zipper proteins form homo- and heterodimers through interactions that generate a stable coiled coil structure between their leucine zippers. A schematic of leucine zipper dimers illustrates this interaction. In general, the formation of leucine zipper dimers is essential for DNA binding and transactivation.

In the fos/jun (AP-1) heterodimer, jun is the most active player. It dimerizes, binds DNA, and weakly activates transcription all by itself. Fos, on the other hand, requires a partner. Even though it has a leucine zipper of its own, fos is incapable of forming homodimers or of specifically binding DNA by itself. It does, however, modulate jun activity: both DNA binding specificity and transactivation activity are appreciably greater for the fos/jun heterodimer than for the jun/jun homodimer. The functional differences between fos and jun appear to be mediated by their respective leucine zippers, since fos acquires the ability to bind DNA and transactivate as a homodimer if its leucine zipper is replaced by the zipper from jun. A number of proteins closely related to fos (e.g., fra-1 and fosB) and jun (e.g., junA, junB, and junD) have recently been identified, raising the possibility of a diverse set of paired heterodimeric regulators (fos/junB, fos/junD, fra-1/jun., etc.) that differ in their binding specificities and transactivation potentials.

Fos *and* jun *are potential early signal transducers in neural cells*

Why should neurobiologists care about nuclear proto-oncogenes such as *fos* and *jun*? Molecular biologists often refer to these genes as **immediate-early genes.** This term is derived from the analysis of the life cycles of animal viruses, which are characterized by the sequential expression of regulatory and structural genes, respectively. In mammalian cells, many of the extracellular stimuli that control proliferation and differentiation result in the rapid but transient induction of nuclear proto-oncogenes. When cultured cells are exposed to polypeptide growth factors, for example, fos and jun mRNAs are almost instantaneously induced, reach peak steady-state levels about 30 minutes after growth factor addition, and by 120 minutes fall back to essentially undetectable levels. The transient induction of nuclear proto-oncogenes is thus among the very first changes in gene activity that can be observed in response to extracellular stimulation.

These signal transduction phenomena are likely to be of direct relevance to the nervous system. Addition of nerve growth factor (NGF) to cultured pheochromocytoma (PC12) cells, for example, initiates an extended cascade of gene expression that ultimately results in the adoption of a neuronal phenotype (see Chapter 11). The first observable nuclear event in this cascade is the induction of fos and jun mRNAs. Remarkably, induction of nuclear proto-oncogene mRNAs is also one of the first observable nuclear consequences of neuronal activation in vivo. Noxious stimulation of primary sensory afferents to the dorsal horn of the rat spinal cord, for example, results in the rapid induction of fos mRNA and protein in post-synaptic neurons. The idea that this induction is important for the subsequent response of these neurons is supported by the observation that these very same cells activate expression of the prodynorphin gene shortly after the fos protein levels rise. The prodynorphin gene is an important component of the pain response, since it encodes a family of opioid peptides (see Box B, Chapter 4) that modulate pain perception. The hypothesis that fos might directly transactivate this gene is in turn supported by the finding that the prodynorphin upstream regulatory region contains a variant TRE site that binds the fos/jun heterodimer. When assayed in vitro, this variant TRE is required for activation of the gene. These and related findings suggest that proto-oncogene transcription factors are likely to function as early nuclear transducers of sensory stimuli in neurons. Distinct neuroanatomical patterns of fos/jun activation have been observed in response to different sensory stimuli, an observation that is being used to map sensory pathways within the nervous system.

CREB is a transcriptional transducer of changes in intracellular cyclic AMP

Many of the hormones and neuromodulators that regulate the synaptic activity of neurons bind to cell surface receptors that are coupled (via G proteins—see Chapter 6) to adenylate cyclase, the enzyme that synthesizes cyclic AMP (cAMP). For these neuromodulators, cAMP is the intracellular second messenger through which their presence is registered. Binding of catecholamines to the β-adrenergic receptors, for example, is directly coupled to cyclase activation and an increase in intracellular cAMP. Changes in the level of cAMP have profound and nearly instantaneous effects on the expression of a variety of neural genes, but until recently the mechanism underlying this rapid regulation was unknown. The first insights into this mechanism came from functional studies of the regulatory regions of cAMP-responsive neuropeptide genes—in particular the somatostatin, proenkephalin, and vasoactive intestinal peptide (VIP) genes. These genes contain either one or both halves of a small palindromic element that is responsible for cAMP induction and that confers cAMP responsiveness when transferred to non–cAMP-responsive genes. This element, analyzed in cell transfection experiments (see Box B), was first recognized in the regulatory region of the rat somatostatin gene, where it occurs as the perfect 8-base pair palindrome 5'TGACGTCA3'. (Note that this element is very similar to the TRE.) It has subsequently been found and studied in many other genes, including the human chorionic gonadotropin, rat corticotropin-releasing hormone, and rat tyrosine hydroxylase genes. Designated the cAMP response element or **CRE,** it has many of the properties of a classical enhancer element.

One of the most powerful features of the analysis of elements such as the CRE and the TRE is the ability to use these small DNA sequences as affinity probes with which to identify, purify, and clone the transcription factors that bind to them. Given their small size, most regulatory elements can easily be synthesized (on an automated oligonucleotide synthesizer) in the large quantities required for the preparation of DNA affinity columns and can thus be used to purify their corresponding transcription factor. They can also be radiolabeled and used as probes in blotting protocols to identify the transcription factor in a heterogeneous mixture, and to clone cDNAs from bacterial expression libraries (see Box C). The first of these approaches was used with the somatostatin CRE to identify, purify, and clone cDNAs for a 43-kD CRE binding protein, or **CREB,** expressed by PC12 cells. Other closely related CREB proteins have subsequently been identified and cloned.

How do CREB and its relatives transduce a change in cAMP levels into a change in gene expression? As described in Chapter 7, most of the biological effects of cAMP are due to its activation of cAMP-dependent protein kinase (protein kinase A, or PKA), which in turns phosphorylates a wide variety of proteins whose activities may be regulated by phos-

Box B Transfection-Based Transcription Assays

Molecular biologists have devised several assays to measure the transcriptional activity of the regulatory region of a cloned gene. Most of these involve the transient expression of hybrid reporter constructs in cultured cells. These constructs are plasmids in which the putative regulatory region is placed upstream of a foreign reporter gene that encodes an easily assayed enzyme. The most popular of these enzymes are bacterial β-galactosidase (β-gal), bacterial chloramphenicol acetyltransferase (CAT), and firefly luciferase. Straightforward biochemical and histochemical procedures have been developed for the quantitative detection of each of these enzymes (see Figure A). Tests of reporter constructs are performed by transfecting these constructs into cultured cells, growing the cells for 1–3 days after transfection, and then assaying extracts of the transfected cells for expression of the reporter enzyme. To a first approximation, the specific activity of the enzyme in each extract is an indicator of the transcriptional strength of the regulatory region under examination.

For the assay of specific regulatory elements (as opposed to a complete regulatory region), the above assays are modified to incorporate a heterologous core promoter sequence into the hybrid constructs. Among the most popular of these core promoters are those of the herpes simplex virus thymidine kinase (HSV TK) gene, and the rabbit β-globin gene. As discussed in the text, these core promoters have minimal transcriptional activity when introduced into cells on their own but provide the basic transcriptional machinery through which the activity of individual enhancer elements may be assayed.

Enzymatic reporter assays are indirect, to the extent that they do not actually measure transcription of reporter gene mRNA, but rather the translation of this mRNA into an assayable protein. A more direct test of tran-

FIGURE A. A chloramphenicol acetyltransferase-based transfection assay of the regulatory region of the rat P_0 gene. (1) In this example, three different fragments of the upstream regulatory region of the rat P_0 gene are linked (in both orientations) to the bacterial chloramphenicol acetyltransferase (CAT) gene. (2) Three days after transfection of these hybrid constructs into cultured Schwann cells, extracts of the transfected cells were assayed for CAT enzymatic activity. The assay was analyzed on a thin layer chromatogram, which resolves the chloramphenicol substrate (radiolabeled with ^3H) from two different forms of acetylated chloramphenicol, the products of the reaction. A transcriptionally active regulatory region results in the production of CAT enzyme, which is signaled by the synthesis of the acetylated forms of chloramphenicol. In (1), the upper P_0 gene map is in exact alignment with the lower construct diagrams. Arrowed regions pointing to the right indicate P_0 DNA in its normal 5' → 3' orientation. Constructs are named according to restriction site boundaries; e.g., HB refers to a HindIII/Bam HI fragment. SVO is a CAT construct that contains no P_0 DNA; constructs HB19 and HB27 are deletions of the P_0 TATA box. (1 and 2 from G. Lemke et al., 1988. Neuron 1: 73–83.)

phorylation. Perhaps not surprisingly, the transcriptional efficacy of CREB depends on PKA activity, and the protein has been demonstrated to contain a critical PKA phosphorylation site (a serine residue in the sequence RRPSY) about 130 residues from the amino terminus. When the cloned protein is mutagenized to eliminate this site, the ability of CREB to stimulate transcription from CRE-containing promoters is drastically reduced.

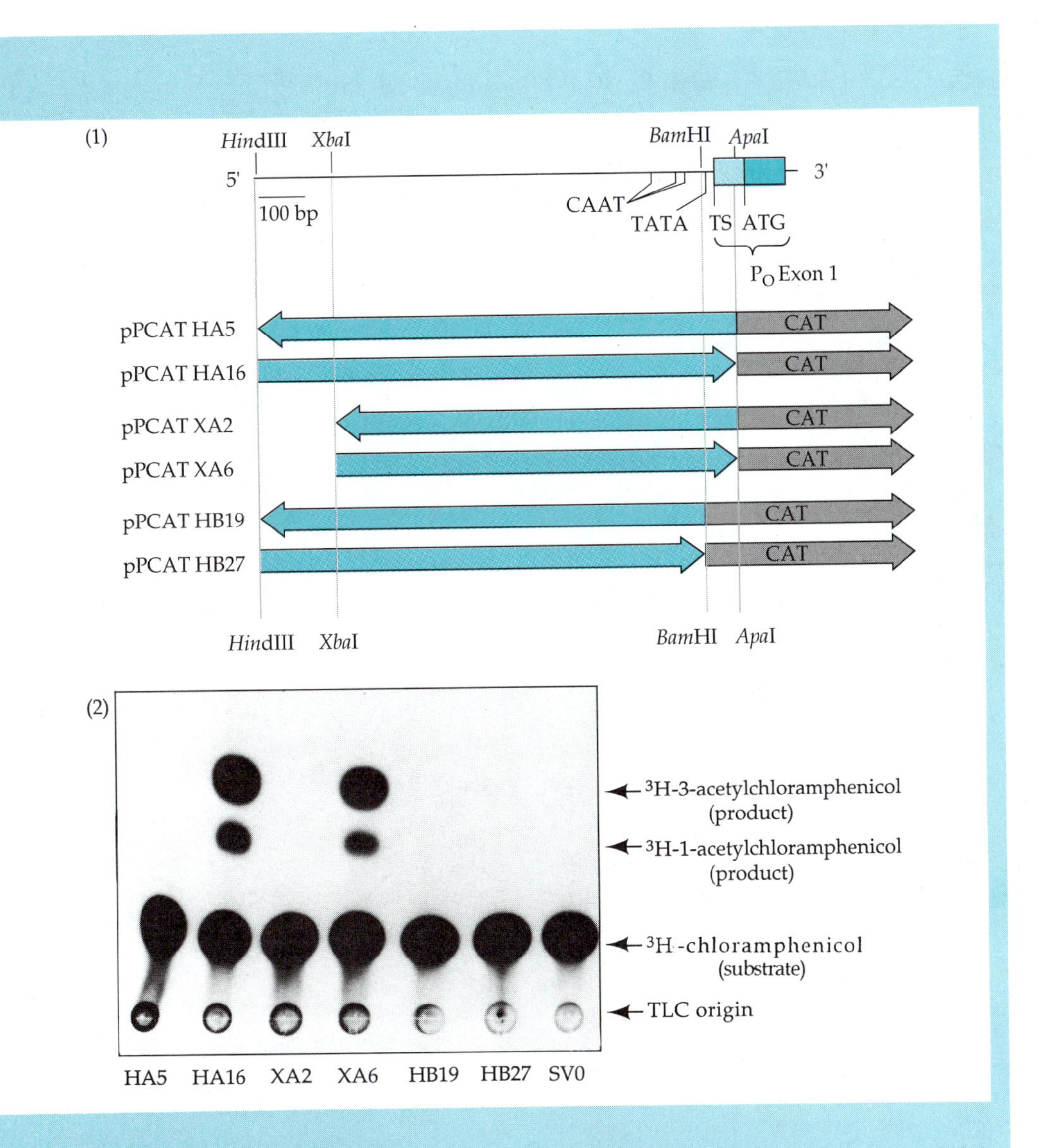

scription can be performed with several sensitive assays that quantitatively measure the levels of transcribed mRNAs in transient transfectants. In these procedures, RNA isolated from transfected cultures is assayed by either primer extension or RNase protection protocols for the specific expression of reporter mRNA rather than reporter protein.

Although the mechanism by which phosphorylation activates CREB is not fully understood in biochemical terms, it appears that the phosphorylated region of the protein interacts with an immediately downstream α helical domain, and that this interaction in turn promotes protein–protein contacts between the amino terminal activation domain of CREB and proteins of the core promoter. The direct sequence of cAMP elevation → PKA

Box C From Enhancer to Transcription Factor

Transfection-based cell culture assays and DNA binding assays make it relatively easy to identify DNA sequence elements important for transcriptional regulation. Biochemical purification of the transcription factors that bind to these elements can be a major undertaking, however. Very often these proteins are present at low concentrations within cells, and their purification therefore requires large quantities of starting material and the application of multiple chromatographic procedures. In an attempt to circumvent these difficulties, several procedures that exploit the sequence elements as affinity probes have been developed. In one such protocol, the double-stranded oligonucleotides that correspond to the binding site are synthesized and attached to a Sepharose or polyacrylamide resin to generate an affinity column. The principle here is exactly the same as that for a column derivatized with either a high affinity ligand for a neurotransmitter receptor (e.g., α-bungarotoxin for the nicotinic acetylcholine receptor) or with the antigen for a specific antibody.

A second protocol permits the identification of transcription factors following their transfer from SDS polyacrylamide gels to a nitrocellulose or nylon membrane. In this variation of a western blot, called a **"southwestern blot,"** the DNA binding protein is detected not with an antibody, but with a radiolabeled oligonucleotide corresponding to its binding site. The southwestern blot allows one to determine the molecular weight of a DNA binding protein and provides a semiquantitative detection assay. One major limitation of the procedure is that the transcription factor under study must retain its DNA binding activity after being subjected to SDS polyacrylamide gel electrophoresis and blotting transfer.

A third protocol permits the direct cloning of transcription factor cDNAs through the screening of λgt11 bacteriophage expression libraries with radiolabeled binding site probes. Again, these probes replace the monospecific antibodies used in conventional λgt11 screens. As noted above, some transcription factors must undergo posttranslational modifications (e.g., phosphorylation) in order to recognize their DNA binding sites. The fact that some of these modifications do not occur in bacteria is one of the limitations of a λgt11 screen. Nonetheless, the procedure has been used to isolate cDNA clones encoding several transcription factors. The end result of a λgt11 screen for the POU domain transcription factor Oct-2 is illustrated in Figure A.

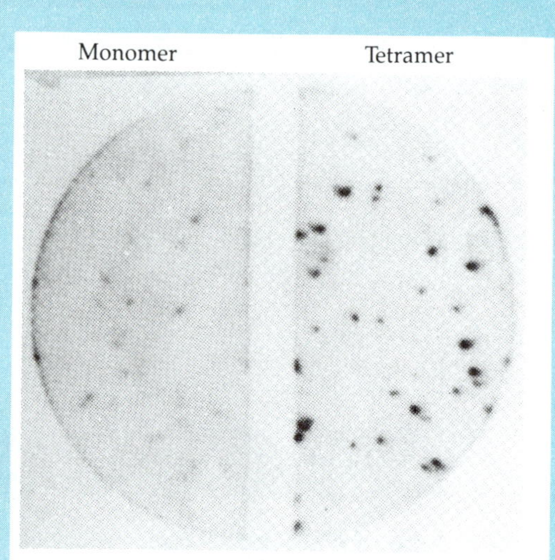

FIGURE A. Oligonucleotide detection of bacteriophage carrying cDNAs for the POU domain transcription factor Oct-2. In this example, filter lifts of λgt11 bacteriophage expression clones have been hybridized with monomeric (left) and tetrameric (right) forms of a radiolabeled double-stranded DNA fragment that contains the octamer motif of immunoglobulin genes. This octamer motif is the normal binding site of Oct-2. Each of the phage clones on the filter are bacterial lysogens that have been induced to express Oct-2 protein from the Oct-2 cDNA inserts that they carry. Multimeric probes are thought to provide a more sensitive detection method due to cooperative binding. (From L. M. Staudt et al., 1988. Science 241: 577–580.)

activation → CREB phosphorylation → CRE-dependent transcription, together with the constitutively high levels of CREB present in most cells, accounts for the very rapid increase in CRE-dependent transcription observed upon elevation of cAMP. Because of cellular phosphatase activity, phosphorylated CREB has a short half-life. A reduction in cAMP levels therefore results relatively quickly in a corresponding reduction in CREB activity.

Although CREB is likely to play a role in the regulation of those genes that are rapidly induced by cAMP elevation, cAMP induction of many genes occurs more slowly. These genes are typically in the middle or at the end of cAMP-initiated cascades (i.e., their expression depends on earlier genes that are rapidly induced by cAMP), and their regulatory regions do not contain CREB binding sites. In addition, recent work has shown that the activity of transcription factors other than CREB is regulated by phosphorylation events mediated by kinases other than PKA.

Like jun and fos, CREB carries a leucine zipper—an amphipathic 21 amino acid α helix, located near its carboxy terminus, in which four leucine residues are each separated by seven intervening charged amino acids. As with the nuclear proto-oncogenes, this leucine zipper is thought to play an essential role in dimerization—in most instances, CREB appears to bind to DNA as a homodimer. Like the fos and jun leucine zippers, the CREB zipper lies immediately downstream of a cluster of basic amino acids that mediate DNA binding.

Steroid and thyroid hormones bind to receptors that are also transcription factors

The differentiation and activity of many neural cells are regulated by precisely timed exposure to steroid hormones such as estrogen, testosterone, cortisol, and aldosterone. Sexually dimorphic nuclei of the vertebrate CNS, for example, are established by exposure to sex steroids during critical periods of development associated first with neurogenesis and early differentiation, as in the spinal nucleus of the bulbocavernosus in mammals, and subsequently with the expression of sex differences in behavior, such as singing behavior in male songbirds (see Chapter 13). Other small hydrophobic compounds—in particular thyroid hormone (an iodinated derivative of tyrosine), and retinoic acid and related molecules (derivatives of vitamin A)—also influence the course of neural development (see Chapter 11 and Figure 5). Although these extracellular signaling compounds affect rapid changes in gene expression, they do not act in a conventional way. Unlike polypeptide growth factors or neuromodulators, these small molecules do not bind to a cell surface receptor that is coupled to a signal transduction system that in turn regulates the synthesis or activity of a transcription factor. Instead, they are lipid soluble, and thus freely diffuse across the plasma membrane. Most importantly, the intracellular receptors to which they bind are themselves transcription factors.

These specialized transcription factors form a large family that includes the receptors for the sex steroids, thyroid hormone, glucocorticoids, retinoic acid, and vitamin D (Figure 5). (Like the *fos* and *jun* genes, the thyroid

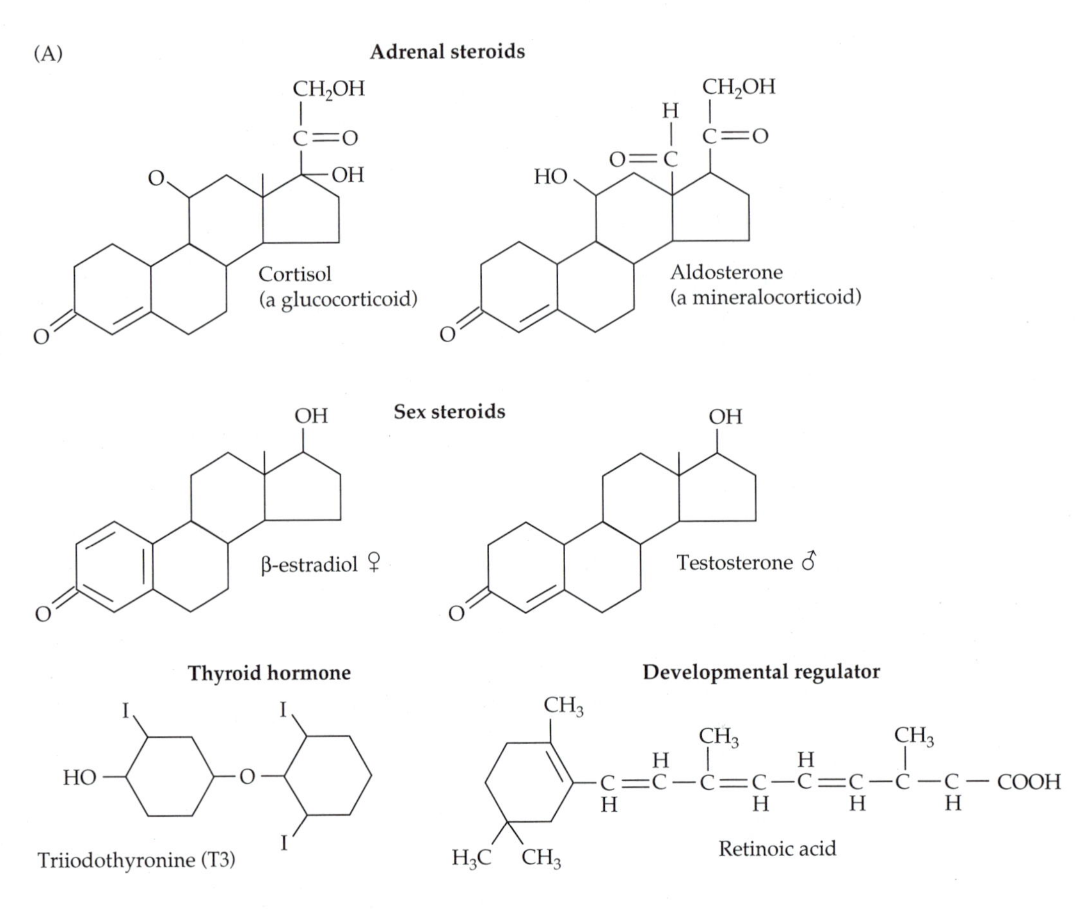

hormone receptor gene is a proto-oncogene.) In spite of this great diversity, each of these receptors represents a variation on the same basic theme. Built from three remarkably interchangeable modules, they contain a DNA binding domain, a hormone binding domain, and a domain responsible for transcriptional regulation (transactivation). The first two of these domains confer specificity of action. A hormone-binding domain that recognizes retinoic acid, for example, will not bind glucocorticoids with high affinity. Similarly, the DNA binding domain of a receptor generally recognizes a specific nucleotide sequence within eukaryotic enhancers. Proteins that recognize very closely related steroids and have nearly identical DNA binding domains, such as the mineralocorticoid and glucocorticoid receptors, are exceptions to this second rule. Functional hybrids can be experimentally generated by swapping domains from one receptor to another. If the DNA binding domain of the estrogen receptor, for example, is replaced by the equivalent domain of the glucocorticoid receptor, genes that were previously sensitive to glucocorticoid induction are rendered estrogen-sensitive (Figure 6). Although modular organization is a general feature of transcription factors, the ease with which domains may be swapped between various members of the steroid receptor family is unusual.

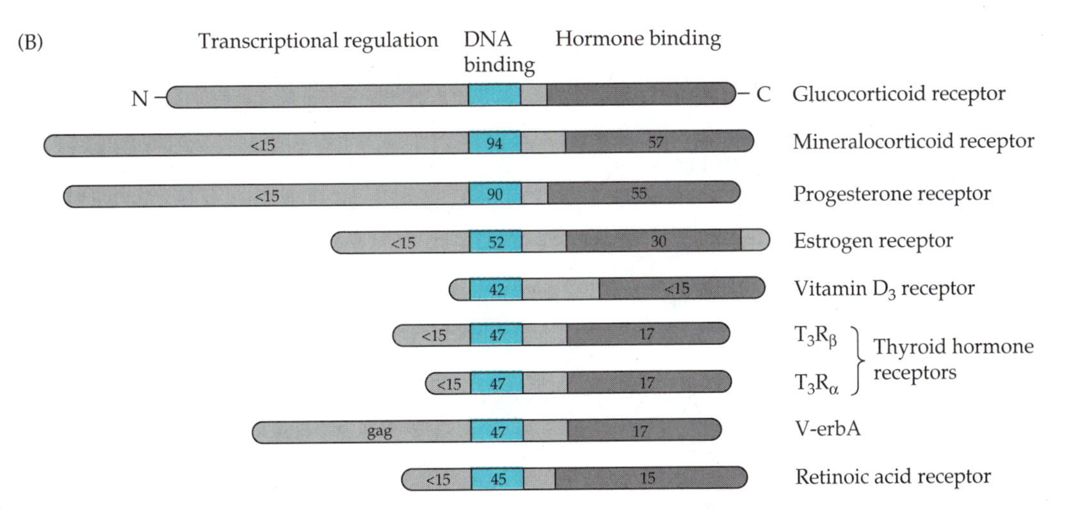

FIGURE 5. Transcription factors of the steroid receptor family. (A) The ligands of the steroid receptor family are small, lipid-soluble molecules that mediate many different biological processes, from basal metabolic regulation (thyroid hormones, mineralocorticoids) to sexual differentiation (sex steroids) to morphogenesis and development (retinoic acid). (B) Related modular structures of transcription factors of the steroid receptor family. The structures of selected members of this family are schematized relative to the human glucocorticoid receptor. Transcriptional regulation, DNA binding, and hormone binding domains are indicated. Numbers within each domain indicate the percentage of amino acid sequence identity relative to the human glucocorticoid receptor. V-erbA, the transforming oncoprotein of avian erythroblastosisvirus, is a retroviral fusion protein with the thyroid hormone receptor. All sequences other than v-erbA are human.

In spite of the fact that the DNA binding and transactivation activities of individual steroid receptors are absolutely dependent on the presence of cognate hormone, mutant receptors in which the hormone binding domain has been deleted are not always inactive. Instead, they are often (depending on receptor type) *constitutively* active. This surprising observation suggests that the hormone binding domain normally inhibits the intrinsic activity of the DNA binding and transactivation domains of steroid receptors. This inhibition is overcome by hormone binding, which presumably induces a conformational change in the receptor that permits DNA binding and transcriptional activation.

The primary structures of the DNA binding domains of steroid receptors are well conserved. These DNA binding domains carry a cysteine-rich motif that is present in a large number of transcription factors, including many that are not directly regulated by hormones. This motif, called the **zinc finger**, was originally observed in a cysteine- and histidine-rich region of the TFIIIA factor that regulates transcription of 5S ribosomal RNA genes. (These are not transcribed by pol II but by another eukaryotic RNA polymerase, pol III.) It consists of an ~20 amino acid sequence in which four (two paired) cysteine residues, or two cysteine plus two histidine residues, are coordinated by a zinc ion, such that the polypeptide between

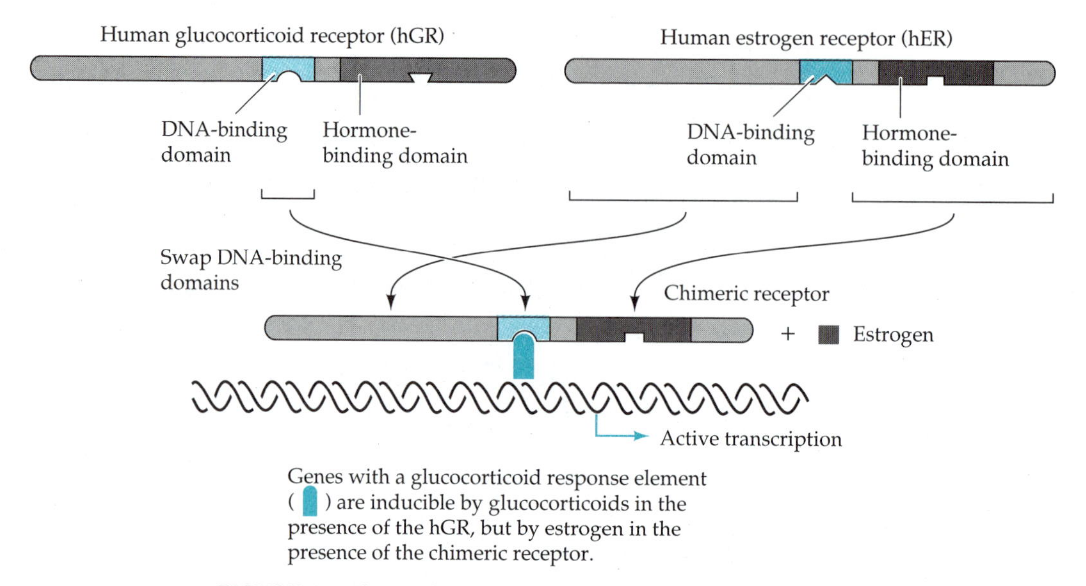

FIGURE 6. The modular organization of transcription factors: domain swapping in the steroid receptor superfamily. When the cDNA segments encoding the DNA binding domains of two different steroid receptors are excised (by restriction digests) and exchanged, a chimeric receptor cDNA is generated. The protein encoded by this chimeric cDNA has the DNA binding properties of one receptor and the hormone binding properties of the other. This procedure has been adapted to characterize the activity of newly cloned receptors whose ligands are unknown. Substituting the DNA binding domain of such a receptor with the corresponding domain of the glucocorticoid receptor (for example) allows one to screen a large panel of potential ligands for their ability to stimulate transcription of genes that are normally induced by glucocorticoids. The retinoic acid receptor was identified using this approach.

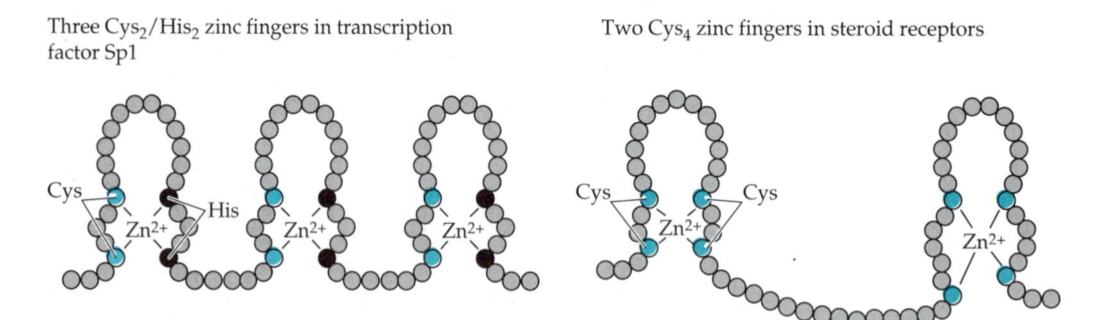

FIGURE 7. Zinc fingers. The zinc finger motif is a feature common to the DNA binding domains of many transcription factors. This motif is generated through the tetrahedral coordination of a zinc ion either by two cysteine and two histidine residues, as occurs in the transcription factor Sp1, or by four cysteine residues, as occurs in proteins of the steroid receptor family. (A number of amino acids other than cysteine and histidine are also conserved within the loops of Cys_2/His_2 and Cys_4 zinc fingers.) As discussed in the text, these two forms of the zinc finger have distinct structures and distinct DNA binding properties.

the paired cysteines (or cysteines plus histidines) "loops out" and folds into a fingered configuration (see Figure 7). Although the Cys_4 and Cys_2His_2 motifs both coordinate zinc, these two kinds of zinc finger have different properties. Whereas Cys_2His_2 fingers appear to bind DNA independently and are found in variable numbers between different transcription factors, the Cys_4 fingers of the steroid family bind DNA cooperatively and are always found in pairs. In several cases, DNA recognition by zinc finger proteins has been demonstrated to require metal coordination. DNA binding by the estrogen receptor, for example, is dependent on the presence of zinc. (Other divalent ions such as Ca^{2+} are ineffective substitutes for zinc). Metal coordination is thought to be essential for the maintenance of appropriate secondary structure within the fingered binding domain. As discussed in Chapter 11, expression of one mammalian member of the zinc finger family, designated *Krox-20,* precisely defines two segments (rhombomeres) early in the development of the mammalian hindbrain.

Transcription Factors and Neural Development

Homeobox proteins are critical determinants of neural development

Just as fos and jun were originally identified as proteins whose mutation results in aberrant proliferation, so another set of transcriptional regulators were first identified as proteins whose mutation results in aberrant development. Among the most bizarre developmental perturbations in *Drosophila* are those associated with **homeotic mutations,** which have the capacity to transform a specific part of one body segment into the homologous part of another. These mutations include rearrangements at the *Antennapedia (Antp)* locus, which (through ectopic expression of Antp protein) transform antennae into legs (Figure 8), and certain loss-of-function mutations at the *bithorax (bx)* locus, which transform thoracic segment 3 into a duplicate of thoracic segment 2, giving mutant flies two pairs of wings. The regulatory genes in which these amazing mutations reside encode a group of nuclear proteins related by the conservation of a highly basic, 60 amino acid domain. This conserved homeotic motif—named the **homeobox**—binds DNA, and the proteins that carry it function as transcription factors when assayed in cell culture or in in vitro transcription reactions.

The homeobox has now been identified in a large number of other regulatory but nonhomeotic genes in *Drosophila,* including several of the segmentation genes discussed in Chapter 11. Highly conserved versions of the homeobox have also been found in the regulatory genes of every higher and lower eukaryotic species examined, including humans. One estimate suggests that nearly 1% of all of the genes expressed in the nematode *C. elegans* encode homeobox-containing proteins! This versatile and widely used DNA binding motif is distantly related to the helix-turn-helix motif present in prokaryotic transcriptional regulators such as the bacteriophage λ repressor. Its structure has been analyzed in solution and bound to DNA, by both two-dimensional NMR and X-ray crystallographic

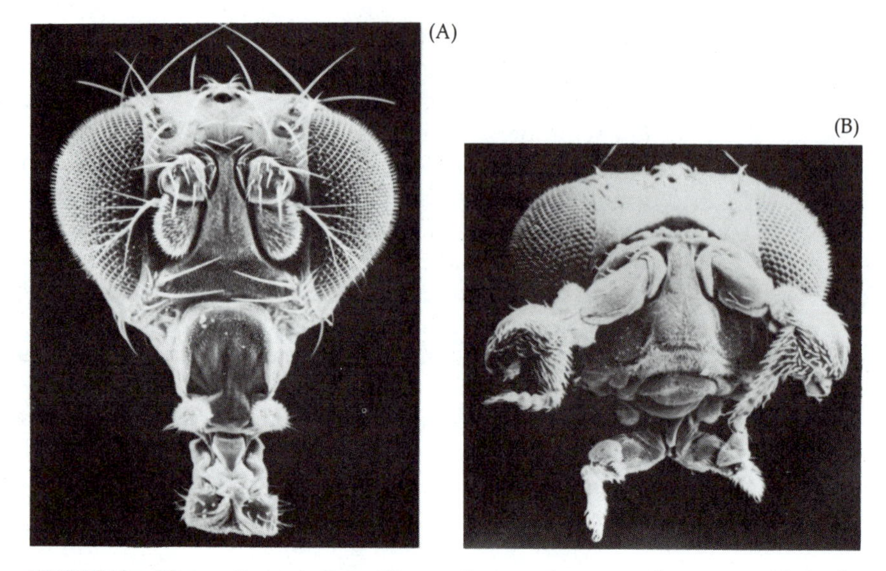

FIGURE 8. Homeotic mutations. Homeotic mutations transform a specific body part of one segment into the homologous part of another segment. (A) The antennae of the wild-type fruit fly on the left are normal; (B) those of the mutant fruit fly on the right have been transformed into legs. This bizarre fly carries an extreme form of a dominant gain-of-function mutation at the *Antennapedia* locus, which results in the ectopic expression of Antennapedia protein. (From J. H. Postlethwait, 1978. *The Genetics and Biology of Drosophila,* Vol. 2C. Academic Press, New York.)

procedures. These studies have indicated that the homeobox consists of three closely linked α helices, the third of which (the "recognition" helix) interacts with DNA at multiple sites. The structure of the homeobox of the engrailed (en) protein of *Drosophila,* as co-crystallized with one of its recognition sites, is illustrated in Figure 9.

Although mutation or aberrant expression of different homeobox proteins results in very different developmental phenotypes, we do not yet have a good biochemical understanding of the mechanisms that underlie this developmental specificity. One feature of many homeobox proteins that distinguishes them from other transcription factors—and that makes them difficult to characterize—is the relatively promiscuous way in which they bind DNA in vitro. Unlike members of the steroid receptor family, for example, many homeobox-containing proteins do not exhibit great selectivity or specificity with regard to the sequence of their binding sites, at least as assayed in isolation (without other DNA binding proteins) in vitro. The products of the *Drosophila engrailed* and *fushi tarazu (ftz)* genes are two such proteins. These proteins each recognize the apparently distinct sequences 5'TCAATTAAATGA3' and 5'TAATAATAATAA3'; other structurally divergent homeodomains also bind these same elements. The equilibrium dissociation constants for these binding reactions are typically 10^{-8}–10^{-9} *M,* as opposed to the 10^{-11}–10^{-12} *M* dissociation constants observed for many other purified transcription factors. In isolation in vitro, most homeobox proteins appear to bind DNA as a monomer, which may account for their reduced template affinities.

In contrast to proteins such as en and ftz, the differing developmental activities of certain other homeobox proteins are correlated with differing specificities of DNA binding. The paired and bicoid proteins, for example, recognize distinct DNA binding sites as a result of a specific difference in the ninth residue of the recognition helix of the homeodomain. Similarly, widespread ectopic expression of the homeobox proteins Deformed (Dfd) and Ultrabithorax (Ubx) in developing flies yields very different developmental phenotypes that seem to be almost entirely a function of the Dfd and Ubx homeoboxes. (Exchanging homeoboxes between these proteins reverses their developmental phenotypes.) The specificity of action of homeobox proteins remains a subject of active research. Studies of a distantly

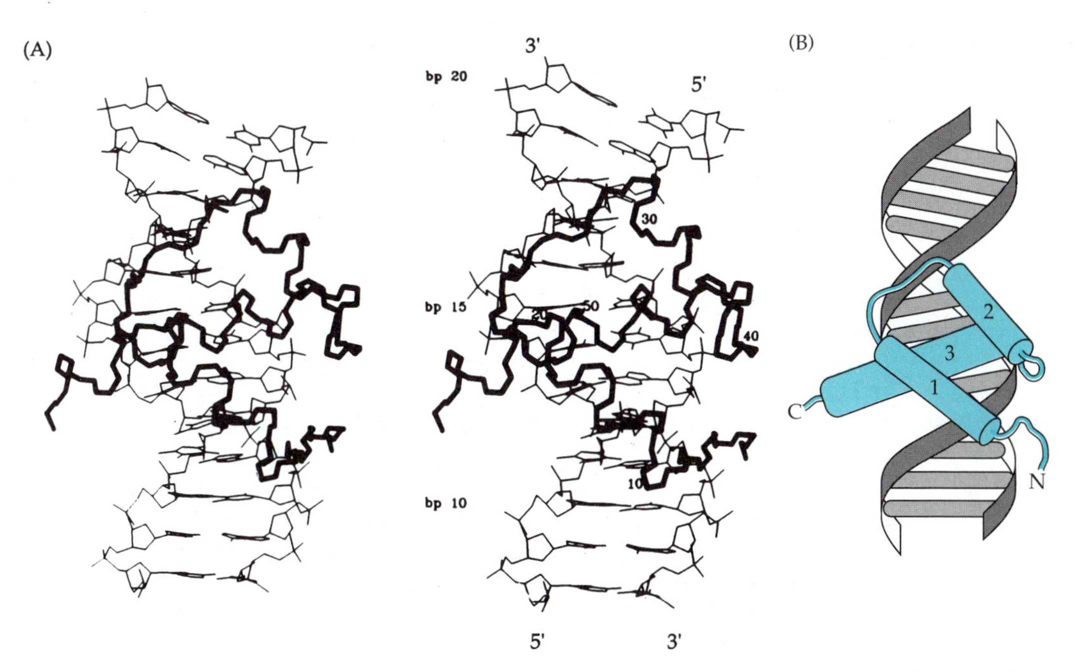

FIGURE 9. The homeobox. Homeotic mutations occur in genes that encode a special class of transcription factors. These proteins are related by a highly conserved DNA binding domain, the homeobox, which is found in every eukaryote from yeast to man. The 60 amino acids of the homeobox are organized into a versatile DNA binding motif of three basic α helices, as illustrated here in a stereo diagram and schematic of the X-ray crystal structure of the homeobox of the *Drosophila* engrailed protein, as co-crystallized with a synthetic binding site. (A) Stereo diagram that shows how the three α helices and the N-terminal arm are arranged in the engrailed homeobox–DNA complex. This diagram shows only the backbone atoms for the protein. Every tenth residue is numbered, as are base pairs 10, 15, and 20 of the engrailed binding site that was co-crystallized with the protein. (B) A schematic that summarizes the relationship of the engrailed α helices and N-terminal arm with respect to the DNA double helix. Cylinders represent α helices, ribbons represent the sugar–phosphate backbone of the DNA, and bars represent base pairs. (From C. R. Kissinger et al., 1990. Cell 63: 579–590.)

related homeobox protein in yeast (the transcriptional repressor α2) suggest that the DNA binding specificity and developmental activity of these proteins may, as for jun, be improved or modified through association with other transcription factors.

Although their specificity of action is not well understood, homeobox proteins are clearly among the central regulators of a complex cascade that ultimately establishes the segmented body plan of *Drosophila.* As summarized in Chapter 11, they control both the early progressive subdivision of the embryo into segments and the subsequent cellular specializations that establish the unique identity of individual segments. As is also described in Chapter 11, a number of experiments in both invertebrates and vertebrates have suggested that these proteins also play a central role in the early developmental segregation of the nervous system, particularly in the specification of the anterior–posterior neural axis and the subsequent specification of individual neural fates. Although mutational analyses in *Drosophila* have partially defined a regulatory hierarchy for the developmental cascade that homeobox-containing proteins generate, in most cases we do not yet have a detailed biochemical understanding of their action in the sense of knowing target sites within the regulatory regions of putative effector genes.

POU proteins are cell-specific regulators that contain a variant of the homeobox

We have a much more satisfying understanding of the target–effector relationship for a set of variant homeobox proteins, which were first identified in mammalian cells. These regulators have been designated **POU** (pronounced "pow") **proteins,** an acronym for *P*it-1/GHF-1, *O*ct-1 and -2, and *u*nc-86, the first four such proteins characterized and cloned. Except for unc-86, a *C. elegans* protein whose mutation leads to selective neuronal loss and the generation of "*unc*oordinated" worms, these proteins were not identified through genetic experiments. Rather, they were uncovered in the course of biochemical studies of cell-specific gene expression. One of the outstanding problems of eukaryotic molecular biology has long been the restricted expression of genes, all of which (to a first approximation) are encoded in all of the cells of the body. Why, for example, is the β-globin gene highly expressed only in mature erythrocytes, immunoglobulin genes only in B lymphocytes, and myelin genes only in Schwann cells and oligodendrocytes?

Several lines of evidence suggest that for certain genes, cell-specific expression is mediated largely by the prior expression of cell-specific transcription factors. Pit-1/GHF-1, for example, was identified, biochemically purified, and cloned as the principal cell-specific activator of the prolactin and growth hormone genes expressed by lactotrophs and somatotrophs of the anterior pituitary, respectively. The target genes for Pit-1/GHF-1 were known prior its cloning, as were the number, position, and relative importance of the Pit-1/GHF-1 binding sites located within the regulatory regions of these genes. Indeed, oligonucleotide probes corresponding to these binding sites were essential reagents in the expression cloning of Pit-1/GHF-1 cDNAs (see Box C). Similarly, Oct-2 was first studied as the principal cell-specific regulator that binds to the "octamer motif"—5'ATGCAAAT3'—contained within the immunoglobulin heavy chain enhancer. (Oct-1 was identified as a more widely expressed protein that also binds to the octamer motif, which is found in genes other than those encoding immunoglobulins.)

These POU proteins define yet another large transcription factor family, which is a spin-off of the homeobox proteins discussed above. Like Antp, Ubx, Dfd, etcetera, POU proteins contain a homeobox that mediates DNA binding. This homeobox is only distantly related to the Antp-class domains, however; about 20 of the 60 amino acids are conserved between the two sets of proteins. The main feature of POU proteins that distinguishes them from Antp-class homeobox proteins is the presence of an additional conserved region: the "POU-specific domain." This ~80 amino acid region is located just upstream of the POU homeobox, is highly conserved among POU proteins, and contributes to the specificity of DNA binding. The POU-specific domain and the POU homeobox, together with an intervening linker region, constitute the POU domain. Unlike most of the Antp-class homeoboxes, the natural DNA binding sites for several POU domains are known with relative certainty. These sites have weak dyad axes of symmetry, and for Pit-1/GHF-1 and Oct-2, there is in vitro evidence that the proteins can bind to them as dimers.

The nearly ubiquitous POU protein Oct-1 provides a compelling example of transcription factor interaction. As noted above, this protein cooperates with the herpes simplex virus transactivator VP16, which is incapable of binding DNA or activating transcription on its own. When complexed with Oct-1, however, VP16 is an exceptionally potent transactivator of a wide variety of genes. The homeobox of the Oct-1 POU domain is required both for binding to DNA and for complex formation with VP16. (Different amino acids of the Oct-1 homeobox α helices mediate the two processes.) VP16 uses an amino terminal domain for Oct-1 binding; a carboxy terminal VP16 acidic domain then apparently contacts proteins of the core promoter and thereby activates transcript initiation. Similar stretches of acidic amino acids—sometimes referred to as "acidic blobs"—are found in many other transcription factors, where they also function as transactivation domains.

Several different lines of experiment suggest that POU proteins can be important regulators of cell-specific gene expression. For example, hybrid reporter constructs containing the prolactin or growth hormone promoters are not transcribed when transfected into HeLa cells *unless* an expression plasmid encoding Pit-1/GHF-1 is simultaneously introduced into the cells. Similar results are obtained with immunoglobulin gene regulatory regions and Oct-2 (see Figure 10). In the mouse, point mutations in the third α helix of the homeodomain of Pit-1/GHF-1 result in dwarf animals. These mutations destroy the ability of the protein to bind to the growth hormone and prolactin promoters, but do not obviously effect the transcription of non-pituitary genes.

Most POU proteins appear to be expressed in only a limited set of

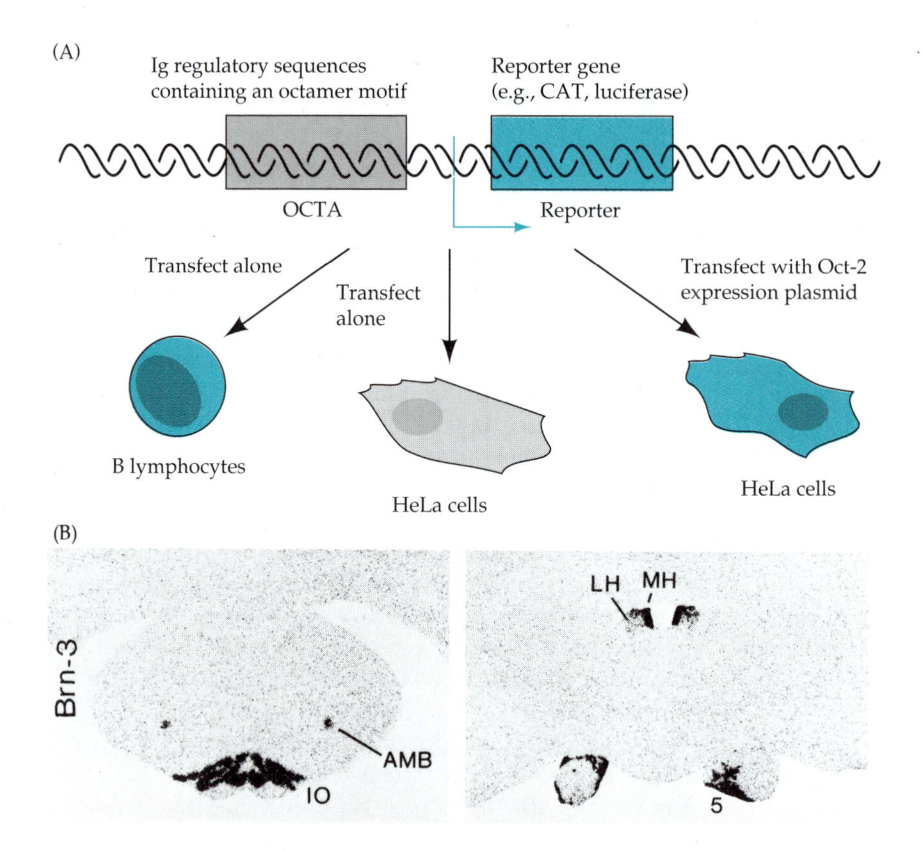

FIGURE 10. The POU proteins. (A) Cell-specific transactivation. POU proteins contain a bipartite DNA binding domain made up of a divergent homeobox and an upstream region called the POU-specific domain. Most of these proteins are expressed only in restricted sets of cells, where they mediate cell-specific gene expression. In the example shown above, hybrid reporter constructs containing immunoglobulin regulatory sequences are expressed when transfected into Oct-2$^+$ lymphoid cells (B lymphocytes). However, these constructs are expressed in non-lymphoid (HeLa) cells *only* if an Oct-2 expression construct is co-transfected. This regulation can be shown to depend on the octamer motif present in the immunoglobulin regulatory sequences: mutation of this motif destroys the ability of co-transfected Oct-2 to activate transcription of the reporter gene. (B) Neural expression. Many recently identified POU proteins are predominantly expressed in subsets of cells in the nervous system. These in situ hybridization profiles through the adult rat brain illustrate the expression pattern of mRNA encoding the POU transcription factor Brain-3 (Brn-3). IO, inferior olive; AMB, nucleus ambiguus; LH, lateral habenula; MH, medial habenula; 5, trigeminal ganglion (B from X. He et al., 1989. Nature 340: 35–41.)

terminally differentiated cells, where they activate transcription of cell-specific genes. This has been demonstrated for Pit-1/GHF-1 and Oct-2, and inferred from the markedly restricted expression of more recently identified POU proteins (see Figure 10). However, some POU proteins appear to function in a limited set of cells earlier in development, prior to terminal differentiation. Oct-3/4, for example, is expressed only by pluripotent cells early in embryogenesis; soon after gastrulation, expression of Oct-3/4 is extinguished in all except germ cells. In the peripheral and central nervous systems, the POU protein SCIP is expressed by rapidly dividing glial precursor cells—proliferative Schwann cells and O-2A cells, respectively (see Chapter 11). When these cells stop dividing and differentiate into myelinating cells (myelin-forming Schwann cells and oligodendrocytes respectively), SCIP expression is dramatically reduced.

Transcription factors of the HLH family are master developmental regulators

Is it possible for one or a small number of transcription factors to directly program the developmental fate of a cell? Remarkably, the answer to this question is yes. In the course of an analysis of muscle differentiation in vitro, expression of a single transcription factor, named **MyoD,** was found to be sufficient to convert a mouse embryo fibroblast cell line (10T1/2) into myoblasts. Transfection-mediated expression of a MyoD cDNA alone initiates a cascade of gene expression that ultimately results in the expression of muscle-specific genes such as the muscle creatine kinase (MCK) gene and the fusion of transfected cells into functional myotubes.

Comparative structural analyses demonstrated that MyoD shares a region of homology with the human κ immunoglobulin enhancer binding proteins E12 and E47, with c-*myc* (like *fos* and *jun,* a nuclear proto-oncogene), and with several neural determination proteins identified in *Drosophila,* including those encoded by the *achaete-scute* complex, *daughterless, hairy,* and *extramacrochaete* genes. These genes, several of which are discussed in Chapter 11, encode a transcription factor family, called the helix-loop-helix, or HLH, family. This name derives from the fact that the conserved regions of these proteins fall into segments that form two amphipathic α-helices of 12–15 amino acids connected by a nonconserved "loop" region of variable length. A consistent feature of the MyoD, c-myc, E12/E47, and *Drosophila* proteins is that the hydrophobic amino acids of their helices (valine, leucine, isoleucine, etc.) all fall along one face of each helix (Figure 11).

This arrangement is of course similar to the alignment of hydrophobic leucine residues in the leucine zippers of fos, jun, and CREB. Perhaps not surprisingly then, members of the HLH family dimerize, and just as dimerization of fos and jun is mediated through their leucine zipper helices, so dimerization of HLH proteins is mediated by the helices of the HLH motif. As for the fos/jun leucine zipper, deletion of the HLH helices blocks protein dimerization, DNA binding, and transcriptional regulation. The HLH proteins therefore provide another example of transcription factors for which dimerization is a prerequisite for transcriptional activation. Consistent with this requirement, the recognition elements to which HLH proteins bind are all at least partially palindromic. In general, HLH heterodimers (between, for example, MyoD and E12/47) bind DNA with a higher affinity than do homodimers.

The dimerization of HLH proteins is also an important feature of the regulation of their transcriptional activity. MyoD, for example, appears to

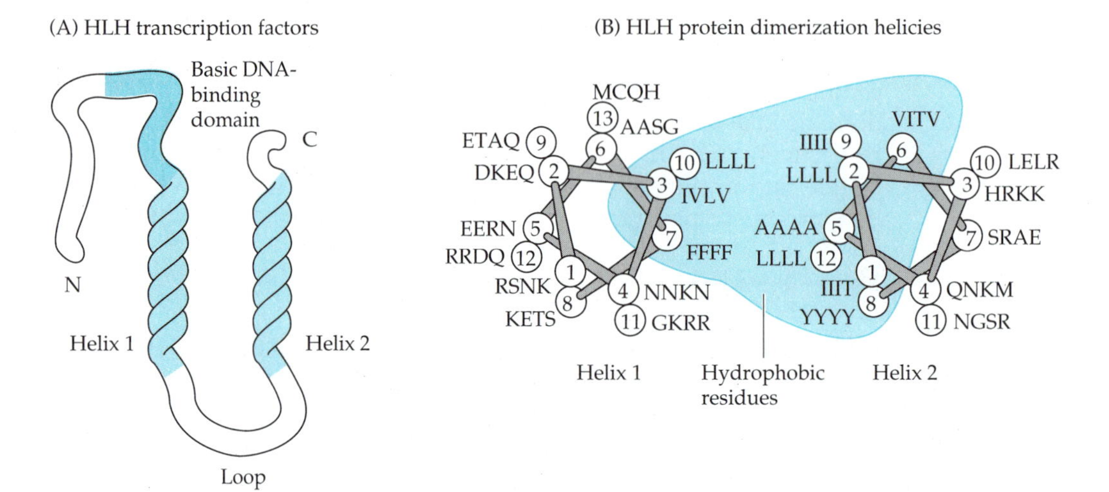

FIGURE 11. The helix-loop-helix motif. The helix-loop-helix family includes a large number of transcription factors that share a protein dimerization motif. This motif consists of two α helices (each of approximately two turns) joined by a loop of variable length. As for proteins of the leucine zipper family (e. g., fos and jun), the dimerization helices of the HLH proteins are immediately preceded by a region of basic amino acids that mediate DNA binding. (A) Structure of a typical HLH protein. An N-terminal regulatory region is followed by a positively charged DNA binding domain of 14 amino acids, an α helix of 15 amino acids, an intervening loop of 10–24 amino acids, a second helix of 15 amino acids, and a very short C-terminal domain. (B) A helical wheel analysis of the two dimerization α helices of the E12, MyoD, c-myc, and T5 achaete-scute HLH proteins. The first 12 amino acids in the helices of these proteins are displayed. (As in Figure 4, the helical wheels spiral into the plane of the page.) Note that the highly conserved hydrophobic residues I, L, V, A, Y, and F align along one face of each helix.

drive the transcription of a wide variety of muscle-specific, differentiation-specific genes (such as the MCK gene) through the direct binding of MyoD-E12/47 heterodimers to the regulatory regions of each of these genes; i.e., all of the genes have MyoD-E12/47 binding sites. Yet in proliferating myoblasts (prior to their fusion into myotubes), MyoD and E12/47 are expressed at high levels at the same time that MCK and other muscle structural genes are transcriptionally silent. When myoblasts in culture are induced to exit the cell cycle, fuse, and differentiate into myotubes, these structural genes are activated. During this terminal differentiation, the levels of MyoD and E12/47 do not change appreciably; they remain high. These data have been reconciled by the discovery that some members of the HLH family function as negative regulators by forming nonfunctional heterodimers with HLH activators such as MyoD and E12/47. These negative regulators include the *Drosophila extramacrochaete* and *hairy* genes, which are important for neural development (see Chapter 11), and a protein—named Id—that blocks the activity of MyoD in muscle development. In general, they lack a region of highly basic amino acids, immediately upstream of the HLH motif, that is known to mediate DNA

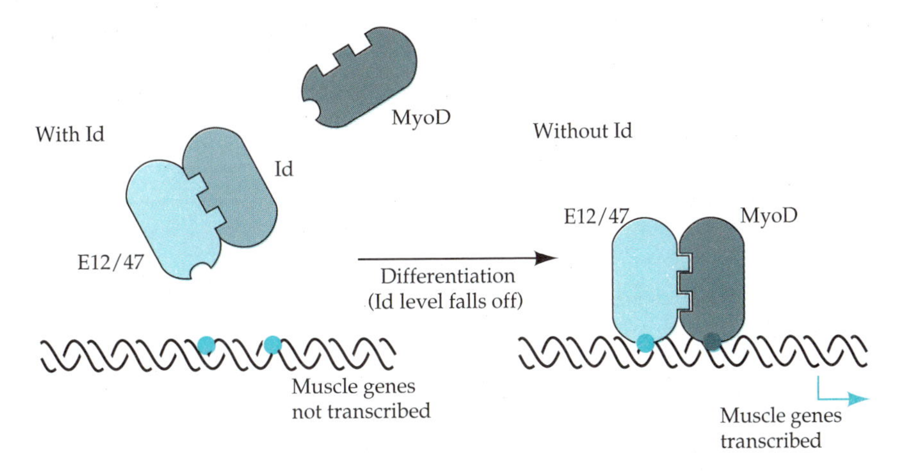

FIGURE 12. Poison partners: the developmental regulation of HLH binding. Several HLH proteins, including the Id and extramachrochaete proteins, appear to act as negative regulators of HLH activity by forming nonproductive heterodimers with HLH activators. In muscle differentiation, for example, it is hypothesized that Id, which lacks a basic DNA binding domain, forms a heterodimer with E12/47 that is incapable of binding DNA or activating transcription. It is further hypothesized that Id effectively titrates out available E12/47 and thereby prevents the productive interaction of E12/47 and MyoD that is required for the activation of muscle-specific genes. Concomitant with muscle differentiation, Id levels fall, allowing for the formation of transcriptionally active E12/47-MyoD heterodimers.

binding. (Note that, just as for fos, jun, and CREB, the highly basic DNA binding domain of HLH activators is located immediately upstream of dimerization α helices.) In effect, Id and its ilk act as "poison partners" (Figure 12). In proliferating myoblasts, expression of Id is extinguished upon terminal differentiation into muscle, and, if expression of the protein is artificially maintained at high levels, muscle differentiation is blocked.

Pre-mRNA Splicing

Pre-mRNA splicing requires a complex multi-protein assembly

Even after all of the remarkable transcriptional regulators discussed above have ensured that a new mRNA molecule is synthesized in the right cells, at the proper time in development, and in response to the appropriate extracellular stimuli, the job of gene regulation is still not finished. This is because the DNA sequence of most of the protein-encoding genes of eukaryotes is not co-linear with the sequence of their mature mRNAs. Instead, these genes are split into segments, such that regions of coding sequence are interrupted by one or more regions of intervening noncoding sequence. These coding and noncoding regions are designated **exons** and **introns**, respectively (Figure 13). The presence of introns means that a newly transcribed pre-mRNA contains sequence information that is irrelevant to the structure of the encoded protein. To generate a mature mRNA, the introns present in the pre-mRNA transcript must therefore be precisely

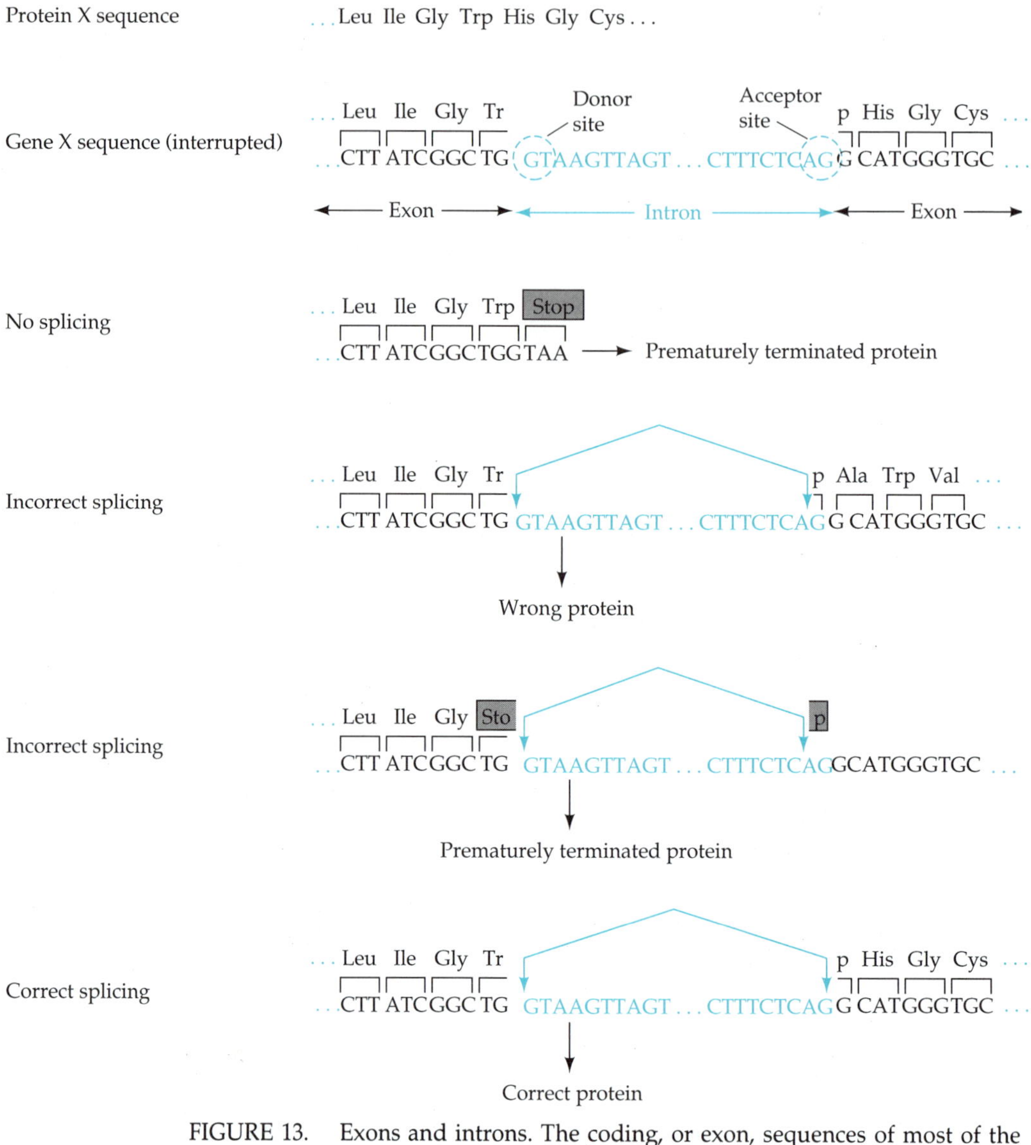

FIGURE 13. Exons and introns. The coding, or exon, sequences of most of the genes of higher eukaryotes are interrupted by noncoding, or intron, sequences. The production of a correct transcript from these interrupted genes requires the precise excision of intron sequences from the precursor messenger RNA. In this example, a region of hypothetical protein X is encoded by interrupted gene X. The intron in this gene occurs between the second and third nucleotides of a tryptophan codon. If the gene X intron is not spliced out of the pre-mRNA, a stop codon is prematurely introduced into the protein X mRNA. If the splicing reaction is not carried out precisely, reading frame errors (that result in an incorrect or prematurely terminated protein sequence) are introduced into the protein X mRNA.

excised, or **spliced** out. There is absolutely no room for error in this splicing operation, since a mistake of even a single nucleotide in the execution of the splice will displace the triplet reading frame of the protein that the mRNA encodes. Although the interruption of coding sequences may seem perverse, there are, as we shall see, good reasons for so doing.

How are splice sites recognized and how is the splicing reaction carried out? Examination of the pre-mRNA sequences of a large number of genes, together with mutational analyses, has identified three essential sequence elements within intron-containing pre-mRNAs (Figure 13). The first of these is the **5' splice site,** or donor site. In mammals, this sequence is (C/A)AG|GURAGU, where | denotes the boundary between exon sequence (left) and intron sequence (right), and R is a purine. Within this consensus sequence, the first two nucleotides of the intron (GU) are highly conserved. Mutation of either of these nucleotides blocks the splicing reaction. The second element required for splicing is the **3' splice site,** or **acceptor site,** which has the sequence Y_nNYAG|G (exon sequence to the *right* of |), where Y is a pyrimidine, N is any nucleotide, and n a number between 10 and 20. In this sequence, only the last two nucleotides of the intron (AG) are absolutely conserved. The splice donor and acceptor sites define the 5' and 3' boundaries of the intron, respectively. A third essential *cis*-acting splicing element, called the **branch point,** is located within the intron. In yeast, this element has the consensus sequence UACUAAC and in mammals the less well-conserved sequence YNYURAY. The branch point is most commonly observed ~30 nucleotides upstream of the 3' splice site, although its exact position is variable. As we shall see, this sequence plays an essential role in one of the intermediates in the multistep splicing reaction, and the conserved A near its 3' end is particularly important for the initial step of the reaction.

The 5' and 3' splice sites and the branch point of the pre-mRNA specifically interact with *trans*-acting elements of the splicing machinery. These include a set of small nuclear RNAs or **snRNAs,** called U1, U2, U4, U5, and U6, some of which are complementary to the sites. The U1 snRNA, for example, hybridizes to the 5' (donor) splice site, and the U2 snRNA recognizes the branch point sequence (see below). The snRNAs associate with a set of specialized nuclear proteins to form ribonucleoprotein complexes or **snRNPs** ("snurps"). These snRNPs assemble with the pre-mRNA and additional proteins to form a large processing structure called the **spliceosome.** The entire splicing reaction is carried out within the spliceosome, which therefore contains the reaction substrate, intermediates, and products. In many respects, the spliceosome is a biochemical machine like the ribosome. Unlike ribosomes, however, spliceosomes are transient structures.

The sequence of events in the splicing reaction (Figure 14) is as follows. The U1 snRNP first associates with and stabilizes the donor (upstream) splice site of the pre-mRNA. Following the binding of the U2 snRNP to the branch point sequence and the subsequent addition of the U4/6 and U5 snRNPs to the spliceosome assembly, the 2' hydroxyl group of the con-

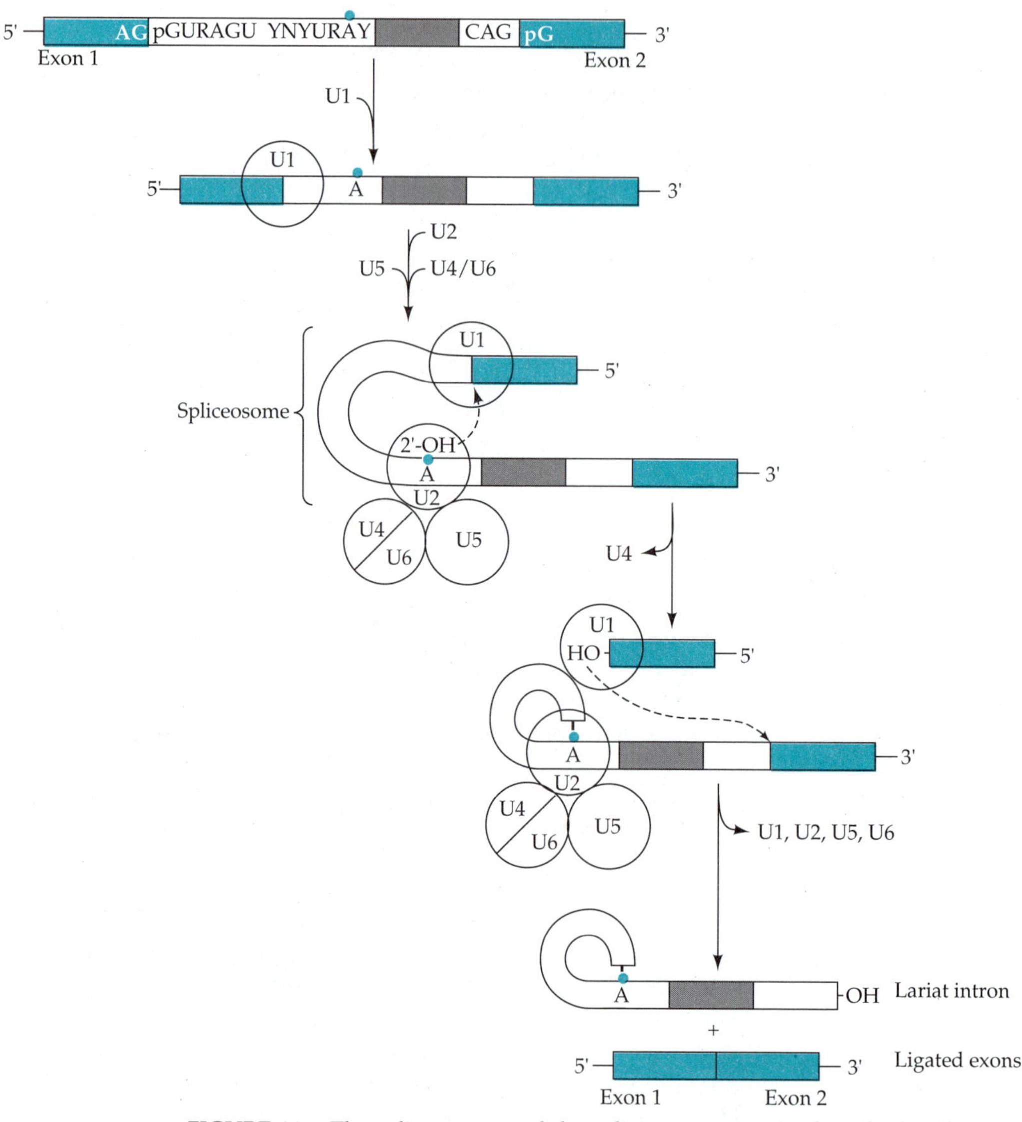

FIGURE 14. The spliceosome and the splicing reaction. As described in the text, the splicing reaction involves a sequence of transesterification reactions that occur in a multiprotein assembly called the spliceosome. In this diagram, exons are indicated by colored boxes, introns by white boxes, the polypyrimidine tract of the splice acceptor site by a gray box, and snRNPs by circles. The sequence of splice donor, splice acceptor, and branch point sites are indicated. The conserved branch point A base, which initiates nucleophilic attack at the 5' (donor) splice site, is marked with a dot. (After C. W. J. Smith et al., 1989. Annu. Rev. Genet. 23: 527–577.)

served A base of the branch point is positioned to attack the donor splice site (see Figure 14). This attack proceeds via a nucleophilic transesterification reaction and cleaves the upstream exon from the remainder of the pre-mRNA. The newly generated free 3' hydroxyl group of the last base of this exon is now free to attack the acceptor (downstream) splice site. Although the mRNA encoded by the upstream exon is temporarily separated from the remainder of the pre-mRNA molecule, it is held in place for this second nucleophilic attack (also a transesterification) by the proteins of the spliceosome. Attack at the acceptor splice site links the upstream and downstream exons and frees the excised intron, which takes the form of a lariat structure (see Figure 14). At this point, the spliceosome dissociates, and the linked exons, lariat intron, U1-6 snRNPs, and remaining spliceosomal proteins are thereby released. The entire process is then reapplied to a new set of splice sites through the sequential reassembly of the spliceosome.

Pre-mRNA splicing is a variable and regulated phenomenon

The main reason that biologists have become preoccupied with RNA splicing is that for many of the genes that contain multiple introns, the pattern of pre-mRNA splicing is not the same in all cells. That is, many pre-mRNA molecules are subject to regulated or **alternative splicing.** Alternative pre-mRNA splicing may take one of several forms and may be specific to certain cells (e.g., neurons) and/or to certain periods of development. The consequences of alternative splicing can be profound, and include alteration of protein localization, modification or loss of protein function, and even alteration of protein identity. In this way, alternative splicing greatly increases the diversity of products that may be encoded in a single gene, a feature of particular importance to the nervous system.

Seven different modes of alternative splicing, schematized in Figure 15, have been described. Perhaps the simplest splicing mechanism involves alternative intron retention, in which an intron is sometimes included in a mature mRNA molecule, and sometimes not. Examples of splice/don't splice decisions are relatively infrequent but are found in the processing of the PDGF A chain and *Drosophila* P element *transposase* genes. A second form of differential splicing involves the use of alternative 5' or 3' splice sites, which is frequently observed in the processing of viral pre-mRNAs. In eukaryotes, a particularly dramatic example of this form of splicing is provided by sex determination in *Drosophila,* a developmental process that is controlled by a regulated cascade of alternative splice donor/acceptor choices (see below). Yet a third set of variations on the alternative splicing theme involves the differential expression of sequences at the extreme 5' and 3' ends of a gene. The former is observed upon the differential use of alternative promoters that are associated with their own exons. The latter is observed upon the use of alternative mRNA termination/polyadenylation sites that are also associated with their own exons. One of the very first examples of alternative splicing to be recognized in the nervous system—the alternative expression of either calcitonin or the calcitonin gene-related peptide (CGRP; see Box C in Chapter 4 and below)—involves the differential use of alternative 3' end exons. Finally, there are two ways in which exons may be selectively excluded from a mature mRNA. The first of these involves the processing of internal, mutually exclusive exons. In this form of alternative splicing, only one member of a pair or set of exons is spliced

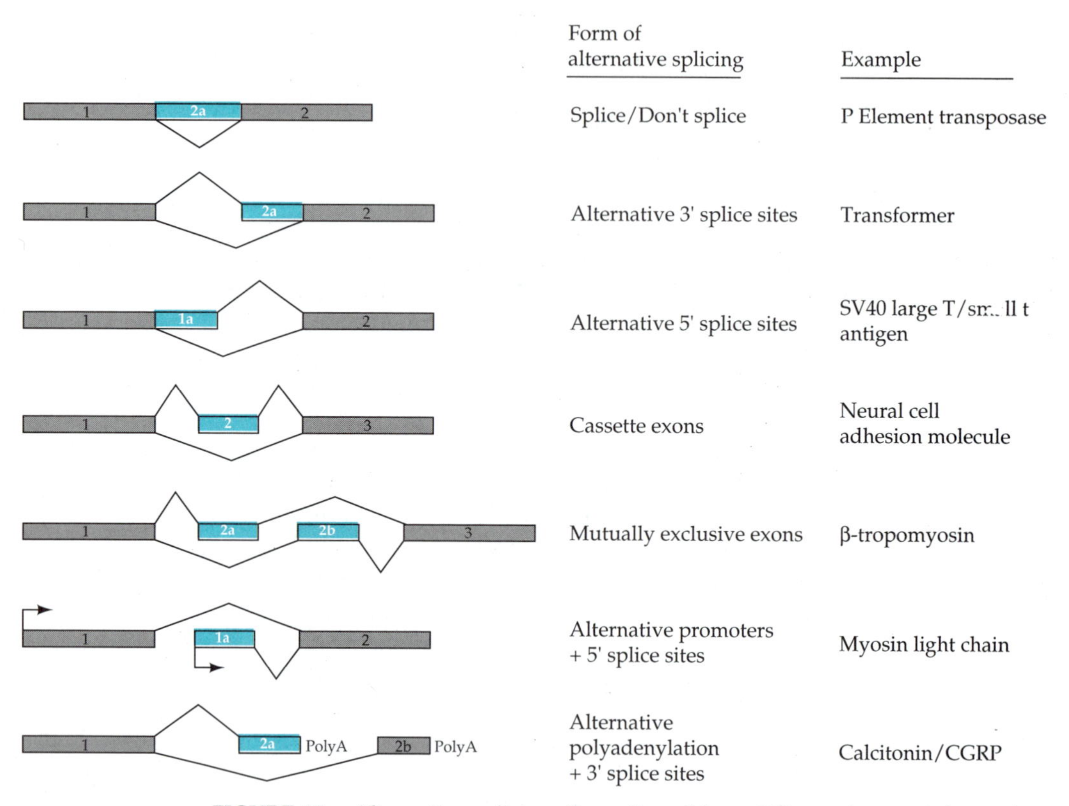

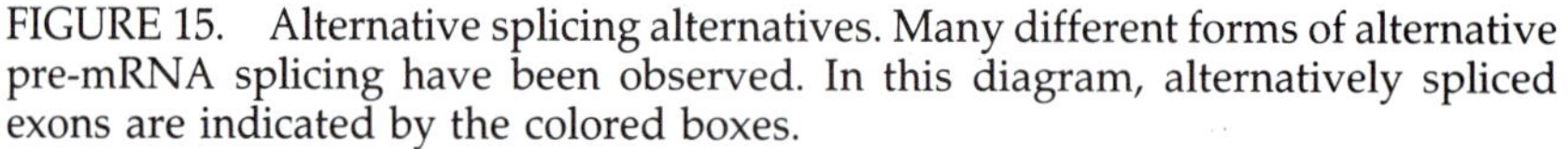
FIGURE 15. Alternative splicing alternatives. Many different forms of alternative pre-mRNA splicing have been observed. In this diagram, alternatively spliced exons are indicated by the colored boxes.

into the mature mRNA. A well-known example of this form of splicing is the alternative processing of pre-mRNAs encoding β-tropomyosin. In fibroblasts, mature β-tropomyosin mRNAs contain sequence encoded in exons 1, 2, 3, 4, 5, 6, 8, 9, and 11, while in skeletal muscle mature mRNAs contain exons 1, 2, 3, 4, 5, 7, 8, 9, and 10. Splicing of the 6/7 and 10/11 exon pairs is thus mutually exclusive. The other form of alternative internal splicing involves the selective use of cassette exons. These may be included or excluded independently of other exons. In such cases, the reading frame of the protein is usually maintained whether an exon is included or not, although frameshift or premature termination has been observed to result from inclusion or exclusion. Two well-known examples of cassette exons in the nervous system occur in the neural cell adhesion molecule (N-CAM) and myelin basic protein (MBP) genes. These genes have multiple cassettes, and are therefore capable of generating a diverse set of mature mRNAs.

What are the consequences of alternative splicing?

As noted above, alternative splicing is used as a means toward multiple ends. One of these is differential protein localization. This form of regulation was first recognized in the immune system. In immature B lymphocytes, the immunoglobulin μ heavy chain (IgM) gene is expressed as a membrane-bound protein. Upon antigen activation and B cell maturation, the amount of membrane-bound IgM dramatically decreases at the same time that a secreted form of IgM increases. The switch between these two forms of IgM is achieved through the alternative use of 3' end exons, one of which encodes a hydrophobic membrane anchor and the other of which does not. Within the nervous system, alternative splicing of the cassette exons of the N-CAM gene results in the production of N-CAM isoforms that differ in their cytoplasmic and membrane attachment domains. One of these isoforms lacks a transmembrane domain entirely, but retains recognition sequences that allow the protein to be attached to the plasma membrane via a phosphatidylinositol linkage. A secreted form of N-CAM is generated by the inclusion of yet another cassette exon that contains a stop codon within sequences that encode the extracellular domain of the protein.

In addition to alteration of protein localization, there are many instances in which alternative splicing is used to modulate or ablate protein function. The fast muscle troponin T (TnT) gene, for example, contains 5 small cassette exons that are spliced into mature mRNAs independently of one another. The multiple troponin T isoforms that are generated from alternative splicing of these exons differ in their ability to associate with tropomyosin, and the muscles that contain them have different activation profiles in response to Ca^{2+} mobilization. In the nervous system, one of the best known examples of the role that alternative splicing may play in modulating protein function is provided by the potassium channels encoded in the *Shaker* gene of *Drosophila.* As a result of alternative splicing, this gene is able to encode a large number (~20) of different A_1-type potassium channel transcripts. As discussed in detail in Chapter 3, the proteins encoded by these transcripts all share a conserved central amphipathic region (the S4 region) that appears to act as the voltage sensor for the channel. They differ, however, as a result of exon choice in both their amino terminal and carboxy terminal regions. These differences in N- and C-terminal sequence in turn translate into differences in channel physiology, most notably into differences in the gating kinetics of the channel.

In a few cases, alternative splicing is capable of generating polypeptides of completely different structure and function. Perhaps the best known of such cases occurs in the nervous system, in the differential processing of pre-mRNAs encoded by the calcitonin/CGRP gene (Chapter 4). Calcitonin is produced in the thyroid, and acts as a circulating regulator of calcium homeostasis, whereas CGRP is produced by neurons, and is locally active as a neuromodulator and neurohormone. In most neurons, the pre-mRNAs transcribed from the calcitonin/CGRP gene are spliced to contain exons 1, 2, 3, 5, and 6, whereas in most non-neural cells in which the gene is expressed (e.g., thyroid cells), the pre-mRNA is spliced to contain exons 1, 2, 3, and 4. Exons 1–3 encode 5' untranslated and common coding sequences, exon 4 encodes the calcitonin peptide sequence and 3' polyA addition site used in nonneural cells, and exons 5 and 6 encode the CGRP coding and 3' polyA addition sequences expressed in neurons. The variable sequences of the mature calcitonin and CGRP peptides are entirely contained within the alternatively spliced exons 4 and 5/6, respectively; alternative splicing is therefore responsible for the generation of completely different peptides.

How is alternative splicing regulated?

Although there has been extensive characterization of differences in pre-mRNA splicing between different cells at different times in development, we know considerably less about the mechanisms that underlie alternative splicing. How is it, for example, that the calcitonin/CGRP gene encodes calcitonin in thyroid cells and CGRP in neurons? The most direct insights into the regulation of alternative splicing have been obtained through molecular and genetic analyses of the somatic sex determination pathway in *Drosophila.* This cascade is initiated by a central control gene, called *Sex-lethal (Sxl),* which activates a downstream gene called *transformer (tra)* (see Figure 16). In conjunction with *transformer-2 (tra-2), tra* acts to establish the female mode of the *double-sex (dsx)* gene. The female form of the *dsx* gene product acts to repress genes responsible for male differentiation. In the absence of tra, *dsx* is expressed in its male form, which acts to repress genes that lead to female differentiation. At each stage of this cascade (except for *dsx*), gene activity is required only for the development of female flies—male development is therefore the default pathway.

Remarkably, the structure and function of the *Sxl, tra,* and *dsx* gene products are controlled by alternative RNA splicing. Early in *Drosophila* development, the initial activity of the *Sxl* gene is set by the X chromosome to autosome ratio. At this time, the activation of a female-specific *Sxl* promoter ensures, by an unknown mechanism, that the *Sxl* pre-mRNA is properly spliced to produce an active Sxl protein (see below). This means that when a second, non-sex-specific promoter is subsequently activated in development, Sxl protein is only present in female flies. This is important because the splicing of the *Sxl* pre-mRNA transcribed from this non-sex-specific promoter is regulated, and this regulation is effected by the Sxl protein itself! In the absence of Sxl (in males), the *Sxl* pre-mRNA is spliced to include exon 3 (Figure 16). This exon contains a stop codon, and its inclusion generates a prematurely terminated, inactive *Sxl* protein. In female flies, on the other hand, exon 2 of the *Sxl* pre-mRNA is spliced directly to exon 4, a process that skips over the stop codon and thereby generates a full-length, functional Sxl protein. In a similar fashion, *Sxl* regulates splicing of the *tra* pre-mRNA. In male flies (which lack Sxl), the *tra* pre-mRNA is spliced to include exon 2a, which (like exon 3 of the *Sxl* gene) contains a stop codon. In female flies, however, an alternative splice to an internal site that defines *tra* exon 2b is performed. This splice site is downstream of the stop codon in exon 2a, and the splicing therefore generates an mRNA that encodes a full-length, active tra protein. Finally, in the presence of tra protein, exon 3 of the *dsx* pre-mRNA is spliced directly to exon 4 to generate a protein that represses expression of male-specific genes. Conversely, in the absence of *tra,* exon 3 is spliced to exons 5 + 6 to generate a dsx protein that represses expression of female-specific genes.

A combination of genetic and biochemical experiments have demon-

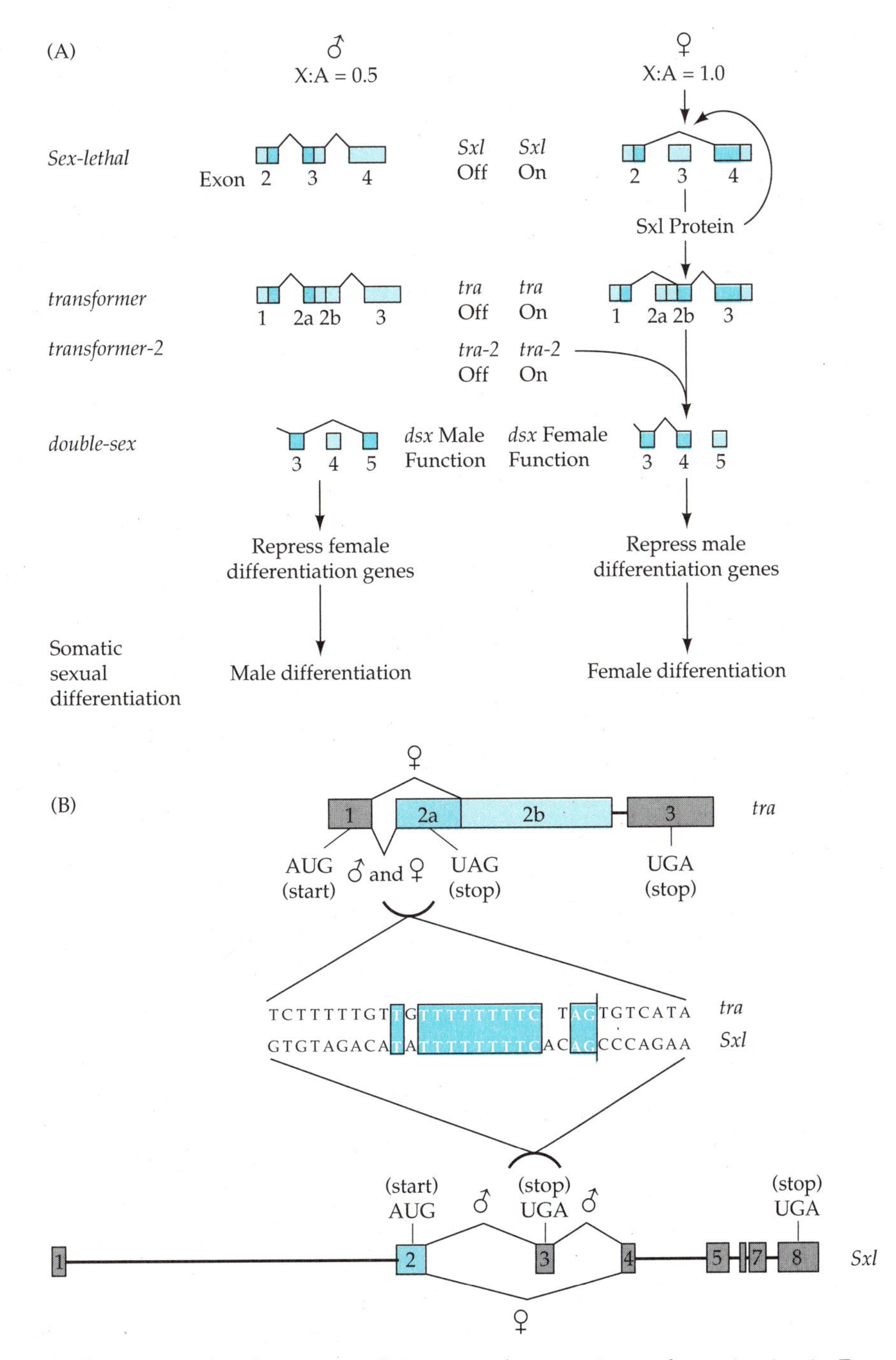

FIGURE 16. An alternative splicing cascade: somatic sex determination in *Drosophila.* (A) The determination of somatic sex in *Drosophila* is controlled initially by the sex chromosome to autosome ratio (X:A) and subsequently by a regulated cascade of alternative pre-mRNA splicing. For most of this cascade, gene activity is required only in female (♀) and not male (♂) flies. See text for details. (B) Conserved nucleotide sequences at the non–sex-specific and male-specific 3' splice sites of the *Drosophila tra* and *Sxl* genes, respectively. Constitutive splicing events in these genes are indicated by straight lines between exons. Conserved splice site nucleotides are boxed, and the border between intron and exon is marked by a vertical line. (A after C. W. J. Smith et al., 1989. Annu. Rev. Genet. 23: 527–577. B after M. McKeown, 1990. Genet. Eng. 12: 139–181.)

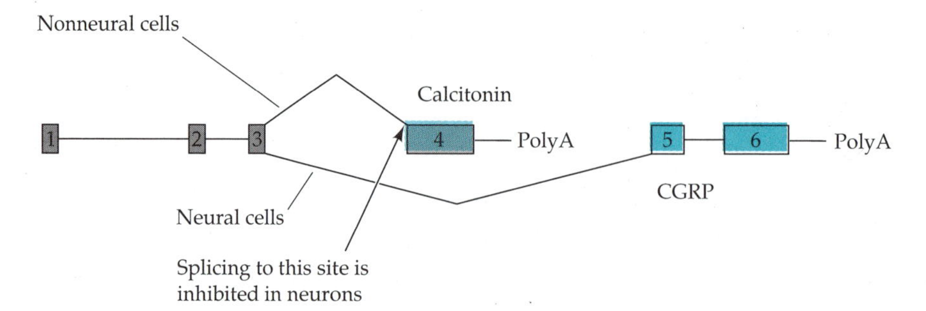

FIGURE 17. Neural-specific splicing: the calcitonin/CGRP gene. The calcitonin/CGRP gene is alternatively spliced to produce calcitonin in non-neural cells and CGRP in neural cells. This regulation is brought about by neural-specific inhibition of splicing to exon 4, which encodes calcitonin.

strated that the Sxl protein exerts its influence by inhibiting splicing to the male-specific splice acceptor site in the *Sxl* gene, and the non-sex-specific site of the *tra* gene. Both intron 2 of the *Sxl* gene and intron 1 of the *tra* gene, for example, carry a conserved sequence immediately upstream of the male-specific exons 3 and 2a, respectively (see Figure 16). Mutational analyses have demonstrated this conserved acceptor sequence is the site of inhibition of male-specific *tra* splicing in female flies. These findings demonstrate that the male versions of *Sxl* and *tra* splicing are constitutive, and offer a biochemical explanation for the fact that maleness is the default pathway of fly development.

Within the nervous system, similar regulatory phenomena have been observed for the neural-specific splicing of the calcitonin/CGRP gene (Figure 17). Splice acceptor sequences immediately upstream of exon 4 (the calcitonin exon) appear to be the regulated sites. When these sequences are replaced with those adjacent to another (unregulated) 3' splice site, exon 4 is spliced into transcripts in both non-neural and neural cells. These and related observations suggest that splicing to the calcitonin-encoding exon 4 is actively inhibited in neural cells, which therefore skip this exon and instead splice to the CGRP-encoding exon 5. They indicate that neural cells are special (at least with regard to this gene) because they inhibit constitutive splicing to non-neural sites. The putative neural-specific splicing inhibitor of the calcitonin/CGRP gene has yet to be identified, and we do not know whether neural-specific inhibition of non-neural splicing extends to the processing of other genes (e.g., the *src* gene) that encode both neural and non-neural transcripts.

Summary

Eukaryotic cells have at their disposal an impressive array of transcriptional and posttranscriptional regulators. These include transcription factors that transduce electrical signals (e.g., nuclear proto-oncogenes); cyclic AMP elevation (e.g., CREB); steroid hormone binding (e.g., the estrogen receptor); transcription factors that regulate development (e.g., Antennapedia and MyoD); and splicing factors that control somatic sex (e.g., Sex lethal). Each of these proteins functions through direct binding to nucleic acids.

The differential activity and expression of transcription factors generates molecular and cellular diversity within multicellular ensembles such as the nervous system. These proteins increase or decrease the rate of transcript initiation by RNA polymerase II. They usually bind, as hetero- or homodimers, to specific nucleotide sequences that have a dyad axis of symmetry, through a DNA binding domain that is rich in positively charged amino acids. Although the total number of different transcription factors is very large, these proteins fall into a handful of overlapping regulatory families that are defined by the conserved structural features of DNA binding, regulatory, transactivation, and protein dimerization domains. Transcription factor families defined by DNA binding domains include the zinc finger and homeobox families; those defined by dimerization motifs include the leucine zipper and helix-loop-helix families; and those defined by regulatory domains include the steroid and retinoic acid receptor families.

Molecular diversity is generated not only by differential cell-specific gene expression, but by differential processing of the same primary transcript in different cells or at different times in development. Alternative pre-mRNA splicing is used to alter the subcellular location and modulate the function of proteins, and, in some cases, to determine which of two structurally distinct proteins (encoded by a single gene) is expressed.

It is easy to be overwhelmed by the sheer number of transcriptional regulators and alternative splicing possibilities. We should bear two things in mind, however. The first is that this number is many orders of magnitude lower than the number of functionally distinct cell types in the vertebrate nervous system. The second is that this number is very much *greater* than the number of regulators expressed by a given (single) neural cell. (A particular neuron in the hippocampus does not express 40 different POU proteins.) These considerations in turn lead to two conclusions: the great cellular diversity of the nervous system is a function of several variables (such as relative cell position) that are not molecular; and the molecular diversity that does exist is likely to be generated through the combinatorial expression of regulators.

References

General references

Berk, A. J. and Schmidt, M. C. 1990. How do transcription factors work? Genes Dev. 4: 151–155.

*McKeown, M. 1990. Regulation of alternative splicing. Genet. Eng. 12: 139–181.

*Mitchell, P. J. and Tjian, R. 1989. Transcriptional regulation in mammalian cells by sequence-specific DNA binding proteins. Science 245: 371–378.

Leucine zipper proteins

Bos, T. J., Bohmann, D., Tsuchie, H., Tjian, R. and Vogt, P. K. 1988. V-jun encodes a nuclear protein with enhancer binding activities of AP-1. Cell 52: 705–712.

Dang, C. V., McGuire, M., Buckmire, M. and Lee, W. M. F. 1989. Involvement of the "leucine zipper" region in the oligomerization and transforming activity of human c-myc protein. Nature 337: 664–666.

Gonzalez, G. A. and Montminy, M. R. 1989. Cyclic AMP stimulates somatostatin gene transcription by phosphorylation of CREB at serine 133. Cell 59: 675–680.

Kouzarides, T. and Ziff, E. 1989. Leucine zippers of fos, jun, and GCN4 dictate dimerization and thereby control DNA binding. Nature 340: 568–571.

*Landschultz, W. H., Johnson, P. F. and McKnight, S. L. 1988. The leucine zipper protein: a hypothetical structure common to a new class of DNA binding proteins. Science 240: 1759–1764.

Montminy, M. R., Sevarino, K. A., Wagner, J. A., Mandel, G. and Goodman, R. H. 1986. Identification of a cAMP-responsive element within the rat somatostatin gene. Proc. Natl. Acad. Sci. USA 83: 6682–6686.

O'Shea, E. K., Rutkowski, R. and Kim, P. S. 1989. Evidence that the leucine zipper is a coiled coil. Science 243: 538–542.

Steroid and related receptors

*Evans, R. M. 1988. The steroid and thyroid hormone superfamily. Science 240: 889–895.

Evans, R. M. and Arriza, J. L. 1989. A molecular framework for the actions of glucocorticoid hormones in the nervous system. Neuron 2: 1105–1112.

Giguere, V., Ong, E. S., Segui, P. and Evans, R. M. 1987. Identification of a receptor for the morphogen retinoic acid. Nature 330: 624–629.

Hollenberg, S. M., Weinberger, C., Ong, E. S., Cerelli, G., Oro, A., Lebo, R., Thompson, E. B., Rosenfeld, M. G. and Evans, R. M. 1985. Identification of a cDNA encoding a functional human glucocorticoid receptor. Nature 318: 635–640.

Homeobox proteins

Gehring, W. 1987. Homeoboxes in the study of development. Science 236: 1245–1252.

Hayashi, S. and Scott, M. P. 1990. What determines the specificity of action of *Drosophila* homeodomain proteins? Cell 63: 883–894.

Kissinger, C. R., Liu, B., Martin-Blanco, E., Kornberg, T. B. and Pabo, C. O. 1990. Crystal structure of an engrailed homeodomain-DNA complex at 2.8 Å resolution. Cell 63: 579–590.

Mann, R. S. and Hogness, D. S. 1990. Functional dissection of Ultrabithorax proteins in *D. melanogaster*. Cell 60: 597–610.

POU domain proteins

He, X., Treacy, M., Simmons, D. M., Ingraham, H. A., Swanson, L. W. and Rosenfeld, M. G. 1989. Expression of a large family of POU domain regulatory genes in mammalian brain development. Nature 340: 35–41.

Monuki, E. S., Kuhn, R., Weinmaster, G., Trapp, B. D. and Lemke, G. 1990. Expression and activity of the POU transcription factor SCIP. Science 249: 1300–1303.

Ruvkun, G. and Finney, M. 1991. Regulation of transcription and cell identity by POU domain proteins. Cell 64: 475–478.

Staudt, L. M., Clerc, R. G., Singh, H., LeBowitz, J. H., Sharp, P. A., and Baltimore, D. 1988. Cloning of a lymphoid-specific cDNA encoding a protein binding the regulatory octamer DNA motif. Science 241: 577–580.

Stern, S., Tanaka, M. and Herr, W. 1989. The Oct-1 homeodomain directs formation of a multiprotein-DNA complex with the HSV transactivator VP16. Nature 341: 624–630.

HLH proteins

*Benezra, R., Davis, R. L., Lockshon, D., Turner, D. L. and Weintraub, H. 1990. The protein Id: A negative regulator of helix-loop-helix proteins. Cell 61: 49–59.

Ellis, H. M., Spann, D. R. and Posakony, J. W. 1990. extramacrochaetae, a negative regulator of sensory organ development in *Drosophila*, defines a new class of helix-loop-helix proteins. Cell 61: 27–38.

Murre, C., Schonleber McCaw, P. and Baltimore, D. 1989. A new DNA binding and dimerization motif in immunoglobulin enhancer binding, daughterless, MyoD, and myc proteins. Cell 56: 777–783.

Murre, C., Schonleber McCaw, P., Vaessin, H., Caudy, M., Jan, L. Y., Jan, Y. N., Cabrera, C. V., Buskin, J. N., Hauschka, S. D., Lassar, A. B., Weintraub, H. and Baltimore, D. 1989. Interactions between heterologous helix-loop-helix proteins generate complexes that bind specifically to a common DNA sequence. Cell 58: 537–544.

Alternative pre-mRNA splicing

Bell, L. R., Maine, E. M., Schedl, P. and Cline, T. W. 1988. *Sex-lethal*, a *Drosophila* sex determination switch gene, inhibits sex-specific RNA splicing. Cell 55: 1037–1046.

Boggs, R. T., Gregor, P., Idriss, S., Belote, J. M. and McKeown, M. 1987. Regulation of sexual differentiation in *D. melanogaster* via alternative splicing of RNA from the transformer gene. Cell 50: 739–747.

Emeson, R. B., Hedjran, F., Yeakley, J. M., Guise, J. W. and Rosenfeld, M. G. 1989. Alternative production of calcitonin and CGRP mRNA is regulated at the calcitonin-specific splice acceptor. Nature 341: 76–80.

Konarska, M. M. and Sharp, P. A. 1987. Interactions between small nuclear ribonucleoprotein particles in formation of spliceosomes. Cell 49: 763–774.

Smith, C. W. J., Patton, J. G. and Nadal-Ginard, B. 1989. Alternative splicing in the control of gene expression. Annu. Rev. Genet. 23: 527–577.

11

Molecular Control of Neural Development

David J. Anderson

THERE ARE PROBABLY more different cell types in the mature nervous system than in all of the other tissues of the body combined. Estimates of the number of different kinds of neurons vary, depending upon the criteria used to distinguish them, from as low as 1000 to as high as 100,000. Broad classes of neurons can be recognized by their distinctive morphologies, as revealed by the brilliant work of the Spanish neurobiologist Ramón y Cajal. Further divisions can be made on the basis of the neurotransmitters and peptides that neurons secrete, and of the receptors and ion channels they contain. More recently, the advent of monoclonal antibodies and cloned DNA probes has revealed that even neurons that appear morphologically identical can be molecularly distinct. The many classes of neurons in the brain are not mixed randomly but are highly organized in three-dimensional patterns. In the cerebral cortex, for example, different types of neurons are arranged in both layers and columns. How are we to begin to understand the molecular mechanisms that generate a structure of such fantastic diversity and complex architecture?

We can divide this very broad question into a series of more manageable and specific questions. First, what mechanisms control the *types* of neurons that arise in the embryonic nervous system? Do developing neuroblasts choose their fate according to intrinsic developmental programs, or according to signals in their local environments? If neuronal precursors have choices of cell fates, how are they limited, and what molecules control the decisions between them? Second, what controls *how many* neurons of a given type are generated? Some types of neurons are more numerous than others. Changes in the relative numbers of different neurons could profoundly affect brain function and might represent an important mechanism for the evolution of behavior. Third, what controls the *timing* of neuronal differentiation? Different classes of neurons must be generated according to a precise schedule to ensure appropriate connectivity and interactions with neighboring cells. Finally, how are different types of neurons generated in the correct place, and organized into an appropriate pattern? Do different neuronal types have a positional identity that helps them know where they are and to which cells they should connect? What is the molecular basis of such positional information, and how is it imparted to cells during development?

The molecular mechanisms that control neuronal development appear to be conserved in evolution and also to be used by many other nonneuronal tissues. This fact has emerged principally from striking similarities in the sequences of regulatory genes that control disparate developmental processes in different systems. Of course, there must be differences in the ways in which these molecular mechanisms are implemented at the cellular level to produce different tissues and species. However, the realization that development can be explained by unifying principles at the molecular level has fueled a rapid progress in understanding neuronal development in a variety of systems. In broad outline, we now understand that cell fate is ultimately controlled by DNA-binding proteins, or **transcription factors,** which activate and repress the expression of specific genes (Chapter 10). The questions of time, place, and magnitude of cell differentiation raised earlier can then be rephrased in terms of the temporal and spatial control of the transcription factors that specify cell fate.

This control is complex and poorly understood. However, it involves a *cascade* of regulatory events in which transcriptional regulators are controlled by yet more transcriptional regulators acting in a hierarchical manner. This control may occur within a single cell or between cells. In the latter case, transcription factors within one cell may cause changes in the expression of intercellular signaling molecules, which in turn act on other cells through specific receptors to alter their complement of transcription factors. In this way, cell–cell interactions and transcriptional regulation are linked into a regulatory network that underlies the developmental program. This linkage requires a complex signal-transduction process that communicates information from the cell membrane to the nucleus and back again (Chapters 6 and 7). Thus, the regulatory molecules that control development include not only cell-surface receptors, their ligands, and transcription factors, but also kinases, phosphatases, GTP-binding proteins and other kinds of molecules involved in signal transduction. In this respect, the mechanisms that control neuronal growth and development are similar to those involved in the main function of the mature nervous system: the transmission and reception of signals between cells. Moreover, other developmental control mechanisms, such as those involving changes in gene expression and morphogenesis, are increasingly seen as contributing to the plasticity of the adult nervous system. These similarities and analogies between the development and function of the nervous system lend credence to the idea that one way to understand how something works is to understand how it is put together.

The Molecular Control of Neuronal Identity

The importance of cell lineage

What is **cell lineage,** and why should it matter? Cell lineage, in its strictest sense, refers to the pattern of cell divisions that leads from a given precursor cell to a particular set of neurons. Knowing a given cell lineage tells us which neurons are sisters, which are first cousins, and so on. A lineage also indicates some of the developmental decisions that precursor cells have to make: does a given precursor give rise to only one kind of neuron, or to many different kinds? The lineage also gives us clues to the mechanisms that may control a particular developmental pathway. For example, in some invertebrates such as the nematode worm *C. elegans*, particular neurons develop through an invariant sequence of cell divisions. This suggests that development in these **determinate** cell lineages is controlled either by a cell-autonomous developmental program, or by highly reproducible cell–cell interactions.

In the vertebrate nervous system, by contrast, cell lineages are usually **indeterminate,** or variable within a species from animal to animal. Developmental fate more often correlates with a cell's position in the embryo than with its ancestry. These facts have emerged from tracing cell lineages using retroviral markers (Box A) or microinjected marker dyes. Individual precursor cells in the vertebrate retina or the neural crest, for example, may give rise to many derivative cell types in no clear genealogical pattern. This suggests that cell fate in these systems is determined either stochastically or by environmental influences. Despite the indeterminacy of vertebrate neurogenesis, cell lineage restrictions do play a role in limiting a cell's choice of fates. Thus, in most organisms cell fates are determined by a combination of ancestry and environmental influence. In molecular terms, this means that development is controlled by regulatory molecules that act both from within and from without the cell. The determination of cell lineage is an important first step in determining what kinds of molecules control a developmental pathway and where and when they may act.

Mutations identify genes that control cell fate in fixed lineages

A powerful method of investigating the mechanisms that control neural development in invertebrates is the isolation of mutants in which a particular step in a given developmental pathway is blocked. The products of the defective genes can then be identified and the mutants used to investigate other steps in the pathway. This approach has been most successful in *Caenorhabditis elegans* and *Drosophila melanogaster,* organisms in which genetic and developmental analyses can be easily performed.

In *C. elegans,* laser ablation and behavioral analysis have been combined with molecular genetics in an elegant but straightforward approach to isolating genes that affect the development of identified neurons. Laser ablation is used to determine the behavioral phenotype of worms lacking particular neurons. Normal worms are then mutagenized and this same phenotype sought among the progeny. The resulting mutants are defective in either the development or the function of the neuron of interest. Direct observation of the transparent developing embryos identifies those mutants in which the development of the neuron is defective. These developmental mutants can then be divided into complementation groups by genetic crosses. Once the mutations have been mapped, the genes affected can be molecularly cloned.

Box A Making Cells Blue

Lineage tracing experiments in vertebrates are often performed by injecting individual cells with large, membrane-impermeant molecules that can later be visualized either directly or indirectly. Such markers include rhodamine-labeled dextran and horseradish peroxidase. A disadvantage of this procedure is that it is often difficult to visualize individual cells to be injected in a vertebrate embryo; also, the injected molecules will eventually be diluted, by multiple cell divisions, to below their thresholds of detectability. These problems can be circumvented by genetically marking cells with a replication-defective, recombinant retrovirus containing the *E. coli* gene β-galactosidase. In this procedure, a suspension of recombinant virus particles is simply injected into the general region of the embryo containing the precursor cells of interest (e.g., a ventricle of the brain). Individual virions then infect cells, so that their genetic material (containing the β-galactosidase gene) is integrated into the host cell's chromosome. The virus suspension is sufficiently dilute that multiple infections, or infections of adjacent cells, do not occur. Once a cell is so marked, it will replicate and transmit the β-galactosidase gene to all its progeny, thus avoiding the problem of dilution. The clonal progeny of such a marked precursor can be visualized at a later time by using the histochemical reaction for β-galactosidase. Marked cells will only be visible if they actually express the β-galactosidase gene.

Retroviruses are used in this procedure because their genetic material normally becomes integrated into the host genome as part of the infection process, and because it is relatively easy to genetically engineer modified viruses that contain heterologous genes, and that lack the capacity to replicate. The retroviral genome consists of an RNA molecule packaged into a virus particle by virally-encoded proteins. Upon infection of a cell, this RNA molecule is converted into double-stranded DNA by reverse transcriptase, then integrated into the host genome. The recombinant virus contains an RNA in which the packaging proteins have been replaced by the β-galactosidase gene. The packaging of this recombinant RNA molecule (in a "packaging mutant" cell line) is accomplished by proteins provided in *trans,* by a "helper genome" (see Figure A). The helper RNA itself lacks packaging sequences, and therefore is not incorporated into virus particles. Recombinant viruses produced in this way can infect a cell once, delivering the β-galactosidase gene, but cannot replicate because they lack the genes encoding the packaging proteins. In this way, the virus serves as a disposable molecular syringe for introducing foreign DNA into a cell of interest.

Such an analysis of neurons that mediate touch sensitivity in *C. elegans* has been especially informative. A set of six **mechanoreceptor,** or touch-sensing, neurons has been identified by their invariant position and morphology (Figure 1A). Laser ablation of these six cells causes the animal to lose its withdrawal response to the light touch of an eyelash (glued to the end of a toothpick). A screen of thousands of animals yielded mutants in three different genes that affect different stages in the development of these neurons (Figure 1C). Mutations in the genes ***lin-32*** and ***unc-86*** prevent the birth of the mechanoreceptor neurons by altering their lineage. By contrast, mutations in a third gene, ***mec-3,*** alter the fate of mechanoreceptor precursors: rather than dividing to generate a single unipolar mechanoreceptor neuron and a bipolar nonmechanoreceptor neuron, they instead gen-

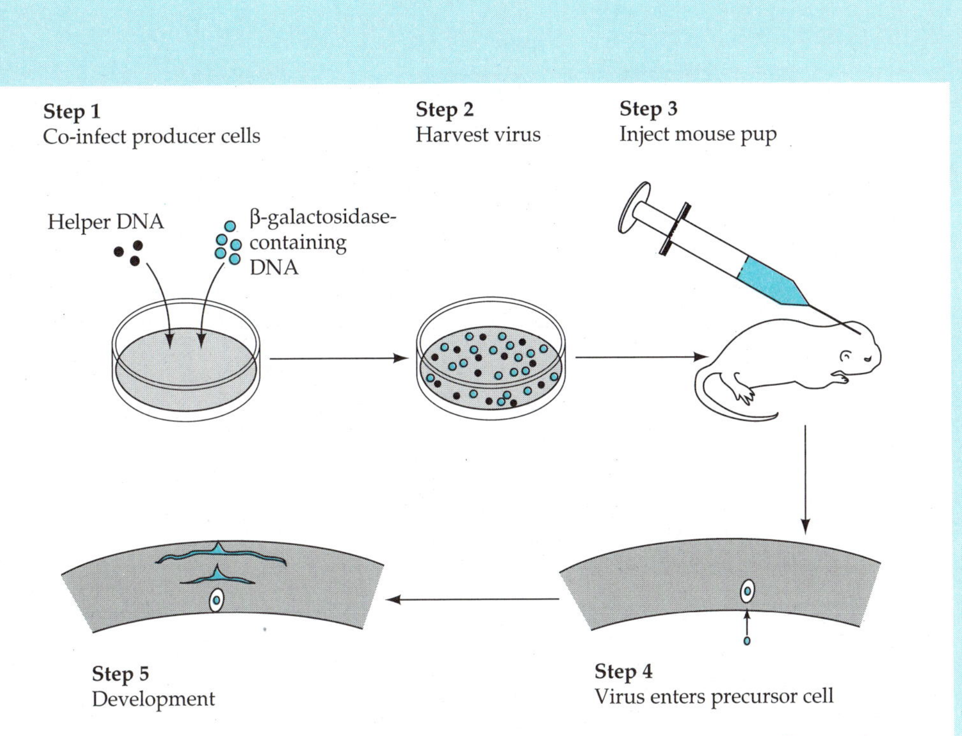

FIGURE A. Recombinant retroviruses are first grown by co-transfecting producer cells with helper DNA (step 1); the β-galactosidase-containing DNA is marked with a blue dot. When secreted virus accumulates in the culture medium (step 2), the virus is harvested, concentrated, and injected into the developing nervous system (step 3). Once injected, the virus suspension is dilute enough so that only one precursor cell in a given region of the ventricular zone will become infected; surrounding cells are uninfected (step 4). The animal is allowed to develop, and the progeny of the infected cell are visualized by staining fixed sections of the brain for β-galactosidase activity; all labeled cells within a confined region are considered to constitute the clonal descendants of a single cell.

erate two bipolar neurons (Figure 1B). It is as though the precursor gave rise to identical rather than to fraternal twins.

The requirement of a single gene, *mec-3*, for the differentiation of mechanoreceptor cells in *C. elegans* suggests that this gene acts as a master regulator specific for this class of neurons. By contrast, *lin-32* and *unc-86* act earlier in the pathway and are required in other lineages as well. Clues to the functions performed by these genes are provided by their sequences. Each of the encoded proteins contains a similar stretch of amino acid sequence that forms a DNA-binding domain called the **homeobox** (Chapter 10). This sequence motif was first identified in homeotic genes that control segment identity during *Drosophila* development (see Box D) and has since been found in regulatory genes in a wide range of organisms. The

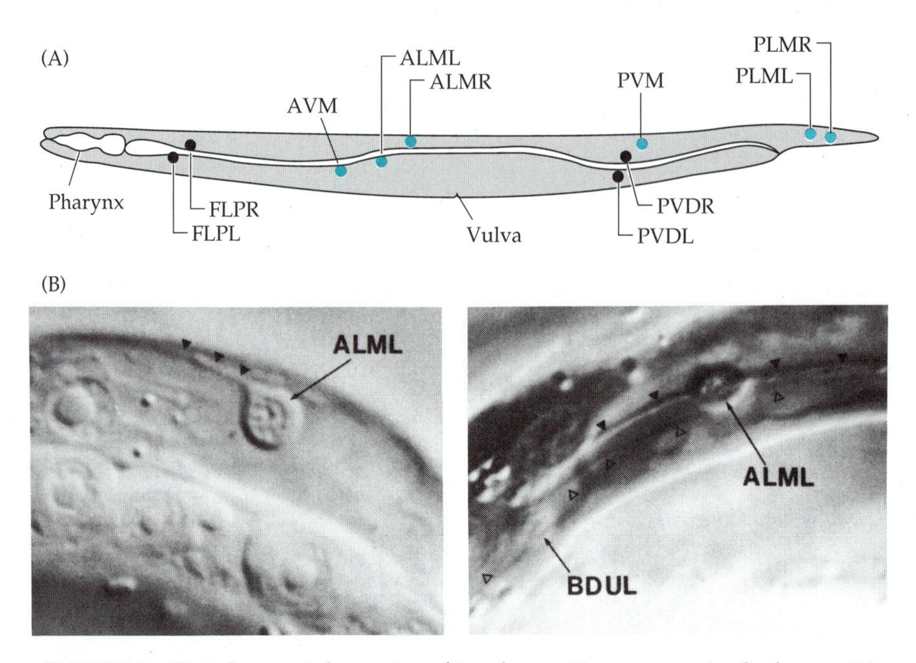

FIGURE 1. Developmental genetics of touch-sensitive neurons in *C. elegans*. (A) The positions of various mechanosensory neurons. The anterior (AVM) and posterior (PVM) touch-sensitive neurons that are produced by the lineage illustrated in (C) are indicated. ALML and ALMR (and PLML and PLMR) are other touch-sensitive neurons. (B) The phenotype of a *mec-3* loss-of-function mutant. Left, a wild-type touch-sensitive neuron (in this case, ALML) is unipolar. Right, in a *mec-3* mutant, the ALML neuron develops a bipolar morphology similar to that of its sister, BDUL. (C) Lineages that generate AVM/PVM and PVD neurons, and mutations that affect them. The defects seen in *mec-3*, *unc-86*, and *lin-32* mutants are seen at progressively earlier stages in both lineages. In a *lin-14* null mutant, the "Q" lineage on the left is converted to one similar to the "V5.pa" lineage on the right, so that PVD-like neurons are generated in place of AVM/PVM neurons. X indicates programmed cell deaths. (D) Hypothetical scheme indicating some of the interactions between the regulatory genes that affect mechanosensory neurons. *lin-32* and *unc-86* are positive activators of *mec-3*, which, once induced, positively feeds back on its own expression. In the Q lineage, the combined actions of the wild-type *mec-3* and *lin-14* gene products are envisioned to activate touch cell–specific genes; in the V5.pa lineage, *lin-14* activity is not required, and the PVD phenotype is produced by the wild-type *mec-3* gene product. (A from M. Chalfie and M. Au, 1989. Science 243: 1027–1033; B from J. Way and M. Chalfie, 1988. Cell 54: 5–16; C from J. Way and M. Chalfie, 1989. Genes Dev. 3: 1823–1833; D from M. Chalfie, personal communication. Photographs courtesy of the authors.)

presence of a homeobox motif in *mec-3*, *unc-86*, and *lin-32* strongly suggests that these genes encode transcription factors that regulate other genes during mechanoreceptor development.

Cascades of interacting regulatory genes control developmental pathways

Unc-86, *lin-32*, and *mec-3* appear to act in a cascade within the mechanoreceptor lineage. For example, *unc-86* is necessary to activate *mec-3* ex-

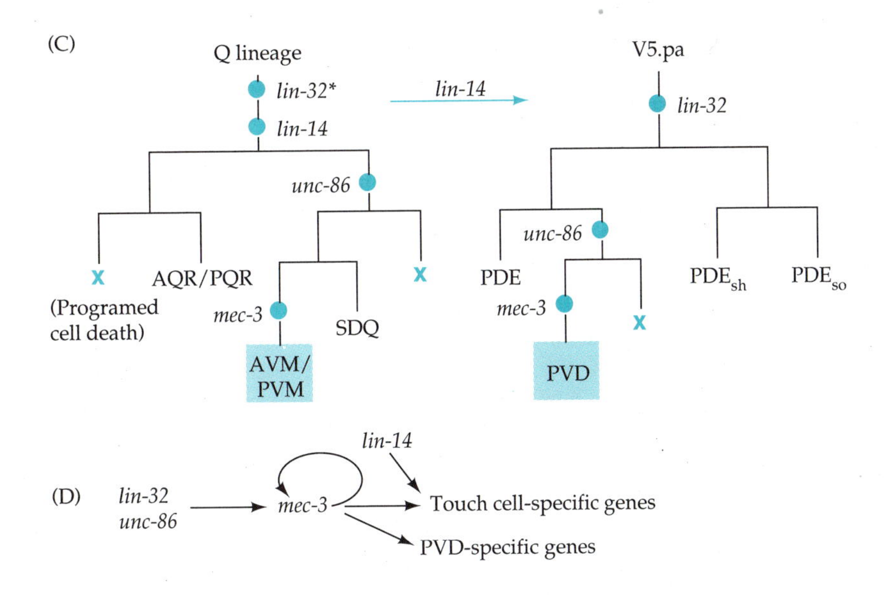

pression in precursor cells (Figure 1D). A mutant lacking unc-86 function fails to express *mec-3*; moreover, the *unc-86* protein has been shown to bind directly to regulatory sequences in the *mec-3* gene. *Lin-32* may also be necessary for *mec-3* activation. Once *mec-3* is turned on, the product of this gene maintains its own synthesis. This property of **autoregulation** permits the differentiated state of the cell to be maintained in the absence of the initial trigger. These features illustrate how developmental pathways can be controlled by a cascade of interacting nuclear regulatory genes. Interestingly, *unc-86* also controls the development of other neuronal lineages, such as one that generates a specific pair of serotonergic neurons called HSN. Thus, the regulatory genes that function early in a cascade may act broadly, but, in concert with other factors, turn on later-acting regulators that are more cell-specific in their function, funneling them, as it were, into the appropriate developmental pathway.

Neuronal diversity may be achieved by the combinatorial action of regulatory genes

As mutations in *mec-3* specifically block the development of touch-sensing neurons, one might have expected that the *mec-3* gene would normally be expressed only in such neurons, or in their immediate precursors. Studies of *mec-3* expression have shown that the gene is indeed expressed in mechanoreceptor precursors but, unexpectedly, is also expressed in other neurons called PVD cells. PVD and mechanoreceptor neurons derive from similar cell lineages (Figure 1C) but differ in their functions: PVD cells mediate a response to a more forceful mechanical stimulus than that sensed by the mechanoreceptor neurons. Since *mec-3* is expressed in two distinct cell types, other regulatory genes must contribute to the difference between the PVD and mechanoreceptor phenotypes. One such candidate is the ***lin-14*** gene. In a *lin-14* mutant, the mechanoreceptor lineage aberrantly generates PVD cells. This result implies that in the absence of *lin-14, mec-3* specifies a PVD phenotype, whereas in the presence of *lin-14, mec-3* specifies a mechanoreceptor phenotype (Figure 1D). In this way, neuronal diversity is generated by different combinations of regulatory genes. Closely related but distinct neurons may be determined by overlapping sets of developmental control genes.

Local cell–cell interactions control developmental fate in the Drosophila *retina*

In invertebrates as well as in vertebrates cell–cell interactions can play an important role in determining cell fates. Study of the compound eye of *Drosophila* has yielded insight into the molecules that mediate such interactions. The eye is a dispensable organ whose sensory neurons are arranged in a repetitive pattern and develop in a stereotyped sequence, facilitating genetic analysis (Figure 2A). Each compound eye is a quasicrystalline array of functional units called **ommatidia,** containing eight **photoreceptors** along with other cell types. The photoreceptors, named R1 through R8, respond to light but display different spectral sensitivities because they express different opsins. They also differ in the specific projections they make into the brain.

The development of photoreceptors in each ommatidium occurs in an invariant sequence. Initially, undifferentiated precursor cells assemble from an epithelial sheet into small clusters of approximately five cells each. These clusters form behind a groove called the **morphogenetic furrow,** which sweeps across the eye imaginal disc epithelium, leading a wave of cellular differentiation that moves in a posterior to anterior direction (Figure 2B and C). Once the ommatidial clusters form, one cell differentiates first as photoreceptor R8, followed by the pairs R2 and R5 and then R3 and R4. After another round of cell division, three more cells join the cluster. These differentiate next as the pair R1 and R6, followed by R7. This stereotyped pattern of photoreceptor differentiation, and the invariant placement of photoreceptors within each ommatidium, initially suggested that the eight photoreceptors might develop as the clonal progeny of a single precursor cell. However, **genetic mosaic analysis** (Box B) indicated that the cells in each ommatidium are not clonally related. This observation suggested instead that the invariant sequence of development reflects a highly reproducible series of cell–cell interactions.

A number of different mutants exist that affect the development of different cells within the ommatidium. The first, and perhaps most dramatic, mutant that was isolated, called ***sevenless* (*sev*),** lacks the **R7** photoreceptor (Figure 3A). This subtle defect, which causes a change in the way the flies respond to ultraviolet light, is the only abnormality evident in mutant animals. In *sev* flies the precursor cell that normally develops into R7 instead migrates apically and differentiates into a cone cell (Figure 3B). Subsequent genetic mosaic analysis has shown that wild-type *sev* (sev^{+}) function is required only in the precursor to R7. Since the fate of the R7 precursor is apparently not determined by cell lineage, this result suggests that the function of the sev^{+} gene product may be to receive or transduce signals from neighboring cells.

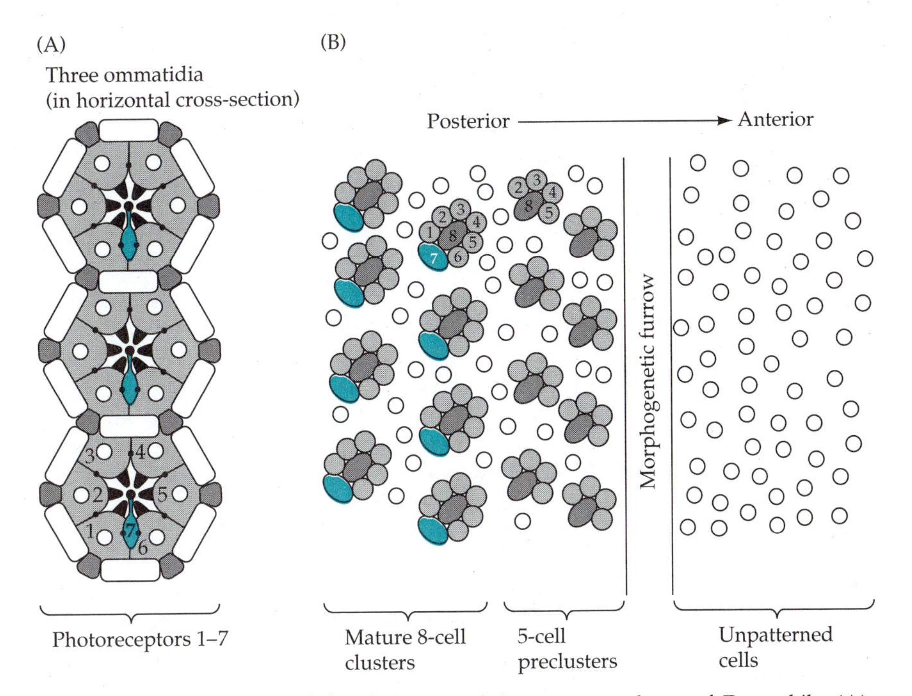

FIGURE 2. Structure and development of the compound eye of *Drosophila*. (A) Schematic cross section through several adult ommatidia. At this plane of section, photoreceptors R1–R6 and R7 can be seen, but not R8. Note the central position occupied by R7 (lower ommatidium). (B) The pattern of cluster formation and cell differentiation in the eye imaginal disc. Development occurs in a posterior to anterior wave, so that the cells to the left are older (and hence more differentiated) than the cells to the right. The "morphogenetic furrow" is a physical groove that sweeps across the disc as development proceeds. To the right (anterior) of this furrow, the cells are undifferentiated and not clustered. Photoreceptors are generated by two successive waves of mitosis posterior to the furrow; the first yields R2–R5 and R8, the second R1, R6 and R7. (After S. L. Zipursky, 1990. Neuron 4: 177–187.)

Consistent with this idea, the sequence of the cloned *sev*$^+$ gene predicts a membrane protein (Figure 3C) with a large extracellular domain and a cytoplasmic domain whose sequence is highly homologous to mammalian **tyrosine kinases** (see Chapter 7). This homology is a tantalizing one because in other cells tyrosine kinases participate in signal transduction pathways mediating growth and differentiation. The receptors for growth factors such as EGF and FGF, for example, contain cytoplasmic domains that have tyrosine kinase activity. This result illustrates how the interpretation of sequences derived from genes isolated by mutation in *Drosophila* or *C. elegans* is aided by information obtained from biochemical studies in nongenetic mammalian systems.

The large size of the sev extracellular domain suggests that it might contain the binding site for a specific ligand. Such a ligand could be provided by cell **R8,** which makes direct contact with the precursor to R7 at the time the latter differentiates. One would predict that mutations eliminating such a ligand would also generate a sevenless phenotype, but

Box B Genetic Mosaic Analysis

Genetic mosaics in *Drosophila* are animals, heterozygous for a particular recessive mutation, in which a single somatic cell has been induced (by X-irradiation) to undergo a mitotic recombination event that makes one daughter of that cell *homozygous* for the mutation (Figure A1). If the recombination event is induced early in development, this cell will then divide to produce a clone of progeny cells that are all homozygous for the mutation and surrounded by heterozygous cells (Figure A2). Such mosaicism has been exploited in several ways by *Drosophila* geneticists. If the homozygous phenotype is visible and distinguishable from the heterozygous phenotype (e.g., by a loss of pigmentation), then the mutation can be used to mark cells for lineage tracing to identify cells that share a common progenitor. In this case, the mutation does not affect the development of the cells, it simply marks them. For example, mosaic analysis in the *Drosophila* retina revealed that individual ommatidia could be composed of mixtures of both pigmented (wild-type) and nonpigmented (mutant) cells (Figure A3). This result indicated that the cells of a single ommatidium were not clonally related.

Genetic mosaics are also used to identify the cell type in which a given mutation acts. For example, consider the case of a mutation that blocks the development of a particular type of neuron. It is important to know whether the mutation acts in that neuron itself (i.e., it is **cell autonomous**), or in some neighboring cell that interacts with it. This can be determined by comparing the phenotype of animals in which either the neuron or the neighboring cell (but not both) is mutant: if development of the neuron is blocked only when the neighboring cell is mutant, and is normal when the neuron is mutant, then the mutation is not cell autonomous and may affect an inductive signal necessary for differentiation, for example. In order to perform such an analysis, it is necessary that the mutant gene of interest be linked (i.e., physically close) to a second, cell-autonomous "marker" gene that can be easily visualized so that the mutant cells can be identified independent of their developmental phenotype. The mutant and marker genes must be close enough on the chromosome so that if one undergoes somatic recombination, the other one does, too. For example, in the case of the *boss* gene that affects photoreceptor R7 development (see text), a nearby gene affecting pigment granules was used as the marker. In mosaics, cells mutant or wild-type for *boss* could be distinguished according to the presence or absence of pigment granules. In this way, it was shown that the *boss* mutation acted in cell R8, although it affected only the development of cell R7.

would act in cell R8 rather than cell R7. Such a mutant has recently been found, called ***bride-of-sevenless* (*boss*).** Mosaic analysis has shown that *boss*$^+$ function is required only in R8 for normal development; all other cells in the ommatidium may be mutant for *boss*, and R7 will still develop normally. The sequence of the *boss*$^+$ gene also predicts a membrane protein, consistent with the idea that *boss*$^+$ could encode a cell-associated ligand of the sev protein (or else a molecule that regulates the expression of the sev ligand). If R8 and R7 can be envisioned as communicating with one another, boss appears to be doing the talking and sev the listening.

Nuclear regulatory genes also influence photoreceptor development

Despite the highly cell-specific phenotype produced by the *sev* mutation, antibody staining has revealed that the sev protein is expressed in most or

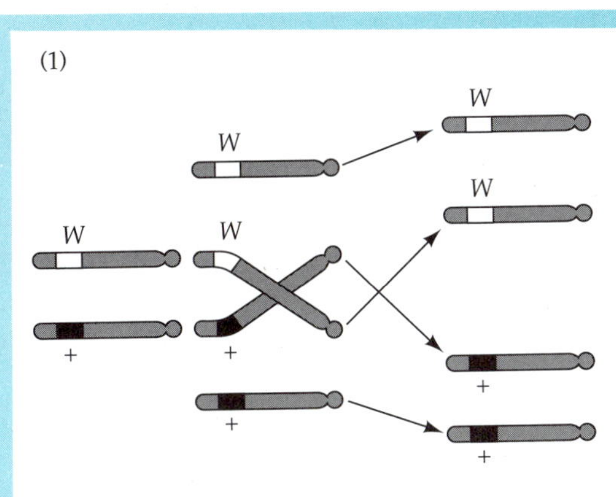

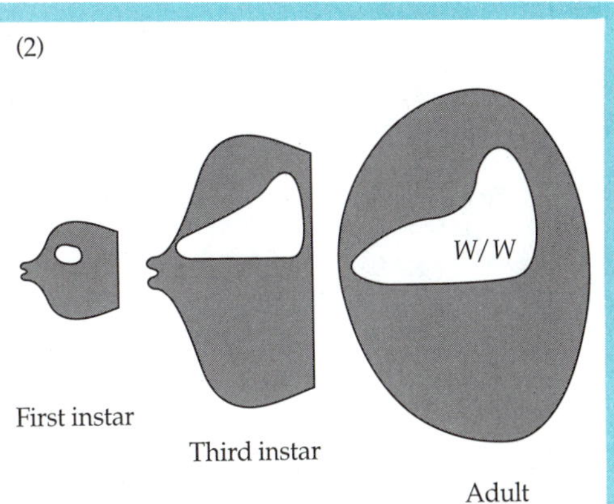

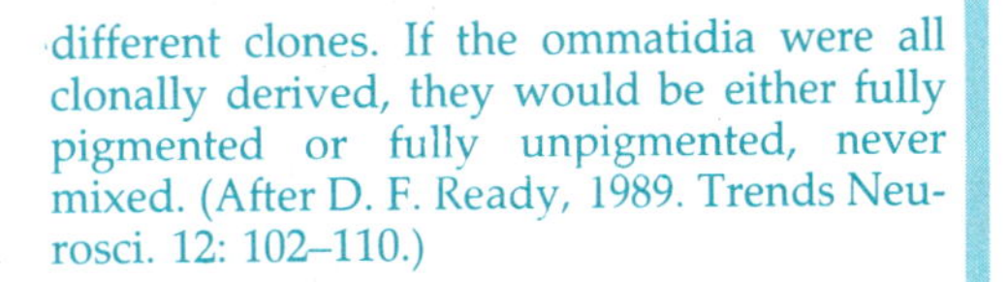

FIGURE A. How mosaic analysis was used to show that cells of a single ommatidium are not clonally related. (1) Flies heterozygous for the recessive white mutation (*W*) were X-irradiated early in eye development, causing somatic recombination in one cell within the eye imaginal disc. One daughter cell received both copies of the mutant *W* gene and therefore this cell and all its progeny can be identified by their lack of pigment. All other cells in the eye are pigmented. (2) As the (*W*/*W*) cell divides, its clonal progeny form a growing unpigmented patch in the developing eye because there is little mingling or migration of the cells. (3) Ommatidia in the eye of the adult fly are examined microscopically along the border of the unpigmented patch. Note that the cells in some ommatidia (hexagonal compartments) are both pigmented and unpigmented. Therefore these mixed ommatidia are formed by cells recruited from different clones. If the ommatidia were all clonally derived, they would be either fully pigmented or fully unpigmented, never mixed. (After D. F. Ready, 1989. Trends Neurosci. 12: 102–110.)

all photoreceptors (and in other parts of the nervous system). Thus, other molecules besides sev must be involved in restricting the function of sev to the R7 precursor. An interesting candidate for such a molecule is the protein encoded by the gene ***seven-up* (*sup*)**. In a *sup* mutant, the precursors of cells R1, R3, R4, and R6 instead develop an R7 phenotype. This result implies that sup^{+} normally functions in those cells to suppress differentiation into an R7 photoreceptor (which might otherwise occur through stimulation of the sev receptor-tyrosine kinase). The sequence of the sup^{+} gene is of interest because it encodes a DNA-binding protein of the steroid-receptor superfamily (Chapter 10). The case of the *sup* gene illustrates how systems whose development requires cell–cell interactions employ nuclear as well as cell-surface regulatory molecules. Many other genes affecting photoreceptor development have recently been isolated,

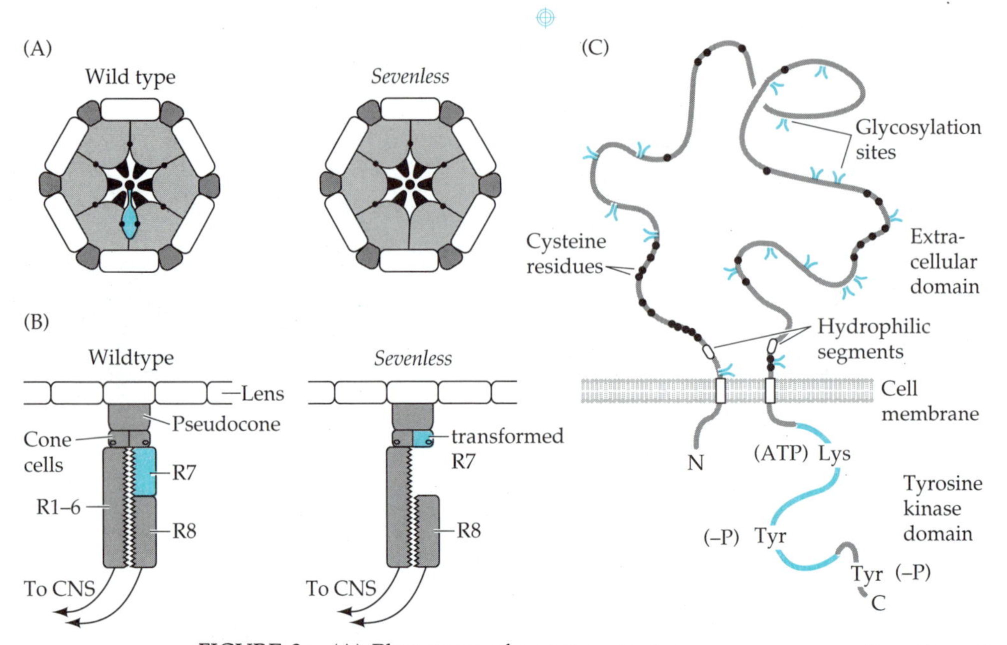

FIGURE 3. (A) Phenotype of a *sev⁻* mutant, seen as a cross section through an individual ommatidium. Note that R7 is absent. (B) Lateral view of wild-type and *sev⁻* ommatidia, showing how the cell that would normally develop into R7 instead has been transformed into a cone cell, which sits atop the ommatidial cluster (and is not seen). (C) Hypothetical model for the structure of the sev protein in the membrane. The tyrosine kinase domain is represented by a colored line. Relative positions of cysteine residues (filled circles) and possible N-linked glycosylation sites (small branches) are indicated in the extracellular domain. Ovals indicate two very hydrophilic segments. (A from K. Basler and E. Hafen, 1989. Development 107: 723–731; B from J. Palka and M. Schubiger, 1988. Trends Neurosci. 11: 515–517; C from K. Basler and E. Hafen, 1988. Cell 54: 299–311.)

but a full discussion is beyond the scope of this chapter. Nevertheless, it should be clear that the developing *Drosophila* ommatidium provides a rich territory that can be mined genetically for molecules that control cell fate.

The fates of migratory progenitor cells can be controlled by diffusible signals and their receptors

The highly ordered arrangement and close packing of the photoreceptor neurons in the *Drosophila* ommatidium is well suited to a developmental pathway that involves an orderly series of cell–cell interactions, mediated by membrane-bound signals and their receptors. In the vertebrate neural crest, by contrast, neuronal precursor cells migrate and disperse widely throughout the embryo before acquiring their final developmental fates. Transplantation experiments have suggested that the differentiation of neural crest cells is influenced by their local environment. The **sympathoadrenal progenitor** illustrates the roles that environmental signals can play. This progenitor arises from a population of initially multipotent neural crest cells that separate from the dorsal neural tube and migrate ventrally (Figure 4A). By an unknown mechanism, some crest cells become committed to the sympathoadrenal lineage during their migration. These

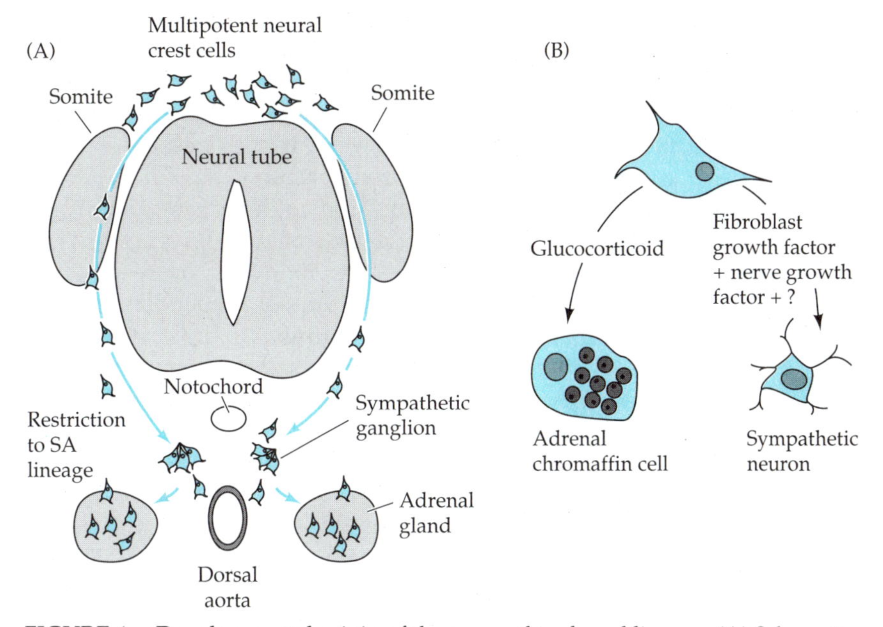

FIGURE 4. Developmental origin of the sympathoadrenal lineage. (A) Schematic cross section through the trunk region of a rat embryo. Initially multipotential neural crest cells migrate ventrally and laterally from the apex of the neural tube, beginning around E9.5–E10 in the rat. The kidney-shaped structures lateral to the neural tube are somites. Some of the crest cells stop migrating near the dorsal aorta, where they aggregate to form sympathetic ganglia, whereas others continue migrating ventrally to invade the developing adrenal gland primordium. (B) Schematic illustration of the bipotential sympathoadrenal progenitor and its choice of cell fates. Cell types are not shown to relative scale.

sympathoadrenal progenitors then differentiate along either of two pathways depending upon their final destination: cells that migrate to the adrenal gland primordium differentiate into endocrine cells called **chromaffin cells,** whereas cells that arrest their migration near the dorsal aorta (a major blood vessel) become **sympathetic neurons** (Figure 4B).

To study the environmental determinants of cell fate in this system, sympathoadrenal progenitor cells have been isolated from rat embryos using specific cell-surface monoclonal antibodies and fluorescence-activated cell sorting and grown in cell culture. Antibody markers are used to follow the fate of the cells as the culture environment is manipulated. Experiments such as these have revealed that individual precursor cells are bipotential, and that **glucocorticoid hormones** are essential for chromaffin cell development (Figure 4B). In the absence of such steroids, the progenitors begin to develop into neurons, a process promoted by specific growth factors (see below and Chapter 13). The requirement for glucocorticoid in vitro is consistent with the fact that, in vivo, chromaffin cells differentiate amidst the glucocorticoid-secreting cells of the adrenal cortex.

The actions of glucocorticoid on sympathoadrenal progenitor cells are mediated through a specific **glucocorticoid receptor.** The receptor is a cytoplasmic molecule that binds glucocorticoids, which diffuse into the cell due to their hydrophobic character. The hormone-receptor complex then enters the nucleus, where it binds to specific genes to regulate their transcription (Chapter 10). The glucocorticoid receptor plays a dual role in the chromaffin-neuron decision. It both induces chromaffin cell–specific genes and represses neuron-specific genes. This example illustrates the general principle that developmental decisions involve both positive and negative regulation. In the case of the glucocorticoid receptor, this dual role can be explained by the ability of the receptor to act either as a transcriptional activator or repressor, depending upon the context of its DNA-binding site (see also Chapter 10). It is interesting to recall that the *Drosophila seven-up* gene, which is also a member of the steroid receptor gene family, plays an inhibitory role in photoreceptor differentiation.

Neuronal differentiation involves a growth factor cascade

At least two different polypeptide growth factors influence the development of sympathoadrenal progenitors along the neuronal pathway of differentiation. One, **fibroblast growth factor (FGF),** is a widely distributed polypeptide that affects the growth and differentiation of many cell types. The other, **nerve growth factor (NGF),** is a neurotrophic factor that functions specifically to support the survival of particular classes of neurons, including those of the sympathetic system. FGF (or a related molecule) acting on sympathoadrenal progenitors within embryonic ganglia promotes proliferation and initial neuronal differentiation but cannot serve as a survival factor. Rather, FGF appears to induce a dependence upon NGF. The timing of NGF dependence is likely to be coincident with the arrival of immature sympathetic axons at the periphery, where they gain access to NGF secreted by target cells. In this way, neuronal development is controlled by a cascade in which one factor promotes the initial differentiation of the progenitor and simultaneously primes the cell to respond to a second factor, which in turn supports further maturation and survival.

The intracellular actions of growth factors are common to many developing systems

FGF and NGF exert their effects on cells by first binding to specific membrane receptors. The FGF and NGF receptors are transmembrane proteins containing an extracellular ligand-binding domain, and a cytoplasmic tyrosine kinase domain analogous to the *Drosophila* sevenless protein mentioned earlier. Ligand binding to these receptors activates tyrosine phosphorylation within seconds by inducing receptor dimerization. In the case of NGF, dimerization may occur with a heterologous receptor polypeptide, p75, which lacks a tyrosine kinase domain. Subsequently, additional "downstream" tyrosine kinases are activated, including the cellular proto-oncogene **pp60-*src*.** Experiments done in PC12 cells, a clonal cell line derived from an adrenal medullary tumor, have shown that activation of src is both necessary and sufficient for neuronal differentiation. Microinjection of anti-src antibodies inhibits FGF or NGF-induced neurite outgrowth, while expression of activated viral src induces neurites in the absence of any growth factor. Similar results have been obtained for the proto-oncogene **c-*ras***, which encodes a GTP-binding protein. However, injection of activated ras can override an antibody blockade of src, implying that *ras* acts downstream of *src*. Interestingly, a receptor-tyrosine kinase and a ras protein function in an analogous pathway mediating vulval development in *C. elegans*. Mutations that inactivate the kinase are suppressed by mutations that constitutively activate ras, implying that ras acts downstream of the receptor kinase in that system as well. Moreover, genetic experiments in the *Drosophila* retina have recently shown that a *ras* gene acts downstream of the sev tyrosine kinase. Such studies illustrate both the complementary nature of genetic and biochemical approaches, and the apparent generality and evolutionary conservation of this transduction mechanism.

Although some of the earliest events in the signaling cascade are becoming clearer, the steps that link these events to changes in gene expression are poorly understood. As discussed in Chapter 10, a number of transcription factors, such as **c-fos** and **c-jun**, are transcriptionally induced and are also activated by phosphorylation in response to growth factor stimulation. These so-called **immediate early genes** are thought to activate or repress in turn the transcription of genes encoding structural proteins or enzymes. In the sympathoadrenal lineage, for example, the transcriptional induction by NGF of tyrosine hydroxylase, a neurotransmitter synthetic enzyme (Chapter 5), requires fos and jun. Such studies begin to provide a link between the earliest nuclear events involved in signal transduction and the expression of lineage-specific genes whose products define cell fate. Although the transcriptional effects of growth factors have been the most intensively studied, these factors also influence cell phenotype at many other levels. For example, NGF causes the rapid assembly of microtubules, important for neurite outgrowth, through phosphorylation of **microtubule-associated proteins** (MAPs) (see Chapter 8). In this way, a single environmental signal orchestrates cell differentiation by coordinating a complex series of cellular responses from transcription to morphogenesis.

Control of Cell Number during Neural Development

In building a nervous system, an embryo has to control the *number* of cells of a given type that differentiate. In principle, there are several ways to do this: by controlling the number of precursor cells initially generated; by controlling the number of times a precursor divides before it differentiates; and by controlling the number of differentiated cells that ultimately survive. In what follows, we will discuss what is known about the first two mechanisms since they operate early in development. The third mechanism operates at later stages of development and is considered in detail in Chapter 13.

Molecules controlling lateral inhibition have been identified by genetic analysis in Drosophila

Neuroblasts in the fruit fly CNS develop within a sheet of epithelial cells in the ventral region of the embryonic ectoderm. A particularly striking series of mutations, called **neurogenic mutations,** produces embryos that have generated too many neuroblasts and thus fail to develop normally (Figure 5A). This suggests that the **neurogenic genes**, whose function is lost in such mutations, normally limit the number of neuroblasts that develop. Experiments performed in the grasshopper have shown that if a newly formed neuroblast is ablated with a laser microbeam, one of the neighboring undifferentiated ectodermal cells is able to replace it. Thus, most or all ectodermal cells have the potential to form neuroblasts but are prevented from doing so if they contact another neuroblast, by a process called **lateral inhibition** (Figure 5B). In neurogenic mutants, ectodermal cells that would normally become epidermoblasts instead become extra neuroblasts, due to a defect in the lateral inhibition mechanism.

The cloning of several neurogenic genes revealed that they encode membrane proteins that could mediate cell–cell interactions important in lateral inhibition. For example, a neurogenic gene called ***big brain* (*bib*)** encodes a hydrophobic integral membrane protein resembling the main intrinsic protein of the vertebrate lens, MP26. As MP26 is thought to function as a channel, by analogy, bib could act to pass an inhibitory signal from one cell into another. A particularly well-studied neurogenic mutant is called ***Notch.*** Mosaic analysis has shown that wild-type Notch function in ectodermal cells is required to prevent the cells from becoming neuroblasts. This suggests that Notch could be important in reception of the postulated lateral inhibition signal.

The *Notch* gene encodes a large transmembrane protein of approximately 300 kD that contains a large cytoplasmic domain (no tyrosine kinase homology!), a single membrane-spanning segment, and a large amino-terminal extracellular domain (Figure 5C). A particularly intriguing aspect of the extracellular domain is the presence of a repeated sequence seen in the epidermal growth factor (EGF) precursor and other proteins. This **EGF repeat,** which occurs 36 times in the Notch protein, is characterized by an invariant spacing of conserved cysteines with intervening residues that are poorly conserved. Many proteins containing this repeat, such as the blood-clotting factors, bind to other proteins, and in some cases the repeat has been demonstrated to be part of a protein-binding site. Since antibody-labeling studies have shown that the Notch protein is present on the cell surface, its EGF repeats could function in binding another protein on an adjacent cell.

In this respect, it is interesting that another neurogenic gene, called ***Delta,*** encodes a smaller homologue of Notch with nine EGF-like repeats in the extracellular domain (Figure 5C). Genetic experiments suggest that *Notch* and *Delta* interact with one another: the phenotype caused by a mutation in *Delta* can be suppressed by an appropriate compensating mutation in *Notch*. Biochemical experiments have now shown that these interactions may be direct. Cultured cells expressing Notch bind, in a calcium-dependent manner, to cells expressing Delta, and the two proteins appear to be concentrated at the points of contact between such cells. Thus Notch and Delta may function as a receptor–ligand system, or as cell adhesion molecules. As in the case of retinal development discussed ear-

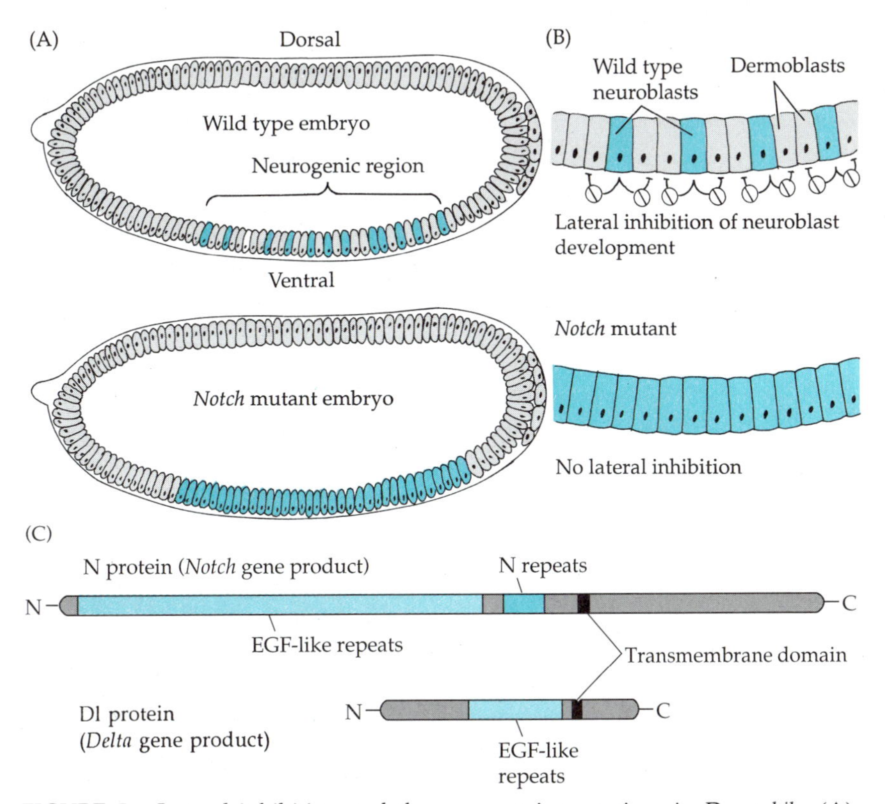

FIGURE 5. Lateral inhibition and the neurogenic mutations in *Drosophila*. (A) Schematic illustration of the effect of a neurogenic mutation such as *Notch, Delta, bib,* or *enhancer of split.* A longitudinal section through a gastrulating embryo is shown, ventral aspect facing down. In a normal embryo (top), cells in the epithelium become either neuroblasts (colored) or dermoblasts (gray). In a loss-of-function neurogenic mutant, such as *Notch* (bottom), all the cells in the neurogenic region become neuroblasts. (B) Lateral inhibition in the neurogenic region. Initially, all cells are equipotent. When cells become committed to the neuroblast fate, they inhibit neighboring cells from acquiring this fate, causing them to become dermoblasts instead. In a *Notch* mutant, lateral inhibition does not occur and all cells become neuroblasts. (C) Schematic illustration of the structure of the *Notch* and *Delta* gene products. The region of EGF-like repeats is shaded. "N repeats" is a different type of repeated motif found in the Notch protein, as well as in the lin-12 protein of *C. elegans.* (A from S. Artavanis-Tsakonas, 1988. Trends Genet. 4: 95–100; C from J. A. Campos-Ortega, 1988. Trends Neurosci. 11: 400–405.)

lier, the cell–cell interactions mediated by Notch and Delta influence cell fate via nuclear regulatory genes; one mutation affecting such genes is *enhancer of split,* a locus containing several related members of the helix-loop-helix class of transcription factors (see below and Chapter 10).

The molecules so far uncovered that limit neuroblast number in *Drosophila* do not directly control the *number* of cells generated. Rather, *Notch, Delta, bib,* and other genes control the cell fate decision of whether to become a neuroblast or an epidermoblast. Thus, the number of neuroblasts generated is indirectly determined by the number of cells that choose to become epidermoblasts. It is important to distinguish the *control* of this process from the machinery necessary for the process to function. For example, lateral inhibition could be controlled by factors that determine the range of action or concentration of the presumed inhibitory signal. Such factors have yet to be identified. Nevertheless, lateral inhibition seems to be a general mechanism for controlling neuronal number, since the neurogenic genes appear to affect neuronal differentiation in many different parts of the *Drosophila* nervous system, including the retina. Interestingly, genes homologous to *Notch* have been found to participate in cell–cell interactions in *C. elegans*. Homologues of the *Notch* gene have also been found in the amphibian *Xenopus,* suggesting that the basic process mediated by this gene may be common to many organisms.

Specific mitogens control progenitor number in a vertebrate glial lineage

In some parts of the nervous system, particular kinds of cells are generated from **committed progenitors,** or "blast" cells. These cells proliferate symmetrically and then differentiate. Examples of such progenitors are the aforementioned sympathoadrenal progenitor and the **O2A progenitor** from optic nerve. Cell culture experiments have shown that O2A progenitors can give rise to either of two glial cell types: oligodendrocytes, or type-2 astrocytes, nonmyelinating glial cells whose role in vivo is not yet clearly established (Chapters 1 and 9) (Figure 6A). Studies of the O2A progenitor have provided some important insights into how cell number can be determined by controlling the extent of proliferation.

Polypeptide growth factors, or mitogens, are important regulators of progenitor cell proliferation. Different mitogens act on different progenitor populations, and the specificity of their action is determined, at least in part, by specific receptors. An important mitogen for the O2A progenitor is called **platelet-derived growth factor (PDGF).** PDGF is produced in the optic nerve by type-1 astrocytes, another glial cell type (Figure 6A). If O2A progenitors are cultured without PDGF, they do not divide but differentiate rapidly into oligodendrocytes. By contrast, PDGF stimulates the progenitors to divide and thereby increases the number of oligodendrocytes that develop. The amount of proliferation that occurs in response to PDGF is limited, however. After a certain number of divisions, O2A progenitors lose the ability to respond to PDGF, and differentiate despite the presence of the mitogen (Figure 6B). The molecular basis for the loss of PDGF-responsiveness is not known; it is not due to a loss of the PDGF receptor. Whatever the explanation, the O2A lineage illustrates how cell number can be controlled by a combination of specific mitogens and inherent limitations on proliferative capacity. Limits on cell proliferation may be controlled by recently discovered **tumor-suppressor genes,** so-called because mutations in them lead to cancer. For example, a tumor-suppressor gene called *retinoblastoma* (see Chapter 15) acts to prevent entry of fibroblasts into S phase of the cell cycle. The role of such genes in neural development remains to be explored.

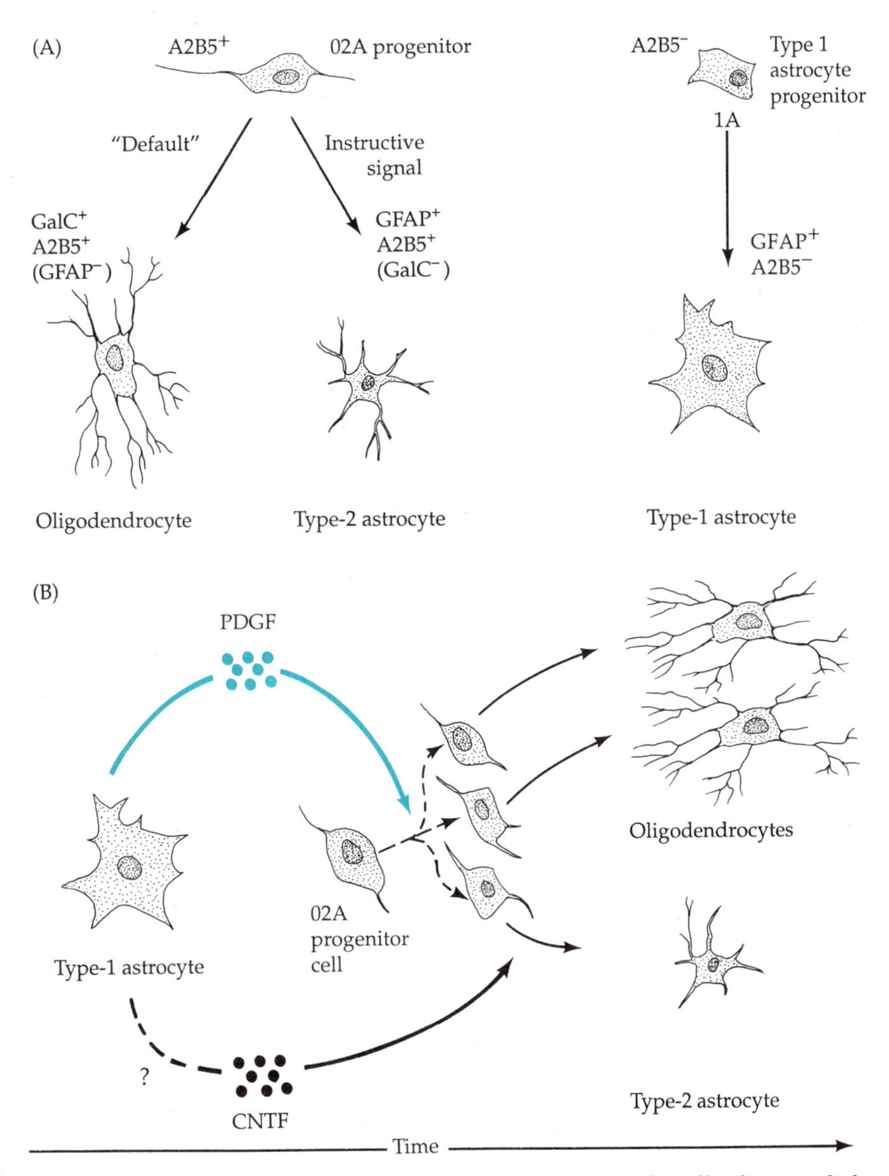

FIGURE 6. Glial cell lineages in the optic nerve. (A) Two lineally distinct glial progenitor cells are found prenatally in the rat optic nerve. Left, the O2A progenitor has the potential to generate either an oligodendrocyte or a type-2 astrocyte. Right, a different progenitor generates only a type-1 astrocyte. The antigenic phenotype of these cells is indicated. A2B5 is a cell-surface ganglioside antigen; GalC is galactocerebroside, a membrane glycolipid; GFAP is the glial-specific intermediate filament protein. Type-1 and type-2 astrocytes are both GFAP$^+$, but only type-2 astrocytes are A2B5$^+$ and have a stellate morphology. (B) Cell–cell interactions that control the direction, number and timing of glial cell differentiation in the optic nerve. Early in development (E17 in the rat), the proliferation of O2A progenitors is driven by PDGF, secreted by type-1 astrocytes. The timing of oligodendrocyte differentiation is determined by the number of cell divisions required for progenitors to lose responsiveness to PDGF (dashed arrows). Beginning in the second postnatal week, an unknown signal causes type-1 astrocytes to begin making CNTF, which (together with additional factors) diverts remaining O2A progenitors to the type-2 astrocyte pathway. (From M. C. Raff, 1989. Science 243: 1450–1455.)

Control of the Timing of Cell Differentiation in the Nervous System

The timing of cell differentiation in the nervous system occurs on a precise schedule, in part to ensure that necessary cell–cell interactions can take place appropriately. In systems where cell fate is controlled by environmental signals, the timing of differentiation could be controlled by the timing of appearance of these signals or of their receptors. Alternatively, development could be under the control of intrinsic cellular clocks. Studies of the O2A lineage and the retina provide glimpses of each of these mechanisms.

O2A progenitors first give rise to oligodendrocytes around the time of birth, whereas they do not begin to generate type-2 astrocytes until the second postnatal week. What is responsible for controlling the precise schedule on which these two types of glial cells develop? If taken from embryonic day 17 (E17) optic nerve and cultured in the absence of mitogens, O2A progenitors differentiate prematurely into oligodendrocytes. If PDGF is provided, however, then differentiation is delayed, so that oligodendrocytes appear on the same schedule in vitro as in vivo. Since O2A progenitors differentiate when they lose responsiveness to PDGF, the schedule of oligodendrocyte production is determined by the time at which responsiveness to PDGF is lost (Figure 6B). It has been suggested that this timing may be controlled by a cellular clock that counts cell divisions, because the progeny of a given progenitor always divide the same number of times and differentiate simultaneously. However, other timing mechanisms, such as the decay of an inhibitor, have not been excluded.

In contrast to the differentiation of oligodendrocytes, which appears to be intrinsically timed, the differentiation of type-2 astrocytes is timed by cell-extrinsic factors. At least one diffusible signal, **ciliary neuronotrophic factor** (**CNTF;** see Chapter 13), together with other signals associated with the extracellular matrix, is required to induce O2A progenitors to develop into type-2 astrocytes. The timing of type-2 astrocyte differentiation is primarily controlled by the time of appearance of CNTF-like activity in the optic nerve, rather than by the time at which O2A progenitor cells acquire responsiveness to this factor (Figure 6B). For example, O2A progenitors from embryonic optic nerve can be made to differentiate prematurely into type-2 astrocytes in vitro by exposing them to purified CNTF. Moreover, CNTF-like activity first appears in the optic nerve in the second postnatal week, coincident with the onset of type-2 astrocyte differentiation. Type-1 astrocytes appear to be the source of CNTF in the optic nerve (Figure 6B), but what controls the timing of CNTF expression is not yet known.

By contrast to the case of the optic nerve, the timing of photoreceptor appearance in the developing vertebrate retina may be controlled by when progenitors become responsive to an inducing signal. Lineage-tracing experiments have shown that the different cell types of the vertebrate retina develop from multipotential precursors, without any apparent genealogical relationship. However, in the mammalian retina they appear on a defined schedule. Most rod cells (photoreceptors) appear at about the time of birth, whereas other cell types appear embryonically. To determine whether rod cells could be made to differentiate prematurely, labeled embryonic retinal precursors were mixed with a large excess of cells from neonatal retina. Under these conditions, the embryonic precursors generated rod cells in greater numbers but at the normal time of development. This suggests that the neonatal retina contains a "rod-inducing signal" but that the schedule of rod cell development may be controlled by the time at which retinal precursors become responsive to this inducer. In this way, the timing of differentiation is controlled by a combination of cell-extrinsic and cell-intrinsic factors.

Pattern Formation and the Spatial Control of Cell Differentiation

The complex architecture of the brain requires that different cell types develop in a precise spatial relationship to one another. To accomplish this, the embryo must not only establish a coordinate system for itself but must also ensure that the appropriate cell types are correctly generated within this coordinate system. Thus, the problems of pattern formation and cell type determination are intimately related. Genetic studies of neuronal development in *Drosophila* have begun to reveal how genes and cells link positional information to differentiation. Moreover, molecular studies of vertebrate neurogenesis have revealed a remarkable evolutionary conservation of these processes.

A family of interacting helix-loop-helix proteins controls neuronal determination in Drosophila

Genetic and molecular study of the *Drosophila* larval peripheral nervous system has yielded insights into the relationship between pattern formation and cell type determination. The larval PNS consists of clusters of sensory neurons that are segmentally repeated. Each cluster contains distinct types of neurons, such as external sensory (es), chordotonal (cd) and multidendritic (md) cells. The development of this system is under the control of a genetic locus called the ***Achaete-Scute Complex*** **(*AS-C*)**. This complex includes at least four related genes, linked in a tandem array: ***lethal of scute, achaete, scute,*** and ***asense***. Genetic deletions within this complex result in the loss of specific subsets of neurons in the larval PNS (as well as in the CNS). Mutations in *achaete* or *scute,* for example, eliminate distinct groups of neurons within each segmentally repeated cluster, whereas mutations in *asense* or *lethal of scute* affect neurons in some segmental clusters but not others. The neurons affected in each mutation appear related by their position rather than by their overt phenotype. Analysis of *AS-C* mutant embryos using specific enhancer-trap markers (Box C) indicates that the mutations act early in development, preventing the appearance of a subset of **sensory mother cells** that are the precursors of sensory neurons (Figure 7 and Box C). These sensory mother cells normally arise at reproducible positions along the embryonic axis.

Box C Enhancer Traps: Cell Markers and Mutations at One Blow

A powerful new method for analyzing the development of specific cell types in *Drosophila* is called the **enhancer trap**. This technique relies on the fact that in higher eukaryotes, enhancers (see Chapter 10) can influence the expression of genes over long distances in chromatin. If an exogenously introduced marker gene that on its own is very weakly expressed integrates into the genome near an enhancer that controls a highly cell-type–specific gene, then the marker will fall under the influence of that enhancer and exhibit the same cell-type–specific expression pattern as the endogenous gene near which it has integrated. If the product of the marker gene is easily visualized, as in the case of β-galactosidase, it now serves as a means of identifying a specific cell type. Moreover, such a marker is inheritable and can be studied in different genetic backgrounds.

A general approach to making such markers, therefore, is to literally riddle the *Drosophila* genome with promoterless β-galactosidase genes in a shotgun manner and then to look for flies expressing the marker in highly specific cell types. Although such an approach is laborious, it is facilitated by the fact that *Drosophila* contains sequences called transposable elements, or **transposons,** which can be made to "jump" from one site in the genome to another if the fly is crossed into an appropriate genetic background. By linking the β-galactosidase marker gene to a transposable element and injecting that marker gene into one "founder" strain, many thousands of fly strains containing different enhancer traps can be generated simply by breeding the founder strain to an appropriate mate, then separating and staining the offspring.

The enhancer trap is a double-barreled shotgun technique. On the one hand, it provides markers to identify cell types or stages in development that would otherwise be unrecognizable, much like a monoclonal antibody. On the other hand, it is a mutagenic agent: if the transposed marker gene lands in the coding region of a cell-specific gene, it may disrupt the coding sequence and inactivate the gene. Since most such mutations are recessive, to determine whether a particular enhancer trap is mutagenic one needs simply to interbreed the strain and examine homozygotes for a mutant phenotype. If an inter-

The sequence of the *AS-C* genes indicates that they belong to a recently discovered family of **helix-loop-helix (HLH)** DNA-binding proteins (Chapter 10). In mammals, an HLH gene called MyoD has been shown to act as a master regulatory gene for **myoblast determination.** DNA-binding by both AS-C gene products and MyoD requires heterodimer formation with another HLH protein, called **E12** (in mammals) or **daughterless (da)** (in *Drosophila*; Chapter 10). This biochemical interaction explains genetic interactions between *AS-C* and *da* which have been observed in the fly. The similarity in structure and biochemical function between *AS-C* and *MyoD* is consistent with the fact that these genes play roles in neurogenic and myogenic determination, respectively.

The requirement that HLH proteins form heterodimers in order to bind to DNA makes them susceptible to a particularly interesting form of negative regulation. This negative regulation appears to be mediated by a subclass of HLH proteins whose members lack the basic region, which is

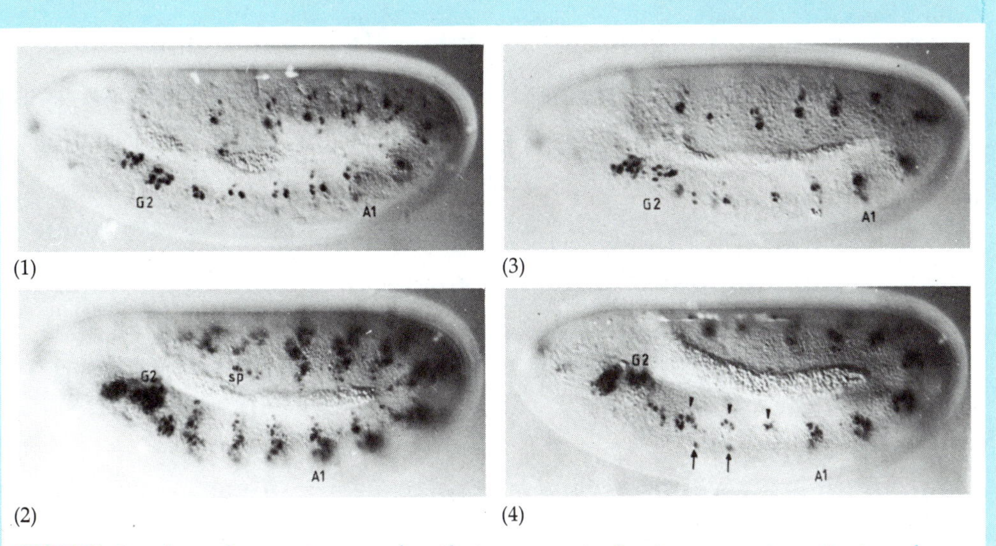

FIGURE A. An enhancer trap marker that labels subpopulations of neural precursors in *Drosophila* embryos. The black dots represent individual cells expressing β-galactosidase enzymatic activity, visualized by a histochemical stain. 1, 2 represent wild-type embryos stained at two different stages of development. 3, 4 represent mutant embryos lacking the *achaete-scute* complex, at the same developmental stages. Note that specific subsets of labeled cells are missing in the mutant embryos, along the ventral (lower) region. (From A. Ghysen and C. O'Kane, 1989. Development 105: 35–52.)

esting phenotype is observed, the inserted marker gene can be used as a handle to rapidly clone the affected surrounding gene. In this way, the enhancer trap method permits one to identify interesting cell types, obtain mutations that affect their development, and clone the endogenous genes affected by the mutation, all in rather short order. A powerful technique, indeed.

required for DNA binding. These proteins, which include **Id** (in mammals) and **extramacrochaete (emc)** (in *Drosophila*), form heterodimers with other HLH proteins, but these heterodimers cannot bind to DNA (Chapter 10). This property makes this subclass of HLH proteins potential competitive inhibitors of MyoD and achaete-scute. An excess of such a protein would dimerize with all of the available E12/da, leaving achaete-scute or MyoD unable to perform its normal function. In mammalian myoblasts, overexpression of Id delays myogenic differentiation. Similarly, chromosomal duplications of *emc* in *Drosophila* reduce the number of neuroblasts that develop; conversely, deletion of this gene increases the number of neuroblasts that form.

The positional control of neuroblast determination appears to be regulated by topographic differences in the relative levels of *AS-C* gene products and inhibitory HLH proteins within the embryonic neuroepithelium. Initially, broad but restricted domains of *AS-C* expression are established

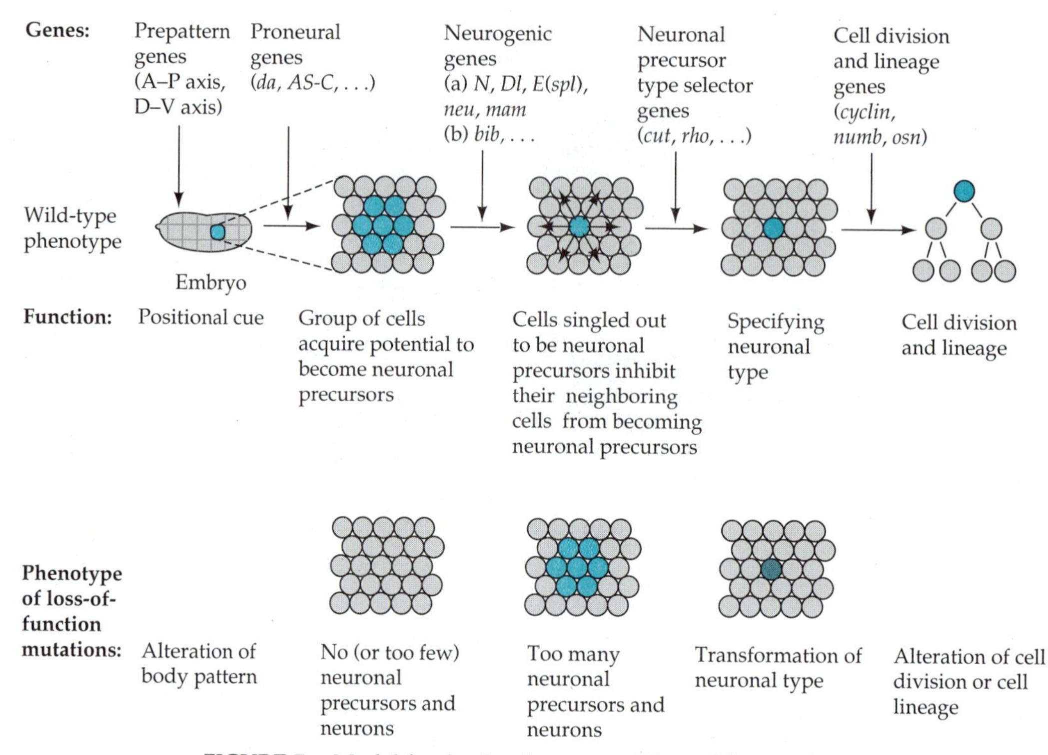

FIGURE 7. Model for the development of *Drosophila* peripheral sensory neurons. Circles indicate a group of cells within a particular region of the ectoderm; many such groups exist at different locations within the embryo. The various genes discussed in the text are shown here to act in a sequential manner, to progressively restrict the fate of the cells. Note that the *AS-C* genes and *da* are referred to as "proneural" genes. The expression of these genes is restricted to a subset of ectodermal cells. Surrounding cells are likely to express inhibitory genes such as *emc,* but this has been omitted for clarity. (After A. Ghysen and C. Dambly-Chaudiere, 1989. Trends Genet. 5: 251–255; modified by Y. N. Jan and L. Y. Jan, 1990. Trends Neurosci. 13: 493–498.)

by regulatory genes acting earlier in development (Figure 7); these domains are later sharpened by competitive interactions between HLH proteins as described above. Sensory mother cells are thought to arise in locations where the levels of *AS-C* gene products titrate out the negative regulators and rise above some threshold critical for neuroblast determination (Figure 7); this process may be accelerated by a positive feedback of *AS-C* genes on their own expression. Because the nuclear regulatory genes in adjacent cells cannot interact directly with one another, cell–cell interactions such as lateral inhibition must be linked to the regulatory network. Ultimately, the domain of *AS-C* function becomes restricted to a single cell within a given region of the epithelium; this cell commits to the neuroblast fate (Figure 7). In this way, a network of interacting regulatory genes *self-organizes* into an appropriate pattern that determines the topographic specificity of cell type determination.

Recently, two rat genes that are homologous to *AS-C* have been isolated.

At least one of these mammalian *AS-C* homologues is expressed in neuronal precursors, i.e., cell types analogous to those in which *AS-C* genes function. This parallel conservation of amino acid sequence and neural-specific expression suggests that at least some of the mechanisms controlling neuronal determination in the fly may be conserved in vertebrates. As we shall see shortly, this conservation of regulatory gene structure between *Drosophila* and mammals is true not only for molecules that control cell type determination but also for those that control pattern formation.

The spatial organization of neuronal development in the vertebrate hindbrain is segmentally controlled

We have discussed mechanisms that determine the local positioning of neuronal differentiation within an epithelial sheet. A more global question concerns how the brain is partitioned into spatially distinct regions, each of which contains many different kinds of neurons. In vertebrates, the central nervous system develops from what is initially a tube of neuroepithelium that runs along the **anterior-posterior (A-P) axis** of the embryo. Eventually, this tube becomes grossly subdivided into regions, corresponding roughly to the future forebrain, midbrain, hindbrain, and spinal cord. Different types of neurons develop in each of these regions. Thus, at the earliest developmental stages, the patterning of cellular differentiation in the brain appears to be foreshadowed by a process of compartmentation. How does this compartmentation of the nervous system occur, and what role does it play in the spatial organization of neuronal differentiation?

Anyone who has ever looked at an earthworm or an insect will have realized that **segmentation** is a fundamental principle of biological organization. A corollary to this rule is that different segments often contain different anatomical structures, depending upon their location along the A-P axis. In some segmented invertebrates, for example, thoracic segments have wings, whereas abdominal segments bear halteres (a vestigial winglike structure). This suggests the fundamental concept that potentially identical segments have acquired differences by being at different positions in the organism, which is referred to as **positional identity.** The principle of segmental organization suggested to neuroanatomists long ago that such a mechanism might be used in some way to organize the nervous system of nonsegmented vertebrates. Such a notion is reinforced, for example, by the fact that the sensory and sympathetic ganglia of the vertebrate peripheral nervous system are organized in a regularly repeated, or **metameric,** array along the spinal cord. In the CNS, evidence of segmental-like structures can be seen in the embryonic **hindbrain,** which becomes transiently subdivided by bulges called **neuromeres** (or, more specifically, **rhombomeres,** because they lie in the rhombencephalon) (Figure 8A). For a long time, however, the developmental significance of rhombomeres was disputed by neurobiologists, some of whom dismissed them as merely morphological curiosities.

The recent spectacular success in understanding the molecular biology and genetics of segmental pattern formation in *Drosophila* (Box D) has

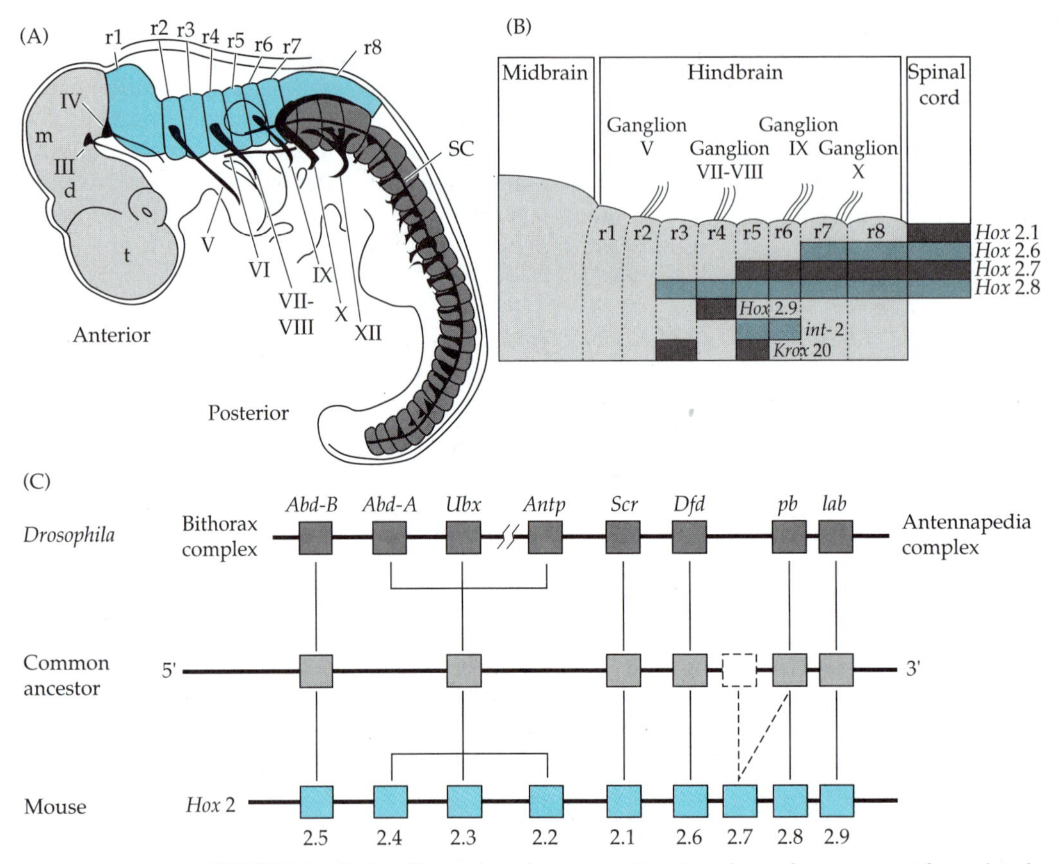

FIGURE 8. Role of *homeobox* genes in patterning the embryonic vertebrate hindbrain. (A) Saggital view of a stage-18 chick embryo. The eight rhombomeres are indicated by r1–r8. III–XII indicate cranial motor nerves; note that nerves V, VII, and IX exit the hindbrain through rhombomeres r2, r4 and r6. sc, spinal cord; m, mesencephalon; d, diencephalon; t, telencephalon. (B) Schematic illustrating the domains of expression of *Hox-2* cluster genes, *Krox-20* and *int-2* (an FGF homologue) with respect to rhombomeres r1–r8. The posterior boundaries of *Hox-2* cluster gene expression extend out of the picture to the right. (C) The mouse *Hox-2* gene complex has a similar arrangement to that of the bithorax complex and antennapedia complex *homeobox* genes in *Drosophila*. Note that the order of related genes in both complexes is identical, suggesting a common origin in an evolutionarily old complex ("common ancestor") antedating the divergence of protostomes and deuterostomes. (A from R. Keynes and A. Lumsden, 1989. Neuron 4: 1–9; B from D. G. Wilkinson and R. Krumlauf, 1990. Trends Neurosci. 13: 335–339.)

prompted a closer re-examination of neuromeres, in the hope that they might reflect underlying developmental mechanisms analogous to those discovered in the fruit fly. Such hopes now appear well on their way to being realized. Neuromeres indeed have been shown to represent metameric domains that correlate with spatial patterns of neuronal differentiation. Underlying this morphological and cytological patterning are restricted spatial domains of expression of genes homologous to those that control segmentation and segmental identity in *Drosophila*. This observation suggests that these genes may play a critical role in establishing and maintaining the positional identity of different neuromeres and hence of the various classes of neurons that develop within them.

Rhombomeres represent metameric domains of cellular and molecular differentiation

During embryonic development, the hindbrain (rhombencephalon) becomes subdivided into eight rhombomeres, called **r1–r8** (Figure 8A). In the hindbrain, different types of neurons develop in different rhombomeres. For example, the cell bodies of the **branchial motor nerves** V, VII, and IX are initially restricted to rhombomeres r2, r4, and r6, respectively (Figure 8A). Similarly, the cranial sensory ganglia V, VII, and IX are located adjacent to r2, r4, and r6, as well. The localization of specific groups of neurons to **alternating rhombomeres** is particularly intriguing. It implies that alternate rhombomeres are somehow more similar than adjacent rhombomeres. This pattern reflects the basic process that underlies segmentation: when adjacent groups of cells become different, segments can form. For example, in *Drosophila*, segments are formed by so-called **pair-rule genes,** such as ***fushi tarazu,*** which are expressed in the blastoderm in stripes that alternate with stripes of expression of other pair-rule genes (Box D).

That rhombomeres are molecularly distinct units has been revealed by staining them with monoclonal antibodies to various neuronal differentiation antigens. For example, an antibody that recognizes most or all developing neurons in the zebrafish embryo labels clusters of cells present in the center of each neuromere so that a periodic pattern of staining is seen. Staining with antibodies to neurofilament protein reveals bundles of axons that lie precisely along the boundaries between rhombomeres in the chick embryo. These patterns of cellular and axonal staining indicate not only that a periodic molecular pattern underlies the periodic morphologic pattern, but also that the boundaries between neuromeres are important elements of the patterning process. This idea is further reinforced by the fact that cell lineages developing within rhombomeres respect the boundaries between these structures. Recent experiments suggest that the lack of mixing between clones in adjacent rhombomeres is not due to a physical barrier to cell migration at the inter-rhombomere border, but rather to intrinsic biochemical differences between the cells in adjacent rhombomeres that prevent their intermixing. Neuromeres are thus repeating, segment-like units of cellular and molecular differentiation.

The boundaries of expression of genes encoding vertebrate homologues of Drosophila *transcription factors coincide with neuromere boundaries*

The segmental patterns of neuronal differentiation seen in rhombomeres could either be a consequence, or a cause, of the physical partitioning of the hindbrain. On the one hand, the physical subdivision of the hindbrain into neuromeres could produce local environments that affect the types of neurons that develop. On the other hand, a latent patterning of the hind-

Box D Pattern Formation in the Drosophila *Embryo*

One of the successes of modern developmental genetics and molecular biology is the understanding of the process of segmentation and pattern formation that occurs during early *Drosophila* embryogenesis. This advance is largely a consequence of a massive screen for embryonic lethal mutations carried out in the early 1980's by Christiane Nusslein-Völlhard and others. Careful analysis of the phenotypes of these mutations, of the interactions between different mutant genes, and, finally, of the molecular structure of these genes, has yielded insight into the mechanisms that build an early fly embryo.

The patterning process in *Drosophila* occurs through a cascade of regulatory interactions that is critically dependent upon an initial asymmetry in the unfertilized egg. This asymmetry is due, at least in part, to the asymmetric distribution of the products of genes such as ***bicoid,*** whose encoded protein is distributed in a relatively steep anterior-posterior gradient (Figure A). Such genes are referred to as *maternal effect* genes, because their products are contributed to the embryo exclusively by the female. Following fertilization, the information contained in such gradients is used to establish the anterior-posterior axis of the embryo. At about the same time, the dorsal-ventral axis of the embryo is generated by a different set of genes. Following the establishment of anterior-posterior and dorsal-ventral axes, the embryo becomes further partitioned by the products of zygotic regulatory genes, called **gap genes,**which are expressed in broad bands along the anterior-posterior axis. The region-specific expression of gap genes is in part controlled by the products of maternal effect genes such as *bicoid.* Some gap genes appear to respond to specific levels of bicoid protein, for example, so that the site of expression is determined by the position of nuclei along the bicoid concentration gradient.

The combined action of the maternal effect and gap genes then initiates the process of segment formation by regulating expression of the **pair-rule genes**. Pair-rule genes are expressed in seven bands of mRNA whose positions correspond to the future locations of alternate segments (more accurately, parasegments). These alternating stripes of pair-rule gene products define the prospective fourteen parasegments that will later physically subdivide the embryo. Within each parasegment, certain **segment polarity** genes are proposed to interact to form segmental boundaries. Along with the basic subdivision

brain, which precedes its overt partitioning, could cause different cell types to develop in different positions; this in turn could generate physical boundaries due to the immiscibility of different cell types. If the latter idea were valid, one might expect to see positional differences in the expression of regulatory genes prior to the formation of rhombomeres. In *Drosophila,* for example, the alternating stripes of pair-rule genes are expressed prior to overt segmentation of the embryo.

At least one putative nuclear regulatory gene shows such an early, spatially restricted pattern of expression in the developing mouse hindbrain. ***Krox-20,*** a gene in the **zinc-finger** class of DNA-binding proteins (Chapter 10), is expressed in patches that coincide precisely with the presumptive positions of alternate rhombomeres 3 and 5 (Figure 8B). This indicates that patches of neuroepithelium can be molecularly distinct *before* they become morphologically distinct. The expression of *Krox-20* in *alternating* prospective rhombomeres is, moreover, reminiscent of the expres-

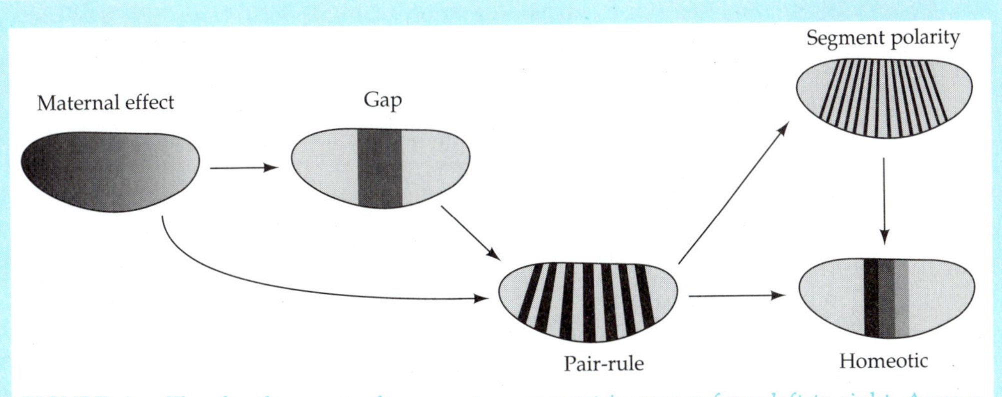

FIGURE A. The development of segments in the *Drosophila* embryo. Oval represents an embryo; shaded regions, zones of gene expression. The different classes of genes active at different stages of development are indicated adjacent to the embryos. Time of development increases from left to right. Arrows indicate that genes active at one stage of development control the expression of other genes at a later stage, or at the same stage. (After D. G. Wilkinson and R. Krumlauff, 1990. Trends Neurosci. 13: 335–339.)

of the embryo into segmental regions, each segment acquires a distinct identity, according to its position along the A-P axis. This is achieved through the differential expression of members of the two clusters of **homeotic genes,** the *Bithorax* and *Antennapedia* gene complexes (see Figure 8C). With the action of these genes, the basic overall body plan of the fly embryo is determined. Many of the genes controlling pattern formation encode transcription factors of one type or another. An important feature of this patterning process is that it fills a pre-existing space, defined by the boundaries of the egg. Initially, nuclei generated within this space exist in a syncitium, so that transcription factors are able to influence adjacent nuclei directly. Subsequently, each nucleus becomes surrounded by a cell membrane. In this way, cells that come to lie in different regions of the embryo fall under the influence of domains of regulatory gene expression that have been set up earlier in the developmental process.

sion of *Drosophila* pair-rule genes in alternating segments. This reinforces the idea that rhombomeres are formed by a patterning mechanism similar to that underlying segmentation in *Drosophila*. Although there is as yet no genetic evidence to illuminate the developmental role of *Krox-20,* the fact that its structure is predictive of DNA-binding activity strongly suggests that it functions to regulate other genes.

Rhombomere boundaries also match the boundaries of expression of a diverse group of homeobox-containing genes. The ***Hox-2*** gene cluster contains a subfamily of nine different homeobox genes (Chapter 10), which are located in a tandem array on mouse chromosome 11 (Figure 8C). These genes are highly homologous to those of the ***Bithorax*** and ***Antennapedia*** gene clusters in *Drosophila* (Figure 8C). In the mouse, different *Hox-2* genes are expressed in overlapping zones within the embryonic neuroepithelium, but each zone of expression has a sharply defined anterior boundary. In the hindbrain, these boundaries coincide with the boundaries between rhombomeres. For example, the anterior limit of *Hox-2.6* expression falls between r6 and r7, while that of *Hox-2.7* falls between r4 and r5 (Figure 8B). In contrast to *Krox-20, Hox-2* genes are expressed after rhombomeres become morphologically visible (about E9–9.5 in the mouse).

The function of *Hox-2* cluster genes in mouse development is not yet understood, but their evolutionary relationship to *Drosophila* genes provides some strong clues. In the fly, the different homeobox genes of the *Antennapedia* complex function to establish different positional identities for different segments along the A-P axis (Box D). The segment in which a given gene functions is related to its position within the *Antennapedia* cluster: genes located toward the "right" (3') end of the cluster tend to affect more anterior segments, while those located at the "left" (5' end) of the cluster affect more posterior segments. Amazingly, this correlation between chromosomal position and domain of expression has been evolutionarily conserved: in mouse, as in *Drosophila*, genes that lie towards the 3' end of the Hox cluster are expressed more anteriorly than those lying towards the 5' end (compare Figure 8B,C). Moreover, the relative order of all nine of the *Hox-2* genes along chromosome 11 corresponds to the relative order of each of their closest *Drosophila* homologues (Figure 8C). This parallel conservation of sequence, chromosomal location and anterior boundary of expression strongly suggests that the *Hox-2* genes play a role in specifying the positional identity of the hindbrain, at least, along the A-P axis.

Although the functions of *Krox-20* and *Hox-2* genes are not yet known, their timing of expression suggests that they could act in a cascade to pattern the vertebrate hindbrain in a manner analogous to their counterparts in *Drosophila*. For example, the early expression of *Krox-20* in alternating patches of hindbrain could cause adjacent groups of cells to become different from one another, leading to rhombomere formation. Subsequently, rhombomeres would acquire distinct positional identities according to which *Hox-2* genes they expressed. Cell-lineage restrictions imposed by rhombomere boundaries would prevent the mingling of cells expressing different *Hox-2* genes, thereby maintaining the positional identity of cells within each rhombomere. In this way, the hindbrain would be "carved up" into distinct spatial domains containing different cell types. Although this scheme is hypothetical, it should be testable using newly developed techniques for manipulating the mouse genome, such as gene disruption by homologous recombination.

Cell lineage, cell fate, and positional identity

We have discussed how the differential expression of homeobox-containing genes along the A-P axis of the hindbrain could impart different positional identities to cells in different regions of the hindbrain. How is positional identity related to the functional properties of a neuron? Positional identity may be specified independently of the properties that determine neuronal type. For example, all dorsal root ganglia neurons may resemble each other in their general physiological properties but differ in their positional identities at different points along the rostral-caudal axis.

Conversely, sensory neurons and their peripheral targets at a given axial level are completely different cell types, but may share positional information that reflects their common axial level of origin. In this respect, positional identity can be considered an aspect of cell phenotype that reflects *location* rather than function: cells must know where they are as well as what they should do.

The imposition of common positional identities upon phenotypically distinct cell types may be one important function provided by cell lineage in systems where lineage does not control cell fate per se. For example, a progenitor cell at a given location could transmit common positional determinants (e.g., patterns of Hox gene expression) to all of its progeny, even if those cells eventually acquired distinct functions. In this way, the cells would share a "memory" of a common positional origin. Such a mechanism could be important in establishing the appropriate connectivity between different cell types. For example, positional determinants could be transcription factors that control the expression of cell-surface molecules important in pathfinding and synaptogenesis (see Chapter 13). The identification of target genes controlled by homeobox proteins in the nervous system should shed light on this issue.

Summary

The molecules that control cell fate determination in the developing nervous system act in a cascade that runs from the cell surface to the nucleus and back again, and from one cell to another cell. These molecules include receptor-tyrosine kinases and EGF-repeat-bearing molecules at the cell surface, which function as receptors for intercellular signals, and mediators of lateral inhibition; a large family of polypeptide growth factors, which serve as extracellular signals for both cell proliferation and differentiation; and in the nucleus, DNA-binding proteins from a variety of multigene families, which control various stages of neuronal determination and differentiation. Among the DNA-binding proteins are homeobox-containing proteins, zinc-finger proteins, steroid-receptors, and helix-loop-helix proteins. Despite this increasingly large cast of regulatory characters, however, we know relatively little about their interactions. For example, the downstream targets of most of the nuclear regulatory genes are still unidentified, as are the ligands of many of the putative receptor-tyrosine kinases, and the molecules that transduce the signal provided by ligand binding from the cell surface to the nucleus. Powerful genetic and biochemical approaches are now available, however, to find these missing links.

Many molecules and mechanisms important for neural development control the development of non-neuronal cells and tissues as well, suggesting that the neural mechanisms evolved from fundamental and universal processes for building an organism from a single cell. What distinguishes the development of the nervous system from that of other tissues, however, is its tremendous complexity. There are an enormous variety of different cell types in the brain; this cellular diversity is reflected moleca great diversity of regulatory molecules, often members of multigene families, which act both inside and outside the cell. Cellular diversity is also generated by the combinatorial action of regulatory factors, as exemplified in studies of identified neuron development in simple invertebrates such as *C. elegans.*

The complexity of nervous system development is also due to intricate spatial organization of the cell types it contains. This degree of organization requires mechanisms for controlling the number of cells of a given type that are generated, and the schedule on which they are born. These mechanisms are only beginning to be understood. Perhaps most intriguing are the mechanisms that generate the overall pattern of cellular differentiation in the brain. These patterning processes provide an underlying topographic map, upon which cell type determination occurs, and are likely to play a key role in determining the development of appropriate patterns of connectivity. Although this process is now only dimly understood, recent insights suggest that it is likely to be based upon evolutionarily conserved mechanisms of profound developmental significance. This further strengthens the imperative of pursuing studies of neuronal development in a variety of organisms, no matter how phylogenetically distant they appear to be. Only through such a concerted and integrated effort will we be able to untangle this most difficult of all problems in developmental biology.

References

General references

Banerjee, U. and Zipursky, S. L. 1990. The role of cell–cell interaction in the development of the *Drosophila* visual system. Neuron 4: 177–187.

Ghysen, A. and Dambly-Chaudiere, C. 1989. Genesis of the *Drosophila* peripheral nervous system. Trends Genet. 5: 251–255.

Lumsden, A. 1990. The cellular basis of segmentation in the developing hindbrain. Trends Neurosci. 13: 329–335.

Raff, M. C. 1989. Glial cell diversification in the rat optic nerve. Science 243: 1450–1455.

Sanes, J. R. 1989. Analyzing cell lineage with a recombinant retrovirus. Trends Neurosci. 12: 21–28.

Sternberg, P. W., Liu, K. and Chamberlin, H. In press. Specification of neuronal identity in *C. elegans*. In: *Determinants of Neuronal Identity*, ed. M. Shankland and E. Macagno. Academic Press, New York.

Wilkinson, D. G. and Krumlauf, R. 1990. Molecular approaches to the segmentation of the hindbrain. Trends Neurosci. 13: 335–339.

Molecular genetics of invariant cell lineages

*Desai, C., Garriga, G., McIntire, S. L. and Horvitz, H. R. 1988. A genetic pathway for the development of the *Caenorhabditis elegans* HSN motor neurons. Nature 336: 638–646.

*Jan, Y. N. and Jan, L. Y. 1990. Genes required for specifying cell fates in *Drosophila* embryonic sensory nervous system. Trends Neurosci. 13: 493–498.

Way, J. C. and Chalfie, M. 1988. *mec-3*, a homeobox-containing gene that specifies differentiation of the touch receptor neurons in *C. elegans*. Cell 54: 5–16.

*Way, J. C. and Chalfie, M. 1989. The *mec-3* gene of *Caenorhabditis elegans* requires its own product for maintained expression and is expressed in three neuronal cell types. Genes Dev. 3: 1823–1833.

Drosophila *eye development*

Hafen, E., Basler, K., Edstroem, H. and Rubin, G. M. 1987. *sevenless*, a cell-specific homeotic gene of Drosophila, encodes a putative transmembrane receptor with a tyrosine kinase domain. Science 236: 55–63.

Mlodzik, M., Hiromi, Y., Weber, U., Goodman, C. S. and Rubin, G. M. 1990. The Drosophila *seven-up* gene, a member of the steroid receptor gene superfamily, controls photoreceptor cell fates. Cell 60: 211–224.

Reinke, R., and Zipursky, S. L. 1988. Cell–cell interaction in the *Drosophila* retina: The *bride of sevenless* gene is required in photoreceptor cell R8 for R7 cell development. Cell 55: 321–330.

Diffusible signals in vertebrate development

Anderson, D. J. 1989. The neural crest cell lineage problem: Neuropoiesis? Neuron 3: 1–12 47.

Doupe, A. J., Patterson, P. H. and Landis, S. C. 1985. Small intensely fluorescent (SIF) cells in culture:

Role of glucocorticoids and growth factors in their development and phenotypic interconversions with other neural crest derivatives. J. Neurosci. 5: 2143–2160.

*Raff, M. C., Miller, R. H. and Noble, M. 1983. A bipotential glial progenitor cell. Nature 303: 390–396.

The Notch *locus*

Artavanis-Tsakonas, S. 1988. The molecular biology of the *Notch* locus and the fine tuning of differentiation in *Drosophila*. Trends Genet. 4: 95–100.

Campos-Ortega, J. A. 1988. Cellular interactions during early neurogenesis of *Drosophila melanogaster*. Trends Neurosci. 11: 400–405.

Doe, C. Q. and Goodman, C. S. 1985. Early events in insect neurogenesis. II. The role of cell interactions and cell lineage in the determination of neuronal precursor cells. Dev. Biol. 111: 206–219.

Fehon, R. G., Kooh, P. J., Rebay, I., Regan, C. L., Xu, T., Muskavitch, M. A. T. and Artavanis-Tsakonas, S. 1990. Molecular interactions between the protein products of the neurogenic loci *Notch* and *Delta*, two EGF-homologous genes in *Drosophila*. Cell 61: 523–534.

Timing of differentiation

*Lillien, L. E. and Raff, M. C. 1990. Differentiation signals in the CNS: Type-2 astrocyte development in vitro as a model system. Neuron 5: 111–119.

Watanabe, T. and Raff, M. C. 1990. Rod photoreceptor development in vitro: Intrinsic properties of proliferating neuroepithelial cells change as development proceeds in the rat retina. Neuron 4: 461–467.

Achaete-scute complex and helix-loop-helix genes

*Davis, R. L., Weintraub, H. and Lassar, A. B. 1987. Expression of a single transfected cDNA converts fibroblasts to myoblasts. Cell 51: 987–1000.

Ellis, H. M., Spann, D. R. and Posakony, J. W. 1990. *extramacrochaete*, a negative regulator of sensory organ development in *Drosophila*, defines a new class of helix-loop-helix proteins. Cell 61: 27–38.

*Ghysen, A. and Dambly-Chaudiere, C. 1988. From DNA to form: The *achaete-scute* complex. Genes Dev. 2: 495–501.

Pattern formation in the hindbrain

*Graham, A., Papalopulu, N. and Krumlauf, R. 1989. The murine and *Drosophila* homeobox gene complexes have common features of organization and expression. Cell 57: 367–378.

Keynes, R. and Lumsden, A. 1990. Segmentation and the origin of regional diversity in the vertebrate central nervous system. Neuron 2: 1–9.

12

Process Outgrowth and the Specificity of Connections

Paul H. Patterson

BECAUSE THE FUNCTION OF THE NERVOUS SYSTEM is based on a highly specific and reproducible pattern of connections between neurons, a central problem in the development of the system is how the wiring diagram is specified. A priori, two extreme models can be envisioned. One model posits that there is little specificity in the outgrowth of axons and dendrites, and that the final pattern of synapses emerges through a selective loss of functionally inappropriate connections. By this hypothesis, random outgrowth is followed by selective stabilization of a wiring pattern that allows proper function of the system. Inappropriate connections would be lost by resorption of axons or by the death of neurons making these errors. In fact, there *is* considerable neuronal death during normal development, but, as will be discussed in more detail in Chapter 13, most of this death is not for the purpose of error correction. Similarly, the idea of an exuberance of initial connections that is selectively pruned by axonal retraction during development is also valid. Moreover, there is strong evidence for selective stabilization of connections by mechanisms involving neuronal activity and competition. As will be discussed below, however, the differential activity in discrete axons that is used to refine the final patterns of connections in several systems does not explain the specificity of the wiring pattern. Initial axonal outgrowth is not, in fact, random, and nerve activity is not responsible for the highly specific patterns of growth that are observed. In addition, highly selective connections can be made in the complete absence of evoked nerve activity; that is, specificity can occur in the absence of functional validation of the connections that are formed.

Another extreme model of specificity posits that the wiring diagram is based on chemical labels that uniquely identify each neuron. Such a system of recognition molecules could, in fact, be feasible for invertebrate nervous systems, where the number of neurons is relatively small, and where particular, individual neurons can be routinely identified from animal to animal, based on their unique shapes, positions, or the molecules expressed on their surfaces. In higher vertebrates, on the other hand, the vast numbers of neurons, their multitudinous interconnections, and the lack of unique features to distinguish among many of them have led many authors to discount the idea of a molecular labeling system for neurospecificity. Other theorists have devised labeling systems involving gradients and combinations of molecules to overcome the numbers problem.

A significant inspiration in the search for recognition molecules has been Roger Sperry's experimental results and his vigorous articulation of the **chemoaffinity hypothesis** (Box A). In fact, recent work has identified a number of candidate recognition molecules within various target areas, as will be discussed below. In addition, molecules have been discovered that are likely cues for the specific guidance of particular axons towards selected target locations. This work, coupled with that on the role of activity and the selective removal of connections, has revealed that the final wiring pattern of the nervous system is not achieved by a single mechanism of specificity but by several, sequential mechanisms that progressively refine the choices made by a developing axon. Initially, growing axons are selectively guided to the target area, they then recognize particular target cells in that area, and, finally, their pattern of connections is refined after the initial synapses are formed. Thus, a number of more or less discrete forces influence connectivity, from the earliest outgrowth of neuronal processes through the modifications of synapses in adulthood. Questions that require answers include: how is the outgrowth of processes initiated, what is the mechanism of elongation, how is the rate of growth controlled, what determines the direction of growth, how are the number and sites of branch points decided, why does growth cease, and how is sprouting at quiescent synapses reinitiated?

Guidance of Outgrowth

Growth cones direct neurite outgrowth

In 1880 Ramón y Cajal observed and named the growth cone (*"cono de crecimiento"*) as the structure leading a new extension of cytoplasm that will become an axon or a dendrite. Although he had only fixed, histological slides on which to make observations, Ramón y Cajal's vivid imagination saw the growth cone in these static images as a "sort of club or battering ram, endowed with exquisite chemical sensitivity, with rapid ameboid movements, and with certain impulsive force, thanks to which it is able to proceed forward and overcome obstacles met in its way, forcing cellular interstices until it arrives at its final destination." He further pondered, "what mysterious forces precede the appearance of these prolongations, promote their growth and ramification, and finally establish the protoplasmic kisses, the intercellular articulations that appear to constitute the final ecstasy of an epic love story."

Growth cones, which were first observed in the living state by Ross Harrison, the inventor of modern tissue culture, have at their center a broad, flattened area at the end of the growing neurite, called the body or palm. Extending from the palm are thin, fingerlike **filopodia**, or a ruffling membrane called a **lamellipodium**, or a combination of a lamellipodium

Box A Sperry's Legacy

Roger Sperry's writings on the retinotectal system from the late 1930s through the early 1960s had a major influence on several generations of biochemists and molecular biologists. Buoyed by their success in unraveling the molecular basis of recognition by enzymes and antibodies, many investigators eagerly applied their skills in the search for the molecules that Sperry (Figure A) postulated to be the basis of neurospecificity. The result has been an exponentially increasing harvest of molecules from the extracellular and surface membrane that are of potential importance for the developing nervous system. Concurrently, a salient theme in the history of science is also on display: new results force the revision of old dogmas.

Santiago Ramón y Cajal, the father of developmental neurobiology, clearly saw the specific nerve growth patterns in embryos, and, as early as 1890, John Langley demonstrated that selective connections can be reestablished during regeneration in adult mammals. These and other early investigators explained their observations with terms such as "chemiotaxis" and "chemotropism," implying that chemical recognition was the basis of the wiring of the nervous system. This fundamental idea came under attack in the 1930s and 1940s, when new theorists proposed that specificity arises from nonselective growth followed by elimination of inappropriate connections. This view became the new orthodoxy until it was challenged by Sperry, along with R. Matthey and L. S. Stone, each working independently. In Sperry's case, his experiments on the retinotectal system overturned dogmas for which both his doctoral and postdoctoral advisors had become famous. His experiments demonstrated that retinal axons persistently innervated their correct targets in the brain even when they were given alternative choices or forced to take novel routes. Sperry interpreted his results by proposing gradients of complementary molecules on the surfaces of retinal and tectal neurons that specifically identified the position of each within a two-dimensional array corresponding to the visual field.

FIGURE A. Roger Sperry at age 68. (Photograph courtesy of California Institute of Technology.)

Although Sperry considered that his experiments had solved the fundamental problem, thus leaving only the identities of the molecules to be determined, present-day revisionists proclaim that the surface recognition molecules envisioned by Sperry, even if found, will prove to play a relatively insignificant role in neurospecificity. According to this view, neuronal activity patterns and competition for trophic support are the major factors in establishing appropriate connections. As the relevant adhesion molecules and surface labels for specific pathways and target cells have been identified only recently, these conjectures are just now being tested. The one certainty is that future Sperrys will try to "prove" the prevailing experts wrong.

with filopodia (Figure 1; see also Chapter 8). Time-lapse studies have confirmed Ramón y Cajal's notion of growth cones as active, mobile structures; moreover, their structures are adaptable and can have different appearances at various stages of development (Figure 2). Enlarged growth cones with filopodia are often found at sites where the neurite is translocating slowly and making choices about which direction to take. In contrast, growth cones have a more narrowed, bullet-like appearance when they move rapidly along a straight course. These observations correlate well with the suggestion that filopodia are part of a sensory apparatus that "tastes" the environment and determines the direction of axonal growth. Indeed, growth cones are often observed to turn in the direction in which filopodia adhere best. In adhering to appropriate surfaces, filopodia can exert measurable tension on the rest of the axon. In one startling experiment in which axons were severed, the growth cones were seen to run on, like dogs trailing their leashes behind them.

The growth cone integrates external signals acting on the growing neurite, and translates them into changes in the rate and/or direction of growth. Before considering how this translation occurs, the behavior of growth cones and the signals that impinge on them will be considered.

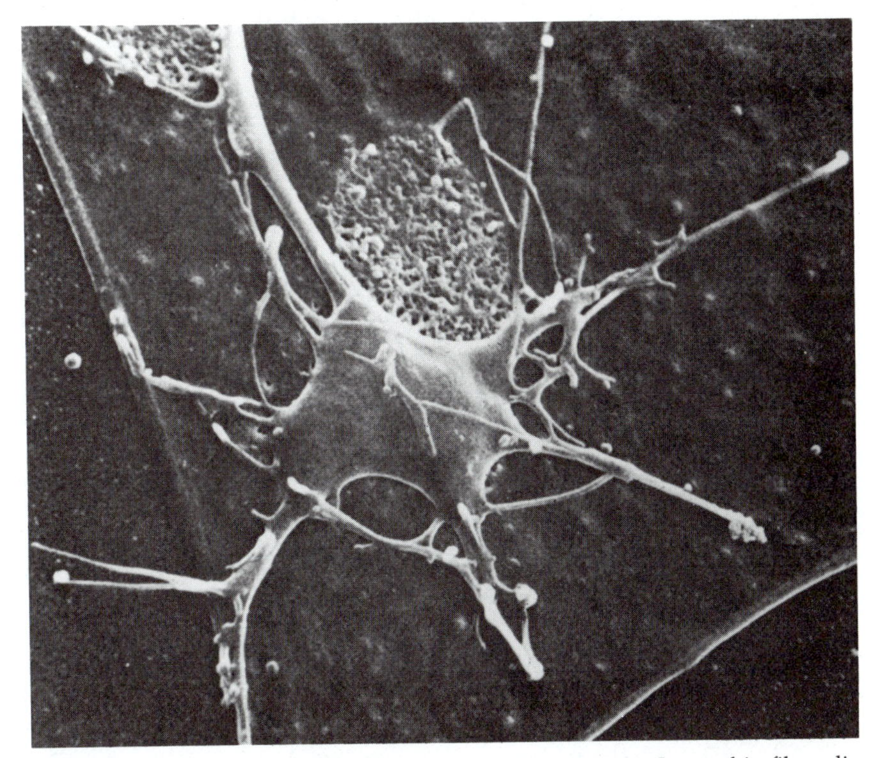

FIGURE 1. Scanning electron micrograph of a growth cone. Long, thin filopodia project out from the body of the structure, contacting a flat cell and the substratum. A ruffling lamellipodium is apparent between the filopodia. (Photograph courtesy of S. C. Landis.)

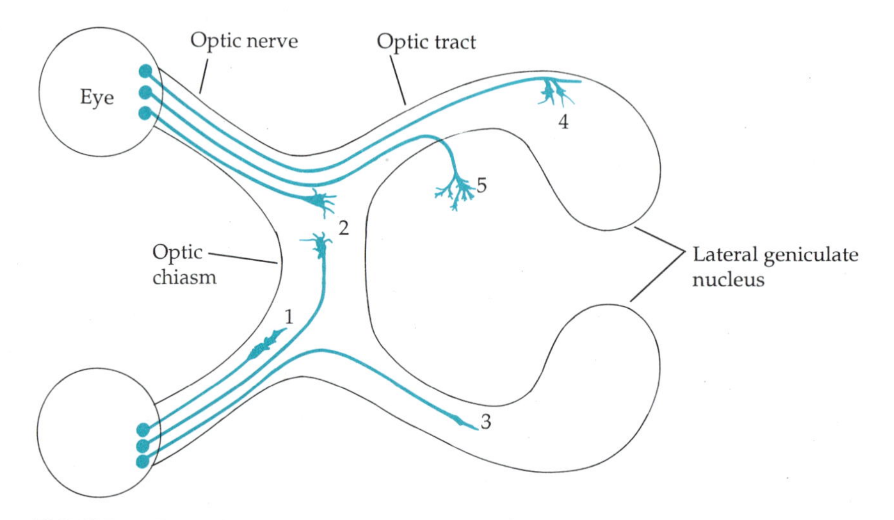

FIGURE 2. Retinal ganglion cell growth cones display different shapes as they traverse the pathway to the tectum. At stage 1, while still in the optic nerve, growth cones are large but lack filopodia. At stage 2, as they pause in the optic chiasm, multiple filopodia appear. While growing rapidly in the optic tract (stage 3), growth cones become quite narrow. Nearing target regions such as the lateral geniculate nucleus (stage 4), filopodia appear to be directed towards the target. Within a target region (stage 5), growth cones display highly branched, large arbors. (From P. Bovolenta and C. Mason, 1987. J. Neurosci. 7: 1447–1460.)

Axons grow to their correct target areas by following environmental cues

The first step in establishing specific connections is the outgrowth of axons to the appropriate part of the nervous system or peripheral tissues of the body. In both vertebrate and invertebrate nervous systems, growing axons appear to follow defined pathways to the target area. Some of the cues along these pathways are provided by other cells, including other axons that have preceded later arriving neurites; other cues are provided by the extracellular matrix.

In the embryonic insect nervous system, the axons and growth cones of individual neurons can be reliably identified in each ganglion and in every animal so that the axonal outgrowth of particular neurons can be followed in repeated experiments. In the central nervous system, axons grow longitudinally along **fascicles**, or bundles, of axons that have preceded them. There are multiple fascicles, however, and a particular neuron must cross over many bundles in order to find the fascicle to which it belongs (Figure 3). At each "choice point," the decision of whether or not to turn defines the pathway that each axon takes to arrive at its appropriate target cell. If a fascicle that is part of the normal pathway of an identified growth cone is eliminated by experimentally ablating the neurons whose axons make up that bundle, the growth cone arrives at the critical point, hesitates for some time, and wanders off in an apparently aimless direction. Such

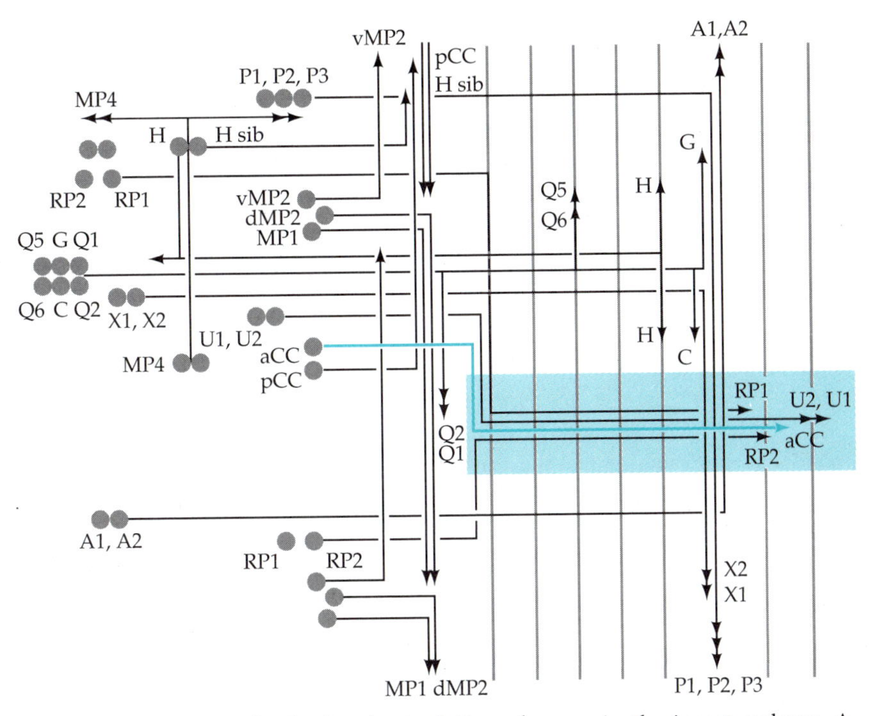

FIGURE 3. Patterns of selective fasciculation of axons in the insect embryo. A fraction of the identified neurons in half of a segmental ganglion are labeled with their distinctive names, and the directions their axons take in the period of initial outgrowth are schematized. The axons form an orthogonal array of bundles by selective fasciculation. A given neuron, such as aCC (colored arrow, center), makes multiple choices as it grows, passing by some bundles (such as that formed by pCC, MP1, and dMP2) and joining up and growing along other bundles (such as that formed by U1, U2, RP1, and RP2). If the development of aCC is delayed so that its axon grows out at a later time, it apparently cannot read these environmental signals, and its growth cone wanders aimlessly in the northeast direction. Note that during neuronal development a given axon may grow along a certain bundle for part of its length, and then diverge and fasciculate with another bundle (see the path of RP1, upper left). Each of these distinctive fascicles are thought to express characteristic surface labels, some of which are illustrated in Figure 9. (From C. S. Goodman et al., 1984. Science 225: 1271–1279.)

observations strongly suggest that fascicles express different molecular labels that can be read by growing axons. As will be described below, several such labels have been identified, some of which have homology to members of the immunoglobulin superfamily of vertebrates.

Growth cones can use non-neuronal cells and the extracellular matrix, as well as other neurons, as cues for directional growth. In insect appendages, identified sensory neurons grow into the CNS along highly stereotyped pathways whose cues include (1) the border formed by the epithelial cells between the segments of the appendage, (2) an apparent distal-to-proximal gradient in the epithelium, and (3) particular epithelial cells that differentiate into neurons (Figure 4). The latter cells have been termed

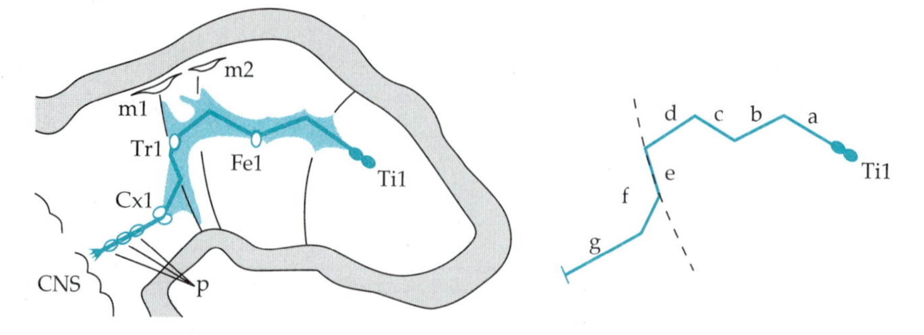

FIGURE 4. Growth cones use many cues along their pathways. The direction of growth cone movements in the insect leg made by the sibling Ti1 neurons as they grow into the CNS indicates several of the cues they use to navigate. Growth cones always emerge in the proximal direction, toward the CNS. Other experiments suggest that this is due in part to a distal-proximal gradient of positional information in the appendage. The initial path of 50 μm , illustrated as segment a on the right, is then reoriented in a proximal-ventral direction as contact is made with the first guidepost cell, neuron Fe1. This forms segment b of the path. At this point, the growth cones orient toward the m1 and m2 cells, forming the c segment. The d segment is formed by a contact with the second guidepost cell, Tr1. From this point, the growth cones turn sharply to follow the segment boundary in the epithelium, forming part e of the pathway. The f segment is formed by a distinct reorientation towards the next guidepost cells, the Cx1 cells. Leaving these cells, the Ti1 growth cones follow a direct line into the CNS along the p cells. The swath that the growth cones sweep as the axons grow is illustrated by the colored region. (From M. Caudy and D. Bentley, 1986. J. Neurosci. 6: 1781–1795.)

"stepping stones" or "**guidepost cells**"; ablating some of these cells with a laser beam results in misdirected growth by the sensory neuron axons. The molecular basis for the cues provided by these cells is not yet understood.

It is also worth noting that the cues in the epithelium just described are primarily used by the initial sensory axons to grow out. Such pathfinding axons are found in many parts of the developing nervous system, and they are often referred to as **pioneer axons**. The large number of axons that grow out after the pioneers tend to track along the prior axons rather than follow the cues on non-neuronal cells. Thus, most axons in the nervous system grow along other axons; the molecules on axons that mediate this behavior are discussed in the following sections. It should be noted, however, that the distinguishing feature of pioneer axons is that they are the first to grow; they are not always qualitatively different from the axons that grow out subsequently. This is shown by experiments in which the pioneer neurons are ablated; subsequent axons can sometimes read the cues provided by the non-neuronal cells, becoming pioneer axons themselves.

In vertebrates, motor neurons of the embryonic chick spinal cord also follow a defined route in innervating muscles of the limb (Figure 5). The neurons that innervate a particular muscle are grouped within the spinal cord to form **motor pools**. When axons grow out, they take a stereotyped pathway directly to the muscle. In the beginning of the pathway, axons

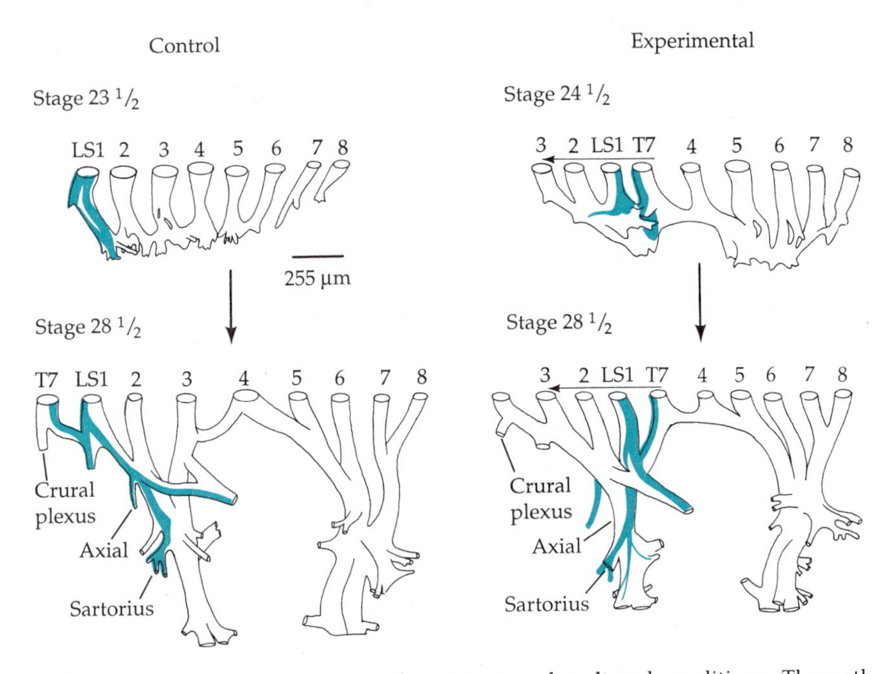

FIGURE 5. Motor axon outgrowth patterns under altered conditions. The pathways taken by chick motor neurons projecting from segments T7 and LS1 were observed by local injections of HRP (color) into the spinal cord before (top) and after (bottom) the period of motor neuron death. In the experimental embryos, four segments of spinal cord were reversed (T7-LS3) at an earlier stage. By stage 28, the injected axons had grown out to their appropriate muscle branches by taking novel routes through the axon plexus. (From C. Lance-Jones and L. Landmesser, 1980. J. Physiol. 302: 581–602.)

from many motor pools grow together, and these become progressively sorted out as the axons near the individual muscles. Although axons always follow a stereotyped pathway, they are not completely dependent on it. If the normal pathway is disrupted, axons can often compensate and find their appropriate muscle using a novel route. For instance, if the limb bud is rotated so that muscles develop in a new position relative to the spinal cord, the appropriate motor neuron axons often find their way to the misplaced muscle. Similar results are seen when the spinal cord is rotated before the axons have grown out. These observations indicate that specific pathways provide a first mechanism for directing growing axons but that their disruption reveals subsidiary mechanisms of specific recognition or guidance between nerve and muscle. These subsidiary mechanisms operate over short distances, however, so that correct pathfinding is observed after relatively small perturbations in the pathway.

Extracellular matrix molecules pave the highways

What are the molecules responsible for these selective patterns of growth? It is useful to begin by distinguishing between broad "highways" of growth, used by many diverse sets of axons, and the more narrowly defined "driveways" to specific subsets of target cells. Many such highways are formed by the extracellular matrix of the basal laminae on epithelia, blood vessels, etc. Several major constituents of this matrix, such as collagen, fibronectin, laminin, and tenascin are large proteins to which many types of cells adhere well (Figure 6). These proteins are also good substrates for neurite outgrowth from many kinds of neurons, **laminin** generally being the most effective (Figure 7). Laminin is composed of several different subunits that can be associated in different combinations

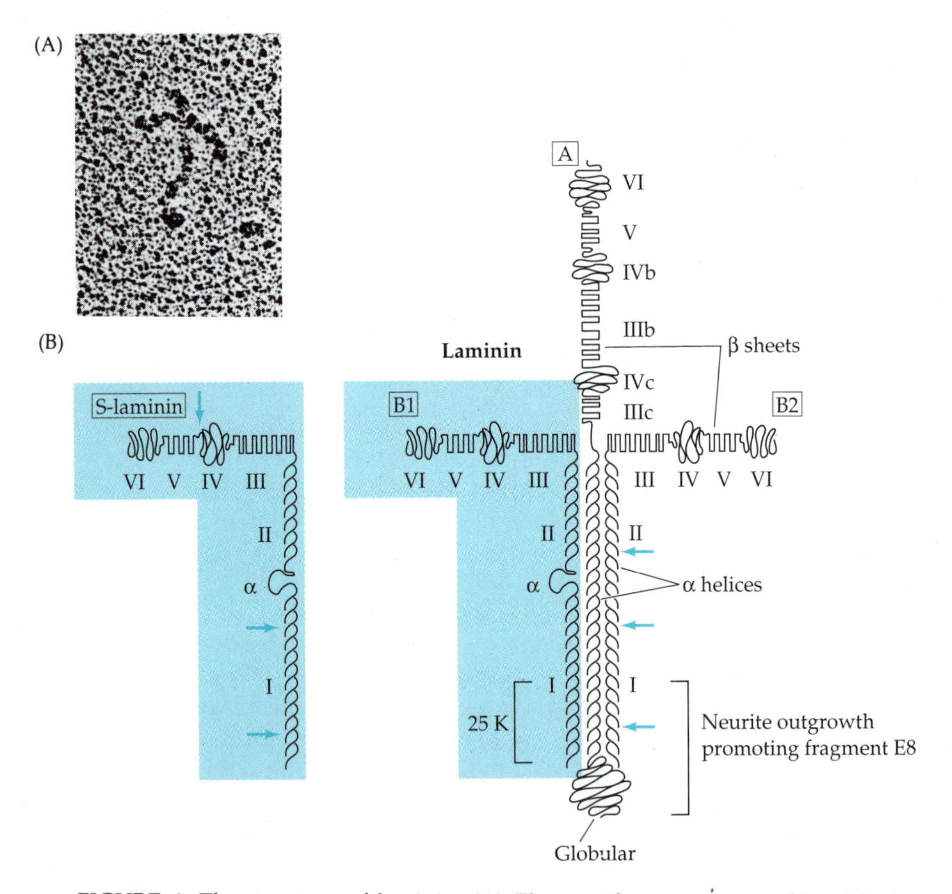

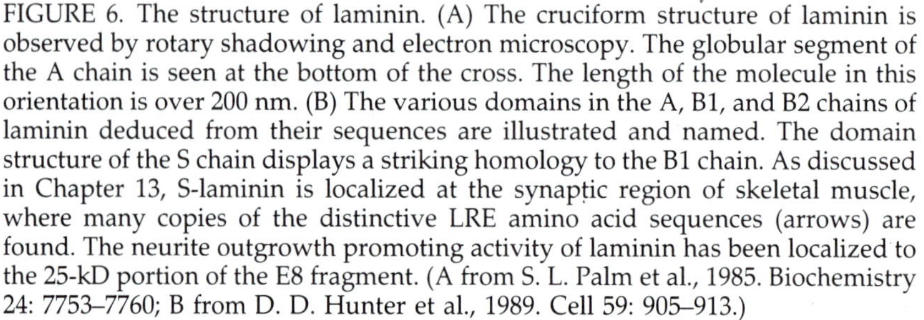

FIGURE 6. The structure of laminin. (A) The cruciform structure of laminin is observed by rotary shadowing and electron microscopy. The globular segment of the A chain is seen at the bottom of the cross. The length of the molecule in this orientation is over 200 nm. (B) The various domains in the A, B1, and B2 chains of laminin deduced from their sequences are illustrated and named. The domain structure of the S chain displays a striking homology to the B1 chain. As discussed in Chapter 13, S-laminin is localized at the synaptic region of skeletal muscle, where many copies of the distinctive LRE amino acid sequences (arrows) are found. The neurite outgrowth promoting activity of laminin has been localized to the 25-kD portion of the E8 fragment. (A from S. L. Palm et al., 1985. Biochemistry 24: 7753–7760; B from D. D. Hunter et al., 1989. Cell 59: 905–913.)

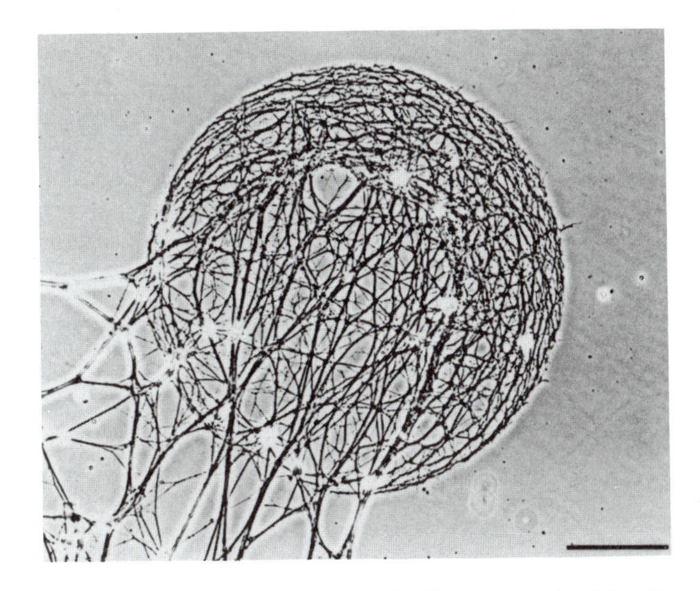

FIGURE 7. Neurites can follow pathways of adhesive proteins. Neurites growing on a collagen-coated culture dish encounter a dot of laminin and branch prolifically, filling the dot with processes. Since laminin is a more favorable substrate than collagen, the neurites do not leave the dot. (From R. W. Gundersen, 1987. Dev. Biol. 121: 423–43.)

to yield distinct forms of the molecule. The forms can have different biological functions, as discussed in the section on the neuromuscular junction in Chapter 13. Molecules like laminin are also extremely large and have many distinct domains. Some of these domains are sites of adhesion for neurons and other cells. Other domains bind and regulate the activities of other proteins such as growth factors and large proteoglycans. Laminin, for instance, binds **heparan sulfate proteoglycan** specifically and with high affinity. There is evidence that the complex with heparan sulfate is an even more effective substrate for outgrowth than laminin alone. Antibodies specific for this complex can inhibit nerve regeneration in vivo as well. Recent work with several mutants of the nematode *C. elegans* has shown that certain neurons display misdirected axonal growth during embryogenesis. One of these mutations, *unc-6*, is in the gene for the worm homologue of the B2 subunit of vertebrate laminin.

Integrins mediate neurite outgrowth in the extracellular matrix

Neurons recognize and adhere to molecules in the extracellular matrix via receptors on the surfaces of growth cones and axons. One large family of receptors for these matrix proteins is called the **integrins**; the name refers to their function of integrating extracellular signals with cytoplasmic responses. The integrins are transmembrane glycoproteins containing two distinct subunits; many different α and β subunits have been described thus far. Various neural and non-neural cells express distinct combinations of these subunits, yielding heterogeneity in binding specificities (Table 1).

TABLE 1 Composition of Various Integrins and Their Preferred Ligands[a]

β Subunit	α Subunit	Ligands
β_1	α_1	Collagen, laminin
	α_2	Collagen, laminin
	α_3	Laminin, fibronectin
	α_4	Fibronectin, V-CAM 1
	α_5	Fibronectin
	α_6	Laminin
	α_7	Laminin
	α_{VN}	Fibronectin, vitronectin
β_2	α_{LFA}	I-CAM 1, I-CAM 2
	$\alpha_{Mac\text{-}1}$	C3bi, fibrinogen
	α_{p150}	?
β_3	α_{IIb}	Fibronectin, vitronectin, von Willebrand factor, fibrinogen
	α_{VN}	Vitronectin, thrombospondin, von Willebrand factor, fibrinogen
β_4	α_6	?
β_5	α_{VN}	Vitronectin, fibronectin
β_P	α_4	?

[a]The integrins are composed of various combinations of α and β subunits, which produce different ligand preferences among the many extracellular matrix proteins. (From L. F. Reichardt and K. J. Tomaselli, 1991. Annu. Rev. Neurosci. 14: 531–570.)

Distinct integrin heterodimers mediate cell attachment to collagen and laminin, for instance, and even to different sites on laminin. Antibodies directed against particular integrin subunits can thus selectively block the interaction of certain types of neurons with specific matrix proteins. The integrins transduce signals across the membrane by regulating second messenger levels, as well as by direct interactions with cytoskeletal proteins. Among the cytoskeletal proteins thought to interact with the cytoplasmic domain of the integrins are talin, vinculin and α-actinin, each of which is found in neuronal growth cones.

In addition to using the biological flexibility of laminin afforded by its multiple forms and multiple sites, its complexes with other molecules, and the variety of laminin receptors, neurons alter their expression of matrix receptors at various stages of development. For example, as retinal ganglion cells grow along a pathway rich in laminin toward the optic tectum, they express high affinity laminin receptors and they respond to laminin-containing surfaces in culture assays by rapidly growing neurites. Upon reaching the laminin-poor tectum, however, the ganglion cells down-regulate these receptors and lose their outgrowth response to laminin surfaces when tested in cell culture. The down-regulation is, in fact, a response to contacting the tectum, because ganglion cells do not decrease their laminin receptors to the same extent in embryos that have had the tectum ablated.

The lower laminin-binding activity is correlated with down-regulation of the expression of the gene encoding the α_6 integrin subunit.

Other matrix receptors that are not members of the integrin family also mediate neurite outgrowth on laminin. One of these is a cell surface galactosyltransferase, an enzyme that binds laminin and modifies its carbohydrate side chains. Blocking the activity of this enzyme, or the glycosyl substrates that it recognizes on laminin, can interfere with neurite outgrowth. The role of carbohydrates in axonal growth and pathfinding will be considered again later in this chapter.

Cadherins promote neurite outgrowth

In addition to interactions with the extracellular matrix, neurons also express membrane proteins that interact directly with integral membrane proteins on other cells. The most widely characterized of these membrane interactions are those mediated by the **cadherin** (**CAD**) family and by the **immunoglobulin (Ig)** superfamily (Figure 8). These molecules promote adhesion between cells by **homophilic interactions** in which the same molecules on different cells bind to each other. That is, a given molecule

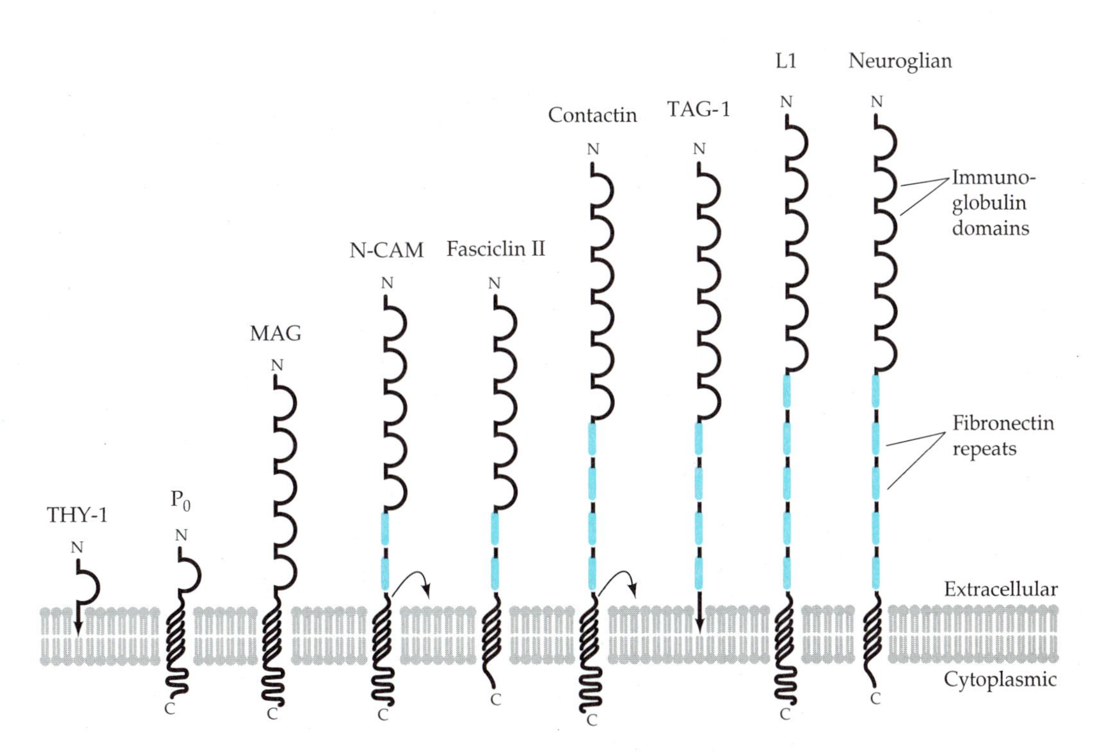

FIGURE 8. A schematic diagram of neural molecules that are members of the immunoglobulin superfamily. Many of these function as cell surface adhesion molecules. The immunoglobulin fold domains are indicated by the half circles, and fibronectin type III repeats are indicated by colored boxes. Thy-1, TAG-1, and variant forms of contactin and N-CAM are attached to the plasma membrane by a phosphoinositol linkage as indicated by arrows.

on one cell, such as neural cadherin (N-CAD), binds to another N-CAD molecule on an adjacent cell instead of binding to a distinct N-CAD receptor. While both families use the homophilic mechanism, CAD-CAD binding requires calcium, and interactions between members of the Ig family do not. As is discussed below, homophilic interactions may be particularly important in mediating fasciculation, a process in which like axons bind to each other.

CADs mediate cell adhesion between a variety of cell types, and six different CADs have been described thus far. N-CAD is found in the nervous system and in mesodermal derivatives, P-CAD in the placenta and a number of other specialized structures, and E-CAD in epithelium and liver. Homophilic binding is selective among members of this family. If P- and E-CAD, for example, are transfected into separate populations of CAD-negative cells and the cells mixed, two types of aggregates form in calcium-containing medium—those containing P-CAD–positive cells and those containing E-CAD–positive cells. Deletion experiments, and experiments in which various domains of one CAD are exchanged for another, have shown that part of the amino terminal, extracellular domain contains the specificity information required for homophilic binding. Surprisingly, these experiments have also shown that the COOH terminal, cytoplasmic domain, which is highly conserved among all CADs, is also necessary for adhesion. The requirement for a cytoplasmic component is intriguing, not only because the binding site is extracellular but because another member of the family, T-CAD, apparently has no cytoplasmic domain. Perhaps T-CAD is normally released from the cell, or acts as an inhibitor of other CADs in the cell's membrane.

Several experiments suggest that CADs play a role in neurite outgrowth. Thus, antibodies against N-CAD inhibit the growth of retinal ganglion cell neurites on astrocytes in culture, and transfection of fibroblasts with a cDNA for N-CAD converts them from cells that do not promote neurite outgrowth to those that do. Other experiments suggest that N-CAD may play a role in the formation of the neural tube. Whether CADs are involved in fasciculation or in mediating specific patterns of outgrowth is not yet known.

Calcium-independent adhesion proteins of the immunoglobulin family promote neurite outgrowth

Many calcium-independent adhesion proteins belong to the Ig superfamily. These proteins can promote neurite outgrowth, and antibodies against them can cause defasciculation of axon bundles. Some of the Ig family members are also expressed by non-neuronal cells such as muscle cells and glia, and these proteins are likely to mediate neuronal interactions with such cells as well. To be considered a member of the Ig family, a protein must contain one or more domains resembling the **Ig fold** in its amino acid sequence (Figure 8). The Ig fold, which is used by antibodies for antigen recognition, is bounded on each side by cysteine residues that form a disulfide link. Members of the family usually share sequence homology on each side of the cysteine residues. The diversity and specificity of antigen-antibody binding is based on the tremendous variability in the amino acid sequences of the Ig fold domain. Since most other members of the family, especially those known to be important in the nervous system, do not display this characteristic sequence heterogeneity in the Ig fold, this region cannot play a variable recognition role in those molecules. Therefore it is not likely that these proteins, such as the **neural cell adhesion molecule (N-CAM),** subserve the same sort of recognition function in the nervous system as do antibodies in the immune system. Nonetheless, the strong conservation of this extracellular domain through evolution suggests that it is well-suited for cell surface interactions of many different types. Besides the Ig fold, other regions of Ig family molecules form strong intermolecular interactions, both homophilic (between constant domains on heavy chains of antibodies, or between N-CAMs) and heterophilic (between heavy and light chains of antibodies, and between the subunits of the T cell receptor), as well as with matrix molecules such as heparan sulfate proteoglycan (N-CAM).

Thy-1 and **P_0**, which contain only one Ig fold domain, are the smallest members of the Ig family (Figure 8). These two proteins are highly concentrated in the nervous system in most species. Recent evidence suggests that Thy-1, which represents a major fraction of the surface membrane protein on long axons, may stabilize neuronal membranes and inhibit neurite sprouting. P_0 is the most abundant protein in PNS myelin, where it acts as a homophilic adhesion molecule, holding the extracellular membranes of myelin together as the Schwann cell wraps around the axon (see Chapter 9). In contrast to these small proteins, N-CAM contains five Ig fold domains. Moreover, unlike Thy-1, P_0, and the CADs, many different forms of N-CAM are generated by alternative splicing of its mRNA, including one form that lacks a cytoplasmic domain and is secreted. Although one of the alternative splicing variants is tissue-specific (found only in muscle), and several are developmentally regulated, the functional importance of this heterogeneity remains to be determined.

Another source of N-CAM heterogeneity is in its glycosylation. In particular, the protein can carry a very distinctive, polysialic acid sidechain that is highly charged. The homophilic binding of N-CAM, which can be seen by the specific aggregation of lipid vesicles containing the protein, is inhibited by the presence of polysialic acid on the N-CAM, suggesting that the bulky polysaccharide prevents protein-protein interaction. Since many axons, as well as non-neuronal cells such as muscle and glia, express N-CAM at various developmental stages, it is possible that this protein could mediate axonal fasciculation and/or nerve-target adhesion. A role for N-CAM in neurite outgrowth is suggested by recent experiments in which N-CAM–negative non-neuronal cells transfected with N-CAM acquire the ability to elicit outgrowth from retinal ganglion cells. Moreover, injection of antibodies against N-CAM into the embryonic eye or tectum disrupts the orderly growth of retinal axons into the tectum. A role for the polysialic acid component of N-CAM in regulating cellular interactions has also been indicated by in vivo experiments. Injection of a specific endoneuraminidase into embryonic chick muscle removes this oligosaccharide from the N-CAM on axons innervating the muscle, resulting in increased axonal fasciculation and decreased nerve branching on the muscle.

The dynamic balance between axonal fasciculation and axon-target interaction is further supported by results obtained with antibodies against another Ig family member, **L1**. This protein is apparently capable of both homophilic and heterophilic binding and is expressed on many axons early in development. Like N-CAM, L1 can promote neurite outgrowth, and antibodies against L1 inhibit axonal fasciculation in culture. In addition, these antibodies, which bind to axons and not to muscle, promote the defasciculation of axons innervating chick embryo muscle. Thus L1 could function to facilitate axon growth along and within fascicles. A number of other, structurally distinct surface membrane proteins—F11, neurofascin, **TAG-1**, and MAG—are also expressed broadly in the nervous system. Each of these proteins has been shown to promote neurite outgrowth, and antibodies against them can cause defasciculation in culture assays.

Since a given neuron can express several Ig family members (as well as calcium-dependent adhesion molecules), the question arises why so many distinct proteins are seemingly involved in fasciculation and axon-target interactions. One answer may be that there is redundancy in the function of some of these proteins (Box B). For instance, application of antibodies specific for N-CAM, N-CAD, or an integrin can fail to inhibit neurite outgrowth unless added simultaneously to cultured neurons. It is also possible that the assays employed thus far are too crude to assess subtle differences in function. An important point is that these proteins are not necessarily expressed on the same axons at the same time. For instance, the expression of TAG-1 on commissural neurons is down-regulated as their axons cross the midline of the spinal cord. At the same time, L1 expression is up regulated in these axons. This result suggests that TAG-1 is involved in the initial commissural axon growth towards the midline of the spinal cord, while L1 is important for fasciculation once the axons cross the midline. These findings also demonstrate that an axon can differentially regulate, in space and time, the expression of several members of the neural Ig family.

Differential expression of axonal surface proteins has been demonstrated dramatically in the embryonic insect CNS. As discussed previously, the patterns of axon growth of identified insect neurons during normal development and after perturbation suggest that the axons of the various fascicles express different labels that can be recognized by growth cones. Some of these labels have now been identified by monoclonal antibodies that bind selectively to certain fascicles (Figure 9). When the genes for several of these protein antigens (**fasciclins I-III**) were cloned, two of them (II and III) were found to be members of the vertebrate Ig superfamily, and all three display homophilic binding. As in the CAD experiments, mixtures of fasciclin-negative cells transfected with different fasciclin genes sort out into separate, homogeneous aggregates based on the particular fasciclin expressed. In addition, a given neuron can express different fasciclins on discrete segments of its axon. As the axon grows during development, different fasciclins appear, corresponding to the fascicle the neurite

Box B Redundancy in the Control of Neurite Growth

A rather startling suggestion has emerged from two very different approaches to the identification of cell surface signals that control axonal growth. One is the perturbation of neurite outgrowth by antibodies to specific surface molecules. This approach requires purified antibodies of known specificity, and the application of many "control" antibodies that bind to the appropriate cells but that do not block the function of the antigen. The best control antibodies are those that bind to the very antigen under study but not to its active site or domain. When antibodies that block the function of cell surface proteins such as N-CAM, L1, and the integrins are added alone to cultured neurons growing on muscle or glial cells, only small effects on neurite outgrowth are seen, but if two antibodies are added simultaneously, a substantial, synergistic inhibition of outgrowth can occur. Control experiments show that other antibodies do not block growth when they bind to the same cells even if added in various combinations. These results suggest that the growth cones are able to use alternative adhesive cues if one set is blocked.

The primary advantage of using *Drosophila* and *C. elegans* as neurobiological preparations is the ability to study the effects of deleting or altering a given protein by mutation of its gene. When growth cone guidance is examined in mutants with a single defective gene for one of several surface molecules, including the fasciclins, no dramatic changes are observed. Disruptions are seen only with double mutants (Figure A). Here again, the interpretation is that alternative or redundant cues are available to growth cones as they make their choices. In this situation, of course, it is important to show that neither mutation causes a generalized disruption of growth or adhesion and that the disruption occur only in locations where both genes are expressed.

While such redundancy may be important for the neurons, it certainly complicates life for the experimenter.

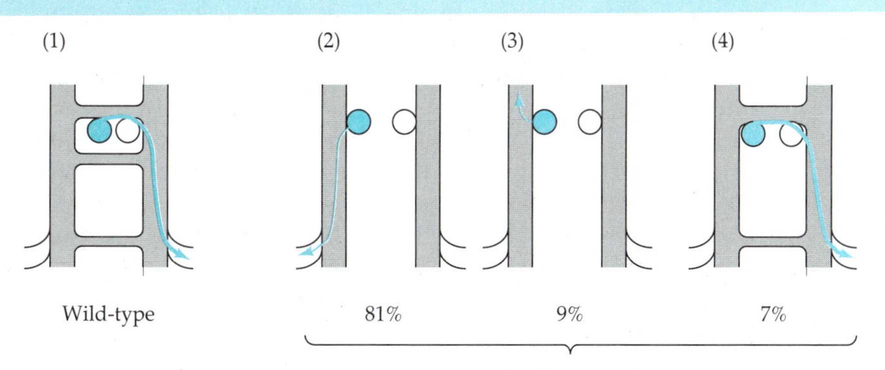

FIGURE A. Redundancy in the control of neurite growth. Schematic diagram of the results of mutating both the fasciclin I gene and a tyrosine kinase gene. (1) The path of the RP1 growth cone is illustrated for wild-type *Drosophila* embryos. The horizontal commissures and longitudinal connectives are shaded. The RP1 axon crosses the midline as it pioneers one of the first bundles in the anterior commissure. After crossing the anterior edge of its contralateral sibling RP1 neuron, it turns posteriorly and then laterally to exit the CNS. (2) In most of the double mutants (81%), the RP1 growth cone grew in the opposite direction, not crossing the midline; it did, however, follow a normal posterior and then lateral course. (3) Other defective paths were observed, and (4) a few of the mutants (7%), appeared to follow a normal pathway. In animals mutant for only one of the two genes, normal RP1 trajectories were observed (not shown). (From T. Elkins et al., 1990. Cell 60: 565–575.)

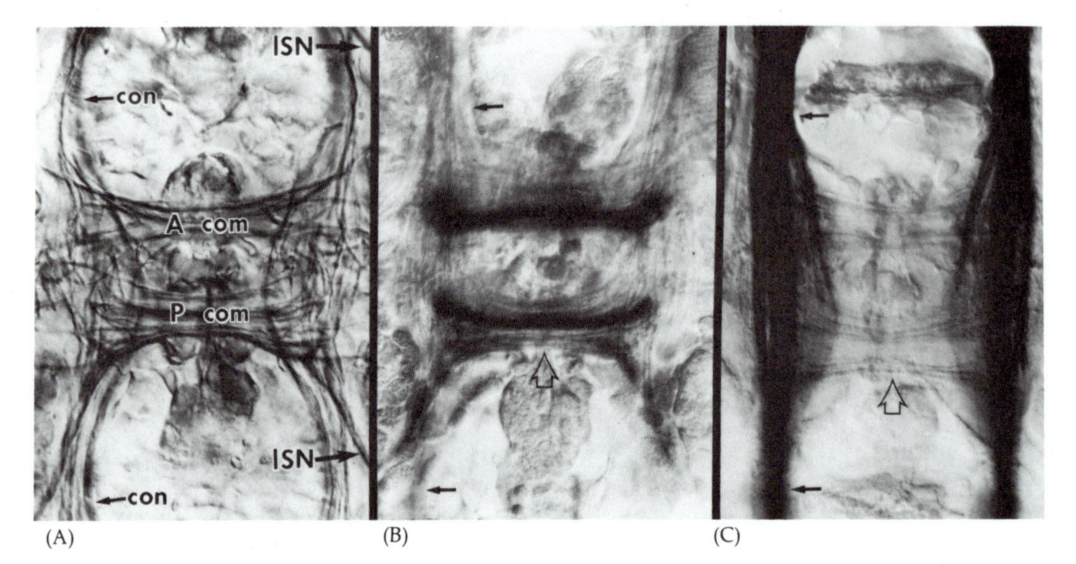

FIGURE 9. Selective staining of axon bundles by fasciclin antibodies. Single segments of embryonic grasshopper CNS are stained with (A) control antibody, which binds all axons, (B) fasciclin I antibody, which primarily stains the anterior and posterior commissures (A and P com) running laterally, or (C) fasciclin II antibody which primarily stains the longitudinal connectives (con). (From M. J. Bastiani et al., 1987. Cell 48: 745–755.)

has joined. Thus neurons are not uniquely identified by a single fasciclin, and expression of these proteins must result from a subtle interplay of neuronal lineage history and environmental cues.

In *Drosophila* the function of proteins of interest can be investigated by producing or finding mutations in the genes encoding the proteins. Surprisingly, null mutations in the fasciclin I gene yield animals without gross abnormalities in neuronal development. When these mutations are combined with mutations in a tyrosine kinase, however, the double mutants display significant, selective defects in the pattern of CNS axonal pathways, as well as in growth cone guidance. The effectiveness of the double mutation may mean that the tyrosine kinase mediates the action of another surface receptor (or is itself that receptor) whose function is redundant with, or partially overlaps, fasciclin I (Box B). If so, deletion of both interactions would be required for a strong defect to be observed.

Diffusible proteins can guide axons

The directed growth of neurites towards their correct targets suggested the possibility to Ramón y Cajal, and many investigators since, that chemotactic signals are produced by targets and diffuse towards receptive neurons, forming gradients. In fact, as discussed in the next chapter, Rita Levi-Montalcini showed that injections of **nerve growth factor** (**NGF**) into the blood of chick embryos caused sympathetic axons to invade the blood vessels. Moreover, NGF injection into the cerebral spinal fluid of the rat caused sympathetic axons to invade the spinal cord. In several different

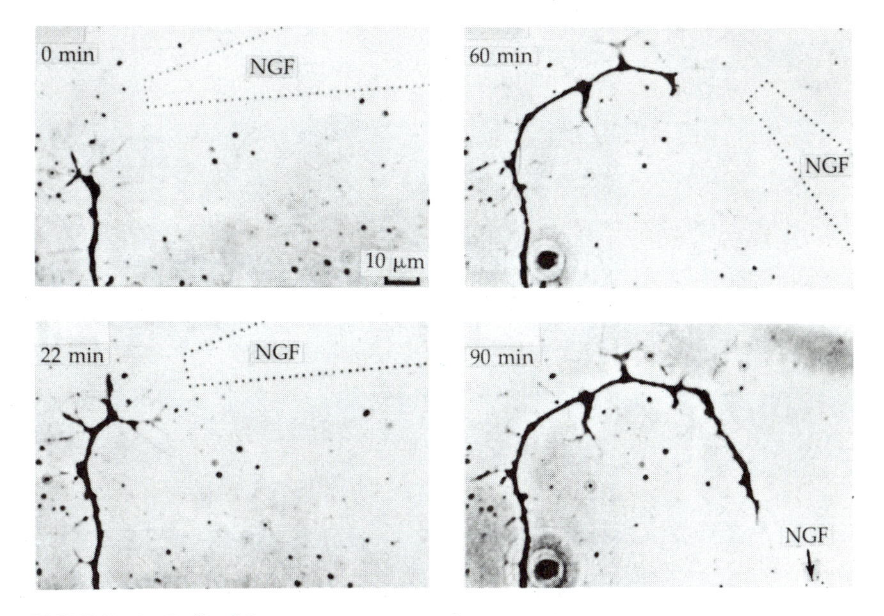

FIGURE 10. Diffusible proteins can guide neurite growth. A micropipette containing NGF (dotted line) was positioned just ahead of a growth cone and gradually moved so as to direct the neurite growth in a 180° turn. The time in minutes is shown in the upper left of each photo. (After R. W. Gundersen and J. N. Barrett, 1979. Science 206: 1079–1080.)

culture paradigms, growth cones have been observed to respond to gradients of NGF by turning sharply so as to move up the gradient (Figure 10). It is not at all clear, however, that NGF gradients do, in fact, guide fibers to their targets in vivo. On the other hand, recent experiments with explants of known targets of particular PNS and CNS neurons have demonstrated preferential growth of the neurons towards the targets. Inappropriate explants from other regions do not evoke this response. It will be important to identity these diffusible factors, and to see how their effects interact with those of the cell surface and matrix-bound adhesive molecules previously considered.

Inhibitory factors can shape axonal pathways

We normally think of positive, adhesive cues in directing the growth of axons. There is, however, no reason to neglect the other likely class of guidance candidates: molecules that inhibit or repulse growth cones. As they grow out the limb, motor and sensory axons appear to avoid developing cartilage; they also avoid growing through the posterior half of each somite (the structure that gives rise to muscle), traversing the anterior half instead. Such observations are echoed in cell culture experiments showing that growth cones are actively repulsed by particular cell surfaces. For example, retinal ganglion cell growth cones rapidly collapse and withdraw upon contact with axons of sympathetic neurons but are unaffected by contact with several other types of axons (Figure 11). Sympathetic growth

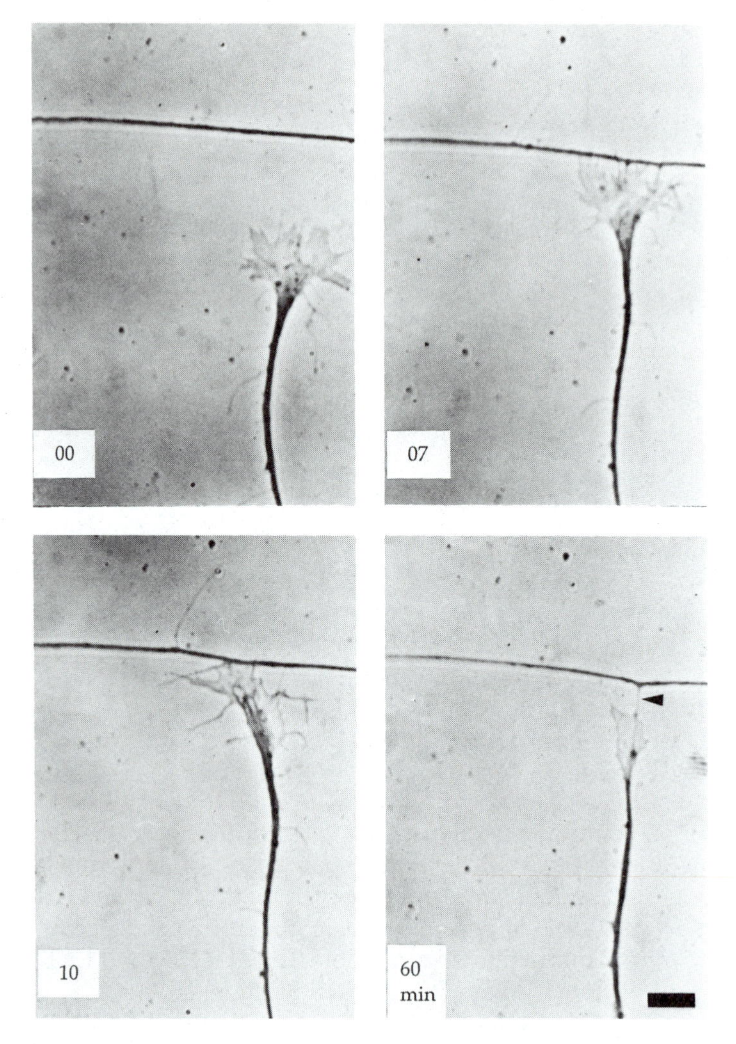

FIGURE 11. Collapse of growth cones upon contact with particular neurites. A retinal growth cone is shown retracting soon after meeting the process of a sympathetic neuron. A similar reaction ensues when growth cones of sympathetic neurons encounter neurites of retinal neurites. Collapse is not seen, however, when retinal growth cones meet other retinal neurites or when sympathetic growth cones meet other sympathetic neurites. The time in minutes is shown in the lower left of each photo. (From J. P. Kapfhammer and J. A. Raper, 1987. J. Neurosci. 7: 201–212.)

cones, on the other hand, are repulsed by contact with retinal, but not sympathetic, axons. The reciprocal cell type preferences in these assays indicate that the effects are specific. Several proteins have recently been isolated that appear to mediate these inhibitory effects. Two proteins that can collapse the growth cones of chick sensory neurons in culture, for example, are preferentially localized in the posterior half of each somite.

The inhibition of growth cones may also be of considerable clinical relevance. The failure of axons in the CNS to regenerate over significant distances (<1 mm) is a long-standing and important problem in human neurology. By contrast, in the PNS, regenerating axons can extend many centimeters. The lack of CNS nerve regeneration is not due to an intrinsic inability of CNS neurons to grow axons, as they can extend axons on appropriate substrates both in culture and in vivo. If an implanted piece of PNS nerve, for instance, is placed next to severed CNS axons, many of the axons will enter the PNS graft and extend long distances. In fact, many of them continue growing until they encounter the CNS environment at the end of the graft. The failure to grow in the CNS environment could be due to a deficiency in factors such as diffusible or matrix-bound proteins that stimulate growth or to inhibitors of growth. Indeed, two inhibitory proteins whose activities in culture are similar to those described above are synthesized by oligodendrocytes and have been isolated from CNS white matter. The key observation is that antibodies against these proteins permit CNS axons to regenerate in the CNS environment, either when added to cultures of CNS tissue or when injected into living animals. Such inhibitory influences may also serve a guidance function in normal development; blockade of oligodendrocyte proliferation and growth in the developing spinal cord can allow axons from the cortex to spread out in a wider array in the cord than they normally do.

Complex oligosaccharides contain information for axonal growth

Although it is natural to think of protein-protein interactions when considering how cells might recognize one another, proteins also interact specifically with complex oligosaccharides. Obvious examples are the recognition of polysaccharide antigens by antibodies, and the enzyme-substrate recognition displayed by the various glycosyltransferases and glycosidases that synthesize and degrade oligosaccharide chains. In addition, there is a growing family of identified **lectins**, proteins that recognize and bind specific carbohydrate residues, some of which mediate specific cell-cell interactions in the immune system. In the nervous system, oligosaccharides mediate at least part of the interaction between growing neurites and laminin, as well as between migrating neural crest cells and laminin. It is also clear that the polysialic acid component of N-CAM can influence fasciculation and nerve-muscle interaction, as discussed above. Furthermore, monoclonal antibodies have delineated the selective expression of particular oligosaccharides on the surfaces of subpopulations of neurons and axons. For instance, two soluble β-galactoside–binding lectins are found in the same subset of sensory neurons that express the oligosaccharides to which the lectins bind. In addition, an antisaccharide antibody displays a gradient of binding in the developing retina. Such results indicate that this is a fruitful area for future research on specificity.

Growth cone motility is regulated by second messengers

In addition to activating receptors that have direct links to the cytoskeleton, environmental signals may also alter growth cone motility via second messengers. One of the most important of these second messengers is calcium, whose concentration influences protein phosphorylation and which is also required by many actin-binding proteins. Calcium appears to be a key regulatory agent in the growth cone, and one of its actions is to regulate the stability of actin filaments. Growth cones of cultured neurons have a higher density of voltage-sensitive calcium channels than do neurite shafts; further, the channels may be clustered to form "hot spots" of calcium entry. High calcium flux at these hot spots could cause local changes in growth cone motility resulting in turning or branching.

Several kinds of experiments indicate that there is an optimal internal concentration of calcium in growth cones and that motility is inhibited both above and below this concentration. Thus, raising intracellular calcium levels by focal application of the ionophore A23187, or by electrical depolarization of the growth cone, causes growth cone collapse and neurite retraction. Lowering extracellular calcium or adding calcium channel blockers also inhibits outgrowth, although in this case the external morphology of the terminal is unchanged.

The effects of calcium have been most thoroughly studied in the large growth cones of cultured invertebrate neurites in which calcium concentrations can be monitored by fluorescent dyes and in which membrane potential can be measured by whole-cell patch clamping. The growth cones of different neurons have various combinations of receptors for neurotransmitters and can respond to them by changes in membrane potential. Application of transmitters to the growth cones often has dramatic effects on their motility (Figure 12). Neurotransmitters that depolarize particular growth cones can inhibit motility, an effect that can be countered by transmitters that hyperpolarize the growth cones. The changes in growth cone motility appear to result, at least in part, from changes in intracellular calcium. Calcium may enter either through voltage-gated channels, or through the ion channels of NMDA receptors; it may be increased by the action of transmitter receptors that are linked to G proteins. The effects of transmitters on growth cone motility raise the interesting possibilities that (1) the synaptic inputs on a neuron's soma or dendrites could control its growth by evoking action potentials and calcium entry in growth cones, and (2) neighboring growth cones could release transmitters on one another and thereby alter the balance of competition for space in a target field. Growth cones release transmitters as they grow, and some neurons may restrict the growth of neighboring cells by such a mechanism in the embryo.

Manipulation of other second messenger systems such as cyclic AMP and the calcium/phospholipid-dependent protein kinase C can also result in altered growth rates. Some of these effects may occur independently of changes in calcium concentration, although this is difficult to establish with certainty. One of the primary substrates for the type C protein kinase in growth cones is **GAP-43** (growth-associated protein, 43 kD apparent molecular weight). GAP-43 levels are highest in embryonic neurons, and are increased during nerve regeneration, suggesting that the protein may be important in nerve growth. In the adult animal, GAP-43 is associated with long-term plastic changes in signaling (see below and Chapter 14).

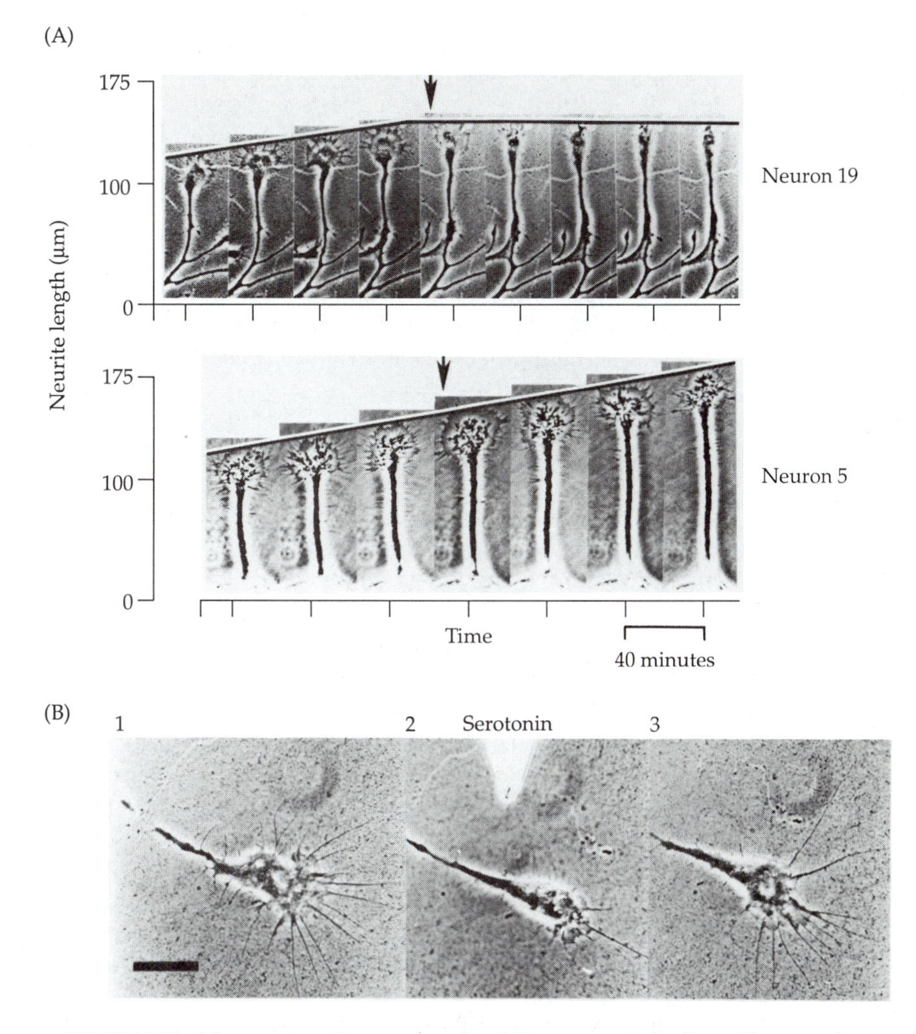

FIGURE 12. Neurotransmitters can control the movement of specific growth cones. Photomicrographs were made of growth cones from identified neurons over time as serotonin was added (arrowheads). (A) While the transmitter has no effect on the rate of growth of neuron 5 (lower panel), it blocks the growth of neuron 19 (upper panel). (B) Application of serotonin to an isolated neuron 19 growth cone reversibly inhibits filopodial extension and translocation. (From P. G. Haydon et al., 1984. Science 226: 561–564.)

In growth cones, GAP-43 is found on the cytoplasmic side of the surface membrane and is especially concentrated in filopodia and lamellipodia. GAP-43 is bound to the membrane via a lipid tail covalently linked to the protein. GAP-43 inhibits phosphatidylinositol phosphate kinase and binds calmodulin, indicating that it may be intimately involved in the regulation of growth cone motility by second messengers.

Growth cones also have a very high concentration of the G protein G_0 (see Chapter 6). This protein mediates the effect of transmitters and other extracellular signaling molecules by changing intracellular second messenger concentrations and by altering channels for calcium and other ions. Its high concentration suggests that growth cone motility is tightly regulated and that the growth cone must integrate many different signals in making its choices. Interestingly, GAP-43 stimulates GTP-γ-S binding to G_0 through an amino-terminal domain in GAP-43 that is homologous to G-linked transmembrane receptors. Although little is known about the precise ways in which extracellular and intracellular messengers determine the decisions that growth cones make and how these decisions are translated into the patterns that axons form, many of the elements of a complex signaling network are clearly present.

Growth cones secrete proteins that alter neurite growth

Growth cones not only sense the extracellular environment and respond accordingly but also actively modify their environment. For example, growth cones secrete a **calcium-dependent metalloprotease** that degrades collagen. This activity may help growth cones penetrate the extracellular matrix. When the enzyme is inhibited, growth cones of cultured neurons are unable to enter three-dimensional collagen gels. A major extracellular matrix component, heparan sulfate proteoglycan, also appears to be cleared away from cultured muscle cell surfaces as growth cones move over these cells.

Another enzyme released by growth cones, **plasminogen activator** (**PA**), also regulates neurite outgrowth (Figure 13). PA is a highly specific protease that, in blood, cleaves plasminogen to yield plasmin, a relatively nonspecific protease involved in blood clotting. Inhibition of PA *increases* the rate of growth cone translocation in culture, a surprising result in light of the role of collagenase. This result takes on additional importance in light of the discovery that muscle cells, glia, and even neurons themselves produce proteins that are natural inhibitors of PA. One of these inhibitors, protease nexin, forms a covalent bond with PA. Why do neurons release PA, only to have it inhibited? One model that seeks to explain these results makes use of the fact that filopodia display contradictory behavior as the growth cone moves forward; they adhere to the substrate and generate force, guiding the direction of growth, and they also detach from the substrate, allowing the growth cone to move on. Without continual detachment, the growth cone might become "stuck" to the surface and slow down. Thus the protease-inhibitor system could play a role in the attach-detach cycle. Such a model would predict differences in the relative PA/inhibitor balance in the microenvironment around filopodia that are attaching ahead of the body of the growth cone (more inhibition?), versus the microenvironment around filopodia that are detaching in the rear of the growth cone (more protease activity?).

Specificity of Connections

Most of the discussion up to this point has concerned the pathways taken by axons towards their target areas. We now turn to the question of how axons distinguish, within the target tissue itself, those cells on which they

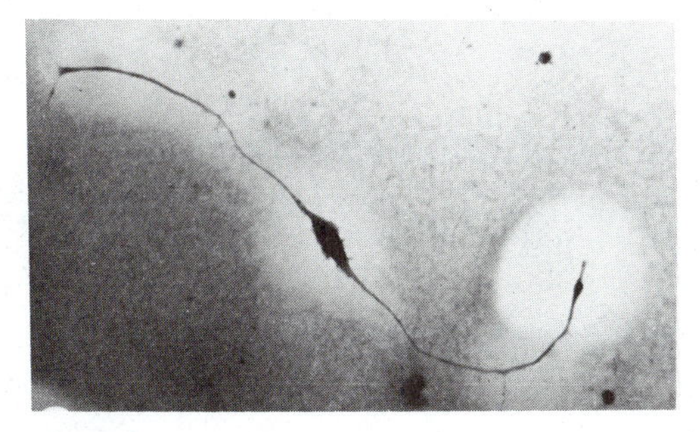

FIGURE 13. Visualization of protease release from a growth cone. Cultured neurons were overlaid with a fibrin clot containing plasminogen. During the incubation, growth cones release the protease, plasminogen activator (PA), which specifically cleaves plasminogen, yielding the enzyme plasmin. The plasmin, in turn, degrades the fibrin, and this local lysis of the clot is seen as an empty space when the clot is stained for total protein, the clear area providing evidence of local release of PA. (From A. Krystosek and N. Seeds, 1981. Science 213: 1532–1534.)

will form synapses. There are three distinct situations that neurons growing towards a target may face. *First*, there may be only one particular cell that is the appropriate target for a neuron to innervate. This is the case for many axons in invertebrate ganglia, which must recognize one neuron out of thousands as their postsynaptic partner. *Second*, incoming axons may be classified according to the information they carry, so that sets of neurons bearing distinct information must synapse with corresponding sets of postsynaptic partners. A common situation in mammals is exemplified by the preganglionic, spinal cord neurons that innervate sympathetic neurons. There are discrete *functional* streams that enable the CNS to separately control blood vessels in the ear and the size of the pupil, for instance. Since the sympathetic neurons that control these many functions are found mixed together in the same ganglia, the incoming preganglionic axons must distinguish among the various sympathetic neurons in order to find those of the appropriate functional stream. *Third*, all of the incoming axons may innervate a single class of postsynaptic cell in the target area but must do so in a way that preserves the *topographical* information that they carry. For example, retinal ganglion cell axons that project to the tectum in the brain all convey the information that light is striking the retina, but each axon originates from a distinct position in the retina. That is, the retinal ganglion cell axons not only convey a functional signal, but also topographical information as to the particular location on the retina that is receiving light. The topographical information conveyed by the entire set of incoming axons represents a map of the retina, and the axons spread out on the tectum in an orderly array to reproduce a map of the retina on the tectum. The coordinates of the map in the tectum correspond to the relative positions of the neurons in the retina.

Molecular gradients may influence the specificity of connections

When the optic nerve of lower vertebrates is cut, retinal axons grow back into the tectum and reestablish the original map of retinal position on the

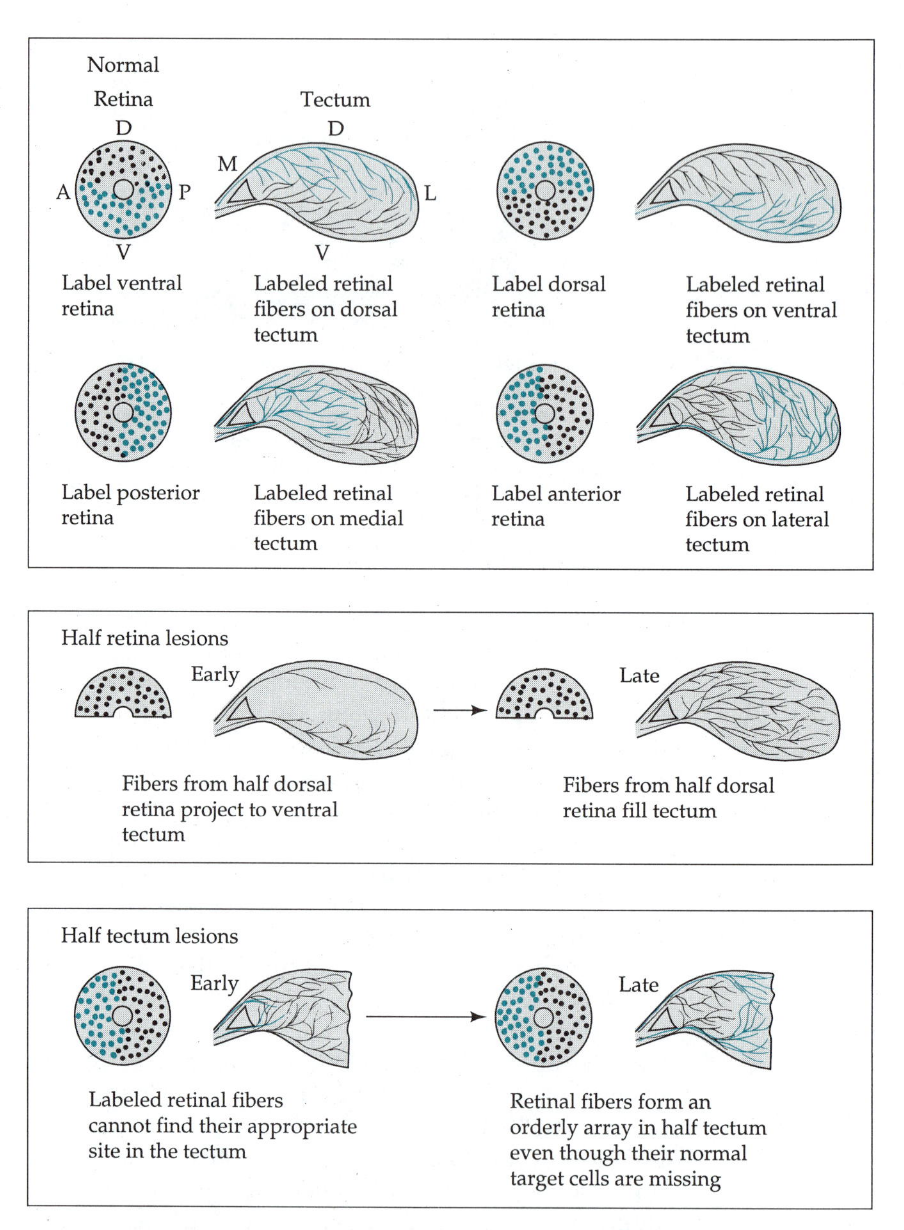

FIGURE 14. The plasticity of the retinotectal projection. Specific regions of the retina are labeled (in color) and the projections of those axons into the tectum are followed. The map of the normal projection is illustrated in the top two rows. In the third row, an experiment is shown in which a half retina is allowed to grow into an intact tectum. At early times after the arrival of the axons (left), the projection is restricted to the region of the tectum appropriate for those retinal axons. Later (right), however, the projection expands to fill the available tectal space. In the bottom row is illustrated an experiment in which an intact retina projects into a half tectum. Initially (left) the retinal axons that normally project to the missing part of the tectum do not innervate the tissue. Over time (right), however, these axons are integrated into a new map in which all regions of the retina are represented on the half tectum and in the appropriate retinotopic positions. (Drawn by N. K. Mahanthappa and P. H. Patterson.)

surface of the tectum (Figure 14 and Box A). In this remarkable example of specific reinnervation, the axons can find their targets even when care is taken to disrupt their normal routes and relative positions. The matching between retinal position and tectal position is not absolute, however. When a disparity is introduced between the size of the retina and that of the tectum (in **size disparity experiments**), the axons adjust to find the best position available. Thus, when most of the retina is removed before regeneration, so that a small piece of retina is projected onto a whole tectum, the retinal fibers occupy the entire retina and maintain correct retinotopic order (Figure 14). Similarly, when the whole retina is projected onto a half tectum, a complete map of the retina is constructed on the remaining portion.

When a whole retina maps onto a half tectum, most of the axons innervate positionally inappropriate parts of the tectum. The same is true for the experiment in which a half retina projects onto a whole tectum. This expansion and compression of the projection does not occur immediately upon reinnervation of the tectum but represents a subsequent adjustment. The original position of the newly arrived fibers from a half retina, for example, is confined to the appropriate half of the tectum (Figure 14) and only later spreads over the entire tectum. Experiments such as these illustrate a general feature of specificity in vertebrates, namely, that it is not a rigid lock-and-key phenomenon but that it depends on the options available. Given the opportunity, axons often make the appropriate choices. If influenced by other forces (such as an abnormally large amount of target space), axons become less selective. Deprived of an opportunity to find their appropriate target, axons often innervate whatever target is available. This is the paradoxical situation that faces the biochemist in searching for specificity molecules.

The relative specificity seen in the size disparity experiments in adult animals is echoed in observations on embryos. In several species, the first retinal fibers to reach the tectum innervate only a small area of what will later become the much larger, adult tectum. New retinal ganglion cells are born at the circumference of the retina, thus representing all areas of the visual field. As the axons of these cells arrive at the enlarging tectum, they must be integrated into the growing retinotopic map. To do this, many of the new fibers must interpolate and displace connections made by the preceding axons. This sequence of new axons bumping the preceding ones is referred to as the **sliding connection** phenomenon.

The changes in position that occur during development are accompanied by a reduction in the terminal arbor of each retinal fiber and by a decreased amount of overlap between the terminals of different retinal ganglion cells. Much of this refinement of the map requires activity in the axons. When action potentials are blocked with tetrodotoxin, or when postsynaptic glutamate receptors are blocked with NMDA antagonists, an increased amount of overlap is seen, with a consequent loss in the precision of the map. This type of activity-dependent rearrangement of connections is also observed in many other parts of the nervous system, as discussed below.

The three-eyed frog has yielded further insight into the rules that govern map formation (Figure 15). In normal frogs, each eye projects only to the contralateral tectum; when an extra eye primordium is implanted into the forebrain of the developing animal, the tectum on each side now receives innervation from the new eye as well as from the eye that normally innervates it. The addition of the extra eye thus causes an increased number of retinal axons to project to each tectum. How do these extra axons map? Do they interdigitate completely with the normal retinal axons, thus maximizing the chances of interacting with the correct part of the tectum, or do they stay together, displacing the normal axons to another part of the tectum? The former result would minimize the associations between axons from the same retina, while the latter would minimize the correct retinotectal associations. Unexpectedly, the third eye fibers appear to compromise between these extremes by forming eye-specific bands to

FIGURE 15. Eye-specific projections in tecta of three-eyed frogs and the role of activity. (A) The primordium for a third eye implanted into the head region at a very early stage of development can grow to essentially normal proportions. (B) When the retinal axons in the third eye are labeled by injection of the eye with HRP, the projection on the tectum is in stripes. That is, the retinal axons from the third eye alternate in bands with the unlabeled axons from the normal, contralateral eye innervating the same tectum. (C) If retinal activity in all eyes is blocked by injections of TTX, the labeled third eye axons spread evenly over the tectum, intermixing with axons from the normal eye. (A from M. Constantine-Paton and M. I. Law, 1978. Science 202: 639–641; B and C from T. A. Reh and M. Constantine-Paton, 1985. J. Neurosci. 5: 1132–1143.)

give stripes of alternating normal and third eye fibers (Figure 15). This pattern is very reminiscent of the **ocular dominance columns** found in normal visual cortex of mammals (see Figure 14 in Chapter 1), as discussed below in the context of synaptic rearrangement. The segregation between the inputs from different eyes in three-eyed frogs and in the normal mammalian cortex both require neuronal activity, as does the refinement of retinotectal connections during normal development. In the case of the retinotectal system, an NMDA receptor appears to be involved the refinement of the map and in the elimination of overlaps between retinal axon terminals.

The flexibility of connections observed in these experiments raises the question of whether any recognition occurs at all between retinal axons and tectal cells during development. Perhaps the map is due primarily to precise ordering of retinal fibers relative to one another, with activity-based adjustments. While such interactions do contribute to the final map, there is convincing data for retinotectal recognition as well. For instance, lesion and size disparity experiments in embryos have shown that retinal axons can innervate their correct tectal location, ignoring inappropriate parts of the tectum, at least initially. Transplant experiments have also shown that retinal fibers can find their appropriate location within the tectum regardless of whether their nearest retinal neighbors are correct. Embryonic retinal fibers can also locate the appropriate part of the tectum using very abnormal pathways to get to the tectum. Finally, there is now good evidence for the existence of surface molecules on both retinal and tectal cells that are related to position in the two-dimensional map of each tissue.

TOP_{DV} is a monoclonal antibody that binds a 47-kD protein in chick retina that is distributed in a dorsal-ventral gradient. A complementary map of TOP_{DV} binding is found in the embryonic tectum. Thus the retinal fibers with the highest levels of TOP_{DV} antigen project to tectal areas with correspondingly high TOP_{DV} levels. If this protein is directly involved in retinotectal interactions, the distribution suggests a homophilic association. Moreover, a second antibody was recently discovered, TOP_{AP}, that binds another protein (40 kD) in chick retina, one that is distributed in an anterior-posterior gradient. As with TOP_{DV}, there is a corresponding, inverted TOP_{AP} gradient in the tectum as well. Thus, these two proteins are distributed in orthogonal gradients in both the retina and the tectum and together could serve as a coordinate system for cell position at all points in the retinotectal map.

How can the function of the TOP proteins be determined? In a cell culture assay, growing retinal axons can discriminate between monolayers of rostral and caudal tectal cells. In the most ingenious form of this experiment, axons from different regions of the retina are grown on lanes of membrane vesicles derived from different parts of the tectum. Given this choice, axons from temporal retina make the correct decision and grow preferentially on lanes of anterior tectal membranes (Figure 16). Surprisingly, the preference for anterior membranes is not due to an increased adhesivity on this surface but rather to an active avoidance of the posterior

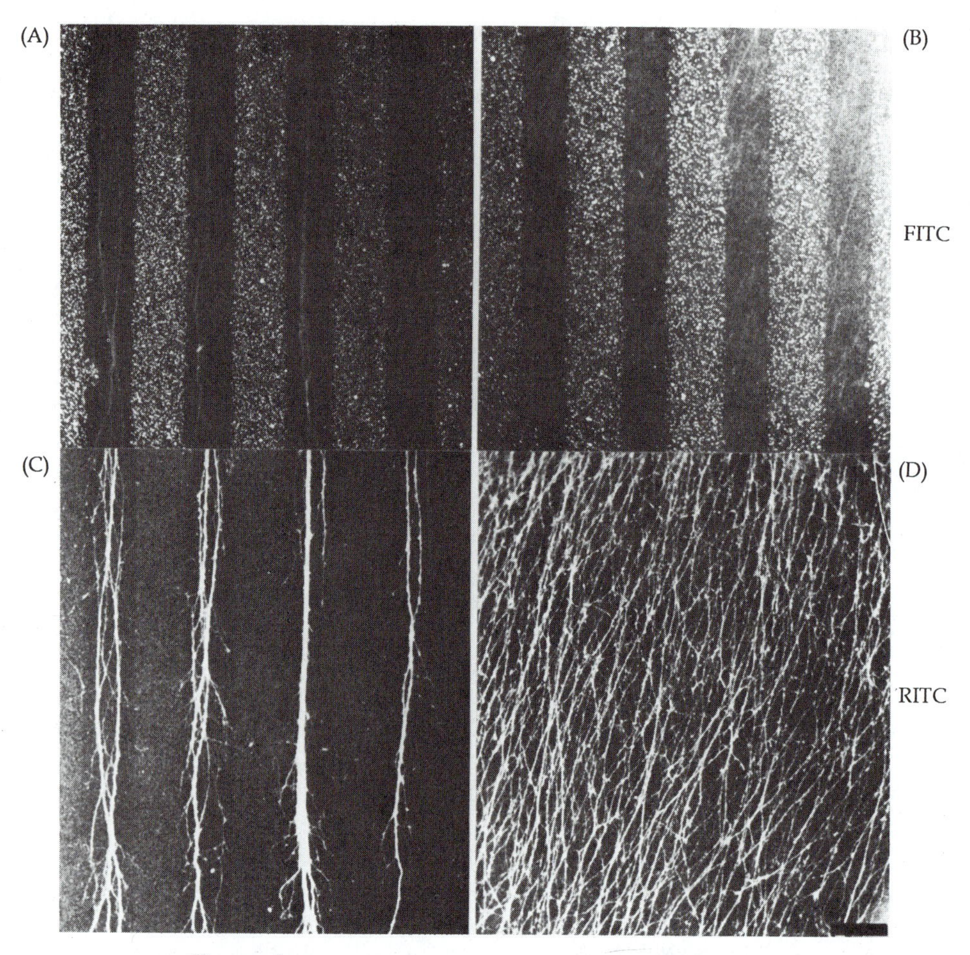

FIGURE 16. Retinal axons discriminate among tectal membranes. Alternating stripes of membranes from anterior and posterior regions of embryonic chick tectum were applied to a filter. Stripes of FITC-labeled posterior tectal membranes are visualized in (A) and (B); the anterior membranes are unlabeled. RITC-labeled retinal explants were then applied to the same filters. Labeled axons from temporal and nasal regions of the retina are visualized in (C) and (D). While axons from nasal retina apparently cannot discriminate between the lanes of tectal membranes, axons from temporal retina grow strictly on on the lanes of anterior tectal membranes. Further experiments indicated that this preference is due to an active avoidance of posterior tectal membranes by the temporal retinal axons. (From J. Walter et al., 1987. Development 101: 685–696.)

tectal membranes. If the posterior membranes are heated to denature the proteins, the axons show no preference for either lane, whereas when the anterior membranes are heated, the preference for anterior membranes is maintained. Thus the active influence is a negative one exerted by the posterior membranes. These results are reminiscent of the experiments discussed above in which growth cones were found to recoil upon contact with inappropriate axons. The molecule responsible for the avoidance of posterior tectal membranes has just been identified as a 33-kD protein. Hence, there are now several candidate molecules that may encode positional information in the retinotectal system.

Growth cones can seek out particular cells as targets

Another form of specificity, one that has a special twist, is seen in studies on the spinal reflex system. For the "knee jerk" reflex to operate properly, the sensory neurons innervating each muscle in the limb must not only find and selectively synapse with motor neurons, but they must do so with those particular motor neurons that innervate the same muscle that the sensory fibers are contacting. Thus there are distinct sensory-motor loops for each muscle. Since the dendrites of the motor neurons that project to various muscles are mixed together in the cord, sensory growth cones must move through a veritable forest of inappropriate processes in order to find the correct target dendrites. This selectivity is achieved at the time of initial innervation; it is not obtained by selective pruning or by activity-dependent rearrangement at later stages. Most interestingly, the target selection made by individual sensory fibers within the spinal cord does not appear to be an intrinsic property of the sensory neurons but is governed by the connections that the sensory neurons make in the periphery. Thoracic dorsal root ganglia contain sensory neurons that normally innervate the skin and make connections with interneurons in the dorsal horn of the spinal cord. Neurons from brachial dorsal root ganglia, on the other hand, normally innervate the forelimb (including the muscle) and make connections with motor neurons in the ventral spinal cord (Figure 17). When thoracic ganglion cells are transplanted to the brachial position, the transplanted sensory neurons grow out into the limb, and some of them innervate muscle, a novel target for these neurons. Subsequently the neurons that innervate muscle project into the spinal cord and make synaptic connections with motor neuron dendrites, a target they would normally never innervate. Moreover, the novel connections in the spinal cord are with those motor neurons that project to the particular muscle that the sensory fibers contact. That is, the connections formed by the transplanted sensory neurons establish the correct reflex arcs.

The fact that the central connections made by transplanted neurons are appropriate for their novel peripheral targets suggests that the immature sensory neurons are plastic and that the peripheral targets that they innervate specify their central connections. A similar phenomenon is found in ganglia of the medicinal leech in which dendrites of identified motor neurons assume a shape characteristic for the peripheral target tissue they innervate. This is discussed in the next chapter in the context of findings that target tissues can also dictate the neurotransmitter and neuropeptide produced by the neurons that innervate them. In this latter case, some of the proteins that direct these transmitter and peptide choices are known. Could the same class of factors dictate the expression of the specificity molecules presumably expressed by motor neurons and recognized by sensory neuron growth cones in the spinal cord?

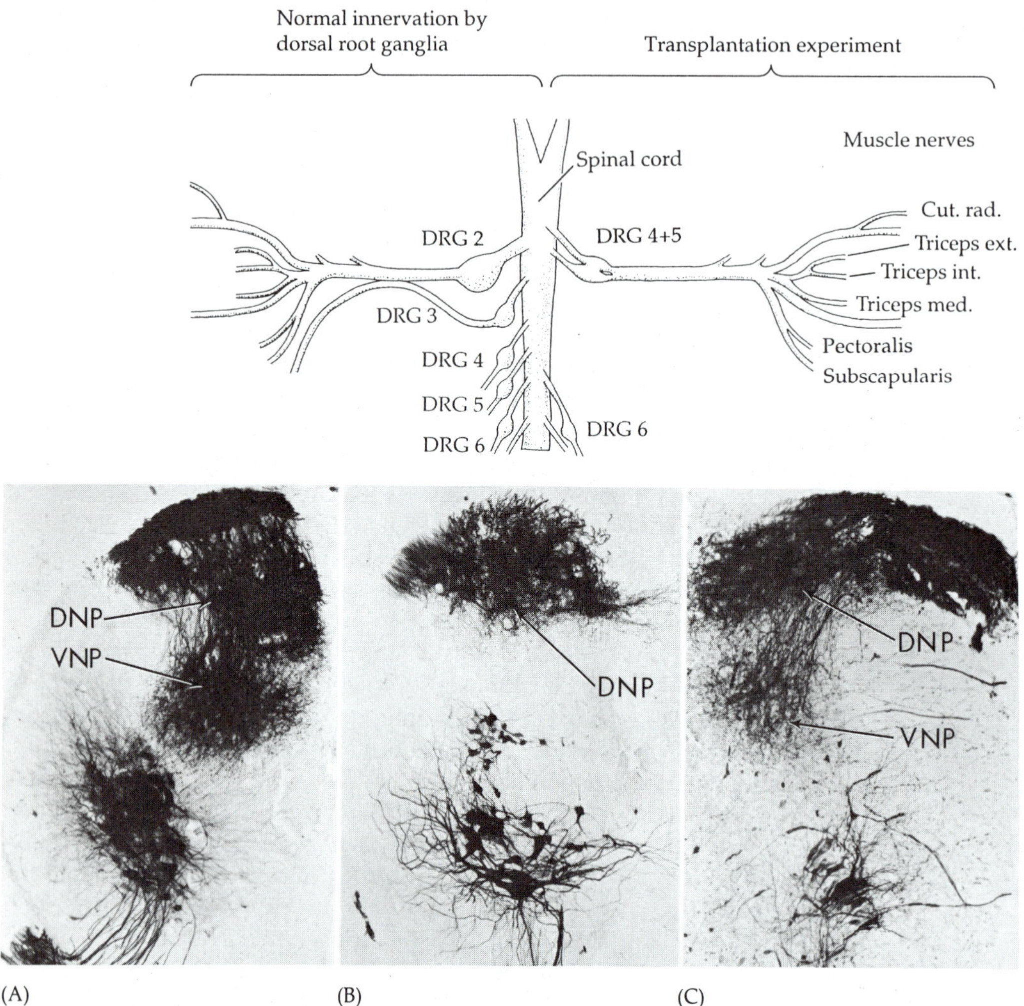

FIGURE 17. Peripheral specification of central targets. Dorsal root sensory ganglia (DRG) were given distinct peripheral targets to innervate and their central connections in the spinal cord were monitored. The upper panel shows the experimental paradigm; the normal pattern of innervation by DRG 2–6 is shown on the left, and the transplantation experiment is shown on the right. DRG 2 and 3 were removed and replaced by ganglia 4 and 5. The effect of this surgery was to take ganglia that would normally innervate body skin and place them in a position to innervate the skin and muscles of the limb. The consequences of this change are illustrated in the lower panel. Both motor and sensory nerves to the periphery were back-labeled so that their processes in the spinal cord could be visualized. The DRG 2 projections are shown in the top part of (A); these neurons, innervating the limb, form both a dorsal and a ventral neuropil (DNP and VNP) in the cord. The dense processes below the VNP are the somas and dendrites of the back-labeled motor neurons. The sensory projection into the VNP corresponds to the innervation of the motor neuron dendrites. In contrast, the normal projection of sensory processes from DRG 4 (B), which does not innervate muscle, includes only the DNP, with no innervation of ventral motor areas of the cord. When DRG 4 is transplanted so that it innervates muscles, however, it forms both a DNP and a VNP projection in the cord (C). (From C. L. Smith and E. Frank, 1987. J. Neurosci. 7: 1537–1549.)

As mentioned at the beginning of this section, the innervation of mammalian sympathetic ganglia by preganglionic axons from the spinal cord displays specificity at the level of individual target cells. For instance, preganglionic neurons that are innervated by higher centers controlling blood pressure send their axons to sympathetic ganglia, where the axons synapse with sympathetic neurons that innervate blood vessels. In addition to this type of specificity, the system displays a bias of connectivity based on position in the rostral-caudal axis. Each ganglion receives innervation from preganglionic neurons located in several different segments of the spinal cord (Figure 18). The different subsets of cord segments that innervate the ganglia are graded along the rostral-caudal axis, such that the more rostral ganglia receive innervation from the more rostral segments of the cord and the more caudal ganglia from caudal cord (Figure 18). This rostral-caudal pattern of ganglionic connections is reestablished with considerable fidelity after cutting the innervation to the ganglia in adult mammals. The functional streams related to peripheral targets such as the ear or the pineal gland, are also regenerated appropriately in the lesioned adult. These findings led to the hypothesis that there are labels on subsets of sympathetic neurons and on preganglionic axons that bias the formation of synaptic connections.

Transplant experiments have reinforced the hypothesis that there are

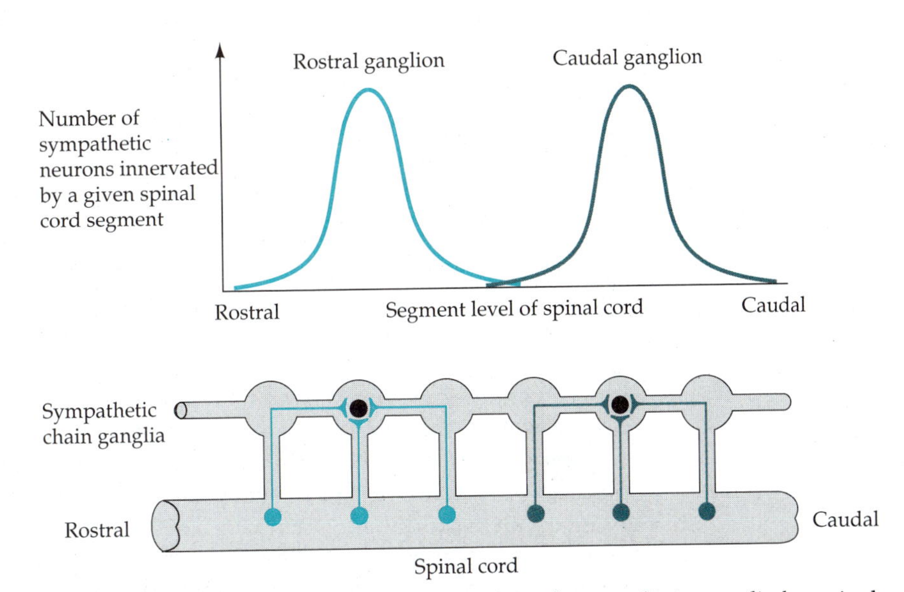

FIGURE 18. Positional bias in the innervation of sympathetic ganglia by spinal axons. Neurons in the spinal cord (color) send axons into the chain of sympathetic ganglia, and the innervation of two sympathetic neurons (black) is illustrated. The lower diagram illustrates that spinal cord neurons from several different segments innervate nearby ganglia. The result of this pattern is that each ganglion receives input from several spinal segments, and that rostral ganglia are preferentially innervated by neurons from rostral cord and caudal ganglia are innervated by neurons from caudal cord. The rostral-caudal gradient in the innervation pattern is illustrated on the upper diagram. (Drawn by N. K. Mahanthappa and P. H. Patterson.)

labels along the sympathetic chain and the spinal cord related to rostral-caudal position. When a caudal sympathetic ganglion is transplanted to a rostral position, the neurons in the caudal ganglion are innervated not by the same set of axons that normally innervate the rostral ganglion but by axons from a more caudal set of cord segments. That is, rostral-caudal positional differences between ganglia are observed even when the ganglia are transplanted to novel locations. Using this same transplant paradigm, a further extraordinary result suggests that the presumptive rostral-caudal positional information may be shared with muscle. When the rostral ganglion is replaced with transplanted rostral or caudal intercostal muscles, the preganglionic spinal cord neurons innervate the skeletal muscle in the absence of their normal ganglionic targets (Figure 19). Most interesting, however, are the relative segmental patterns of innervation of the muscles; rostral muscle is innervated by a more rostral set of spinal cord axons than is caudal muscle. Thus, the apparent ability of the preganglionic axons to distinguish cells on the basis of their rostral-caudal position may hold for both neuronal and muscle targets even though these axons normally never contact muscle. Although little biochemical information is as yet available on the molecular basis for this positional bias, a monoclonal antibody was recently produced that can distinguish rostral vs. caudal sympathetic ganglia as well as rostral vs. caudal nerves that innervate intercostal muscles. This ROCA1 (for ROstral-CAudal) antibody binds a 26-kD protein that shows a graded, rostral-caudal distribution in staining of sections of these tissues.

In sum, neurospecificity still retains its wonder and mystery, but neurochemists are beginning to find some promising clues. Fascinating times are ahead.

Synaptic connections are remodeled after being formed

After axon guidance and selective synapse formation takes place, a process of synaptic rearrangement begins. This involves the loss of many of the initial synapses and the formation of new ones. The phenomenon is widespread in the nervous system and is not the result of neuronal cell death, which occurs earlier in development. The end result is that each neuron has stronger connections with fewer target cells, a process called **convergence**. In skeletal muscle, for example, each muscle fiber initially receives synapses from five or six different motor axons. After the rearrangement process (often referred to as "**synapse elimination**"), each fiber is strongly innervated, but by just one axon. This process is competitive in that surgically removing some of the axons allows the maintenance of multiple innervation of muscle fibers by the remaining axons. The competition depends at least in part on activity. If action potentials in the axons are blocked by application of tetrodotoxin, rearrangement does not occur as quickly.

The mammalian visual system undergoes a similar rearrangement; the inputs from both eyes initially overlap in the visual cortex and then segregate into ocular dominance columns (Chapter 1). The phenomenon is similar to that illustrated in Figure 15 for the frog, where two eyes were

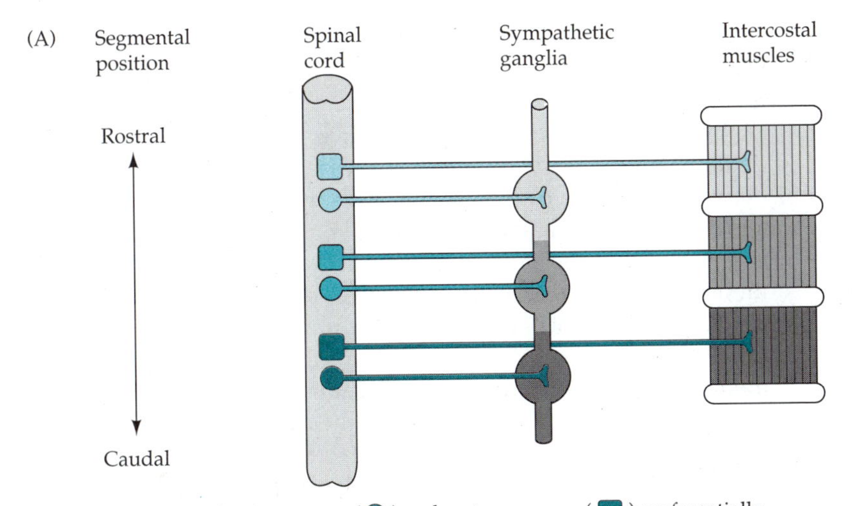

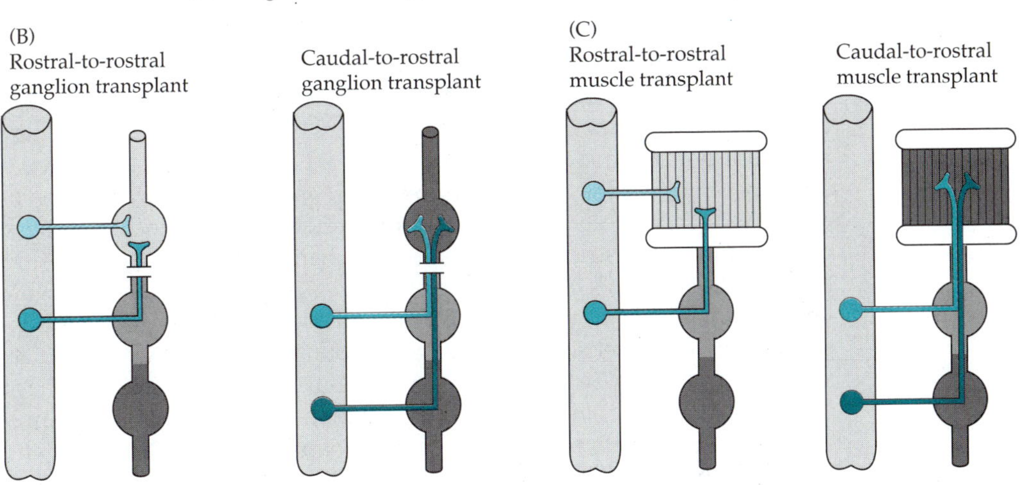

FIGURE 19. The specificity of innervation of transplanted ganglia and muscles depends on rostral–caudal position. The findings suggest that rostral–caudal positional information is shared between nerve and muscle. (From D. J. Wigston and J. R. Sanes, 1985. J. Neurosci. 5: 1208–1221.)

made to innervate the same tectum. Rearrangement in the visual system, like that in muscle, depends on electrical activity. There is also a role for the NMDA receptor; activation or blockade of this receptor can lead to changes in the rate of synapse elimination. Although the biochemical basis of the rearrangement phenomenon is not understood, models include activity regulation of target-derived neuronal trophic factors (discussed in Chapter 13), or inhibitors like those previously mentioned that repulse growth cones.

Recent work on the neuromuscular junction and autonomic ganglia has demonstrated that synaptic rearrangement continues throughout life. In a mature rodent the overall shape of an individual neuron's dendritic tree can change significantly over a period of several weeks in vivo (Figure 20), suggesting that the presynaptic inputs must also be remodeled. Such remodeling has been directly confirmed by staining presynaptic nerve terminals on the surfaces of identified, autonomic ganglion neurons at several time points in the living, adult mouse. Observations over a 1–3 week interval reveal that the pattern of terminals changes gradually. Thus, there is nothing immutable or fixed about the elaborate synaptic specializations described in the next chapter. The lability of dendrites and synapses is consistent with another body of work on sprouting and reinnervation of adult ganglia and muscles. Those data also support the contention that synaptic relations in the mature organism represent a dynamic balance of the same forces driving synapse formation and elimination that are more dramatically manifest during development.

What regulates remodeling in the mature system? There are many hints that, in addition to the short-term modifications that are made in ion channel function during learning (see Chapter 14), long-term learning may involve structural modifications in the relevant circuits. It has been known for some time that long-term learning can require RNA and protein synthesis, and recent experiments on a sensory-motor synapse in *Aplysia* have demonstrated that striking morphological changes accompany long-term learning. Long-term sensitization yields larger responses to stimuli, while long-term habituation yields smaller responses to stimuli. Electron microscopic analysis of sensory neuron presynaptic terminals (the site of plasticity for short-term learning) was carried out on animals that had been trained for four days in long-term sensitization or habituation paradigms. The result was that the number of presynaptic sensory terminals and synaptic active zones were larger in animals showing sensitization than in control animals, and smaller in animals showing habituation (Figure 21). These morphological changes persisted for several weeks, as did the sensitization itself. Since application of the putative sensory transmitter serotonin mimics the behavioral and electrophysiological effects of long-term sensitization, the effects of serotonin on the structural changes have also been recently studied. Application of the transmitter to sensory neurons co-cultured with motor neurons produces a long-term enhancement of synaptic efficacy and a long-lasting stimulation of axonal growth from the sensory neuron. A further provocative finding is that these changes induced by serotonin require the presence of the motor neuron; sensory neuron growth or sprouting is not enhanced by serotonin in the absence of the target cell or in the presence of a nontarget neuron. This raises the testable hypothesis that the transmitter stimulates the motor neuron to produce a factor that influences sprouting by the sensory neurons. This type of interaction between developing pre- and postsynaptic partners is the topic of the next chapter.

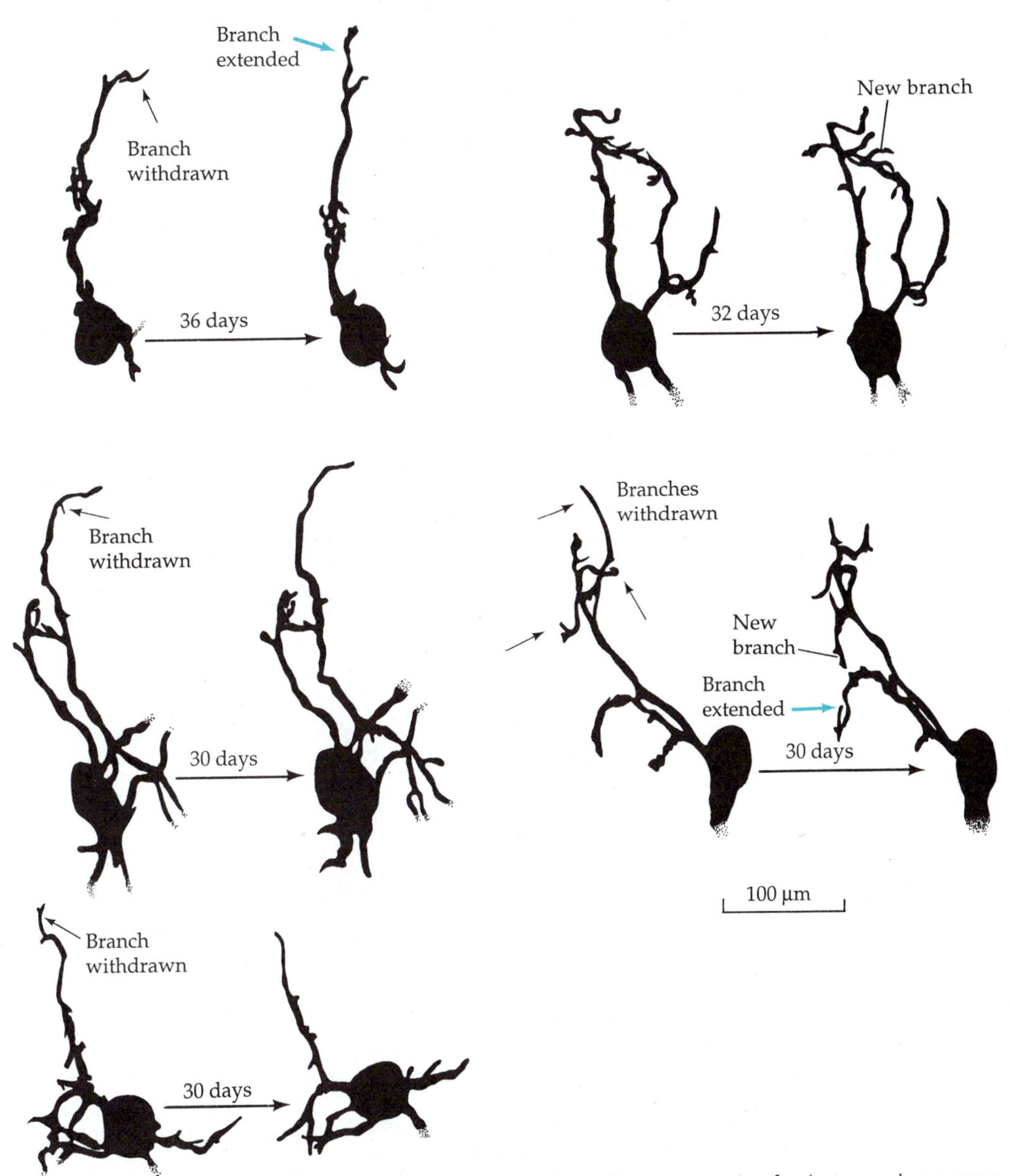

FIGURE 20. Dendritic structure is plastic in mature animals. Autonomic neurons were visualized and their dendrites drawn at two different times, 30 to 36 days apart. The arrows indicate branches that were withdrawn, extended, and new branches formed. That the visualization process itself did not induce the changes was suggested by the finding that dendritic changes were more marked at longer time intervals between observations. (From D. Purves et al., 1986. J. Neurosci. 6: 1051–1060.)

Summary

There are several levels of specificity in the wiring of the embryonic nervous system. Growing axons can recognize particular target cells as

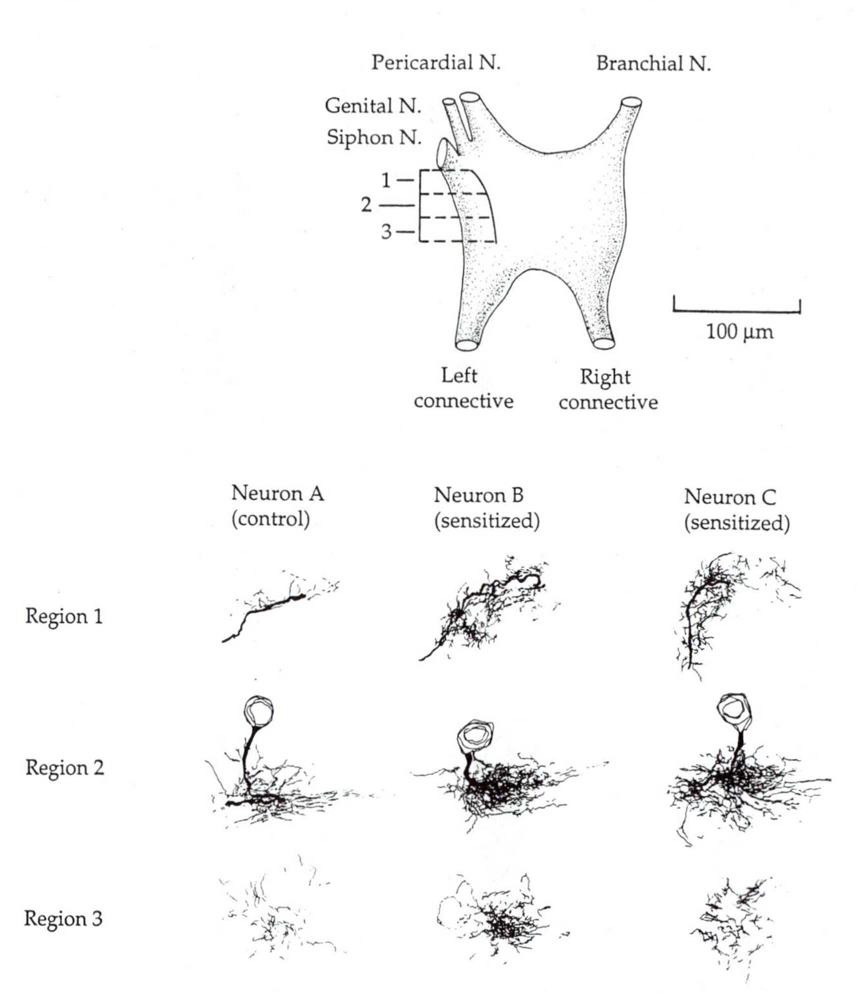

FIGURE 21. Structure of individual neurons change during learning. Sensory neurons from *Aplysia* subjected to long-term sensitization and from control animals were reconstructed and representative cells illustrated. The ganglion is shown in the upper panel, and the three areas of the neurons taken for illustration are indicated. A control neuron is depicted on the left (A); the region of its soma is shown in panel 2, the region of its process near the siphon nerve is shown in panel 1, and the region of its processes in the other direction is shown in panel 3. The same regions of two sensory neurons subjected to sensitization are illustrated to the right of the control cell (B,C). Training resulted in the growth of many more fine branches in these latter neurons. (From S. H. Bailey and M. Chen, 1988. Proc. Natl. Acad. Sci. USA 85: 2373–2377.)

unique individuals, or as classes of target cells that share a common functional modality, or as cells in a certain position in a two- or three-dimensional array. Moreover, several such levels of specificity are often employed simultaneously, so that a growing axon can recognize the appropriate class of target cells in a mixture and, at the same time, select a cell in a distinct position in the array. This extraordinary selectivity is built up in stages: initial outgrowth on certain "highways" containing many thousands of other axons, leaving these highways at characteristic exits, recognition of particular streets, and focusing on certain cellular addresses. The latter stages involve not only the formation of selective contacts but the selective withdrawal of processes. The selection of particular growth pathways and connections is relative and depends on the choices available. Axons deprived of their specific partners, for example, will form connections with the next most appropriate targets available.

Decisions as to the rate and direction of outgrowth are made by the growth cone. This highly motile structure uses its filopodia and lamellipodium to detect molecular heterogeneity in its environment. Relevant cues are provided by small molecules such as neurotransmitters, diffusible proteins such as growth factors, membrane-bound proteins such as members of the Ig and CAD families, proteoglycans of the extracellular matrix such as laminin, and distinctive oligosaccharide sequences on glycoproteins and glycolipids. Growth cones of different neurons adhere preferentially to discrete subsets of these molecules and thereby either continue their progress in the same direction, turn towards a new pathway, or form a branch. They can also be actively repulsed and withdraw upon contact with certain cells and molecules in their paths. The intracellular computations that control these decisions, and the coordination of cytoskeletal and membrane molecules that mediate growth, turning, and branching are poorly understood. Rounding up the usual suspects has provided evidence that calcium levels in the growth cone can regulate the rate of translocation, and that the expression of several proteins and their phosphorylation status change with the state of process outgrowth.

Not only do growth cones receive, compute, and act on signals from the environment, they also send out signals of their own to modify their surroundings. Growth cones release neurotransmitters that can control growth by neighboring processes and organize molecules in the postsynaptic cell. Growth cones also secrete several types of proteases; these enzymes may participate in the mechanism of movement itself and can enable the growth cone to penetrate extracellular matrices.

Highways for axonal growth can consist of matrix proteins such as laminin that are permissive for neurite growth by many types of neurons. Many highways consist of fascicles of other axons. In this case, the permissive substrates are axonal surface proteins such as the Ig family members L1, N-CAM, and TAG-1, or the calcium-dependent CAD adhesion proteins. Since many of these proteins utilize homophilic binding, their selective expression could be a first stage in choosing the direction of outgrowth. Several mechanisms are used to enhance the specificity of these major adhesive proteins. A number of different forms of N-CAM have been found, and the protein can be glycosylated in ways that profoundly alter its adhesive function. The proteins may also be used in combinations that could yield further specificity. Axons up-regulate or down-regulate these proteins or the receptors for these proteins during development, thereby altering their adhesive preferences. A clear example of selective highway labeling is the fasciclin family. These homophilic adhesion proteins are specifically expressed in certain axonal bundles, and neurons can localize particular fasciclins to distinct regions of their axons. Despite this precise labeling, there is evidence for redundancy among the adhesion molecules. If one of these proteins is blocked by antibodies or deleted by mutation, another set of proteins can sometimes be used as a substrate, with only minor effects on the pattern of growth.

Exiting from a highway depends in part on the available choices; growth cones compare the relative affinities of the pathway they are following with that of the surrounding environment. When a more attractive substrate becomes available, or if the highway is modified during development by changes in the proteins or oligosaccharides expressed, the growth cones may choose to exit and grow into or towards a particular target tissue. Growth cones can also be induced to turn off a highway by diffusible factors produced by target cells.

Once in the target area, growth cones are clearly able distinguish among potential postsynaptic partners. This discrimination can be based on the functional modality the target cells subserve or the position the cells occupy within the target field. While little is known about the molecular basis of the first category of specificity, there are a number of candidate proteins for the second, positional category. TOP_{DV} and TOP_{AP} are distributed in orthogonal gradients in the retinotectal system, and a protein that inhibits the growth of certain retinal axons is found in a gradient on tectal membranes as well. Another protein is found in a rostral-caudal gradient in sympathetic ganglia and intercostal nerves. The roles that these and other candidate molecules play in the specificity of axon guidance and/or synapse formation is an area of active investigation.

Following the initial formation of connections, a process of synaptic rearrangement ensues. This phenomenon generally sharpens the focus of each axon, narrowing the convergence. In the case of the visual cortex, the inputs corresponding to each eye segregate into nonoverlapping columns. In striated muscle, fibers that are initially innervated by five to six axons end up being innervated by a single axon. This rearrangement is competitive; it is enhanced by activity in the nerves, but the nature of the competition at the molecular level is not understood.

Rearrangement of connections continues through adulthood; changes in inputs to identified cells can be observed in the living animal. Many lines of evidence corroborate that this rearrangement exemplifies a state of dynamic equilibrium in the connections of the mature system. Turnover of connections at steady state may be useful in repairing damage to the system and it may play a role in the synaptic plasticity associated with learning and memory. The number of synapses, as well as their relative strengths, can be modified by various learning paradigms. Thus, the mechanisms used to form the initial wiring diagram are used to modify it according to experience in adulthood.

References

General

*Purves, D. and Lichtman, J. W. 1985. *Principles of Neural Development.* Sinauer Associates, Sunderland, MA.

Growth cones and motility

Bray, D. and Hollenbeck, P. J. 1988. Growth cone motility and guidance. Annu. Rev. Cell Biol. 4: 43–62.

Mills, L. R. and Kater, S. B. 1990. Neuron-specific and state-specific differences in calcium homeostasis regulate the generation and degeneration of neuronal architecture. Neuron 2: 149–163.

Pittman, R. N. 1990. Developmental roles of proteases and inhibitors. Semin. Dev. Biol. 1: 65–74.

Strittmatter, S. M., Valenzuela, D., Kennedy, T. E., Neer, E. J. and Fishman, M. C. 1990. G_0 is a major growth cone protein subject to regulation by GAP-43. Nature 344: 836–841.

Molecules controlling neurite adhesion and outgrowth

Caudy, M. and Bentley, D. 1986. Pioneer growth cone steering along a series of neuronal and non-neuronal cues of different affinities. J. Neurosci. 6: 1781–1795.

Doe, C. Q., Bastiani, M. J. and Goodman, C. S. 1986. Guidance of neuronal growth cones in the grasshopper embryo. IV. Temporal delay experiments. J. Neurosci. 6: 3552–3563.

*Elkins, T., Zinn, K., McAllister, L., Hoffman, F. M. and Goodman, C. S. 1990. Genetic analysis of a *Drosophila* neural cell adhesion molecule: Interaction of fasciclin I and Abelson tyrosine kinase mutations. Cell 60: 565–575.

Hedgecock, E. M., Culotti, J. G. and Hall, D. H. 1990. The *unc-5*, *unc-6*, and *unc-40* genes guide circumferential migrations of pioneer axons and mesodermal cells on the epidermis in *C. elegans.* Neuron 2: 61–65.

Kapfhammer, J. P. and Raper, J. A. 1987. Collapse of growth cone structure on contact with specific neurites in culture. J. Neurosci. 7: 201–212.

Landmesser, L., Dahm, L., Tang, J. and Rutishauser, U. 1990. Polysialic acid as a regulator of intramuscular nerve branching during embryonic development. Neuron 4: 655–667.

*Reichardt, L. F. and Tomaselli, K. J. 1991. Extracellular matrix molecules and their receptors: Functions in neural development. Annu. Rev. Neurosci. 14: 531–570.

Rutishauser, U. and Jessell, T. M. 1988. Cell adhesion molecules in vertebrate neural development. Physiol. Rev. 68: 819–857.

Schnell, L. and Schwab, M. E. 1990. Axonal regeneration in the rat spinal cord produced by an antibody against myelin-associated neurite growth inhibitors. Nature 343: 269–272.

*Takeichi, M. 1990. Cadherins: A molecular family important in selective cell-cell adhesion. Annu. Rev. Biochem. 59: 237–52.

The specificity of connections

Landmesser, L. 1984. The development of specific motor pathways in the chick embryo. Trends Neurosci. 7: 336–339.

Smith, C. L. and Frank, E. 1987. Peripheral specification of sensory neurons transplanted to novel locations along the neuroaxis. J. Neurosci. 7: 1537–1549.

Sperry, R. W. 1963. Chemoaffinity in the orderly growth of nerve fiber patterns and connections. Proc. Natl. Acad. Sci. USA 50: 703–710.

Trisler, D. 1990. Cell recognition and pattern formation in the developing nervous system. J. Exp. Biol. 153: 11–27.

Udin, S. B. and Fawcett, J. W. 1988. Formation of topographic maps. Annu. Rev. Neurosci. 11: 289–327.

Walter, J., Kern-Veits, B., Huf, J., Stolze, B. and Bonhoeffer, F. 1987. Recognition of position-specific properties of tectal cell membranes by retinal axons *in vitro.* Development 101: 685–696.

Wigston, D. J. and Sanes, J. R. 1985. Selective reinnervation of intercostal muscles transplanted from different segmental levels to a common site. J. Neurosci. 5: 1208–1221.

Remodeling of connections

Glanzman, D. L., Kandel, E. R. and Schacher, S. 1990. Target-dependent structural changes accompanying long-term synaptic facilitation in *Aplysia* neurons. Science 249: 799–802.

*Greenough, W. T. and Bailey, C. H. 1988. The anatomy of a memory: Convergence of results across a diversity of tests. Trends Neurosci. 11: 142–147.

Jansen, J. K. S. and Fladby, T. 1990. The perinatal reorganization of the innervation of skeletal muscle in mammals. Prog. Neurobiol. 34: 39–90.

Purves, D., Hadley, R. D. and Voyvodic, J. T. 1986. Dynamic changes in the dendritic geometry of individual neurons visualized over periods of up to three months in the superior cervical ganglion of living mice. J. Neurosci. 6: 1051–1060.

Shatz, C. J. 1990. Impulse activity and the patterning of connections during CNS development. Neuron 5: 745–756.

13

Neuron-Target Interactions

Paul H. Patterson

THE PREVIOUS CHAPTER considered the initial stages of neuronal differentiation, including the outgrowth of processes and the control of growth cone movement, the various influences on the direction of growth, and the selection of target cells with which synapses are formed. Once target cell selection is made, the actual process of constructing these highly specialized connections begins. Synapse formation is, in fact, even more of an "epic love story" than Ramón y Cajal imagined when he used that phrase to describe the formation of these "protoplasmic kisses" in the late 19th century. As in a delicate romance, the pre- and postsynaptic partners engage in a protracted dialogue, exchanging many signals essential for the further advances and differentiation of each participant. Some of these signals have now been identified and their roles at least partially defined. Such molecules not only mediate the assembly of the junction, but they can control the chemistry, the morphology, and even the survival of the synaptic partners. Thus, synapse formation can be viewed as a matching of partners in two stages: an initial recognition (romantic) phase and an adjustment and adaptation (marital) phase.

Synapse Formation

The synapse is biochemically specialized

Because there is great variability in structure among synapses in various parts of the nervous system, there is no truly prototypical synapse. There is, however, one case for which we have a significant biochemical, anatomical, and physiological description of synaptogenesis, and that is the vertebrate neuromuscular junction. Only three primary cells types interact at this relatively simple synapse: the motor neuron, the skeletal muscle, and the Schwann cell (Figure 1).

Both nerve and muscle cells are highly specialized at the mature synapse, as is the extracellular matrix in the cleft between them. The most prominent feature of the nerve terminal is the large number of synaptic vesicles. These are clustered near active zones, the presynaptic sites of transmitter release that are directly opposite the postsynaptic folds in the muscle fiber. These deep folds, which are found only underneath the nerve

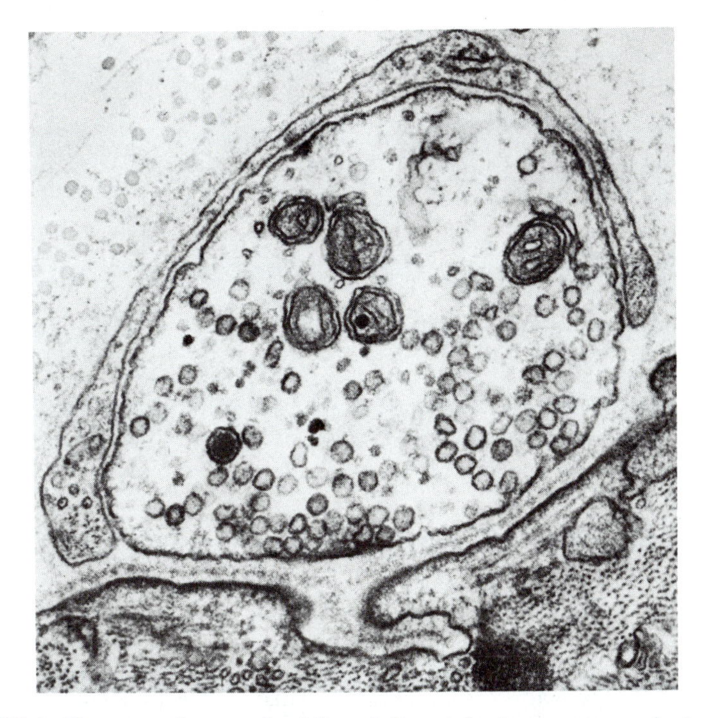

FIGURE 1. Electron micrograph of the adult vertebrate neuromuscular junction. A Schwann cell caps the presynaptic nerve terminal, which contains mitochondria and synaptic vesicles. The latter are clustered in active zones opposite the postsynaptic folds in the muscle membrane. The extracellular matrix, or basal lamina, surrounds the muscle and penetrates the folds as well. (Micrograph by J. R. Sanes, from J. R. Sanes et al., 1980. In *Nerve Repair and Regeneration,* D. L. Jewett and H. R. McCarroll, Eds. Mosby, St. Louis.

terminal, are composed of two domains—the crests of the folds, which contain the highly concentrated acetylcholine (ACh) receptor and an associated 43-kD protein, and the sides and bottoms, which contain sodium channels and ankyrin (see Chapter 8).

Other molecules, many of them related to known cytoskeletal proteins, are associated with the postsynaptic membrane. The basal lamina, which passes between the nerve and muscle and ensheathes each muscle fiber, also contains components that are specifically concentrated at the synapse. These include acetylcholinesterase (AChE), the A and S subunits of laminin, and the $\alpha3$ and $\alpha4$ chains of type IV collagen.

Synaptogenesis is a complex and extended event

The various morphological and biochemical specializations of the neuromuscular junction can be used to mark the progress of synaptogenesis (Figure 2). Two important points emerge. First, not all specializations appear at once, indicating that development of the mature synapse occurs in multiple stages. Second, the process is quite prolonged, requiring two to three weeks in rodents and birds, a period equivalent in these species to the entire period of prenatal development.

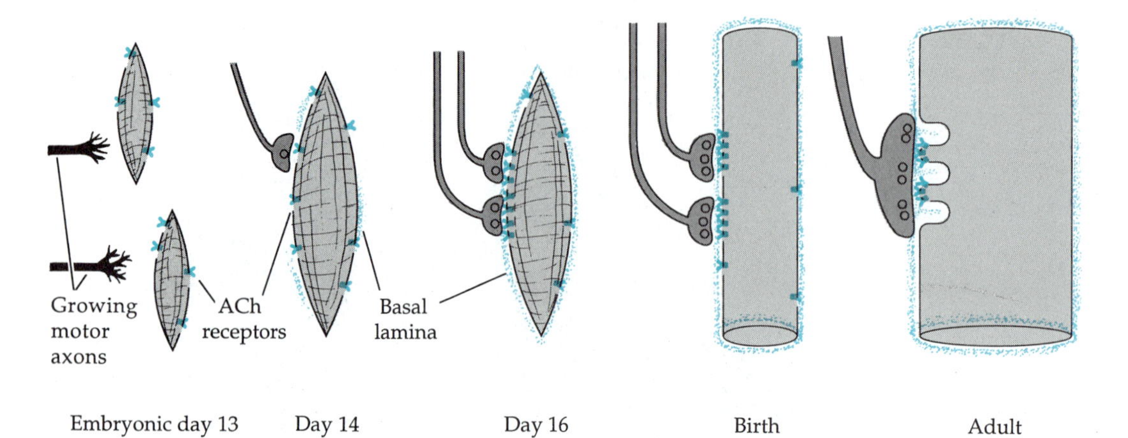

FIGURE 2. Schematic drawing of the major stages of neuromuscular junction formation in the intercostal muscles of the rat. (Embryonic day 13) Motor nerves releasing ACh enter the muscle where myoblasts are fusing to form myotubes; the myotubes express ACh receptors on their surfaces. (E.D. 14) The nerves contact the myotubes; patches of basal lamina form on the muscle surface. (E.D. 16) ACh receptors cluster under the nerve terminals; myotubes become multiply innervated; the basal lamina becomes continuous. (Birth) The density of extrajunctional receptors begins to decline; ACh receptor turnover time changes and the ACh receptors become resistant to dispersal (not illustrated). At this stage, each muscle fiber is innervated by several motor neurons. (Adult) Extrajunctional ACh receptors disappear; the junctional folds appear; nerve terminals from all axons except one are withdrawn; the subunits and gating properties of the ACh receptors change (not shown). See Figure 4 for change in subunits.

In the rat, motor axons first enter the developing muscle masses of the trunk and limbs at the same time that myotubes are beginning to form by fusion of myoblasts (embryonic day 13, or E13). The first functional contact between nerve and muscle is detected on E14, when low-frequency, spontaneous, miniature endplate potentials are seen and when endplate potentials can be elicited by nerve stimulation. Both nerve and muscle cells appear to be prepared for their first encounter. Growth cones of the extending axons release ACh, and the myotubes have ACh receptors (AChRs) over their entire surface. Morphological specializations are not observed in either nerve or muscle at the time of first contact.

During E15–20, major changes occur. The appearance of AChRs clustered at high density near the nerve terminal is the first sign of postsynaptic specialization. Synaptic vesicles accumulate in the nerve terminal, and levels of choline acetyltransferase, the enzyme that synthesizes ACh, increase significantly. At about E16, AChE becomes concentrated in the synaptic cleft, and other specialized basal lamina components appear. At E18–20, the clustered AChRs become metabolically stabilized, so that their half-life in the membrane increases from about one day to 10 days. The half-life of AChRs in the nonsynaptic, extrajunctional muscle membrane remains short.

Further changes occur during the first two postnatal weeks. The synaptic folds appear, AChRs in the extrasynaptic membrane disappear, and there is a change in the subunit composition of the endplate AChRs, as the ε-subunit replaces the γ-subunit, thus altering the channel properties of the endplate receptors (see below and Chapter 3).

Multiple innervation is lost postnatally

An important synaptic rearrangement, similar to those discussed in the previous chapter, also takes place postnatally. Individual myotubes initially receive immature synaptic contacts from several motor neurons; most of these are eliminated during the first and second postnatal weeks, leaving each mature muscle fiber with a single motor axon. The elimination of multiple innervation appears to be the result of competition between the axons for synaptic contact on the myotube, with each axon trying to increase its area of contact by terminal arborization. As with the retinotectal system (see Chapter 12), this rearrangement is affected by the level of activity in the nerves. Elimination of synaptic terminals is blocked by complete paralysis of nerve-muscle activity, slowed by reduction of nerve activity, and accelerated by enhanced activity. While the effects of activity may be mediated by factors produced by the muscle, the identity of these factors is not yet known.

Apparently there is an upper limit on the number of synapses that can be supported by each motor neuron. That is, even in the absence of competition, the neuron cannot maintain an unlimited number of axonal branches contacting muscle fibers. Moreover, in the competition for synaptic contact with muscle cells, the terminals of "over-extended" neurons compete less well; the competitive vigor of a particular terminal may be inversely related to the number of terminals that its axon has formed. The limit on the number of terminals and the competition between axons combine to ensure that each muscle fiber is innervated by only a single axon, and that there is an appropriate distribution of muscle fibers among the pool of motor neurons.

Muscle activity is required for some, but not all, steps in synaptic development

How do signals exchanged between the nerve and muscle mediate or modulate the protracted process of synaptic development? The activation of muscles by nerves through the secretion of ACh is the most obvious and continuous line of communication between the cells. Disruption of this line by blockade of neuromuscular transmission does not affect the initial events of synapse formation, however. Thus, both in vivo and in nerve-muscle co-cultures, the clustering of AChRs occurs even in the presence of curare or α-bungarotoxin. Several subsequent events do, however, depend on the electrical and/or mechanical activity evoked in muscle cells by the nerve. One example is the accumulation of synaptic AChE in chick muscle, which can be prevented by neuromuscular blockade. In blocked muscles AChE is restored by direct electrical stimulation of the muscle.

Muscle activity also controls the density of AChRs in the extra-synaptic membrane of both developing and adult muscles. Postsynaptic blockade of neuromuscular transmission prevents the decline in density of extra-synaptic receptors that occurs during early postnatal life. The receptors' normal disappearance is thus likely to be the result of activity in the newly innervated myofibers. In adult muscle, denervation causes the reappearance of AChRs in the extra-synaptic membrane (a phenomenon called "denervation supersensitivity"). The increase in density of extra-synaptic AChRs can also be elicited by postsynaptic blockade and can be prevented by electrical stimulation of the muscle. In addition to its effect on receptors, denervation of adult muscle results in widespread changes, including muscle atrophy. Some of these changes can be partially prevented by direct electrical stimulation of the muscle.

Neurons secrete factors that direct the local expression and organization of synaptic molecules

Because action potentials depolarize the entire muscle fiber, raising the calcium concentration everywhere, they cannot by themselves provide an adequate signal for local specializations at the junctional site. How could muscle activity alone, for example, cause the discrete synaptic localization of AChE? Clearly there must also be local signals that arise from the nerve that stimulate or guide synaptic development. Strong evidence now supports the idea that factors secreted from the nerve direct the accumulation of AChRs, AChE, and other macromolecular components at the synapse; they locally stimulate AChR synthesis; and they regulate the subunit composition of the AChR. Some of these activities can now be attributed to specific proteins.

When axons make contact with muscle cells, they induce the ACh receptors already on the surface of myotubes to form high-density clusters in the muscle membrane near the site of contact. This was originally demonstrated by using fluorescently labeled α-bungarotoxin to prelabel AChRs in muscle cultures to which nerves were subsequently added (Figure 3). The clusters that formed along the path of nerve growth contained labeled receptors, indicating that the nerve had caused a redistribution of AChRs already in the membrane. Synthesis of new receptors is thus not required to produce clusters. AChR clustering is induced specifically by motor neurons (not by sensory neurons), but is unrelated to synaptic activity, as it occurs in the presence of curare, as well as α-bungarotoxin.

These observations suggest that a factor released by the nerve is responsible for inducing clusters. A protein, **agrin**, purified from *Torpedo* electric organ (a rich source of cholinergic synapses, see Chapter 3), has the properties expected of such a factor. When agrin is added to cultures of chick myotubes, it causes the formation of multiple clusters of AChRs in the absence of nerves and of protein synthesis. Moreover, a protein immunologically related to agrin is found in the basal lamina at the synapse. The AChR clusters induced by agrin are associated with patches of several other synaptic components including the 43-kD protein, AChE, and heparan sulfate proteoglycan. Thus agrin appears to organize many components of the postsynaptic specialization. Although the mechanism by

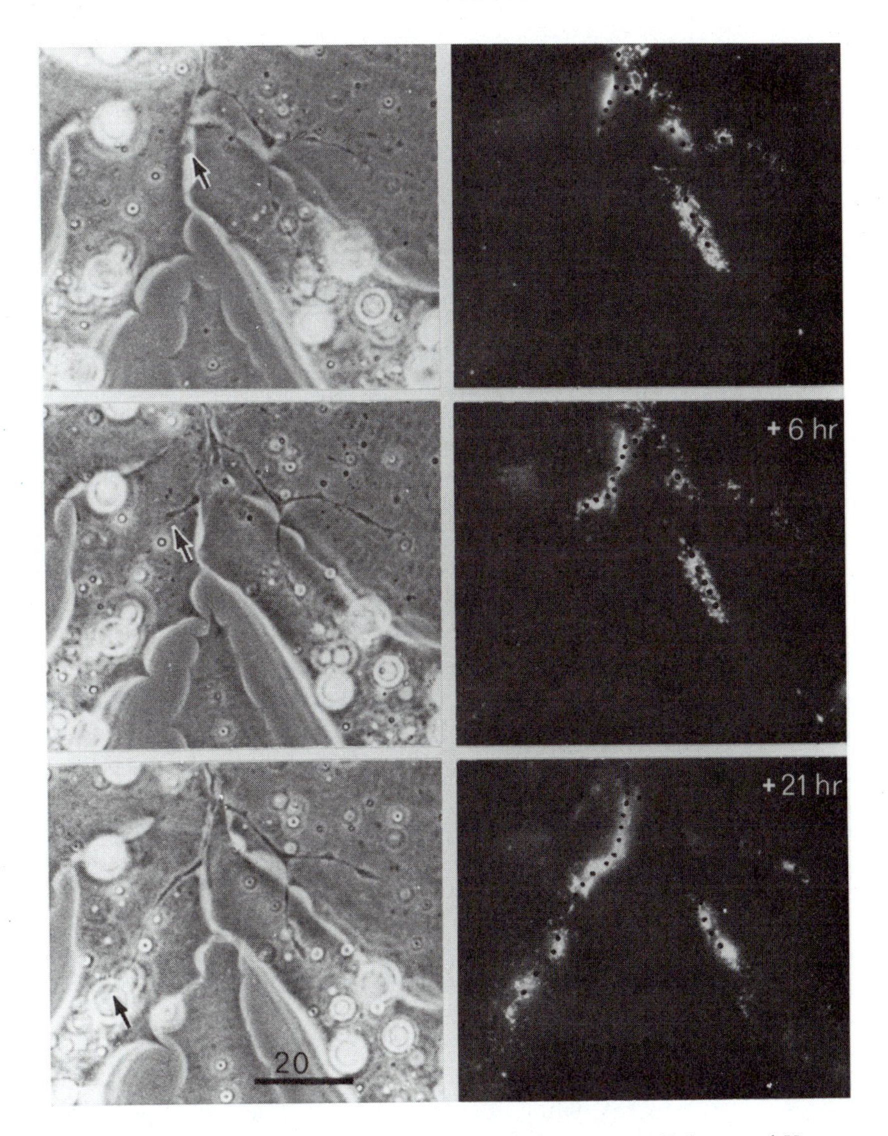

FIGURE 3. ACh receptors cluster under growing neurites. Cultures of *Xenopus* myocytes were incubated with fluorescent-labeled α-bungarotoxin and then washed to localize the receptors on the surfaces of the cells, as neurites (arrows) from spinal cord explants grow across the myocytes. The same field was photographed by phase contrast optics to visualize the cells (left) and by fluorescence optics to visualize the labeled toxin (right). The large, refractile images are fat droplets in the myocytes. At 6 and 21 hours, the left branch of the neurite can be seen to grow significantly, and the fluorescent label extends along this branch. Note also the loss of fluorescent stain along other points of nerve-muscle contact. Since the label was removed from the medium at time zero, only those receptors on the surface at that time are labeled; that is, new sites of staining represent movement of receptors in the membrane rather than the insertion of newly synthesized receptors. Black dots were added to the fluorescence micrographs to illustrate the sites of nerve-muscle contact. (From M. Cohen et al., 1979. Prog. Brain Res. 49: 335–349.)

which agrin induces AChR clusters is not known, it has recently been shown to cause tyrosine phosphorylation of the β-subunit of the AChR.

Immunostaining for agrin protein and in situ hybridization analysis of agrin mRNA show that motor neurons produce agrin and transport it to their terminals. Although the release of agrin from nerves has not been demonstrated, anti-agrin antibodies block the clustering of AChRs at sites of nerve contact in nerve-muscle co-cultures. These experiments suggest that agrin may play an important role both in inducing the concentration of postsynaptic components at the newly formed synapse and in maintaining these components in the adult.

Motor neurons contain peptides that regulate AChR subunit synthesis

Motor nerves also secrete factors that increase the local synthesis of the AChR. In adult muscle fibers, electrical activity suppresses synthesis of AChR subunit mRNA over most of the fiber (Figure 4). Nuclei near the endplate, however, continue to produce AChR mRNA in response to a local signal from the nerve. At least two polypeptides found in motor neurons increase the synthesis of mRNA for the α-subunit of the AChR in muscle fibers. One of these is the neuropeptide **CGRP (calcitonin-gene–related peptide**) (see Chapter 4), which is found in large, dense-cored vesicles in nerve terminals. CGRP also increases cAMP in the muscle. The protein **ARIA (AChR inducing activity),** purified from embryonic brain, increases synthesis of the α-subunit of the AChR, but by a mechanism that does not involve cAMP. The precise roles of CGRP and ARIA in the formation and maintenance of the postsynaptic membrane are not known, but they, or factors like them, could function both to increase the concentration of AChR at the site of initial nerve-muscle contact, and to maintain AChR synthesis at the adult synapse.

The switch of the γ- and ε-subunits of the AChR during development (Figure 4) may also be regulated by similar factors. In adult muscle, in addition to α-, β-, and δ-subunit mRNA, ε-subunit mRNA is made by nuclei near the endplate, and γ-subunit mRNA is not. After denervation, nuclei near the old endplate continue to make ε-, and not γ-, subunit mRNA, while other nuclei in the fiber, in response to muscle inactivity, make mRNA for α, β, γ, and δ. The induction of ε, and the suppression of γ, mRNA synthesis in nuclei near the endplate thus appears to be a permanent alteration, enduring after the nerve is lost. The detailed mechanisms by which the transcription of AChR genes are controlled are not known but are likely to be similar in principle to those seen for other genes that are subject to multiple regulatory factors (see Chapter 10).

Our present understanding of the control of AChR expression and distribution can be summarized as follows. Myotubes are preprogrammed to produce AChR, even in the absence of nerves. These preexisting receptors can also form randomly scattered clusters on the surfaces of muscle cells in the absence of neurons. When a motor axon is present, the clusters organize preferentially under the nerve and cluster formation elsewhere on the muscle is suppressed. There are several good candidates for mol-

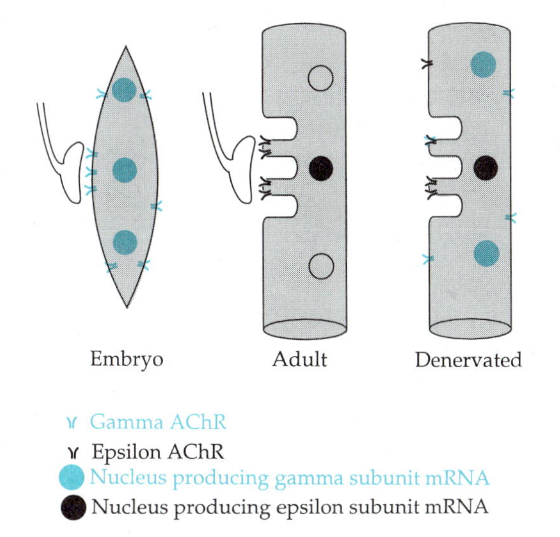

FIGURE 4. Distribution of embryonic and adult forms of the ACh receptor in embryonic, adult, and denervated muscle. The ACh receptor made in embryonic muscle (color) contains α, β, γ, and δ subunits, whereas the adult form (black) contains α, β, ε, and δ subunits. In embryonic muscle, the ACh receptor is distributed over the entire muscle surface, and nuclei throughout the muscle make receptor mRNA. In adult muscle, the receptor is confined to the postsynaptic membrane, and only nuclei near the synapse (black) make receptor mRNA. The synaptic nuclei make ε mRNA but not γ mRNA. After denervation, the embryonic form (γ) of the ACh receptor is expressed throughout the muscle, while the synaptic nuclei continue to make mRNA for the adult form (ε).

ecules that motor neurons secrete to cause receptor clustering at developing synapses. Nerve-evoked activity in the myotube down-regulates receptor expression in extra-junctional areas. In addition to the changes in receptor distribution, there is a switch in expression of two of the receptor subunits that begins after the nerve contacts the muscle. The adult-specific ε-subunit is substituted for the embryo-specific γ-subunit. The initial appearance of the ε-subunit is induced only in the myotube nuclei at the site of nerve-muscle contact, but its continued expression does not require nerve activity.

There are many other examples of nerve-induced target cell differentiation in which we do not yet know the molecular basis of the interaction. Some classical cases include the induction of sense organs, such as taste buds of the tongue, and muscle spindle organs in skeletal muscle, by the sensory nerves innervating them. In cases such as these, the sense organ does not differentiate until the neurites arrive, and differentiation does not proceed if the innervating axons are prevented from arriving. There are also experiments suggesting that substituting other axons for the normal sensory processes will not induce organ differentiation. These classical phenomena await renewed investigation with current techniques.

The basal lamina regulates presynaptic differentiation

In addition to the communication from nerve to muscle, muscle cells provide signals that can direct the differentiation of the nerve terminal. The best evidence for such signals comes from a series of experiments on the reinnervation of denervated muscle. In the frog, when the cut motor nerve grows back after denervation, the axons locate the sites of the previous neuromuscular junctions on the muscle surface and form new synapses at these sites with unerring accuracy. The information that identifies the old sites is contained in the basal lamina. If the muscle is damaged at the time of denervation, and irradiated to prevent muscle fiber regeneration, all that is present when the nerves grow back are the empty sheaths of basal lamina (Figure 5). Nevertheless, even in the absence of muscle cells, the nerves reinnervate the old synaptic sites, identified by AChE. Moreover, the new nerve terminals accumulate synaptic vesicles and form active zones precisely across from the preexisting folds in the basal lamina.

A molecule has recently been isolated from muscle basal lamina that could serve as a signal to terminate axonal growth at the old synaptic sites. Synaptic basal lamina contains a special form of laminin in which **s-laminin** (Figure 6 in Chapter 12) substitutes for the B1 subunit, the other subunits being B2 and either the merosin or A subunits. When tested in culture, s-laminin, like laminin, promotes the adhesion of neurites. But while laminin indiscriminately supports the adhesion of neurites from many neurons, s-laminin preferentially promotes the adhesion of motor neurons rather than sensory or CNS neurons. Adhesion to s-laminin can be prevented by a simple tripeptide, Leu-Arg-Glu (LRE in the single-letter code), suggesting that this sequence in the intact molecule may function as the neuron attachment site, as the sequence RGD does in laminin and fibronectin. Interestingly, the LRE sequence occurs at higher-than-expected frequency in two other synapse-specific proteins, the tailed form of AChE and agrin, suggesting that the synapse contains a high concentration of this attachment signal. The LRE peptide and antibodies against it should provide powerful tools for future investigations of synapse formation.

The A subunit of laminin is also highly enriched in the synaptic basal lamina, as are two collagen type IV chains, α3 and α4. Laminin containing the A subunit is known to promote neurite outgrowth from many types of neurons, as does collagen type IV. Thus these proteins could also play synapse-specific roles as well.

Other molecules are likely to promote the innervation of muscle

Other proteins, not localized specifically at synaptic sites, may also contribute to motor axon growth on muscle. The neural cell adhesion molecule (N-CAM) and neural cadherin (N-CAD), for instance, are expressed at high levels on embryonic muscle, and antibodies against these proteins can perturb axonal growth on myotubes. Recall also the discussion of the role of the polysialic acid component of N-CAM, and the conflicting attractions of axonal fasciculation and nerve-muscle contact (Chapter 12). As previously discussed, the protein tenascin and the INO epitope formed by the complex of laminin and heparan sulfate proteoglycan can also promote

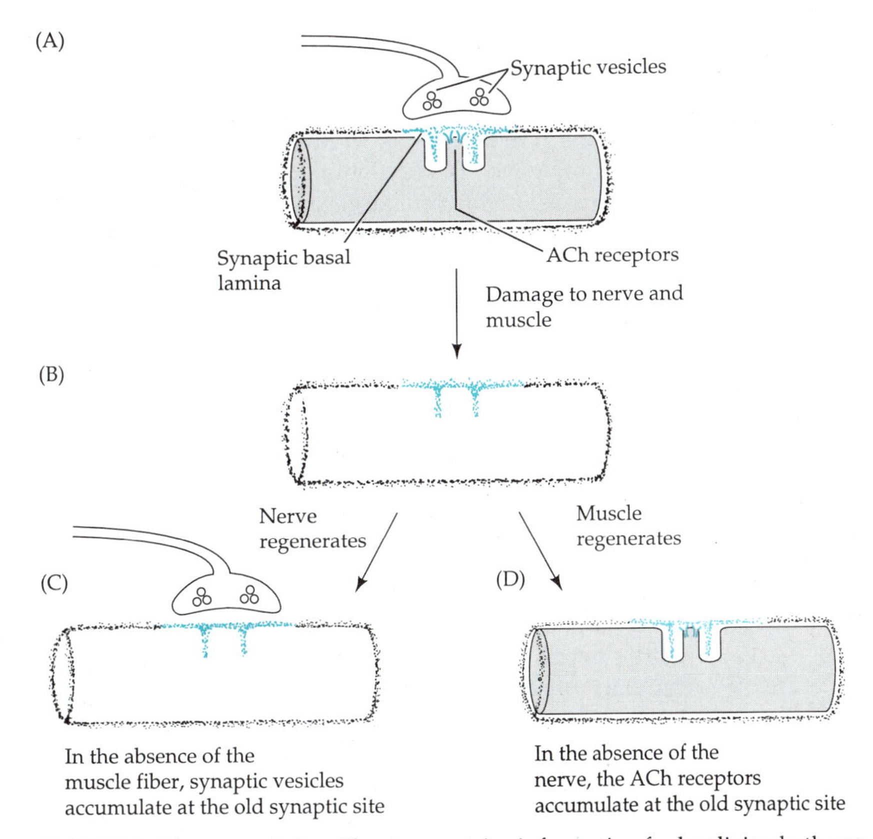

FIGURE 5. The synaptic basal lamina contains information for localizing both pre- and postsynaptic specializations at the neuromuscular junction. (A) In adult frog muscle the synaptic basal lamina (color) differs from the extrasynaptic basal lamina (black). ACh receptors are concentrated in the postsynaptic membrane and synaptic vesicles are clustered opposite the projections of the basal lamina that extend into the folds of the postsynaptic membrane. (B) When both the motor nerve axons and the muscle fibers are damaged, the basal lamina remains intact. (C) When the nerve reinnervates the empty basal lamina sheath in the absence of muscle regeneration, nerve terminals innervate the old synaptic sites, and vesicles cluster opposite the synaptic projections of the basal lamina. (D) When the muscle regenerates in the absence of the nerve, myotubes form inside the basal lamina sheaths and the ACh receptors are concentrated at the old synaptic site.

axonal growth. Like N-CAM, the expression of these molecules is high in embryonic muscle, decreasing after the muscle is innervated and increasing significantly when muscle is denervated. Embryonic and denervated muscles also contain soluble proteins that can enhance motor neuron survival and growth. Some or all of these various components could aid regenerating axons in growing within the muscle.

In addition to specifying the site of nerve terminal differentiation, target cells can also specify the particular type of synaptic terminal that is formed. For instance, when retinal ganglion cells innervate their normal target cells in the lateral geniculate nucleus, they make characteristic morphological specializations. If the embryonic retinal axons are diverted into the somatosensory nucleus, however, they form presynaptic terminals with a morphology characteristic of the normal innervation of that nucleus. This phenomenon is also likely to occur during normal development; apparently homogeneous populations of neurons form morphologically distinct synapses in different target areas. At the single-cell level, an identified neuron in the marine mollusc *Aplysia* has been found to send different neuropeptides, in distinct vesicles, down its two axons, each projecting to a different location (see Chapter 4). Thus, target influences may regulate the differentiation of presynaptic components in two axons of the same neuron, and the cytoskeleton (Chapter 8) presumably routes the various molecules specifically to those target areas.

Control of Neuronal Phenotype

Target and glial cells can control neuronal shape

Although still somewhat mysterious, the induction of presynaptic specializations by a target cell may be viewed as part of the matching of pre- and postsynaptic partners that occurs locally at the point of apposition. The nature of this interaction, however, extends far beyond this site. Not only can the morphology of the presynaptic *terminal* be influenced by the target cell, but the morphology of the entire presynaptic *neuron* can be altered in highly specific ways. This instructive influence is well illustrated by identified neurons in developing invertebrates. Retzius cells are serotonergic neurons found in the same position in each of the reiterated abdominal ganglia of the leech (see Figure 10 in Chapter 1). These neurons send their axons to the adjacent body wall to innervate the skin, and they also send axons to neighboring ganglia via the connectives that run between the ganglia. Within the ganglion itself, each Retzius cell produces a rich arbor of dendritic processes. Retzius neurons that innervate the reproductive organs in the tail segments differ from their counterparts in other ganglia in that they have much smaller dendritic arbors and do not send axons into the next ganglia. During early embryogenesis, however, all Retzius cells look alike; it is only upon innervating skin versus reproductive tissue that their shapes become different. The decisive influence of the target organs is shown in experiments in which they are ablated or transplanted. After ablation of the reproductive organs the neurons that would have innervated them develop more like Retzius cells in other segments. When the reproductive organs are transplanted ectopically, nearby Retzius cells innervate them and change their axonal and dendritic projections to resemble the Retzius cells that normally innervate these organs.

Target organs not only influence the shape of the dendritic arbor, but also the presynaptic connections on the arbor. After removal of the reproductive tissue, the Retzius cells in that segment receive synaptic inputs normally found on Retzius cells in other segments that innervate skin. Such specification of central connections by peripheral targets is reminiscent of similar findings with dorsal root ganglion neurons that innervate skin versus muscle, resulting in distinct projections in the vertebrate spinal cord, as discussed in Chapter 12.

Dendritic morphology can also be influenced by the glial cells that surround neuronal somas. Sympathetic neurons cultured alone or with various non-neuronal cells, such as heart cells, develop very long axons, but few if any dendrites. When grown with glial cells, however, the neurons also display robust dendritic arbors (Figure 6). Thus, axonal and dendritic growth can be differentially controlled. The signal for dendritic growth in this case may reside in the extracellular matrix, because an extract of basement membrane can mimic the effect of glial cells. For embryonic mesencephalic neurons, astrocytes from different brain regions have different effects. The neurons form both axons and dendrites when cultured with astrocytes from the mesencephalon (where the neuronal somas would normally be located); they only form axons when cultured with astrocytes from the striatum (a target area for these neurons). As with sympathetic neurons, the extracellular matrix is thought to play a role in controlling the morphology of the neurons.

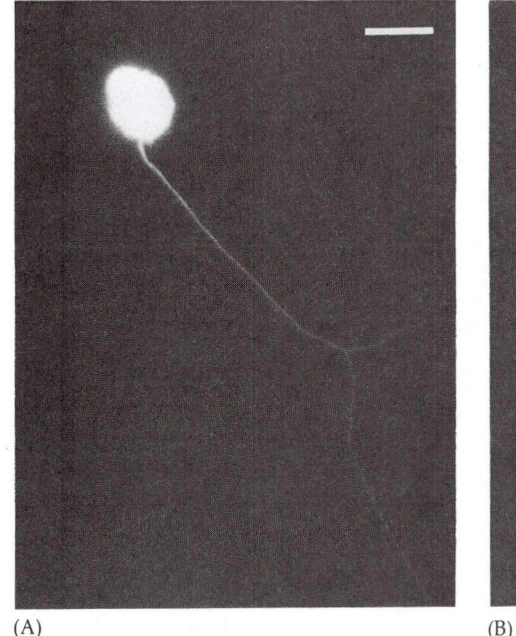

(A)

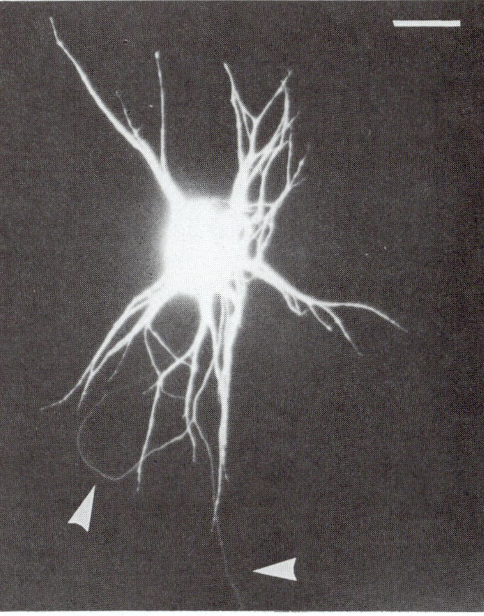

(B)

FIGURE 6. The growth of dendrites can be controlled by glial cells. The morphology of cultured sympathetic neurons is visualized by injection of the dye Lucifer Yellow and observation with fluorescence optics. When grown in the absence of glia (A), the neurons exhibit long axons and few, if any, dendrites. When co-cultured with glial cells from sympathetic ganglia, in contrast, the neurons produce many dendrites as well as an axon (B). The latter has branched (B, arrowheads) in both cases. Dendrites are identified as being much shorter and thicker than axons, and gradually tapering rather than having a constant, narrow bore like axons. The dendritic nature of these processes has also been confirmed using antibody markers and by electron microscopy. (From M. Tropea et al., 1988. Glia 1: 380–392.)

Instructive factors can control the chemical identity of neurons

Neurotransmitter and neuropeptide phenotype is another fundamental aspect of neuronal identity that can be influenced by environmental signals. As described in Chapter 4, each neuron produces and secretes several chemical messengers, chosen from a group of about a dozen "classical" neurotransmitters and over 75 neuropeptides. How do neurons "decide" which of these molecules to produce? As discussed in Chapter 11, neuronal phenotype is determined by an interplay between the cell's lineage history and its environment. Such environmental signals can come from other cells near the neuronal soma, or they can be produced by target tissues. In the latter case, a neuron's phenotype would not be finally determined until its axons reached the target area.

Most neurons in sympathetic ganglia, such as those innervating the salivary glands, produce norepinephrine and neuropeptide Y (NPY). There are exceptions, however, such as the neurons that innervate the sweat glands in the rat foot that produce ACh and vasoactive intestinal polypeptide (VIP). During development, the sympathetic neurons that are destined to innervate the sweat glands initially produce norepinephrine, just like the other neurons in the ganglion. Only after their axons reach the sweat glands do they switch their phenotype, down-regulating the production of norepinephrine and initiating the expression of ACh. This correlation between target and transmitter/peptide identity is reinforced by a series of transplant experiments (Figure 7). When a piece of salivary gland is substituted for the sweat glands in the foot, sympathetic axons innervating this ectopic target do not switch their phenotype and become cholinergic as they would normally do. Conversely, when sweat glands are transplanted into the hairy skin of the rat, noradrenergic sympathetic axons that would normally innervate nearby hair papillae and blood vessels now innervate the transplant. These neurons become cholinergic and produce VIP. Thus, postmitotic, fully differentiated neurons are plastic with respect to their transmitter/peptide identity, and the particular target tissue they innervate can control that identity. This idea is confirmed by culture experiments that demonstrate unambiguously that individual sympathetic neurons can switch their phenotype from noradrenergic to cholinergic. This plasticity is not restricted to the developmental period; sympathetic neurons cultured from adult rats as well as from embryos can be converted from a noradrenergic to a cholinergic phenotype.

Instructive factors can have overlapping specificities

What signals do target cells use to effect these changes? Several proteins that instruct neurons in their choice of transmitter and neuropeptide expression have been purified using neuronal cultures as assay systems. Media conditioned by cell cultures from various tissues, as well as extracts of target and nervous tissues, have yielded a growing family of these **instructive factors** (Table I). Interestingly, while many of these proteins have distinct effects on transmitter and neuropeptide gene expression, their effects are often partially overlapping, or redundant. That is, several different proteins can induce ACh production as well as expression of VIP

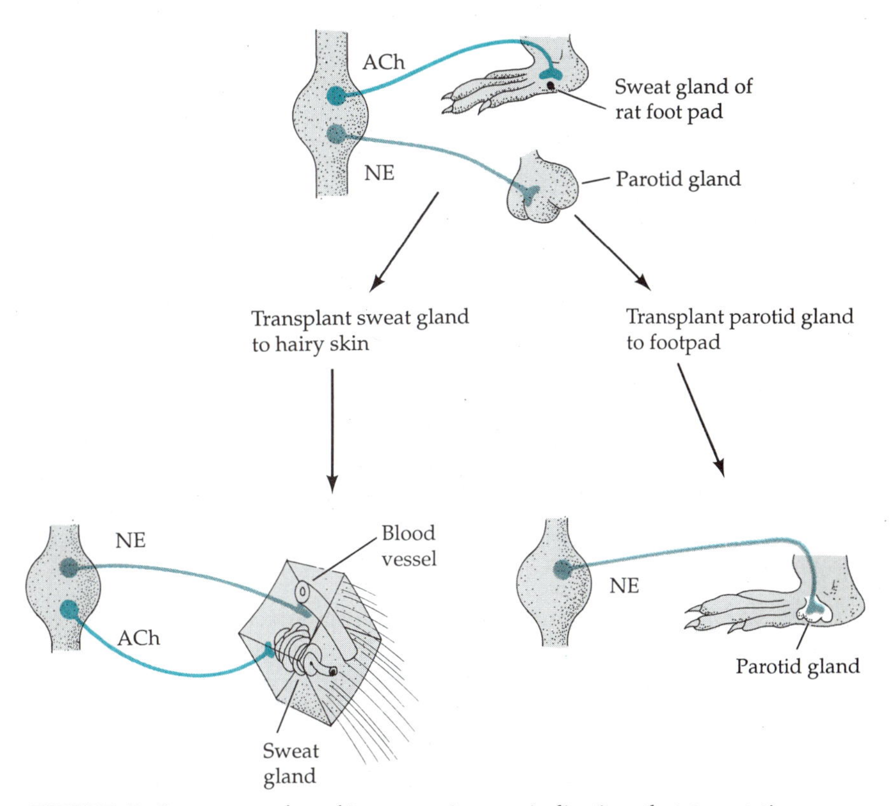

FIGURE 7. Summary of grafting experiments indicating that target tissue can control the phenotype of the neurons that innervate them. Neurons in sympathetic ganglia innervate many different target tissues, such as the parotid glands, and the sweat glands of the footpad in the rat. The transmitter phenotype of neurons innervating the former is noradrenergic (NE), while neurons innervating the latter become cholinergic (ACh) after contacting the sweat glands. Sweat glands transplanted to hairy skin, where the sympathetic neurons innervating blood vessels and hair papillae are noradrenergic, attract sympathetic innervation that becomes cholinergic upon contact with the glands. In contrast, if the sweat glands of the footpad are removed and replaced by parotid glands, the innervation remains noradrenergic. (Drawn by N. K. Mahanthappa and P. H. Patterson.)

and somatostatin in sympathetic neurons; another protein induces just VIP; and a third induces only somatostatin. These proteins have distinct biochemical properties, and those whose cDNAs have been cloned show no significant sequence homologies. The emerging picture of neuronal differentiation factors has a striking parallel to the control of phenotypic decisions in the hematopoietic (blood forming) system. Here too, biochemically diverse proteins induce distinct differentiation responses, but with partially overlapping effects. Thus, while four hematopoietic regulators elicit diverse arrays of derivatives from multipotential stem cells, all four promote the granulocyte and/or macrophage differentiation pathway. Recently, this analogy between the nervous and hematopoietic systems has been carried a step further, with the demonstration that the same

TABLE 1. Comparison of Neurotransmitter/Neuropeptide Specifying Factors[a]

	ACh	CA	VIP	NPY	SOM	SP	mENK	Molecular Weight (kD)
I (CDF)	↑↑↑	↓↓	↑	↓	↑	↑↑↑	—	45
II			↑↑					85
III					↑↑			45
CNTF	↑↑↑	↓↓	↑	↓	↑	↑↑↑	—	23

[a]The table compares the biological and biochemical properties of several proteins that affect the expression of neurotransmitters and neuropeptides of cultured sympathetic neurons. Upward arrows indicate the induction of a neurotransmitter synthesizing enzyme or neuropeptide and its mRNA; downward arrows indicate the suppression of such expression. Dashes indicate no effect. ACh, acetylcholine synthesis and choline acetyltransferase expression; CA, catecholamine synthesis and tyrosine hydroxylase expression; VIP, vasoactive intestinal polypeptide expression; NPY, neuropeptide Y expression; SOM, somatostatin expression; SP, substance P expression; mENK, metenkephalin expression. CDF, cholinergic differentiation factor, also known as leukemia inhibitory factor; CNTF, ciliary neurotrophic factor.

protein can influence differentiation choices in both systems. A **cholinergic differentiation factor** (**CDF**) from heart cells was shown to be identical to **leukemia inhibitory factor** (**LIF**), a protein that inhibits the proliferation of a myeloid cell line and induces the differentiation of macrophage characteristics. In fact, CDF/LIF can affect the proliferation and differentiation of embryonic stem cells, adult liver cells, and embryonic and adult bone cells. In addition, CDF/LIF is produced in both membrane-bound and diffusible forms, suggesting that it can act locally as well as at a distance.

The tissues that produce these instructive factors must now be determined and their potential functions perturbed with immunological and molecular genetic methods. It is possible, however, that the redundancy in their effects may mitigate the effects of perturbation, as we saw in the previous chapter with the neurite outgrowth and fasciculation molecules. It will also be of considerable interest to localize the receptors for these signals on various populations of neurons. The intracellular mechanisms by which these factors induce their distinct, but partially overlapping, effects on neuronal gene expression could reveal interesting combinatorial effects. A neuron's phenotype could be decided not only by the presence of particular instructive factors in its environment but also by its combination of receptors for the factors and by the status of its transcription machinery. Supporting this notion is the recent finding that cultured sensory neurons, like their sympathetic counterparts, are induced by CDF/LIF to produce ACh and VIP but, unlike the sympathetic neurons, the sensory neurons are not induced to express either somatostatin or substance P.

The last several sections make it clear that target cells can control many aspects of a neuron's development. These include determination of which processes become axons and dendrites, how the dendrites are shaped, where synapses are elaborated, which transmitters are secreted, and even where to localize the sites of secretion within the nerve terminal. Control of its dendritic properties leads, in turn, to an influence on the connections that the neuron receives. Indeed, the extremely elaborate web of interactions that comprise the process of synapse formation can be seen as a cascade in which the specific connections in the wiring pattern influence the nature of the units in that pattern.

Neuronal Life and Death: Trophic Factors

Targets and presynaptic inputs can control neuronal survival and growth

One of the most dramatic, and in some respects still puzzling, phenomena in the development of the nervous system is neuronal cell death. In many different parts of the PNS and CNS, half or more of the neurons die as part of normal ontogeny. While cell death is routinely found in developing vertebrates, its extent varies greatly between species. In cat retina, for instance, 80 percent of retinal ganglion cells die, whereas the corresponding figure is 40 percent for the chick retina, and none for the retinas of fish and amphibia. In general, neuronal death is least pronounced in those species that continue to grow throughout life.

In most cases, death is not due to an obvious defect in the neurons; like their surviving counterparts, neurons that die differentiate, receive presynaptic input, and successfully grow their axons into the appropriate target area. The major role of death in these cases is thought to be in matching the sizes of neuronal and target cell populations, connected at some distance to each other. Because they are generated separately, a mechanism is required to adjust the ratios of pre- and postsynaptic partners once they are connected. It has also been argued that the generation of a large excess of neurons could be useful in evolution as a source for experiments in the formation of new pathways.

How is the matching of neuronal populations accomplished? There is compelling evidence that the size of the target field plays a key role in regulating the survival of the neurons that innervate it. First, if the sole target of a given neuronal population is removed before their axons reach it, 85–90 percent of the neurons die, as opposed to the 50–60 percent that normally die during development. All of these neurons can be shown to survive under appropriate culture conditions, demonstrating that they are not irreversibly preprogrammed to die. Second, if extra target tissue is provided, 20–70 percent of the neurons that would normally die survive. The fact that experiments designed to rescue all of the neurons in situ have not succeeded suggests that factors other than those derived from target cells may be involved. Third, several types of genetic manipulations have been used to vary the numbers of postsynaptic cells, and a direct correlation between the sizes of the pre- and the postsynaptic populations has been found. Fourth, a number of target-derived **neurotrophic factors** (*trophikos* is Greek for "nourishment") have been isolated, as described in subsequent sections. These proteins specifically support the survival of the neurons innervating the targets producing the proteins. Thus, vertebrate neuronal survival is dependent, in part, on success in obtaining a target-derived substance present in limiting amounts.

Competition between neurons influences their *growth* as well as their survival. Neurons can be experimentally provided with extra target tissue in a number of ways, including grafting additional tissue, or removing other neurons that compete for the target. These manipulations often result in larger neuronal somas, axons, and dendrites. If additional irides are grafted into the anterior chamber of the eye, for example, sympathetic neurons can grow axonal arbors more than eightfold larger than normal. Such transplantation experiments also reveal a cascade effect. Larger peripheral targets yield greater numbers of larger neurons innervating them. These neurons, in turn, provide a larger target population for their presynaptic partners, and so on. Thus, a change in one population can have significant repercussions for other parts of the system, and possibly for selection during evolution.

Neuronal survival is also influenced by the presynaptic input neurons receive. As discussed previously, postsynaptic cells deprived of their input often atrophy. This atrophy can result in death, as seen in the accessory optic nucleus (AON)–ciliary ganglion–iris pathway. The AON provides the only input to the parasympathetic neurons of the ciliary ganglion. The latter neurons, lying behind the eye, innervate the iris, and about 50 percent of them die during normal development. If the AON is ablated before its axons innervate the ganglion, 85–90 percent of the ciliary neurons die. If the iris is ablated before the ciliary axons reach it, a similar increase in ciliary neuronal death occurs. If both the pre- and the postsynaptic partners of the ciliary neurons are ablated, 100 percent of them die. Although both denervation and target deprivation cause cell death, the mechanisms by which they do so are different, as discussed below.

Is electrical activity of the target involved in the competition for survival? Perhaps surprisingly, if synaptic transmission at the neuromuscular junction is blocked by α-bungarotoxin or curare, the number of motor neurons *increases*. In fact, all the neurons that would normally die can be rescued, as in the culture experiments. If the neuromuscular blockade is removed, all of the "excess" neurons die. Conversely, if motor nerves are stimulated with action potentials during the time that they innervate the muscle, *more* neurons die than is normally the case. One hypothesis to explain these results involves regulation of putative neurotrophic factors by muscle activity. If the synthesis or secretion of the trophic factors were down-regulated when the muscle is active, such as with normal, functional innervation, the factors might become limiting, resulting in the death of some neurons. If functional transmission were prevented, factor levels might remain high enough to support greater numbers of neurons. On the other hand, if transmission is enhanced, factor levels might fall prematurely or fall below normal levels, enhancing neuronal death. Evidence that trophic-factor production is linked to innervation or activity is weak at this time, however. A related notion is that *access* to the factor is limiting. For instance, when activity is blocked, the motor axons are seen to branch more profusely, and this could provide many more uptake sites for the trophic factor.

In addition to the matching of population sizes, neuronal death can also serve as a mechanism for correcting errors. That is, neurons whose axons do not locate the appropriate target area could be eliminated by death, possibly by lack of access to the appropriate trophic factor. Although not thought to be a general rule, neuronal death does appear to serve this purpose in several cases (see Chapter 12).

NGF: The paradigmatic neurotrophic factor

Nerve growth factor (**NGF**) was the first neurotrophic factor to be discovered (see Box A), and it remains the factor whose biological function is best understood. A series of decisive experiments demonstrates that this protein has the trophic characteristics predicted from the biological results described in the previous section. First, injection of NGF into embryos rescues sympathetic and certain sensory neurons that would normally die, and it elicits an extensive outgrowth of neurites from these cells. Thus, this protein can mimic a target-derived neurotrophic factor. Second, injection of anti-NGF antibodies into a developing animal (or induction of the antibodies in pregnant mothers) results in an almost complete loss of sympathetic neurons (Figure 8). Thus, NGF is likely to be the factor that acts during normal development. Third, pure cultures of sympathetic and

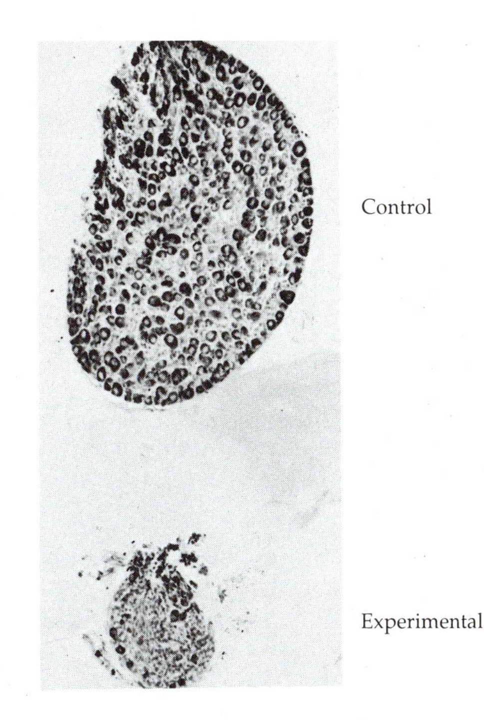

FIGURE 8. Administration of anti-NGF antibodies to young mice results in the death of sympathetic neurons. Mice were injected daily for 9 days after birth with antisera generated against NGF. Superior cervical sympathetic ganglia were removed and sectioned to reveal the number of surviving neurons. (From R. Levi-Montalcini and B. Booker, 1960. Proc. Natl. Acad. Sci. USA 46: 384–391.)

Box A Rita Levi-Montalcini and the Discovery of NGF

From the beginning of her career, Rita Levi-Montalcini, the discoverer of NGF, has demonstrated an unreserved passion for science, and this commitment has been essential for her success. An Italian M.D. specializing in neurology and psychiatry, her medical career was aborted at the start because of Mussolini's exclusion of "non-Aryans" in the professions. Although new to experimental neuroembryology, she set up a laboratory in her bedroom and began to extend Viktor Hamburger's work on the effects of target tissues on the development of the CNS of chick embryos. Throughout the course of anti-Semitic campaigns and mass killings, as well as the Allied bombing of her hometown, Turin, she persevered and produced enough work for several papers. In one of these, Levi-Montalcini came to a different conclusion than had Hamburger. After the war, he invited her to work with him on the problem in St. Louis (Figure A), and thus began a very fruitful collaboration.

FIGURE A. Rita Levi-Montalcini. (Photograph courtesy of V. Hamburger.)

The key lead was provided by experiments of Ernst Bueker, a former student of Hamburger's who had attempted to increase the amount of target tissue for neurons to innervate by injecting a rapidly growing mouse tumor into chick embryos. The injected tissue had not only been innervated by motor neurons, but had elicited a massive outgrowth of fibers from sympathetic ganglia, even from ganglia at some distance from the tumor. Neurites from these distant ganglia filled the viscera and invaded blood vessels, a most salient result that Levi-Montalcini says she and Hamburger did not fully appreciate for some time. A decisive experiment then revealed that tumor cells placed on the chorioallantoic membrane of the egg released something into the circulation so powerful that it could stim-

sensory neurons are completely dependent on added NGF. Thus, NGF acts directly on the neurons and not via another cell type. Fourth, cutting the axons ("axotomy") of young neurons results in death that can be prevented by application of NGF. Fifth, NGF mRNA and protein are present in the appropriate target tissues, i.e., those innervated by sympathetic and sensory neurons. These results also support the supposition that NGF is limiting during normal development; there is a linear relation between the amount of NGF and the amount of sympathetic innervation in various target tissues.

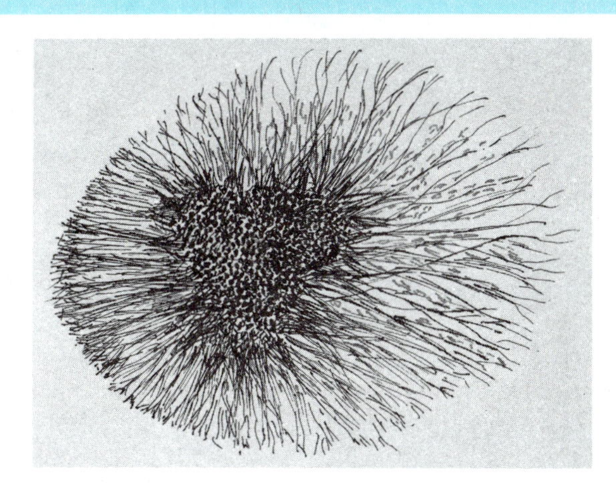

FIGURE B. Original drawing made by Levi-Montalcini in 1951 of the "halo effect." Nerve fibers from an explant of a sympathetic ganglion were stimulated to grow out upon coculture with sarcoma cells. The active principal released by the sarcoma cells was subsequently shown to be NGF. (Courtesy of R. Levi-Montalcini and V. Hamburger.)

ulate sympathetic and sensory neurites to grow into the blood vessels.

Not content with describing this unexpected phenomenon, Levi-Montalcini realized the need for a convenient culture assay to use in the purification of the active agent. She flew to Brazil to learn the necessary culture methods, and after the first successful experiment excitedly sent Hamburger a drawing of the result, the famous "halo" of fibers that grows out of a ganglion in the presence of NGF (Figure B). She then recruited the young biochemist Stanley Cohen to purify the factor. This task was aided greatly by the serendipitous discovery of an extremely rich source of the activity. In the initial stages of characterizing the factor, Arthur Kornberg suggested incubating it with snake venom, which contains an enzyme that can degrade nucleic acids. Degradation of the nerve outgrowth activity would suggest that the factor contained nucleic acid. Far from degrading the activity, however, snake-venom–treated factor increased neurite outgrowth many fold. Thus it was found that snake venom (and later, mouse salivary glands) contains high levels of NGF, an observation whose significance is still not fully understood. After the purification was achieved, Cohen played a key role in producing anti-NGF antiserum, a crucial reagent for showing the biological role of NGF in vivo.

Subsequent work has identified the neuronal receptor for NGF and revealed that it is regulated along with NGF. During development of the sensory innervation of the mouse whisker pads neither NGF synthesis by the target nor expression of NGF receptors in the nerves is significant until the sensory fibers enter the target area. The findings argue that, while NGF can guide growing neurites towards a source of NGF, it probably does not do so in this case. The guidance response could, however, play a role in keeping neurites from growing *out* of the target area. While there is a correlation between the onset of NGF production and the arrival of sensory fibers, NGF mRNA expression is maintained even in the absence of sensory innervation. As discussed earlier, blockade of synaptic transmission between motor neurons and skeletal muscle prevents naturally occurring neuronal death, prompting the notion that neurotrophic factor production might be regulated by innervation. Blockade of synaptic transmission in an NGF-producing target would test this hypothesis further.

The amino acid sequence of NGF was determined in 1971, and the cDNA cloned in 1983. The active protein is a small homodimer, each chain consisting of 118 amino acids, with three disulfide bridges. The sequence is highly conserved in snake, frog, chicken, and mammals, including man. Differential splicing yields at least four different mRNA transcripts. Several peptides, corresponding to amino acids 26–40, can inhibit the neurite-promoting activity of the protein, presumably by binding to the NGF receptor. The inhibition is specific for this sequence and the peptides do not block neuronal responses to other trophic agents.

The mechanism by which NGF exerts its effects is still not well understood. It is clear, however, that NGF binds to a receptor on the surface membrane of axons and is internalized and retrogradely transported to the neuronal cell body. At some stage in this sequence, a series of changes in several second messenger systems is induced, and these lead to alterations in gene expression, and enhanced neurite outgrowth and neuronal survival.

The initiation of the process involves NGF receptors, of which there are low- and high-affinity forms. It is the high-affinity form that is critical for NGF's neuronal survival and neurite outgrowth activities. The cDNA for this protein was recently cloned and the sequence identifies the receptor as the proto-oncogene, *trk*. This 140-kD transmembrane tyrosine kinase presumably initiates the cascade of events that follows NGF binding by phosphorylating one or more protein substrates. While NGF clearly affects the transcription of numerous genes, it appears that NGF itself does not enter the nucleus to alter gene expression. After internalization and retrograde transport of labeled NGF, most of the label is localized over lysosomes where the protein is degraded. Anti-NGF antibodies do not block NGF action when injected into the soma, nor is NGF active when injected into neurons. Therefore, it seems likely that retrograde transport primarily serves to degrade NGF (lysosomes are not present within nerve terminals or axons). The second messenger signals induced by NGF binding to its receptor may also be retrogradely transported in order to affect transcription.

NGF was thought for many years to act only in the PNS. Recently, however, clear evidence of its selective action in the CNS has been obtained. One responsive population is the cholinergic neurons found predominantly in the basal forebrain nuclei that project to the cortex and hippocampus. The action of NGF on these neurons appears to be much the same as in the periphery. The targets of the cholinergic projection, the neurons of the hippocampus and cortex, contain both NGF and its mRNA, and the cholinergic neurons have NGF receptors and retrogradely transport NGF back to their somas. Studies of cultured cholinergic neurons indicate that they respond to NGF; also, anti-NGF injections into the brain lower choline acetyltransferase levels and significantly reduce the number of neurons expressing this enzyme. Axotomy experiments also support a target-derived trophic factor role in the CNS. As in the periphery, young central cholinergic neurons die if their axons are cut, but they can be rescued by injections of NGF. Although the action of NGF in the CNS appears to be selective, it is likely that not all subpopulations of neurons that require this protein for survival during development have been identified. Resolving this problem is not entirely straightforward, given that the NGF requirement for sympathetic and sensory neurons appears only at certain critical periods in development.

The effect of NGF on damaged cholinergic neurons could have clinical relevance for Alzheimer's disease and possibly other memory disorders. An important deficit in this disease is a loss of cholinergic neurons in the basal forebrain (see Chapter 15). One intriguing experiment has demonstrated that injections of NGF can have positive effects on aged rats with learning and memory deficits. The performance of the injected rats on learning tasks is restored to that of their normal, aged siblings, and there is considerable sprouting of neurites by the central cholinergic neurons. Thus far, however, postmortem analysis of the brains of Alzheimer's patients has not revealed striking deficits in NGF or NGF receptors.

NGF: The family

The paradigm of a target-related neurotrophic factor so beautifully exemplified by NGF may have broad application throughout the nervous system. Neuronal death is widespread during development, and many types of neurons die when they are separated from their targets. In fact, several new neurotrophic factors have been identified. Two of these resemble NGF in their sequence and in their interaction with receptors.

New neurotrophic factor activities are assayed by adding tissue extracts to neural explants and looking for "halos" of fiber outgrowth, or by adding extracts to dissociated, embryonic neurons and looking for increased survival. In spite of the simplicity of these assays, identification and purification of neurotrophic factors other than NGF has been difficult, in part because they are present at very low concentrations. Because purified populations of neuronal targets are readily available, factors directed toward cells from the PNS have been the first to be isolated. Like NGF, however, the other neurotrophic factors are active not only on cells in the PNS, but also on selected target neurons in the CNS. These factors are thus of great potential interest in understanding disease and regeneration in the brain.

Two of the factors, **brain-derived neurotrophic factor** (**BDNF**) and **neurotrophin-3** (**NT-3**), are closely related to NGF. BDNF was isolated from brain on the basis of its ability to enhance the survival of dorsal root ganglion neurons. cDNA cloning revealed it to be similar in amino acid sequence to NGF, suggesting that the two might be members of a larger family. Using the polymerase chain reaction with oligonucleotides from conserved regions in NGF and BDNF as primers, it has now been possible to isolate other cDNAs for related proteins. The best characterized of these is NT-3. The three proteins NGF, BDNF, and NT-3 are closely related, each about 120 amino acids in length, and any two of the three sharing about 60 percent amino acid identity. The greatest homology is around the three disulfide bonds, whose positions are conserved in all three.

The three factors have distinct but overlapping cellular specificities. Both BDNF and NT-3, like NGF, support the survival of cultured sensory neurons, and BDNF also prevents the naturally occurring cell death of dorsal root ganglion neurons in developing ganglia. Neither BDNF nor NT-3, however, increases the survival of sympathetic neurons in culture. BDNF is also active on retinal ganglion neurons and on central dopaminergic and cholinergic neurons.

There also appears to be overlap in binding of the three neurotrophic factors to their receptors. NGF and BDNF each bind to specific, high-affinity receptors that mediate their biological activities. BDNF, and perhaps NT-3 as well, also binds to the low-affinity form of the NGF receptor. These observations suggest that the receptors for the three factors are related, and that understanding the relation between the high- and low-affinity states of the NGF receptor may yield information about how specificity is achieved for each of the neurotrophic factors.

Other candidate neurotrophic proteins may provide new twists in the story

Fibroblast growth factor (FGF), which is best known for its ability to promote proliferation of fibroblasts, also promotes the survival of many types of neurons in culture. FGF is a small protein (16 kD) originally described as having two forms, acidic and basic (aFGF and bFGF, respectively). It is now known to be a member of a larger family that includes at least seven proteins that share greater than 40 percent homology and that are thought to be descended from a common ancestral gene. Under a new system of nomenclature, several members of the family belong to a group of heparin-binding growth factors that are divided into class 1 (anionic) and class 2 (cationic) factors, called HBGF-1 and HBGF-2, respectively. Members of the entire family of proteins not only share a common structure but have broadly similar activities in a wide range of biological processes that include wound healing, carcinogenesis, vascularization, mesoderm induction, liver regeneration, and ovarian function. The FGF family acts through a corresponding family of related receptors that have specificity either for HBGF-1 or HBGF-2.

In vitro, bFGF, the best studied of the factors, promotes the survival of some neurons (hippocampus, and ciliary and dorsal root ganglion), and stimulates neurite outgrowth from others (PC12 and adrenal chromaffin cells; embryonic sympathoadrenal progenitors; and the progenitor cell line, MAH). Its most novel role, however, is in the embryonic sympathoadrenal progenitor cell line, in which it induces the expression of the NGF receptor and renders the cells dependent on NGF for survival. This process, which is described in more detail in Chapter 11, is an early step of neuronal differentiation in the sympathoadrenal lineage. Interestingly, when administered in the eye, bFGF binds to retinal ganglion cells and is transported in the anterograde direction to the brain. This is the reverse of the NGF paradigm and suggests that FGF could be used as a trophic signal by presynaptic inputs. Application of bFGF can also reverse neuronal death in a mutant mouse model of retinal degeneration.

If FGF is a conventional neurotrophic factor, it is presumably secreted by the cells that produce it. Indeed, cultured cardiac muscle cells, which synthesize bFGF, do not retain it but secrete it into the medium. In contrast to most secreted proteins, however, both aFGF and bFGF lack signal sequences and consensus glycosylation sites, leading some to postulate that these proteins do not play a role that is similar to NGF, but are accumulated inside the cell and released by cell damage. There is also evidence that FGF is transported to the cell nucleus and that this transport is required for some of its actions.

Other observations suggest yet another, also unorthodox, role for FGF. FGF binds heparin, an extracellular matrix component, and can be localized to the matrix in vivo by immunohistochemistry. As heparin-bound FGF can stimulate neurite outgrowth and survival, extracellular FGF could serve as a depot of trophic activity. According to this hypothesis, the limiting element in naturally occurring cell death would be *access* to, rather than *production* of, the factor. One clear difference between FGF and members of the NGF family is that its concentration in target tissues is more than 100-fold higher than those of NGF or BDNF, suggesting that the action of the latter factors may not be the appropriate model for members of the FGF family.

Studies on another intriguing candidate, **ciliary neurotrophic factor (CNTF)**, have raised many of the same questions considered for the FGFs. Like FGF, CNTF exerts potent survival effects on many peripheral neurons. Thus, there is an even greater overlap of neurotrophic factors for many neurons, including, for dorsal root ganglion neurons for instance, NGF, BDNF, NT-3, FGF, and CNTF, and for central cholinergic neurons, NGF, BDNF, FGF, insulin-like growth factor, interleukin 3, and possibly others. The specter of redundancy, previously raised in the context of neurospecificity and guidance molecules (Box B in Chapter 12), as well as with phenotype-specifying proteins, now looms large in this field. On the other hand, factors may act at different times during the development of a neuron, as when FGF induces NGF receptors. While application of CNTF can rescue embryonic motor neurons in vivo and in culture, it is not produced by skeletal muscle. Schwann cells, on the other hand, produce high levels of CNTF and could use this trophic factor in the rescue of damaged motor neurons postnatally. Multiple factors could also act simultaneously but on separate parts of a neuron, as in the case of BDNF acting on the central projections of dorsal root ganglion neurons and NGF acting on the peripheral projections. A major focus for the future will be to clarify where the various factors are produced in the embryo, to localize their receptors, and to determine the effects of lowering the levels of the factors in the developing and adult animal.

Some surprises in the mechanisms of neuronal death

How do neurons die? A number of theories have been suggested: first degree murder (the neuron is actively killed by another cell); second degree murder (many neurons are killed by a diffusible, broadly acting agent); manslaughter (the neuron dies when essential resources are consumed by other neurons); or suicide (the neuron turns on its own lytic enzymes). Very few examples of first type of killing are known in normal development. In the nematode, laser ablation of particular cells that normally engulf their neighbors allows the neighbors to survive. There are, of course, a number of well-described instances of killing in various pathological circumstances. A clinically relevant case involves the so-called **excitotoxins**, excitatory amino acid transmitters that, when present in abnormally high concentrations, can kill sensitive neurons (see Chapter 15). In some cases at least, this killing is caused by an abnormal ion flux elicited by the transmitter, leading to osmotic swelling and death. In the second category of cell death, an agent that acts at a distance to cause the death of particular neurons is found in insect metamorphosis. The hormone ecdysterone triggers many changes in the neuromuscular system of larvae, including the death of certain motor neurons.

Most frequently, ontogenetic neuronal death in vertebrates appears to be a combination of manslaughter and suicide; external events trigger the expression of degradative events within the neuron. There is, however, morphological evidence for a diversity of mechanisms. For instance, death in some classes of neurons involves condensation of the nucleus and cytoplasm; other dying neurons display prominent autophagic vacuoles; while death in still other neurons is marked by swelling of cellular organelles. Interestingly, in one case, neuronal death caused by target removal results in an autophagic morphology, while death caused by deafferentation is nonautophagic in appearance. When the two operations are combined, autophagy is not apparent.

Curiously, inhibitors of protein and RNA synthesis do not enhance the rate or extent of the pathology observed in dying neurons, as might be expected, but can actually *prevent* cell death. Both in culture, and in vivo during normal ontogenetic death or death induced by target removal, neurons are rescued by these inhibitors. These results support the suicide model (as opposed to starvation), and suggest that targets may serve to suppress an endogenous program of neuronal cell death genes. Rescue by these inhibitors has been observed with many types of dying non-neuronal cells as well, so it is possible that useful generalizations may arise from this work. A number of genes involved in cell death and degeneration have been cloned in the nematode, and it will be interesting to see if the expression of the vertebrate homologues of these genes are influenced by cellular interactions.

A number of fundamental issues remain to be explored. What is the nature of the afferent neurotrophic signals? What is the role of glial cells in these events? Why do half of the neurons die during development, rather than all of the neurons becoming half as large? That is, when the level of trophic factor is reduced by half with cultured neurons, the growth of all neurons is reduced; half of them do not die allowing the other half to remain unaffected. This question may be closely related to the problem of what is meant by access to trophic factors. As mentioned above, there are several well-described cases in which neurons that die can be shown to have their axons in the correct target area and to have received presynaptic input. In what critical way are they different from their siblings?

Steroids alter neuronal circuits

Despite their broad distribution and limited diversity (relative to proteins), hormones can control the development of specific subpopulations of neurons. Two examples mentioned elsewhere are corticosteroids, which govern phenotypic choices in the sympathoadrenal lineage (Chapter 11), and ecdysterone, which kills particular neurons during insect metamorphosis (see above). Sex steroids are another class of hormones that exert a critical influence on neuronal development. These hormones profoundly affect areas of the CNS that are different between the sexes (sexually dimorphic areas), as well as the motor and autonomic innervation of vertebrate sex organs. They can act directly on neurons, or they can exert their actions indirectly, as on the development of sexually dimorphic muscles.

The effects of steroids on the avian song system is a particularly fascinating example, illustrating the interplay between cellular mechanisms and behavior. In songbirds such as canaries and zebra finches, males produce species-specific songs. Females vocalize, but do not produce the phrases characteristic of song. Birdsong is a learned behavior; the young male zebra finch hears and memorizes his father's song and then later reproduces it by trial and error. The young bird must hear his own vocalizations as he gradually improves performance, matching his vocalizations to his memory, or template, of his father's song. The neurophysiological basis of this behavior is in an early stage of analysis, but enticing clues are emerging. Neurons that respond well only to the bird's own song or to discrete phrases in that particular song can be found in the adult male. Thus, the learning process results in auditory neurons that have highly selective response properties. These properties must arise because the early auditory input has caused previously existing synaptic connections to be modified or new ones to be made.

Male song behavior is hormone-dependent in two ways. As adults, the birds require androgen to sing; during development, estrogen is required in order for the androgen to have its effect in maturity. To understand this relay or transition between the effects of androgen and estrogen, it is useful to recall that two actions of sex hormones, organizational and activational, are commonly distinguished. The former occurs during development, when the hormones permanently affect the organization of the brain. Later, in maturity, hormones can have acute effects, activating systems that were laid down in development. In the case of birdsong, estrogen acts as the organizing hormone, while androgen activates the male behavior in adults. While it is curious that the female hormone is needed for the development of a male behavior, organizational effects of estrogen have been observed on a number of male-specific behaviors in mammals as well.

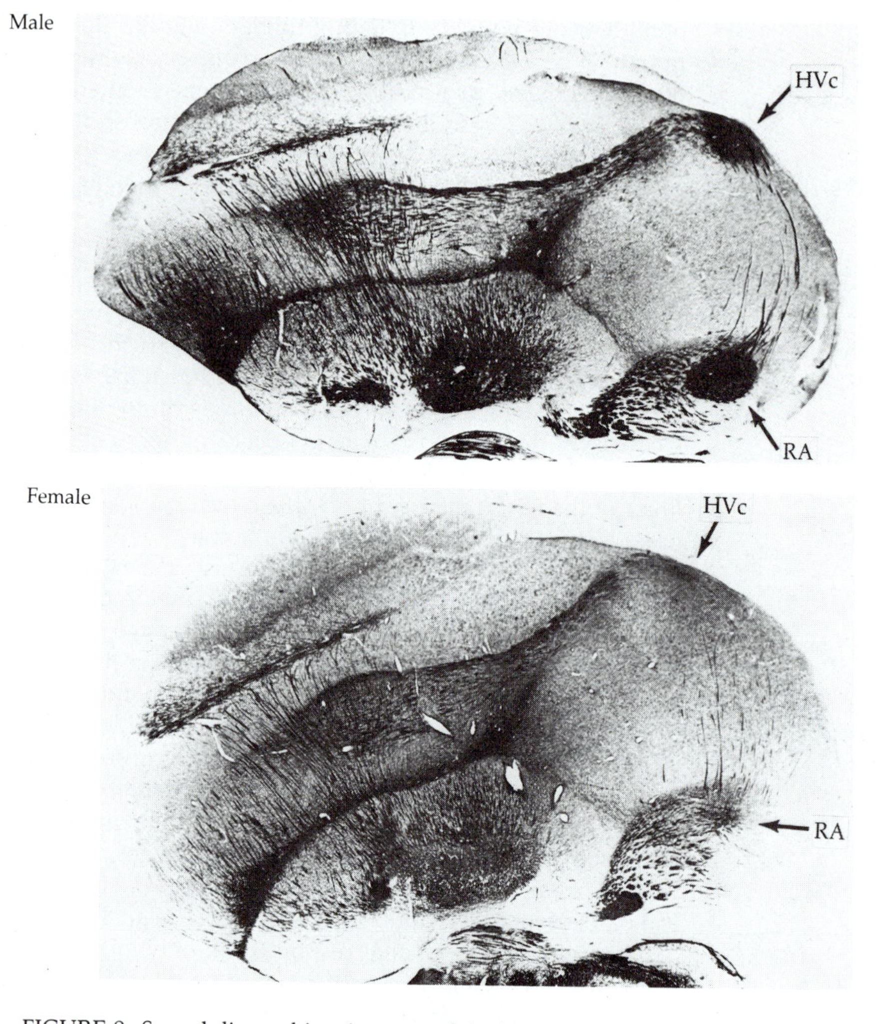

FIGURE 9. Sexual dimorphism in areas of the brain that control birdsong. Brains of adult male (top) and female (bottom) zebra finches were sectioned and stained to reveal nerve fiber tracts. The arrows highlight two areas that are part of the song control system. (M. Konishi, unpublished.)

Estrogen prevents cell death in birdsong nuclei

What is the role of sex steroids in setting up the song system? The areas in the brain that receive auditory input and produce motor output resulting in birdsong are much larger in the male (Figure 9), both in terms of the numbers of neurons in each area, and in the size of the neurons and their dendritic arbors (Figure 10). This **sexual dimorphism** could arise from hormone effects on the initial genesis of the neurons, on their survival, and/or on the specification of the types of neurons that develop.

The rate of production of neurons born in each vocal control area can be measured by using ^{3}H-thymidine to label dividing neuronal precursors,

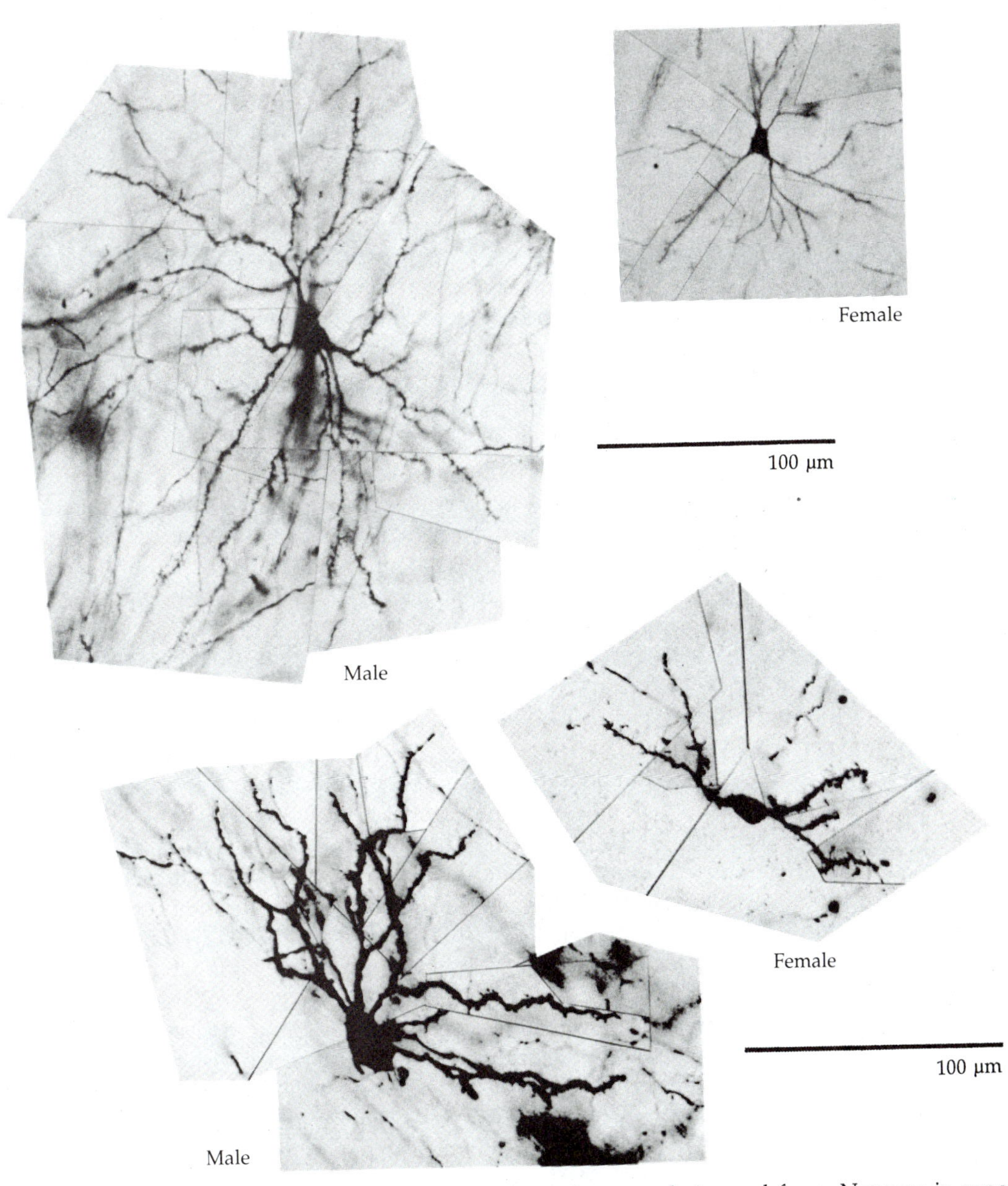

FIGURE 10. Sex steroids can control neuronal size and form. Neurons in song control areas of zebra finch brains were visualized with the Golgi method, and cells of similar type from males and females compared. (From M. Gurney, 1981. J. Neurosci. 1: 658–673.)

and cell death can be quantitated by counts of healthy and dying neurons at various ages. The results of such experiments reveal a significant difference in neuronal death between males and females; estrogen prevents or slows the death that occurs during the first weeks after hatching. In addition, there are differences between the sexes in the number of newly generated neurons in certain song areas. These differences could be due to differential survival or to sexually dimorphic neurogenesis. Dimorphism in the phenotypic specification of neurons has been postulated but not definitively proved. It is probably not a coincidence that the birth of new neurons, as well as their differentiation and death, occurs during the first two months after hatching in the zebra finch, just at the time the new song is being learned.

Many of the song system neurons accumulate labeled androgens, suggesting a direct action of the sex steroids on these cells. Some song system neurons do not accumulate androgens, however, and must be influenced indirectly. Most strikingly, only one song nucleus in the young bird appears to contain estrogen receptors, even though this hormone has potent effects on all of the song nuclei. The issue of indirect hormone action is critical for understanding another steroid effect; in addition to the sexual dimorphism in number and sizes of neurons, there are gender differences in the connections of the song system. For instance, there is a major projection from area HVc to area RA (see Figure 9) in male (or steroid-treated female) zebra finches, but little or no projection in normal females. It is not clear how the steroid directs this connectivity because the projecting neurons are not selectively spared from death by hormone treatment. Estrogen could act on certain neurons by affecting their pre- and/or postsynaptic partners through a cascade as discussed earlier for other parts of the nervous system.

Some species of songbirds, such as the canary, learn new songs each year. Canaries generate large numbers of new neurons throughout adulthood, some of which become integrated into the song system. This phenomenon does not appear to be related to the sexual dimorphism of the song system, however. Although the genesis and integration of new neurons into the circuitry of the mature brain is not a feature of the mammalian CNS, grafts of immature brain tissue can become well integrated into the adult CNS.

Clearly, the song system offers rich material for molecular studies, encompassing as it does learning, memory, neurogenesis, cell migration and synapse formation, and neuronal death, as well as the steroid control of many of these phenomena. The linking of molecules and behavior will be fascinating to watch as the study of this system progresses.

Summary

The function of the nervous system is based on exquisitely controlled communication between cells. It may not be surprising, then, to find that the mature circuitry is built during development by an enormously complex series of intercellular interactions. What may be more difficult to comprehend is the scope of the events dependent on these interactions and the staggering number of different molecules that must be used as signals to direct growth and differentiation.

Although detailed biochemical information is available on only one synapse, the vertebrate neuromuscular junction, the data reveal a highly specialized structure, containing a number of molecules that are highly enriched at the junction. These include surface membrane proteins such as the acetylcholine receptor (AChR) and the associated 43-kD protein, as well as basal lamina components such as the S and A chains of laminin and the α3 and α4 chains of type IV collagen. These molecules, and the synaptic structural specializations such as the deep folds in the muscle membrane and the clusters of synaptic vesicles at release sites, are produced at distinct times over an extended developmental period. The timing of these events suggests that distinct mechanisms underly each stage. While myotubes produce AChR in the absence of nerves, the AChR clusters form preferentially under motor axons. The protein agrin is a good candidate for the clustering factor released by motor neurons. Nerve-evoked activity in the myotube down-regulates AChR expression in extra-junctional areas. In addition to the changes in receptor distribution, there is a switch in expression of two of the receptor subunits that begins after the nerve contacts the muscle. The adult-specific ε-subunit is substituted for the embryo-specific γ-subunit. The initial appearance of the ε-subunit is induced only in the myotube nuclei at the site of nerve-muscle contact, but its continued expression does not require nerve activity. The peptides CGRP and ARIA are produced by neurons and can up-regulate AChR synthesis.

Differentiation signals also flow in the retrograde direction. The synaptic basal lamina contains the signal(s) that causes regenerating motor axons to cease growth and form structural specializations characteristic of the presynaptic nerve terminal. A leading candidate for the differentiation signal in the basal lamina is the S subunit of laminin, which is highly enriched at the synapse and has motor-neuron–specific attachment activity. A number of other molecules produced by muscle may serve as growth promoting and/or guidance cues that enhance nerve regeneration. Soluble growth factors and membrane/matrix proteins such as N-CAM, N-CAD, and INO also have biological activities relevant for such a role.

Target cells also produce signals that influence the morphological and chemical phenotype of neurons innervating them. The shape of the dendritic arbor as well as the synaptic connections made on the dendrites can be regulated by particular target tissues. Glia also exert striking control on the differentiation of neurites into axons versus dendrites. Target cells can control the choice of neurotransmitter and neuropeptide produced by neurons through instructive differentiation factors such as the proteins CDF/LIF and CNTF. Such factors have partially overlapping effects on neuronal gene expression, as do the factors that control differentiation decisions in the hematopoietic system.

Target effects extend to the very survival of the neurons innervating them. Neuronal death is a quantitatively significant and widespread feature of normal vertebrate development. Depriving neurons of their innervation during critical periods can cause death, as can depriving them of access to their target tissues. Target-related death involves competition between neurons for a neurotrophic factor provided by the target. NGF is such a factor; this protein is produced by targets of peripheral and central neurons that require it for survival. NGF appears to be present in limiting quantities and it stimulates neuronal growth and differentiation as well as preventing death. The high-affinity NGF receptor has been identified as the tyrosine kinase *trk*; upon binding, NGF is internalized and retrogradely transported to the soma. NGF activates a number of second messenger systems, as well as immediate early genes. Since inhibition of mRNA or protein synthesis can prevent neuronal death during NGF deprivation, it is thought that NGF also acts to suppress the synthesis of a protein that is part of a death program.

Two other neurotrophic factors with strong sequence homology to NGF have been discovered, and the three related factors act on distinct but partially overlapping sets of neurons. There are also proteins that have neurotrophic activity that are not related in sequence to the NGF family. In addition, steroids are trophic factors for certain neurons. Sex steroids prevent neuronal death in sexually dimorphic brain areas involved in bird song. Estrogen also stimulates the growth of sensitive neurons and enhances the formation of the circuits involved in the song behavior.

Target cells therefore can control the number and nature of the neurons that innervate them. Innervating axons, in turn, influence the growth and differentiation of the target cells on which they form synapses. These influences extend transynaptically, in both anterograde and retrograde directions, beyond the synapse under consideration. The competition between neurons for survival and synapse formation, coupled with the transynaptic reverberations of these effects, would appear to provide fertile ground for evolutionary change in neural systems.

References

General References

Purves, D. and Lichtman, J. W. 1985. *Principles of Neural Development.* Sinauer Associates, Sunderland, MA.

Synapse formation

Brehm, P. and Henderson, L. 1988. Regulation of receptor channel function during development of skeletal muscle. Dev. Biol. 129: 1–11.

Cohen, M. W., Anderson, M. J., Zorychta, E. and Weldon, P. R. 1979. Accumulation of acetylcholine receptors at nerve-muscle contacts in culture. Prog. Brain Res. 49: 335–349.

*Dahm, L. M. and Landmesser, L. T. 1991. The regulation of synaptogenesis during normal development and following activity blockade. J. Neurosci. 11: 238–255.

Dennis, M. J., Ziskind-Conhaim, L. and Harris, A. J. 1981. Development of neuromuscular junctions in rat embryos. Dev. Biol. 81: 266–279.

Hunter, D. D., Porter, B. E., Bulock, J. W., Adams, S. P., Merlie, J. P. and Sanes, J. R. 1989. Primary sequence of a motor neuron-selective adhesive site in the synaptic basal lamina protein s-laminin. Cell 59: 905–913.

Jansen, J. K. S. and Fladby, T. 1990. The perinatal reorganization of the innervation of skeletal muscle in mammals. Prog. Neurobiol. 34: 39–90.

*Nitkin, R. M., Smith, M. A., Magill, C., Fallon, J. R., Yao, Y. -M. M., Wallace, B. G. and McMahan, U. J. 1987. Identification of agrin, a synaptic organizing protein from *Torpedo* electric organ. J. Cell Biol. 105: 2471–2478.

Sanes, J. R., Marshall, L. M. and McMahan, U. J. 1978. Reinnervation of muscle fiber basal lamina after removal of myofibers. J. Cell Biol. 78: 176–198.

Control of neuronal phenotype

Loer, C. M. and Kristan, W. B. 1989. Central synaptic inputs to identified leech neurons determined by peripheral targets. Science 244: 64–66.

*Nawa, H. and Patterson, P. H. 1990. Separation and partial characterization of neuropeptide-inducing factors in heart cell conditioned medium. Neuron 4: 269–277.

Schotzinger, R. J. and Landis, S. C. 1990. Acquisition of cholinergic and peptidergic properties by sympathetic innervation of rat sweat glands requires interaction with normal target. Neuron 5: 91–100.

Tropea, M., Johnson, M. I. and Higgins, D. 1989. Glial cells promote dendritic development in rat sympathetic neurons in vitro. Glia 1: 380–392.

Yamamori, T., Fukada, K., Aebersold, R., Korsching, S., Fann, M.-J. and Patterson, P. H. 1989. The cholinergic neuronal differentiation factor from heart cells is identical to leukemia inhibitory factor. Science 246: 1412–1416.

Neuronal trophic factors

*Barde, Y.-A. 1989. Trophic factors and neuronal survival. Neuron 2: 1525–1534.

*Davies, A. M., Bandtlow, C., Heumann, R., Korsching, S., Rohrer, H. and Thoenen, H. 1987. Timing and site of nerve growth factor synthesis in developing skin in relation to innervation and expression of the receptor. Nature 326: 353–358.

Ernfors, P., Ibanez, C. F., Ebendal, T., Olson, L. and Persson, H. 1990. Molecular cloning and neurotrophic activities of a protein with structural similarities to nerve growth factor: Developmental and topographical expression in the brain. Proc. Natl. Acad. Sci. USA 87: 5454–5458.

Fallon, J. H. and Loughlin, S. E. (Eds.) *Neurotrophic Factors.* Academic Press, San Diego, CA, in press.

Konishi, M. 1989. Birdsong for neurobiologists. Neuron 3: 541–549.

Levi-Montalcini, R. 1987. The nerve growth factor 35 years later. Science 237: 1154–1162.

Oppenheim, R. W., Prevette, D., Tytell, M. and Homma, S. 1990. Naturally occurring and induced neuronal death in the chick embryo *in vivo* requires protein and RNA synthesis: Evidence for the role of cell death genes. Dev. Biol. 138: 104–113.

Rodriguez-Tebar, A., Dechant, G. and Barde, Y.-A. 1990. Binding of brain-derived neurotrophic factor to the nerve growth factor receptor. Neuron 4: 487–492.

Sengelaub, D. R., Nordeen, E. J., Nordeen, K. W. and Arnold, A. P. 1989. Hormonal control of neuron number in sexually dimorphic spinal nuclei of the rat: III. Differential effects of the androgen dihydrotestosterone. J. Comp. Neurol. 280: 637–644.

Williams, R. W. and Herrup, K. 1988. The control of neuron number. Annu. Rev. Neurosci. 11: 423–453.

Part Four

Complex Interactions of Neurons and Neuronal Disorders

14

Cellular and Molecular Mechanisms of Neuronal Plasticity

Mary B. Kennedy and Eve Marder

One goal of neurobiological research is to understand how the nervous system adapts the behavior of organisms to their external environment. Most of the chapters in this textbook have described molecular and cellular mechanisms that are used by neurons to process and transmit information. Ultimately, we want to understand how these mechanisms function within the context of neural networks that control behavior. In the case of the mammalian brain, we still know too little about individual neurons and the synapses underlying functional neural circuits to understand the properties of networks in terms of the characteristics of their constituent neurons. Until quite recently, many of the experimental preparations used for the study of neural systems have been difficult to examine with cellular and molecular techniques. However, in a few favorable systems we are beginning to understand the relationship between molecular and cellular properties of individual neurons and the functioning of the networks to which they belong. In this chapter, we describe three examples of such systems.

Some general principles have already emerged from the study of these and other systems. The intrinsic membrane properties of particular neurons, defined by the complement of ion channels in their membranes, control the patterns of electrical activity that they display. Neural systems adapt to the changing demands of their environment by modulating both the intrinsic membrane properties of neurons and the strength of the synaptic connections among them. Such modulation can occur at all levels within a nervous system from sensory input to motor output.

The Motor Rhythms of the Crustacean Stomatogastric Ganglion

The stomatogastric ganglion controls the timing of movements of the stomachs of lobsters and crabs

Integrative physiologists have long been attracted to the study of motor systems that generate rhythmic behaviors. In humans these systems control such important actions as walking, breathing, and speech. The networks underlying such behaviors are relatively easy to study because of the repetitive and often stereotyped nature of their output. In both vertebrate and invertebrate nervous systems rhythmic movements are generated by neuronal networks in the central nervous system called **central pattern generators**. To understand how these networks function it is important to know which neurons are part of the network, how they are connected to each other, and how the specialized molecular properties of each neuron contribute to the properties of the network as a whole.

Considerable progress has been made on most of these questions in the stomatogastric nervous system, a small motor system that controls the movements of parts of the crustacean stomach. The stomatogastric nervous system contains a very small number of elements, yet it is capable of producing rich and diverse behaviors. Much of this richness is a consequence of the presence of multiple modulatory inputs. Each of these can act in a unique way on a different group of target neurons to modulate channels that control their electrical properties and to change the strength of their synaptic relationships.

The foregut (esophagus and stomach) of lobsters and crabs is quite different from the equivalent parts of the mammalian digestive system. It is a complex mechanical structure that may resemble the mammalian mouth more than the stomach (Figure 1). Food enters the esophagus, and is moved through it by alternating contractions of striated muscles that constrict and dilate the esophageal walls. From the esophagus, food travels into the cardiac sac, a storage area that also has muscles that alternately constrict and dilate it. From there food moves into the gastric mill, a chamber containing three teeth that chew and grind the food. Movements of the teeth are controlled by a specialized set of gastric mill muscles. After it is broken down into small pieces, food passes into the pylorus, a sieving and filtering device. In the pylorus, a series of muscles alternately dilates and constricts the region to move food forward into the hepatopancreas and intestine.

The stomatogastric ganglion (STG) controls and coordinates movements of the muscular chambers and bony teeth of the gastric mill and pylorus. It contains 30 large neurons, most of which are motor neurons directly innervating the gastric mill and pyloric muscles. Unlike most other central pattern generators, which consist of interneurons that are difficult to locate, the central pattern generating circuits in the STG are formed primarily from the motor neurons themselves. As a result, recordings from the motor neurons simultaneously provide information about intrinsic membrane properties, patterns of motor output, and the synaptic relationships that give rise to these patterns. The rhythmic output of the ganglion is responsible for the alternating pattern of contraction by dilator and constrictor muscles that moves food through the gastric mill and the pylorus. The motor patterns generated by the ganglion are modulated by sensory input and by modulatory hormones secreted in the vicinity of the ganglionic neurons. This system provides a rich illustration of the many roles that cellular and molecular processes can play in the operation of a

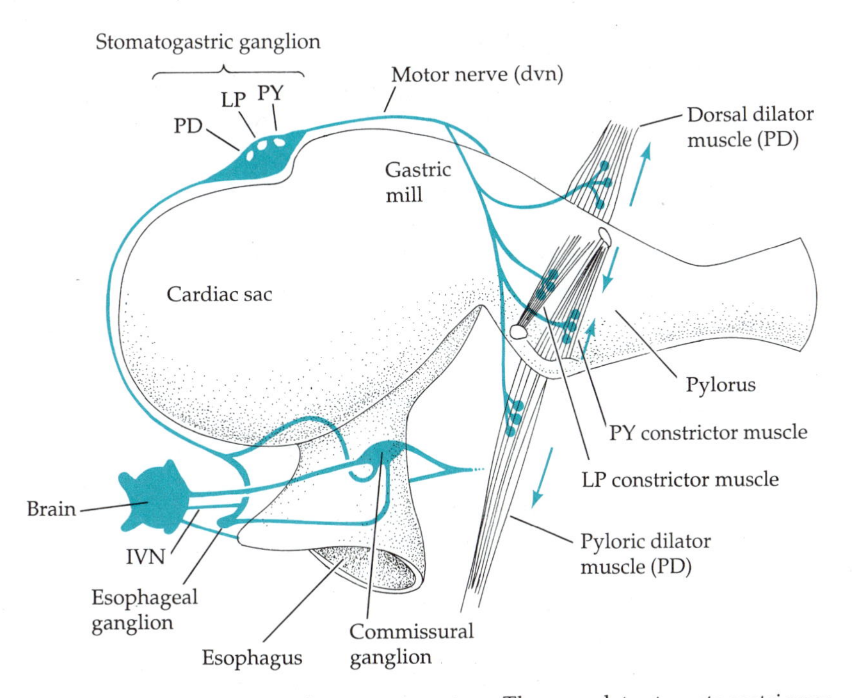

FIGURE 1. The stomatogastric nervous system. The complete stomatogastric nervous system includes four ganglia: the stomatogastric ganglion on the dorsal surface of the stomach, the esophageal ganglion on the anterior surface of the esophagus, and the paired commissural ganglia lateral to the esophagus. The ganglia and the nerves that connect them are shown in color. Muscles that move the esophagus, cardiac sac, gastric mill, and pylorus are controlled by the stomatogastric nervous system. The position of the two sets of muscles that dilate and constrict the pylorus are shown in color. Contraction of the dilator muscles opens the pylorus, whereas contraction of the constrictor muscles narrows it. The dilator muscles are innervated by the two pyloric dilator (PD) motor neurons, the anterior constrictor muscles by the lateral pyloric (LP) neuron, and the more posterior constrictor muscles by the pyloric (PY) neurons.

neural network by tailoring its output to conform to the changing needs of the animal.

A network formed by 14 neurons generates the pyloric rhythm

Thirteen motor neurons and one interneuron within the STG are interconnected to form a central pattern generator that controls the pyloric rhythm. Among these are the two pyloric dilator neurons (PD), eight pyloric neurons (PY), a lateral pyloric neuron (LP), a ventricular dilator (VD), an inferior cardiac (IC) neuron, and one interneuron, the anterior burster (AB). The pyloric motor pattern is illustrated in the recordings in Figure 2. The top trace is an extracellular recording made from the motor nerve which contains axons from three classes of motor neurons, PD, LP, and PY. The recording illustrates the triphasic pyloric rhythm, which consists of sequential bursts of action potentials in the LP, PY, and PD neurons.

The pyloric rhythm can be seen more clearly in the bottom three traces

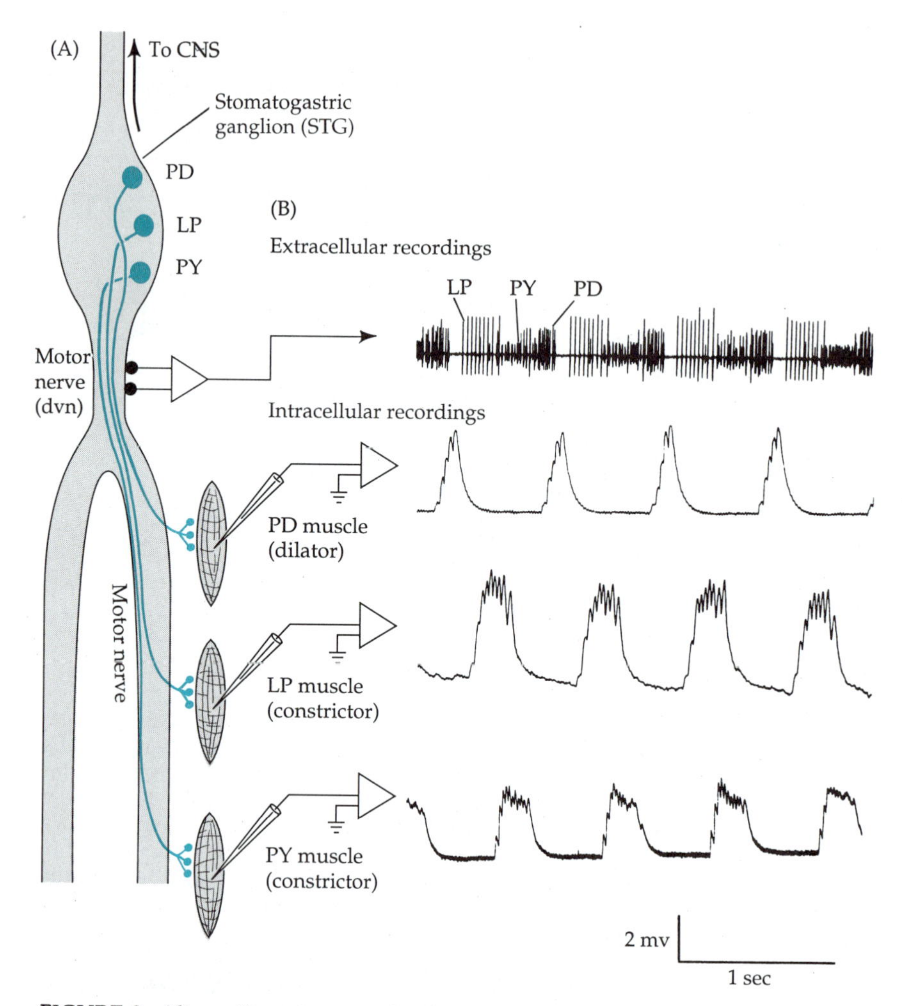

FIGURE 2. Alternating patterns of activity in the pyloric muscles. (A) Schematic diagram of the positions of electrodes used for the recordings shown in (B). Note that the PD, LP and PY neurons are located in the STG and send axons into the motor nerves that branch to innervate their muscles. (B) An extracellular, multiunit recording from the motor nerve made with wire electrodes placed as shown in (A). The recording shows the activity of axons from the LP, PY, and PD neurons. The LP neuron, which has the largest diameter axon, has the largest action potential in this recording. The three lower records are simultaneous intracellular recordings from muscles innervated by each of the three motor neurons. Summed excitatory junctional potentials (EJPs) rhythmically depolarize each muscle in turn, causing it to contract. (After S. L. Hooper et al., 1986. J. Comp. Physiol. A 159: 227–240.)

in Figure 2, which are simultaneous intracellular recordings from three different muscles, one innervated by the LP neuron, one innervated by the two PD neurons, and one innervated by several of the PY neurons. These muscles, like many crustacean muscles, do not fire action potentials. Instead, action potentials in the motor axons release a neurotransmitter that depolarizes the muscle fibers in a graded fashion. The amplitude of depolarization depends on the number and rate of action potentials in the motor neuron. Tension and movement are a direct function of the extent of depolarization and muscle load. Therefore, even relatively minor changes in the frequency and timing of bursts of action potentials in a motor neuron can produce significantly different movements. The triphasic rhythm seen in the recording from the motor nerve produces a triphasic pattern of muscle depolarization (Figure 2). The muscles innervated by the PD neurons dilate the pylorus, while the muscles innervated by the LP and PY neurons constrict the pylorus. Therefore the pattern of motor neuron and muscle activation produces repeating dilations and constrictions of the pylorus. The constriction phase takes place in two stages, because the LP and PY neurons are activated sequentially.

Two questions are important for understanding the pyloric rhythm: (1) What generates the essential rhythm? and (2) How are appropriate phase relationships in the activity of the pyloric neurons produced and maintained? The answers to these questions come from experiments in which simultaneous intracellular recordings are made from several pyloric neurons and from experiments in which specific neurons are deleted from the network. Neurons are selectively killed by intensely illuminating them after filling them with a dye such as Lucifer Yellow that forms toxic biproducts upon photolysis. When simultaneous intracellular recordings from the single interneuron, AB, a PD neuron, and the LP neuron are compared (Figure 3B), the AB and PD neurons depolarize at the same time. As they do so, the LP neuron is hyperpolarized by a compound, inhibitory postsynaptic potential (ipsp) evoked by the PD and AB neurons. Each action potential in LP in turn evokes an ipsp in PD, and a smaller, slower hyperpolarization in AB. PD and LP thus show **reciprocal inhibition.**

The synchronous depolarization of AB and PD suggests that they are electrically coupled. Indeed, if electrodes are placed in each of these neurons, current passed into one produces voltage deflections in the other. Although AB and PD are electrically coupled and depolarize synchronously, they have quite different properties. AB is an **endogenous burster**, a neuron that fires rhythmic bursts of action potentials in the absence of timed synaptic input. When the PD neurons are deleted from the network, AB still generates bursts of action potentials that resemble those produced in the intact network. If AB is deleted, however, the PD neurons lose their ability to generate bursts; they either fire tonically, or generate long, slow plateau potentials. AB and PD also differ in their synaptic connections. AB makes inhibitory glutamatergic synapses in the ganglion, whereas the PD neurons make inhibitory cholinergic connections in the ganglion (remember that PD neurons also make excitatory neuromuscular connections). LP inhibits the PD neurons through direct synaptic contacts, but it inhibits AB only indirectly through the electrical synapse between AB and the PD neurons.

Figure 3C summarizes the connections among the pyloric neurons and helps to explain how the pyloric rhythm is generated. The source of the rhythm is the bursts of action potentials produced by AB. The PD neurons

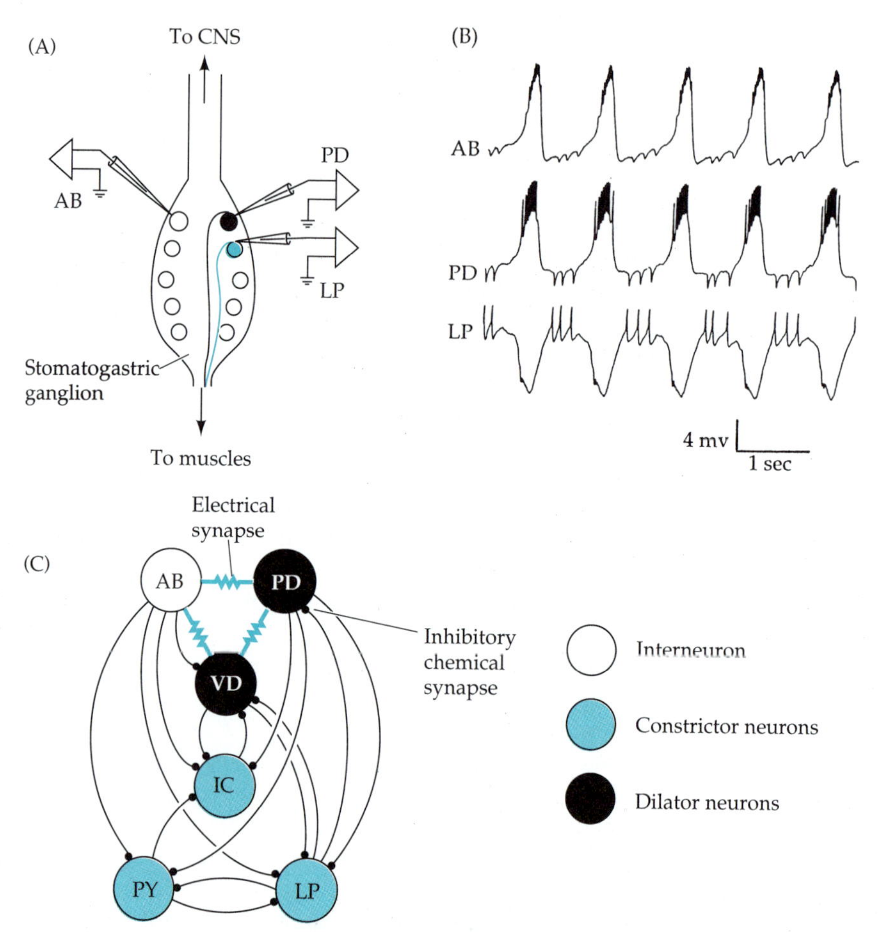

FIGURE 3. The pyloric neural network. (A) Schematic diagram showing the position of the electrodes during the recordings in (B). (B) Simultaneous intracellular recordings from the AB, PD, and LP neurons of the stomatogastric ganglion. The AB and PD neurons depolarize simultaneously, and cause an inhibitory potential (ipsp) in LP. Each action potential in LP produces an ipsp in PD (and in AB because the AB and PD neurons are electrically coupled). (C) Schematic diagram of connections among the neurons of the pyloric network established using intracellular recordings like those in (B). Electrical coupling is represented by the resistor symbols. Filled circles denote chemical synapses, all of which are inhibitory. Neurons innervating antagonistic muscles often make reciprocal inhibitory connections. (B after J. S. Eisen and E. Marder, 1982. J. Neurophysiol. 48: 1392–1415; C after E. Marder and J. S. Eisen, 1984. J. Neurophysiol. 51: 1345–1361.)

depolarize and burst in synchrony with AB because of the electrical connection between them. AB and the PD neurons both inhibit LP and PY. The reciprocal inhibition between the LP and the PD neurons forces them to fire out of phase with each other. This causes the dilator muscles innervated by the PD neurons, and the constrictor muscles innervated by the LP and PY neurons, to contract in an alternating rhythm. The PY neurons fire later than LP, as shown in Figure 2, because LP recovers faster from the phasic inhibition than do the PY neurons. These two features, a **pacemaker neuron** such as AB that provides intrinsic timing information, and reciprocal inhibition between functional antagonists, are key elements in the network, and are found often in pattern-generating networks in both vertebrates and invertebrates.

The pyloric central pattern generator, like many neural networks, is modulated by an array of hormones and transmitters

Rhythmic behaviors, and therefore the neural networks that generate them, must be flexible so that animals can adapt to different environmental conditions. The circuits that control locomotion must be able to change speed and patterns of activity to allow the animal to move over widely differing terrains, or to carry different loads. Many mechanisms are used by animals to modulate the output patterns of rhythm-generating systems. Some of these are based on conventional synaptic mechanisms; in others, substances such as biogenic amines and neuropeptides are released more diffusely than conventional transmitters and influence many cells simultaneously, modulating cellular properties of pattern-generating neurons and their synaptic connections.

The stomatogastric nervous system illustrates the remarkable flexibility that modulatory agents can confer on a pattern-generating circuit. Modulators can have many different effects on rhythmic networks. They can activate silent rhythms or speed up their pace; they can change the pattern of a particular network; they can also change the synaptic relationships between networks and thereby alter the way that networks work together.

The 30 neurons located in the STG are either cholinergic or glutamatergic. However, more than 10 different modulating agents are released by fibers that come from other parts of the nervous system and enter the central neuropil of the ganglion where neurons of the ganglion form all their synaptic connections (Figure 4). The fibers ramify widely throughout the neuropil, where they terminate. We now know the locations of the cell

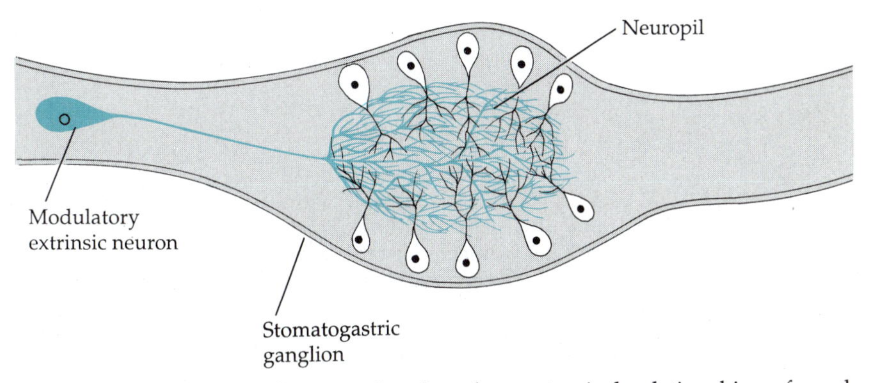

FIGURE 4. Schematic diagram showing the anatomical relationships of modulatory input neurons and the neurons of the STG. Note STG somata form a circle, or outer rind, around the central area, or neuropil, where synaptic contacts are made.

bodies of some of the modulatory neurons, many of which project to the STG from other ganglia. These neurons are thought to respond to various environmental conditions, firing action potentials to release their modulatory transmitters when a change in motor output is required.

When all extrinsic modulatory and synaptic inputs to the STG are removed, the pattern generators in the STG become less active: they stop firing, or fire slowly. This occurs because the intrinsic membrane properties that are crucial to rhythmic firing (bursting and plateau potentials) depend on tonic modulatory inputs. Indeed, many of the modulators in fibers entering the STG can activate the central pattern generators. Application of proctolin, a neuropeptide found in crustaceans, or stimulation of a proctolin-containing neuron strongly activates a nonrhythmical or slowly cycling preparation (Figures 5A and B). The proctolin-activated pyloric rhythm is characterized by long, high-frequency bursts of LP neuron activity, which produces strong, long-lasting constrictions of the pyloric sac

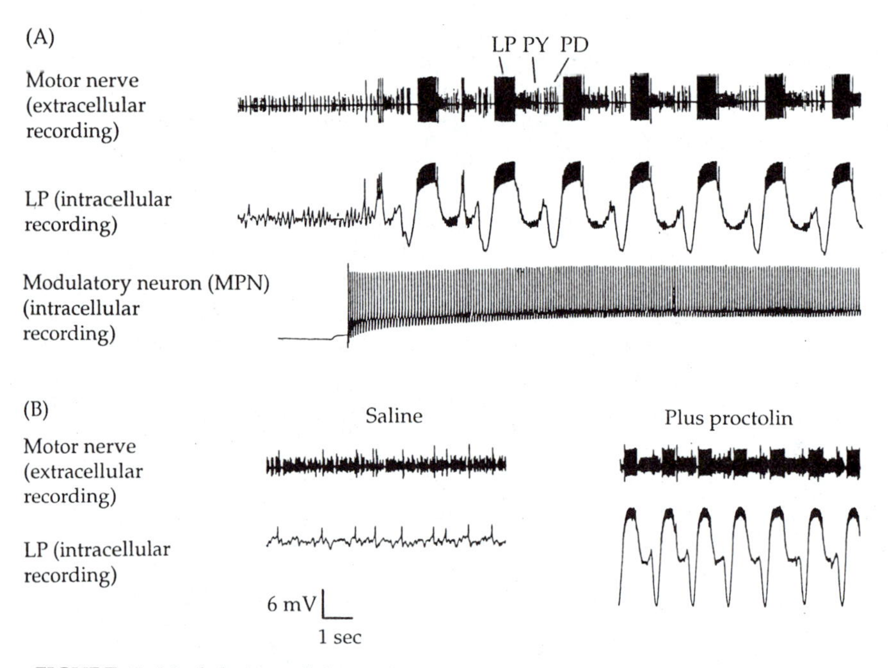

FIGURE 5. Modulation of the pyloric rhythm. (A) Intracellular stimulation of MPN, a modulatory proctolin-containing neuron, activates the pyloric rhythm. As in Figure 2, an extracellular recording from the motor nerve shows activity of the LP, PY, and PD neurons (top trace). The bottom traces (LP and MPN) are intracellular recordings from LP and MPN, respectively. MPN action potentials were evoked by depolarizing the cell body. (B) Control recordings on the left show activity when the preparation is bathed in saline alone. Recordings on the right show activity of the same preparation with 10^{-6} *M* proctolin in the bath. Both stimulation of MPN (shown in A) and bath application of proctolin result in long, high-frequency bursts of action potentials in LP. (A, B after E. Marder and M. P. Nusbaum, 1989. *Perspectives in Neural Systems and Behavior*, T. C. Carew and D. Kelley, eds., Liss, New York.)

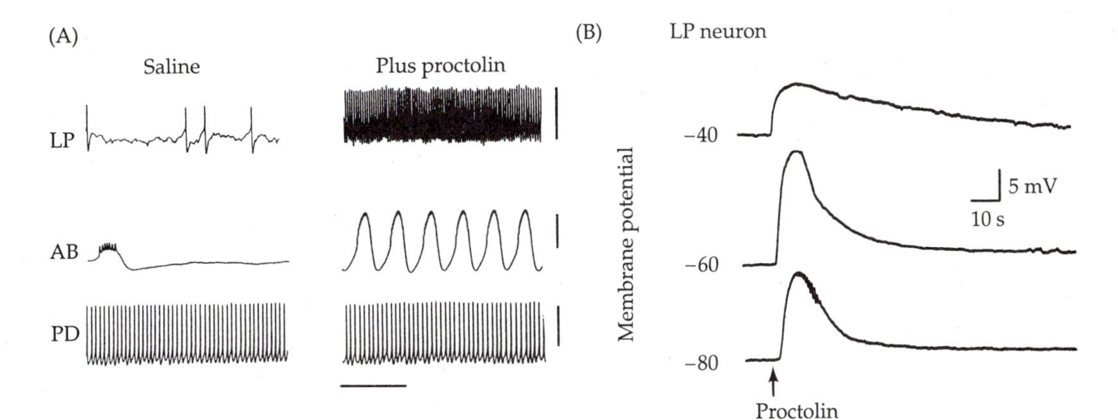

FIGURE 6. Proctolin actions on individual target neurons. (A) The top recording is from an isolated LP neuron in control saline, and in saline containing proctolin. Proctolin produces a slight depolarization and high frequency action potentials. The middle recording is from an isolated AB neuron. In control saline the neuron displays low amplitude and low-frequency bursts of action potentials. Proctolin enhances the amplitude and frequency of the bursts. The isolated PD neuron is not affected by proctolin. Vertical bars, 5 mV; horizontal bar, 3 sec for LP, 4 sec for PD, 1 sec for AB. (B) The response of the LP neuron to proctolin is voltage-dependent. Proctolin applied in a puff from a micropipette at the time indicated by the arrow produces depolarization of LP that is maximal at −60 mV. (A after S. L. Hooper and E. Marder, 1987. J. Neurosci. 7:2097–2112; B after J. Golowasch and E. Marder, 1991. J. Neurosci., in press.)

alternating with dilations. Stimulation of serotonin-containing sensory neurons located in the motor nerve can also activate the pyloric rhythm. The rhythm that serotonin activates is characterized by long bursts in LP and short bursts in PD.

Proctolin acts directly on both the AB and the LP neurons. Proctolin increases the frequency and amplitude of the AB neuron bursts (Figure 6A), thus activating the inactive pyloric rhythm. Extended bursts produced in the LP neuron by proctolin occur because proctolin depolarizes the LP neuron by activating an inward current, carried mainly by Na^+, that shows marked outward rectification. This current is largest close to the action potential threshold of the LP neuron (Figure 6B), so that even a small activation of the current markedly increases excitability.

Modulators can change the patterns of rhythmic movements

Each of the modulators of the STG evokes distinct forms of the pyloric rhythm. Three of these characteristic pyloric rhythms, produced by a muscarinic agonist (pilocarpine), serotonin, and proctolins are illustrated in Figure 7. In this experiment, an isolated STG preparation with no ongoing rhythm was bathed in each agent and then extensively washed to return it to its baseline activity before the next application. The different rhythms activated by each agent are easily seen by comparing the balance of activity in LP and the PD neurons. Pilocarpine produces strong alternating constrictions (LP) and dilations (PD), whereas other agents produce slower

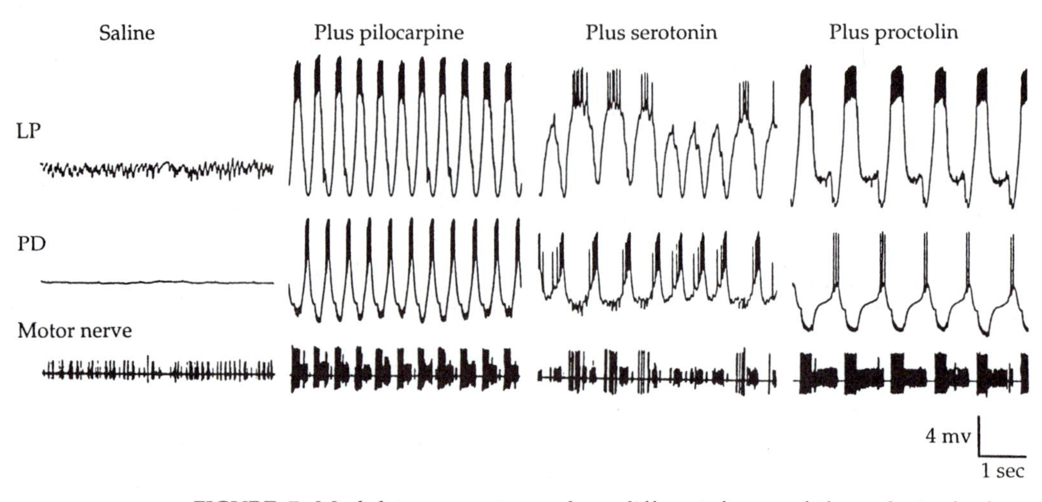

FIGURE 7. Modulatory agents produce different forms of the pyloric rhythm. Each panel shows intracellular recording from the LP neuron and the PD neuron, respectively, and the bottom trace is an extracellular recording from the motor nerve. In control saline the pyloric rhythm is not active. Bath application of pilocarpine (a muscarinic agonist), serotonin, and proctolin turns on the pyloric rhythm. However, the form of the rhythm in the presence of each agonist is different. (After E. Marder and J. M. Weimann, 1991. *Neurobiology of Motor Programme Selection: New Approaches to Mechanisms of Behavioral Choice,* J. Kien, C. McCrohan and W. Winlow, eds., Manchester University Press, in press.)

rhythms, some biased toward constriction. The pyloric rhythm includes the output of six different neuronal types, and each modulator is likely to act on a different subset of membrane channels located on distinct sets of target neurons. Thus, it is easy to imagine how the same small network can produce a wide variety of behaviors.

The same agents also modulate the gastric mill rhythm. In this case, different rhythms produce quite different coordinated movements of the lateral and medial teeth, corresponding to different chewing patterns.

Modulators alter the interactions between different central pattern generators

In addition to activating and altering motor patterns in single neuronal circuits, modulatory agents can create new operational networks. One result is that individual neurons can participate in more than one functional network. The effects of the peptide red pigment-concentrating hormone (RPCH) on the gastric mill and cardiac sac central pattern generators provide an example. RPCH was first characterized in crustaceans as a circulating hormone that controls the animal's color by regulating the extent of dispersion or contraction of red pigment granules. However, like many circulating hormones, RPCH is also found in the nervous system and is contained in fibers that project into the neuropil of the STG.

When RPCH is applied to the STG, a novel rhythm is produced that employs neurons from both the cardiac sac and gastric mill networks

(Figure 8). During spontaneous activity (Figure 8A), the cardiac sac rhythm produces slow, irregular bursts in the cardiac sac dilator 2 neuron (CD2), while the gastric mill rhythm results in more rapid bursts in the lateral posterior gastric (LPG) neurons. The two rhythms are not linked, and their intrinsic periods differ by nearly an order of magnitude. After application of RCPH (Figure 8A), a new rhythm is produced in which CD2 and LPG fire in strict alternation. The cellular mechanisms responsible for this new, fused rhythm include a strong potentiation of the synaptic connections made by the inferior ventricular (iv) neurons onto both CD2 and LPG (Figure 8B). Application of the peptide increases the postsynaptic poten-

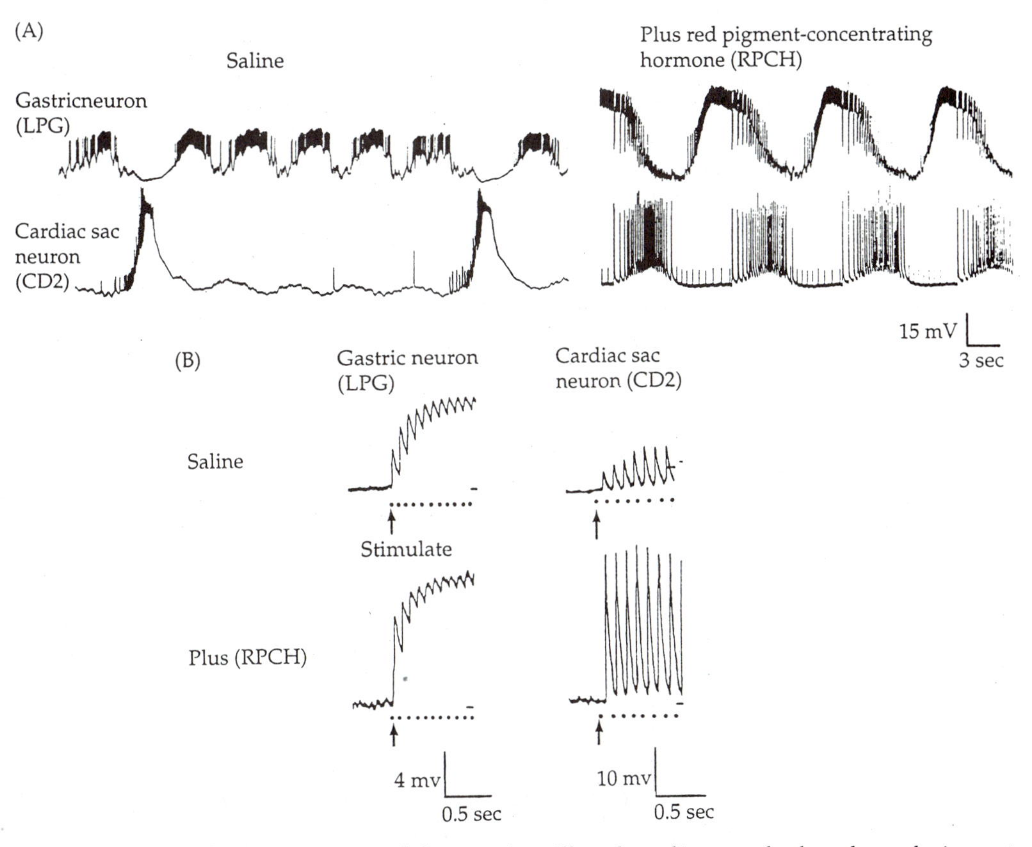

FIGURE 8. Fusion of the gastric mill and cardiac sac rhythms by red pigment concentrating hormone. (A) Simultaneous intracellular recordings from the lateral posterior gastric (LPG) neuron and the cardiac sac motor neuron CD2. In this recording, the period of the CD2 burst of action potentials is much longer than that of bursts in neurons of the gastric rhythm. Application of red pigment concentrating hormone (RPCH) evokes a novel rhythm in which LPG and CD2 fire in alternating bursts. (B) The novel fused rhythm evoked by RPCH in (A) occurs because RPCH strongly potentiates synaptic connections from the inferior ventricular neurons onto LPG and CD2. Top traces show excitatory postsynaptic potentials (epsp's) evoked in LPG and CD2 by stimulating iv neurons in control saline; bottom traces show epsp's evoked in the presence of RPCH. The dots represent repeated stimuli. (After P. S. Dickinson et al., 1990. Nature 334: 155–158.)

tials produced by the iv neurons in CD2 and LPG three- to tenfold. The RPCH-potentiated postsynaptic potentials are now strong enough to control the activity of CD2 and LPG, thus coordinating their activity.

The stomatogastric nervous system illustrates the rich variety of ways that neuronal plasticity can contribute to the regulation of behavior

Although the stomatogastric nervous system is capable of producing a variety of behaviors under modulatory influences, not all features of the behaviors or of the networks are plastic. For example, the strong reciprocal inhibition that forces the PD and LP neurons to fire alternately is retained whenever they are both activated. Some features of neural networks are relatively constant despite modulatory influences, and others are subject to alteration.

It is interesting to imagine how the principles illustrated by the stomatogastric nervous system could scale up within the large computing networks of vertebrate brains. Many of the cellular and molecular mechanisms and modulatory agents used by invertebrate neurons are present in the vertebrate brain, but we have much to learn about how they are employed.

Short-Term and Long-Term Synaptic Modulation in a Reflex Pathway of the Sea Hare, *Aplysia*

Aplysia *offers many advantages for the study of cellular neurobiology*

Aplysia californica and related species of gastropod molluscs that are found in shallow ocean waters have a small repertoire of stereotyped behaviors that includes feeding, egg-laying, and a variety of protective maneuvers. The central nervous system of these organisms is relatively simple, consisting of eight paired ganglia arranged around the esophagus, and a large, fused ganglion in the abdomen (Figure 9). As is the case in most invertebrate ganglia, such as the stomatogastric ganglion, neuronal cell bodies are located in a rind around the outside of each ganglion and send processes into the interior ganglionic neuropil, where synaptic connections are located. Axons leave the ganglia in discrete nerve bundles. The larger ganglia contain approximately 2000 neurons that are easy to identify and impale with microelectrodes because of their large diameters (50 μm to nearly 1 mm). Many individual neurons that contribute to specific behaviors have been identified. Studies of their cellular properties have contributed immensely to our understanding of neuronal modulation and its relationship to behavior.

The gill- and siphon-withdrawal reflexes are altered by prior experience

In response to a threatening stimulus, such as a tap on the mantle or siphon, *Aplysia* reflexively withdraws the siphon into the parapodium, and the gill toward the mantle shelf (Figure 10A). These defensive reflexes are mediated by neurons in the abdominal ganglion, including sensory neurons that carry tactile information from the siphon and mantle, motor

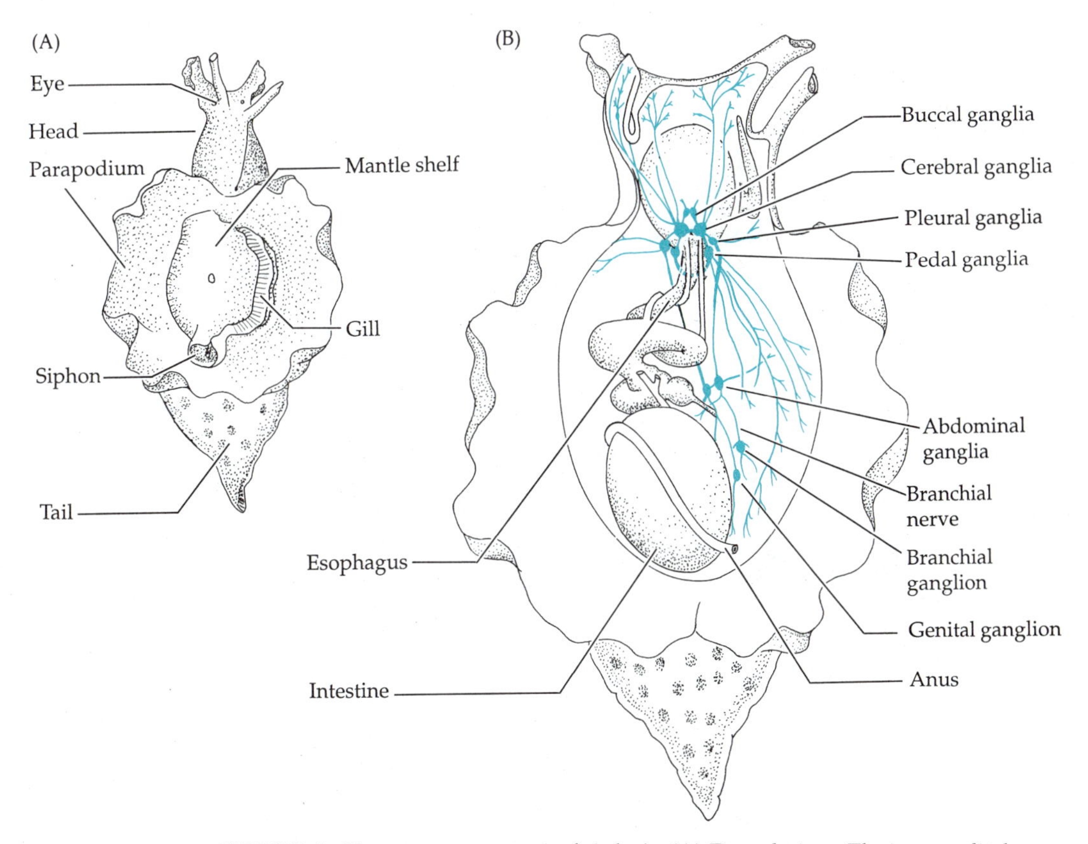

FIGURE 9. The nervous system of *Aplysia*. (A) Dorsal view. The parapodia have been peeled back to partially expose the mantle, gill, and siphon. (B) Plan of the nervous system of the animal in relation to internal body organs. Four of the major ganglia form a ring around the esophagus. The fifth, the abdominal ganglion, is located near the digestive system. (After E. Kandel, 1976. *Cellular Basis of Behavior*, W. H. Freeman, San Francisco.)

neurons that innervate the muscles of the gill and siphon, and a number of interneurons (Figure 10B). A schematic diagram of some of the synaptic interactions among these neurons is shown in Figure 11. The sensory neurons that innervate the siphon make direct, **monosynaptic** contact with the siphon and gill motor neurons. In addition, the sensory neurons synapse onto interneurons in the ganglion that carry secondary, or **polysynaptic** sensory input to the motor neurons. The diagram suggests that the strength of the reflex response could be altered by changing the strength of synaptic transmission at any or all of these synapses.

When an *Aplysia* is tapped repeatedly on the mantle, each touch elicits the **withdrawal reflex**, but the strength of the response gradually decreases, or **habituates**. Conversely, when the animal experiences an unpleasant stimulus, such as a sharp blow to the mantle or a mild electric shock to the tail, the reflex is **sensitized** so that for many minutes after the unpleasant stimulus, the animal withdraws its gill and siphon more vigorously in response to each subsequent touch.

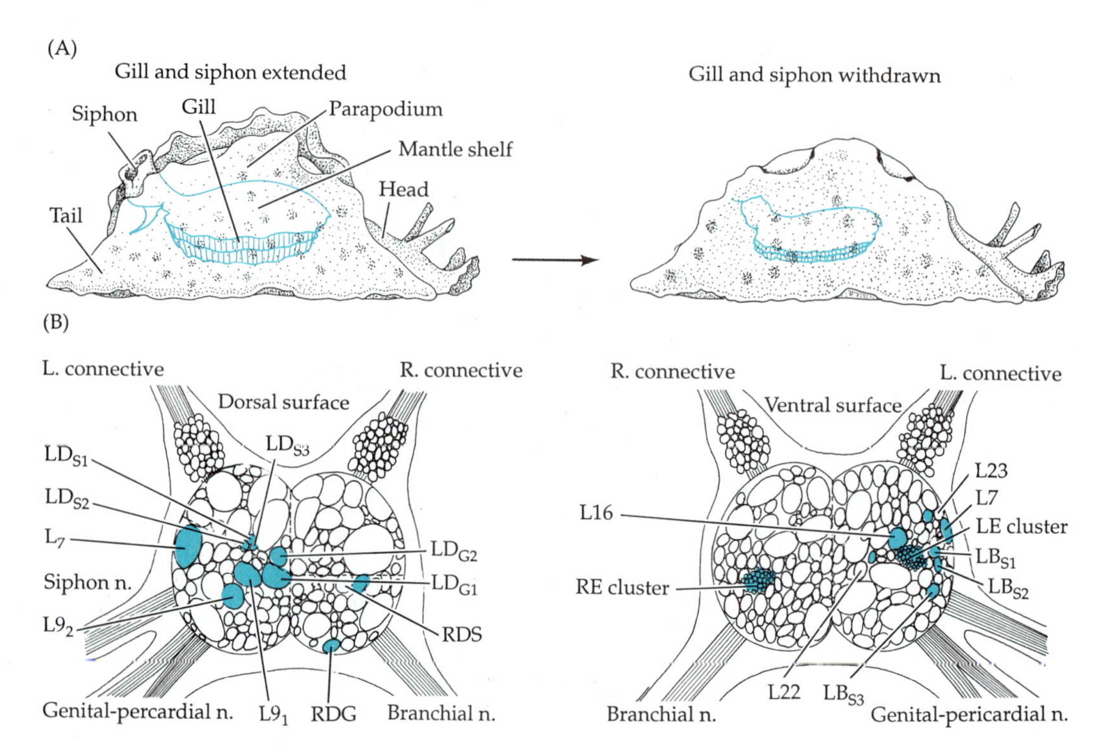

FIGURE 10. The gill and siphon withdrawal reflex. (A) Tactile stimulation of the mantle or siphon of an *Aplysia* causes it to contract the gill and withdraw the siphon into the parapodium. Noxious stimulation of the siphon, mantle, head, or tail sensitizes the reflex. (B) Location of identified neurons and clusters of neurons that participate in the gill and siphon withdrawal reflex. Sensory neurons are located in the LE cluster; L22 and L23 are excitatory interneurons; L16 is an inhibitory interneuron; most of the other neurons are motor neurons. The somata of the facilitating interneurons are located in other ganglia. (After E. Kandel, 1976. *Cellular Basis of Behavior*, W. H. Freeman, San Francisco.)

These behavioral changes are mediated by changes in excitability of neurons in the reflex circuit and by changes in the strength, or **efficacy**, of synaptic transmission at several points within the circuit. While the polysynaptic pathway contributes substantially to the reflex, the monosynaptic pathway is easier to study, and, in the gill-withdrawal reflex, it shows changes in synaptic transmission that often correlate with the changes in behavior. Habituation is associated with **depression** of the synapses from sensory neurons onto gill and siphon motor neurons produced by reduction in transmitter release from the presynaptic terminals following each action potential (Figure 12A,B). Sensitization of the reflex is usually associated with **facilitation**, an increase in release of transmitter from sensory terminals. In this case, the increase in synaptic efficacy is called **heterosynaptic facilitation**, because it is produced by release of modulatory transmitters by interneurons (Figure 12C). It is important to distinguish between behavioral terms, such as "habituation" and "sensitization," and

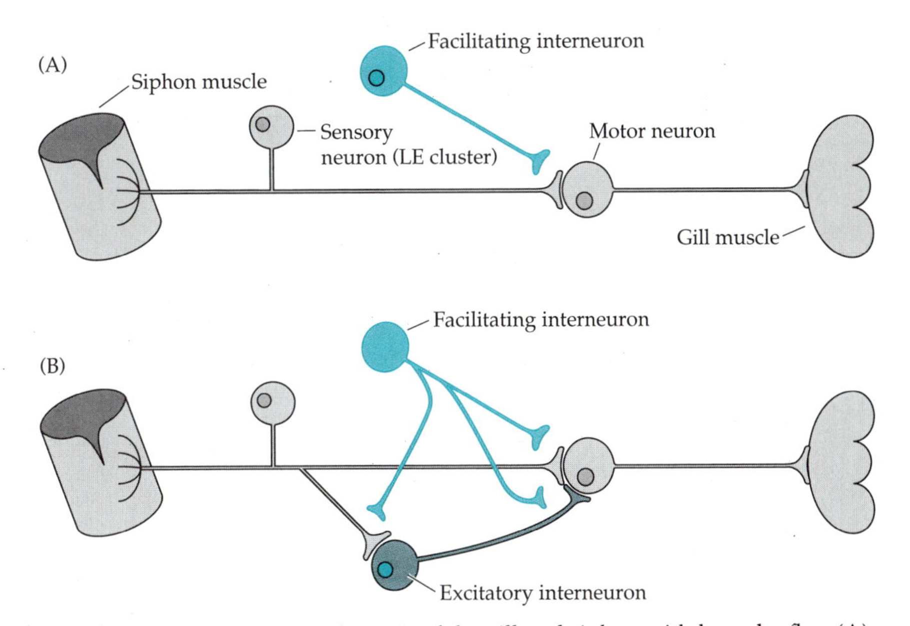

FIGURE 11. Schematic neural circuit of the gill and siphon withdrawal reflex. (A) Direct monosynaptic connections are made between sensory neurons in the LE cluster and the motor neurons that retract the gill and siphon. Facilitating interneurons, whose somata are located in other ganglia, release facilitating transmitter in the vicinity of the sensory neuron synapses in the abdominal ganglion. (B) Sensory neurons from the siphon and mantle also synapse onto interneurons that excite the motor neurons, forming a polysynaptic reflex. Facilitating interneurons can influence the strength of these synapses. Thus, the polysynaptic reflex sometimes plays an important role in behavioral habituation and sensitization.

synaptic terms, such as "depression" and "facilitation." This is because under some circumstances a behavioral change such as habituation could result from facilitation of an inhibitory synaptic pathway; conversely, sensitization might sometimes result from depression of an inhibitory pathway.

Serotonin and other modulatory compounds facilitate the gill-withdrawal reflex

The nervous systems of molluscs, like those of crustaceans, contain modulatory neurons that release a variety of peptide hormones and neurotransmitters into the blood and into the ganglionic neuropil. Release of serotonin from modulatory neurons, called **facilitator neurons**, is triggered by noxious stimulation of the head or tail and produces sensitization of the gill- and siphon-withdrawal reflex. A pair of serotonergic facilitator neurons are located in the cerebral ganglion and release serotonin near the sensory neurons innervating the gill motor neurons. Other modulatory agents are released by the same stimuli, but the molecular bases of their actions are less well understood.

The effects of serotonin on sensory synapses have been studied in the

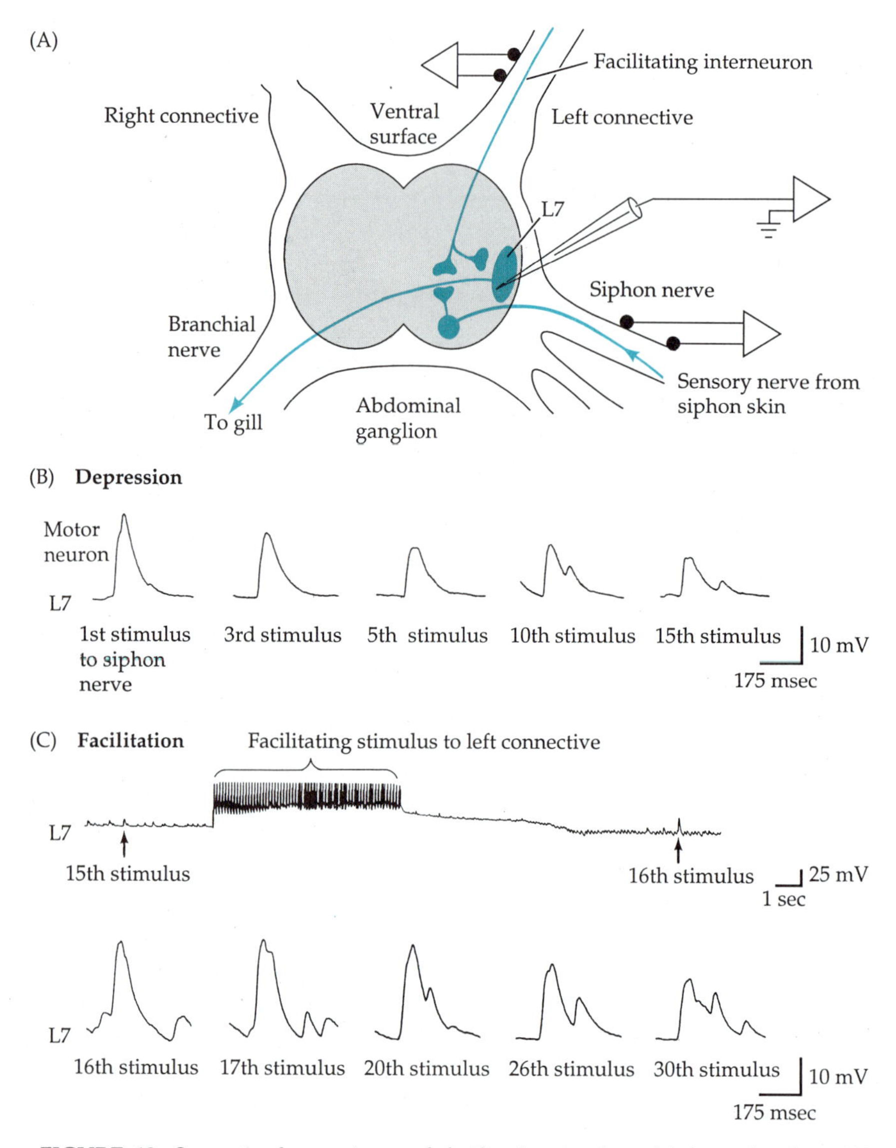

FIGURE 12. Synaptic depression and facilitation in the withdrawal reflex. (A) Schematic diagram illustrating experimental arrangement for recording from the gill motor neuron, L7. (B) Synaptic depression at the sensory synapse onto L7. When the sensory nerve is repeatedly stimulated at a rate that produces habituation of the reflex in the intact animal (once every 10 s to once every 3 min), the excitatory potential in L7 becomes depressed. (C) Synaptic facilitation produced after depression of the sensory synapse onto L7. Facilitation was initiated after the fifteenth stimulus by stimulation of the left connective, which contains axons of facilitatory interneurons. The sixteenth excitatory potential is larger than the fifteenth. (Also note the polysynaptic excitatory potentials following the first, monosynaptic, potential.) Continuing low-frequency stimulation of the sensory nerve depresses the synapse again. (Adapted from E. Kandel, 1976. *Cellular Basis of Behavior,* W. H. Freeman, San Francisco.)

intact abdominal ganglion, in individual neurons removed from the ganglion, and in neurons maintained in culture. Application of serotonin to sensory neurons triggers two parallel but distinct cellular mechanisms that cause enhanced release of transmitter. First, the amount of transmitter released per impulse is increased by prolonging the action potential. Even slight lengthening of the action potential allows significantly more Ca^{2+} to flow into the neuron through voltage-dependent calcium channels, thus prolonging transmitter release (see Chapter 5). At synapses that have not been depressed, about two-thirds of the facilitation produced by serotonin is accounted for by this mechanism. Second, a process triggered by serotonin, which is not understood at the molecular level, enhances transmitter release by improving the efficiency of Ca^{2+}-dependent release processes. Greater efficiency may result from an increase in the number of vesicles available for release, an increase in intracellular Ca^{2+} caused by a change in cytosolic Ca^{2+} metabolism, increased efficiency of the release machinery itself, or a combination of these mechanisms. When serotonin is applied to synapses that have been depressed by repeated stimulation, about two-thirds of the resulting facilitation is by this mechanism.
itation is by this mechanism.

Serotonin activates cAMP-dependent protein kinase, which inactivates a specific potassium channel in sensory neurons

Activation of serotonin receptors on *Aplysia* sensory neurons stimulates adenylyl cyclase (Chapter 6), which increases synthesis of the second messenger cAMP. The increased concentration of cAMP activates the enzyme cAMP-dependent protein kinase (Chapter 7). Phosphorylation by the kinase inactivates a potassium channel, called K_S, that is otherwise active at resting membrane potentials (see Box A). Because there are fewer active potassium channels in the membrane, the repolarization phase of the action potential is slowed and the action potential is prolonged. The delayed repolarization causes voltage-dependent Ca^{2+} channels activated by the action potential to remain open slightly longer, permitting a greater flux of calcium into the neuron. The ultimate effect is enhancement of the amount of transmitter released (Figure 12).

Inactivation of K_S is the best understood action of cAMP in the sensory neurons, but it is not the only one. Cyclic AMP also participates in the enhancement of transmitter release by the second mechanism described above. The molecular pathway that underlies the second process is unknown but may involve phosphorylation of the synaptic vesicle protein synapsin I (see Chapters 5 and 7).

Long-term facilitation requires protein synthesis

Short-term sensitization of the gill withdrawal reflex lasts for several minutes after an *Aplysia* receives a single sensitizing stimulus. In the isolated *Aplysia* nervous system, five minutes of exposure to serotonin produces facilitation of sensory synapses that also lasts several minutes. In the isolated system, the duration of short-term facilitation is determined by the

Box A Modulation of a Potassium Channel by the cAMP-Dependent Protein Kinase

A series of experiments by Eric Kandel and colleagues provided critical evidence that cAMP-dependent protein kinase is involved in the modulation of transmitter release associated with behavioral sensitization.

In one of the first studies that provided direct evidence for modulation of ionic currents by protein phosphorylation, Vincent Castellucci and collaborators injected purified catalytic subunits of the cAMP-dependent protein kinase directly into sensory neurons in isolated *Aplysia* nervous systems while monitoring the electrical properties of the sensory neurons themselves and the synaptic potentials that they evoked in motor neurons. Injection of the catalytic subunits produced action potential broadening and facilitation of synaptic potentials similar to that produced by firing of facilitator neurons, application of serotonin, or injection of cAMP directly into the neurons (Figure A). Injection of heat-inactivated catalytic subunits had no effect.

To determine which ion channels were modulated to produce the spike broadening,

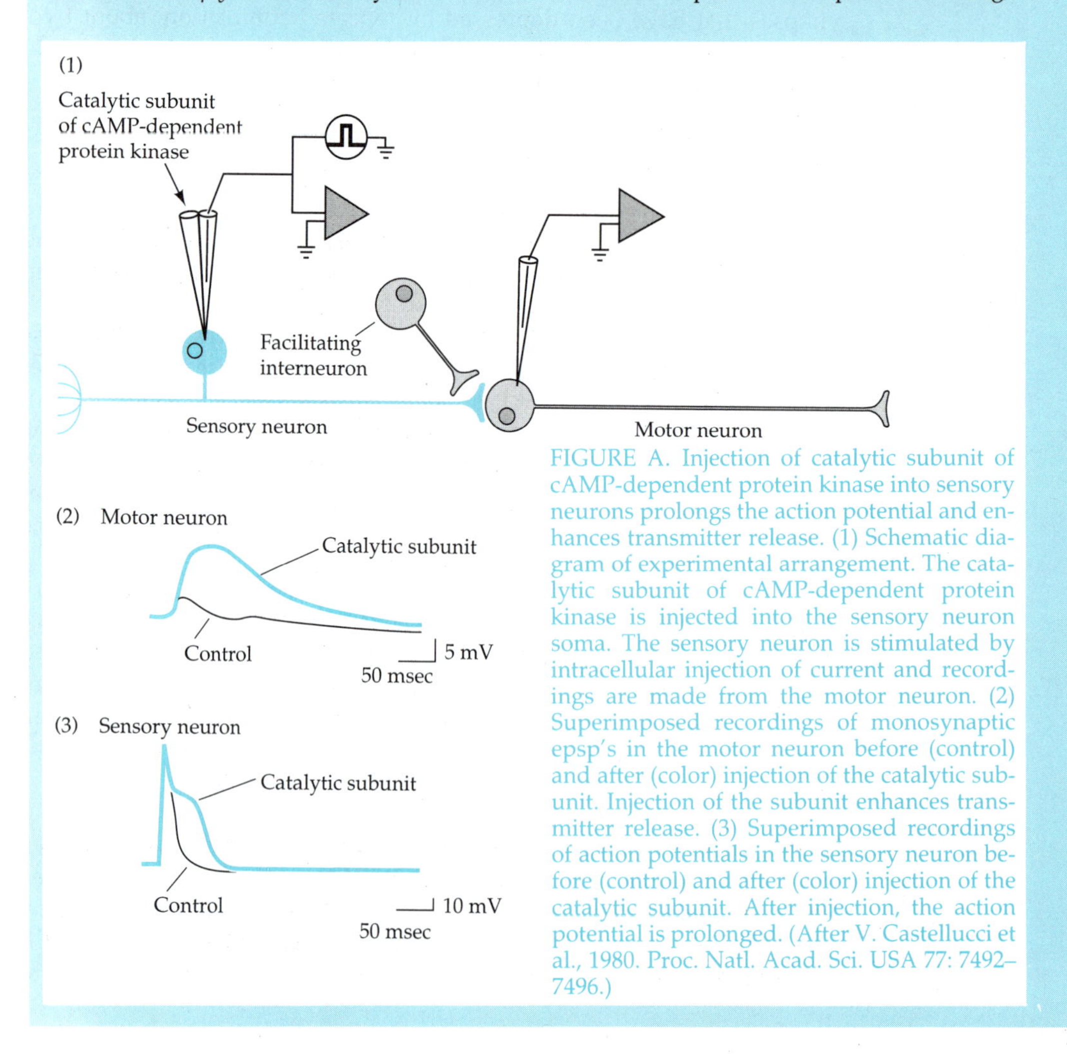

FIGURE A. Injection of catalytic subunit of cAMP-dependent protein kinase into sensory neurons prolongs the action potential and enhances transmitter release. (1) Schematic diagram of experimental arrangement. The catalytic subunit of cAMP-dependent protein kinase is injected into the sensory neuron soma. The sensory neuron is stimulated by intracellular injection of current and recordings are made from the motor neuron. (2) Superimposed recordings of monosynaptic epsp's in the motor neuron before (control) and after (color) injection of the catalytic subunit. Injection of the subunit enhances transmitter release. (3) Superimposed recordings of action potentials in the sensory neuron before (control) and after (color) injection of the catalytic subunit. After injection, the action potential is prolonged. (After V. Castellucci et al., 1980. Proc. Natl. Acad. Sci. USA 77: 7492–7496.)

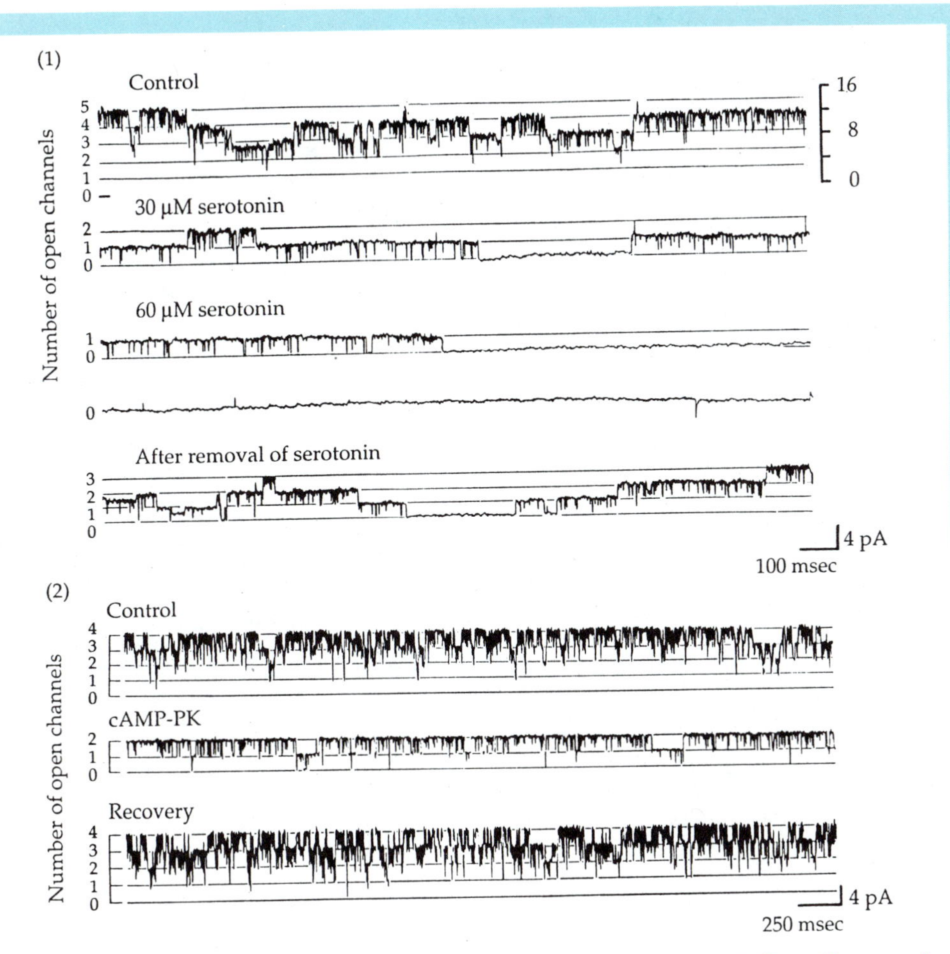

FIGURE B. Single-channel recordings of K_S. (1) Recordings made with a patch electrode attached to an *Aplysia* sensory neuron from the abdominal ganglion. Currents were recorded in response to steady depolarization of the membrane to 0 mV. (Control) Single-channel recording from five K_S channels prior to application of serotonin to the sensory neuron. Individual current steps are 2.6 pA. (30-μ*M* serotonin) Currents 2 min after addition of serotonin to the bathing medium. (60-μ*M* serotonin) Currents 1 min after an additional dose of serotonin, raising the final concentration to 60 μ*M*. The two traces are a continuous recording. (After removal of serotonin) Currents 5 min after superfusion with serotonin-free bathing medium. Channels have begun to reopen. (2) Recordings made with a patch electrode attached to an inside-out membrane patch from an *Aplysia* sensory neuron. The membrane was held at 0 mV throughout the experiment. (Control) Single-channel recordings from four K_S channels before addition of the catalytic subunit of cAMP-dependent protein kinase (cAMP-PK) to the bath. Mg-ATP was present. Channels open and close randomly, but are open most of the time. (cAMP-PK) Recordings from the same channels after addition of catalytic subunit and ATP to the bath. (Recovery) After recovery, the channels are open again. (1 from S. Siegelbaum et al., 1982. Nature 299: 413–417; 2 from M. J. Shuster et al., 1985. Nature 313: 392–395.)

(*Box A continues*)

(Box A continued)

Steven Siegelbaum and colleagues made patch clamp recordings from membranes of intact sensory neurons. At resting membrane potential, they observed potassium channels whose conductance was reduced by application of serotonin to the outside of the neurons or by injection of cAMP into the cytosol (Figure B1). This experiment provided clear evidence for modulation of a potassium channel protein by the second messenger cAMP. The potassium channel was not sensitive to internal Ca^{2+} or to external tetraethylammonium (TEA). These traits, and its weak sensitivity to voltage, distinguished it from the voltage-sensitive potassium channel previously defined by Alan Hodgkin and Andrew Huxley, and from other known potassium channels. The new potassium channel was named the S-channel, or $\mathbf{K_S}$, to signify its sensitivity to serotonin. K_S contributes to repolarization of the action potential, an effect that is seen most clearly when the voltage-sensitive potassium channel is blocked by TEA. Under these conditions, serotonin produces a substantial broadening of the action potential that enhances Ca^{2+} influx and facilitates transmitter release. In the absence of tetraethylammonium, the broadening produced by application of serotonin is smaller but still sufficient to account for about two thirds of the facilitation produced when serotonin is applied to nondepressed synapses.

To determine whether the cAMP-dependent protein kinase directly phosphorylates proteins in the membrane, the Kandel group studied inside-out patches of membrane pulled away from sensory neurons after formation of a gigaohm seal (Chapter 2). Individual openings of S-current channels were recorded in the membrane patches before and after their exposure to a solution containing free, purified catalytic subunits of cAMP-dependent protein kinase and ATP (Figure B2). In most of the experiments, exposure to both agents caused prolonged closure of 40 to 50% of the channels in a patch whereas the effect of either agent alone was minimal. Thus, either the K_S channel itself, or a protein closely associated with it, is phosphorylated by the cAMP-dependent protein kinase, resulting in channel closure.

time course of the elevation of cAMP produced by serotonin. A much longer-lasting sensitization of the reflex can be produced in the animal by repeated exposure to the unpleasant stimulus. For example, several closely spaced shocks applied to the tail over a few hours each day for several days produces sensitization of the withdrawal reflex that lasts for days or even weeks. Similarly, in the isolated *Aplysia* nervous system, or in cultures of *Aplysia* neurons, a two-hour exposure to serotonin results in facilitation of sensory synapses that lasts for 24 hours or more. The molecular mechanism of the transition from short-term facilitation to long-term facilitation has been studied in these reduced preparations. This transition is often considered a model for the transition from short-term to long-term behavioral sensitization in the animal.

In some ways, the cellular mechanisms underlying short-term and long-term facilitation are similar. For example, in both processes the K_S current is suppressed, elevating transmitter release. In short-term facilitation, the current reappears after a few minutes; in long-term facilitation, it remains suppressed for at least 24 hours. However, long-term facilitation differs from short-term facilitation in two additional ways: it lasts much longer than the serotonin-induced elevation of cAMP, and it requires protein synthesis. Thus, the synthesis of new proteins, and perhaps the expression of new genes, are required to produce facilitation that endures beyond the transient increase in cAMP concentration.

Long-term facilitation may be maintained by a persistent activation of cAMP-dependent protein kinase

Recall that the cAMP-dependent protein kinase is a tetramer of two regulatory subunits and two catalytic subunits (Chapter 7). When cAMP concentration is low, the regulatory subunits are bound to the catalytic subunits, inhibiting their kinase activity. When cAMP concentration is increased, it binds to the regulatory subunits, releasing active catalytic subunits. The prolonged increase in cAMP caused either by repeated sensitizing stimuli to the animal, or by a two-hour application of serotonin to reduced experimental systems, results in a small, persistent reduction in the steady state level of regulatory subunits in the sensory neurons. This reduction shifts the binding equilibrium in the cytosol toward free catalytic subunits (Figure 13) causing a higher steady state level of protein phosphorylation by catalytic subunits throughout the sensory neuron. The enhanced protein phosphorylation presumably reduces the number of open K_S channels, resulting in a facilitation that lasts until the original ratio of the two subunits is restored.

The reduction in regulatory subunits has two critical characteristics in common with long-term facilitation: it persists for at least 24 hours after the sensitizing stimuli, although the cAMP concentration falls a few minutes after the stimuli end, and it does not occur if protein synthesis is inhibited during the sensitizing stimuli. Furthermore, the reduction in regulatory subunits does not occur after a stimulation that produces only short-term facilitation. The requirement for protein synthesis during the sensitizing stimuli suggests that a protease, or perhaps a repressor of transcription of the regulatory subunit gene, may be synthesized during the stimuli, initiat-

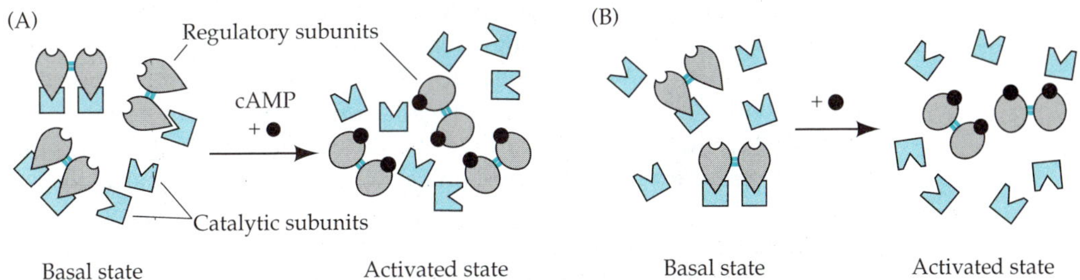

FIGURE 13. Schematic diagram of the change in ratio of regulatory and catalytic subunits of the cAMP-dependent protein kinase during long-term facilitation in *Aplysia*. (A) Before facilitation, the regulatory and catalytic subunits are present in approximately equal amounts. Therefore the number of free, active catalytic subunits is small at basal cAMP concentrations. When cAMP concentration rises, the catalytic subunits are released. (B) After long-term facilitation, the number of regulatory subunits is reduced in proportion to catalytic subunits, shifting the steady state toward more free catalytic subunits at basal cAMP concentration.

ing a series of events that results in persistent reduction in the steady state level of regulatory subunits.

Facilitation of transmitter release at the synapses of sensory neurons onto gill motor neurons now appears to be only one regulatory event in an intricate and interdependent network of regulatory events underlying modulation of the withdrawal reflex. However, the mechanism of facilitation illustrates two molecular themes that reappear in other neuronal modulatory pathways. First, short-term changes in function are controlled by rapidly reversible posttranslational modifications of key proteins, in this case by phosphorylation of a potassium channel or a protein closely associated with it. Second, the transition to long-term changes involves modification of the same key protein(s), but new protein synthesis is required to make the change persistent.

Long-Term Potentiation in the Mammalian Hippocampus

The hippocampus is necessary for early stages of memory formation

Enduring memories, such as the names and faces of our immediate family, or the locations of our grammar and high schools, are stored in a stable form that survives seizures, most injuries to the brain, and coma. Such memories appear to be stored, at least in part, as structural changes at specific synapses within neural networks that encode and connect the sensory impressions associated with each memory. These structural changes are thought to occur in the neocortex. Loss of neurons in a particular area of cortex never erases one particular set of memories but can degrade specific aspects of many memories. Thus, memories seem to be stored in a highly distributed fashion throughout many neocortical areas.

Some of the processes underlying the early stages of memory formation are more localized. For example, a set of cortical structures including the **entorhinal cortex**, the **subiculum**, and the **hippocampus** are necessary to transform short-term **declarative memories** (memories of persons, places, and things) into a more permanent form. The role of the hippocampus, a cashew-shaped structure that lies beneath the back (caudal) edge of the neocortex (Figure 14A), has been most thoroughly examined. Studies of primates with brain lesions and clinical studies of human patients have demonstrated that the hippocampus is a crucial link in the formation of long-term declarative memories. Clinical studies have demonstrated that destruction of parts of the hippocampus, for example by ischemia, is sufficient to produce severe anterograde amnesia, an inability to form new memories. Anterograde amnesia can occur even in patients that retain normal intelligence and a normal ability to recall events that occurred in the years before their brain injury.

In one controlled study, monkeys were trained to perform a declarative memory task 2 to 16 weeks before surgical lesioning of the hippocampal formation. After their recovery, monkeys that had been trained 4 weeks or less before surgery were severely impaired in their ability to recall the task when compared to control monkeys. In contrast, monkeys trained 12 to 16

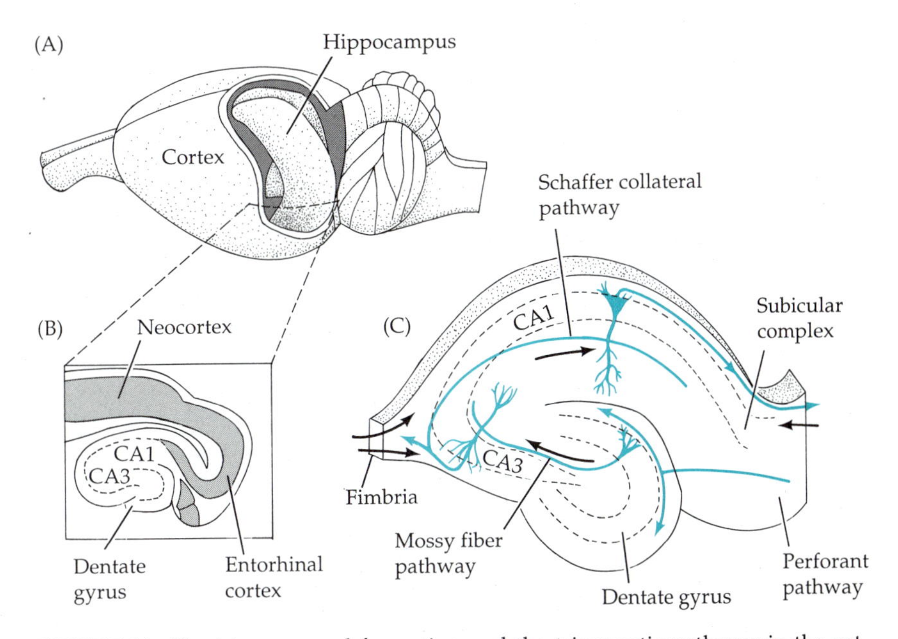

FIGURE 14. The hippocampal formation and the trisynaptic pathway in the rat. (A) Schematic diagram of the rat brain showing the position of the hippocampal formation. The dashed line shows the position of the horizontal section shown in (B). (B) Schematic diagram of a horizontal section through the hippocampal formation and overlying neocortex. Dashed lines and shaded spaces represent layers of neuronal cell bodies. In the CA fields and dentate gyrus, cell bodies are packed into narrow layers. CA1 and CA3 are areas of the hippocampus. (C) Schematic diagram of the trisynaptic circuit in a hippocampal slice. Dashed lines represent outlines of tightly packed layers of neuronal cell bodies. Areas outside the dashed lines contain axons, dendrites, and numerous synaptic contacts. Additional axons enter the hippocampus through the fimbria or through the subicular complex and entorhinal cortex (A and B after M. B. Kennedy, 1989. Cell 59: 777–787; C after D. G. Amaral, 1987. *Handbook of Physiology: The Nervous System*, vol. V, part 1, J. M. Brookhard and V. B. Mountcastle, eds., American Physiological Society, Bethesda, MD.)

weeks before the hippocampal lesion were not deficient in their memory of the task. These experiments suggest that short-term declarative memories are stored in a temporary form in the hippocampal formation and associated structures for about 4 weeks. After 4 weeks, memories become encoded in a more permanent form in the neocortex by an unknown process.

The hippocampus sends and receives connections from many parts of the neocortex

The anatomy of the hippocampal formation and its connections to the rest of the brain suggest how it might be involved in forming memories. The hippocampal formation is composed of two sheets of neurons that are highly interconnected with each other (Figure 14B). The first sheet consists of three contiguous areas of cortex called the **CA fields** (cornu Ammonis or Ammon's Horn), the subicular complex, and the entorhinal cortex, which is continuous with the neocortex. The CA fields are often referred to as the hippocampus proper. They form a large flap of tightly packed pyramid-shaped neurons that is folded and tucked under the edge of the neocortex. The second sheet of neurons in the hippocampal formation is the **dentate gyrus**, which contains round, tightly packed neurons called granule cells. The dentate gyrus is not continuous with the sheet of pyramidal neurons. Rather it bends around the end of the sheet and resembles the top of a tooth in cross section, hence its name.

In primates, the axons of pyramidal neurons from many **neocortical association areas** project into the entorhinal cortex. Information from these areas is relayed via a fan of axons called the perforant pathway through the subiculum and into the dentate gyrus. This cortical information is processed in the hippocampus through a network of synaptic connections that is not yet completely understood. One major part of the network, called the **trisynaptic loop**, has been extensively studied in vitro because it is preserved in thin slices of hippocampus (Figure 14C). The first limb of this loop is formed by the synapses from **perforant path** axons onto dentate granule cells. Axons of the granule cells, called **mossy fibers**, leave the dentate gyrus and make the second set of synapses on dendrites of pyramidal cells in area CA3 in the hippocampus proper. The pyramidal cells of area CA3 produce branched axons. One branch leaves the hippocampus; the other makes a loop to form synapses with dendrites of pyramidal cells in area CA1. This third projection in the trisynaptic loop is called the **Schaffer collateral pathway**. Axons of the CA1 neurons project to dendrites of neurons in the subicular complex and, to complete the circuit, these neurons send axons to the entorhinal cortex. All areas of the hippocampal formation except the dentate gyrus contain neurons that project to various other parts of the brain. The apparent flow of information from neocortical association areas, through the hippocampal formation, and back out to the rest of the brain suggests that a function of the hippocampus may be to "bind" activity from areas of the brain encoding information from many different sensory modalities into a record of a specific object or event. Such widespread temporary associations may constitute a short-term memory, which can be slowly transformed into a permanent memory encoded in the neocortex.

Synapses in the trisynaptic loop display long-term potentiation

In 1966, synapses from the perforant pathway onto dentate granule cells were found to have an unusual property. In intact animals, brief (one second or less), high-frequency stimulation of the perforant path resulted in enhanced transmission at synapses in the dentate gyrus for as long as hours to weeks. This phenomenon is now called **long-term potentiation**, or **LTP**. When studied in vitro in slices of hippocampus, LTP lasts for two to three hours (Figure 15). Much is now known about the mechanisms underlying LTP in vitro; additional mechanisms may be required in the animal to produce LTP that lasts for days or weeks.

Although all synapses in the trisynaptic loop display LTP in vitro, there

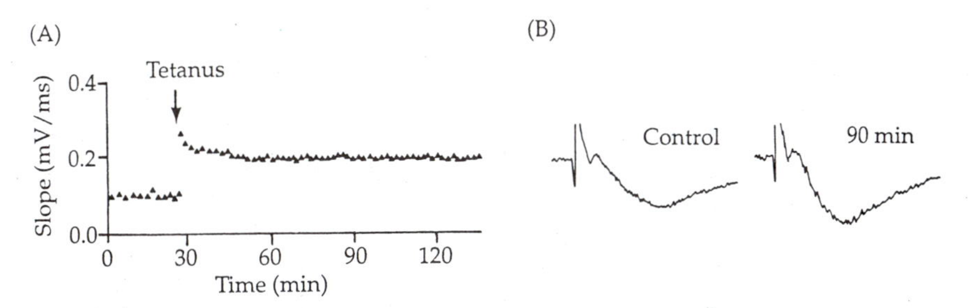

FIGURE 15. Electrophysiological measurement of long-term potentiation. (A) In hippocampal slices, LTP (long-term potentiation) is recorded in small populations of neurons through an extracellular electrode. Single test stimuli are delivered at regular intervals to a bundle of axons through a stimulating electrode, and the slope of the population excitatory postsynaptic potential (epsp) is recorded. After a tetanic stimulus is delivered to induce LTP, the slope of the epsp increases and remains larger for several hours. The amplitude of the epsp increases as well, but the slope is considered a more reliable measure of LTP. (B) Examples of individual population epsp's recorded extracellularly before and 90 min after a tetanic stimulus. The duration of the epsp is approximately 20 ms. Its amplitude depends on the position of the extracellular electrode and ranges from 0.5 to 1 mV. (From R. Nicoll et al., 1988. Neuron 1: 97–103; M. B. Kennedy, 1989. Cell 59: 777–787.)

are at least two distinct mechanisms by which LTP is produced at different sites. At synapses from the mossy fibers onto pyramidal neurons in area CA3, LTP can be induced only by high-frequency stimulation of presynaptic terminals. The strength of stimulation required is not influenced by depolarization of the postsynaptic membrane. This type of LTP is called **homosynaptic**, or **nonassociative**, because it is independent of events occurring at other, neighboring synapses. A second type of LTP occurs at the first and third synapses of the trisynaptic loop. In these two synapses, LTP is induced by strong, high-frequency presynaptic stimulation and also by weak stimulation that occurs at the same time that the postsynaptic membrane is depolarized, either by injection of current or by stimulation of nearby excitatory synapses. This form of LTP is called **heterosynaptic** or **associative** LTP because stimulation of one set of synapses increases the probability that other synapses on the same neuron will be potentiated if the synapses are activated at the same time or a few milliseconds later. This second form of LTP provides a possible cellular mechanism for encoding associations between different events or sensations that happen at the same time, as is seen during formation of memories.

A subset of glutamate receptors, called N-methyl-D-aspartate (NMDA) receptors, is responsible for the associative property of LTP

Glutamate, or a similar compound, is the principal excitatory transmitter in the hippocampus as well as in most other parts of the CNS. Three pharmacologically distinct glutamate receptors have been defined based on their selective responses to three specific agonists: **quisqualate, kainate** and **N-methyl-D-aspartate (NMDA)** (see Chapter 3). The quisqualate and kainate receptors may not be distinct entities, and they are often collectively called Q/K receptors. The receptor subtypes have different functions in the hippocampus. Antagonists of Q/K receptors, such as CNQX, completely block the excitatory postsynaptic potential at hippocampal synapses, indicating that their activation is essential for normal synaptic transmission. In contrast, antagonists of NMDA receptors, such as APV (5-phosphono-aminovaleric acid), have little effect on the excitatory postsynaptic potential. However, NMDA antagonists completely block the ability to induce LTP in synapses of the perforant path and the Schaffer collateral pathway. Thus, activation of a specific subset of receptors distinct from those involved in normal transmission is required to induce LTP.

The unusual voltage-dependent properties of NMDA receptors underlie the associative property of LTP in the perforant and Schaffer collateral pathways (Figure 16). At resting membrane potentials, the ion channels linked to NMDA receptors are partially blocked by the normal extracellular concentration of Mg^{2+}. However, when the membrane is depolarized, either by high-frequency stimulation or by simultaneous stimulation of neighboring synapses, the affinity of the channels for Mg^{2+} decreases and the Mg^{2+} blockade is relieved. Under these circumstances, binding of glutamate to the NMDA receptors opens the channels. NMDA receptors are highly concentrated at synapses in the perforant and Schaffer collateral pathways, but they are sparse at mossy fiber synapses in area CA3. This accounts for the observation that induction of LTP in mossy fiber synapses does not depend upon NMDA-receptor activation and is not associative.

The discovery that at many synapses in the hippocampus, NMDA receptor activation is required for induction of LTP but not for normal transmission led to a pharmacological test of the role of LTP in memory formation. Infusion of the hippocampi of laboratory rats with APV disrupts their ability to perform tasks requiring spatial memory without disrupting their ability to perform tasks that do not require spatial memory. This result is consistent with the idea that LTP is important for formation of memories.

The increase in postsynaptic calcium concentration produced by activation of NMDA receptors is required to induce associative LTP

In addition to their different gating mechanisms, the Q/K and NMDA receptors differ in the selectivity of their ion channels. The channel gated by Q/K-type glutamate receptors is highly selective for Na^+ and K^+, whereas the channel gated by NMDA-type receptors is permeable to Ca^{2+} as well as to Na^+ and K^+. The increased concentration of Ca^{2+} in the neuronal cytosol produced by activation of NMDA receptors can be detected by a change in absorbance of the calcium-sensitive dye Arsenazo III, or by changes in the fluorescence of Fura-2 (see Chapter 7). The entry of Ca^{2+} into the postsynaptic neuron produced by activation of NMDA receptors is necessary for induction of LTP. If Ca^{2+} chelators such as EGTA or Nitr-5 are injected into a hippocampal neuron, normal synaptic transmission is not altered, but LTP can no longer be induced in synapses on that neuron.

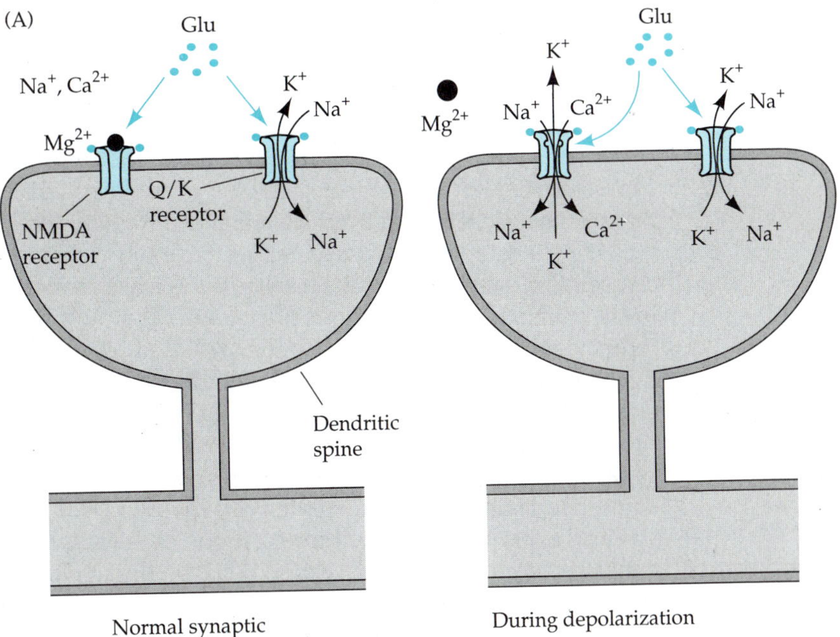

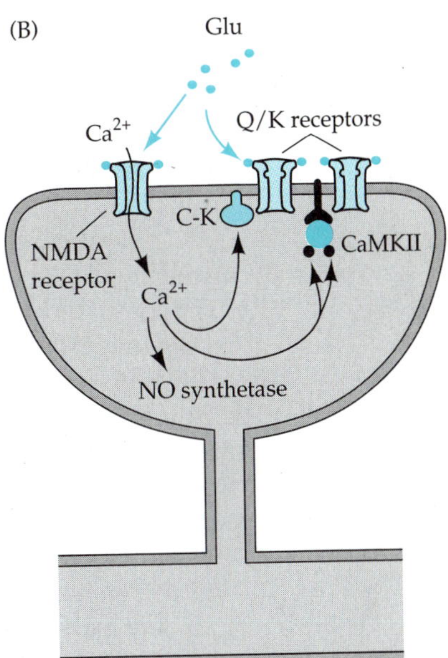

FIGURE 16. Induction of LTP. (A) Schematic diagram of the role of NMDA receptors. Glutamate is released from the presynaptic terminal and binds to both the N-methyl-D-aspartate (NMDA) and quisqualate/kainate (Q/K) receptors. Na^+ and K^+ flow through the Q/K receptor channel but not through the NMDA receptor channel because Mg^{2+} blocks the channel at potentials near the resting membrane potential (left). When the terminal is depolarized (right), as during a high-frequency tetanus, the Mg^{2+} block is relieved, allowing Na^+, K^+, and Ca^{2+} to flow through the NMDA channel. The rise in Ca^{2+} produced in the dendritic spine provides a necessary trigger for subsequent events leading to LTP. (B) Schematic diagram of potential postsynaptic targets for Ca^{2+} that may contribute to induction of LTP. CaM kinase II is concentrated in postsynaptic densities, whereas C kinase moves to the membrane upon activation (see Chapter 7). Activity of both kinases is necessary to generate LTP; the kinases may phosphorylate membrane proteins leading to generation of LTP. Other possible targets of Ca^{2+} include phospholipase A, calpain (the Ca^{2+}-dependent protease), and nitric oxide synthetase. (A after R. Nicoll et al., 1988. Neuron 1: 97–103; B after M. B. Kennedy, 1989. Trends Neurosci. 12: 417–419.)

Although activation of NMDA receptors is necessary to induce LTP in area CA1, it is not sufficient. Application of large, depolarizing doses of NMDA or glutamate to hippocampal slices potentiates the synaptic response, but the potentiation decays with a half-life of 10 to 15 minutes. The failure to induce LTP by application of NMDA implies that, in addition to the activation of NMDA receptors, induction of LTP requires another condition provided by stimulation of presynaptic terminals. What this element might be remains unknown.

The enduring change that maintains expression of LTP may be either presynaptic or postsynaptic

Although the induction of associative LTP clearly requires postsynaptic events, the location of the final molecular modifications that underlie expression of LTP remains unclear. One possibility is that modifications in the presynaptic terminal produce enhanced transmitter release, as seen in the *Aplysia* synapses discussed earlier. Several mechanisms, including increased Ca^{2+} sensitivity of the transmitter release apparatus or an increase in the number or sensitivity of presynaptic channels that control calcium influx, could increase the transmitter released by each stimulus. Alternatively, new receptors could be added to the postsynaptic membrane or the existing receptors could be potentiated to increase the voltage change produced by each quanta of transmitter. Because excitatory hippocampal synapses occur on postsynaptic spines, enlargement of the neck of the spine could facilitate spread of the postsynaptic potential into the dendrite.

Early experiments in the hippocampus of intact animals demonstrated enhanced accumulation of released, extracellular glutamate after induction of LTP, thus implicating the presynaptic terminal as the site of expression. More recently, quantal analysis of hippocampal synapses before and after induction of LTP made with whole-cell voltage-clamp recording also seem to imply that expression of LTP involves primarily an enduring increase in release of transmitter.

Not all investigators agree, however, that a presynaptic change in release of transmitter is the principal mechanism of expression of LTP. Two laboratories have found that Q/K receptors appear to be potentiated during LTP. Using drugs that specifically block either NMDA receptors or Q/K receptors, these investigators found that synaptic potentials produced by Q/K receptors alone are strongly enhanced after induction of LTP, compared to those produced by NMDA receptors. Control experiments suggested that this difference could not be explained by saturation of NMDA-receptors after induction of LTP. These results seem to imply that increased sensitivity of Q/K receptors, rather than increased transmitter release, is the important change underlying LTP. Thus, it appears that at least two mechanisms, increased transmitter release and enhanced sensitivity of Q/K receptors, may underlie expression of LTP. The contradictory results from different investigators need to be reconciled before we will understand the mechanism of LTP at the molecular level.

If any part of the expression of associative LTP involves enhanced release of transmitter, the induction process, which clearly requires an increase in postsynaptic Ca^{2+}, must depend upon retrograde communication from postsynaptic spine to presynaptic terminal. No examples of such retrograde synaptic communication are presently known; however, eicosanoids, metabolites of the fatty acid arachidonic acid, and the gas nitric oxide (Chapter 7) could diffuse across the two synaptic membranes. Eicosanoids mediate a form of presynaptic inhibition by the peptide FMRF amide in *Aplysia* through direct activation of a potassium channel. Activation of NMDA receptors in central nervous system neurons stimulates the

Ca^{2+}-requiring enzyme phospholipase A_2, causing release of arachidonic acid from neurons. Nitric oxide synthetase, the enzyme responsible for synthesis of nitric oxide, is activated by Ca^{2+}/calmodulin. Thus, at hippocampal synapses, arachidonic acid or its metabolites, or nitric oxide, could be generated at a postsynaptic site during induction of LTP and diffuse into the presynaptic terminal to trigger potentiation of transmitter release.

Induction of LTP requires protein kinase activity

Very little is known about the molecular events triggered by postsynaptic Ca^{2+} that culminate in expression of LTP. Before they can be fully understood it will be necessary to understand more clearly the molecular structure of central nervous system synapses and the organization of their biochemical regulatory networks (Chapter 7). Nevertheless, experiments employing pharmacological interventions suggest that activity of protein kinases is necessary for induction of LTP, although continuous kinase activity does not appear to be required for its maintenance.

C kinase is activated synergistically by Ca^{2+} and by diacylglycerol, which is normally produced in the membrane by hydrolysis of phosphatidylinositol bisphosphate. During high-frequency stimulation of the perforant pathway to induce LTP, a portion of C kinase moves from the cytosol into membranes, suggesting that the kinase is activated by the stimulation (Chapter 7). Furthermore, pharmacological activation of C kinase alters synaptic function. Application to hippocampal slices of phorbol esters, a class of chemicals that dissolve in the membrane and mimic diacylglycerol, causes a potentiation of synaptic transmission that decays in about 30 minutes. Direct injection of C kinase into CA1 neurons also potentiates their postsynaptic response, but this potentiation may be due to modulation of potassium or chloride currents unrelated to those altered in LTP.

A second protein kinase, type II Ca^{2+}/calmodulin-dependent (CaM) protein kinase has also been implicated in induction of LTP. This enzyme is present at unusually high concentrations in hippocampal neurons and is a major constituent of postsynaptic densities, where it seems likely to be exposed to the calcium influx mediated by NMDA receptors. Moreover, its switchlike activation mechanism (Chapter 7) may be suited for induction or maintenance of LTP. Because pharmacological agents that specifically activate the CaM kinase in intact cells are not known, there is no experimental data regarding the effect of its activation on neuronal physiology. However, injection of peptides that bind calmodulin into hippocampal neurons prevents induction of LTP in synapses that communicate with the injected neurons.

Strong evidence that activity of both of these protein kinases is necessary for induction of LTP comes from experiments with specific peptide pseudosubstrate inhibitors of both C kinase and type II CaM kinase (Chapter 7). Injection of the inhibitors into hippocampal neurons prevents induction of LTP by tetanic stimulation of synaptic terminals on the neurons (Figure 17). Injection of similar peptides that do not inhibit either of the kinases does not block induction. These experiments are not conclusive

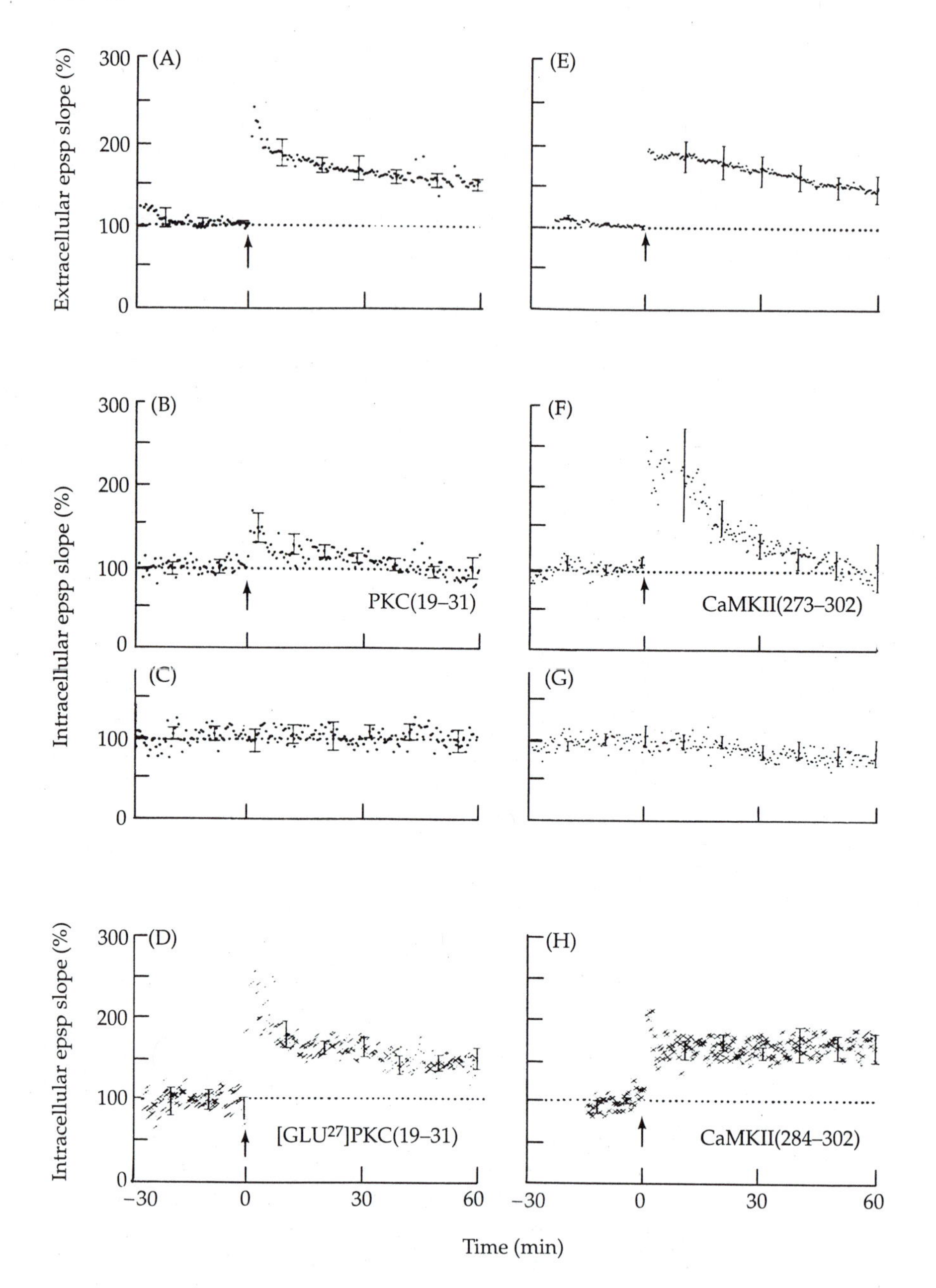

because the intracellular concentrations of the peptides after injection cannot be measured, and, at high concentrations, either peptide can inhibit both kinases. Nevertheless, the results suggest that both C kinase and CaM kinase are required for induction of LTP. The critical proteins phosphorylated by these protein kinases during induction remain to be discovered.

In addition to direct activation of the two protein kinases by Ca^{2+}/diacylglycerol or Ca^{2+}/calmodulin, they could be indirectly activated by the

FIGURE 17. Selective postsynaptic block of protein kinase C or CaM kinase II by pseudosubstrate inhibitory peptides prevents induction of LTP. (A) and (E) Extracellular recordings from hippocampal slices demonstrate induction of LTP in a population of neurons after tetanus (arrow). (B) Recording of synaptic potentials in a single neuron from slice shown in (A) with an intracellular microelectrode containing the peptide inhibitor of protein kinase C (PKC(19–36); see Table 3, Chapter 7). In this cell, the tetanus does not produce potentiation. (C) Transmission through a nontetanized pathway, measured in the same cell through the same peptide-containing microelectrode, is constant throughout the experiment. (D) Transmission measured in a different set of slices with an inactive derivative of the inhibitory peptide in the microelectrode. In this cell a tetanus produces LTP. (F) Recording of synaptic potentials in a single neuron from slice shown in (E) with an intracellular microelectrode containing the inhibitor of CaM kinase II (CaMKII(273–302); see Table 3, Chapter 7). In this cell, the tetanus does not produce potentiation. (G) Transmission in a nontetanized pathway, measured in the same cell as in (F) through the same peptide-containing microelectrode, is constant throughout the experiment. (H) Transmission measured in a different set of slices with an inactive derivative of the CaMKII inhibitory peptide in the microelectrode. In this cell a tetanus produces LTP. (After R. Malinow et al., 1989. Science 245: 862–866.)

Ca^{2+}-activated protease, calpain I (Chapter 7). In the presence of Ca^{2+}, purified calpain can hydrolyze the regulatory domains of both C kinase and type II CaM kinase to generate constitutively active catalytic fragments. Therefore, calpain activation in hippocampal synapses might lead to an increase in constitutive protein kinase activity. An appealing aspect of this mechanism of activation is its irreversibility; only further degradation of the constitutively active fragments and synthesis of new, regulated kinase molecules could reverse activation of the kinases by proteolysis. Therefore, specific local proteolysis by calpain could generate long-lasting changes in the rate of protein phosphorylation. Calpain I can be activated in neurons by intense stimulation with excitatory agents, but it is not known whether it is located at postsynaptic sites where it could be easily activated by the increases in Ca^{2+} concentration that lead to induction of LTP.

Long-lasting adaptation of the nervous system and memory storage require specific regulation of gene expression in mature neurons

In slices of hippocampus, LTP begins to decay within a few hours. In the animal, however, LTP can last for days to weeks. The longer lasting forms of LTP, as well as memory storage in the neocortex, are widely believed to require alterations in gene expression that lead ultimately to enduring or even permanent changes in cellular and synaptic physiology. Changes in gene expression in response to environmental stimuli must initially be triggered by events at the synapses themselves. The ways that synaptic events give rise to signals that are transmitted to the nucleus are only beginning to be understood (Chapter 10) and are an active area of study in the *Aplysia* nervous system and in the mammalian brain. The answer to this puzzle seems likely to provide us with an understanding at the molecular level of the most stable forms of memory storage at synapses.

Summary

The plasticity of the nervous system determines the ability of an organism to change its behavior in response to stimuli from the external and internal environments. Recent experiments in three favorable preparations reveal cellular and molecular changes that influence behavior. In the stomatogastric ganglion, which controls the movements of the crustacean stomach, a defined network of neurons generates a rhythmic motor output based on the intrinsic activity of a pacemaker cell and on a pattern of reciprocal inhibition between neurons that innervate antagonistic muscles. Amines and peptides secreted by modulatory neurons whose endings branch throughout the ganglion alter the rhythmic output to produce a rich variety of motor patterns. The modulating agents act by changing the membrane properties of individual neurons and by changing the strength of synaptic relations between them.

In the defensive gill- and siphon-withdrawal reflexes of the sea hare, *Aplysia,* specific patterns of stimuli to the skin cause well-defined and sometimes long-lasting changes in the behavior of the reflex that can be related to changes in synaptic transmission between identified cells. Habituation of the gill-withdrawal reflex is often associated with depression of synaptic transmission between sensory and motor neurons mediating the reflex, whereas sensitization of the reflex is associated with facilitation of transmission. One mechanism for facilitating transmission is initiated by the action of serotonin on presynaptic endings. Serotonin increases cAMP, which, acting through the cAMP-dependent protein kinase, inactivates a potassium channel, thus broadening the action potential. Increased calcium entry into the terminal then increases transmitter release. Repeated noxious stimulation of the skin causes a long-lasting facilitation of synaptic transmission (up to 24 hours) that requires protein synthesis and is associated with persistent activation of the cAMP-dependent kinase caused by a prolonged decrease in the concentration of kinase regulatory subunits.

In the hippocampus, a part of the mammalian brain that is required for early stages of memory formation, synaptic transmission can be enhanced for hours or weeks following a brief period of high-frequency stimulation. Furthermore, weak stimuli at one set of synapses on a cell can become more effective in causing the enhancement by pairing them with stronger stimuli at other nearby synapses. This phenomenon, called long-term potentiation, provides a possible cellular model for associative memory. The pairing of presynaptic activity with strong postsynaptic depolarization increases the entry of calcium into postsynaptic sites through the channels of NMDA-type glutamate receptors. The increased calcium induces long-term potentiation by a mechanism that is not fully understood, but appears to involve activation of postsynaptic protein kinases. The persistent change in synaptic strength may be caused by biochemical alterations in the presynaptic transmitter release mechanism, in postsynaptic receptors, or both.

The examples illustrate that neural systems adapt to the changing demands of their environment by modulating both the intrinsic membrane properties of neurons and the strength of their synaptic connections. These adaptations can occur at all levels including sensory input, association areas, and motor output.

References

Crustacean stomatogastric system

Dickinson, P. S., Mecsas, C. and Marder, E. 1990. Neuropeptide fusion of two motor-pattern generator circuits. Nature 344: 155–158.

*Harris-Warrick, R. M. and Marder, E. 1991. Modulation of neural networks for behavior. Annu. Rev. Neurosci. 14: 39–57.

Katz, P. S. and Harris-Warrick, R. M. 1990. Actions of identified neuromodulatory neurons in a simple motor system. Trends Neurosci. 13: 367–372.

Marder, E. 1987. Neurotransmitters and neuromodulators. In A. I. Selverston and M. Moulins, eds. *The Crustacean Stomatogastric System*. Springer-Verlag, Berlin, pp. 263–300.

Modulation of Aplysia *sensory neurons.*

Carew, T. J. 1989. Developmental assembly of learning in *Aplysia*. Trends Neurosci. 12: 389–394.

*Goelet, P., Castellucci, V. F., Schacher, S. and Kandel, E. R. 1986. The long and short of long-term memory—a molecular framework. Nature 322: 419–422.

Greenberg, S. M., Castellucci, V. F., Bayley, H. and Schwartz, J. H. 1987. A molecular mechanism for long-term sensitization in *Aplysia*. Nature 329: 62–65.

Kandel, E. R. 1976. *Cellular Basis of Behavior*. W. H. Freeman, San Francisco.

*Shuster, M. J., Camardo, J. S., Siegelbaum, S. A. and Kandel, E. R. 1985. Cyclic AMP-dependent protein kinase closes the serotonin-sensitive K^+ channels of *Aplysia* sensory neurons in cell-free membrane patches. Nature 313: 392–395.

Long-term potentiation

Bliss, T. V. P. 1990. Maintenance is presynaptic. Nature 346: 698–699.

*Kennedy, M. B. 1989. Regulation of synaptic transmission in the central nervous system. Long-term potentiation. Cell 59: 777–787.

*Madison, D. V., Malenka, R. C., and Nicoll, R. A. 1991. Mechanisms underlying long-term potentiation of synaptic transmission. Annu. Rev. Neurosci. 14: 379–397.

Malinow, R., Schulman, H. and Tsien, R. W. 1989. Inhibition of postsynaptic PKC or CaMKII blocks induction but not expression of LTP. Science 245: 862–866.

Morris, R. G. M., Anderson, E., Lynch, G. S. and Baudry, M. 1986. Selective impairment of learning and blockade of long-term potentiation by an N-methyl-D-aspartate receptor antagonist, AP5. Nature 319: 774–776.

Zola-Morgan, S. M., and Squire, L. R. 1990. The primate hippocampal formation: evidence for a time-limited role in memory storage. Science 250: 288–290.

15

Molecular Approaches to Diseases of the Nervous System

Xandra O. Breakefield

One of the most exciting dimensions of modern neurobiology is understanding the molecular basis of diseases of the nervous system. Several factors have made neurological and psychiatric diseases particularly difficult to understand, including the cellular and functional complexity of the mammalian brain, its limited accessibility, and our incomplete knowledge of the molecules important for neuronal functions. Nevertheless, with new tools of molecular neurobiology and genetics, we appear to be on the threshold of great advances in this area. Eludicating the molecular and cellular basis of disease will be helpful not only in devising therapies, but also in understanding normal functions in the nervous system. Human diseases are particularly important in the latter context as they cover a broad spectrum of disabilities, including those involving mood, learning, and speech, that are not easily evaluated in animals. Further, affected individuals themselves can monitor subtle changes in behavior that would be hard to measure directly.

A disease state can be caused by a defective gene or an environmental insult. Understanding the primary cause is critical to unraveling the cascade of events producing the symptoms and neuropathology. The progression, type, and severity of symptoms are, in turn, influenced by the interaction of additional genetic and environmental factors. Some diseases result from abnormal development of the nervous system, such as the generation of too few or too many neurons, improper neuronal migration, formation of inappropriate synapses, or poor myelination. In other diseases, the structure of the nervous system is normal, but neuronal function is altered by external agents such as drugs or viruses, or internal factors such as hormones, that change the physiological state of neurons. In many of these diseases, damage and death of specific neurons occurs over time. In order to prevent or ameliorate a disease, it is thus necessary to understand the inciting event, the development and physiology of the neurons involved, and the genetic and environmental factors that influence these neurons.

The application of molecular techniques has led to rapid progress in investigation of inherited diseases. In muscular dystrophy and cystic fibrosis, for example, molecular genetic analysis has resulted in the identification of new gene products that, when defective, cause disease. In other cases, such as Huntington disease and Alzheimer disease, genetic loci for the diseases have been mapped to specific regions of chromosomes. This information, in turn, provides a means to identify the responsible genes. A wide range of diseases in neurology, as well as several important psychiatric diseases such as manic-depressive illness and schizophrenia, have strong genetic components and are amenable to similar analyses. The study of diseases through genetics will allow us to discover new neural proteins, to understand functional domains of their structure, and to unravel their interaction with other proteins in affecting physiological functions.

Genetic Diseases of the Nervous System

About 30,000 genes are expressed specifically in the nervous system, but the gene products for only a few thousand of these are known. Although dysfunction of at least 1000 genes can cause neurologic disease, the responsible genes for only about 30 inherited neurological diseases have been identified. Thus, the study of genetic diseases of the nervous system is in its infancy. Current molecular biological techniques make it possible to elucidate the molecular basis of many of these diseases and to create animal models for them. Study of these models will likely lead not only to therapies, but also to an understanding of the normal function of these genes in the nervous system.

The position of genes on chromosomes can be mapped

Even when the pathophysiology of a hereditary disease is not understood, the responsible genes can often be isolated based on their location at a specific "address" in the human genome. This address is found by locating the position of the genes relative to nearby genes on the same chromosome.

The total human genome consists of about 3×10^9 nucleotides (or base pairs, abbreviated bp), and contains about 100,000 genes. These genes comprise 10–30% of the DNA sequence; the remainder consists of large expanses of repeat elements and other elements of unknown function. Genes range in size from 1 to 200,000 kilobases (kb), with an average size estimated at 10 kb. Within each gene, the DNA that encodes proteins is fragmented, with bits and pieces of coding sequences (exons) interspersed within noncoding sequences (introns) and flanking regions that control gene expression. In humans, the genes are divided among 22 autosomes and the X and Y chromosomes, which determine sex. Within each chromosome, the genes are arranged in a linear sequence.

Genes that are near each other on a chromosome tend to be inherited together in future generations. In the formation of gametes, homologous chromosomes pair during meiosis, and exchange information through

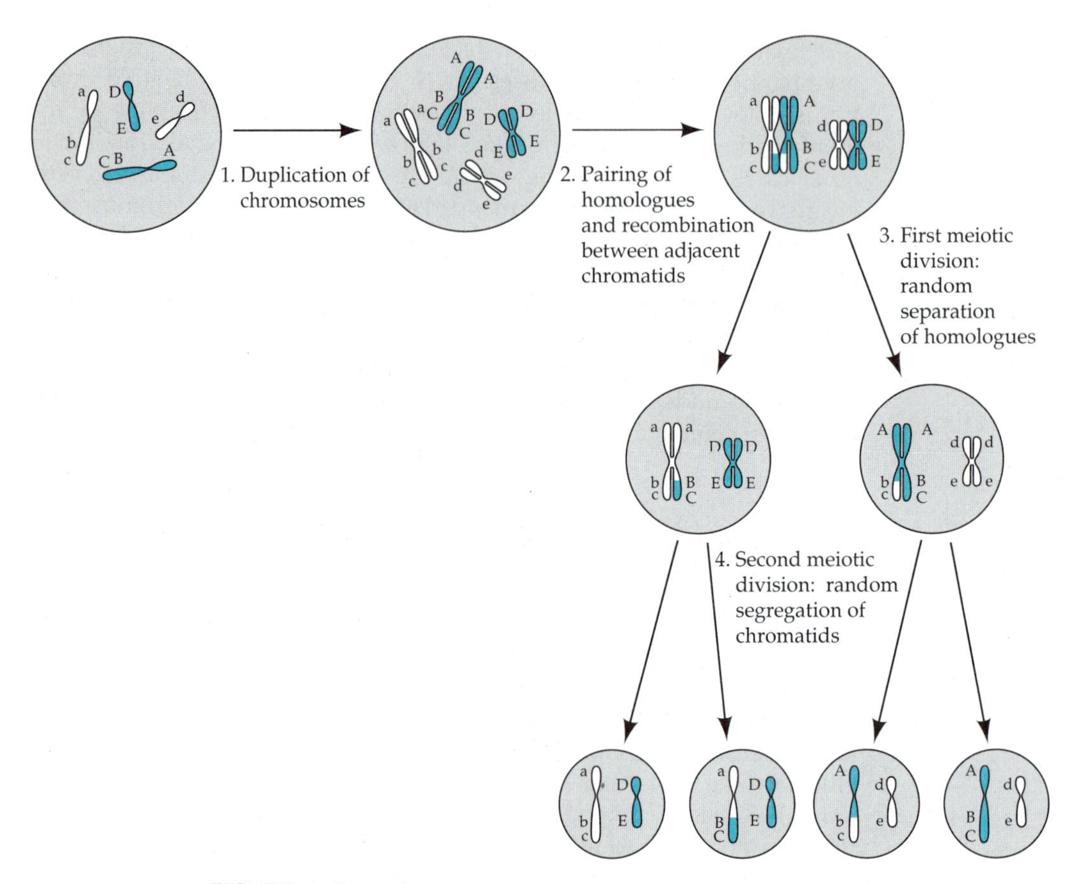

FIGURE 1. Recombination at meiosis. During the formation of gametes, each chromosome first duplicates (forming two chromatids) and condenses (1). Then two homologous chromosomes (same size, one colored and one white), pair with each other, and crossover events (recombinations) occur between homologous chromatids (2). This results in a recombined array of genetic information on individual chromatids, shown here by the realignment of three loci represented as alleles *a, b,* and *c* on one chromosome and as *A, B,* and *C* on its homologue, which recombine to *a, B, C* and *A, b, c,* respectively. At the first meiotic division, homologous chromosomes segregate (3). At the second meiotic division, duplicate chromatids segregate, yielding cells that contain only one copy of each homologous chromosome (4). (After A. Blumenfeld, personal communication.)

breakage and recombination (Figure 1). The distribution of genetic information to gametes can be followed by distinguishing the two **alleles** or variants for the same gene present on homologous chromosomes. Alleles for genes that are far apart on a chromosome are more likely to be separated by recombination events and thus to end up in different gametes than are those that are close together, or linked. The frequency with which recombination events separate two genes is related to the amount of DNA between them. A recombination frequency of 1%, which is equivalent to 1 **centiMorgan** (1 cM; or a recombination unit (θ) = 0.01), means that two genes recombine on average once in every 100 meiotic events. One cM is equivalent to about 1000 kb of DNA sequence. Analysis of recombination frequencies among genes establishes their linear order and the distances between them. Such a linkage analysis, extended over an entire chromosome, gives a **genetic map** of the chromosome. Different regions of the genome vary dramatically in their tendency to recombine, and, thus, genetic and physical maps have the same gene order but show different relative distances between genes (Figure 2).

Genes can be mapped using either phenotypic traits or DNA sequence polymorphisms

In order to map genetic loci on chromosomes, one must be able to distinguish different variants of these loci on homologous chromosomes. Dif-

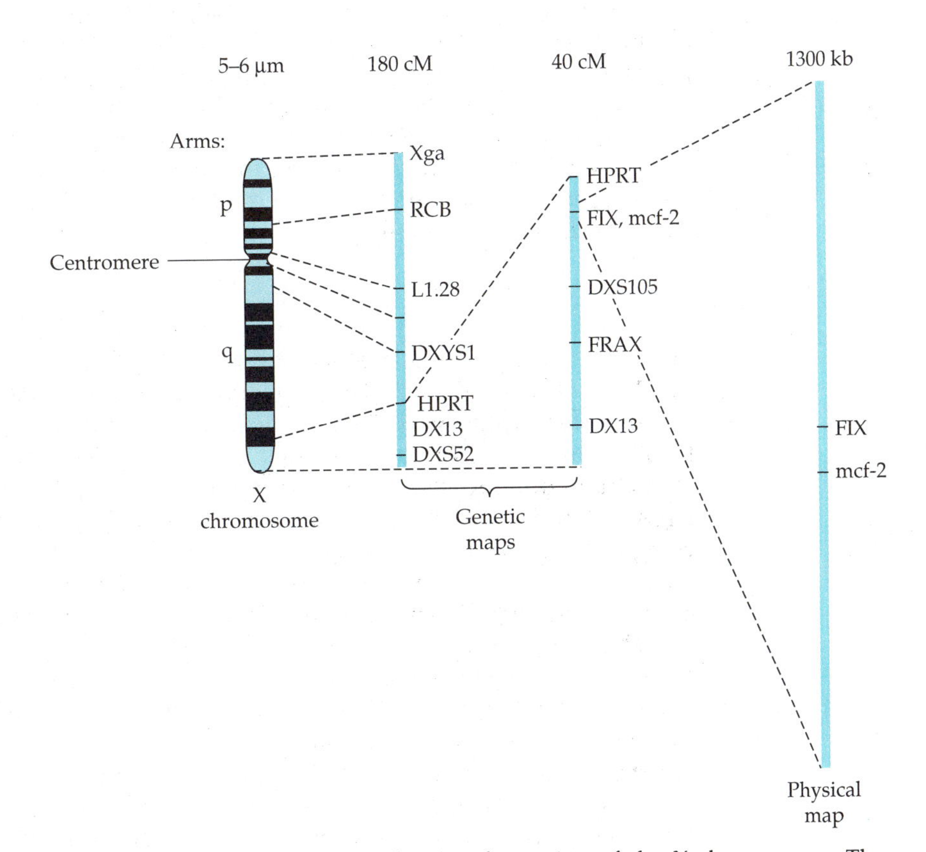

FIGURE 2. Genetic and physical map of a region of the X chromosome. The relationship between loci is measured in genetic maps as the frequency of recombinations between them, and in physical maps as the distance in nucleotides. These maps have the same linear sequence, but there is no direct correlation of distances. The human X chromosome is shown on the left in metaphase with its characteristic Giemsa banding pattern. To the right of it are two genetic maps giving the distance between loci in cMs, with progressively finer detail in the telomeric region of Xq. Two loci, *FIX* (factor 9) and *mcf-2*, are tightly linked. The physical map of large genomic fragments was resolved by digestion of these regions with restriction enzymes, which recognize rare sites (e.g., *Bss*HII and *Eag*I), and hybridization to probes for these loci. These procedures generate a long-range restriction map and place these loci near each other in the same 1300-kb fragment. (After C. Nguyen et al., 1987. EMBO J. 6: 3285–3289.)

ferent alleles can underlie discernible physical traits, different physical or biochemical properties of a protein, or detectable variations in DNA sequence. These normal variations, termed **polymorphisms** if they occur in at least 1% of alleles, result from errors in DNA replication or repair and are passed from generation to generation. These variations are phenotypically silent if they cause no change in amino acid sequence or produce a change that has no consequence to protein function, or if they fall outside of the coding and critical regulatory sequences of a gene.

Even though these sequence variations may have no phenotypic effect, they can be used for linkage studies in the same way as are traditional **genetic markers** (any gene or locus that is used as a reference point). Because sequence variations occur frequently throughout the genome and can be detected at the DNA level, they have expanded enormously the power of linkage analysis and made it possible, in principle, to have markers anywhere on the chromosomes. With about 200 markers for DNA polymorphisms spaced evenly across the genome, any unknown gene would lie within 10 cM of a marker. In other words, at least 90% of the time, a specific marker would be inherited together with the disease gene and thus define its location.

Polymorphisms can be detected by restriction enzymes or by analysis of repetitive elements

There are two types of normal variations in DNA sequence that are commonly used to mark loci. The first type, termed **restriction fragment length polymorphisms (RFLPs)**, is detected by variations in the size of DNA fragments generated following digestion of genomic DNA with **restriction endonucleases**. These bacterial enzymes cut double-stranded DNA at specific 4–8 bp sequences. A restriction enzyme with a 6-bp recognition site will cut the genome into about one million fragments of varying lengths. The marker fragments can be resolved on the basis of size by gel electrophoresis and identified, following denaturation to a single-stranded state, by hybridization to a labeled, homologous DNA probe (Figure 3). On average, two alleles differ in one nucleotide per 100–500 nucleotides in noncoding DNA sequences. Some of these differences determine whether the sequences are cut by a particular restriction enzyme and thus influence the size of DNA fragments between cutting sites (Figure 4). The second type of polymorphism involves variations in the number of times short DNA sequences are repeated in particular regions of the genome. These **repeat elements** almost always occur in noncoding regions and can involve from 2 to 30 nucleotides, tandemly repeated from two to 500 times. Some of the common repeat elements occur frequently throughout the genome, as, for example, the Alu repeat, which occurs about once every 5 kb, or the dinucleotide (dGdT or dCdA), which occurs about every 50 kb. The latter are frequent enough that they can be used to mark essentially any gene of interest (Figure 5). Repetitious sequences are usually inherited stably across generations and yet are so highly variable among individuals that they form the basis of DNA fingerprinting, allowing individuals to be uniquely identified by repetitive elements at a num-

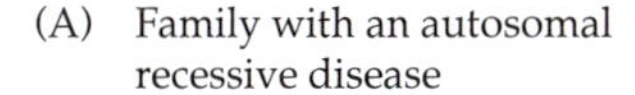

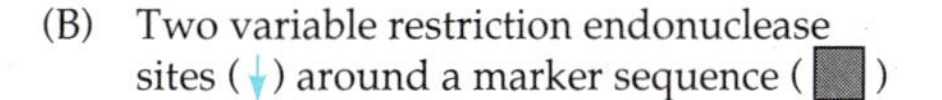

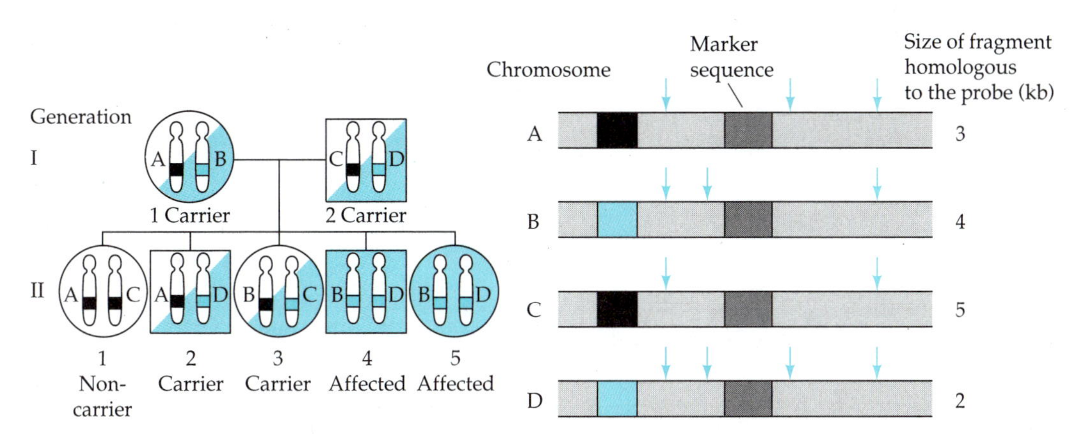

(C) Autoradiogram

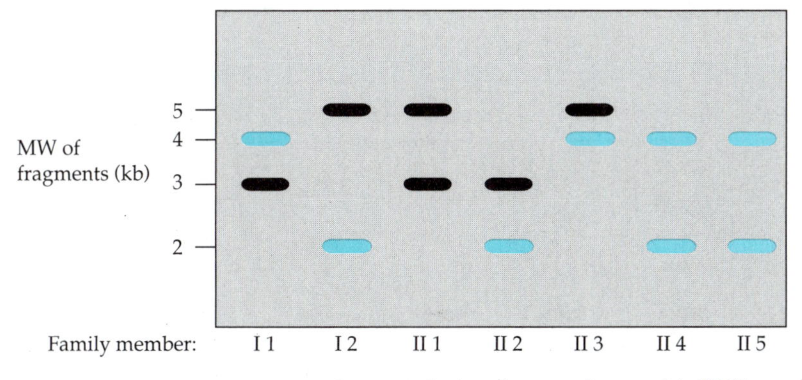

FIGURE 3. Linkage analysis using a polymorphic DNA marker. This is a theoretical diagram of (A) the inheritance of a mutant allele that causes an autosomal recessive disease and (B) and (C) the use of restriction-fragment–length polymorphisms near the mutant allele to identify its presence in family members. (A) A family in which two children (II4 and II5) have the disease and both parents (I1 and I2) and two children (II2 and II3) are carriers of the mutant allele. Marker alleles near the gene on each homologous parental chromosome are indicated by *A, B, C* and *D* (disease gene, colored square; normal gene, black square). (B) Theoretical positions of specific restriction endonuclease cleavage sites in the DNA from different chromosomes around the region of the marker are shown (↓), along with sizes of DNA fragments generated by cleavage and digested with an endonuclease. (C) DNA fragments are separated on the basis of molecular weight in agarose gels. The position of the fragments that contain the responsible gene is determined by hybridization to a labeled probe that is homologous to the marker gene, followed by autoradiography. The marker allele contribution from each chromosome of each parent can be distinguished on the basis of the size fragment within which it is contained. In this example, both affected children inherit the same fragment sizes, and none of the unaffected children has both these same fragments, indicating that the marker is likely linked to the disease gene. (After M. B. Rosenberg et al., 1985. Prog. Neurobiol. 24: 95–140.)

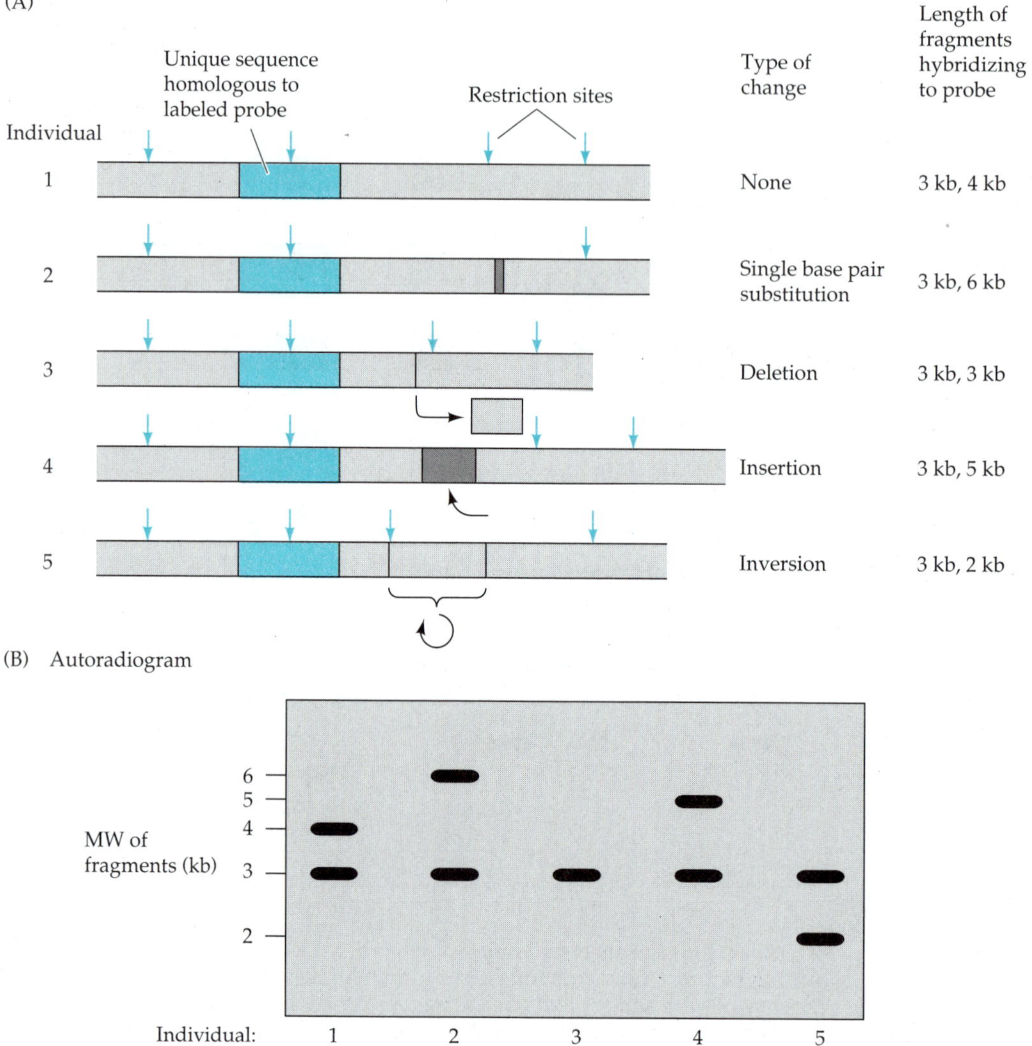

FIGURE 4. Variations in DNA sequence underlying RFLPs. The size of DNA fragments in a particular region generated by digestion with restriction enzymes varies as a result of single nucleotide differences in the enzyme recognition site, as well as of other more extensive changes in DNA in the region, such as loss, inversion, or insertion of DNA sequences. (After M. B. Rosenberg et al., 1985, Prog. Neurobiol. 24: 95–140.)

ber of different loci. In addition to these two types of variations, small differences in DNA sequences among alleles can also affect the mobility of fragments containing the sequences in an electric field, their sensitivity to denaturation, and their ability to be cleaved by chemicals. Thus, polymorphisms at almost any locus can be detected.

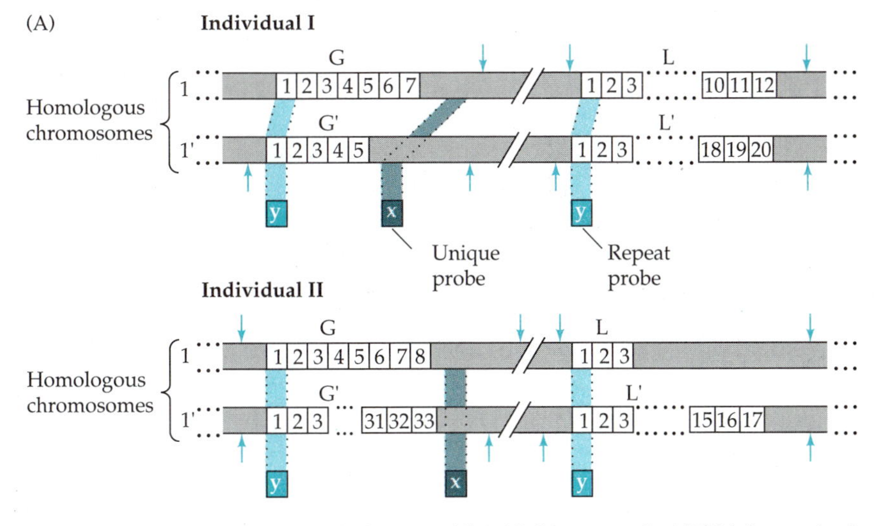

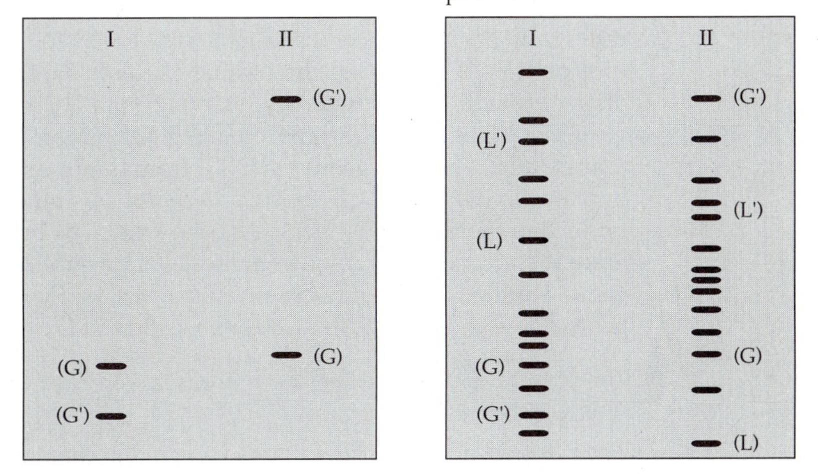

FIGURE 5. Use of repeat elements in the genome to distinguish alleles at one or more loci. Repeat elements are scattered through the genome and contain tandem copies of the same DNA sequences repeated a variable number of times in different positions. (A) The same repeat unit (open box) repeated a variable number of times (numbers in boxes) is present in alleles at two loci (*G* and *L*) on two homologous chromosomes (1 and 1′) from two individuals (I and II). By using a restriction enzyme that does not have a recognition site (arrows) within the repeat unit, one can obtain fragments whose length varies proportionally to the number of times the unit is repeated. (B) Fragments resolved by gel electrophoresis (see Figure 4) are hybridized to a cloned probe (x) that recognizes a unique sequence flanking one of the repeat loci (*G*). This analysis reveals a polymorphism at a single locus, which is highly informative due to high variability in the number of unit repeats among alleles. (C) If the same genomic fragments are hybridized to a cloned probe that recognizes the repeated unit itself (y), multiple loci in the genome will be visualized simultaneously, each of which will likely have two distinguishable alleles. The pattern of fragment sizes bearing the same repeat element can be unique for an individual and thus represents a DNA "fingerprint." Bands corresponding to loci G and L are indicated; other bands correspond to other loci scattered throughout the genome.

Disease may provide clues to the location of the gene

In some cases, certain aspects of a disease help to identify the chromosomal location of the responsible gene. For example, if the disease almost always affects males and is not passed from father to son, it is likely to be caused by a gene on the X chromosome. In some cases the disease state may be associated with cytogenetically visible alterations in chromosome structure, such as a **translocation** that joins different chromosomes to each other, or the loss (**deletion**) of part of a chromosome. Because such changes in DNA structure can cause a disease by disrupting the responsible gene, they can be used to locate the disease gene. Almost all disease genes found to date have been marked by such gross alterations in chromosome structure. Somatic mutations can also give clues to gene location. For example, hereditary cancers are frequently caused by a germ-line mutation in one allele of an autosomal gene that normally acts to inhibit cell growth (a tumor suppressor gene), with subsequent loss of the normal allele in a susceptible cell type during somatic growth. In these cases, tumor cells differ from normal cells of the same individual by loss of the chromosome, or a portion of it, bearing the normal allele. If the individual bears two distinguishable alleles (**heterozygosity**) at a marker locus near the disease gene, loss of the critical chromosomal region will result in retention of only a single allele (**homozygosity**) in DNA from tumor cells. In addition, some human diseases have genetic counterparts in mutant mice, where the position of the responsible gene has been located by breeding experiments. Since the mouse and human genomes contain similar subsets of ordered genes, shuffled in different chromosomal arrays, the location of a human gene can often be predicted from its location in the mouse genome.

The chromosomal region containing the disease gene can be defined by linkage analysis

The position of an unknown gene can be determined by finding marker loci that tend to co-inherit with it through meioses. The segregation of markers and phenotypic traits is assessed in families by **linkage analysis**. This search requires determining whether transmission is **autosomal** or **X-linked**, and whether it is **dominant** (in which case affected individuals have one mutant copy and one normal copy of the responsible gene) or **recessive** (wherein affected individuals have no normal copy of the gene). Other factors that must be considered in the analysis include the frequency in the population of the disease gene and marker alleles, the age of onset of the disease phenotype, and the **penetrance** of the disease gene (the probability that an individual carrying the mutant gene will have the disease). Since co-inheritance of a disease gene and a marker is a probabilistic event, the significance of linkage results must be evaluated statistically. This is done by calculating the **lod score**, or the logarithm of the ratio of the likelihood that two loci are linked at a given distance to the likelihood that they are not linked (Box A). The advantage of using logarithmic values is that they can be summed among families in which the same genetic trait is being followed. Lod scores are calculated at a series

Box A Lod Score Analysis

The purpose of calculating the lod score is to statistically evaluate a set of linkage data between two markers by comparing the probability of obtaining a particular pedigree at a given recombination distance (θ) between the two markers with that of obtaining the same pedigree when there is no linkage between the two markers ($\theta = 0.5$). These probabilities depend on two experimental values: n, the total number of offspring in the F_1 generation; and r, the number of offspring showing recombination between the two markers. Lod scores from different pedigrees can be summed.

The probability P of obtaining a pedigree of r recombinants among n offspring in the F_1 generation at recombination distance θ is calculated as

$$P(\theta) = \tfrac{1}{2}[\theta^r(1 - \theta)^{n-r} + \theta^{n-r}(1 - \theta^r)]$$

The lod score, Z, is the log of the ratio $P(\theta)/P(0.5)$:

$$Z = \log \frac{P(\theta)}{P(0.5)}$$

Z is then calculated for a series of values of θ. A plot of lod score versus θ for three hypothetical pedigrees a, b, and c appears as follows:

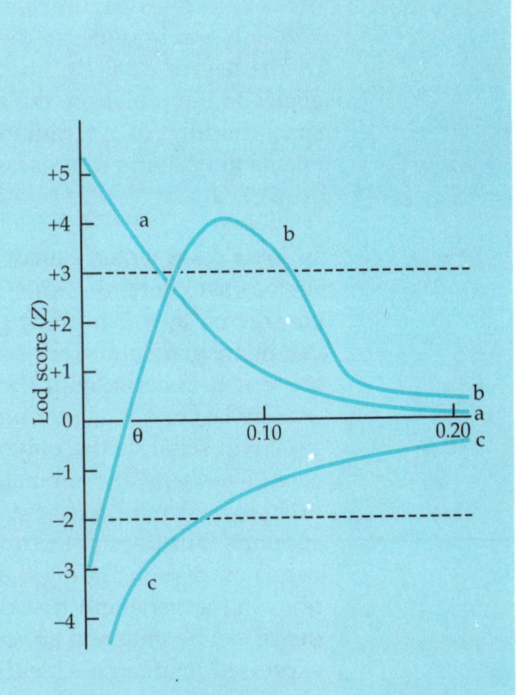

A value of $Z = +3$ (upper dashed line) is considered significant for linkage; a value of $Z = -2$ (lower dashed line) is considered significant against linkage. Thus in case a, the two loci are likely to be within 2 cM of each other; in case b, the two loci are likely to be within 5–12 cM of each other; and in case c, the two loci tested are likely to be more than 6 cM apart.

of recombination distances between loci; scores above +3 (1000:1 odds) indicate linkage, while those less than −2 are taken as proof against linkage.

Once a marker linked to a disease gene is found, other markers in the region are tested to confirm the location. Eventually the position of the disease gene is defined to a 1–2 cM area delimited by flanking markers that show recombination with the disease gene. This is about the smallest area that can be resolved by linkage analysis, as over 100 meiotic events must be analyzed to determine 1% recombination (1 cM). A 2-cM interval corresponds on average to 1000–2000 kb and contains 30–60 genes. In some diseases the location of the gene can be further restricted by demonstrating the association of specific marker alleles with the disease gene. This phenomenon, termed **allele association**, or **linkage disequilibrium**, results from the fact that the mutation originally took place within a particular set of marker alleles (Figure 6). Those alleles that are very close to the mutation will only rarely undergo recombination events that dissociate them from it. The degree of allele association varies with respect to the nearness of alleles to the mutation, the number of meiotic events that have occurred since the time of the mutation, and the number of different mutational events underlying the disease state.

Identification of disease genes involves multiple strategies

In some cases, linkage analysis places the disease gene in a chromosomal region that contains a known gene that, by virtue of its function, can be considered as a **candidate gene**. A simple way to include or exclude a candidate gene in further analysis is to mark it with a polymorphism and see if it co-inherits with the disease state. If the disease gene is not unusually large, it is highly unlikely that a recombination event within the gene will separate the polymorphism from the mutation causing the disease. Thus, any recombination event observed in affected families will exclude the candidate gene. If no recombination events are observed, this supports, but does not prove, the identity of the disease gene. Proof depends on demonstrating that affected individuals have damaging mutations in the gene and that control individuals do not. Incriminating evidence can be obtained either at the mRNA or gene level. If the mRNA is expressed in an accessible tissue, such as lymphocytes or skin fibroblasts, altered expression can be evaluated in patients by Northern blot analysis. Many mutations do not disrupt expression of the gene, however, but lead to structural alterations in the message that affects the function or stability

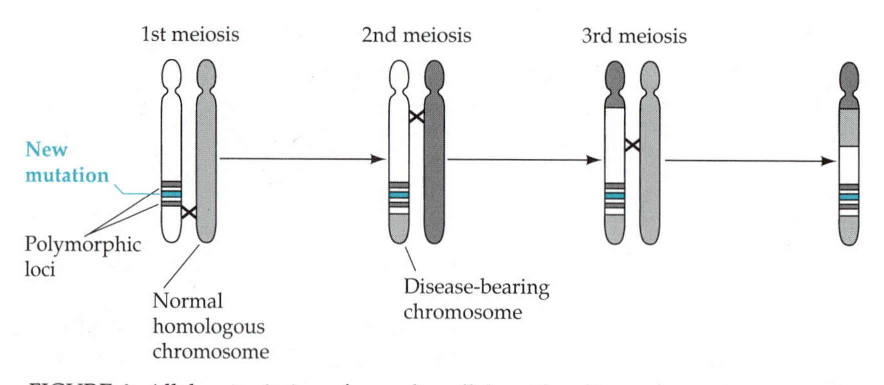

FIGURE 6. Allele association of a marker allele with a disease gene. At some point in evolution a mutation (blue line) can take place that produces a disease gene. This mutation is a specific event that occurs on a particular DNA background of allelic polymorphisms. Through subsequent meiotic events recombinations will occur throughout the chromosome. Allelic markers that are further away from the disease gene will tend to segregate away from it, while those close to it will tend to remain with it. Thus there is an association between the disease gene and particular polymorphic markers near it. The size of this region of association will depend on how many meiotic events have occurred between the mutational event and the evaluation, and the probability with which DNA in this particular region undergoes recombination events.

of the gene product. These mutations can be elucidated by sequencing cDNA generated from mRNA.

Unfortunately, many genes important for neural function are not expressed in cells outside of the nervous system, and intact RNA from brain is difficult to obtain at autopsy. In order to evaluate mutations in neural-specific genes, exons from them must be sequenced using genomic DNA, which can be obtained from peripheral tissues. This sequence analysis requires knowing the structure of the normal gene. Some sequence differences found in genes from patient and control genes may clearly alter function of the encoded protein; others may represent normal polymorphisms. Mutations can also occur in flanking and intron sequences and disrupt expression or processing of mRNA. Changes in these regions are harder to evaluate. Disease-specific alterations in candidate genes have been found in several inherited neurologic diseases (Table 1).

Given our limited knowledge of neural genes, genetic analysis of most diseases of the nervous system is likely to lead to the description of new genes (see Table 1 for examples). Identification of an unknown gene in a large region of DNA (500–2000 kb), delimited only by linkage analysis, is a formidable task. Cloned pieces of genomic DNA must be obtained throughout the region and oriented in a physical map. Although tradi-

TABLE 1. Identification of Human Disease Genes

Disease	Defective	Function
Diseases caused by previously known proteins:		
Pelizaeus–Merzbacher	Proteolipoprotein	Component of CNS myelin
Gerstmann–Straussler[a]	Prion protein	Unknown
Malignant hyperthermia	Ryanodine receptor	Calcium release channel in muscle
Retinitis pigmentosa[b]	Rhodopsin	Photoreceptor pigment
Hyperkalemic periodic paralysis	Muscle-specific sodium channel	Generation of action potential
Dutch cerebral amyloidosis	Beta amyloid	Unknown
Alzheimer disease[c]	Beta amyloid	Unknown
Diseases caused by previously unknown proteins:		
Retinoblastoma	RB	Nuclear phosphoprotein, "tumor suppressor"
Duchenne muscular dystrophy	Dystrophin	Structural protein associated with plasma membrane
Cystic fibrosis	CFTR	Chloride channel or channel regulator
Neurofibromatosis type 1	NF1 peptide	GTPase activating protein

[a] In some familial forms.
[b] In some autosomal dominant forms.
[c] In some familial cases with early onset.

tional cloning and electrophoretic techniques are useful for manipulating DNA fragments in the range of 1–50 kb, new methods have been devised to clone and analyze larger DNA fragments for these gene searches. New cloning vectors include P1 phage and yeast artificial chromosomes (**YACs**), which can carry up to 150 or 800 kb, respectively, of foreign DNA sequence. Restriction enzymes that recognize rare sequences, e.g., 8 bp, can be used to cut genomic DNA into large DNA fragments (50–800 kb), and electrophoretic techniques using large pore agarose gels and alternating electric fields, e.g., pulsed field gel electrophoresis, can be used to resolve these fragments. Large DNA fragments can be aligned in overlapping, contiguous segments (**contigs**) based on homologous sequences, common repeat elements, and similar patterns of restriction enzyme sites (Figure 7). Isolation of discrete regions of human chromosomes is possible by X-irradiation of human donor cells to break up the chromosomes, followed by fusion of these cells with rodent cells to generate **somatic cell hybrids**. Discrete fragments of human chromosomes become translocated onto rodent chromosomes and are propagated along with them in these hybrid cell lines. These hybrids can be used to rapidly assess the genomic location of new clones and provide a source of human genomic DNA from a particular region. Human clones can be identified in genomic libraries prepared from these hybrids by screening for the presence of human-specific DNA repeat elements.

Once a substantial amount of genomic DNA in the critical region has been cloned, the next step is to define the position of genes within it. Several strategies have been used successfully. First, coding sequences tend to be conserved across species, while noncoding sequences are not. Therefore, clones containing exon sequences will hybridize to discrete fragments of genomic DNA from a number of different species (**Zoo blots**). Second, many genes contain CpG-rich elements (islands) in their 5′ regulatory regions that are recognized by certain restriction enzymes, e.g., HpaI and BssHI (Figure 2). Digestion of these regions yields small pieces of DNA, termed HpaI tiny fragments (HTF, hence the term **HTF islands**) which mark the 5′ end of genes. Third, genomic clones from the region of interest can be used to screen **cDNA libraries** prepared from the normal counterpart of the tissues affected in the disease state. Once coding sequences have been identified, they can be used to obtain full-length cDNA clones. Isolation of a particular cDNA can be more or less difficult depending on the size, abundance, and tissue distribution of the corresponding mRNA.

Every gene found in the critical genetic region must be characterized and evaluated as a candidate gene. Characterization includes evaluation of the tissue distribution of the mRNA, as well as its expression and sequence in patients and controls. The size and cellular localization of the encoded protein can be determined by generating antibodies to peptide sequences deduced from the cDNA. Sequence information itself can provide some clues as to protein function through domains of homology with other known proteins. Ultimately the identification of the disease gene depends on finding **mutations** in it associated with the disease state. The more

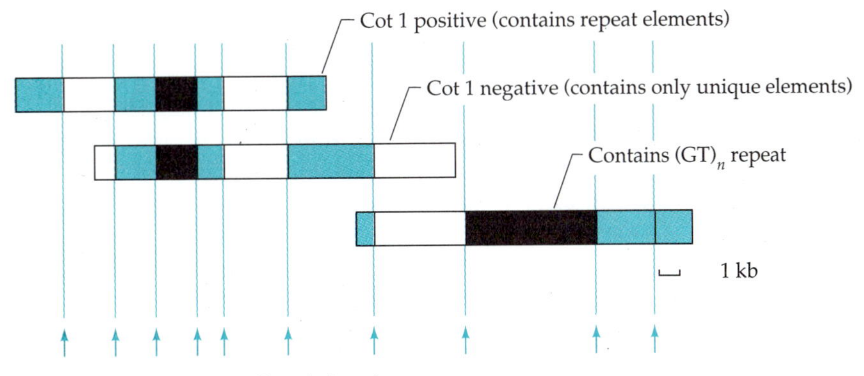

FIGURE 7. Contig map of a part of the human genome. Large segments of the human genome can be cloned in cosmid and yeast artificial chromosome (YAC) vectors (200–800 kb inserts) and aligned by common restriction fragments and repeat elements. Cot1 refers to rehybridization kinetics, with Cot1-positive (color) indicating the presence of repeat elements, which hybridize rapidly, and Cot1-negative (white) representing unique elements, which hybridize slowly. GT-positive means the fragment contains a $(GT)n$ repeat (black). (After R. L. Stallings et al., 1990. Proc. Natl. Acad. Sci. USA 87: 6218–6222.)

different mutations that underlie the same disease state, the more likely it is that one of them will incriminate the gene. In general, the larger the gene the higher the frequency of mutations within it. In recessive diseases, mutations causing **loss of gene function** can involve deletions or rearrangements of substantial amounts of DNA and can thus be readily apparent even when large fragments of DNA are analyzed. In dominant diseases, one abnormal copy of the gene causes the disease. In some cases the abnormality may involve loss of gene function, as in recessive mutations, with half the remaining amount of a protein being too little for normal function. In other dominant diseases, an abnormal protein may be produced that disrupts function of the normal protein. These **gain-of-function** mutations may involve discrete changes in gene structure. The difficulty in detecting these small alterations, plus the fact that affected individuals express both normal and abnormal gene products, can make resolution of the defect difficult.

Disease genes have many different guises

Even when the disease gene has been identified, the function of its normal counterpart and its role in pathogenesis may not be clear. Sequence information obtained through the gene's identification, however, provides a new tool to further elucidate these molecular mechanisms. Several examples of successful gene "hunting" illustrate how such information can be used. **Duchenne muscular dystrophy** is a recessive, X-linked disease characterized by the gradual degeneration of skeletal muscle with eventual paraplegia and death of affected boys in adolescence. The responsible gene was located on the p arm (see Figure 2) of the X chromosome by the observations that some affected males manifest chromosomal deletions in this region and that some rare affected females bear chromosome translocations involving this region. The very large gene (2000 kb) responsible for this disease was identified by conserved exon regions within it and by the many different mutations that disrupt it. Most patients show loss of the mRNA encoding a large skeletal muscle protein, dystrophin. In some cases, deletions yield a smaller mRNA; if the mRNA still codes for the correct amino acid sequences downstream from the mutation, patients have a milder form of the disease, called Becker muscular dystrophy. Dystrophin shares domains with spectrins and alpha-actinin, and, like them, is thought to function as a part of the cytoskeleton associated with the plasma membrane of skeletal muscle fibers. The exact function of dystrophin is not known. It is not needed for muscle function, as move-

Box B Molecular Diagnostics Leads to Cancer Cure

Retinoblastoma is a highly malignant cancer that begins during development of the eye through loss of a gene critical to differentiation of neuroepithelial precursor cells. The disease occurs in young children, and malignant cells are first seen as a white opacity in the eye (Figure A). From there they spread down the optic tract, into the brain and throughout the body. Once the disease has spread, it is universally fatal. Retinoblastoma was first recognized by John Waldrop in 1809. He tracked the origin of the tumor to the retina and recommended removal of the eye as soon as it was detected, a procedure that was extremely painful, as there was no anesthesia at the time. Because there were also no ophthalmoscopes, detection was frequently too late.

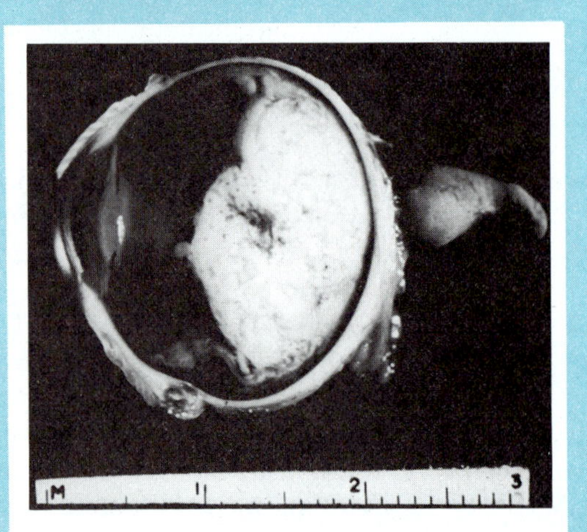

FIGURE A. Retinoblastoma tumor. A cross section of the eye shows a large tumor mass filling the vitreous space. (Photograph courtesy of T. Dryja.)

Early on, this tumor was recognized as a familial disease. In about 65% of cases the tumor arises sporadically in one eye; in another 35% of hereditary cases tumors form in both eyes. From this pattern, John Knudson deduced in 1971 a two-hit theory of cancer development which has provided major insight into how some cancer genes act. This theory states that loss of some genes can lead to uncontrolled cell growth. For autosomal genes this requires two mutational events. In the case of sporadic tumors, two mutational events in the same cell are needed to knock out both homologous genes. In hereditary tumors, all cells carry a mutation in one copy; a second somatic mutation occurs in a susceptible cell type. In fact, a large percentage (10–30%) of mutations predisposing to retinoblastoma are deletions, and a large percentage of the second event represents loss of the chromosome bearing the normal allele. These deletions were critical in locating and identifying the responsible *RB* gene on chromosome 13q (Figure B).

ment is relatively normal in affected boys until about two years of age. Further, in other species, disruption of the equivalent gene does not lead to muscle wasting. In humans, absence of dystrophin appears to trigger secondary responses that cause muscle degeneration. The dystrophin protein is also expressed in brain where it is concentrated at synaptic terminals; loss of this protein then may underlie cognitive impairment found in some patients, again through an as-yet-undefined mechanism.

A very interesting example of a disease gene that is recessive at the cellular level but dominant at the organismal level is hereditary **retinoblastoma**. In this disease patients develop bilateral tumors of the retina, which initially appear in the first few years of life and are highly malignant (Box B). Here again the location of the disease gene was found through rare patients bearing a deletion on the q arm of chromosome 13. Further, tumor

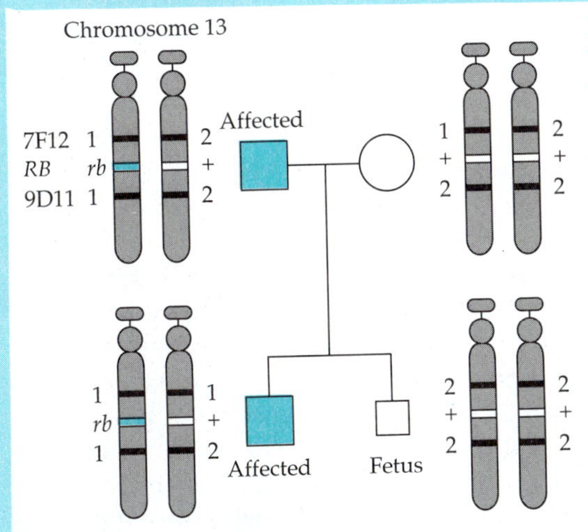

FIGURE B. Molecular diagnosis in family with hereditary retinoblastoma. Allelic status of two markers (7F12 and 9D11) flanking the *RB* gene was assessed in members of a family in which the father and a son had retinoblastomas (color squares). The defective RB gene (*rb*) is flanked by "1" alleles for marker loci in the two affected members. The fetus (arrow) is determined to be a male homozygous for allele 2 at both markers; thus he is not at risk for developing retinoblastoma and would not need to undergo frequent eye exams. (After D. H. Ledbetter and W. K. Cavenee, 1989. *The Metabolic Basis of Inherited Disease*, 6th ed., C. R. Scriver, A. L. Beander, W. S. Sly and D. Valle, eds. New York, McGraw-Hill.)

Once one member of a family is affected with retinoblastoma, others are deemed to be at risk. Using polymorphic DNA markers the mutant gene can be tracked through the family (Figure B). Individuals can be identified in utero or at birth as being susceptible to formation of this cancer. These children are now followed by ophthalmoscopic exams under anesthesia every three months for the first few years of life. This allows tumors to be detected early when they can be killed using focal photocoagulation, laser therapy, or cryotherapy, thus sparing vision. After 4 years of age the risk of retinoblastoma formation is very small and these individuals go on to lead a normal life, although some develop sarcoma tumors through the same mechanism later in life.

This same tumor suppressor gene appears to play a role in many other types of cancers. Several DNA tumor viruses, including adenovirus, simian virus 40, and human papilloma virus, encode proteins that bind and block the action of the normal RB protein, an ability that is critical to their transforming potential. Also, loss of the RB genes is common in small lung cell carcinomas and bladder carcinomas. This protein thus holds secrets to the mechanism of retinal differentiation, and to the means of controlling cell proliferation in many cell types.

tissue from patients who were constitutionally heterozygous for markers in this chromosomal region was frequently found to be homozygous, suggesting somatic loss of alleles in this region. Again, the gene was identified through conserved sequences that were missing in tumor tissue from patients due to very small deletions in this region. The relevant mRNA was absent in retinoblastomas but present in almost every other cell type examined. The protein encoded by this gene is a nuclear, DNA-binding protein that is phosphorylated and appears to be involved in the regulation of cell growth. Although the exact function of this protein is unknown, it is defined as a tumor suppressor; interference with its function has been found in many forms of cancer. Interestingly, in gene carriers a mutation in one copy of this gene has no consequences to most cells of the body. If, however, at a critical time in the development of the eye a neuroepithelial precursor cell loses the normal gene copy, it will fail to differentiate and instead continue to proliferate, forming a highly malignant tumor.

Only a few dominant disease genes have been identified in humans. One can speculate that in order for a mutant protein to interfere with the function of its normal counterpart, it must interact with it to "alter" its structure or change its regulation. A possible example of the former mechanism is a mutant form of the eye pigment protein, rhodopsin, found in individuals with an autosomal dominant form of **retinitis pigmentosa**. Here, the disease gene was located by linkage analysis in a subset of affected families to the q arm of chromosome 3, the same region that bears the rhodopsin gene. Rhodopsin is a photosensitive eye pigment found exclusively in rods of the eye, which function in dim light (Chapter 6). Patients with retinitis pigmentosa have night blindness in adolescence followed by complete loss of vision in adulthood. Sequencing of the rhodopsin gene in these patients revealed single nucleotide changes in regions encoding critical parts of the protein. The mutant gene underlies abnormal light-evoked electrical responses from the retina in otherwise presymptomatic individuals and eventually leads to progressive degeneration of both rod and cone photoreceptor cells. The precise mechanism of degeneration is unknown, but it may result from the gradual accumulation of undegraded mutant rhodopsin and abnormal membranous discs in the rod cells, with secondary responses of the retina to this malformation.

Recent understanding of the defect in **hyperkalemic periodic paralysis** illustrates an example of a dominant mutant gene that causes episodic symptoms triggered by environmental factors. In patients with this disease, elevated levels of potassium in the serum cause severe muscle weakness. Linkage analysis placed this disease gene and the gene for the subunit of a muscle-specific sodium channel in the same region on the long arm of chromosome 17, thereby suggesting a role for this channel in the disease. Subsequent single channel recordings from myotubes cultured from patients demonstrated prolonged sodium channel opening in the presence of elevated potassium concentrations, indicating that the mutant channel did not inactivate normally (see Chapter 2). The sustained influx of sodium through the mutant channels leads to membrane depolarization, which in turn inactivates normal sodium channels. Muscle fibers are thus temporarily unable to contract, as they cannot generate an action potential until extracellular potassium levels decrease. At normal serum potassium levels the mutant channel functions normally. This type of conditional mutation offers a possible model for episodic behavioral disorders that have a strong genetic component, such as manic-depressive illness.

Animal Models

Two classes of animal models can help to unravel the molecular basis of neurological diseases. In one class, the disease phenotype is created in normal animals by exposure to drugs or other environmental conditions. In the other, mutant genes, which disrupt particular functions, are introduced into the genome of animals.

Neurotoxins can cause death of certain neuronal populations

Several drug-induced animal models mimic degenerative diseases in human brain. **Parkinson disease** is a fairly common movement disorder characterized by difficulty in producing voluntary movement. Extensive, age-related loss of dopaminergic neurons in the substantia nigra causes the disease (Figure 8). These neurons project to the striatum, which is the part of the basal ganglia that regulates subconscious aspects of movement. Although in most cases Parkinson disease does not appear to be hereditary, genetic factors may make some individuals more susceptible to it than others. Some motor function can be regained in patients by administration of L-DOPA, a precursor of dopamine (Chapter 5), which presumably acts by increasing the amount of dopamine that is released by surviving neurons. An animal model of Parkinson disease can be created by administration of the compound 1-methyl-4-phenyl-1,2,3,6-tetrahydropyridine (MPTP), which is metabolized in the brain by monoamine oxidase (MAO) to a neurotoxin, MPP^+ (Box C). MAO has two isoforms, A and B, which are together the primary enzymes involved in degradation of amine transmitters in the brain (Chapter 5). The selective death of dopamine neurons induced by MPTP is apparently due to their ability to concentrate MPP^+ through a high-affinity uptake process in nerve terminals normally used for dopamine recovery. MPP^+ in turn, inhibits oxidative phosphorylation in mitochondria and thus compromises cellular energy metabolism. Based on this animal model, Parkinson disease could be caused by exogenous or endogenous MPP^+-like compounds; if so, inhibition of MAO activity could prevent or slow the progression of the disease.

Huntington disease is a relatively rare, autosomal dominant disease characterized by midlife onset of movement disabilities, behavioral disturbances, and eventually dementia. Although the defective gene has been localized to the p arm of chromosome 4 (Box D), it has not yet been identified. Patients have an accelerated loss of specific neuronal populations throughout the brain, with greatest attrition in the caudate and

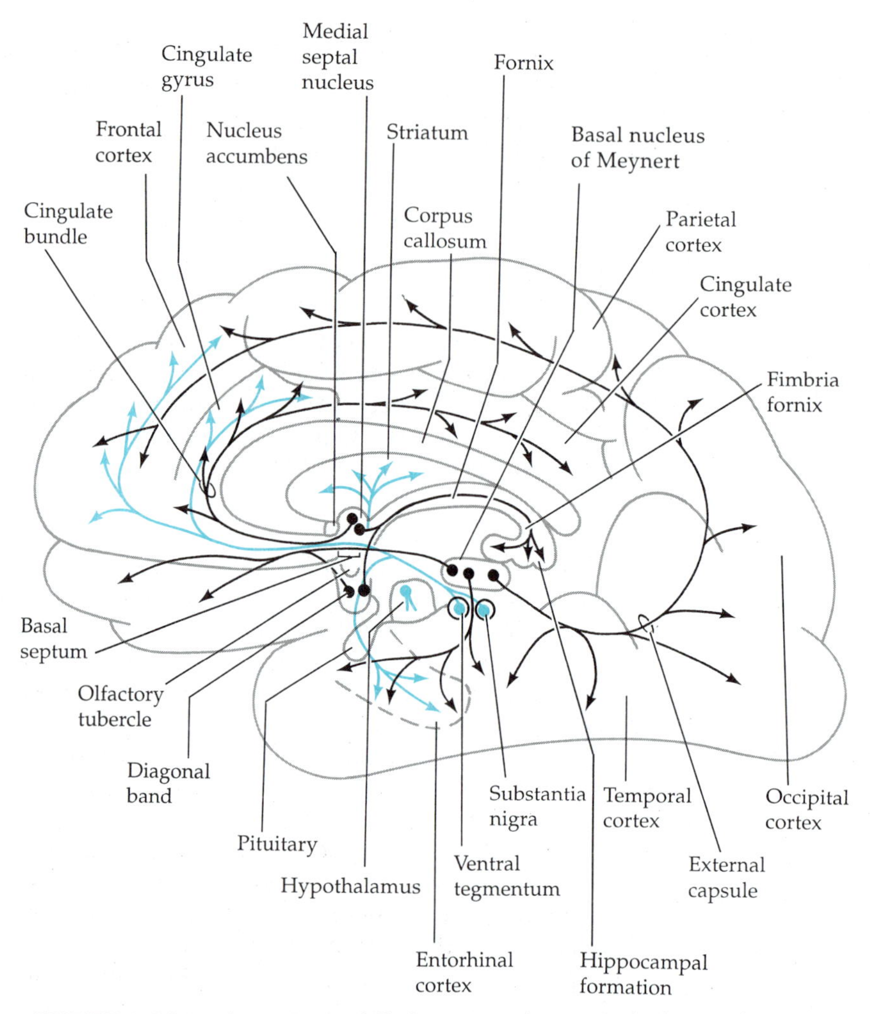

FIGURE 8. Major dopamine and cholinergic pathways in the human brain. Dopamine neurons (color) projecting from the substantia nigra to the striatum are involved in control of movement. These neurons are lost in Parkinson disease. The major cholinergic nuclei (black) are concentrated in the basal forebrain and project to the hippocampus. Some of these neurons are lost in Alzheimer disease. (From Q. Huang, after C. B. Saper, 1987. In *Handbook of Physiology*, F. Plum, ed., American Physiological Society, Bethesda, MD.)

putamen of the basal ganglia. In advanced cases this brain region is reduced to less than 10% of its original volume with concomitant enlargement of the adjacent ventricles (Figure 9A). There are few clues as to why neurons die in Huntington disease. One attractive hypothesis implicates abnormal stimulation of cells by excitatory amino acid transmitters. Glutamate, which is the principle excitatory transmitter in the brain, causes influx of calcium through binding to a class of glutamate receptors called NMDA receptors (Chapter 3). Excessive influx of calcium can cause cell death, possibly through the action of calcium-dependent proteases or ki-

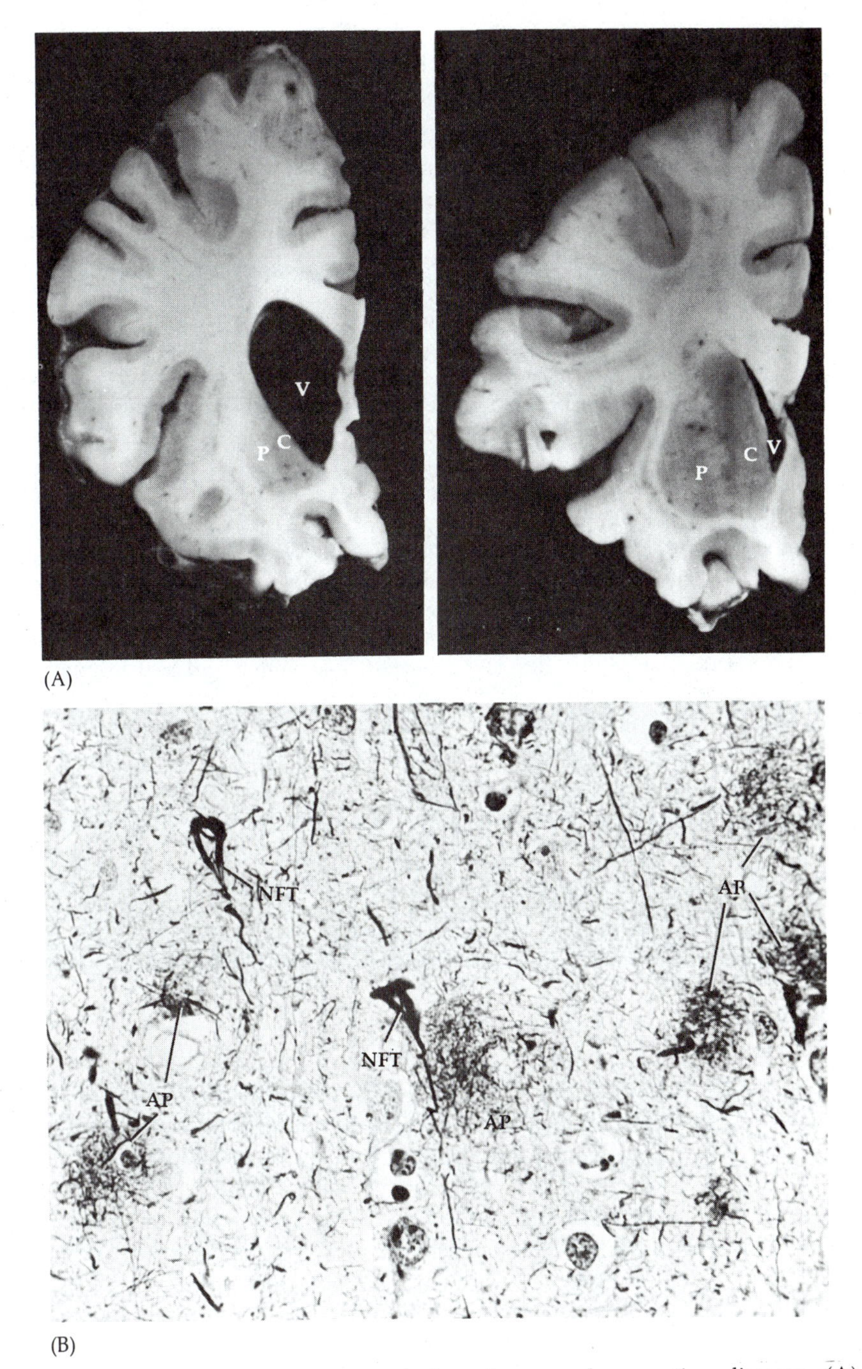

FIGURE 9. Characteristic neuropathology in neurodegenerative diseases. (A) Huntington disease. Coronal sections of the left brain hemisphere from a patient (left) and the left brain hemisphere from a control individual (right). Extensive cell loss in the caudate (C) and putamen (P) of the patient's basal ganglia results in an enlarged ventricle (V). (B) Alzheimer disease. Sections of affected brain regions shows neurofibrillary tangles (NFT) within neurons and associated extracellular amyloid plaques (AP). (Photographs courtesy of J.-P. Vonsattel.)

Box C Street Drugs Provide Clues to Parkinson Disease

Why do particular neurons die in neurodegenerative diseases? A model for what may happen in Parkinson disease comes from an unexpected and unhappy source: the recent development of Parkinson-like symptoms in several young drug users.

The first clue to the cellular and molecular basis of Parkinson disease was the finding by H. Kornycywicz in Austria in 1960 of a dramatic and selective loss of dopaminergic neurons in the substantia nigra. Kornycywicz's observation led to an attempt to improve the function of the remaining neurons by supplying them with the precursor to dopamine, L-DOPA (see Chapter 5). Although the small doses initially tried produced only nausea and vomiting, massive doses (10 gm/day) led to a dramatic improvement in motor function, a finding which is the basis of current L-DOPA therapy for Parkinson disease.

In 1982 physicians in the San Francisco Bay area were baffled by several young patients with neurological symptoms closely resembling those seen in Parkinson disease, which usually affects only the elderly. The symptoms of the patients, who were all heroin addicts, were relieved by L-DOPA. One such patient, who later died of a drug overdose, was found on autopsy to have extensive loss of neurons in the substantia nigra. The responsible agent turned out to be a contaminant in a "designer drug" preparation, 1-methyl-4-phenyl-1,2,3,6-tetrahydropyridine (MPTP). The critical experiment was the finding that administration of MPTP to monkeys produced symptoms similar to those seen in humans with Parkinson disease.

Further investigation showed that MPTP, which is lipophilic and easily gains access to the brain, is not itself toxic, but is rendered so by oxidation to 1-methyl-4-phenylpyridinium (MPP^+). The conversion is catalyzed most efficiently by monoamine oxidase B, a form of the enzyme found in glia (Figure A). Monoamine oxidase A, which is the form of the enzyme found in the dopaminergic cells that die, has a different substrate specificity and catalyzes the reaction less efficiently. Selective inhibitors of MAO-B, but not MAO-A, block the toxic effects of MPTP. How does the toxic MPP^+ get into dopaminergic neurons? Current evidence favors the idea that MPP^+ is selectively concentrated by the high-affinity uptake system for dopamine (Chapter 5) in the nerve terminals of these neurons. Within the

nases. Chronic stimulation of a second class of glutamate receptors, the kainate receptors, leads to influx of sodium into cells, which also causes cell damage by osmotic swelling. Lesions in the basal ganglia resembling those seen in Huntington disease can be produced by stereotactic infusion into this region of kainate or other glutamate analogues, such as quinolinic acid, a normal metabolite of tryptophan. Huntington disease thus may be caused by a genetic defect in the metabolism of compounds that bind to glutamate receptors or in the receptors themselves. There is as yet no proof, however, that excitotoxins are the mode of cell death in this or any other degenerative disease.

Protein deposits can cause neuronal degeneration

Several neurodegenerative diseases are caused by transmissible agents that cause slow, progressive neuronal loss, but without signs of virus or immune response to them. These diseases include **kuru, Creutzfeld–Jakob disease,** and **Gerstmann–Straussler syndrome** in humans, as well as **scra-**

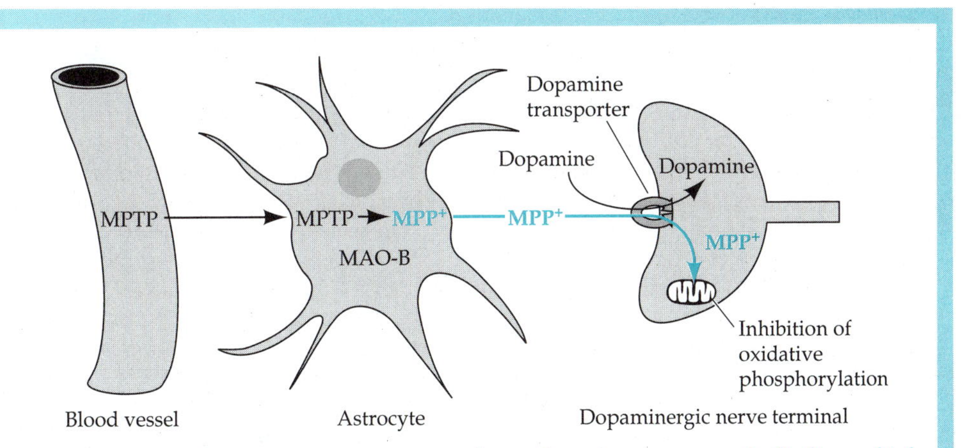

FIGURE A. Toxic pathway of MPTP in the CNS. MPTP crosses the blood-brain barrier and enters neurons, astrocytes, and other cells. It is converted to MPP^+ primarily by MAO-B in astrocytes. MPP^+ is taken up into dopaminergic nerve terminals by a high-affinity dopamine transporter. MPP^+ appears to be concentrated by mitochondria and can inhibit oxidative phosphorylation.

cells, MPP^+ is concentrated in the mitochondria, where it blocks NADH-linked oxidative phosphorylation.

The action of MPTP raises the possibility that Parkinson disease is caused by toxins in the environment. Pyridines related to MPTP are found in the environment both as industrial pollutants and in foods. Low-level exposures over a lifetime could lead to a slow and steady loss of dopaminergic cells that becomes critical late in life when few are left. In laboratory mice raised in a controlled environment, dopaminergic neurons appear to be particularly sensitive to aging, suggesting that oxidation of dopamine or other endogenous compounds may be toxic through production of hydrogen peroxide and free radicals.

pie in sheep and cattle. The neuropathology is characterized by formation of **amyloid plaques** (insoluble protein deposits), spongiform encephalopathy (the appearance of prominent vacuoles in cells), and gliosis (reactive proliferation of glia). Even after extensive efforts, no viral DNA or RNA has been demonstrated in infectious material. A protein has been purified, however, from the scrapie-associated fibrils that constitute the major component of the plaques. Following sequencing and cloning, this protein has been demonstrated to be a normal cell-surface component of many cells, with highest levels in neurons. No function has yet been assigned to this "prion" protein, but a posttranslationally altered form of it is hypothesized to be the neurodegenerative factor in these diseases. This possibility is supported by the demonstration of disease-specific mutations in the prion gene in familial cases of Gerstmann-Straussler and Creutzfeld-Jakob diseases. Further, transgenic mice that overexpress the mutant form of the prion protein spontaneously develop spongiform encephalopathy. It is still not clear how the altered form of this protein could increase the propensity

Box D A Heroic Woman and a Horrible Disease.

Hereditary diseases are so common that they affect almost every family. When such a disease is diagnosed, the family frequently has a great sense of powerlessness. How the human psyche responds to such an overwhelming blow is the stuff heroes are made of. Nancy Wexler discovered at the age of 20 that her family, and possibly she herself, carried the deadly Huntington disease gene, which manifests itself in midlife. Her reaction was to "move mountains" to uncover the identity of this disease gene. She and her father, Milton Wexler, organized the Hereditary Disease Foundation, which since 1968 has been organizing catalytic workshops that bring together some of the most creative scientists in the country to look for the cause of this mysterious disease. In 1980 Raymond White and David Botstein cracked the facade when they published a paper describing a theoretical means to find any gene in the genome using DNA polymorphisms. This approach required no understanding of the nature of the disease gene, just the ability to follow its inheritance through large pedigrees. To locate the Huntington gene—the first effort of its kind—a young molecular biologist, James Gusella, and several distinguished neurologists, Joseph Martin, Anne Young, Jack Penny, and Ira Shoulson, teamed up with Nancy Wexler, a psychologist. Together with Dr. Americo Negrette, they located the largest family in the world with Huntington disease, in the Lake Maracaibo region of northwestern Venezuela. This family has now been traced to include almost 12,000 individuals—of which 258 have the disease, 1227 are at 50% risk, and 2885 are at 25% risk—all the descendants of one woman who lived in the early 1800s. The arrival by boat into this remote area of South America of Dr. Wexler and a team of neurologists bearing videocameras, test tubes, and medical equipment was met with initial distrust by the inhabitants. In spite of the difficulties, which included extreme heat and parasites of every description, the investigators persevered to make friends, help people with medical problems, collect blood samples, and describe the full clinical spectrum of the disease. Their scientific efforts paid off. Using the samples that were sent back, Gusella and his colleagues found that the disease gene was located on a small region of chromosome 4p. Although the gene is not yet identified, genetic markers for molecular diagnosis of carrier status are now available. Moreover, the initial success of this effort has stimulated others who have subsequently identified genes for muscular dystrophy, cystic fibrosis, and other disorders.

of the normal protein to assume an altered state to become the transmissible agent. The possibility is still open that the mutant protein increases cellular susceptibility to an atypical viral infection.

There is increasing evidence that degradation of some proteins can yield peptide fragments that form insoluble amyloid deposits that, in turn, lead to progressive neuronal degeneration. Some familial forms of early-onset **Alzheimer disease** appear to be caused by just such an "**amyloid precursor protein.**" Alzheimer disease is the most common form of dementia in the elderly. Progressive loss of memory and other intellectual functions is accompanied by the appearance of **amyloid plaques** and **neurofibrillary tangles** (Figure 9B). Many different types of neurons die in Alzheimer disease, with greatest losses in the cholinergic neurons of the basal forebrain (Figure 8). These neurons innervate the hippocampus and cortex and are believed to have a major role in information processing and storage.

FIGURE A. Nancy Wexler, who led a team of neurologists and geneticists in a search for the Huntington disease gene. (Photograph courtesy of R. Gumpel and S. Uzzell.)

The peptide present in the amyloid plaques of Alzheimer patients has been purified, cloned, and found to be generated from a larger normal protein of unknown function (different from the prion protein) that is expressed in neurons and other cell types. Some, but not all, familial cases of Alzheimer disease appear to have a disease-specific mutation in this protein, which presumably predisposes it to amyloid formation. A different mutation in the same protein also underlies another familial neurological disease, Dutch cerebral hemorrhage, in which amyloid deposits disrupt blood vessels in the brain. The phenotype of these two diseases is quite distinct, however. As noted above, in Alzheimer disease there is an accumulation of neurofibrillary tangles composed of neurofilaments and microtubule-associated proteins (Chapter 8), presumably reflecting disrupted axonal transport and slow death of neurons. In Dutch cerebral hemorrhage, no neurofibrillary tangles appear, and neurons seem unaffected until death occurs through internal bleeding. Clearly, it will be important to introduce the mutant forms of these amyloid precursor proteins into transgenic mice to verify their roles in these diseases.

Other animal models of neurodegenerative diseases include lesion models in which neuronal processes are severed, thus depriving neurons of neurotrophic substances received from their postsynaptic targets. For example, lesion of the fibria fornix leads to death of cholinergic neurons in the basal septum. Such lesions deprive these cholinergic neurons of an essential supply of nerve growth factor (NGF) produced by the hippocampal neurons that they innervate. In fact supplementation with NGF can prevent death of lesioned cholinergic neurons. This same population of neurons is especially vulnerable in Alzheimer disease, a parallel that suggests that Alzheimer disease may result from a trophic deficit.

Yet other animal models include induction of **autoimmune diseases**. For example, peripheral injections of antigenic portions of myelin basic protein, a component of brain-specific myelin, can lead to antibody-mediated demyelination in the CNS. This model phenotypically resembles **multiple sclerosis,** in which focal demyelination in discrete regions of the brain leads to compromise of neural transmission. Defects in neuronal development can be produced by a variety of means. For example, injection of heterologous NGF into animals leads to production of antibodies that block functions of endogenous NGF. In pregnant animals these antibodies cross the placenta and interfere with development of sensory and sympathetic neurons. Developmental loss of these neurons parallels neuropathological findings in familial dysautonomia, an autosomal recessive condition in humans characterized by poor sensory and sympathetic function. The animal model suggests that a mutation interfering with the action of NGF or a related neurotrophic protein may underlie this disease.

Some spontaneous mouse mutants mimic human diseases

A long-term systematic search for mouse mutants has yielded many with neurological diseases. Most show abnormal movements and have altered cerebellar development, defective myelination, and/or loss of specific neuronal populations. Some of these mutants are authentic genetic models for human diseases. For example, defects in the X-linked gene encoding the proteolipoprotein of CNS myelin (discussed in Chapter 9) result in failure of myelination in *jimpy* mice and in human males with Pelizaeus-Merzbacher disease. Deficiency of the lysosomal enzyme, glucocerebrosidase, results in accumulation of undegraded glycolipids in lysosomes and eventual death of neurons in the *twitcher* mouse and in humans with Krabbe disease. In some cases, however, disruption of a common gene in mice and humans does not lead to the same phenotype. As mentioned above, mutations in the dystrophin gene in humans causes Duchenne muscular dystrophy, a progressive, muscle-wasting syndrome, while similar mutations in mice and cats yield only a mild compromise of muscle function. In another example, loss of an enzyme involved in purine salvage, hypoxanthine phosphoribosyl transferase (HPRT), leads to compulsive self-mutilation and loss of motor coordination—the Lesch Nyhan syndrome—in humans, while mice with the same lesion appear normal.

New mouse mutants can be created

Current advances in molecular biology now make it possible to modify the mouse genome by introducing new genes into it or by disrupting endogenous genes. The mutants thus created can serve as animal models of human diseases where the gene defect is known, or as a means of creating new diseases in order to assess the function of specific genes during the development and life span of mice. DNA can be introduced into the genome by infection of early mouse embryos with retroviruses, or by microinjection of cloned DNA fragments into the pronucleus of the fertilized egg. Exogenous DNA sequences usually integrate at random sites in the mouse genome. The resulting mice are screened for the presence of these foreign sequences and for manifestion of mutant phenotypes. If a disease phenotype is produced by disruption of an endogenous gene, the disrupted gene can be identified by virtue of the foreign sequences embedded within it. Random insertion of retrovirally encoded sequences have led, for example, to the generation of "dystonic" mice. This autosomal recessive condition leads to sustained contractures of skeletal muscle groups, which apparently results from the loss of sensory neurons and hence of peripheral sensory input. Similar movement disturbances are observed in humans with hereditary **dystonia**, but in this case the cause is a dominantly acting gene associated with dysfunction of neurons in the basal ganglia, with apparently normal function of sensory neurons.

Alternatively, the function of a novel gene of interest can be elucidated by inserting it into the mouse genome under the control of a promoter element that confers either ubiquitous or cell-specific expression. For example, viral *onc* genes introduced into mouse embryos can lead to the development of tumors whose type depends on the particular *onc* gene and the cell specificity of the promoter. The simian virus type 40, transforming T antigen expressed under the control of a promoter active during eye development, for example, produces hereditary retinoblastoma in mice that is phenotypically similar to the human condition but is genetically different, as the former is a gain-of-function, dominant mutation, and the latter a loss-of-function, recessive mutation.

Specific genes can be disrupted by homologous recombination with exogenous DNA sequences. This new technology allows site-directed mutagenesis of genes in living animals (see Box E). The resulting mutations may act in a recessive or dominant manner to cause developmental lethality or a disease state. In other cases there is no apparent phenotype. Homologous recombination was initially used to create the $HPRT^-$ "Lesch-Nyhan" mouse. This technique has now been used to evaluate a number of genes thought to be involved in neuronal development. For example, disruption of the engrailed 2 locus, which contains a homeobox domain thought to encode a transcriptional regulatory element, produces apparently normal mice with subtle alterations in cerebellar morphogenesis. Alterations in cerebellar development have also been noted in mice deficient in the *int-1* proto-oncogene, which is believed to function in cell-to-cell signaling.

Box E Homologous Recombination in Yeast and Mice

Part of the power of modern yeast genetics is the ability to make specific changes in a targeted gene. A modified gene transfected into a yeast cell will recombine with, and replace, its endogenous homologue with over 90% likelihood. Thus, the functional consequences of disrupting any cloned gene can be determined. The relatively small size of the yeast chromosome and its simple structure facilitate homologous recombination. The mammalian genome was, until recently, believed to be too large and complex to be amenable to this approach. Over the past few years, Oliver Smithies and Mario Capecchi have dramatically changed this assessment by finding ways to increase the frequency of homologous recombination in mammalian cells and by devising techniques to select the rare homologous recombinants.

The genetic manipulations are most easily done with embryonic stem (ES) cells, which can be grown in culture and yet retain full developmental potential. Cloned DNA is transfected into the cells, and the homologous recombinant cells identified and cloned. The difficulties are twofold: first, only some of the cells integrate the transfected DNA into their genome; second, most integration events occur at heterologous sites. Thus in the transfected population of cells only a few will have undergone homologous recombination. By clever design of the transfecting DNA, however, both positive and negative selection can be used to fish out the rare homologous recombinants.

For homologous recombination, cloned gene sequences from mice are modified by mutation or by the insertion of foreign DNA. A convenient method is to insert a gene that confers resistance to an antibiotic (e.g., neo^R) into an exon, thus disrupting it, and providing a means of positive selection (Figure A). After transfection, only those cells which express the foreign gene will grow in the antibiotic. The population can be further enriched for homologous recombinants by the addition

FIGURE A. Design of DNA elements for homologous recombination into cellular genes. DNA elements can be engineered to select for successful homologous recombination events and against nonhomologous recombination events in several ways. (1) An endogenous gene (host DNA) with a 5′ promoter element (colored box), five exons (gray boxes), and introns or flanking sequences (open boxes). On top is a cloned DNA element containing sequences from the endogenous cellular gene and two foreign sequences (colored boxes). One foreign sequence bears a constitutive promoter driving a positively acting selection marker (S^+) that has been inserted into one of the exons so as to disrupt it. S^+ markers can include prokaryotic genes that confer resistance to drugs such as neomycin and hygromycin. The other foreign sequence has a constitutive promoter driving a negatively acting selection marker (S^-) attached to the end of the DNA element—for example, the herpes simplex virus thymidine kinase gene, which confers sensitivity to nucleoside analogues such as ganciclovir. Following transfection of this element into cells, some portion of it will recombine into the homologous gene (brackets) or insert into other regions of the genome. Recombinant cells are selected for by growth in the presence of the appropriate drugs (S^+ selection). Nonhomologous events are selected against by exposure of cells to ganciclovir (S^- selection). (2) Recombination using DNA elements without a second selectable marker. A cloned DNA element containing sequences from the endogenous gene and a foreign sequence encoding a positively selectable drug resistance marker without a promoter attached to it have been inserted inframe with coding sequences of the first exon. Expression of the selectable marker will occur only when the foreign gene is inserted downstream of an active cellular promoter element; this occurs most commonly as a result of homologous recombination. (After S. Mansour et al., 1988. Nature 336: 348–352.)

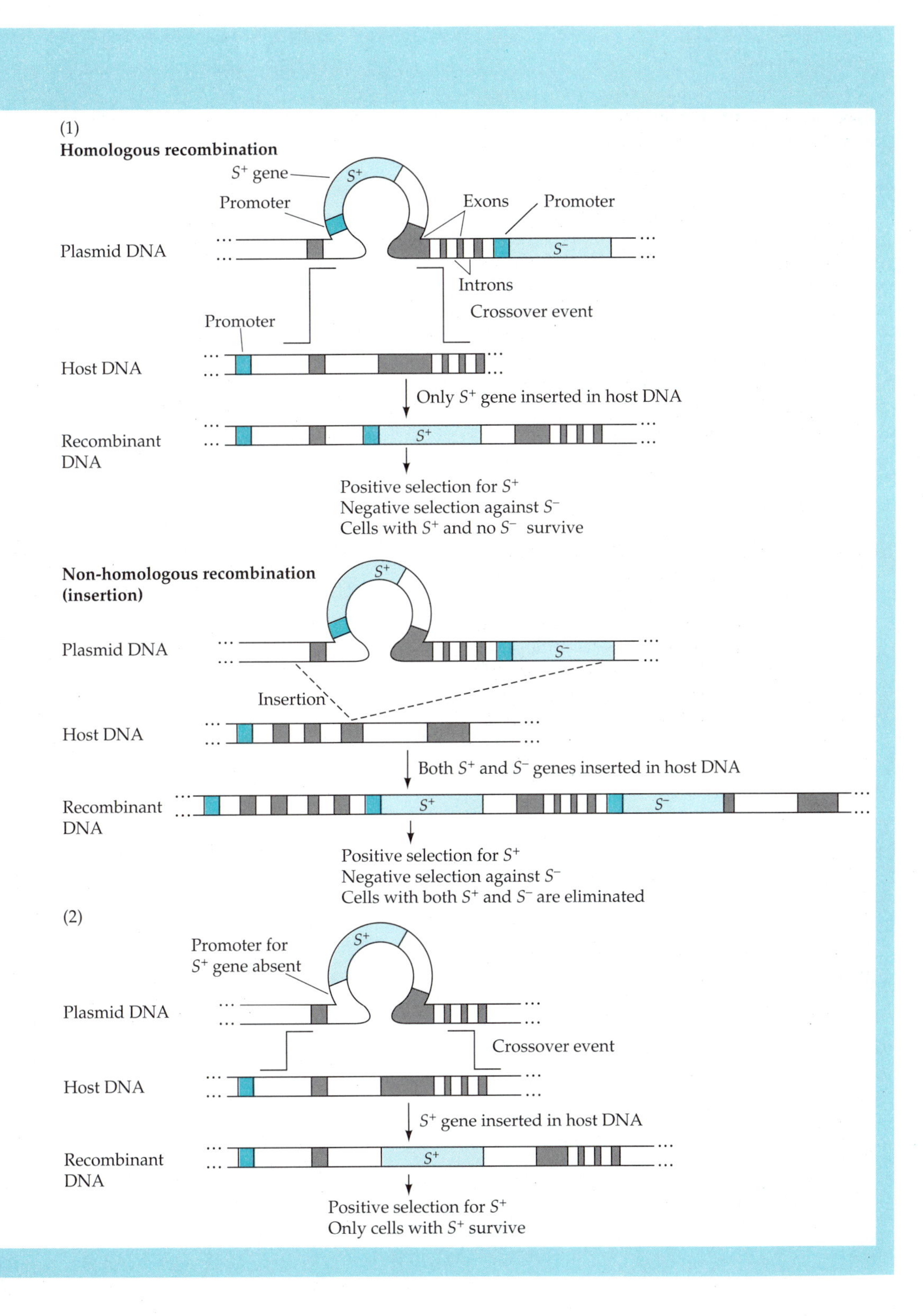

(*Box E continues*)

(Box E continued)

of the thymidine kinase gene from herpes simplex virus to one end of the transfecting DNA. Because it is at the end, this gene is lost by homologous recombination, but retained after integration at heterologous sites. As the enzyme encoded by this gene converts certain nucleoside analogues to toxic compounds, all cells expressing the gene can be eliminated. Homologous recombinants can also be identified by screening clones of the transfected cells, using the polymerase chain reaction (PCR) to identify cells bearing sequences that span the junction between the foreign DNA and the endogenous gene.

After recombinant ES cells are obtained in which the target gene is disrupted, they are mixed with normal embryonic cells to produce chimeric embryos (Figure B). Ideally these yield offspring that produce gametes carrying the new mutation. Breeding the chimeric mice generates heterozygous animals, which can be bred further to yield animals in which both copies of an autosomal gene are mutated. The potential of this technology is extraordinary. Rather than waiting for the spontaneous appearance of a mutated gene, one can alter genes at will and evaluate the effects on function of the cell and the organism.

FIGURE B. Creation of new mouse mutants for an autosomal recessive condition involving a cloned gene. Cultured embryonic stem cells [derived from a strain of mice with one coat color (white)] are transfected with a disrupted gene sequence (colored box) homologous to an autosomal gene (white box). Cells are selected that have successfully undergone homologous recombination into one copy of the endogenous gene, but still retain the normal gene on the homologous chromosome. Cloned cells are combined with embryonic cells from a normal zygote derived from a strain of mice marked by another coat color (black) and implanted into a pseudopregnant female. Some offspring will bear cells derived from embryonic stem cells and thus have a patchy coat color, and some of these will generate gametes of the mutant phenotype. Breeding of these latter offspring with normal mice will yield some heterozygotes for the mutation. When these F_1 offspring are bred to each other they will yield mice homozygous and heterozygous for the mutation, as well as wild-type offspring.

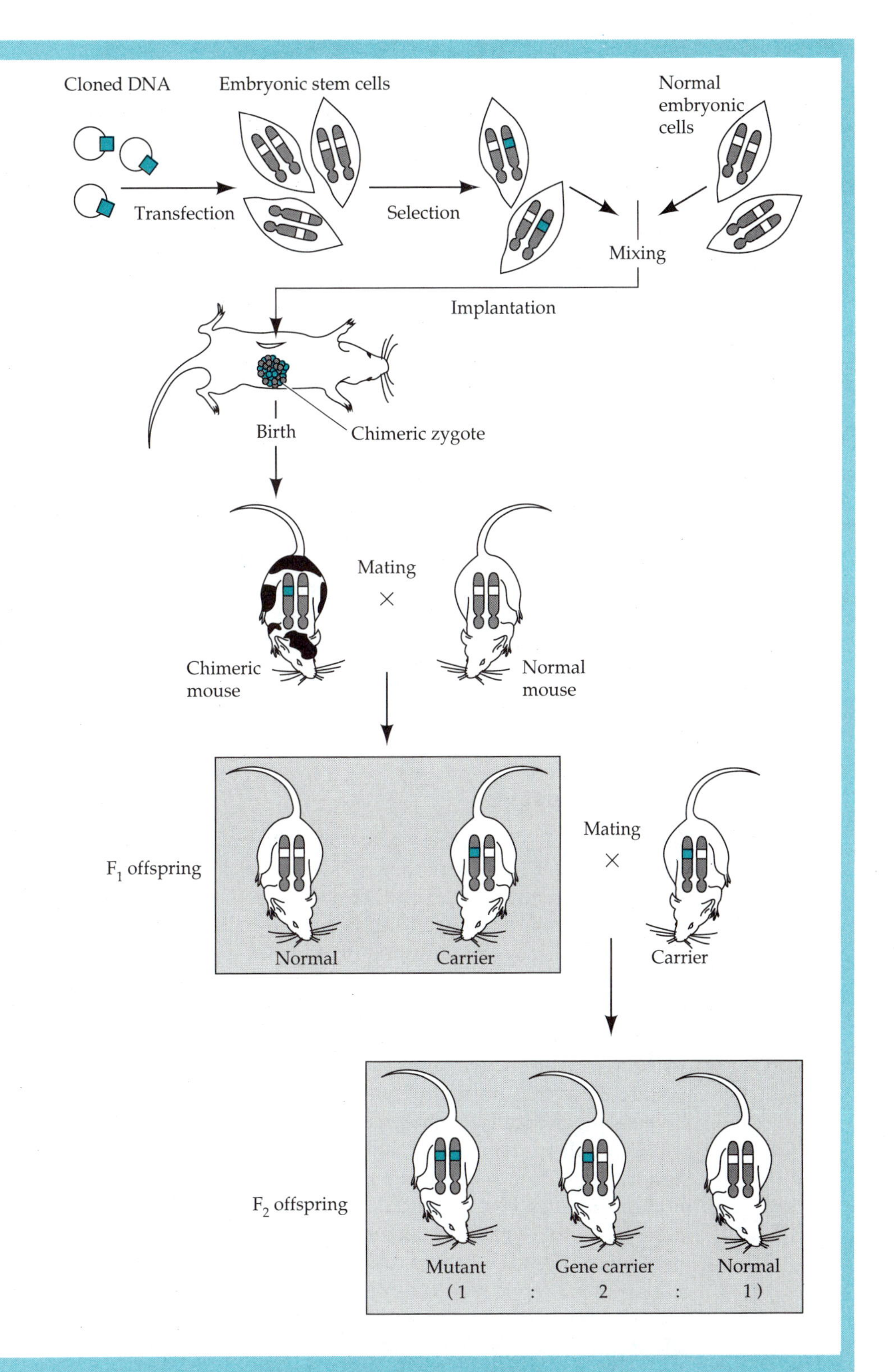
Cloned DNA
Embryonic stem cells
Normal embryonic cells
Transfection
Selection
Mixing
Implantation
Birth
Chimeric zygote
Mating
×
Chimeric mouse
Normal mouse
F1 offspring
Normal
Carrier
Mating
×
Carrier
F2 offspring
Mutant
Gene carrier
Normal
(1 : 2 : 1)

Other methods are being developed to block gene expression

In addition to direct disruption of a gene, its function can also be blocked by inhibiting RNA transcription, processing, and/or translation. One means is to introduce into cells an antisense RNA that hybridizes to specific mRNA species. Cells can be programmed to produce an antisense RNA by transfection with a recombinant gene in which inverted gene sequences are driven by a strong promoter element. Alternatively, cells can be incubated with high concentrations of short, antisense oligonucleotides, which are taken up by the cells, hybridize to the mRNA, and prevent its translation. In either case there must be a large excess of antisense RNA to sense RNA to decrease the amount of gene product translated. Even then, inhibition is usually not complete. A new method involves transfection of genes encoding **ribozymes** (RNA enzymes) into cells. Ribozymes have antisense sequences that recognize specific mRNAs coupled to a catalytic sequence that cleaves the corresponding message, thus resulting in its irreversible inactivation. Gene function can also be disrupted at the protein level by engineering genes with dominantly acting mutations that interfere with the function of the normal protein. Alternatively, antibodies can be introduced into cells that recognize antigenic determinants of endogenous proteins and block their action. These antibodies have a finite half-life when microinjected into cells, but stable expression can be achieved by transfecting into cells genes engineered to encode the recognition site of the antibody. The development of these new techniques will be crucial in attempts to understand the function of new neural proteins.

Genetic Therapies

Recombinant DNA technologies can potentially be used to ameliorate diseases caused by genetic or environmental factors. These technologies include pharmacological production of proteins and peptides in vitro, either in their natural state or in genetically altered forms with increased stability or efficacy. In addition, methods for gene transfer to the nervous system are being developed; these include grafting of genetically modified cells or direct viral delivery of genes to endogenous cells (Figure 10).

Cells grafted into the brain can be an effective means of long-term delivery of neuroactive substances to focal regions. Possible donor cells include those that normally make a compound of interest and those that have been genetically engineered to produce a particular substance. Fetal cells and cells from histocompatible donors tend to survive longer as grafts than do cells from older animals or those from different strains. Genetic modification can be carried out on cultured cells by DNA transfection or infection with retrovirus vectors. Retrovirus vectors can deliver genes more efficiently to cells than can most transfection procedures but require cellular DNA replication for integration into the genome. Repeated infection with retrovirus vectors can be used to deliver genes to most primary cells, such as skin fibroblasts or astrocytes, which divide slowly and have a finite life span in culture. Cells can thus be modified to release neuroac-

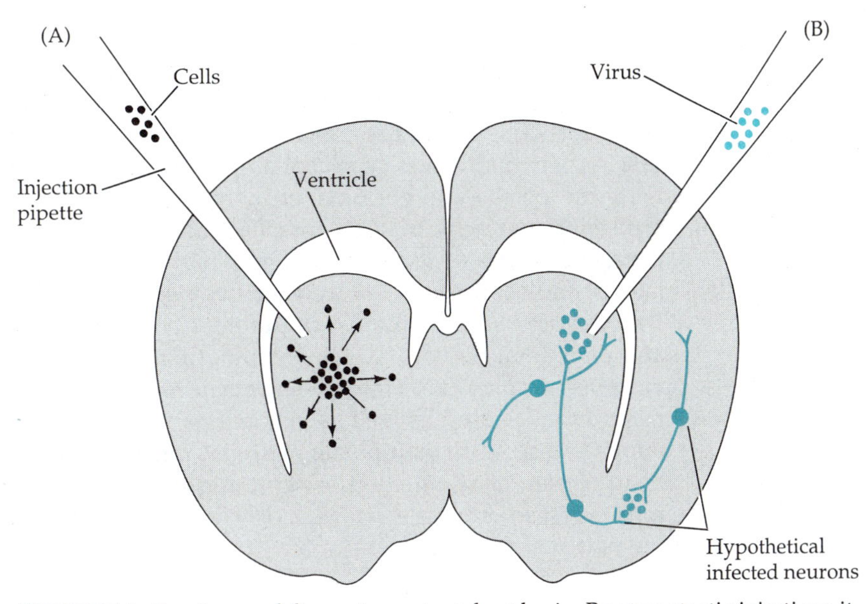

FIGURE 10. Focal gene delivery to postnatal rat brain. By stereotactic injections it is possible to introduce either cells (A) or virus particles (B) into specific regions of the brain. Grafted cells may migrate and can be genetically modified so as to release diffusible neuroactive substances or metabolize neurotoxic compounds. Virus vectors can be introduced directly. Some types of virus, e.g., herpes, will infect a number of different cell types and yield stable gene delivery through latency in at least some neurons. The virus can be transported within the brain by means of retrograde and anterograde transport and may be passed across synapses.

tive compounds, such as neurotransmitters or growth factors, or to metabolize toxic compounds. For example, skin fibroblasts transfected with the catecholamine synthetic enzyme tyrosine hydroxylase synthesize L-DOPA and, when grafted into the striatum, can deliver it to surviving dopaminergic neurons in animal models of Parkinson disease. Similarly, cells modified to produce NGF have proved effective in supporting the survival of several types of CNS neurons following toxic insults, such as lesion of neuronal processes and exposure to excitotoxins. The types of cells that can be grafted and the gene products that can be delivered are still being explored, but they potentially include many that are appropriate for a number of disease conditions.

Viral vectors can deliver genes directly to endogenous brain cells. Retrovirus vectors are inefficient at delivering genes to postnatal neurons as most are postmitotic. Possibly other cell types, such as astrocytes and oligodendrocytes, which do divide after birth in some circumstances, could serve as gene recipients. These retrovirus vectors can be used to deliver toxic genes selectively to tumor cells in the brain as these cells are an actively dividing population within a background of nondividing cells. Vectors derived from herpes simplex virus can be used for gene delivery to neurons. These virus particles are taken up intact at synaptic terminals and carried by rapid retrograde transport to the cell nucleus. Within the nucleus, the virus either replicates or enters a state of latency. Replication leads to death of the host neuron with the production of more infectious virus particles, which are passed transynaptically through the nervous system, eventually causing encephalitis and death of the animal. In latency the virus exists as an episomal element in the cell nucleus, retains some transcriptional activity, and does not interfere with neuronal function. Defective herpes vectors, which cannot replicate in neurons, can still enter latency and express genes. Foreign genes inserted into these vectors can be stably expressed in a number of different types of neurons in the central and peripheral nervous systems without pathogenicity. These and other virus vectors may eventually provide gene replacement therapy for certain hereditary diseases, as well as in some nonhereditary disease conditions where one seeks to manipulate neuronal function—for example, in treatment of pain or stimulation of nerve regeneration. Gene transfer to neurons may also provide a means of defining neural promoters and identifying new neural genes.

Summary

Molecular medicine has begun to unravel some of the mechanisms underlying inherited neurologic diseases, which, in turn, provide a means for understanding complex functions of the human nervous system. By assessing the co-inheritance of DNA markers and a disease phenotype the position of a disease gene can be located within the human genome. This positional information provides a segment of DNA containing the gene that can be cloned. New molecular biological methods facilitate this process by increasing the usefulness of markers through repeat sequence polymorphisms, by expanding the means to handle large DNA fragments, and by devising means to find genes within genomic fragments. The "reverse genetics" approach has been used successfully in identifying several disease genes, including those for Duchenne muscular dystrophy, a form of retinitis pigmentosa, neurofibromatosis type 1, and retinoblastoma. Identification of the disease gene has, in turn, provided new insights into the role of the normal protein and into the importance of reactions secondary to the primary defect. Altered channels, receptors, DNA regulatory elements and cytoskeletal components have all been implicated in disease states. In addition, accumulation of insoluble protein deposits formed from otherwise normal proteins can cause neuronal degeneration, as in Alzheimer disease.

Molecular genetic techniques are also being used to further understand disease states through the use of animal models. These studies include analysis of spontaneous mouse mutants with genetic defects and creation of new mutant mouse strains through random insertional mutagenesis, atypical expression of normal or mutant proteins, generation of antisense RNA, and targeted mutagenesis through homologous recombination. These techniques, which involve manipulation of embryos and embryonic stem cells, provide a powerful means for analyzing the structure and activity of proteins critical to the development and function of the nervous system.

Other efforts are being developed to deliver genes to the nervous system. Retrovirus vectors have proved useful in delivering genes to neuroblasts, with the result that all their clonal progeny carry the gene. These vectors are not effective postnatally as their integration into the genome requires DNA replication, and most cells in the nervous system, including neurons, are postmitotic after birth. Other means of delivering genes to the brain include grafting of genetically modified cells, and infection with replication-defective herpes simplex virus vectors, which can achieve stable expression of foreign genes in latency.

References

General references

Breakefield, X. O. and Cambi, R. 1987. Molecular genetic insights into neurologic diseases. Annu. Rev. Neurosci. 10: 535–594.

*Capecchi, M. R. 1989. Altering the genome by homologous recombination. Science 244: 1288–1292.

Gage, F. H. and Fisher, L. J. 1991. Intracerebral grafting: A tool for the neurobiologist. Neuron 6: 1–12.

*Rossant, J. 1990. Manipulating the mouse genome: Implications for neurobiology. Neuron 2: 323–334.

Watkins, P. C. 1988. Restriction fragment length polymorphism (RFLP): Applications in human chromosome mapping and genetic disease research. Biotechniques 6: 310–320.

*Wexler, N. S., Rose, E. A. and Housman, D. E. 1991. Molecular approaches to hereditary diseases of the nervous system: Huntington's disease as a paradigm. Annu. Rev. Neurosci. 14: 503–529.

Genetic mapping

Burke, D. T., Carle, G. F. and Olson, M. V. 1987. Cloning of large segments of exogenous DNA into yeast by means of artificial chromosome vectors. Science 236: 806–812.

Evans, G. A. and Lewis, K. A. 1989. Physical mapping of complex genomes by cosmid multiplex analysis. Proc. Natl. Acad. Sci. USA 86: 5030–5034.

Feener, C. A., Boyce, R. M. and Kunkel, L. M. 1991. Rapid detection of CA polymorphisms in cloned DNA: Application of the 5′ region of the dystrophin gene. Am. J. Hum. Genet. 48: 621–627.

*Gill, P., Jeffreys, A. J. and Werrett, D. J. 1985. Forensic application of DNA 'fingerprints,' Nature 318: 577–579.

Nguyen, C., Pontarotti, P., Birnbaum, D., Chimini, G., Rey, J. A., Mattei, J. F. and Jordan, B. R. 1987. Large scale physical mapping in the q27 region of the human X chromosome: The coagulation factor IX gene and the *mcf*.2 transforming sequence are separated by at most 270 kilobase pairs and are surrounded by several "HTF islands." EMBO J. 6: 3285–3289.

Smith, C. L., Lawrence, S. K., Gillespie, G. A., Cantor, C. R., Weissman, S. M. and Collins, F. S. 1987. Strategies for mapping and cloning macroregions of mammalian genomes. Methods Enzymol. 151: 461–489.

Stallings, R. L., Torney, D. C., Hildebrand, C. E., Longmire, J. L., Deaven, L. L., Jett, J. H., Doggett, N. A. and Moyzis, R. K. 1990. Physical mapping of human chromosome by repetitive sequence fingerprinting. Proc. Natl. Acad. Sci. USA 87: 6218–6222.

Genetic diseases of the nervous system

Cannon, S. C., Brown, R. H. Jr. and Corey, D. P. 1991. A sodium channel defect in hyperkalemic periodic paralysis: Potassium-induced failure of inactivation. Neuron 4: 619–626.

Cawthon, R. M., Weiss, R., Xu, G., Viskochil, D., Culver, M., Stevens, J., Robertson, M., Dunn, D., Gesteland, R., O'Connell, P. and White, R. 1990. A major segment of the neurofibromatosis type 1 gene: cDNA sequence, genomic structure, and point mutations. Cell 62: 193–201.

*Dryja, T. P., McGee, T. L., Reichel, E., Hahn, L. B., Cowley G. S., Yandell, D. W., Sandberg, M. A. and Berson, E. L. 1990. A point mutation of the rhodopsin gene in one form of retinitis pigmentosa. Nature 343: 364–366.

Kang, J., Lemaire, H.-G., Unterbeck, A., Salbaum, J. M., Masters, C. L., Grzeschik, K.-H., Multhaup, G., Beyreuther, K. and Muller-Hill, B. 1987. The precursor of Alzheimer's disease amyloid A4 protein resembles a cell-surface receptor. Nature 325: 733–736.

Lee, W.-H., Bookstein, R., Hong, F., Young, L.-J., Shew, J.-Y. and Lee, E. Y.-H. P. 1987. Human retinoblastoma susceptibility gene: Cloning, identification, and sequence. Science 235: 1394–1399.

*Monaco, A. P., Neve, R. L., Colletti-Feener, C., Bertelson C. J., Kurnit, D. M. and Kunkel, L. M. 1986. Isolation of candidate cDNAs for portions of the Duchenne muscular dystrophy gene. Nature 323: 646–650.

Orita, M., Suzuki, Y., Sekiya, T. and Hayashi, K. 1989. Rapid and sensitive detection of point mutations and DNA polymorphisms using the polymerase chain reaction. Genomics 5: 874–879.

Saiki, R. K., Gelfand, D. H., Stoffel, S., Scharf, S. J., Higuchi, R., Horn, G. T., Mullis, K. B. and Erlich, H. A. 1988. Primer-directed enzymatic amplification of DNA with a thermostable DNA polymerase. Science 239: 487–491.

Animal models

Hsiao, K. K., Scott, M., Foster, D., Groth, D. F., DeArmond, S. J. and Prusiner, S. B. 1990. Spontaneous neurodegeneration in transgenic mice with mutant prion protein. Science 250: 1587–1590.

Mansour, S. L., Thomas, K. R. and Capecchi, M. R. 1988. Disruption of the proto-oncogene *int-2* in mouse embryo-derived stem cells: A general strategy for targeting mutations to non-selectable genes. Nature 336: 348–352.

McMahon, A. and Bradley, A. 1990. The *Wnt-1 (int-1)* protooncogene is required for development of a large region of the mouse brain. Cell 62: 1073–1085.

Sarver, N., Cantin, E. M., Chang, P. S., Zaia, J. A., Ladne, P. A., Stephens, D. A. and Rossi, J. J. 1990. Ribozymes as potential anti-HIV-1 therapeutic agents. Science 247: 1222–1225.

Singer, T. P., Castagnoli, N. Jr., Ramsay, R. R. and Trevor, A. J. 1987. Biochemical events in the development of Parkinsonism induced by 1-methyl-4-phenyl-1,2,3,6-tetrahydropyridine. J. Neurochem. 49: 1–8.

Genetic therapies

Breakefield, X. O. and DeLuca, N. A. 1991. Herpes simplex virus for gene delivery to neurons. New Biologist, 3: 203–218.

Cepko, C. 1988. Retrovirus vectors and their applications in neurobiology. Neuron 1: 345–353.

Dobson, A. T., Sedarati, F., Devi-Rao, G., Flanagan, M., Farrell, M. J., Stevens, J. G., Wagner, D. K. and Feldman, L. T. 1989. Identification of the latency-associated transcript promoter by expression of rabbit beta-globin mRNA in mouse sensory nerve ganglia latently infected with a recombinant herpes simplex virus. J. Virol. 63: 3844–3851.

Wictorin, K., Brundin, P., Gustavii, B., Lindvall, O. and Björklund, A. 1990. Reformation of long axon pathways in adult rat central nervous system by human forebrain neuroblasts. Nature 347: 556–558.

INDEX

ABOUT THE BOOK

Editor Andrew D. Sinauer

Project Editor Carol J. Wigg

Production Manager Joseph J. Vesely

Book Production Janice Holabird

Illustrations J/B Woolsey Associates

Book and Cover Design Joseph J. Vesely

Composition Ampersand Publisher Services, Inc.

Cover Manufacture New England Book Components

Book Manufacture Hamilton Printing Company